TRIBOLOGICAL TECHNOLOGY – VOLUME II

NATO ADVANCED STUDY INSTITUTES SERIES

Proceedings of the Advanced Study Institute Programme, which aims at the dissemination of advanced knowledge and the formation of contacts among scientists from different countries.

The series is published by an international board of publishers in conjunction with NATO Scientific Affairs Division

A	Life Sciences	Plenum Publishing Corporation
B	Physics	London and New York
C	Mathematical and Physical Sciences	D. Reidel Publishing Company Dordrecht and Boston
D	Behavioural and Social Sciences	Martinus Nijhoff Publishers The Hague, Boston and London
E	Applied Sciences	

Series E: Applied Sciences – No. 57

TRIBOLOGICAL TECHNOLOGY VOLUME II

Proceedings of the NATO Advanced Study Institute on Tribological Technology,
Maratea, Italy, September 13 – 26, 1981

edited by

Peter B. Senholzi
Mechanical Technology Incorporated
1656 Homewood Landing Road
Annapolis, Maryland, U.S.A.

1982

Springer-Science+Business Media, B.V.

Distributors:

for the United States and Canada
Kluwer Boston, Inc.
190 Old Derby Street
Hingham, MA 02043
USA

for all other countries
Kluwer Academic Publishers Group
Distribution Center
P.O.Box 322
3300 AH Dordrecht
The Netherlands

ISBN 978-94-011-9809-7   ISBN 978-94-011-9807-3 (eBook)
DOI 10.1007/978-94-011-9807-3

Copyright © 1982 by **Springer Science+Business Media Dordrecht**
Originally published by Martinus Nijhoff Publishers in 1982
Softcover reprint of the hardcover 1st edition 1982

All rights reserved. No part of this publication may be reproduced, stored in a retrieval system, or transmitted, in any form or by any means, mechanical, photocopying, recording, or otherwise, without the prior written permission of the publishers, Springer-Science+Business Media, B.V.

CONTENTS OF VOLUME II

Chapter VI
W.O. Winer: Lubrication   347

Chapter VII
W.O. Winer: Lubricants   407

Chapter VIII
E.C. Fitch: Contamination in Fluid Systems   469

Chapter IX
M.B. Petersen: Trobological Failures and Mechanical Design

Chapter X
M. Godet: Tribo-Testing   535

Chapter XI
D. Scott: Monitoring   611

Chapter XI
B.R. Reason: The Multidisciplinary Approach   635

Appendices

Appendix A
D. Koshal, W.B. Rowe: Pressured Bearings   713

Appendix B
A. Kumar, B. Reason: Tribological Investigations of the
Contact Mechanics in a Rotary Positive Displacement Machine   747

Appendix C
B. Snaith, M.J. Edmonds, S.D. Probert: Three Dimensional
Topographical Descriptions of Solid Surfaces   759

LUBRICATION

W. O. Winer
Georgia Institute of Technology

1. INTRODUCTION

The objective of lubrication is to separate two surfaces of a mechanical system which are moving relative to one another so that the energy dissipation and the surface degradation are held at acceptably low levels consistent with the engineering design objectives for the mechanism. There are many possible ways of meeting these objectives including full film lubrication, boundary lubrication, solid film lubrication, and mixed film lubrication. In this chapter, the function and regimes of lubrication are discussed as well as several specific mechanisms. Emphasis in the material presented, is on introducing the concepts with physical understanding, and directing the reader to the literature available for in-depth study.

The types of mechanical components that require lubrication include slider bearings, rolling bearings, gears, cams, guideways, and others. These components have the common functional requirements of transmitting a force and/or guidance from one component to the other. They all involve the relative motion of one surface to another. The role of lubrication is to provide for minimum surface degradation and minimum energy dissipation at the interface between the two surfaces. The nature of this solution depends on the many defining characteristics and design requirements for the system. These considerations include the kinematics and dynamics of the system, the nature of the environment in which it must function, the composition of the bodies involved, and the design life criteria to be employed.

The objective of lubrication as stated above can lead to solutions which are not normallly though of as lubrication. For example, Halling suggests that the question should be "What is the best solution to the problem of carrying load across the interface with acceptable friction and wear?[1]" When asked in this fashion, we realize that the support of load with minimum friction and wear can be accomplished in some cases without the presence of a lubricant or lubrication. For example, in Figure 1 from Halling, eight different solutions to tribological problems are presented. At least three of these, (e, f, and h) would not normally be considered solutions involving lubrication. In those cases where the relative motion between the two surfaces has very small amplitude, the load could be carried by an elastomer, flexible strips, or a magnetic field, all of which are capable of accommodating the kinematic motion between the surfaces. The solutions shown in (a & g), namely dry contact and rolling elements, although not necessarily involving lubricants, would be considered to be traditional lubrication solutions to tribological problems. The three remaining examples, (b) chemical films, (c) laminar solid films, and (d) pressurized lubricant films are in the traditional sense lubrication solutions to the tribological problems. The chemical film solution (b) and the lamellar or solid film solution (c) can be thought of as similar mechanisms in the sense that they both involve solid boundary films on the substrate surfaces which permit the two solids to come together causing attrition and shear of the solid films between the surfaces. Within solution type (d) (pressure lubricant films) are included many classical types of lubrication where a pressurized film of a fluid (gas or liquid) is formed between the two surfaces causing them to be separated by virtue of the fact that the integrated value of the pressure over the surface is equal to the load being applied. The lubricant film in this case, is readily deformed and sheared with relatively low energy dissipation. Because, in this case, the surfaces are completely separated, there is little or no wear occurring on the surface. The pressurized lubricant film that exists can be formed in many ways. The film can be the result of an externally pressurized lubricant being injected between the two surfaces, in which case it is referred to as hydrostatic lubrication or it can be formed as a result of the relative motion of the two surfaces and the geometry of the film resulting in a self-acting bearing.

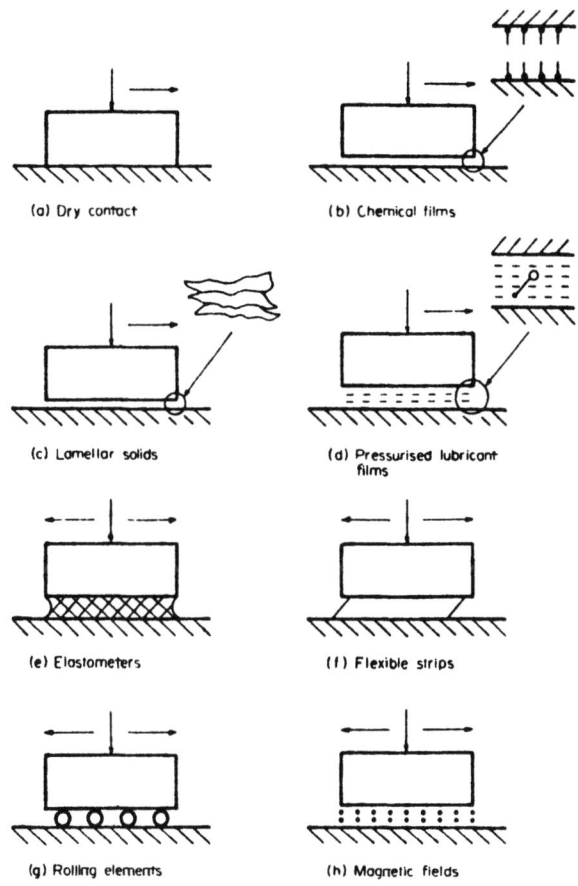

FIGURE 1 METHODS OF SOLUTION OF TRIBOLOGICAL PROBLEMS (HALLING, 1)

In many mechanical mechanisms to be lubricated, more than one type of lubrication can exist in the operating life of the device. For example, in self-acting fluid film lubrication, the relative motion of the two surfaces is necessary to generate the film pressure. However, in such mechanisms at some time during the operating cycle the kinematics may change, for example when the machine comes to rest, or the loads may change in such a fashion that the pressurized film is no longer generated. Under such conditions, the surfaces are permitted to come together resulting in the boundary lubrication or chemical film lubrication mechanism referred to in Figure 1 (b) or (c).

2. REGIMES OF LUBRICATION

Regimes of lubrication are normally associated with the dominate lubrication mechanism for the mechanical component being studied. In the present context, it is useful to divide the lubrication mechanisms into three broad regimes. These three regimes can be referred to as full film lubrication, boundary lubrication, and mixed film lubrication. In the literature it is not uncommon for other authors to divide the regimes of lubrication more finely than this. For example, to divide full film lubrication into hydrodynamic lubrication, hydrostatic lubrication, and elastohydrodynamic lubrication. The concept of regimes of lubrication is also used within these various categories to refer to the dominant mechanisms operating for particular contact. For example, as we will see in the regimes herein referred to as full film lubrication , there is a sub-category of thin film lubrication in which there are four possible dominant mechanisms, depending on the kinematics and material properties of the contact. These four sub-categories will also be referred to as regimes of thin film lubrication. However, for introductory purposes, it is helpful to think in terms of full film lubrication being that lubrication mechanism where the surfaces are completely separated by a film of lubricant, which may be a fluid, a gas, or a grease. Within this category, we find hydrodynamic lubrication, gas bearings, hydrostatic lubrication, and elastohydrodynamics.

The second broad category referred to here is boundary lubrication. In this regime, lubrication is accomplished by the presence of surface layers on the solid which are sacrificially worn away as a result of the sliding motion between the two surfaces. No hydrodynamic action is present or necessary. In virtually all situations, there is some sort of an adsorbed layer formed on the surface which inhibits adhesion between the surfaces and is sheared during relative motion, hence acting as a lubricant. Within this category of boundary lubrication, we have the sub-categories referred to as EP (extreme pressure) lubrication, solid film lubrication, and unlubricated mechanisms such as those cases where the solid component is made out of a self-lubricating material.

The third general regime of interest is mixed film lubrication, which is, as the name implies, a combination of the full film type lubrication and boundary lubrication. This regime occurs when the hydrodynamic action generates a pressure which is insufficient to completely separate the two surfaces. This separation can be viewed in the context of the ratio of the film generated by the full film mechanisms to the combined surface roughness of the two surfaces. In virtually all hydrodynamic lubrication analysis, it is assumed that the surfaces are completely smooth when a film thickness is predicted from the analysis. If that film thickness is not greater than the composite surface roughness of the two mating surfaces, then one would expect some surface contact to occur between the high spots of the two surfaces. At those contacts the lubrication mechanism must be of the boundary lubrication type and, therefore, the overall lubrication mechanism for the device is a mixed film type of lubrication.

The relative scales of the solid components, the lubricant films, and the surface roughnesses involved in these contacts are shown in Figures 2(a) and 2(b). Figure 2(a) is attributed to Archard in the Wear Control Handbook[2]. It shows on the logarithmic scales, the relative sizes of the typical lubricant molecules, and typical oxide film thicknesses, as well as surface roughnesses and engineering tolerances for typical engineering objects. For comparison, it also shows the relative sizes of objects that are readily understood by the reader. Figure 2(b), which is attributed to Halling, shows the relative sizes of the different lubricant layers[1]. Chemisorbed gas, lubricant monolayers, and oxide film are typical boundary lubricant films on the surface. They can be seen in relation to the surface roughness of a lapped engineering surface that is a very smooth engineering surface and elastohydrodynamic or normal full film lubricant films which are typical of the full film lubrication regime. In Figure 2(b), the relationship can also be seen between these two types of lubricant films and the solid surface structure of the surface to be lubricated. This surface consists of a rather ill-defined surface layer referred to as the Beilby layer, a heavily deformed region resulting from production of the surface, and the more lightly deformed material below that, before the solid substrate is reached. The relative scales for the two figures can be easily appreciated by noting that one Angstrom is 10^{-7} mm.

The ratio of the full film lubrication film thickness to the combined surface roughness has long been recognized as an important parameter in lubrication. The early work of Stribeck on journal bearings showed that for small values of ZN/P (Z being the viscosity, N the shaft rotational speed, and P the load on the bearing), hydrodynamic lubrication would be less likely to occur resulting in higher values of the coefficient of fiction and thus greater possibilities of wear occurring. As the parameter of ZN/P increased, the coefficient of friction decreased, went through a minimum,, and then slowly increased again as shown in Figure 3, from Bowden and Tabor[3]. The increase in coefficient of friction for low ZN/P, is the result of increasing solid surface

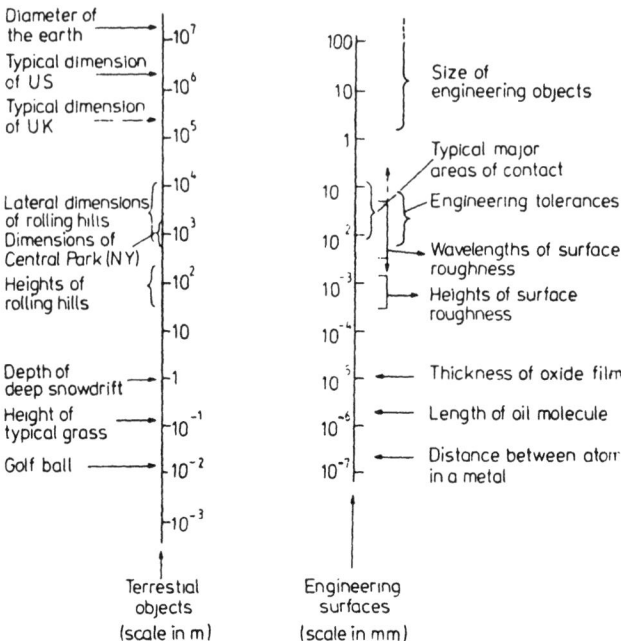

FIGURE 2(A) COMPARISON BETWEEN SCALES OF SIZE OF ENGINEERING & TERRESTRIAL OBJECTS (REF. 2)

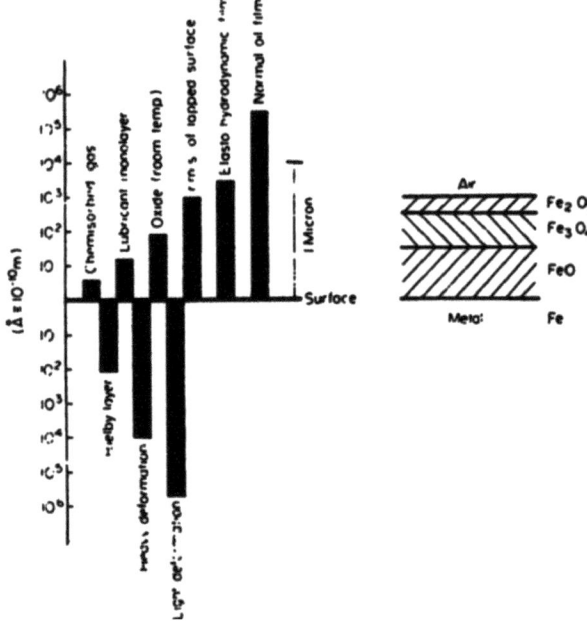

FIGURE 2(B)   ORDER OF MAGNITUDE OF SURFACE FEATURES (REF. 1)

contact as the hydrodynamic load carrying ability of the film
decreases with decreasing ZN/P. This coefficient of friction
is the result of a mixed film mechanism combining the viscous
shear of the hydrodynamic film and the solid contact shearing
of the boundary lubrication films. The boundary lubrication
films typically have higher resistance to shear resulting in
higher coefficients of friction. It is also in this regime of
low ZN/P, where wear will occur while at the higher ZN/P
region where full fluid film lubrication dominates, little
wear is expected.

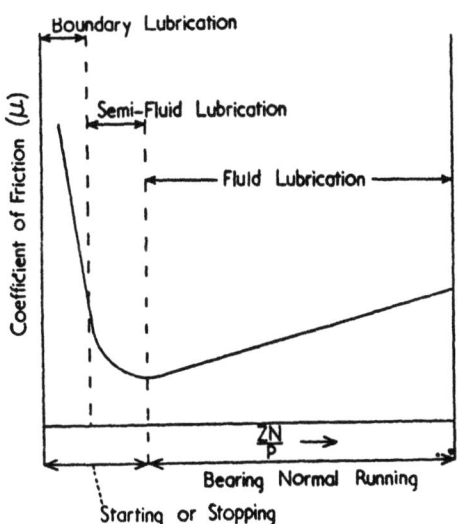

FIGURE 3 FRICTIONAL BEHAVIOR OF A JOURNAL AND BEARING AS THE
VISCOSITY Z, NUMBER OF REVS. PER MINUTE N, AND NOMINAL
PRESSURE P, ARE VARIED. UNDER CONDITIONS OF PURE
FLUID LUBRICATION THE COEFFICIENT OF FRICTION μ IS LOW
AND PROPORTIONAL TO AN/P. UNDER LESS FAVORABLE CONDI-
TIONS THE LUBRICANT FILM BREAKS DOWN AND THERE IS A
MARKED RISE IN FRICTION (REF. 3)

The fact that the increase in coefieicnt of friction for
low ZN/P is the result of contact between the solids is
clearly demonstrated by Czichos in Figure 4[4]. In this figure,
the coefficient of friction and the electrical contact time
between the surfaces are each plotted as a function of sliding
velocity in the left hand part of Figure 4. In each case of a
different temperature or different surface roughness, it can
be seen that as the time during which no contact occurs is
increasing, the coefficient of friction drops and reaches a
minimum when the no-contact time reaches 100%. On the
right-hand portion of Figure 4, the coefficient of friction
and no-contact time fraction are plotted as functions of the

λ ratio which is the ratio of the hydrodynamically generated film thickness to the composite surface roughness. The composite surface roughness is defined as the square root of the sum of the squares of the two RMS surface roughnesses. When the results are rationalized in this fashion, it becomes clear that as the λ ratio increases, the coefficient of friction decreases and the amount of surface interaction decreases. When the λ ratio is less than one, the amount of surface interaction and the coefficient of friction increase and expected wear would also increase. In this particular case, λ ratios less than one appear to indicate that boundary lubrication is the dominant mode, λ ratios from about 1 to 2-1/2 represent mixed film lubrication and for λ ratios greater than about 2-1/2 the mode of lubrication seems to be completely full film lubrication. Consequently, from a design standpoint one can expect, for low λ ratios in the neighborhood of 1 or 2 or less, that the energy dissipated in the contact and the wear occurring in the contact will be greater than that occurring at higher λ ratios where the surfaces are completely separated by a hydrodynamic lubrication film.

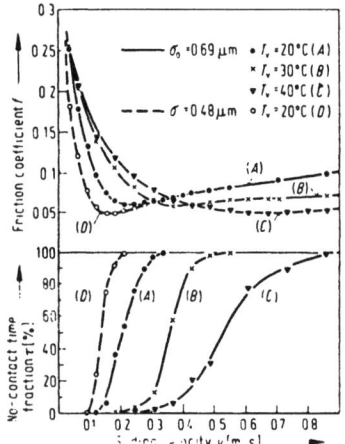
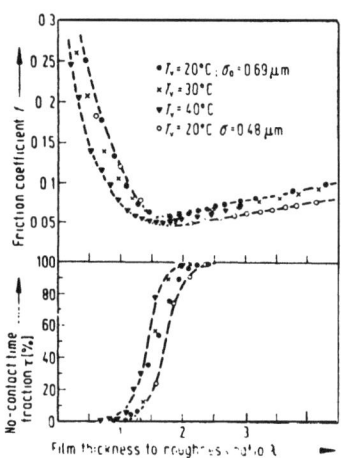

FIGURE 4 (REF. 4) STRIBECK'S CURVE AND THE NO-CONTACT TIME FRACTION FOR A ROUGH LINE CONTACT (F_N=9N; p_H-32.3 MN m^{-2}).

FRICTION AND NO-CONTACT TIME FRACTION VERSUS FILM THICKNESS-TO-ROUGHNESS RATIO (F_N=9N; p_H=32.3 MN m^{-2}).

The λ ratio in concentrated contacts such as cams, gears, and rolling element bearings also has an influence on the life

of the component. In these concentrated contacts if the
surfaces are completely separated by a hydrodynamic lubricant
film, the mode of failure would be expected to be one of
fatigue failure of the surfaces. However, if the surfaces are
interacting as a result of a low λ ratio, permitting the
asperities to contact one another from the two surfaces, the
life of the component will be reduced due to this interaction.
Although the details of the relationship between life and
lambda ratio are not completely agreed upon by all those
active in the field, it can be schematically shown in Figure
5. As a general concept to qualitatively guide design, this
is a useful idea; however in quantitative detail, it must be
used with some caution. The standard life referred to here is
the catalog life in the case of rolling element bearings, but
the amount of increase or decrease expected is not clearly
defined. In the case of the definition of the λ ratio, there
is some uncertainty associated with the film thickness
developed and also in the definition of the composite surface
roughness. As discussed elsewhere in this text, the details
of measuring the composite surface roughness are debatable.
For example, cutoff length for sampling wavelength range to be
considered and the relationship of these to the size of the
lubricated contact, all influence the surface roughness
measurement.

Recent work by Winer and Bair has shown, Figure 6, that
for concentrated contacts the coefficient of friction (or
traction) in the transition region between full film and
boundary lubrication is more complicated than in the case of
the conformal contacts such as journal bearings[5]. In the case
of conformal contacts, the pressures are relatively low and
the viscous drag causing the friction in the full film portion
of the behavior is dominated by the viscosity of the
lubricant. However, in concentrated contacts where the
pressures can be quite high, the limiting shear stress
behavior of the lubricant plays a role in determining the
friction. This property will be discussed more fully in the
following subsection on "Lubricants".

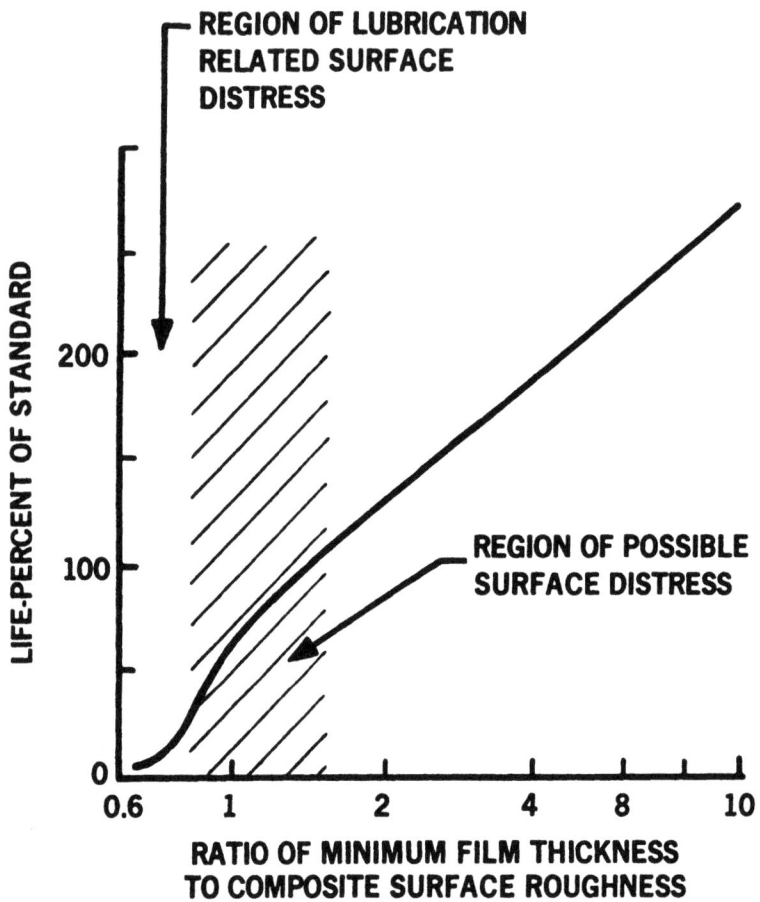

FIGURE 5 FATIGUE LIFE AS A FUNCTION OF THE MINIMUM LUBRICANT FILM THICKNESS/ROUGHNESS RATIO

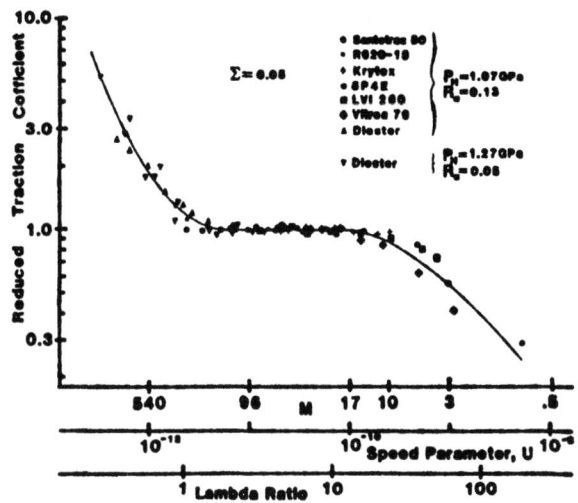

FIGURE 6 REDUCED TRACTION COEFFICIENT IN CONCENTRATED CONTACTS FOR A NUMBER OF LUBRICANTS AS A FUNCTION OF VARIOUS PARAMETERS (REF. 5)

3. BOUNDARY LUBRICATION

In the context of this discussion of lubrication, boundary lubrication includes all cases where the lubricant is a surface film or surface layer of a solid or near solid material which is sacrificially sheared between the two surfaces in order to maintain low friction and wear. In more complete discussions found elsewhere in the literature, a finer division is made in terminology and various types of what is herein referred to as boundary lubrication which may be discussed as separate categories under different names such as solid lubricants, E.P. lubricants, or oxide film lubrication. Boundary lubrication may exist as the only mode of lubrication between two surfaces, during part of the operating cycle, or over part of the geometry as in the mixed film lubrication mode.

The boundary lubrication films can be formed by several mechanisms. They may be mechanically put in place, such as burnishing on solid lubricants, for example, MoS_2 or graphite. Boundary lubricant films may be formed as a result of physical adsorption of materials in the environment onto the surface. They may be the result of chemical adsorption of materials in the immediate environment reacting with materials already on the surface, or they may even be the result of self-lubricating materials as the substrate structural material when the component is made out of a composite material.

Boundary lubrication is provided by a sacrificial, low shear strength solid layer which adheres to the surface and is replenished, as it is worn away, by adsorption, reaction or insitu formation mechanisms in the temperature range required by the application. A highly simplified example which is attributed to Bowden and Tabor is shown in Figure 7 from Czichos[3,4]. This is a hypothetical schematic of what might be expected of a system with a lubricating oil present and requiring boundary lubrication. If the operating conditions are such that no hydrodynamic film is developed as a result of the oil present, then the boundary lubrication characteristics of a nonpolar or nonreactive base oil would be rather poor resulting in a high coefficient of friction over the temperature range of interest as indicated by curve I. If a small percentage (a fraction of one percent) of a fatty acid which melts at some temperature T_m, is added to the base oil, the fatty acid will preferentially adsorb on the surface forming a low shear strength solid fatty acid layer which acts as a boundary lubricant at temperatures below its melting point. This fatty acid layer will result in low friction as the temperature is increased until the melting temperature is reached, at which point the fatty acid layer is desorbed leaving the surfaces again without a boundary lubricant and thus the coefficient of friction will rise to that which occurred with the base oil alone. If instead of the fatty acid one were to add an E.P. (extreme pressure) additive which chemically reacts with the surface of the solid at some temperture, T_r, then at lower temperatures there would be no solid boundary film on the surface and the friction would be high until the temperature of the surface reached the reaction temperature for the extreme pressure additive. At this temperature and above, the reaction which takes place results in the formation of a new material which is a solid layer on the surface and has a low shear strength resulting in low friction. Therefore, the E.P. additive acts as a good boundary lubricant at temperatures above its reaction temperature. Although not shown in this figure, the reaction products of the E.P. additive would of course melt or desorb at some higher temperature, again resulting in no lubrication and high friction. The E.P. additive is shown in curve III. Hypothetically one could formulate a lubricant consisting of both a small amount of the fatty acid and a small amount of the E.P. additive and resulting in low friction and good lubrication over the entire temperature range as shown by curve IV, the hypothetical composite lubricant behavior. In theory, with this composite lubricant the fatty acid would act

as the lubricant at low temperatures and would not desorb
until temperatures above those at which the E.P. additive
reacted to form another film on the surface. At the higher
temperatures, the E.P. additive reaction products would
function as the lubricant. At still higher temperatures, the
reaction products of the E.P. additive would also be desorbed,
as previously stated.

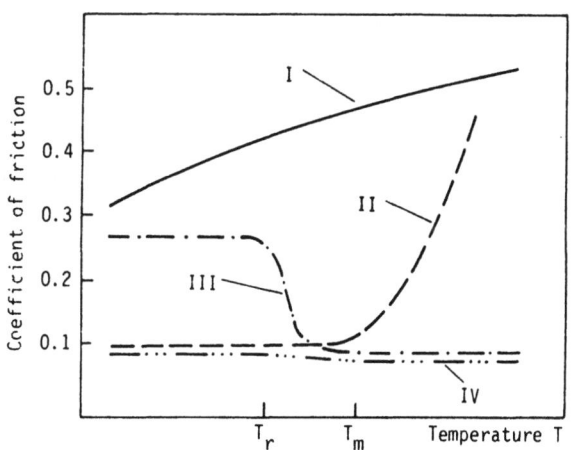

FIGURE 7 FRICTION BEHAVIOR OF BOUNDARY-LUBRICATION SYSTEMS
(SCHEMATIC) (REF. 4)
I.   NON-POLAR BASE OIL (B.O.)
II. B.O. PLUS FATTY ACID WHICH MELTS AT T_M
III. B.O. PLUS E.P. ADDITIVE WHICH REACTS AT T_R
IV. HYPOTHETICAL COMPOSITE

Although the E.P. additive characteristics described
above are frequently referred to in the literature, it is
generally recognized that such behavior is incorrectly named.
The term "extreme pressure" (E.P.) was the result of early
studies of the behavior of certain boundary lubrication
additives to lubricating oils. In those studies it was
recognized that these additives became active at high loads,
that is high contact pressures, hence the name "extreme
pressure additives" was given to them. However, it is now
recognized that it is not the pressure that is important but

the temperature which resulted from the high pressure which actually causes or permits the reaction to take place. Therefore, these additives would more correctly be termed "extreme temperature additives."

In practice, of course, boundary lubrication is much more complicated than the simple description given above. Boundary lubrication is an area of considerable practical importance. However, it is an area in which relatively little fundamental research has been done and relatively little fundamental understanding exists, at least when compared to the existing understanding and efforts put into hydrodynamic or elastohydrodynamic lubrication. Boundary lubrication is very much an interdisciplinary field involving not only mechanical engineering but also very significant portions of chemistry and physical chemistry. The details of the reactions and behavior of boundary lubricants, which in general are proportionately small quantities added to bulk lubricants, are very complex. Additive interaction and additive competition for reactions with the surface play a significant role in the practical functioning of lubricants in the boundary lubrication regime. The limits of lubrication which are associated with gross surface disruption and sometimes referred to as scoring, scuffing, or severe wear, are also very much dependent upon the limits of boundary lubrication and are extensively studied but poorly understood.

For a more complete understanding of the state of the knowledge on boundary lubrication, the reader is directed to the vast literature on the subject. The books of Bowden and Tabor, Czhichos, Braithwaite, Cameron, and the _Standard Handbook of Lubrication Engineering_ have extensive discussions of boundary lubrication[3, 4, 6, 7, 8]. The ASME publication, Boundary Lubrication: An Appraisal of the World Literature, published in 1969 and edited by Ling, Klaus, and Fein is a very excellent source of the review of the world literature through 1969[9]. A more recent review by Sakurai entitiled, "The Role of Chemistry in the Lubrication of Concentrated Contacts," is also a very thorough review of the work of one of the leaders in the field of boundary lubrication[10].

4. MECHANISMS AND TYPES OF FULL FILM LUBRICATION

Full film lubrication exists when the two surfaces moving relative to one another are completely separated by a lubricant film. The lubricant might be either a liquid or a gas and the surface motion might be steady or unsteady. Whatever the material and the surface kinematics are, the operating conditions are such that the lubricant has a pressure developed in it which when integrated over the surface area is equal to the load being applied to the two surfaces. The behavior of the film can be analyzed and described in terms of fundamental principles of mechanics. The fundamental principles involved are continuity (the conservation of mass) and the conservation of momentum (Newton's Second Law). These two fundamental concepts are combined with a number of appropriate simplifying assumptions into a single equation which is referred to as the Reynolds' Equation. Reynolds first developed and presented this equation governing hydrodynamic lubrication in 1886. If the system can be assumed to have constant properties, then the Reynolds' Equation with appropriate boundary conditions, is all that is required to analyze the behavior of the full film. However, if variable properties, as a result of temperature rise, are to be included, then one must also use the energy equation to predict the temperatures in the film and some appropriate constitutive equations describing the change of the properties with temperature. If property variations are permitted with pressure, as in the case of elastohydrodynamic lubrication or gas bearing lubrication, then additional constitutive equations relating the properties to temperature and/or pressure are required.

A detailed derivation of the Reynolds' Equation can be found in many texts in the literature. These derivations can be found at whatever level of sophistication suits the reader from the relatively elementary presentation of Halling, Trumpler, or Dowson and Higginson to the more advanced treatments of Gross, Constantinescu, Pinkus and Sternlicht, and Tipei [1, 11, 12, 13, 14, 15, 16].

Fundamentally, one analyzes the volume between the surfaces and requires continuity of flow, i.e., conservation of mass of the flow through this volume, and obtains the velocity profiles from the momentum equation. These velocity profiles are integrated across the film resulting in a single equation of the form shown in Equation (1) for the geometry shown in Figure 8.

$$\frac{\partial}{\partial x}\left(\frac{\delta h^3}{\mu}\frac{\partial p}{\partial x}\right) + \frac{\partial}{\partial y}\left(\frac{\delta h^3}{\mu}\frac{\partial p}{\partial y}\right) = 6[U_1 - U_2]\frac{\partial(\delta h)}{\partial x} + 6\delta h \frac{\partial}{\partial x}[U_1 + U_2] + 12\frac{\partial(\delta h)}{\partial t} \qquad (1)$$

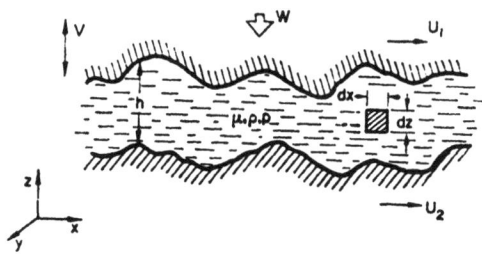

FIGURE 8   HYDRODYNAMIC LUBRICANT FILM GEOMETRY

The major assumptions that were used to develop Equation (1) are the neglect of gravitational and inertial force terms, the assumption of Newtonian lubricant behavior, the assumption that the film thickness is small compared to the lateral dimensions of the film, and the no-slip condition of the liquid at the liquid-solid or gas-solid boundaries of the film. Equation (1) is a rather complicated nonlinear partial differential equation in three independent variables, x and y being the lateral dimensions of the film, and t as time. U_1 and U_2 are the surface velocities of the two surfaces as shown in Figure 8; ρ, p and μ are the lubricant density, pressure, and viscosity, respectively, and are in general treated as functions of space and time; h is the lubricant film thickness, also a function of space and time. Physically, the left hand side of Equation (1) can be thought of as being related to the load being carried by the film because it is related to the pressure distribution which when integrated over the contact area will give the load being carried. The right hand side of Equation (1) can be thought of as related to the mechanisms generating the pressure and therefore generating the load being carried. The first term on the right hand side is the "geometric wedge" term which involves the product of the velocity difference between the two surfaces and the wedge shape or slope of the film resulting from the fact that the two surfaces are not parallel. This is the most common mechanism of generating the pressure in a hydrodynamic lubrication film. The second term on the right hand side is sometimes referred to as the stretching term. It involves the product of the film thickness and the change of surface velocity with distance along the surface. This is a much less common mechanism for generating pressure in the film than the first term.

The third term on the right hand side is referred to as the squeeze film term because it involves the time rate of change of film thickness between the two surfaces. That is, it represents the normal motion between the two films as the gap is closed. The squeeze film term is common in time varying load or time varying film thickness applications, such as in connecting rod bearings. It is a time unsteady phenomena.

If we further assume that the lubricant properties of density and viscosity are constant and the lower surface is not moving ($U_2 = 0$), then Equation (1) becomes Equation (2) which is a much more commonly recognized form of the Reynolds' Equation. All three basic terms described above in Equation (1) are still present on the right hand side of the equation.

$$\frac{\partial}{\partial x}\left(h^3 \frac{\partial p}{\partial x}\right) + \frac{\partial}{\partial y}\left(h^3 \frac{\partial p}{\partial y}\right) = 6\bar{\mu}\, U_1 \frac{\partial h}{\partial x} + 6\bar{\mu} h \frac{\partial U_1}{\partial x} + 12\bar{\mu} V \qquad (2)$$

In Equation (2) the new variable of μ represents the average viscosity in the film; U_1 represents the sliding velocity between the two surfaces; and V represents the relative velocity between the two surfaces in the direction perpendicular to the plane, that is, the squeeze film velocity. Equation (2) is a partial differential equation with the pressure as the dependent variable for which we are solving. The film thickness, surface velocities, and viscosity are assumed to be known parameters. The solution of this equation requires boundary conditions which will be discussed subsequently. The solution results in the pressure distribution being a function of x, y and t' as well as being inherently a function of the film thickness, surface velocities, and viscosity.

The surface motions associated with the wedge and squeeze film mechanisms of film generation are shown in Figure 9[1]. The wedge action is shown in Figure 9(a) and the squeeze film action is shown in Figure 9(b). Figure 9(c) combines the squeeze film and wedging action which exists in the case of lubricated rollers or in a journal bearing application.

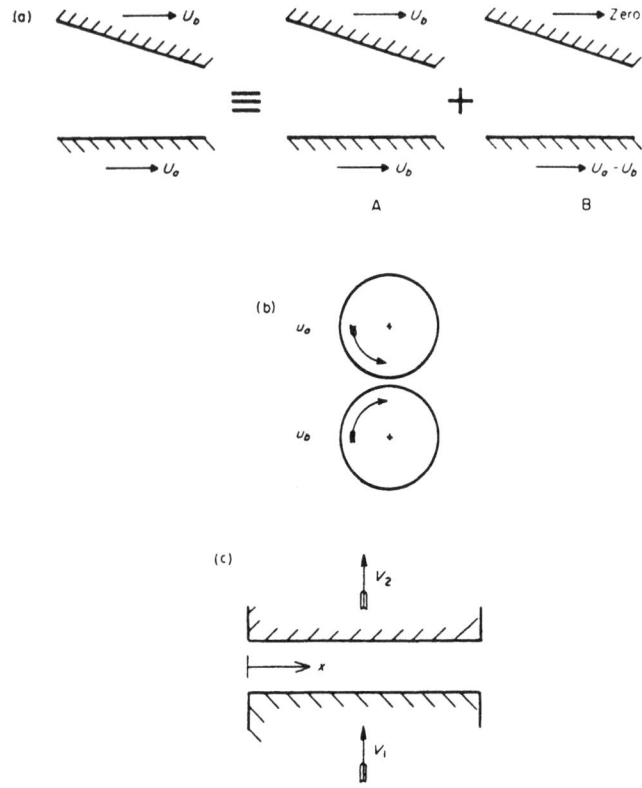

FIGURE 9   SCHEMATIC OF THE MECHANISMS OF HYDRODYNAMIC LUBRICATION (REF. 1)

The boundary conditions used in solving the Reynolds' Equation fall into three basic types: (a) the full Sommerfeld, (b) the half Sommerfeld, and (c) the Reynolds' boundary conditions. The choice of boundary conditions used depends on the particular system of interest and the applications of interest. All solutions to the Reynolds' Equation that exist in the literature and from which design curves have been made, have assumed one of these different types of boundary conditions. It is important to understand the difference in the boundary conditions, the applicability of the boundary conditions, and to make the correct choice when using design charts.

In slider bearings, the boundary conditions are less ambiguous because the pressure can be assumed to be the ambient pressure around the edge of the slider bearing pad.

However in journal bearings, the boundary conditions become more complicated and the three different choices are available and have been used. The full Sommerfeld solution was first used by Arnold Sommerfeld shortly after the turn of the century and consists of assuming that the pressure is equal to some ambient pressure, p_o, at $\theta = 0$, and 2π where θ represents the angular coordinate around the circumference of the bearing. The load vector passes through $\theta = 0$ and π with $\theta = \pi$ at the point of minimum film thickness. It is reasonable to expect that the pressure at the location chosen as 0 is equal to that at the same location which also represents 2π around the bearing. The difficulty with the full Sommerfeld boundary condition assumption is the pressure distribution that results from it when applying that boundary condition to the Reynolds' Equation.

Figure 10 shows the pressure distribution that results in the case of each of these three boundary conditions[1]. The full Sommerfeld solution is shown to result in a sub-ambient pressure distribution in the diverging region of the contact. The angular coordinate, θ, is selected in such a way that $\theta = 0$ represents the location of maximum film thickness, and $\theta = \pi$ represents the location of minumum film thickness in the journal bearing. Therefore, for $\theta = 0$ to π, the film is converging while from π to 2π the film is diverging in thickness. Depending on the operating conditions and the magnitude of p_o, the ambient pressure, the sub-ambient pressure in the diverging region could be large enough to result in cavitation or rupture of the film. If the film ruptures there will be no pressure in that film other than the ambient pressure, p_o. However, the sub-ambient pressure in the diverging region can exist if the p_o is sufficiently high that the pressure reduction in the diverging region does not result in a low enough pressure to cause cavitation in the film. In that case, the full Sommerfeld boundary condition and solution would be appropriate and the load would be the integral of the curve shown. However, in the vast majority of applications, the ambient pressure, p_o, is essentially atmospheric pressure and the pressures developed in the film are large compared to the ambient pressure. In this case, the sub-ambient pressure of the diverging region predicted by the Sommerfeld solution, would result in a near absolute zero pressure or even negative pressure (that is, tensile stress in the film) which, except in very special circumstances, the film could not withstand. The film in the diverging region would therefore rupture. In this much more common situation, the remaining two types of boundary conditions are more appropriate. The most elementary of the remaining two types of boundary conditions is the half Sommerfeld boundary condition, which simply takes the full Sommerfeld solution and acknowledges that the sub-ambient pressure predicted in the diverging region will not occur, and therefore sets the pressure in the diverging region (i.e., from $\pi \leq \theta \leq 2\pi$) to be the ambient pressure p_o. In this case only the positive portion of the pressure distribution in the converging region is integrated to obtain the load being carried by the film.

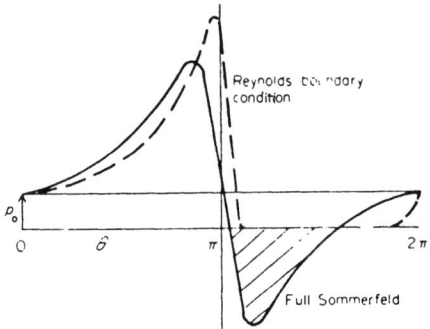

FIGURE 10   PRESSURE DISTRIBUTIONS IN JOURNAL BEARING RESULTING FROM CHOICE OF DIFFERENT BOUNDARY CONDITIONS (REF. 1)

The Reynolds' boundary condition is for a similar operating bearing to that used for the half Sommerfeld solution, but is somewhat more complicated and considered to be more correct for the fundamental behavior of the lubricant. The Reynolds' boundary condition states that the pressure must be equal to the ambient pressure at $\theta = 0$, but at the end of the positive portion of the pressure generated at some unknown location, θ_a, both the pressure must be equal to the ambient pressure and the pressure gradient, $dp/d\theta$, must be equal to 0. The difference between the results from the half Sommerfeld and the Reynolds' boundary conditions is in most cases small. The difference between the solutions obtained with the full Sommerfeld boundary conditions compared to the half Sommerfeld or Reynolds' conditions can be significant depending on the operating regime of the bearing.

The half Sommerfeld and Reynolds' boundary conditions should be used in most applications, particularly those where the ambient pressure of the bearing is atmospheric or near atmospheric pressure. However, if the bearing is operating in a pressurized environment, it would be appropriate to consider the full Sommerfeld solution and to examine the pressure generated in the diverging region relative to the ambient pressure. In the case of gas bearings, the full Sommerfeld boundary condition is the appropriate one to use, but because the density is not constant the pressure distribution developed will not be antisymmetrical about the minimum film thickness as shown in Figure 10, for an incompressible liquid lubricant.

The designer employing hydrodynamic lubrication solutions would prefer to have as a solution to the Reynolds' Equation the minumum film thickness as a function of the viscosity, surface velocity, and time. The designer would likewise prefer to have the viscous drag, flow rates, and energy dissipation also as a function of the independent variables available, namely the viscosity, speeds, and load. This is referred to as the inverse hydrodynamic problem. However, what is obtained from the Reynolds' Equation is the pressure distribution which then must be integrated to find the load as a function of the film thickness as well as the other variables. These solutions are then plotted in dimensionless fashion to obtain design curves which can be used.

From the above discussion of the Reynolds' Equation, it becomes apparent that the only property of the liquid lubricant that is important in determining its ability to be a hydrodynamic lubricant is the viscosity of the lubricant. Therefore the lubricant can be any liquid material and, depending on the surface speeds, geometry, and load, any viscosity could be used to develop a hydrodynamic lubrication film. The viscosity of liquids is a strong function of temperature. Figures 11(a) and (b) show the viscosity of typical ASTM grade hydrocarbon lubricants and a wide variety of liquids, respectively, as a function of temperature. Additional discussion of lubrica nt viscosity behavior can be found in the following chapter on lubricants.

4.1 Hydrodynamic Lubrication With Liquids in Journal Bearings

Solutions of the Reynolds' Equation for journal bearings can be found for design purposes in several locations in the literature. The early work of Ramondi and Boyd is readily available in the Standard Handbook of Lubrication Engineering [8]. Other sources include those already mentioned; References 1, 7, 11 - 16. Additional more recent work with special emphasis can be found in the publications of Rohde, et al and Szeri [17, 18, 19]. In these publications, the results are presented as design charts relating the dimensionless groups of important variables for the operation of the journal bearing. The general journal bearing geometry considered is shown in Figure 12. The primary independent variables employed include the Sommerfeld number, S:

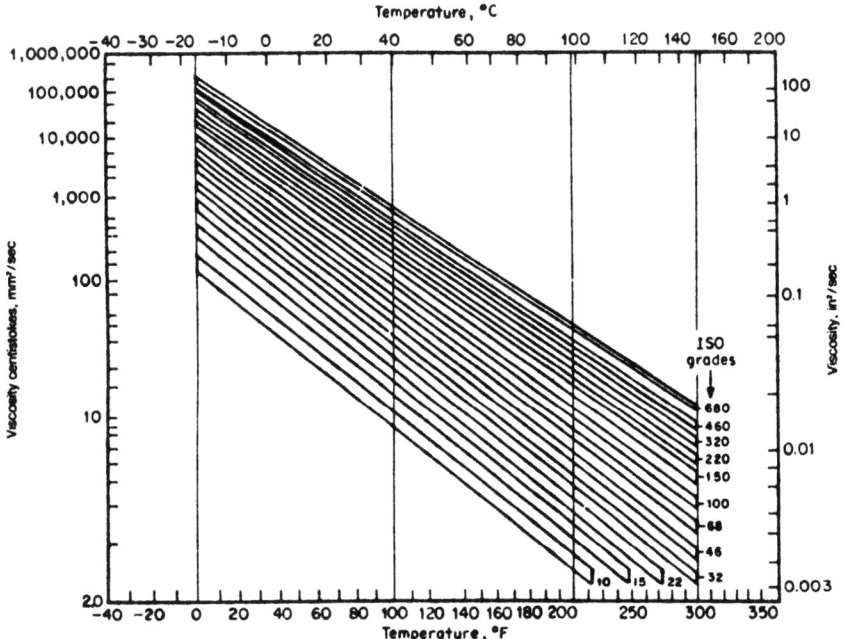

FIGURE 11(A) VISCOSITY-TEMPERATURE BEHAVIOR OF TYPICAL 100 VI LUBRICANT IN ISO VISCOSITY GRADES (REF. 2)

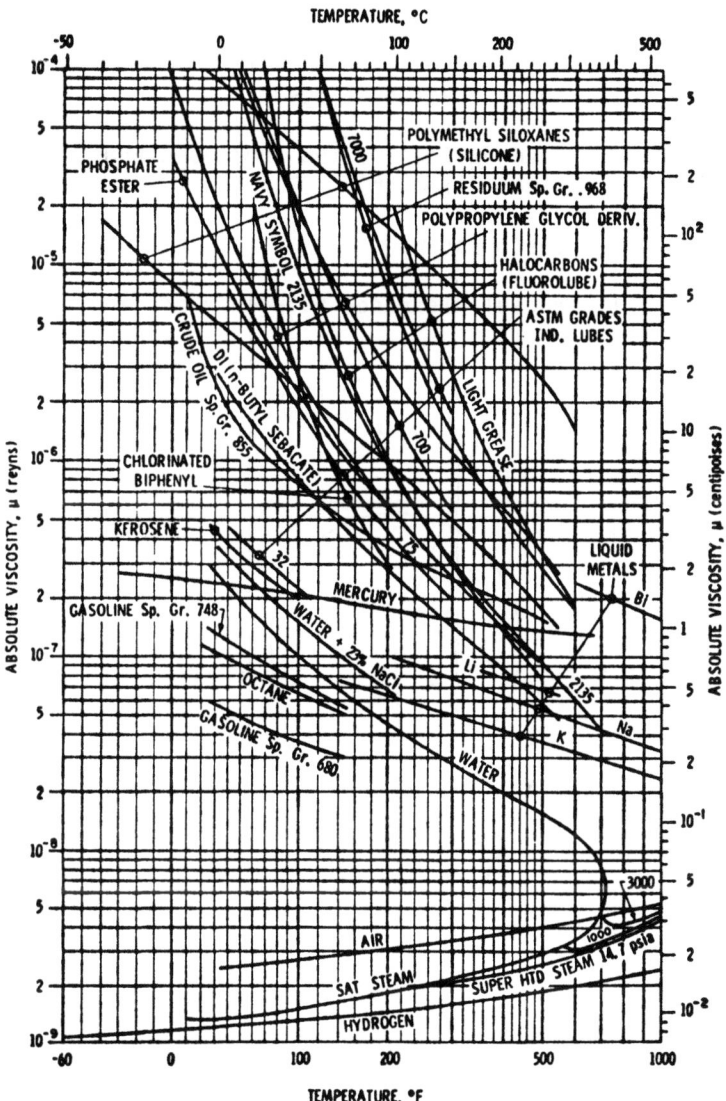

FIGURE 11(B)   VISCOSITY-TEMPERATURE CURVES FOR A NUMBER OF FLUIDS (REF. 8)

$$S \equiv \frac{\mu N}{P}\left(\frac{r}{c}\right)^2$$

where μ = lubricant viscosity

N = shaft rotational speed

P = load divided by the projected area (length times diameter)

r = shaft radius

c = radial clearance between shaft and bearing

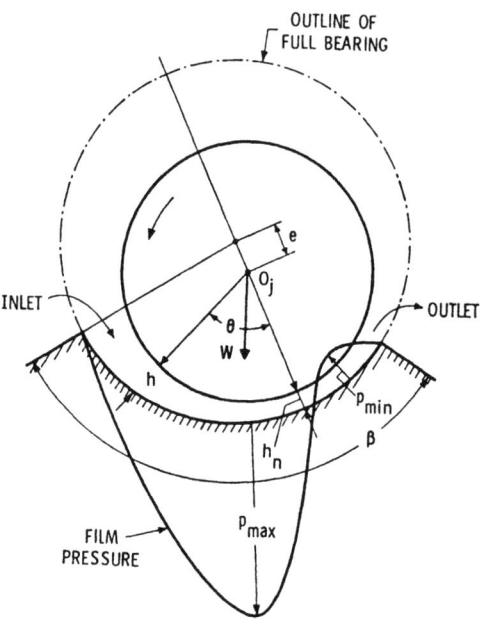

FIGURE 12   JOURNAL BEARING NOMENCLATURE (REF. 8)

The length-to-diameter ratio of the bearing is L/D' and the bearing angle is β. A bearing angle of 2π represents a full journal bearing surrounding the shaft and a bearing angle less than 2π represents a bearing shell which only surrounds part of the shaft.

In the design charts available, a number of dependent dimensionless groups are considered. These groups include: the dimensionless film thickness variable which is the minimum film thickness divided by the bearing radial clearance; ho/C; the eccentricity, E, which is one minus the film thickness variable

$$\varepsilon = 1 - h_o/C;$$

the temperature rise variable

$$\delta C_p \Delta t/P$$

where Δt is the average temperature rise of the lubricant and C the specific heat;

the total flow variable

$$Q/rcNL$$

representing the lubricant flow rate required to maintain the film, Q, the side flow variable

$$Q_s/Q$$

representing the portion of lubricant that flows out the end of the bearing; Q_s,

the friction variable

$$f\frac{r}{c}$$

which is a measure of the effective coefficient of friction, f, resulting from the viscous shearing action of the film; and the attitude angle of the bearing, ϕ, which is the angle between the load vector and the location of minimum film thickness.

The attitude angle with the eccentricity ratio gives a measure of the relative positions of the shaft center and the bearing center. In some solutions, the location and value of the maximum pressure and the zero pressure of the film are also given.

For the purposes of illustration and introducing the ideas involoved, the film thickness chart and the temperature rise chart for the full 360° journal bearing in a range of length-to-diameter ratios are presented in Figures 13 and 14. In each of these cases, it is seen that the abscissa is the Sommerfeld number or bearing characteristic number, which includes the viscosity, the rotational speed, load, and the

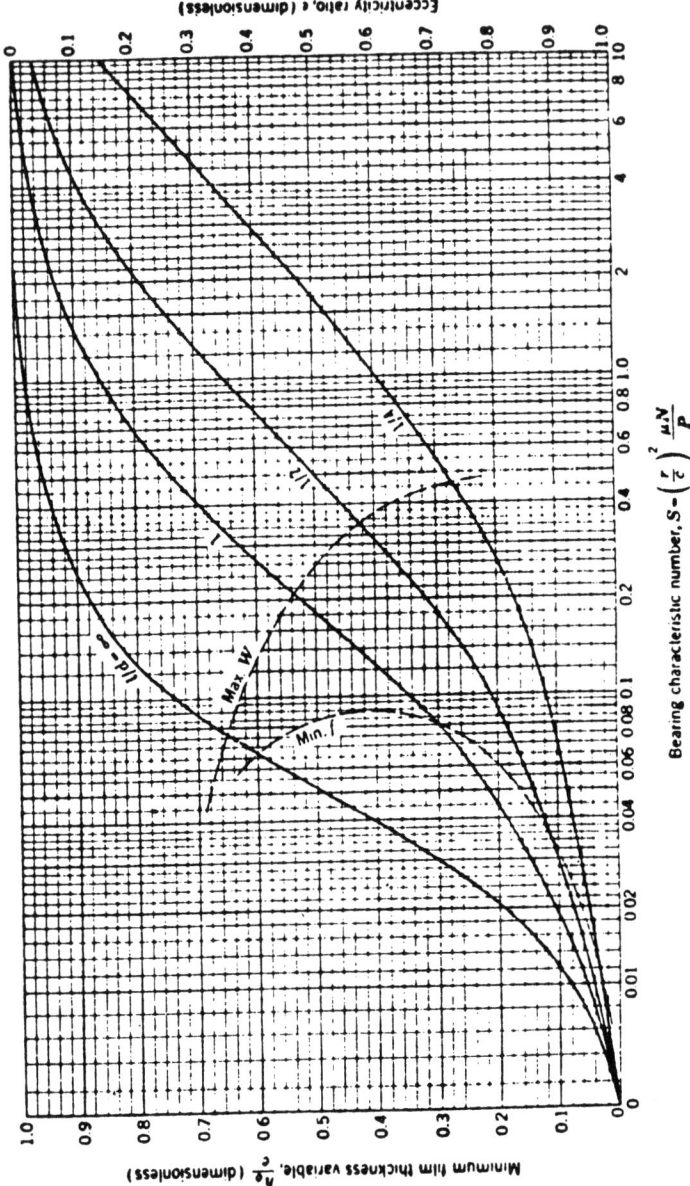

FIGURE 13 CHART FOR MINIMUM-FILM THICKNESS VARIABLE AND ECCENTRICITY RATIO. THE LEFT BOUNDARY OF THE ZONE BETWEEN THE MAX. W AND MIN. F CURVES DEFINE THE OPTIMUM h_o FOR THE MINIMUM FRICTION; THE RIGHT BOUNDARY IS THE OPTIMUM h_o FOR MAXIMUM LOAD (REF. 2). AS ADAPTED FROM RAIMONDI AND BOYD (REF. 8)

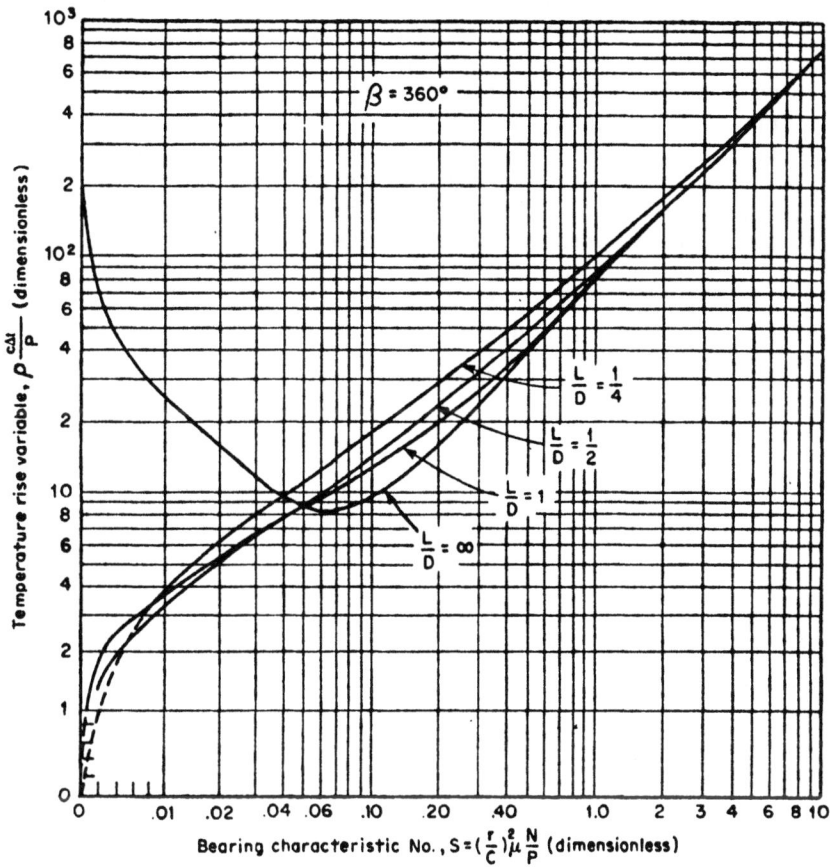

FIGURE 14 TEMPERATURE RISE VARIABLE FOR FULL JOURNAL BEARINGS (REF. 2) (ADAPTED FROM RAIMONDI AND BOYD REF. 8)

clearance ratio for the bearing. The dependent variable or ordinate on Figure 13 is the minimum film thickness ratio of the eccentricity ratio while on Figure 14 the ordinate is the dimensionless temperature rise variable. The temperature rise variable is based on the assumption that all of the energy dissipated by viscous shearing of the film is carried away by the flow of lubricant through the film and none of that energy is conducted to the solid surfaces. This is a conservative estimate in most applications. The temperature rise variable is important because the temperature rise in the film causes a reduction in the lubricant viscosity and the viscosity enters into the bearing characteristic number. Therefore, an iterative solution is required if one is to consider the influence of temperature rise on the bearing operation.

To illustrate the use of these charts consider the following operating conditions for a full journal bearing;

$r = 25.4$ mm

$c = 38$ μm

$L = 51$ mm

$W = 4480$ N

$N = 30$ rps

$\mu = 200$ mPa.sec (e.g. a typical SAE 20 lubricant at 58C)

The average pressure P in the bearing is

$$P = \frac{W}{2rL} = 1.73 \text{ MPa}$$

The Sommerfeld number, also known as the bearing characteristic number S, needed in Figure 13 is;

$$S = \left(\frac{r}{c}\right)^2 \frac{\mu N}{P} = \left(\frac{0.0254}{38 \times 10^{-6}}\right)^2 \frac{20 \times 10^{-3} \times 30}{1.73 \times 10^{-6}} = 0.155$$

From Figure 13 for $S = 0.155$ and $L/d = 1$,

$$\frac{h_o}{c} = 0.46$$

Therefore, the minimum film thickness is

$$h_o = 0.46 \times 38 \times 10^{-6} = 17.5 \text{ μm}$$

The accuracy of h_o using the above simple procedure depends on how accurately one can estimate the effective viscosity, μ, based on an estimated effective film temperature. Because of the strong dependence of viscosity on temperature, the influence of increased effective temperature on film thickness should be checked. Assuming the heat generated by viscous action is available for increasing the lubricant temperature and that the mean temperature of the lubricant leaving the sides of the bearing is equal to the average of the inlet and exit temperatures, the temperature rise of the lubricant is a function of the Sommerfeld number, S, and L/d. The dimensionless temperature rise, $\rho c \Delta t /P$, is presented in Figure 14.

The relationship between the effective film temperature (T_{eff}), used to determine the effective film viscosity, and the temperature rise (Δt) depends on the size of the bearing, the method of lubricant feed, and the thermal characteristics of the bearing. Small bearings (D $\leq$ 75 mm = 3 in.) tend to be nonadiabatic and a recommended relationship between the temperature rise (Δt) and the effective film temperature (T_{eff}) is

$$T_{eff} = T_i + \frac{\Delta t}{2} \quad (D \leq 75 \text{ mm})$$

which is the relationship assumed in constructing Figure 14. However, larger bearings (typically D > 75 mm = 3 in.) fed with an adequate supply of lubricant tend to operate nearly adiabatically and therefore the recommended effective working temperature (T_{eff}) is taken as

$$T_{eff} = T_i + \Delta t \quad (D > 75 \text{ mm})$$

In both cases, T_i is the inlet temperature of the lubricant supplied to the bearings.

The effective temperature is used to determine the effective film viscosity from a viscosity temperature plot for the lubricant. This viscosity is then used to calculate the Sommerfeld number, S, to determine the thermally corrected minumum film thickness, h_o, from Figure 13.

The determination of the effective operating temperature requires an iterative approach. The approach suggested is to plot the effective temperature of the bearing, as a function of lubricant viscosity, on the same graph as the viscosity temperature data for the lubricant. Where this bearing characteristic curve crosses the lubricant data curve will give the effective film temperature and effective viscosity for that bearing-lubricant combination.

The lubricant property, ρc, the volumetric specific heat, does not vary greatly with temperature within a given chemical class of lubricants. For hydrocarbon mineral oil lubricants, which are the most common, this property is 1.36 MPa/C, 1.36 x $10^6 N/m^2/C$ (107 lb/in.^{2}F).

For the above journal bearing example (e.g. D = 51 mm, L/D = 1, therefore use Figure 14 for the effective lubricant temperature)

$$S = \frac{\mu N}{P} \left(\frac{r}{C}\right)^2 = 7.74 \mu$$

if the viscosity is in units of Pa.sec. By selecting values of viscosity, the above expression along with Figure 14, can be used to calculate the effective temperature as a function of viscosity. If the inlet temperature is 55C for this example, the values of this calculation are shown below:

μ Pa.sec	S	$\frac{\rho c \Delta t}{P}$	$\frac{\Delta t}{2}$/C	T_{eff}/C
7 x 10^{-3}	0.054	8.7	5.5	60.5
14 x 10^{-3}	0.108	13.0	8.3	63.0
28 x 10^{-3}	0.217	21.0	13.4	68.0
56 x 10^{-3}	0.433	36.0	23.0	78.0
100 x 10^{-3}	0.774	65.0	41.0	96.0

If these are plotted on a viscosity temperature curve for the lubricant where the bearing curve crosses, the lubricant curve indicates the effective lubricant viscosity and temperature for that combination of bearing and lubricant. If the lubricant were a typical SAE 10, the effective viscosity and temperature would be 14 mPa.sec and 62°C. Substituting this viscosity into the Sommerfeld Number for this example gives S = 0.11 and from Figure 13, h_o/c = 0.37 so the thermally corrected h_o is 14 μm for a reduction of 20% compared to the isothermal calculation. The above method assumes perfect axial alignment between the shaft and bearing. Once this average effective bearing characteristic is obtained, one can go to the other design charts that are available and determine the viscous torque on the shaft, the flow rate of lubricant required, the shaft center location relative to the bearing center location, as well as the location and magnitude of the maximum pressure and the location of the beginning of film rupture.

4.2 Slider Bearings

Dimensionless design charts obtained from solutions of the Reynolds' Equation are also available for slider geometry. They exist for both the fixed geometry slider pad and the tilting pad bearing. Complete sets of the design charts can be found in the literature, i.e., the solutions of Ramondi and Boyd[8]. For the purposes of introductory example, the fixed pad slider will be introduced in Figure 15. In this case, the dimensionless independent parameters are the bearing characteristic number,

$$K_f = \frac{1}{m^2} \frac{\mu U}{PB}$$

and the length to width ratio of the bearing represented by $\frac{L}{B}$.

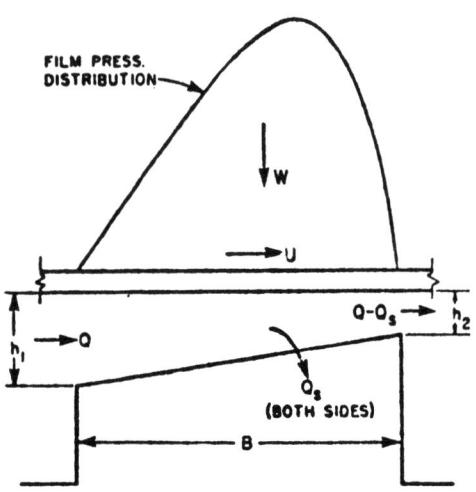

FIGURE 15 DIAGRAMMATIC SKETCH OF FIXED-PAD BEARING (REF. 2)

The bearing characteristic number, K_f, and the width to length ratio, L/B, are dimensionless where

B = length of pad in direction of motion, (m)

L = width of pad perpendicular to direction of motion, (m)

m = slope of pad surface = $(h_1 - h_2)/B$, (dimensionless)

h_1 = inlet film thickness, (m)

h_2 = outlet or minimum film thickness, (m)

U = velocity of sliding surface, (m/sec)

P = load per unit area = W/LB, (N/m^2)
W = load on pad, (N or lb)

μ = average lubricant viscosity, (N.sec/m^2)

As in the journal bearing case, slider bearing design charts are available for several independent dimensionless groups including the minimum film thickness variable and the temperature rise variable, as shown in Figures 16 and 17

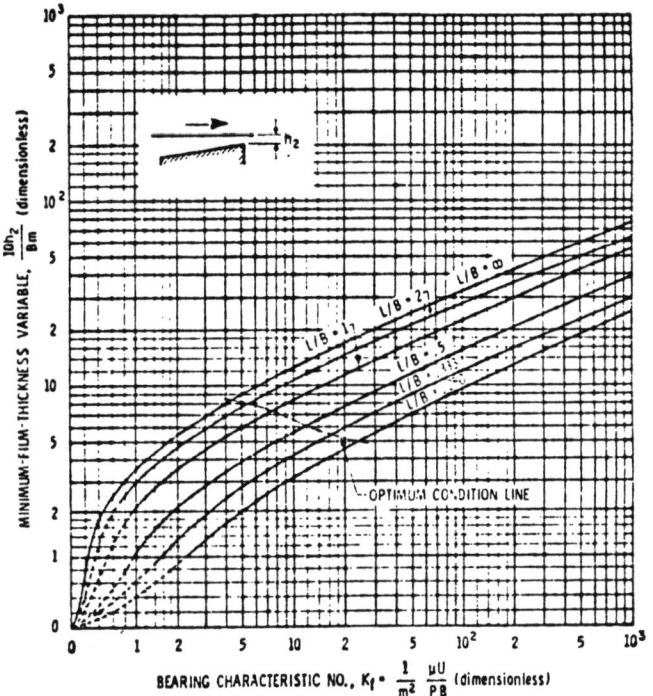

FIGURE 16   CHART FOR DETERMINING MINIMUM FILM THICKNESS (REF. 8)

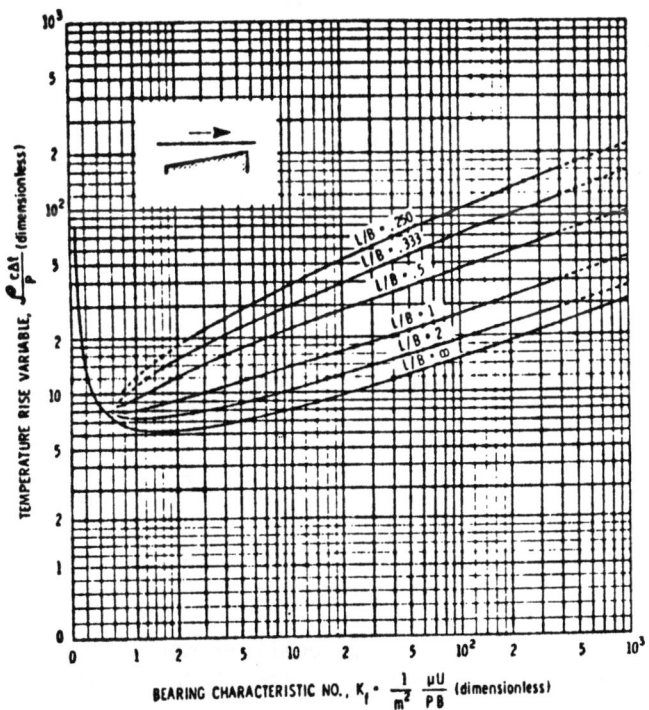

FIGURE 17   CHART FOR DETERMINING TEMPERATURE RISE (REF. 8)

respectively. In the reference literature, design charts for the dimensionless flow rate, side flow, and friction variable are also available.

The use of these charts can best be seen by the following example involving the analysis of a simple single slider with temperature dependent viscosity as the lubricant.

For example, consider the following case:

W = 1780 N

U = 30.5 m/sec

B = 0.0254 m

L = 0.0254 m

m = 0.001 = $(h_1 - h_2)/B$ (dimensionless)

and a typical mineral oil supplied at 70 °C.

As in the preceeding example with the journal bearing, the bearing behavior is calculated for a range of viscosities. These calculated results are shown as follows:

$\mu / \frac{N \cdot sec}{m^2}$	K_f	$\frac{\rho c \Delta t}{P}$	$\frac{\Delta t}{2}/C$	T_{ave}/C
10^2	4.35	10	10	80
2×10^{-2}	8.70	11.8	12	82
4×10^2	17.40	14.0	14	84
8×10^2	34.8	17.0	17	87
10^3	43.5	18.0	18	88

If these were plotted on a viscosity-temperature curve for various SAE oils, the resulting operating points would be found:

SAE Grade	T_{ave}	K_f	$\dfrac{10 \text{ hz}}{m}$	h_2
30	81°C	5.66	6.5	16.5 μm
10	79°C	3.39	5.0	12.7 μm

4.3 Squeeze Films

As mentioned in the introductory portion of this section, squeeze films result from the time unsteady behavior of lubricating films. If the load is fluctuating or the surface velocities perpendicular to the plane of the film change, the lubricant must be squeezed out from between the films resulting in squeeze film action. During the time of the squeezing, a pressure is developed in the film which carries a transient load applied to the surface. This mechanism arises from the last term on the right hand of the Reynolds' Equation. Squeeze film behavior in its simplest form has been demonstrated for well over a hundred years and was analyzed and published prior to the Reynolds' Equation case.

In applications of journal bearings or slider bearings with fluctuating loads, such as connecting rod bearings and reciprocating devices, the load histroy is complex and the solution of the problem is complicated. This field is highly specialized and has relatively few persons contributing to it. Recent examples of some of these developments can be found in the references of Booker and co-worker[20, 21]

5. HYDROSTATIC LUBRICATION

Hydrostatic lubrication results from the introduction of a pressurized lubricant into the oil film. This is usually done with a recessed pocket in the oil film and the pressure distribution, resulting from this oil passing out through the oil film, carries the load. Unlike the hydrodynamic lubrication described above, no surface velocity is required to develop the film. The pressure is developed from an external source of a hydraulic pump. From the mathematical standpoint, this means that the right-hand side of the Reynolds' Equation is or can be zero and the boundary conditions are different in that a particular pressure is imposed upon the system. Hydrostatic bearings are particularly advantageous for carrying very high loads with little or no velocity between the two surfaces. A typical hydrostatic bearing configuration is shown in Figures 18 and 19[1].

The hydrostatic bearing system consists of a reservoir and associated filtering system, a hydraulic pump, and a compensating element (or restrictor) between the pump and the pad recess. The lubricant may be fed to several pads from a single pipe through a manifold system with restrictors between the manifold and the individual pads.

As in the case of journal bearings and slider bearings, hydrostatic bearings are also a highly developed field in which design curves can be obtained in the literature[1,8,22].

The design of a hydrostatic bearing involves the behavior of a single pad and the interaction of several single pads and compensator systems. Both the relative size of the recess in the pad and the pressure drop across the compensator influence the performance of the bearing. The performance is measured in terms of flow rate, power consumption, and stiffness. The stiffness is the derivative of the load with respect to change in film thickness. The behavior of these quantities is given in terms of dimesionless pressure, flow, and power factors for a given geometry as illustrated in Figure 20 for a single circular pad with a circular recess. One can see from Figure 20 that the minimum power consumption, that is pump power required, will occur with a dimensionless pocket radius of about 0.5 in this particular case. In the literature cited, similar dimensionless design curves can be found for a variety of different pad and pocket geometries.

As illustrated in Figure 21 the pads can be arranged around a shaft to result in a hydrostatic journal bearing as well as the linear slider bearings implied above[1].

6. GAS BEARINGS

Gases can be viewed as fluids with very low viscosity and variable density. There is no reason why the previously discussed Reynolds' Equation cannot be applied to gases as lubricants. Because of the very low viscosity of the gases, the operating conditions suitable for gas film lubrication are different from those associated with liquid film lubrication. However, because of the very low viscosity, one would expect low friction, low energy dissipation, and low temperature rise. Gases afford a wide range of temperature for applications because of the high thermal stability of gases at elevated temperatures as well as low temperature capabilities beyond that normally thought of with most liquids. However,

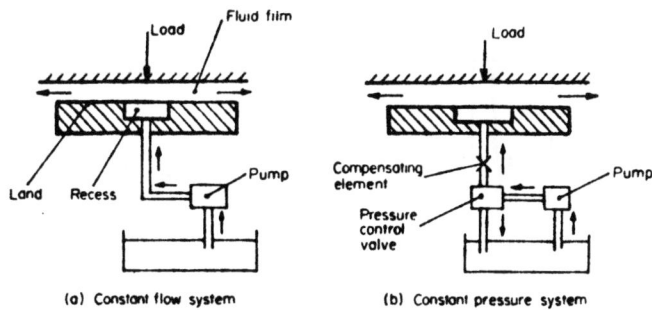

FIGURE 18   TYPICAL ARRANGEMENTS OF HYDROSTATIC BEARING SYSTEMS (REF. 1)

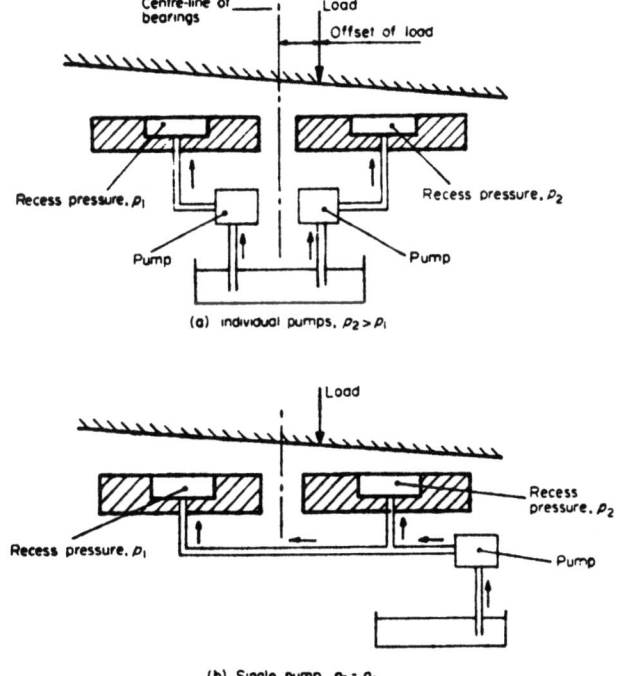

FIGURE 19   MULTIBEARING SYSTEMS (REF. 1)

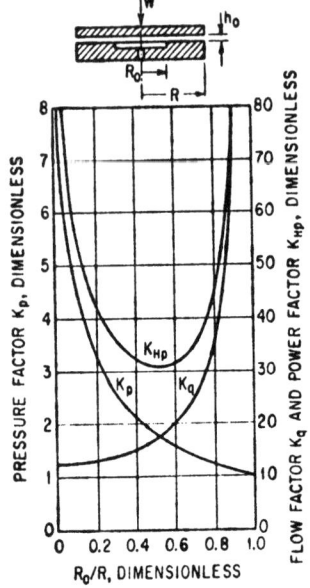

$$p_\bullet = H_p \frac{W}{\pi R^2}$$

$$Q = K_q \frac{W}{\pi R^2} \frac{h^3}{12\mu}$$

$$Hp = K_{Hp} \left[\frac{W}{\pi R^2}\right]^2 \frac{h^3}{12\mu}$$

FIGURE 20   PERFORMANCE FACTORS FOR CIRCULAR STEP BEARING (REF. 8)

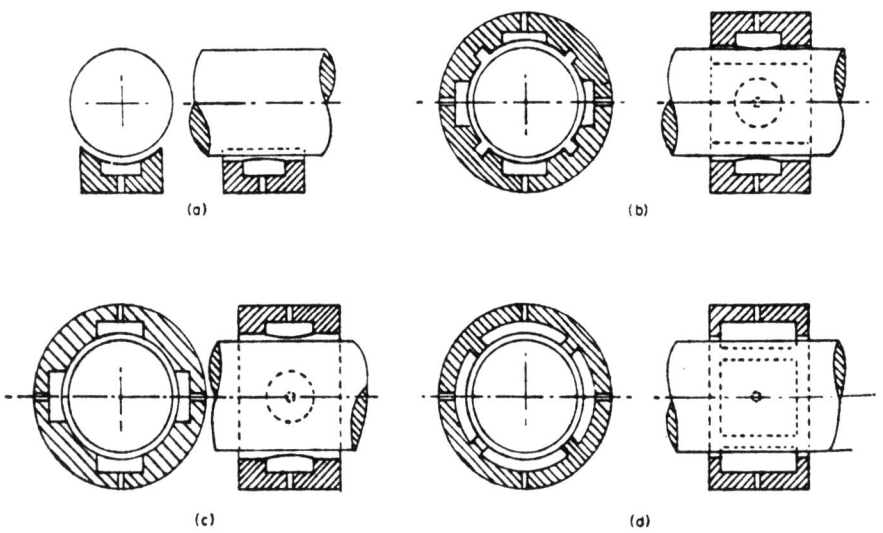

FIGURE 21   EVOLUTION OF HYDROSTATIC JOURNAL BEARINGS (REF. 1)

the disadvantages of gas film lubrication are also associated with low viscosity, which requires either thin films or high velocities and low loads to find suitable operating conditions. Also in the case of a gas as a lubricant, there is no boundary lubricant or extreme pressure lubricant present that can act as a backup lubricant when surface contact occurs. Because of the compressibility of the lubricant and the very low dissipation associated with low viscosity, gas bearings have more of a tendency to become dynamically unstable than liquid film bearings. The dynamic instabilities associated with gas bearings are for more important and more complicated than those associated with liquid bearings. Because of the very small film thickness at which gas bearings operate, the manufacturing of gas bearings is difficult requiring high precision in both the manufacturing techniques and the alignment of the bearing. Comparisons of gas, oil and water bearings, and their range of minimum film thickness operation and typical unit loading ranges are shown in Table 1. Recent developments in compliant gas film bearings have permitted the unit loading to be increased from that shown in Table 1 to as much as 50 psi. Design solutions for gas bearings are also available in the literature[13, 23].

Table 1.

Practical Comparison of Bearings Using Various Fluid Lubricants

Fluid	Viscosity, reyns	Typical min film thickness in bearing applications, in.	Typical unit load* P, psi
Oil	3×10^{-6}	0.002 - 0.004	200 - 500
Water	1×10^{-7}	0.0004 - 0.001	25 - 75
Air	0.3×10^{-8}	0.00005 - 0.0004	1 - 10

* Hydrodynamic bearings

The pressure distribution in a gas bearing is different from that in a liquid bearing because of the absence of cavitation in a diverging region. The pressure distribution in a typical journal bearing is shown in Figures 22 and 23[8].

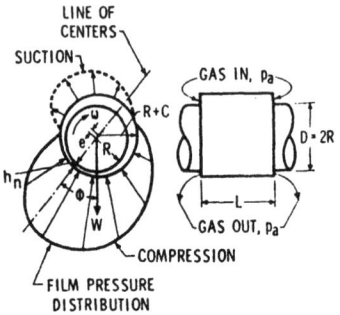

FIGURE 22 FULL JOURNAL BEARING OR PLAIN CYLINDRICAL BEARING (REF.8)

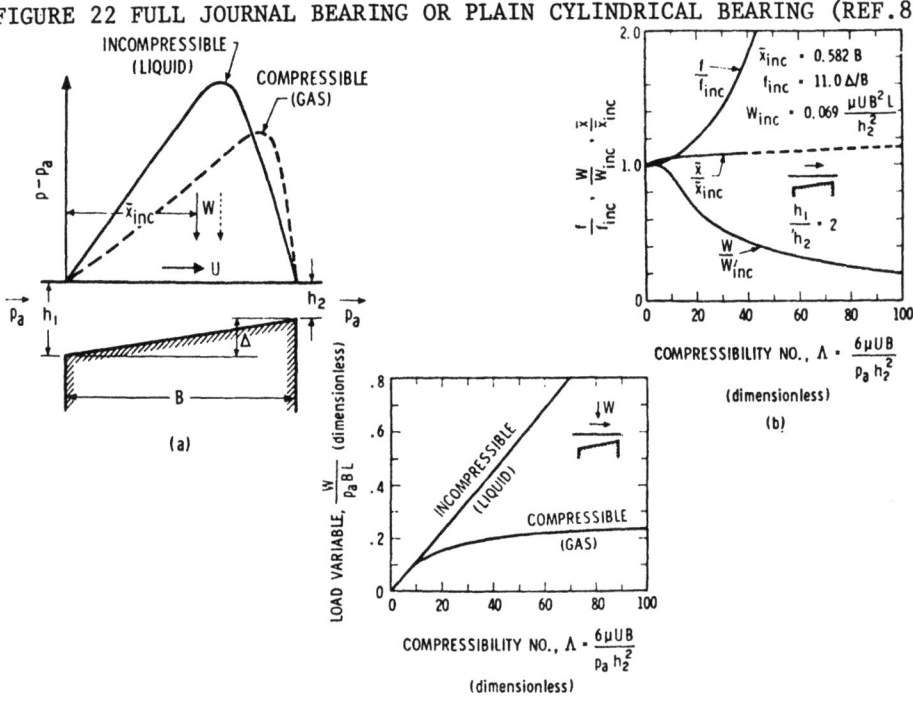

FIGURE 23 TYPICAL COMPRESSIBILITY EFFECTS IN HYDRODYNAMIC BEARINGS (SLIDER BEARING, L/B=1, $h_1/h_2=2$). (A) EFFECT ON PRESSURE DISTRIBUTION. (B) EFFECT ON COEFFICIENT OF FRICTION f= RATIO OF FRICTION FORCE PER POUND OF LOAD CARRIED = RECIPROCAL OF EFFICIENCY; EFFECT ON CENTER OF PRESSURE (OR PIVOT POSITION) $\bar{x}$; AND EFFECT ON LOAD CAPACITY W. (SUBSCRIPT "INC." DENOTES VALUE FOR LIQUID LUBRICATED BEARING.) NOTE THAT BOTH EFFICIENCY AND LOAD CAPACITY OF GAS BEARING DECREASE RELATIVE TO LIQUID BEARING BECAUSE OF COMPRESSIBILITY; CENTER OF PRESSURE MOVES TO REAR. ALSO NOTE THAT FOR SMALL Λ ($\Lambda < 5$, SAY) PERFORMANCE OF GAS AND LIQUID BEARINGS IS IDENTICAL. (C) EFFECT OF LOAD CAPACITY (REF. 8)

The importance of the boundary conditions discussed above in the case of liquid bearings is not as important as gas bearings. However, the density must be assumed to vary with the gas pressure in the bearing, although the viscosity is normally assumed to be constant because of the low energy dissipation and therefore the low temperature rise. In the case of gas bearings, the dimensionless groups are obtained for plotting the design relations for the important operating variables. The primary independent variables in this case are the length-diameter ratio and the compressibility number, Λ, which replaces the Sommerfeld number or the bearing characteristic number in the cases of the journal bearing or the slider bearing respectively. Figure 23 shows a comparison of incompressible liquid and compressible gas in a slider bearing. The difference in pressure distribution, shown in Figure 23(a), indicates that the peak pressure moves towards the exit position of the pad relative to that of the incompressible bearing. In Figure 23(b), the friction load ratio and the center of pressure for compressible versus incompressible bearings, are plotted as a function of compressibility number. The pivot position for a tilting pad bearing is only slightly influenced by the compressibility with the pivot position moving towards the rear of the pad compared to that of the incompressible case. However, the load of the compressible bearing decreases substantially with increasing compressibility number. This influence of the load is shown not only in Figure 23(b) but also in Figure 23(c), which shows that as a compressibility number increases, the load being carried asymptotically approaches a value related to the atmospheric pressure, Pa, for the gas bearing. Figure 24 shows a typical design curve for gas journal bearings in which the two independent parameters are the length-to-diameter ratio and the compressibility number. In this case, the eccentricity ratio is the additional parameter in plotting the results of a load variable versus compressibility number. For each eccentricity ratio, the load being carried approaches an asymptotic value for large values of the compressibility parameter.

The following example gives some indication of the magnitude of variables involved in a gas journal bearing. Consider the following operating conditions for a journal bearing operating on air at one atmosphere pressure and 50° C;

$D = L = 100$ mm

$W = 10^3$ N

$C = 5$ μm

Pa = 100 kPa

W = 100 rad/s.

μ = 2 x 10^{-5} Pas

The load variable is

$$\frac{W}{2RLPa} = \frac{10^3}{0.1 \times 0.1 \times 10^5} = 1.0$$

and the compressibility number is

$$\frac{\Delta}{6} = \frac{\mu W}{Pa}\left(\frac{R}{C}\right)^2 = \frac{2 \times 10^{-5} \times 100}{10^5}\left(\frac{5 \times 10^{-2}}{5 \times 10^{-6}}\right)^2 = 2$$

From Figure 24, for these conditions, the eccentricity is found to be about 0.53. Therefore, the minimum film thickness is h_o = ε x C = 0.53 x 5 = 2.6 μm.

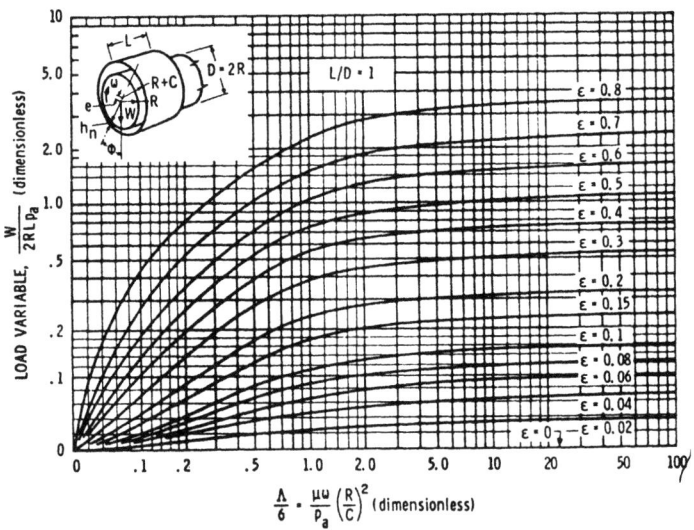

FIGURE 24 LOAD CAPACITY FOR FULL BEARING, L/D=1 (REF. 8)

7. ELASTOHYDRODYNAMIC LUBRICATION

Elastohydrodynamic lubrication occurs in concentrated contacts such as cams, gears, and rolling bearings. The origin of the name comes from the fact that in these cases the load is concentrated over a very small region because the surfaces are nonconformal in nature. These concentrated loads result in sufficiently high local pressures that the surfaces deform elastically in amounts significant relative to the size of the generated hydrodynamic lubrication film thickness. In addition, because of the high pressure the lubricant properties, primarily the viscosity, will also change. Discussions of elastohydrodynamic lubrication can be found in the traditional lubrication texts such as Halling and Cameron[1, 7]. More thorough discussions of elastohydrodynamic lubrication will be found in the works of Dowson and Higginson and Hamrock and Dowson[12, 24].

The geometries associated with elastohydrodynamic lubrication are usually classified as either line contact or elliptical contacts. Sometimes elliptical contacts are also referred to as point contacts which are actually a special case of the elliptical contact. Line contact is that which occurs when two cylinders are loaded against one another as might be expected in the case of gears, cams, or cylindrical roller bearings. The fundamental geometry for line contact is shown in Figure 25[2]. In the highly loaded line contact case, the pressure distribution in the film is assumed to be Hertzian in nature even if the surfaces are separated by a thin film. The film thickness distribution and the pressure distribution are also shown in Figure 25. The most commonly accepted expression for film thickness distribution in line contacts, is that attributed to Dowson and Higginson and shown in Equation (3) in which the dimensionless film thickness, h_{min}/R, is shown to be a power law function of a dimensionless material parameter, G, the dimensionless speed parameter, U, and a dimensionless load parameter, W.

$$\frac{h_{min}}{R} = 2.65 \frac{G^{0.54} U^{0.7}}{W^{0.13}}$$

where $G = \alpha E$ dimensionless material parameter

$U = \dfrac{\mu_o (u_1 + u_2)}{2ER}$ dimensionless speed parameter

$W = \dfrac{w}{ERL}$ dimensionless load parameter

α = pressure viscosity coefficient based on the piezo-viscous relation $\mu = \mu_o e^{\alpha p}$

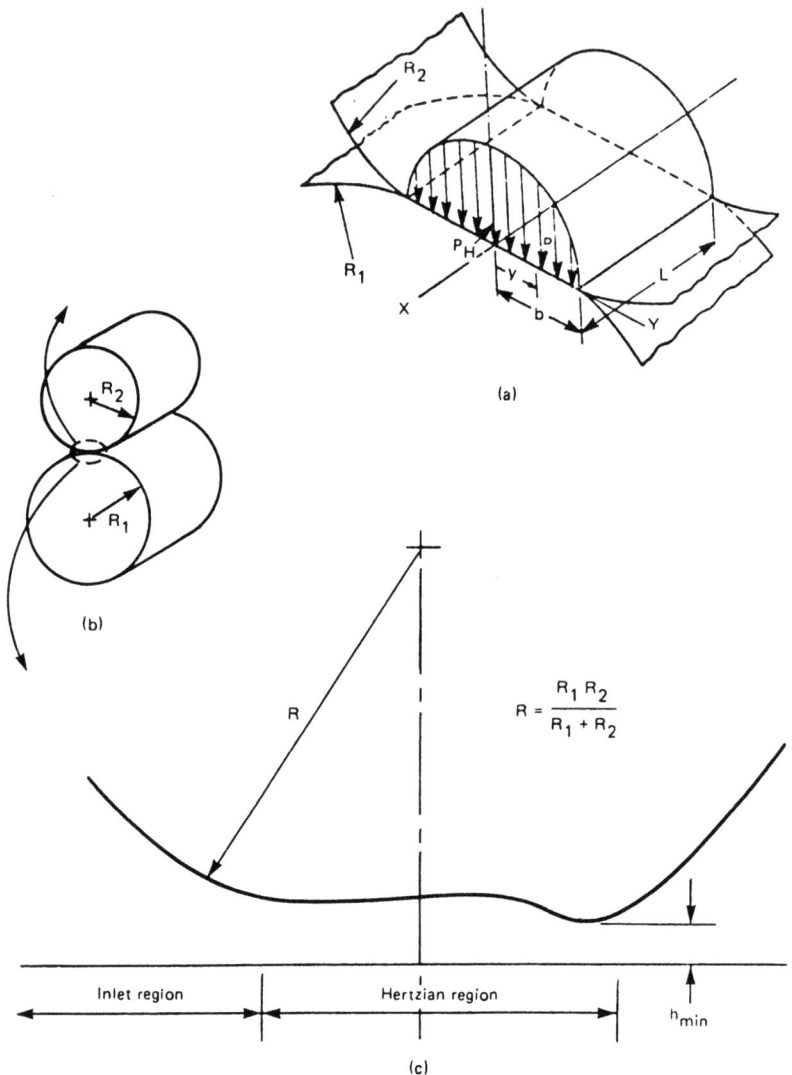

FIGURE 25 GEOMETRY OF A LINE CONTACT (REF. 2)
  (A) HERTZIAN PRESSURE DISTRIBUTION
  (B) CONTACTING CYLINDERS
  (C) EHD FILM THICKNESS DISTRIBUTION

$$= \frac{1}{\mu}\frac{\partial \mu}{\partial P}\bigg|_{p=0} \quad (m^2/N) \quad (e.g., \alpha_{OT})$$

or

$$= \alpha^* = \left[\int_0^{p-\infty} \frac{\mu_o}{\mu(P)} dp\right]^{-1} \quad (m^2/N)$$

$$\frac{1}{E} = \frac{1}{2}\left(\frac{1-v_1^2}{E_1} + \frac{1-v_2^2}{E_2}\right) \quad (m^2/N)$$

μ_o = lubricant viscosity (Pa.sec, N.sec/m) at inlet surface temperature

u_1, u_2 = surface velocities relative to contact region (m/sec)

$$R = \frac{R_1 R_2}{R_1 \pm R_2} \quad (m)$$

where the plus assumes external contact (both surfaces convex) and the minus internal contact (the surface with the larger radius of curvature is concave).

w = total load on the cylinder (N)

L = length of the cylinder (m)

One of the more interesting aspects of elastohydrodynamic film thickness is the rather small dependence of the film thickness on the load (e.g., to the -0.13 power). This expression shows that for a given system the film thickness is primarily a function of the speed parameter which is primarily dependent upon the viscosity and the average surface velocity.

In elastohydrodynamic lubrication, two physical lubricant properties are important in determining the film thickness. They are the viscosity in the inlet region and the pressure viscosity coefficient of the lubricant. Both the viscosity and the pressure viscosity coefficient are functions of temperature and are discussed in more detail in the following chapter entitled, "Lubricants."

The applicability of Equation (3) can be seen from an example of two steel rollers in external contact with a typical SAE 10 oil at about 40°C:

$L = 100$ mm

$R_1 = R_2 = 76$ mm

$w = 35.3$ kN

$E_1 = E_2 = 207$ GPa

$v_1 = v_2 = 1/3$

$u_1 = u_2 = 12.7$ m/s

$\mu_o = 27.6$ mPas

$\alpha = 14.5$ GPa^{-1}

Therefore

$R = 38$ mm

$E = 233$ GPa

$G = 3379$

$U = 3.96 \times 10^{-11}$

$W = 3.987 \times 10^{-11}$

and

$$\frac{h_{min}}{R} = 1.58 \; \mu m$$

At high surface speeds the lubricant viscous energy dissipation in the inlet region is sufficient to cause an increase in the temperature of the lubricant resulting in a decrease in the inlet viscosity. This phenomena is referred to as the thermal film thickness reduction and can be related to the energy dissipation in the inlet and the thermal conductivity of the lubricant. A thermal reduction factor, which is the ratio of the film thickness resulting when energy dissipation occurs in the inlet, divided by the isothermal film thickness as a function of a thermal reduction factor is attributed to Cheng and shown in Figure 26[2].

The thermal reduction factor Q_m is defined as follows:

$$Q_m = \frac{\mu_o u^2 \delta}{k_f} \text{ (dimensionless)}$$

where

μ_o = lubricant vicosity at atmospheric pressure and the inlet surface temperature ($\frac{N \cdot sec}{m^2}$)

$\mu = \frac{\mu_1 + \mu_2}{2}$ the average surface velocity (m/sec)

δ = lubricant viscosity temperature collection $\frac{1}{\mu}\frac{\Delta\mu}{\Delta T}(C^{-1})$

k_f = lubricant thermal conductivity ($\frac{N}{sec\ K}$ which is $\frac{W\ m}{m^2 K}$)

In the above example, with a typical SAE 10 mineral oil for which the thermal conductivity is about 0.13 W/mK and the viscosity-temperature coefficient is about 0.045 C^{-1}, the Q_m = 1.6 and the thermal reduction is ϕ_T = 0.76. Therefore, the thermally corrected film thickness would be $h_{oT} = h\phi_T$ = 1.2 μm.

Concentrated contacts are not always elastohydrodynamic in the sense that both the elasticity and the viscosity variation of the lubricant with pressure is important. In the case of line contacts, there are three other possible regimes of operation. They are: 1) the case where neither elastic deformation nor viscosity change are important, which is referred to as the rigid isoviscous case, (R-I); 2) where the surfaces can be assumed to be rigid and the viscosity varies with pressure (R-V); and 3) where the elasticity of the surfaces results in significant deformation of the surface but the pressures are low enough that the viscosity can be treated as an isoviscous material (E-I). The equations governing the film thickness in these different regimes are quite different in their nature. The regimes can be divided in terms of two dimensionless variables, g_1 and g_3, which are shown in Figure 27[2]. The governing film thickness equations for the different regimes are presented as follows:

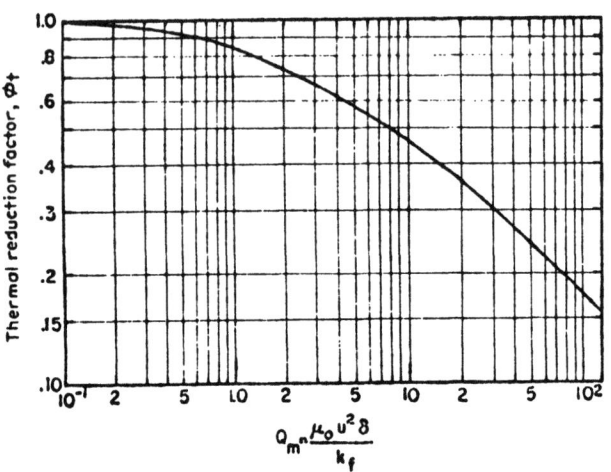

FIGURE 26   THERMAL REDUCTION FACTOR (REF. 2)

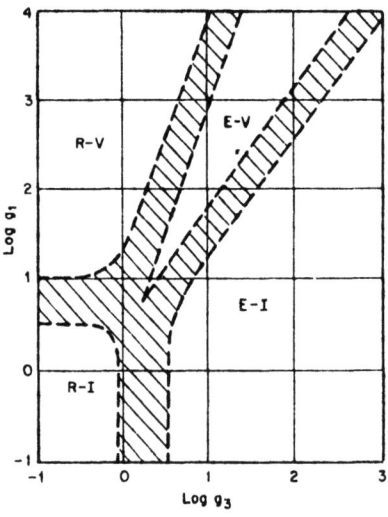

FIGURE 27   REGIMES OF LUBRICATION FOR LINE CONTACTS (REF. 2)

E-V: the elastic solid-variable viscosity lubricant regime which is assumed for the film thickness Equation (3). In terms of the variables used here Equation (3) becomes

$$h' = 2.65 \, g_1^{0.54} g_3^{0.06}$$

E-I: elastic solid-isoviscous lubricant regime where the film thickness equation is

$$h' = 3.0 \, g_3^0$$

R-V: rigid solid-variable lubricant regime where the film thickness equation is

$$h' = 1.66 \, g_1^{2/3}$$

R-I: rigid solid-isoviscous lubricant regime where the film thickness equation is

$$h' = 4.9$$

In Figure 27 and the above film thickness equations, the following definitions have been used,

$$h' \equiv \frac{2h_{min} \, w}{\mu_o R(u_1 + u_2)} \quad \text{dimensionless minimum film thickness}$$

$$g_1 \equiv \left[\frac{2\alpha^2 w^3}{\mu_o R^2 (u_1 + u_2)} \right]^{1/2}$$

$$g_3 \equiv \left[\frac{2w^2}{\mu_o ER(u_1 + u_2)} \right]^{1/2}$$

$$w \equiv \frac{\text{load}}{\text{length of contact}}$$

Elliptical point contacts result when two curve surfaces with curvatures in both directions come in contact, such as a ball bearing or many gear configurations. The generalized geometry for elliptical contact is shown in Figure 28. The governing equations appropriate for the elliptical contact attributed to Hamrock and Dowson are presented as follows[24];

$$H_{c.F} = 2.69 \ U^{0.67} G^{0.53} W^{-0.067} (1 - 0.61 e^{-0.73k}) \quad (4a)$$

$$H_{min.F} = 3.63 \ U^{0.68} G^{0.49} W^{-0.073} (1 - e^{-0.68k}) \quad (4b)$$

where

$H_{c.F} = h_{c.F}/R_x$ (dimensionless)

$h_{c.F}$ = central film thickness for flooded contacts

$H_{min.F} = h_{min.F}/R_x$ (dimensionless)

$h_{min.F}$ = minimum film thickness for flooded contacts

$k = \dfrac{a}{b}$ = elliptically parameter

$R_x = R_{x1} R_{x2}/(R_{x1} + R_{x2})$ (m)

$R_y = R_{y1} R_{y2}/(R_{y1} + R_{y2})$ (m)

a = semi-axis of contact ellipse perpendicular to the motion (m)

b = semi-axis of contact ellipse parallel to the motion (m)

$W = \dfrac{w}{ER_x^2}$ (dimensionless)

w = the total load (N)

$U = \dfrac{\mu_o (u_1 + u_2)}{2 \ ER_x}$ (dimensionless)

$G = \alpha E$ (dimensionless)

$\dfrac{1}{E} = \dfrac{1}{2} \left(\dfrac{1 - v_1^2}{E_1} + \dfrac{1 - v_2^2}{E_2} \right)$ (m^2/N)

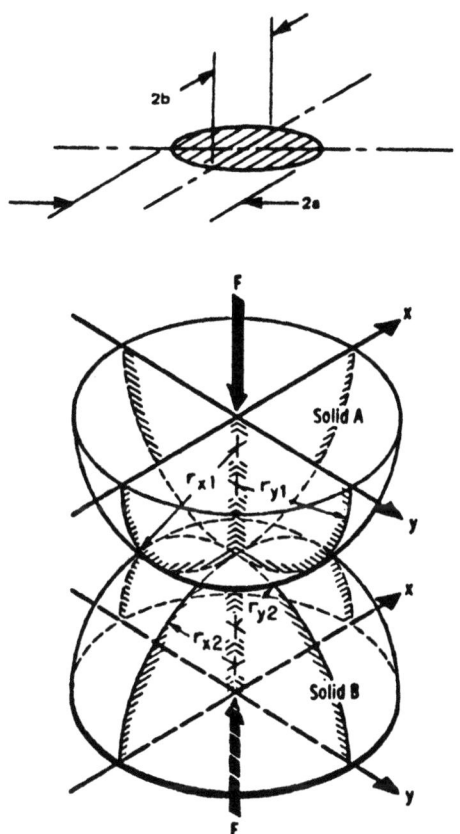

FIGURE 28   GEOMETRY OF CONTACTING ELASTIC SOLIDS (REF. 24)

E_1, E_2, v_1, v_2 = moduli of elasticity and Poission's ratio of body 1 and 2. The surface velocities, u_1 and u_2, lubricant inlet viscosity, μ_o, and pressure viscosity coefficient, α, are defined for line contacts.

The radii of curvature are assumed positive if the line surface is convex and negative if concave. The radius of curvature for a flat surface is infinite.

For example, consider the following elliptical contact. The principal radii of an EHD point contact between a 25.4 mm diameter steel roller with a 51 mm crown radius (surface 2) and a 76.2 mm diameter steel cylinder with no crown radius (surface 1) are

$R_{x1} = 38$ mm

$R_{x2} = 12$ mm

$R_{y1} = \infty$ m

$R_{y2} = 51$ mm

The load and speeds are

$w = 445$ N

$u_1 = 12.7$ m/sec

$u_2 = 12.7$ m/sec

$\mu_o = 27.6$ mPa.sec

$\alpha = 14.5$ GPa^{-1}

$R_x = \dfrac{R_{x1} R_{x2}}{R_{x1}+R_{x2}} = \dfrac{38 \times 12.7}{38 + 12.7} = 9.52$ mm

$R_y = \dfrac{R_{y1} R_{y2}}{R_{y1}+R_{y2}} = 51$ mm

$E = 233$ GPa

$G = \alpha E = 3373$

$U = \dfrac{\mu_o(u_1 + u_2)}{2 \times E \times R_x} = \dfrac{27.6 \times 10^{-3}(12.7+12.7)}{2 \times 233 \times 10^9 \times 9.52 \times 10^{-3}} = 1.58 \times 10^{-10}$

$W = \dfrac{w}{ER_x^2} = \dfrac{445}{233 \times 10^9 \times (9.52 \times 10^{-3})^2} = 2.11 \times 10^{-5}$

$$k = \frac{a}{b} = 1.03 \, (R_y/R_x)^{0.64} \quad \text{[See section 3.3, Equation 16 of Reference (2)]}$$

$$= 1.03 \, (51 \times 9.52)^{0.64}$$

$$= 3.015$$

Then Equation (4a) gives

$$H_{c.F} = 2.69 \times (1.58 \times 10^{-10})^{0.67} \, (3373)^{0.53}$$

$$(2.11 \times 10^{-5})^{-0.067}$$

$$\times (1 - 0.61 \times e^{-0.73 \times 3.015})$$

$$= 1.037 \times 10^{-4}$$

$$h_{c.F} = 9.52 \times 10^{-3} \times 10.7 \times 10^{-4} = 0.98 \; \mu m$$

and Equation (4b) gives

$$H_{min.f} = 3.63 \times (1.58 \times 10^{-10})^{0.68} \, (3373)^{0.49}$$

$$(2.11 \times 10^{-5})^{-0.073}$$

$$\times (1 - e^{-0.68 \times 3.015})$$

$$= 8.04 \times 10^{-5}$$

$$h_{min.F} = 8.04 \times 10^{-5} \times 9.52 \times 10^{-3}$$

$$= 0.76 \; \mu m$$

As in the case of line contacts, there are different regimes of operation for the elliptical point contacts. These regimes have been developed by Hamrock and Dowson and are shown for the case of an ellipticity ratio of one in Figure 29[24]. The four different regimes involved are the same as those in the case of the line contact, however, in the case of point contact there is the added variable of the ellipticity ratio of the contact. Additional plots for other ellipticity ratios can be found in Reference 24. The appropriate

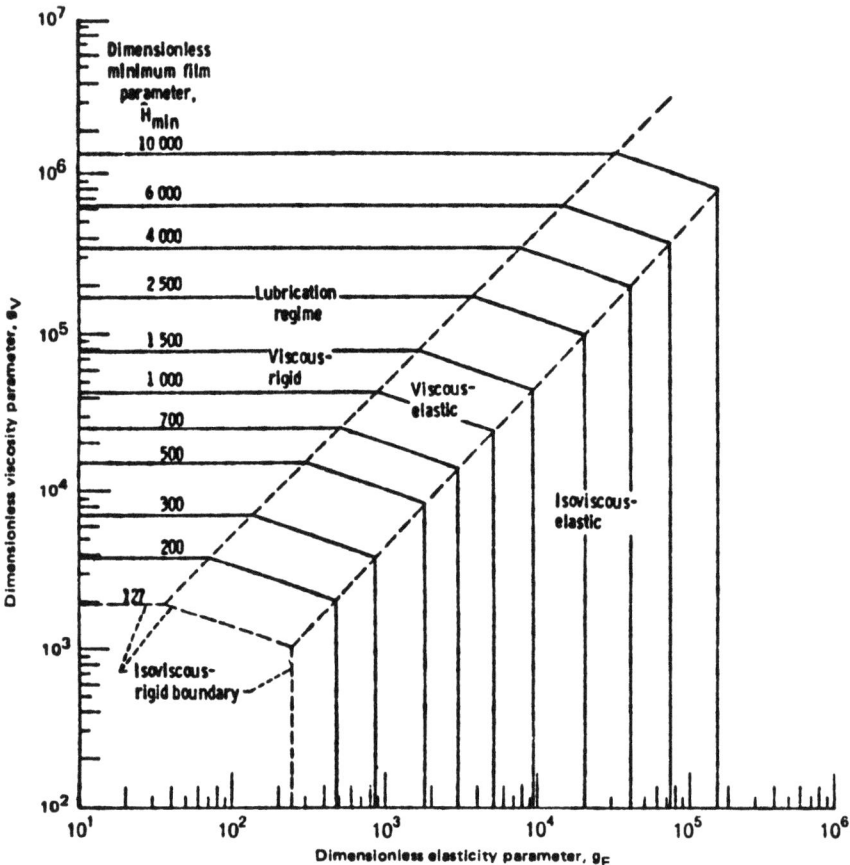

FIGURE 29 MAP OF LUBRICATION REGIMES WITH DIMENSIONLESS-FILM-PARAMETER CONTOURS ON LOG-LOG GRID OF DIMENSIONLESS VISCOSITY AND ELASTICITY PARAMETERS FOR ELLIPTICITY PARAMETERS OF 1 (k = 1) (REF. 24)

equations for the different regimes of the elliptic contact are given below:

IR: (Isoviscous-rigid regime)

$$\{H_{min}\}_{IR} = 128\beta \, \Phi^2 \, [0.13 \, \tan^{-1}(\tfrac{\beta}{2}) + 1.68]^2$$

where

$$\beta = \frac{R_y}{R_x} \simeq (\frac{k}{1.03})^{1/0.64}$$

$$\Phi = (1 + \frac{2}{3\beta})^{-1}$$

VR (Viscous-rigid regime):

$$\{H_{min}\}_{VR} = 1.66 \, g_v^{2/3} \, (1 - e^{-0.68k})$$

IE (Isoviscous-elastic regime):

$$\{H_{min}\}_{IE} = 8.70 \, g_E^{0/67} \, (1 - 0.85 \, e^{-0.31k})$$

VE (Viscous-elastic regime):

$$\{H_{min}\}_{VE} = 3.45 \, g_V^{0.49} \, g_E^{0.17} \, (1 - e^{-0.68k})$$

where all variables are defined as previously in this section except

$$\{\overline{H}_{min}\} = H_{min} \, (\frac{W}{U})^2 = \frac{4w^2 h_{min}}{(u_1 + u_2)^2 \, \mu_o^2 \, R_x^3}$$

where

$$g_v = \frac{GW^3}{U^2}$$

$$g_e = \frac{W^{8/3}}{U^2}$$

The above discussed methods of determining film thickness in elastohydrodynamic lubrication are not the only ones available in the literature. A very useful and simplified approach for determining the elastohydrodynamic film thickness in standard machine components has been developed and presented in the Mobil EHL Design Guide[25]. In this design guide, component materials are assumed to be steel. Certain standard geometries and surface roughnesses are assumed and the operating parameters are given in terms of quantities more familiar to the designers, such as rotational speed of the shaft and load on the shaft. The lubricant properties are lumped into a lubricant parameter which is the product of the viscosity and the pressure viscosity coefficient. When compared with the above equation for elastohydrodynamic lubrication, it is noted that these two parameters are in those equations to two different powers. However, because of the relatively limited range of variation of the pressure viscosity coefficient and the practically small difference between the power on the viscosity and the power on the pressure viscosity coefficient the lumping of these two parameters seems to be quite satisfactory for determining design information.

8. MIXED FILM LUBRICATION

Mixed film lubrication exists when the film thickness developed by the above hydrodynamic mechanisms is inadequate to separate the two surfaces being lubricated. In this case, some surface asperities interact with resulting boundary lubrication at those interactions. Wear, surface distress, and an increase in the energy dissipation compared to the complete hydrodynamic film can be expected. A clear understanding of the nature of mixed film lubrication is not available. All component run-in or wear-in is under mixed film lubrication. A better understanding of run-in is desirable because the difference between run-in and failure in new components is often unpredictable and certainly not well understood. Mixed film lubrication is an area of current active interest and one in which there is a great deal of descriptive information in the literature. A recent symposium held by ASME gives and indication of the current state-of-the-art in this field[26].

9. CONCLUSION

Lubrication is a complex and broad subject difficult to summarize in one brief review chapter. The various modes of lubrication have been introduced and the reader is directed to the appropriate literature for further study.

10. REFERENCES

1. Halling, J., Ed., "Principles of Tribology," MacMillan Press, Ltd., London (1975).

2. Peterson, M.B., and Winer, W.O., Eds., "Wear Control Handbook," ASME, New York (1980).

3. Bowden, F. P. and Tabor, D., "The Friction and Lubrication of Solids," Oxford University Press, New York (1954), and "Friction and Lubrication of Solids, II," Oxford University Press (1964).

4. Czichos, H., "Tribology," Elsevier Scientific Publishing Company, New York (1978).

5. Bair, S., and Winer, W. O., "Regimes of Traction in Concentrated Contact Lubrication," ASME Paper No. 81-LUB-16 (to be published in Trans. ASME, Journal of Lubrication Technology, 1982).

6. Braithwaite, E. R., "Solid Lubricants and Surfaces," Pergamon Press, New York (1964).

7. Cameron, A., "Basic Lubrication Theory," Longman, London (1966), and "Principles of Lubrication," J. Wiley and Sons, New York (1966).

8. O'Connor, J. J. and Boyd, J., "Standard Handbook of Lubrication Engineering," McGraw-Hill, New York (1968).

9. Ling, F. F., Klaus, E.E. and Fein, R. S., Ed., "Boundary Lubrication: An Appraisal of World Literature," ASME (1969).

10. Sakurai, T., "Role of Chemistry in the Lubrication of Concentrated Contacts," Trans. ASME, Journal of Lubrication Technology, 103, No. 4 (1981) 473 - 485.

11. Trumpler, P.R., "Design of Film Bearings," MacMillan, New York (1966).

12. Dowson, D., and Higginson, G. R., "Elasto-hydrodynamic Lubrication," Pergamon Press, New York (1966).

13. Gross, W. A., et. al, "Fluid Film Lubrication," John Wiley and Sons, New York (1980).

14. Constantinescu, V. N., "Gas Lubrication," ASME, New York (1969).

15. Pinkus, O., and Sternlicht, B., "Theory of Hydrodynamic Lubrication," McGraw-Hill, Inc., New York (1961).

16. Tipei, N., "Theory of Lubrication with Application to Liquid-Gas-Film Lubrication," Sanford University Press (1962).

17. Rohde, S. M., Maday, C. J., and Allaire, P. E., Eds., "Fundamentals of the Design of Fluid Film Bearings," American Society of Mechanical Engineers, New York City (1979).

18. Rohde, S. M., Wilcock, D. F., and Cheng, H. S., "Energy Conservation through Fluid Film Lubrication Technology: Frontiers in Research and Design," American Society of Mechanical Engineers, New York City (1979).

19. Szeri, A. Z., "Tribology," Hemisphere Publishing Company, New York (1980).

20. Booker, J. F., "Design of Dynamically Loaded Journal Bearings," (in Reference [17]) (1979).

21. Goenka, P.K., and Booker, J. R., "Effect of Surface Ellipticity on Dynamically Loaded Cylindrical Bearings and Joints," ASME Paper No. 81-LUB-1 (to be published in Trans. ASME, Journal of Lubrication Technology, 1982).

22. Rippel, H. C., "Cast Bronze Hydrostatic Bearings Design Manual," Cast Bronze Bearing Institute, Inc. (1964).

23. Grassam, N. S., and Powell, J. W., Eds., "Gas Lubricated Bearings," Butterworths, London (1964).

24. Hamrock, B. J., and Dowson, D., "Ball Bearing Lubrication: The Elastohydrodynamics of Elliptical Contacts," J. Wiley and Sons, (1981).

25. Mobil EHL Design Guide, Mobil Oil Co.

26. Rohde, S. M., and Cheng, H. S., "Surface Roughness Effects in Hydrodynamics and Mixed Lubrication," American Society of Mechanical Engineers, New York (1980).

11. ACKNOWLEDGEMENTS

Figure Number	Source (Chapter Reference)/Citation
1, 2(B)	Ref. 1, with permission of the Editor, J. Halling
3	Ref. 3, with permission of D. Tabor, F.R.S.
4, 7	Ref. 4, with permission of H. Czichos
9, 10	Ref. 1, with permission of the Editor, J. Halling
11(B), 12, 16, 17	Ref. 8, Standard Handbook of Lubrication Engineering, O'Connor and Boyd, Ed., 1968 McGraw-Hill Book Co., with permission
18, 19	Ref. 1, with permission of the Editor, J. Halling
20	With permission of the American Society of Lubrication Engineers, publishers of Lubrication Engineering and ASLE Transactions
21	Ref. 1, with permission of the Editor, J. Halling
22, 23, 24	Ref. 8, Standard Handbook of Lubrication Engineering, O'Connor and Boyd, Ed., 1968 McGraw-Hill Book Co., with permission

LUBRICANTS

W.O. WINER
GEORGIA INSTITUTE OF TECHNOLOGY

1. INTRODUCTION

A lubricant is any material used to separate two surfaces in relative motion which can be readily sheared while adhering to the surfaces. In the process of acting as a lubricant, the material protects the solid surface from unacceptable wear while dissipating an accepatably small amount of energy. Any deformable media is a potential lubricant. The lubricant might be a deformable media readily available in the manufacturing process (that is, a process fluid lubricant) or a material specifically selected and introduced for its lubricating quality.

This chapter will discuss the range of materials used as generic lubricants (solids, liquids, gases, and greases), relevant physical and chemical properties, typical lubricants, major selection criteria, and an overview of specification procedures. Chapter emphasis will concern itself with an introduction to lubricants and with the literature on lubricants.

2. FUNCTIONS OF A LUBRICANT

Lubricants are functional materials in the mechanical system. The primary functions to be performed by the lubricant are to separate the surfaces, control wear, reduce friction, and reduce pitting fatigue. Frequently the function of separating surfaces is referred to as the load-carrying capacity or ability of the lubricant. Most of these primary functions of the lubricant are really not functions simply of the lubricant itself, but of the complete mechanism, and therefore depend not only on the lubricant properties and characteristics but also on the kinematics and dynamics of the system being lubricated.

Secondary, yet very important functions of the lubricant in tribological systems, are the scavenging of heat, dirt, and wear debris from the contact region, the prevention of corrosion throughout the system, and sealing, when mechanical seals as opposed to bearing surfaces are being lubricated.

These primary and secondary functions of the lubricant are related to a number of physical and chemical properties of the lubricant and the lubricant bearing material system. Table 1 consists of a list of a number of the lubricant and system/lubricant properties which are frequently discussed when considering a particular selection.

TABLE 1

LUBRICANT/SYSTEM PROPERTIES AFFECTING LUBRICANT FUNCTIONAL PERFORMANCE

DENSITY	SOLVENCY	VOLATILITY
VISCOSITY	PITTING	PAINTABILITY
POURABILITY	CORROSION	MISCIBILITY
FLAMMABILITY	RUST	EMULSIBILITY
FILTERABILITY	WETTABILITY	SURFACE TENSION
COMPATIBILITY	THERMAL COND.	SLUMPABILITY
STABILITY	ELECTRICAL COND.	DROPPING POINT
THERMAL	PENETRATION	VISCOELASTICITY
OXIDATIVE		FOAM
HYDROLYTIC		AIR RELEASE
BULK MODULUS		
HEAT CAPACITY		

The selection of a lubricant for a particular application is frequently very complex. The decision requires balancing the physical property requirements of the system, the chemical property requirements of the system, and the life requirements for the particular application, as well as the considerations of availability and cost. The Tribology Handbook presents a very useful but somewhat arbitrary initial selection procedure for lubricants based on bearings, loads and speeds as shown in Figure 1[1]. These selection limits are related to the thermal limits of materials as well as to considerations of feeding and traditional lubrication mechanisms used for the different range for the variables plotted. The first estimate

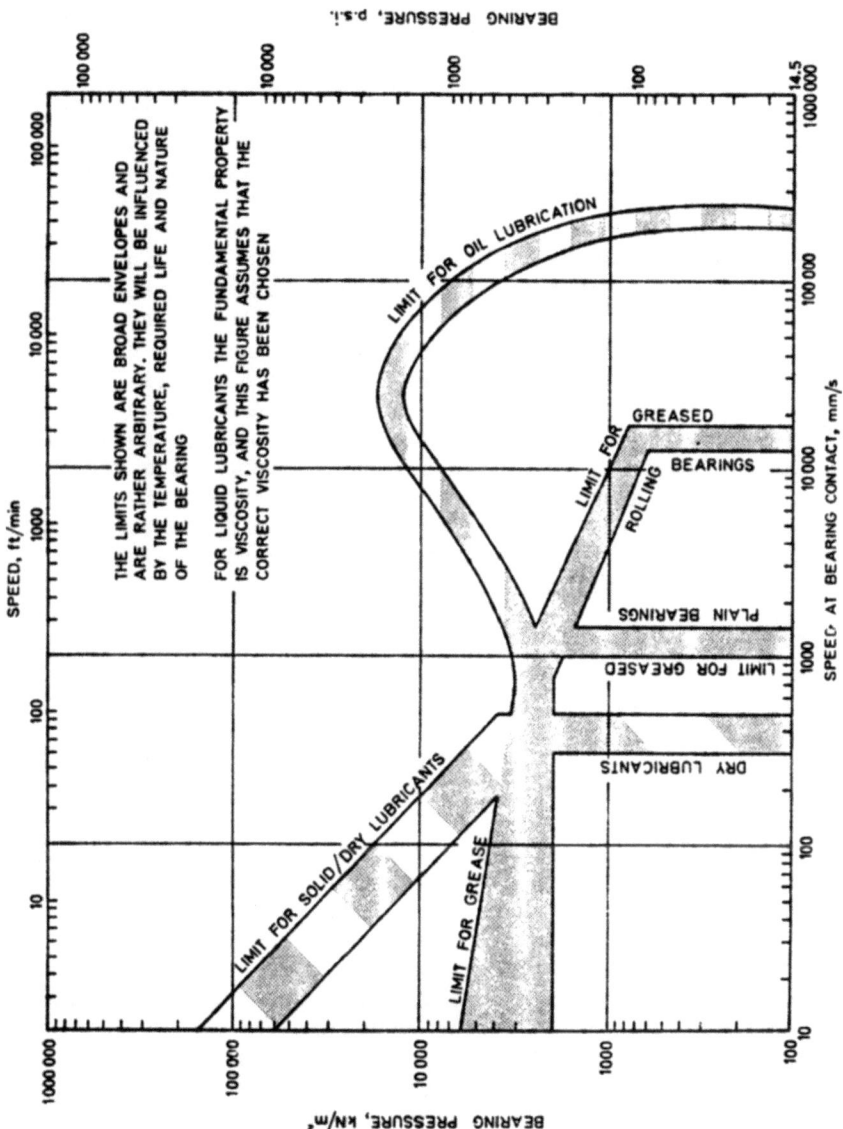

FIGURE 1 SPEED/LOAD LIMITATIONS FOR DIFFERENT TYPES OF LUBRICANT (REF. 1)

selection procedure shown in Figure 1 is to be modified and qualified by additional considerations presented throughout this chapter.

3. PHYSICAL AND CHEMICAL PROPERTIES OF LUBRICANTS (NON-RHEOLOGICAL)

Among the many physical properties of a lubricant that are of interest, the rheological properties are the most significant and will be discussed separately. In general, the remaining physical properties of interest vary over a relatively narrow range when considering a given chemical class of lubricants. This can be readily seen in Table 2 from the Tribology Handbook which presents the typical physical properties for highly refined mineral oils[1]. As we see from examining Table 2, the physical properties such as density, bulk modulus, thermal capacity, thermal conductivity, and vapor pressure, each have a relatively narrow range of property values for all of the mineral oils presented in the table. However, the viscosity has a rather wide range of values both among the different materials and at different temperatures for the same material. The different oils in the table represent a range of average molecular weights. The molecular weight has a strong influence on the viscosity as well as the pour point, vapor pressure, and the flash point for the material. Similar generalizations concerning the variation of physical properties within classes of synthetic lubricants can also be made. Properties for synthetic lubricants similar to those for mineral oils shown in Table 2, can be found in the literature, Tribology Handbook, Lubrication Handbook, and Synthetic Lubricants[1,2,3].

Of the chemical properties mentioned in Table 1, by far the most important is the oxidation stability of the lubricant which normally determines the life/temperature limitations of the lubricant in a particular application. In most lubrication applications there is adequate oxygen present in the system such that the life/temperature limitation is determined by the oxidation deterioration of the lubricant. In those special situations, where oxygen is not present, the thermal stability or the chemical reactivity of the lubricant at elevated temperature is the limiting characteristic.

Oxidation or thermal decomposition of the lubricant results in changing reactivity of the lubricant towards the system as well as changing rheological properties of the lubricant. In most situations, the result of oxidation or thermal degradation is an increase in viscosity which can inhibit the functioning of the lubricant as well as the circulation of the lubricant in the system.

Typical physical properties of highly refined mineral oils (*courtesy*: Institution of Mechanical Engineers)

		Naphthenic oils			Paraffinic oils		
		Spindle	Light machine	Heavy machine	Light machine	Heavy machine	Cylinder
Density (kg/m³) at	25°C	862	880	897	862	875	891
Viscosity (mNs/m²) at	30°C	18.6	45.0	171	42.0	153	810
	60°C	6.3	12.0	31	13.5	34	135
	100°C	2.4	3.9	7.5	4.3	9.1	27
Dynamic viscosity index		92	68	38	109	96	96
Kinematic viscosity index		45	45	43	98	95	95
Pour point, °C		−43	−40	−29	−9	−9	−9
Pressure—viscosity coefficient (m²/N × 10⁸) at	30°C	2.1	2.6	2.8	2.2	2.4	3.4
	60°C	1.6	2.0	2.3	1.9	2.1	2.8
	100°C	1.3	1.6	1.8	1.4	1.6	2.2
Isentropic secant bulk modulus at 35 MN/m² and	30°C	—	—	—	198	206	—
	60°C	—	—	—	172	177	—
	100°C	—	—	—	141	149	—
Thermal capacity (J/kg °C) at	30°C	1880	1860	1850	1960	1910	1880
	60°C	1990	1960	1910	2020	2010	1990
	100°C	2120	2100	2080	2170	2150	2120
Thermal conductivity (Wm/m² °C) at	30°C	0.132	0.130	0.128	0.133	0.131	0.128
	60°C	0.131	0.128	0.126	0.131	0.129	0.126
	100°C	0.127	0.125	0.123	0.127	0.126	0.123
Temperature (°C) for vapour pressure of 0.001 mmHg		35	60	95	95	110	125
Flash point, open, °C		163	175	210	227	257	300

TABLE 2   TYPICAL PHYSICAL PROPERTIES OF HIGHLY REFINED MINERAL OILS (COURTESY: INSTITUTION OF MECHANICAL ENGINEERS)

The rate at which oxidation degradation occurs in the lubricant is influenced by a number of factors. The temperature level is a major contributing factor with the oxidation rate doubling with approximating every 8° to 10°C temperature of the lubricant system. The oxidation rate is also a function of the access to oxygen which is related to the degree of agitation of the oil with air and the presence of oxygen in the ambient. The presence of catalytic materials, particularly iron and copper with large surface areas exposed to the oil, also contribute to the oxidation. The type of oil, e.g., the chemical makeup of the base lubricant itself, as well as the presence of various additives including antioxidant additives, are also factors which determine the life limiting behavior of the oil.

A final influencing factor frequently overlooked is the rate of addition of lubricant to the system. The addition of fresh lubricant replenishes oxidation inhibitors as well as dilutes the buildup of oxidation products in the bulk material. Figure 2 presents some general guidelines concerning the time and temperature limitations of mineral oils[1]. As indicated in this figure, the lower limit of temperature application is primarily the result of rheological property changes at low temperature resulting in too high a viscosity or pour point difficulties inhibiting the flow of the lubricant. These low temperature rheological properties can also be influenced by additives which will be discussed.

Similar data to that shown in Figure 2 for mineral oils can be found in the literature for synthetic oils and greases, Reference 1, Tribology Handbook.

4. ADDITIVES

Although the rheological and boundary lubrication properties of a lubricant are crucial to its primary functions as a lubricant, the total and rather complex additive package used in lubricants is necessary for the successful functioning of lubricants in many applications. When examining the literature on additive technology, one finds at least fourteen different additives, named on the basis of their function, which are used in various lubricants. Probably the largest number of additives used in any lubricant system is that used in internal combustion engine applications where up to ten different additives can be found in a single lubricant.

Table 3 lists fourteen different additives presented by function performed and indicates in which type of machinery applications additives are utilized. In two cases, namely

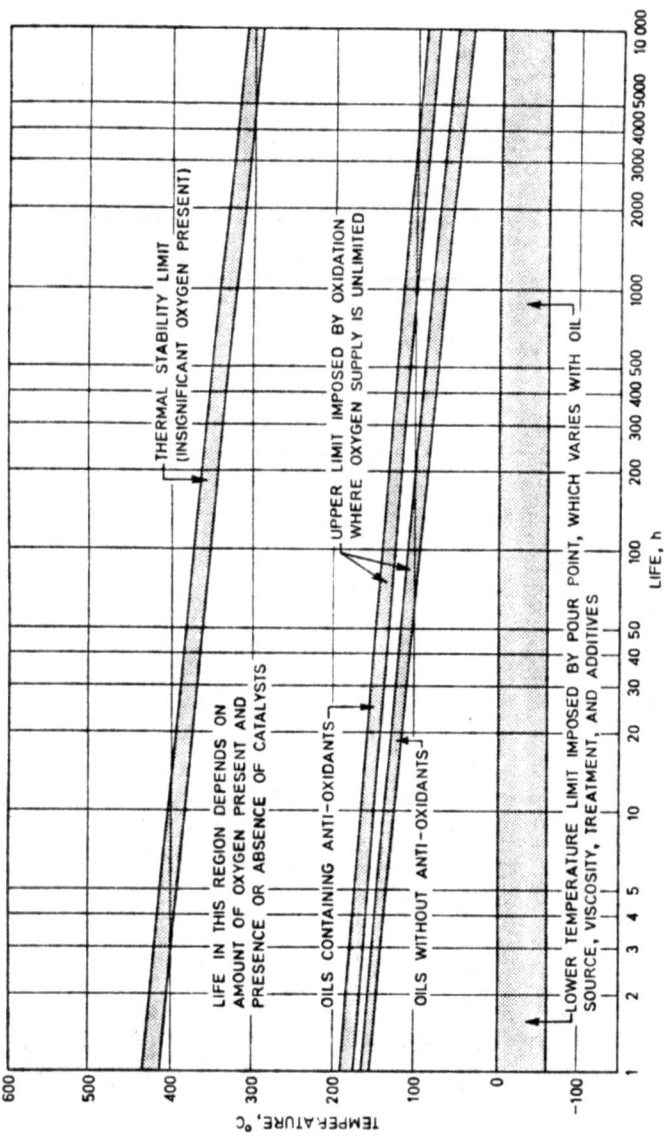

FIGURE 2 TEMPERATURE LIMITS FOR MINERAL OILS (REF. 1)

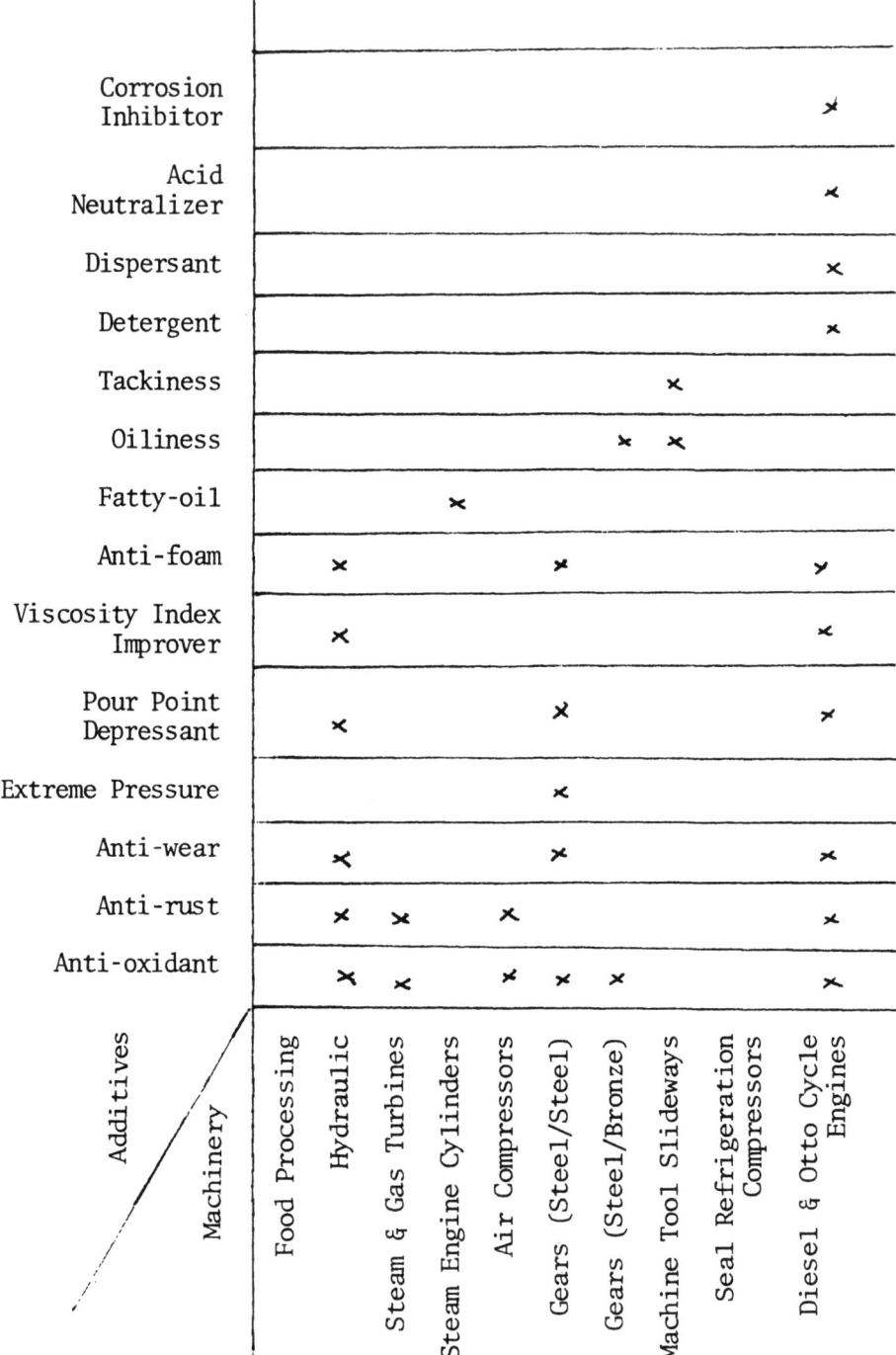

TABLE 3 TYPICAL ADDITIVE TYPES FOR MINERAL OIL LUBRICANTS IN VARIOUS TYPES OF MACHINERY

food processing machinery and sealed refrigeration compressors, it is likely that no additives will be found. This situation is prevalent in the food processing industry because the primary consideration is to avoid contamination of the product. In the case of the sealed refrigeration compressors, the lack of oxygen in the system and the design of the compressors permits operating in many cases, without additives.

In some cases, the additives listed are not particularly well defined and in fact the terminology used for some of them, e.g., oiliness and tackiness additives, is frequently objected to by some people in the field. These additives, however, will still be found in promotional literature. Oiliness additives are somtimes referred to as boundary lubrication additives where their function is to reduce friction under boundary lubrication conditions. These additives may increase load-carrying capacity by forming a solid lubricant film on the surface. Tackiness additives, on the other hand, are materials used to prevent the drainage of lubricant from surfaces either due to gravitational or centrifugal force fields.

Table 4, taken from Lubrication Magazine (a Texaco publication), shows examples of additives used in lubricating oils with typical chemical compositions of materials used for the additive type as well as a summary of the purpose of the particular additive. The chemical compositions listed should be considered only as typical and not all inclusive. For example, under viscosity index improvers, many other types of polymers are used in addition to the methacrylate and butylene polymers listed in the table.

Additive chemistry and additive formulation is a complex art in the field of lubrication. Many of the additives are chemically reactive under the conditions of application and several of them are surface reactive as well. The competition among additives for functioning is a major consideration in the formulating of lubricants. In addition, a seldom recognized complication is that there are many additives employed in the earlier stages of lubricant production as well as in other parts of a lubricating system which can interact with the additives in the lubricant itself.

There are appproximately twelve additive types used in the production of crude oil and refinery processing which might be found as carry-overs into the final product of a mineral oil lubricant. The behavior of these additives, in conjunction with those added to the lubricant for lubricating

Additive Type	Chemical Composition	Purpose
Viscosity index improver	Methacrylate polymers, Butylene polymers	Lower the rate of change of viscosity with temperature.
Pour point depressant	Alkylated naphthalene	Decrease pour point of oil.
Detergent-dispersant	Alkyl P_2S_5 products, metal sulfonates, alkylpolyamide, metal alkyl phenolates	Keep insolubles in suspension and maintain cleanliness.
Oxidation inhibitor	Zinc dialkyldithiophosphate	Retard oxidation of oils.
Rust inhibitor	Alkylamines	Prevent rusting of ferrous metals.
Corrosion inhibitor	Basic metal sulfonates	Prevent acidic materials from attaching to metal surfaces.
Extreme pressure agent	Sulfurized olefins, chlorinated paraffins	Prevent seizure of metal surfaces.
Foam inhibitor	Silicone polymers	Decrease tendency to foam.
Anti scuff-wear agent	Metal salts of alkyl acid phosphates.	Provide chemical polishing and reduce wear.

TABLE 4   EXAMPLES OF ADDITIVES USED IN LUBRICATING OILS (COURTESY TEXACO MAGAZINE LUBRICATION)

characteristics, can be a complicating factor in the final performance of the lubricant.

In addition, in the application of internal combustion engines there may be up to ten additives used in the engine fuel which, as a result of piston blow-by of the combustion gases into the engine sump, may end up in the lubricating oil. The presence of these fuel additives in the lubricant can be either detrimental or beneficial depending on the additive, its interaction with the lubricant, and the environmental conditions. For example, if tetraethyl lead antiknock compound is used in the fuel, lead will be found in very high concentrations in engine oils after a significant number of miles have been accumulated on the oil. These lead compounds frequently act as lubricants in highly loaded contact areas such as valve-trained mechanisms. The removal of tetraethyl lead from the fuel of vehicles in the United States put additional requirements on the engine oil additives to make up for the loss of the lubricating performance of the lead compounds in the lubricating oil.

Lubricating greases are also likely to contain a substantial number of additives to improve functional performance. Table 5, taken from Lubrication Magazine, is a typical example of the list of types of additives that may be found in lubricating greases. Like the base oils themselves, the additive life and degradation is also a function of system temperature. Additive degradation or depletion is frequently the major cause for lubricant change requirements.

5. SOLID LUBRICANTS AND BOUNDARY LUBRICANTS

As discussed in the previous chapter on Lubrication, solid lubricants and boundary lubricants can be classified conceptually in the same category because they both function as the result of the formation of a solid film surface. That solid film on the surface, to function acceptably, must adhere to the surface and be readily sheared by the motion of the two solid surfaces.

Solid lubricants are typically solid materials which are directly applied to the surface by a number of processes or put into solution in the oil or as a filler in the grease in order to be attached to the surface during the lubrication process. Boundary lubricants, on the other hand, are generally liquids or chemicals which are soluble in the lubricant as an additive and form a solid film on the surface by one of a number of possible interactions with the surface.

Additive Type	Chemical Composition	Purpose
Thickening agent	Metal soaps	Hold fluid by adsorption.
Fillers	Metal oxides	Add bulk to grease.
Oxidation inhibitor	Phenyl-beta-naphthylamine	Inhibit oxidation.
Metal deactivator	Mercaptobenzothiazole	Prevent catalytic effect of metals.
Corrosion inhibitor	Ammonium dinonyl naphthalene sulfonate	Arrest corrosion.
Anti-wear agent	Dibenzyl disulfide	Reduce wear.
Extreme pressure agent	Chlorinated wax Lead naphthenate	Reduce friction.
Dropping point improver	Fatty soaps	Increase dropping point.
Stabilizer	Fatty acid esters	Increase usable temperature.
Tackiness agent	Polybutylenes	Provide adhesiveness on metal surface.

TABLE 5 EXAMPLES OF ADDITIVES USED IN LUBRICATING GREASES
(COURTESY TEXACO MAGAZINE LUBRICATION)

The solid film of the boundary lubricant can be formed on the surface by physical adsorption, which is a relatively weakly bonded film somewhat similar to the condensation process of a gas onto a surface, or by chemical adsorption which is a somewhat more strongly bonded film on the surface resulting from a chemical bonding between the boundary lubricant additive and molecules in the surface. It can also be formed by chemical reaction which is similar to a mild form of corrosion in which surface molecules are removed to form new chemical compounds with the boundary lubricant. The resulting compound material remains on the surface. The films formed by reaction are generally the more strongly adhered films. The mechanism of film formation is dependent upon the composition of both the boundary lubricant additive and the surfaces being lubricated as well as the temperature and pressure of the environment.

Solid lubricants are frequently thought of as being lamellar solids, such as molybdemum disulfide and graphite, the structures of which are shown in Figure 3. These two materials are thought to function well as solid lubricants because of the relatively low shear strength planes separating the lamellar structure of the molecule. Although this concept has some basis, it is a gross over-simplification of the performance of these molecules. In particular, the graphite performance is very much dependent upon the atmosphere in which lubrication takes place because the low shear strength of the graphite crystal structure is apparently dependent upon the presence of condensable materials at the edges of the crystal structure. The graphite is not a suitable lubricant under very high vacuum conditions where there are no condensable materials present, whereas the molybdemum disulfide is more suitable under these conditions.

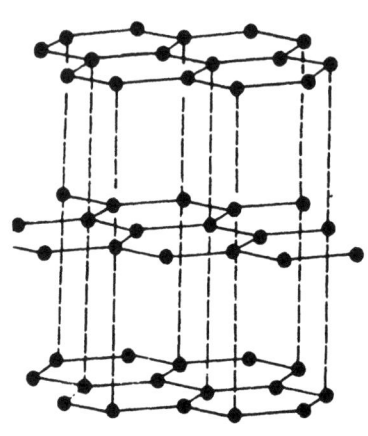
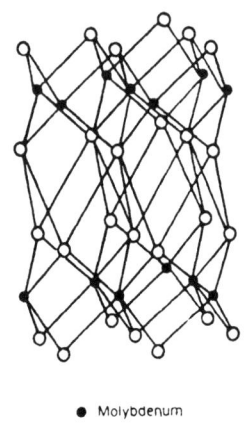

THE CRYSTAL STRUCTURE　　　　THE CRYSTAL STRUCTURE OF
OF GRAPHITE　　　　　　　　　　MOLYBDENUM DISULPHIDE

FIGURE 3

Although solid lubricants are frequently thought to be lamellar solids, all lamellar solids are not good solid lubricants, and conversely many good solid lubricants are not lamellar crystal structure materials. In fact, many amorphous solid materials possess the necessary characteristics of low shear strength and good adhesion to solid surfaces. In the case of the lamellar solids, orientation of the crystal structure on the surface influences the resulting friction and wear when these materials are used.

A typical schematic of the way boundary lubricants are thought to function is shown in Figure 4. These chemically formed boundary lubricant films are thought to perform several functions. They are thought to fill the valleys and cushion the surfaces through the smoothing of the surfaces permitting micro-elastohydrodynamic lubrication to occur at the asperities. They may also change the mechanical properties of the surface layers of the solids through such phenomena as the Rehbinder effect. The boundary lubricant fillms can range in thickness from a few angstroms to 1 μm or more.

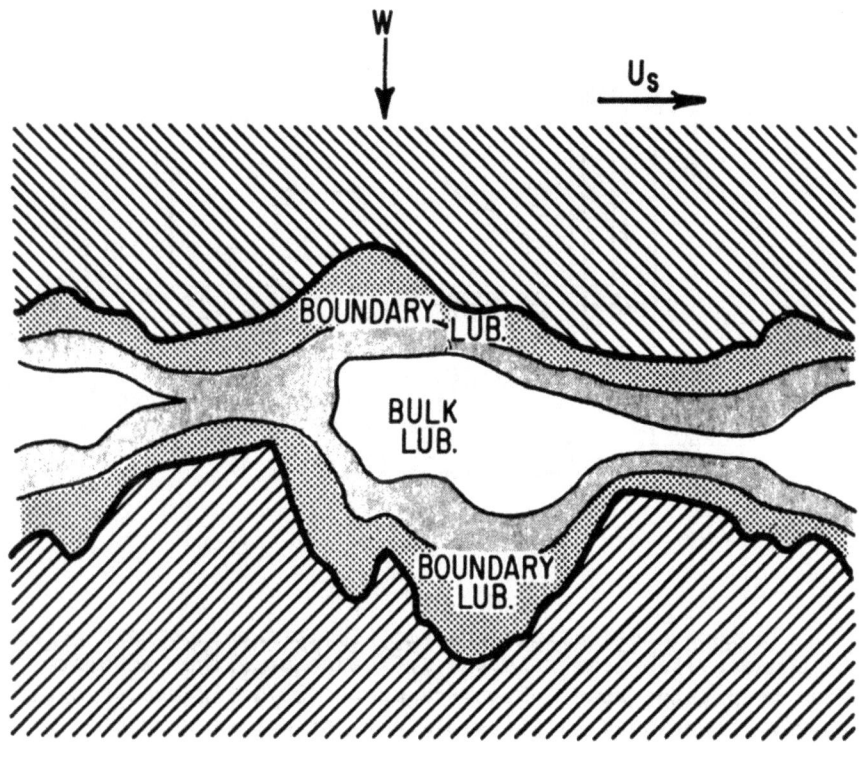

FIGURE 4 SCHEMATIC OF BOUNDARY LUBRICANT FILMS

Even gases which condense on and/or react with the surface can act as boundary lubricants in the sense that they reduce or inhibit seizure or adhesion between the surfaces and can reduce friction compared to the absence of any film on the surface. Figure 5 is an example from Bowden and Tabor of early studies on the presence of gases (H_2S and Cl_2) in vacuum environments on the friction of clean iron surfaces[4]. At room temperature with clean surfaces, there is a seizure between the two iron surfaces resulting in coefficients of friction in excess of two. When gas is present, the coefficient of friction drops to about 1.2 in case of H_2S and to about 0.4 in the case of Cl_2. As the temperature rises, the gas film is eventually desorbed resulting in seizure.

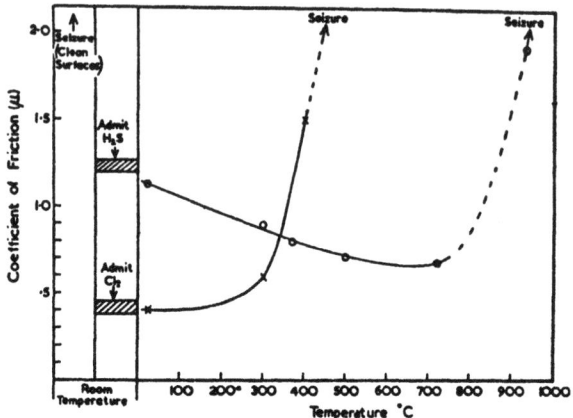

FIGURE 5 EFFECT OF TEMPERATURE ON FRICTION OF CHLORIDE AND SULPHIDE FILMS FORMED ON CLEAN IRON SURFACES. THE CHLORIDE GIVES THE LOWER FRICTION, BUT BREAKS DOWN AT A MUCH LOWER TEMPERATURE (REF. 4)

In most applications however, the primary concern related to boundary lubricants is the influence on either adhesive wear or pitting wear. Figure 6 shows the influence of typical boundary lubricant additives within various classes of chemicals, on the wear coefficient in four ball tests. The four ball tests consist of one rotating ball loaded against three stationary balls and is a common lubricant test device.

Figure 6 shows that some additives can reduce wear by up to three orders of magnitude as compared to that exhibited by the base oil, and in some cases additives can increase the wear compared to the base oil wear. Table 6 shows a range of wear coefficient values for four ball tests on 52100 steel with various lubricants and boundary lubrication additives. The ambient atmosphere, as well as the lubricant type and additive, have significant influence on the wear coefficient. This influence, for the conditions shown in Table 6, can vary by more than eight orders of magnitude from that existing in a dry argon environment to that existing with a fully formulated engine oil in an air environment.

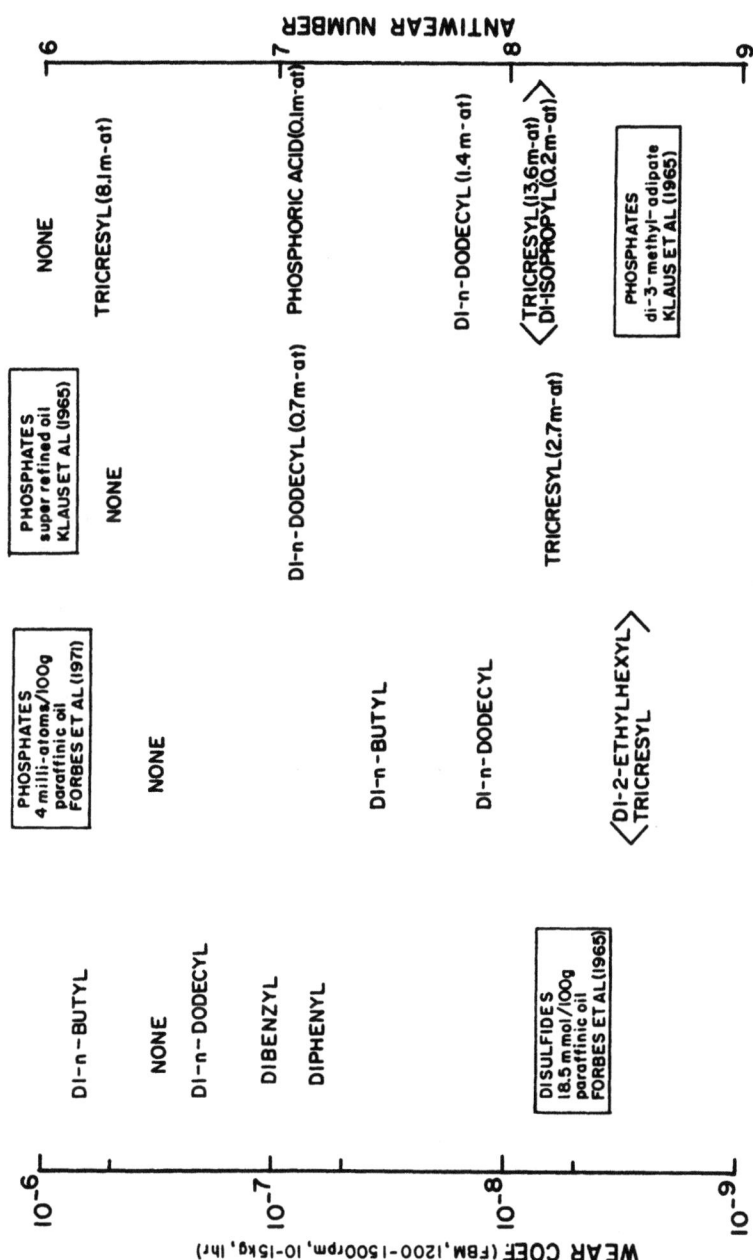

FIGURE 6 INFLUENCE OF TYPICAL BOUNDARY LUBRICANTS ON WEAR IN FOUR BALL MACHINE (FEIN, REF. 5)

AISI 52100 STEEL

Lubricant	Atmosphere	K
None	Dry Argon	1.0×10^{-2}
None	Dry Air	1.0×10^{-3}
Cyclohexane	Air	8.4×10^{-6}
Paraffinic Oil	Air	3.2×10^{-7}
Tricresylphosphate/ Paraffinic Oil	Air	3.3×10^{-9}
Engine Oil	Air	$< 2.0 \times 10^{-10}$

TABLE 6 COMPARISON OF K VALUES (REF. 5)

For any given type of boundary lubrication additive, the chemical reactivity with the surface is important to the performance of the additive and is heavily dependent upon the surface temperature. Figure 7 shows the influence of surface temperature on the wear in a four ball machine for a one percent organophosphite additive in mineral oil[6]. The different surface temperatures shown in the curve resulted from changes in not only ambient temperature but also load and speed in the four ball test.

Another measure of the performance of boundary lubrication additives is the load-carrying capacity of the resulting lubricant in a mechanism. A load-carrying capacity, although not a well-defined property, can be thought of as that load at which severe surface distress occurs resulting in the inoperability of the tribo-contact. Figure 8 shows a plot of wear versus work transmitted in an FZG gear test rig for a base oil, a base oil plus a mild EP additive, and a base oil plus a strong EP additive. The stronger the EP additive, e.g., the greater the reactivity between the EP additive and the surface, the more boundary lubricant film that is formed and thus the greater the amount of energy that can be transmitted through the gear mesh.

Figure 9 contains data showing the relative improvement of load-carrying capacity as a result of different chemical reactivity in two types of EP additives. Not only can the difference be large between classes of lubricants, but the range of load-carrying capacity is also large within a given lubricant class for different chemical structures.

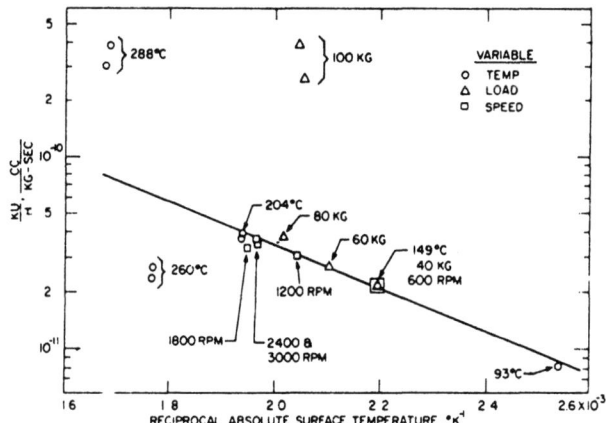

FIGURE 7 CORRELATION OF WEAR WITH SURFACE TEMPERATURES: FOUR BALL MACHINE, 1% ORGANOPHOSPHITE ADDITIVE IN MINERAL OIL (REF. 6)

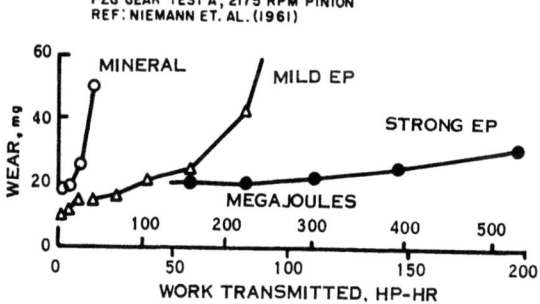

FIGURE 8 LOAD CARRYING LIMIT FOR VARIOUS LUBRICANTS

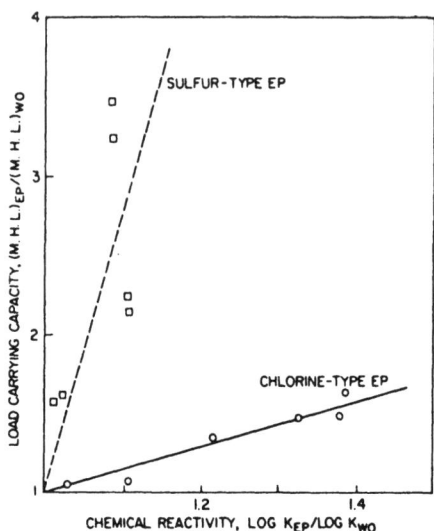

FIGURE 9   DEPENDENCE OF LOAD CARRYING CAPACITY ON CHEMICAL REACTIVITY (REF. 7)

In addition to their influence on adhesive wear coefficients and load-carrying capacity, boundary lubrication additives will also influence fatigue life of bearings. Even the presence of small quantities of water can change the fatigue life of bearings as shown in Figures 10 and 11.

The additives discussed above may be present in lubricants from as low a concentration as a few parts per million up to approximately ten percent by weight. The very low concentration in the parts per million level might be the intentionally added anti-foam agents or the unintentionally present water, both of which have significant effects on the performance of the lubricant. At the higher concentration levels are such additives as detergents, dispersants, and V.I. improvers in automotive engine oils. The more common concentration level for most additives is in the range of one percent or less by volume.

6. RHEOLOGY OF LUBRICANTS

Rheology is the study of the flow and deformation of materials. The rheology of lubricants is primarily concerned with the shear stress shear rate behavior of lubricants or more commonly the viscosity of lubricants. The viscosity or effective viscosity of lubricants is probably the most

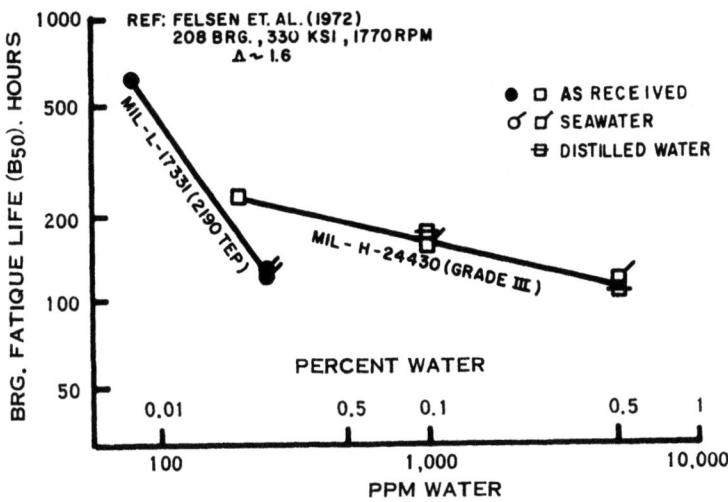

FIGURE 10   INFLUENCE OF WATER ON BEARING FATIGUE LIFE

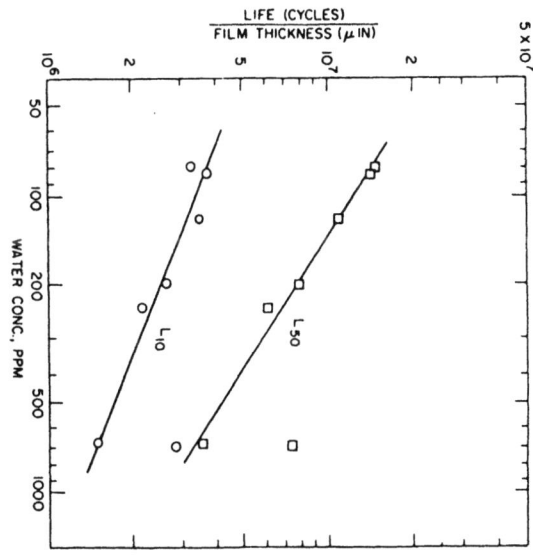

FIGURE 11   DEPENDENCE OF FATIGUE LIFE ON WATER CONTENT OF HYDRAULIC OILS AS PURCHASED (REF. 8)

commonly used property for specifying lubricants. The viscosity of the lubricant can be a function of temperature, pressure, and rate of shearing of the material. In addition to viscosity, the shear properties related to liquid-amorphous-solid transitions and the yield shear properties of lubricants in the amorphous solid regime are also important and will be discussed.

The viscosity is defined as the material property which is the ratio of shear stress to the rate of shearing strain in the material. An idealized experiment for measuring the viscosity is shown schematically in Figure 12. This figure depicts the relative motion between two parallel plates enclosing the fluid whose viscosity is being measured. The shear stress occurring is equivalent to the force required to move the plate divided by the area of the plate, and the rate of shearing strain is equivalent to the velocity gradient of the fluid between the two plates, which in steady flow for a Newtonian fluid, is simply the ratio of the relative velocity divided by the thickness of the film between the two surfaces. This viscosity is sometimes referred to as the absolute or dynamic viscosity and, given its definition, has dimensions of force times time divided by a length squared, which in the S.I. system of units is

$$\frac{N}{m^2} s \text{ or Pas}$$

In the more traditional English units, this might be in units of pounds per force seconds per square inch.

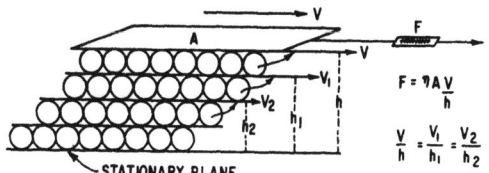

F = FORCE IN N
η = ABSOLUTE VISCOSITY, $\frac{NS}{m^2}$
A = AREA, m^2
V = VELOCITY, m/s
h = PLANE SEPARATION, m

FIGURE 12   SCHEMATIC FOR DEFINING VISCOSITY

Because of convenience in measuring techniques and usefulness in other fields, such as heat transfer and fluid mechanics, a second measure of viscosity is frequently reported. This second measure is referred to as the kinematic viscosity and is simply the ratio of the viscosity described above to the density of the material. The kinematic viscosity has dimensions of length squared per unit time which in the S.I. unit system is usually

$\frac{m^2}{s}$ or $\frac{mm^2}{s}$

Both the absolute viscosity and the kinematic viscosity can be reported in many different units which can be quite confusing even to the seasoned worker in the field. Probably the most common additional units used for these properties are the Poise or centipoise for absolute viscosity where the centipoise is a mPas/sec. In the case of the kinematic viscosity, a common unit is the centistoke which is the same as mm^2/s in the S.I. system.

The rheological behavior of lubricants, more particularly the viscous behavior of lubricants, is frequently divided into two broad categories referred to as Newtonian and non-Newtonian behavior. The non-Newtonian behavior is subsequently divided into a number of special cases many of which are appropriate for descriptions of lubricants.

Newtonian behavior is very common and is the simplest type of viscous behavior of materials. This behavior is shown schematically in Figure 13(a) and is simply the case where the viscosity is not a function of the rate of shearing of the material but solely a function of temperature and/or pressure. The two most common ways of presenting the viscous behavior of lubricants is to plot the viscosity versus rate of shear or the shear stress versus the rate of shear, the latter being referred to as a flow curve. For a Newtonian fluid, these two plots appear as (A) and (B) respectively in Figure 13(a).

Non-Newtonian behavior simply refers to any shear behavior which does not have a viscosity which is constant with shear rate. In Figure 13(b), the most common type of non-Newtonian behavior in lubricants is shown. This behavior is the so-called pseudo-plastic behavior. For a non-Newtonian fluid the viscosity, sometimes referred to as apparent viscosity, is a function of shear rate being applied to the material. The non-Newtonian case shown in Figure 13(b) is the more common type of non-Newtonian flow found with lubricants, that is, a decreasing viscosity with increasing shear rate. However, many other types of non-Newtonian flow also exist.

The non-Newtonian behavior might be either time independent or time dependent. Time independent non-Newtonian flow is divided into the categories of plastic or pseudo-plastic and dilatant depending on whether the apparent viscosity decreases or increases with increasing shear rate. The pseudo-plastic behavior, that is the decreasing apparent viscosity with increasing shear rate, is the most common type of non-Newtonian flow among lubricating materials. Figure 14 shows the typical flow curve and apparent viscosity curve for pseudo-plastic non-Newtonian lubricants. In this case it is common to speak of an apparent viscosity which is the ratio of the shear stress to the shear rate at a particular shear rate. As a general rule low molecular weight materials (< 1000 amu) tend to be Newtonian materials, while very high molecular weight materials (in excess of 10,000 or 20,000 amu) tend to be non-Newtonian materials. This means that unblended mineral oils could be expected to be Newtonian while those containing high polymers, particularly viscosity index improvers, would be expected to be non-Newtonian.

Non-Newtonian behavior introduces one of the major difficulties in the understanding of the relationship of the viscosity to lubricant performance. As shown in Figure 15, taken from the Tribology Handbook, the viscosity of a number of the typical lubricants as a function of shear rate might be Newtonian or non-Newtonian in the range of shear rates from those typically occurring in viscosity measurement apparatus

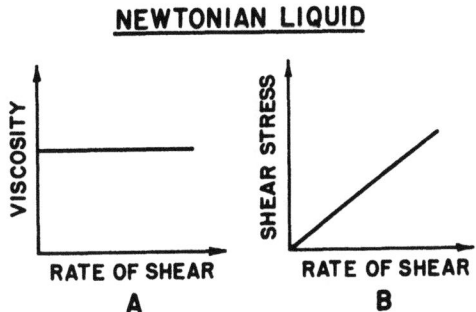

FIGURE 13(A) CHARACTERISTICS OF NEWTONIAN LIQUIDS. CURVE A: VISCOSITY IS INDEPENDENT OF RATE OF SHEAR. CURVE B: SHEAR STRESS IS DIRECTLY PROPORTIONAL TO RATE OF SHEAR

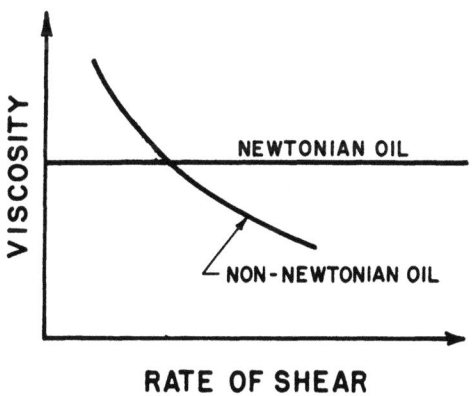

FIGURE 13(B) VISCOSITY VERSUS RATE OF SHEAR FOR NEWTONIAN AND NON-NEWTONIAN OILS

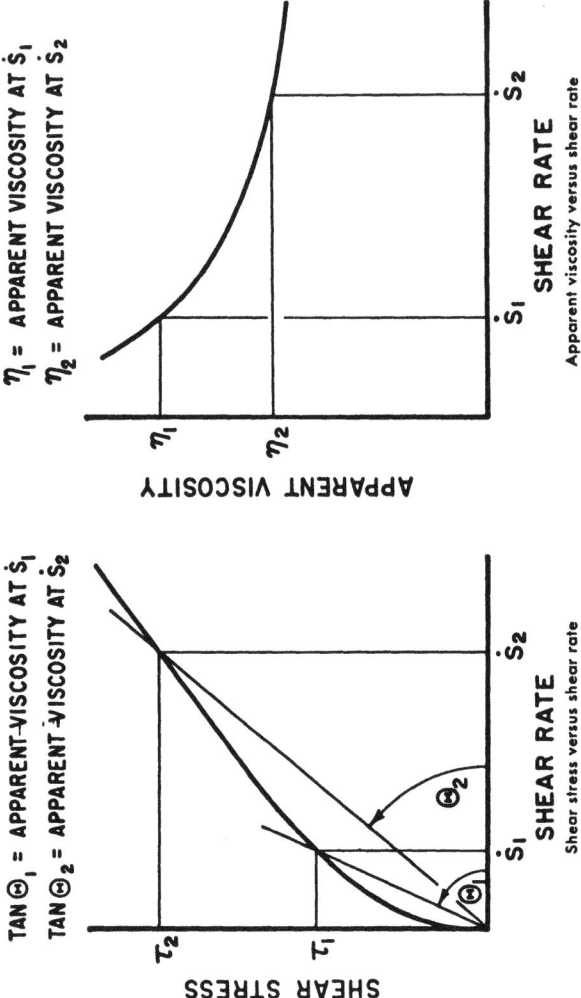

FIGURE 14 TYPICAL VISCOUS CHARACTERISTICS OF A NON-NEWTONIAN FLUID

to the very high values occurring in bearing applications[1]. Most viscosity measurements are made at very low shear rates, one reciprocal second or less. However, the shear rates occurring in most bearings are quite high, usually 10^5 and sometimes in excess of 10^6 reciprocal seconds. Depending on the polymeric material involved, this large difference in shear rate can cause the measured viscosity to be many times greater than the effective viscosity in a bearing.

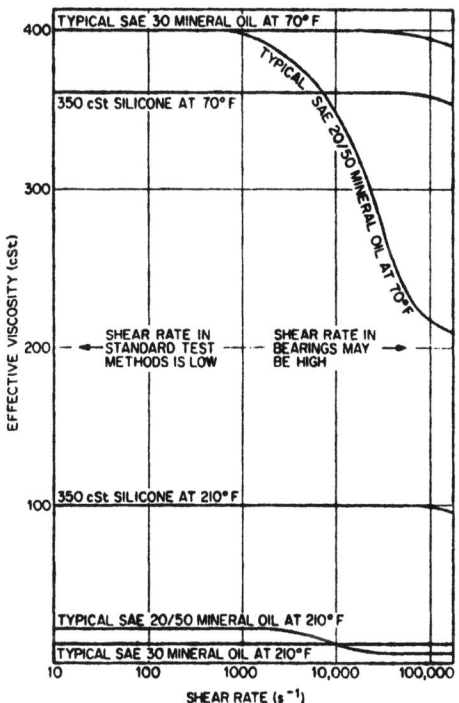

FIGURE 15   VARIATION OF VISCOSITY WITH SHEAR RATE (REF. 1)

In addition to time steady non-Newtonian behavior as a function of shear rate described above, it is also possible to have time dependent viscosity behavior. If the viscosity is a function of time at constant shear rate, the behavior is divided into two categories referred to as rheopectic and thixotropic depending on whether the viscosity is increasing or decreasing with time respectively. This time dependent behavior is shown schematically in Figure 16. Time dependent behavior is less common than time independent behavior among lubricants, but it does occur. For example, in the case of automotive engine lubricants, both thixotropic and rheopectic behavior can be observed over the drain interval for the engine. The thixotropic behavior is related to polymer degradation. In this situation the high molecular weight polymers will slowly be mechanically degraded into smaller molecules resulting in a lower viscosity with time. The thixotropic behavior in automotive engine lubricants can be observed as a result of combustion product blow-by and oxidation causing the oil to thicken with time over the drain interval. These two effects tend to counteract each other with the result that the viscosity will increase or decrease in a particular application depending on which effect predominates.

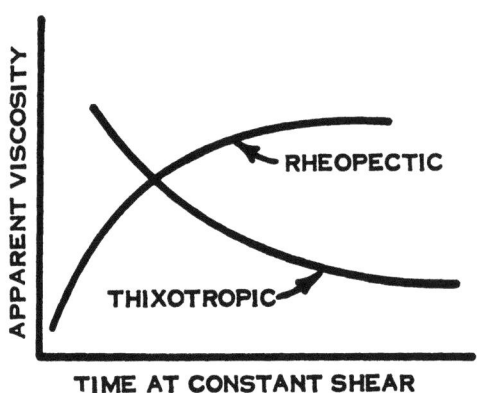

FIGURE 16   DIFFERENT TYPES OF NON-NEWTONIAN BEHAVIOR

There are many different types of apparatus for measuring viscosity of lubricants. Although the viscosity is defined as the ratio of shear stress to shear rate, it can also be viewed as a lubricant property which relates the way that mechanical energy is dissipated into thermal energy in the material. Therefore, measurements can be made by determining the time required for a given mass of material to be lowered from one height to another while passing through some flow constriction. Measurements can also be made by rotational devices in which a continuously rotating member, with a torque applied, is transmitting energy to the fluid film to be dissipated.

By far the most common method of measuring lubricant viscosity is by measuring the flow of a fixed quantity of material through a capillary constriction. If the volume of fluid, the distance to travel, and the geometry of the orifice are known, the measurement is simply a measurement of time for the flow of a given volume to occur. The Saybolt viscometer is a common device used for this type of viscosity measurement. In the Saybolt viscometer, a fixed quantity of oil (60 ml) is allowed to pass from a constant temperature bath through an orifice into a measuring flask. The measurment consists simply of the time required for the 60 ml to flow through the orifice. In this case, the density influences the potential to drive the fluid through the orifice and therefore the time is a function of the ratio of the absolute viscosity to the density of the material. The flow time is actually a measure of the kinematic viscosity of the oil. It is traditional to report viscosities measured in a Saybolt viscometer in terms of Saybolt's seconds, e.g., the time required for the flow to occur in that particular device. Unfortunately, the units of seconds bear no direct relation to kinematic viscosity or absolute viscosity in the terms most logically associated with the definition of the property as discussed above.

Other devices which are similar in principle and commonly used are the Redwood viscometer and the Engler viscometer. In each case the principle is the same as the Saybolt viscometer, but the geometry and volume of flow are different and therfore the generated numbers will be different. In the case of the Redwood viscometer, the viscosity is reported in Redwood seconds; while in the case of the Engler viscometer the viscosity, is reported in Engler degrees. In both the Saybolt viscometer and the Redwood viscometer different orifices are available for measuring viscosities in different ranges and therefore the times are usually qualified by which orifice has been employed.

Glass capillary viscometers are similar in function to the Saybolt viscometer described above. The glass capillary viscometers are immersed in constant temperature baths and permit the flow of a fixed quantity of fluid through a prescribed capillary in the glass. Again the measurement is the time required for the flow to occur. However, in the case of the glass capillaries, the results are usually reported in terms of traditional units of viscosity. The Saybolt viscometer and the glass capillary viscometers are both used in ASTM Standards for viscosity measurement .

Figure 17 is a nomograph which permits the conversation of kinematic viscosities from one set of units to another.

The viscosity of lubricants is a strong function of temperature, and whenever viscosity is quoted it is important that the temperature at which the measurement was made is also

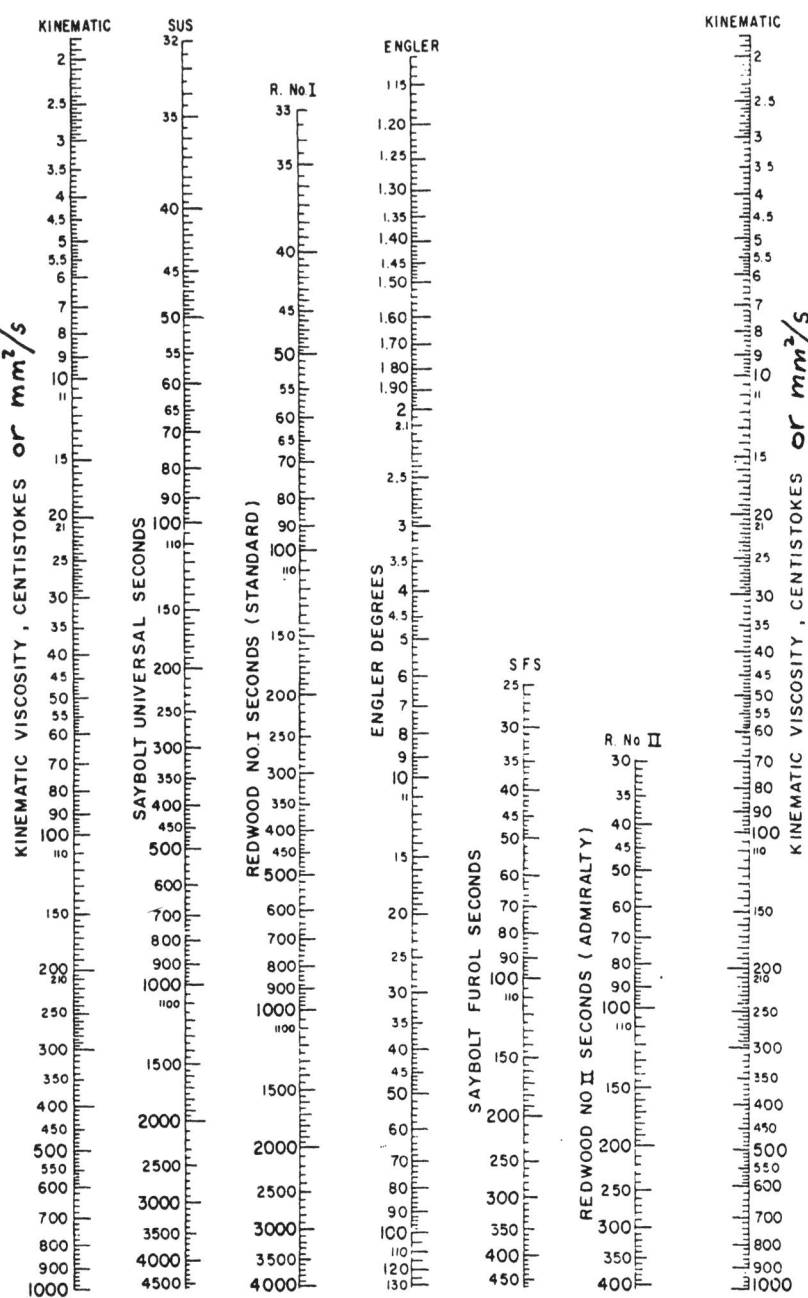

FIGURE 17 VISCOSITY CONVERSION NOMOGRAPH
(COURTESY TEXACO LUBRICATION MAGAZINE)

quoted. The most common method of presenting viscosity as a function of temperature is done with the ASTM Viscosity Temperature Chart as shown in Figure 18. The ASTM Viscosity Temperature Chart is based on the Walther Equation for describing the viscosity change with temperature. The scales on this chart can be quite misleading if not viewed with care. The kinematic viscosity range is a doubled log scale while the temperature range is a logarithmic scale. As a result, subsequent plotting of the two variables results in the viscosity temperature behavior of most materials being a straight line on the graph. This permits the convenience of measuring the viscosity at only two temperatures, plotting them on the curve and making possible the extrapolation and interpolation over other temperatures.

The double logarithmic scale on viscosity results in a very distorted scale for the kinematic viscosity. The major increments at the lower portion of the kinematic viscosity scale are only from two to three centistokes while the major divisions at the top of the scale are from ten million to twenty million centistokes. Also shown in Figure 18 are relative temperature viscosity behavior of some typical lubricating materials.

Although very important, the temperature viscosity behavior of lubricants is difficult to conveniently quantify. The most common method of specifying the temperature viscosity characteristics is to use the ASTM Dean and Davis Viscosity Index Scale.9 This scale is a relative scale and shown schematically in Figure 19. It is based on two standard categories of mineral oil, one of which exhibits a relatively low change of viscosity with temperature which is arbitrarily set at 100 on the VI scale. Each of these groups of lubricants, the zero and the 100 VI standards, have a range of viscosities at 210°F. Viscosity index of an unknown fluid is determined by comparing its viscosity at 100°F with the 100°F viscosities of the standard fluids, which have the same viscosity at 210°F, as shown in Figure 19 and by using the Equation

$$VI_u = \frac{U - H}{L - H}$$ (variables as presented in Figure 19).

Tables for this purpose are in the Appendix to ASTM Method D-2270-77.

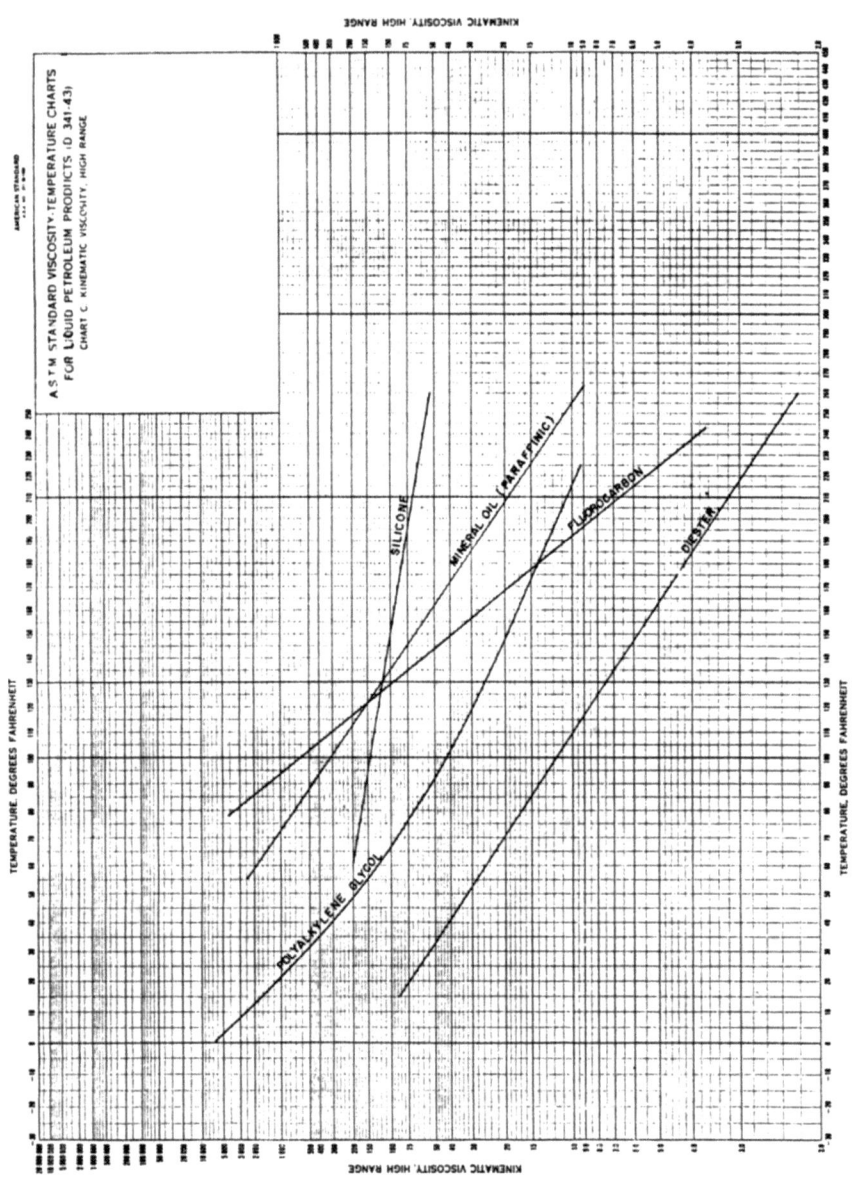

FIGURE 18 AN ASTM VISCOSITY-TEMPERATURE CHART WITH CURVES FOR SEVERAL FLUIDS

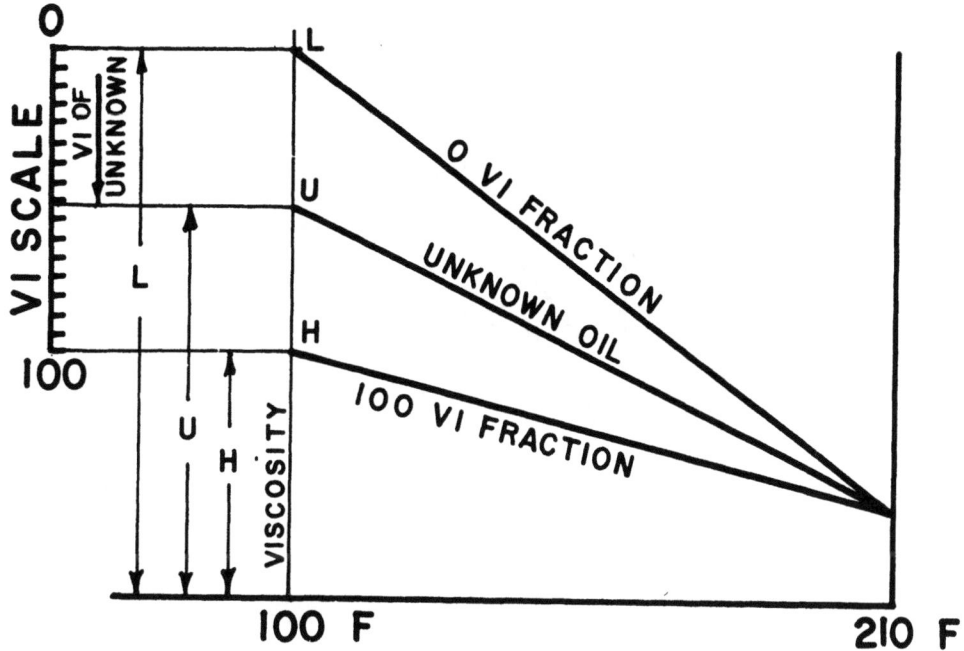

FIGURE 19   A SCHEMATIC ILLUSTRATION OF THE VISCOSITY INDEX SYSTEM

The viscosity index scale was established over forty years ago and has worked quite satisfactorily for some time. However, when considering today's lubricants, it is possible to have viscosity indices in excess of 100. The behavior of the viscosity index scale becomes somewhat different in the high VI range as compared to the lower 0 to 100 range for which it was originally established. Figure 20(a) represents a plot of kinematic viscosity at 100°F versus kinematic viscosity at 210°F with lines of constant viscosity index. Figure 20(b) shows the viscosity index of a hypothetical ideal material which has a constant viscosity versus the 210°F viscosity of the material. As can be seen from this figure, rather large changes in viscosity index occur for low viscosity materials. The viscosity index of a material can be obtained from extensive tables existing in the ASTM Standards.

Several other measures of viscosity temperature change can be found in the literature. These include simply reporting the ratio of the viscosity at 100°C to the viscosity at 40°C, the percent change in viscosity from 40°C to 100°C, or the logarithmic derivative of the viscosity at any particular temperature.

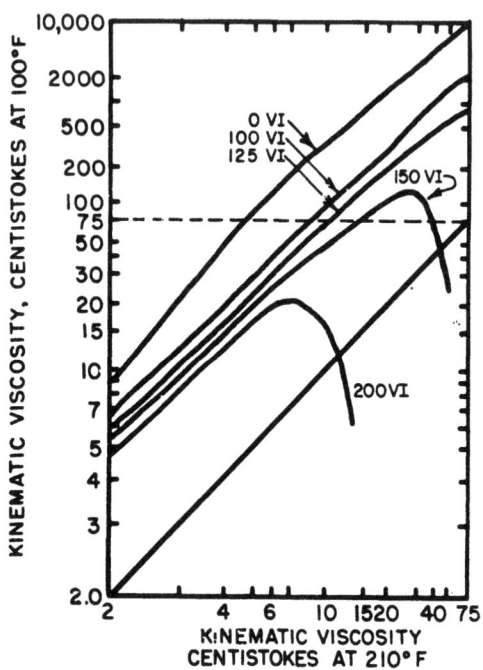

FIGURE 20(A) PLOT OF LINES OF CONSTANT ASTM VISCOSITY INDEX
(COURTESY OF TEXACO LUBRICATION MAGAZINE)

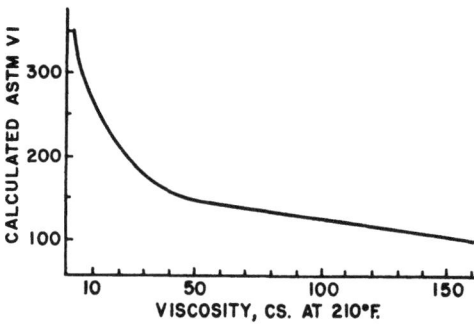

FIGURE 20(B) CALCULATED VISCOSITY INDEX VALUES FOR IDEAL OILS
(WHOSE VISCOSITIES DO NOT VARY WITH TEMPERATURE)
VERSUS THEIR VISCOSITIES AT 210°F
(COURTESY TEXACO LUBRICATION MAGAZINE)

The viscosity of lubricants also changes with pressure. As a first approximation, the viscosity increases exponentially with pressure such that the viscosity pressure isotherm plotted on a semilog plot (log viscosity versus pressure) will be nearly a straight line. The viscosity change is among the largest of all physical properties to change with pressure. However, in lubrication applications this change does not normally become important enough to consider unless the pressures are in excess of 10,000 psi or approximately 100 MPas. Therefore, the variation of viscosity with pressure is generally not considered important in thick film hydrodynamic lubrication, but does become important in the relatively thin film elastohydrodynamic lubrication.

Some typical results of the pressure viscosity dependence of lubricants are shown in Figures 21(a) and (b) taken from the ASME Pressure Viscosity Report[10]. In the case of mineral oil shown in Figure 21(a), for the same base viscosity, the slope of the pressure viscosity curve tends to increase with increasing naphthenic content. Also for the same base viscosity, the temperature viscosity dependence also tends to increase with increasing naphthenic content of the oil. Figure 21(b) shows the relative behavior of a number of synthetic lubricants.

Figure 22 shows the pressure viscosity isotherms for a synthetic lubricant (Polyphenol Ether) and shows the rather large change of viscosity pressure dependence with temperature. The slope of the pressure viscosity curve is the pressure viscosity coefficient which is seen to decrease substantially with increasing temperature. As discussed in the previous chapter on lubrication, the pressure viscosity coefficient is the luricant property which, along with the viscosity, determines the film thickness in elastohydrodynamic lubrication. The pressure viscosity coefficient is frequently presented in two different forms. These forms are either the slope of the logarithmic plot of viscosity versus pressure for an isotherm at atmospheric pressure ($\alpha 0T$) or an integrated relation between the viscosity dependence on pressure which is referred to as α^*.

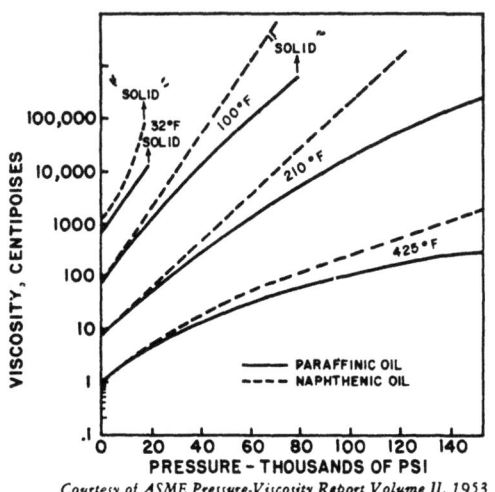

FIGURE 21(A)   PRESSURE-VISCOSITY CURVES FOR PARAFFINIC AND NAPHTHENIC OILS AT FOUR TEMPERATURES

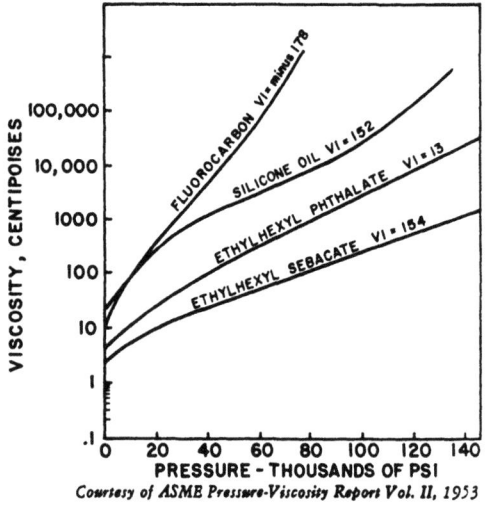

FIGURE 21(B)   THE EFFECT OF PRESSURE ON THE VISCOSITY AT $210°F$ OF SEVERAL SYNTHETIC OILS

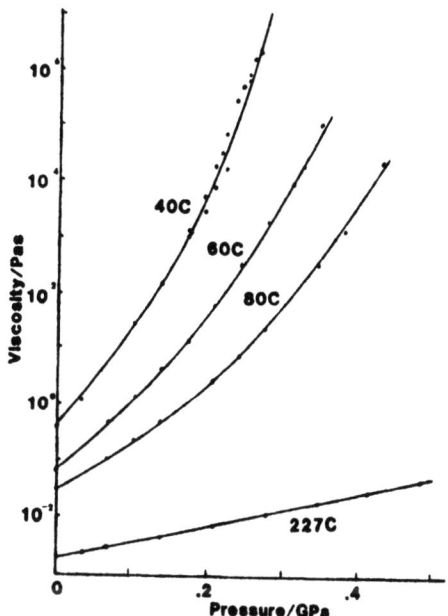

FIGURE 22   PRESSURE-VISCOSITY ISOTHERMS FOR 5PAE (POLYPHENOL ETHER), (REF. 11)

Table 7 presents the pressure viscosity coefficients, α_{OT} and α^* as a function of temperature for a number of typical lubricants. As decribed in the previous chapter, the lubricant property determining the film thickness and concentrated contacts or elastohydrodynamic lubrication is frequently presented as a product of the viscosity and the pressure viscosity, α^*, and referred to as a lubricant parameter. This lubricant parameter is presented as a function of temperature for a number of typical synthetic materials in Figure 23.

Fluid	T/C	α_{OT}	α_T^*
R620-15	26	27.4	27.4
	40	21.9	21.9
	99	15.4	14.8
	149	10.7	11.0
	227	12.0	8.85
R620-16	26	35.6	35.8
	99	19.8	19.8
	227	10.8	10.6
R620-15 + 4523	26	25.5	25.7
	99	17.1	15.0
	227	16.8	10.3
R620-15 + 4521	26	24.2	24.9
	99	15.0	15.3
	227	13.8	9.8
5P4E	40	40.6	41.2
	60	27.6	29.3
	80	20.0	20.1
	227	6.8	6.8
LVI 260	30	31.9	34.8
VITREA 79	30	22.6	22.6
	40	22.6	22.6
Turbo 33	30	19.6	19.6
	40	16.7	16.7

$$\alpha_{OT} \equiv \frac{d\ln\mu}{dp}\bigg|_{T,p=0} \qquad \alpha_T^* \equiv \left[\int_0^{p\to\infty} \frac{\mu(T,p=0)}{\mu(T,p)} dp\right]^{-1}\bigg|_T$$

Experimental fluids

Name	Type
R620-15	naphthenic mineral oil
R620-16	naphthenic mineral oil
R620-15 plus 4 wt. percent PL4520	Polymer blend
R620-15 plus 4 wt. percent PL4521	Polymer blend
R620-15 plus 4 wt. percent PL4522	Polymer blend
LVI 260	mineral oil
Vitrea 79	mineral oil
Turbo 33	mineral oil
5P4E	5-ring polyphenylether
Santotrac 50	Traction fluid
Krytox	Perfluorinated polyether
XRM 177F	Synthetic paraffinic hydrocarbon

TABLE 7   PRESSURE-VISCOSITY COEFFICIENTS (REF. 11)

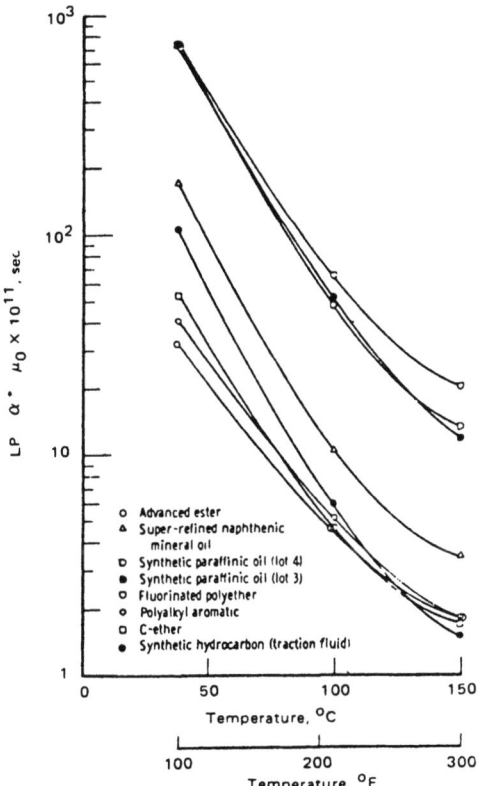

FIGURE 23   EHD FILM FORMING CAPABILITY AS FUNCTION OF TEMPERATURE FOR UNFORMULATED FLUIDS (REF. 12)

7. AMORPHOUS SOLID BEHAVIOR OF LUBRICANTS

In concentrated contacts with elastohydrodynamic lubrication, the pressures can vary from 600 MPa to 2 or 3 GPa. At these pressures, the viscosity of the lubricant becomes extremely high and the lubricant may no longer be able to respond by deforming viscously to the stress applied to it by the sliding surfaces. Therefore, the lubricant behavior may become similar to that of a relatively low shear strength amorphous solid. The liquid-amorphous transition in the lubricant is a function of temperature, pressure, and rate of stress application. This behavior is related to the glass transition of the material which is normally measured at atmospheric pressure. The glass transition is related to a very low rate of stress application and, for most lubricants at atmospheric pressure, occurs at temperatures below -20°C.

However, as the pressure is increased, the glass transition temperature also increases and at the very high pressures occurring in concentrated contacts, the glass transition temperature can be equal to or above the ambient temperature depending on the lubricant employed. Figure 24 is a temperature-pressure plot schematically showing a typical elastohydrodynamic contact and the glass transition behavior of three lubricants; 5P4E, a polyphenol ether which has the highest glass transition temperature of the lubricants shown; N1, a naphthenic mineral oil; and XRM 177, a low molecular weight synthetic material with very good low temperature viscosity. The curve for N1 is typical of many mineral oils.

Also shown schematically on Figure 24 are the approximate ranges of temperature and pressure in parts of an elastohydrodynamic contact. The inlet zone is at low pressure and the Hertzian zone is the higher pressure range. The temperatures in a rolling contact with little energy dissipation will be relatively low and near ambient temperature, while increased sliding in the contact will cause higher temperatures to be present. For a given lubricant whose curve is shown on Figure 24, the behavior in the concentrated contact would be expected to be liquid-like if the temperatures and pressures occurring are above and to the left of the curve shown. If the temperatures and pressures occurring in the contact for that lubricant are below and to the right of the curve, the material behavior would be expected to be that of an amorphous solid. If the material behavior is liquid-like, the lubricant viscosity should be the determining rheological property. If the behavior is amorphous solid in nature, properties such as elastic shear modulus and maximum shear stress should be the controlling rheological properties. To a first approximation, one would expect the inlet zones of virtually all elastohydrodynamic contacts to be in the liquid-like region. Therefore, because film thicknesses are determined by behavior in the inlet zone, film thicknesses in elastohydrodynamic contacts would be determined by the viscous behavior of the lubricant. This logic is consistent with what has been found experimentally. However, most of the Hertzian zones in concentrated contacts will be in the amorphous solid region for many lubricants. The friction in the contact is primarily determined by behavior in the Hertzian zone. Therefore, the friction or traction behavior will be determined by properties in the amorphous solid regime for the lubricant. Similar transition curves for a number of other lubricants can be found in the literature[13].

The transition curves shown in Figure 24 are for very low rates of stress application. If the rate of stress

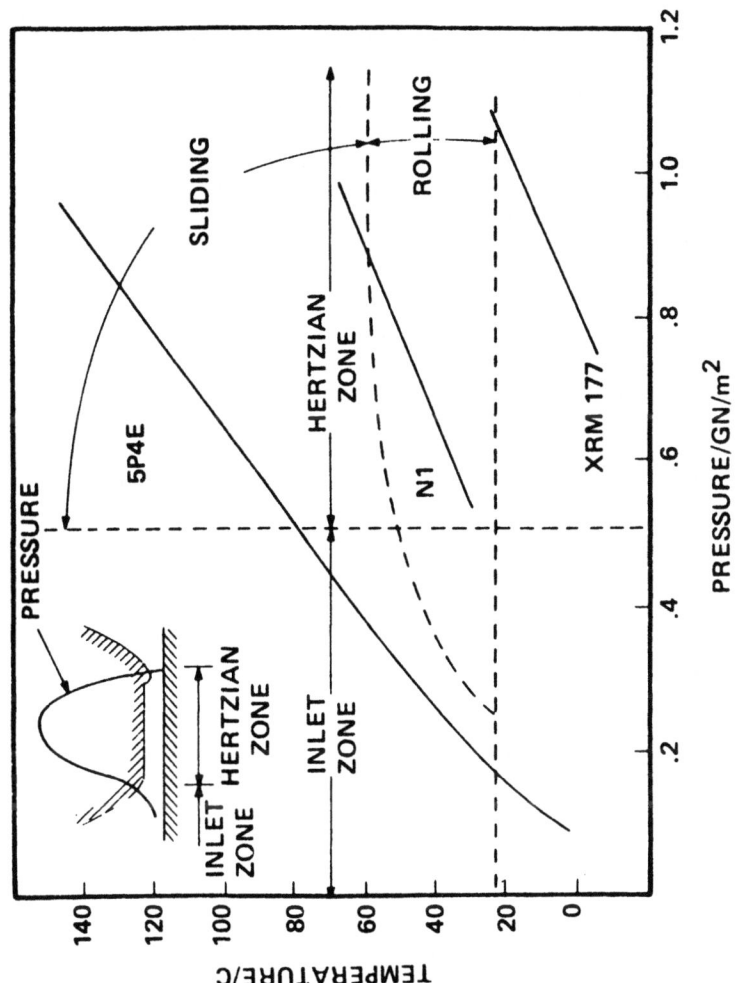

FIGURE 24 HEURISTIC ESTIMATES OF THE RELATIONSHIP BETWEEN CONDITIONS IN AN EHD CONTACT AND GLASS-LIQUID TRANSITION DIAGRAM OF SOME LUBRICANTS (LUBRICANT SUPPLY TEMPERATURE ABOUT 20°C) (REF. 13)

application is increased, these transition curves will shift up and to the left, e.g., to lower pressures for the same temperature or higher temperatures for the same pressure. This results in an expansion of the amorphous solid regime and a contraction of the liquid-like regime. Figure 25 shows the transition curves which result at a constant pressure, in this case atmospheric pressure, as the frequency of stress application, is changed. As the frequency of stress application increases, the temperature at which the transition occurs increases. The relative slopes of different materials is clear. For some materials, i.e., the polyphenol ether, the change of transition temperature with frequency is small as compared to the change that occurs with the mineral oil, N1. This change of transition with frequency can also be shown on the temperature pressure plot as shown in Figure 26.

It is reasonable to argue that the constant frequency relates to a constant relaxation time, which in turn is approximately equal to a constant viscosity for the material[14]. Therefore, lines of constant frequency of stress application and constant viscosity will plot as nearly parallel lines on the temperature-pressure transition curve. They should be parallel to the limiting low rate stress application performed by dilatometry. Such curves are shown in Figure 26 for polyphenol ether.

8. SHEAR STRESS STRAIN BEHAVIOR IN THE AMORPHOUS SOLID REGION

In the amorphous solid region the material will behave like an elastic-plastic solid. This elastic-plastic shear stress shear strain behavior will be the determining material property for the friction or traction in the concentrated contact. Figure 27 shows the stress strain behavior measured for a polyphenol ether at 40 kpsi pressure at the temperatures indicated. As the temperature is decreased, starting at $38°$C and going down to -27°C, the elastic-plastic behavior of the material begins to become apparent as the temperature passes through the amorphous solid transition temperature as shown in Figure 26. The rate of stress application for the data in Figure 27 is very slow. As the temperature decreases the initial slope of the curve, which is the elastic shear modulus, slowly increases to a nearly constant value as one goes into the amorphous solid region. The maximum shear stress the material can withstand also increases as the temperature is lowered and the material is taken further into the amorphous solid region. The maximum values of the elastic shear modulus and the shear stress are referred to as the limiting values in each case. Similar behavior would be observed if the temperature and the strain rate were held

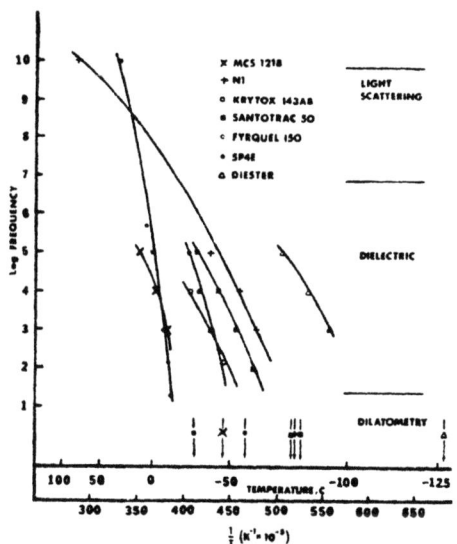

FIGURE 25   DIELECTRIC ($10 \leq$ Hz $\leq 10^6$), LOW RATE DILATOMETRY (ARROWS) AND LIGHT SCATTERING (Hz=10^{10}) TRANSITION DATA AT ATMOSPHERIC PRESSURE, (REF. 14)

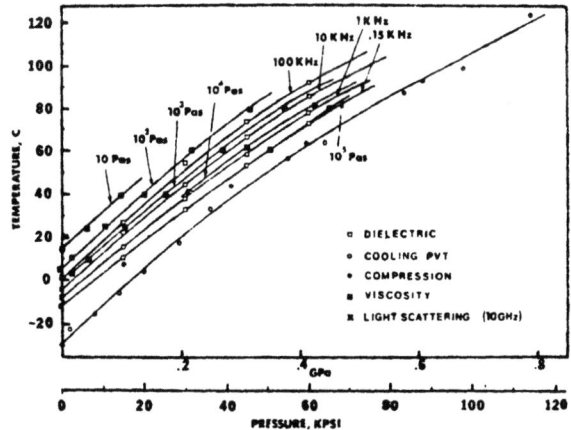

FIGURE 26   TRANSITION DIAGRAM FOR POLYPHENYL ETHER (5P4) AND SEVERAL METHODS (REF. 14)

constant while the pressure was increased. Similar behavior would also be observed if the temperature and pressure were held constant while the rate of stress application was increased.

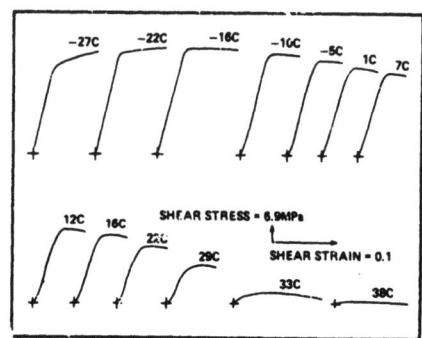

FIGURE 27 RECORDER PLOT OF SHEAR STRESS VERSUS SHEAR STRAIN FOR POLYPHENYL ETHER (5P4E) AT 0.275 GPa (40 kpsi) AND INDICATED TEMPERATURES (REF. 13)

Because in the concentrated contact the rate of stress application is very rapid, the limiting values of the elastic shear modulus and the maximum shear stress will be the controlling properties determining the traction in elastohydrodynamic contacts. Accompanying the maximum shear stress is a maximum recoverable elastic strain which can be seen from Figure 27 to be in the neighborhood of 0.02 to 0.03. If the strain in the concentrated contact at very low slide-roll ratios is less than this value of limiting recoverable strain, the traction will be determined by the elastic shear modulus of the material. However, at higher slide-roll ratios, the traction will be determined by the maximum shear stress the material can withstand.

Figure 28(a) and (b) contain the traditional format of viscosity versus shear rate or shear stress versus shear rate respectively for 5P4E lubricant. The entire range of behavior from liquid-like behavior on one side of the amorphous transition curve to limiting shear stress behavior on the other side of the transition curve is presented in a single flow curve. The data here presented represents measurements from three quite different types of measuring apparatus[15]. In Figure 28(b) the Newtonian behavior is represented by a line of slope 1 and the limiting shear stress is represented by a line of slope zero. As the shear rate is increased, the

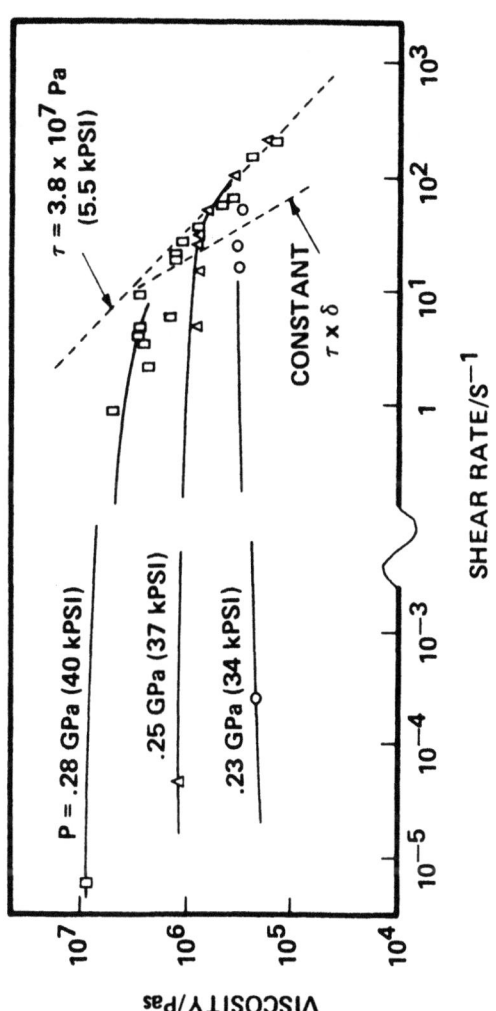

FIGURE 28(A)   VISCOSITY OF 5P4E VERSUS SHEAR RATE SHOWING THE LIMITING SHEAR STRESS AT 40°C (REF. 15)

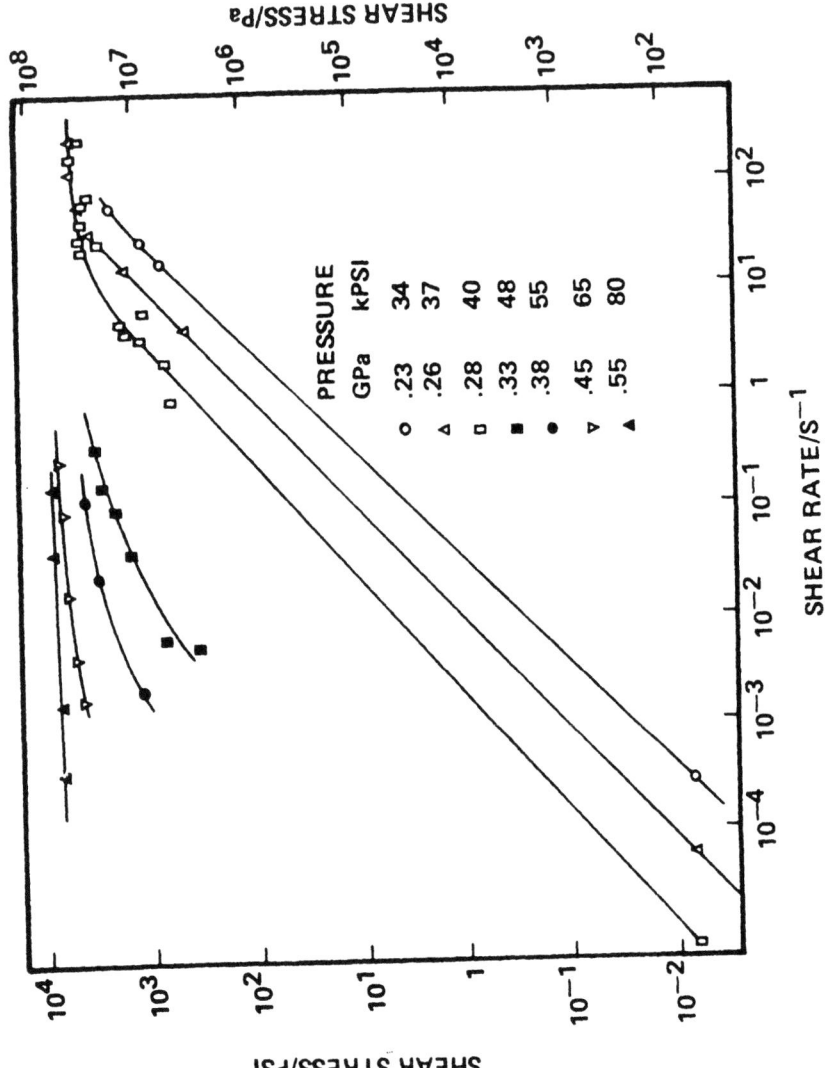

FIGURE 28(B) SHEAR STRESS-SHEAR STRAIN RATE FOR 5P4E AT 40°C AND INDICATED PRESSURE (THREE DIFFERENT METHODS) (REF.15)

behavior is Newtonian up to some nearly maximum value where the shear stress reaches a maximum.

The limiting shear stress is the property primarily responsible for traction in highly loaded concentrated contacts. Limiting shear stress for a number of materials as a function of temperature is shown in Figure 29(a) and (b). The limiting shear stress decreases with increasing temperature and increases with increasing pressure. Also note on Figure 29(a), the influence of the presence of a polymer in a mineral oil by comparing N1, which is a naphthenic mineral oil, and N1 + 4% PAMA polymer. This influence has been found to be representative for PAMA polymers of several molecular weights. In all cases the limiting shear stress with the polymer present is approximately 15% less than that of the base oil alone[11]. This is somewhat unexpected in light of the fact that the presence of the polymer can increase the viscosity by a factor of five to eight times.

The limiting shear stress of some solid lubricating materials is very similar to that of liquid lubricants. In Figure 30 there are two solid plastic materials, PVC and Teflon, which are commonly used as solid lubricating materials. This similar behavior of limiting shear stress for both solid lubricating materials and typical liquid lubricants is consistent with the fact that in concentrated contacts with these materials, the coefficients of friction are similar. In the limiting case, the coefficient of friction should simply be the ratio of the limiting shear stress to the pressure on the contact.

Table 8 shows a comparison of the average shear stress based on traction measurements at low slide-roll ratios for three materials and the limiting shear stress measurement at the same pressure and temperature[11]. The agreement between the property measurement and the tribological traction measurement is extremely good.

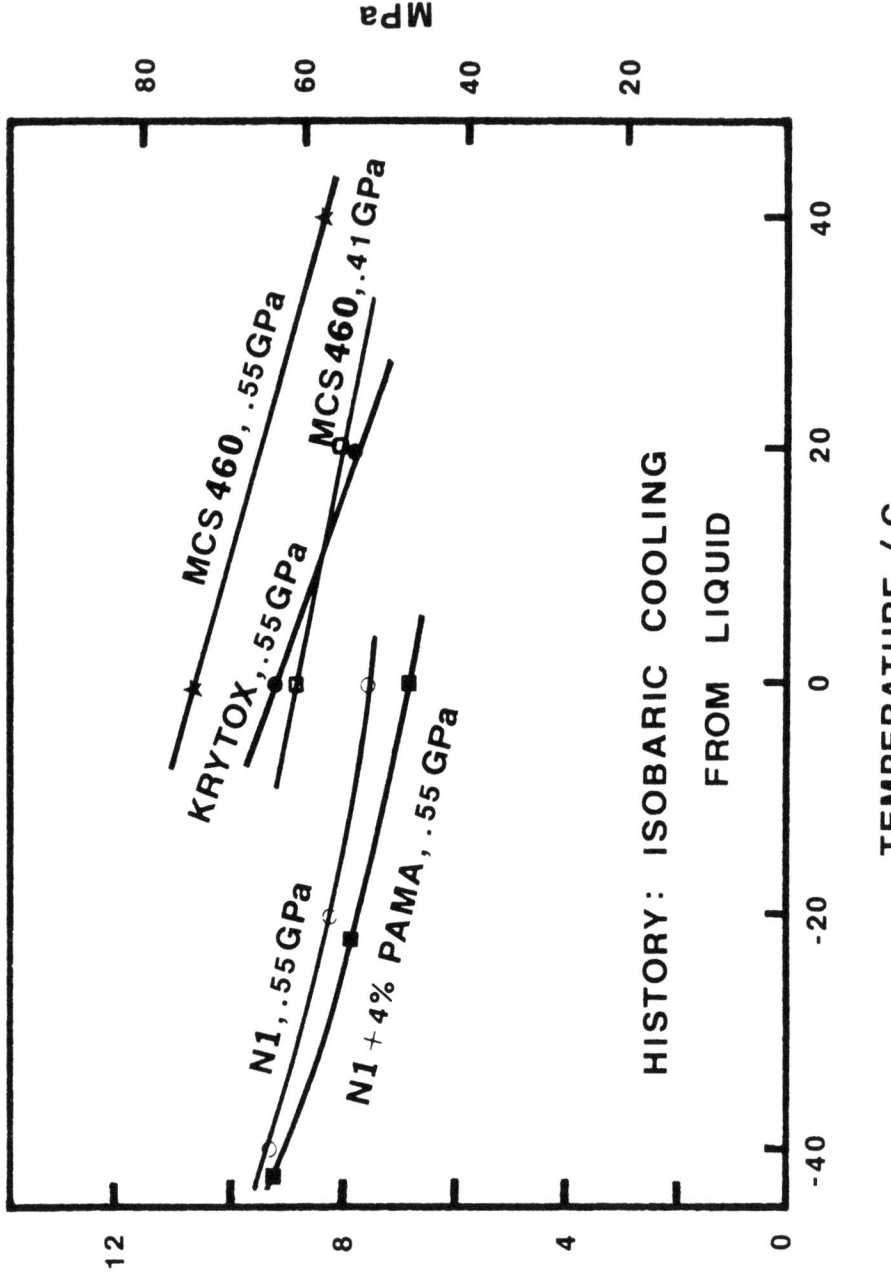

FIGURE 29(A) LIMITING SHEAR STRESS VERSUS TEMPERATURE FOR MCS 460, KRYTOX, N1, AND N1 + 4 PERCENT PAMA (COLLING HISTORY)

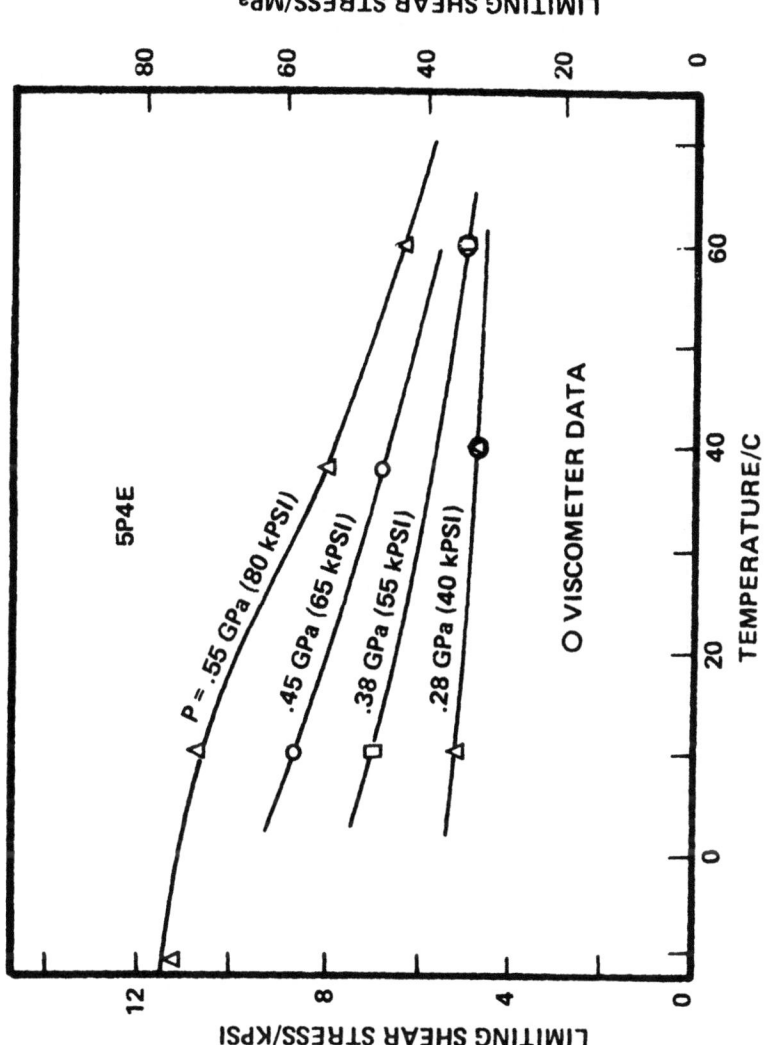

FIGURE 29(B) LIMITING SHEAR STRESS FOR 5P4E AS A FUNCTION OF TEMPERATURE AT INDICATED PRESSURES. CIRCLE AROUND DATA POINT INDICATES IT WAS OBTAINED WITH THE HIGH STRESS VISCOMETER (REF. 15)

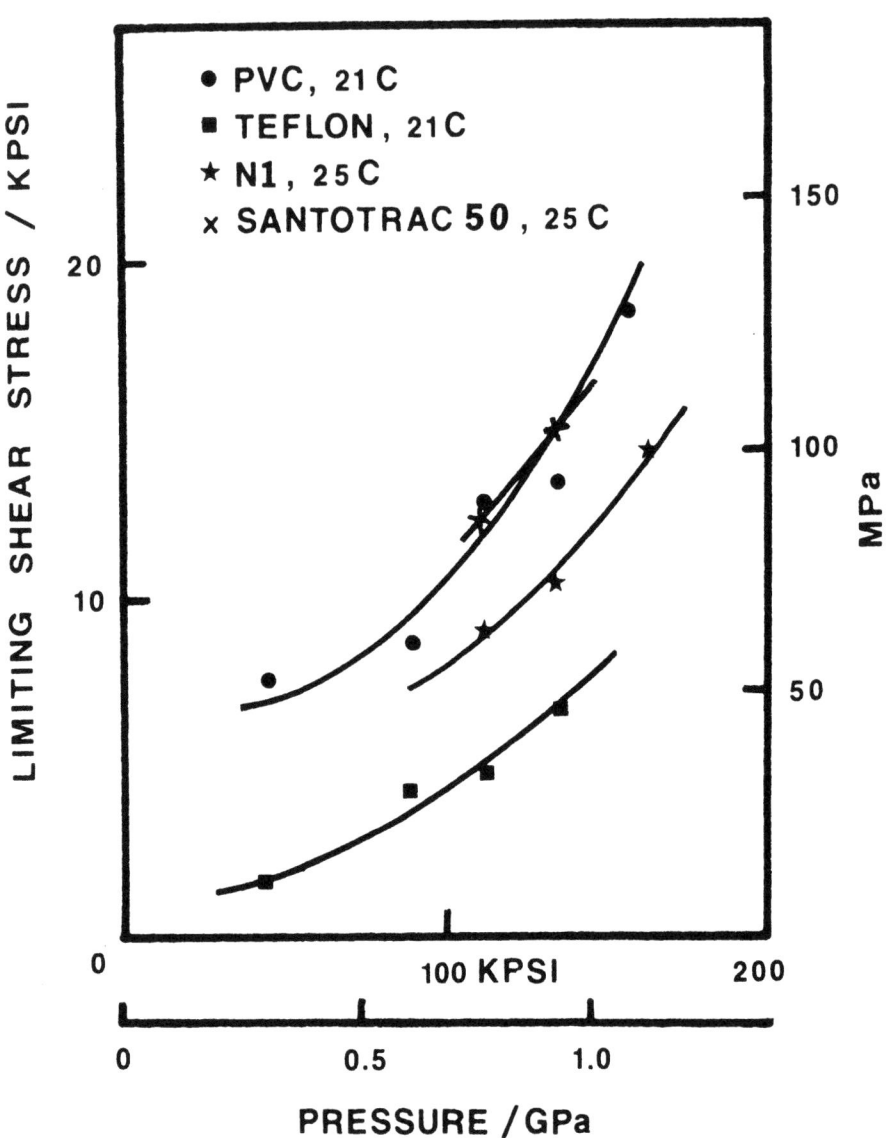

FIGURE 30 SHEAR STRENGTH OR LIMITING STRESS VERSUS PRESSURE FOR N1 (NAPHTHENIC BASE OIL), SANTOTRAC 50 (CYCLOALIPHATIC HYDROCARBONO, POLYVINYL CHLORIDE, AND TEFLON

Fluid	Limiting shear stress, τ_L, at $P = 0.67$ GPa	Average shear stress based on traction measurement at a slide-roll ratio of 10^{-1}, average pressure of 0.67 GPa [9]
5P4E	58 MPa (38°C)	55 MPa (35°C)
Vitrea 79	31 MPa (26°C)	33 MPa (30°C)
LVI 260	49 MPa (35°C)	51 MPa (35°C)

TABLE 8   COMPARISON OF LIMITING SHEAR STRESS WITH EHD TRACTION SHEAR STRESS, (REF. 11)

Figure 31 shows the typical log viscosity versus pressure isotherm for a material (5P4E) for different rates of stress application. At very low shear stress, the effective viscosity behavior is the traditional logarithmic dependence upon pressure. However, as the pressure increases the limiting shear stress is reached resulting in a nearly constant effective viscosity. A shear rheological model suitable for the entire range of behavior can be found in Reference 16.

9. SPECIFICATIONS OF LUBRICANTS

As indicated in the previous discussion on lubricants, lubricant behavior is dependent upon rheological properties and chemical properties. The specification of lubricants is a very complex consideration based on a number of criteria. The criteria include both readily measurable rheological properties, primarily viscosity at various temperatures, and performance criteria based on accepted application tests.

Probably the most common specification for lubricant selection is the viscosity of the material at the anticipated temperature of application. In this case there are ISO grades of lubricant viscosity, as indicated in Figure 32, which are approximately the kinematic viscosity in centistokes (mm^2/sec) at 40°C 9. The ISO specifications are commonly used in industrial lubricants. There are also the SAE viscosity grades for automotive engine lubricants and automotive transmission and axle lubricants[17].

The current SAE viscosity grades for engine oils are shown in Table 9(a) and (b). These viscosity grades are based on measurements at high temperature by kinematic viscosity (e.g., low shear stress viscosity) as well as low temperature viscosities based on a pumpability consideration and a non-Newtonian increase of viscosity at low temperature. Very

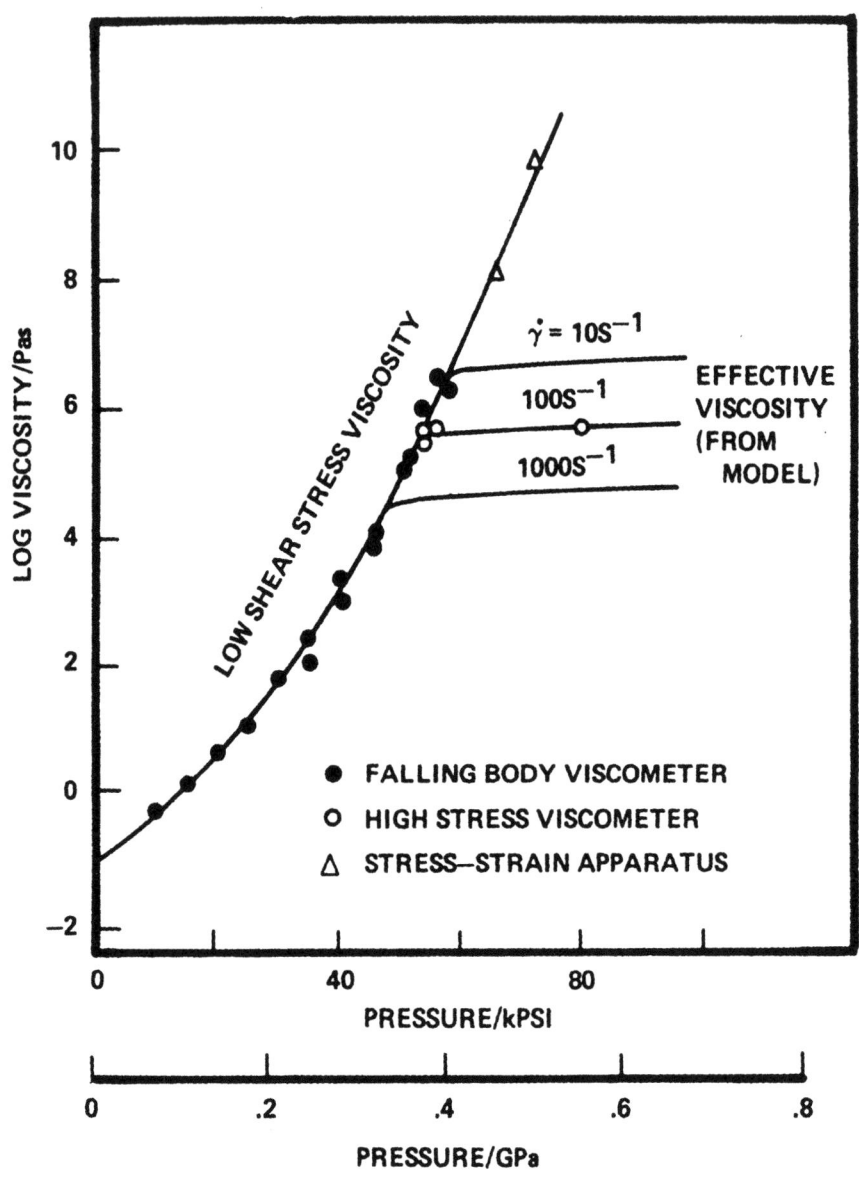

FIGURE 31 VISCOSITY PRESSURE ISOTHERM (60°C) FOR 5P4E BY INDICATED METHODS OF MEASUREMENT. LINES OF CONSTANT SHEAR RATE PREDICTED FORM MODEL (REF. 16)

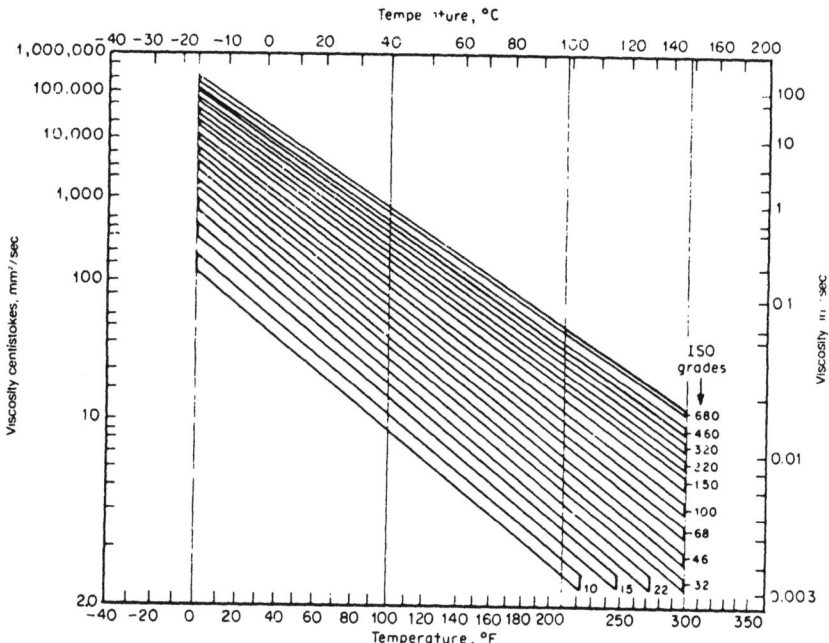

FIGURE 32 VISCOSITY-TEMPERATURE BEHAVIOR OF TYPICAL 100 VI LUBRICANT IN ISO VISCOSITY GRADES (REF. 8)

SAE Viscosity Grade	Viscosity Range		
	Centipoises (cP) at −18°C (ASTM D 2602)	Centistokes (cSt) at 100°C (ASTM D 445)	
	Max	Min	Max
5W	1 250	3.8	—
10W	2 500	4.1	—
20Wª	10 000	5.6	—
20	—	5.6	less than 9.3
30	—	9.3	less than 12.5
40	—	12.5	less than 16.3
50	—	16.3	less than 21.9

Note: 1 cP = 1 mPa·s; 1 cSt = 1 mm²/s
ªSAE 15W may be used to identify SAE 20W oils which have a maximum viscosity at −18°C of 5 000 cP.

TABLE 9(A) SAE VISCOSITY GRADES FOR ENGINE OILS (REF. 17)

SAE Viscosity Grade	Viscosity* (cP) at Temperature (°C)	Borderline Pumping Temperature** (°C)	Viscosity*** (cSt) at 100°C	
	Max	Max	Min	Max
0W	3 250 at −30	−35	3.8	−
5W	3 500 at −25	−30	3.8	−
10W	3 500 at −20	−25	4.1	−
15W	3 500 at −15	−20	5.6	−
20W	4 500 at −10	−15	5.6	−
25W	6 000 at − 5	−10	9.3	−

* By proposed modification of ASTM D 2602,

** ASTM D 3829,   *** ASTM D 445

SAE Information Report

TABLE 9(B) PROPOSED NEW DEFINITIONS FOR "W" GRADES

likely in the near future, additional criteria will be added to the SAE engine oil viscosity specifications for high temperature (150°C) and high shear rate viscosity of the oil.

The SAE viscosity classifications for axle and transmission lubricants are shown in Table 10 and Figure 33, taken from the SAE Handbook[17]. Similarly in the case of greases, a rheological specification is provided by National Lubricating Grease Institute, referred to as NLGI consistency numbers and shown in Table 11. These classifications are also based on measurements according to an ASTM Standard.

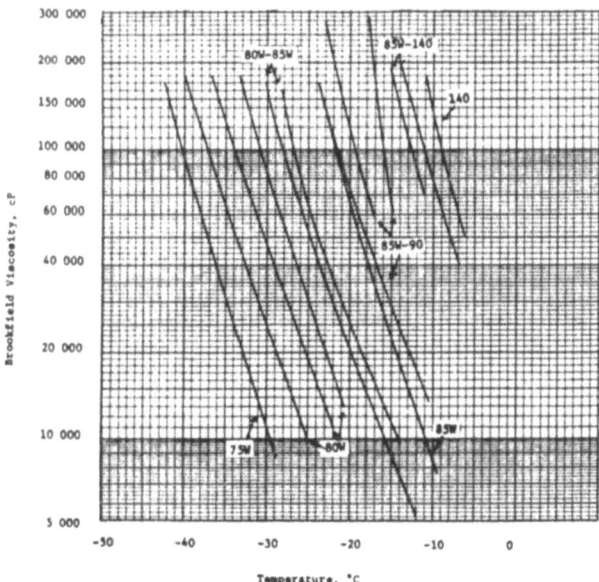

FIGURE 33   BROOKFIELD VISCOSITY VERSUS TEMPERATURE FOR TYPICAL GEAR LUBRICANTS (SAE VISCOSITY GRADES INDICATED) (REF. 17)

The performance specifications on lubricants vary with the particular application. Probably the most highly developed of these is the performance applications utilized by the American Petroleum Institute (API) concerned with automotive engine oil performance. A general description of the performance designations taken from the SAE Handbook is shown in Table 12 [17]. More detailed discussion of the

SAE Viscosity Grade	Maximum Temperature for Viscosity of 150 000 cP[a] °C	Viscosity at 100°C[b] cSt	
		Minimum	Maximum
75W	−40	4.1	—
80W	−26	7.0	—
85W	−12	11.0	—
90	—	13.5	<24.0
140	—	24.0	<41.0
250	—	41.0	—

[a] Centipoise (cP) is the customary absolute viscosity unit and is numerically equal to the corresponding SI unit of millipascal-second (mPa·s).

[b] Centistokes (cSt) is the customary kinematic viscosity unit and is numerically equal to the corresponding SI unit of square millimetre per second (mm² s).

[c] The new viscosity classification represents a conversion to international SI units using degrees Celsius and with a minimum change in viscosity limits relative to prior practice. By early 1982, it is the aim to define the low temperature requirements at suitable multiples of 5°C while retaining 100°C for the high temperature range. The proposed revision will necessitate considering changes of the viscosity limits for the high and/or low temperatures used to define the new system.

TABLE 10   AXLE AND MANUAL TRANSMISSION LUBRICANT VISCOSITY CLASSIFICATION (REF. 17)

NLGI Consistency No.	ASTM Worked (60 Strokes) Penetration at 25°C (77°F) tenths of a millimetre	NLGI Consistency No.	ASTM Worked (60 Strokes) Penetration at 25°C (77°F) tenths of a millimetre
000	445 to 475	3	220 to 250
00	400 to 430	4	175 to 205
0	355 to 385	5	130 to 160
1	310 to 340	6	85 to 115
2	265 to 295		

[a] National Lubricating Grease Institute, 4635 Wyandotte St., Kansas City, Missouri 64112

TABLE 11   NLGI* CONSISTENCY NUMBERS

Letter Designation	API Engine Service Description	ASTM Engine Oil Description
SA	**Formerly for Utility Gasoline and Diesel Engine Service** Service typical of older engines operated under such mild conditions that the protection afforded by compounded oils is not required. This category has no performance requirements and oils in this category should not be used in any engine unless specifically recommended by the equipment manufacturer.	Oil without additive except that it may contain pour and or foam depressants.
SB	**Minimum Duty Gasoline Engine Service** Service typical of older gasoline engines operated under such mild conditions that only minimum protection afforded by compounding is desired. Oils designed for this service have been used since the 1930s and provide only antiscuff capability and resistance to oil oxidation and bearing corrosion. They should not be used in any engine unless specifically recommended by the equipment manufacturer.	Provides some antioxidant and antiscuff capabilities.
SC	**1964 Gasoline Engine Warranty Service** Service typical of gasoline engines in 1964-1967 models of passenger cars and trucks operating under engine manufacturers' warranties in effect during those model years. Oils designed for this service provide control of high and low temperature deposits, wear, rust, and corrosion in gasoline engines.	Oil meeting the 1964-1967 requirements of the automobile manufacturers. Intended primarily for use in passenger cars. Provides low temperature antisludge and antirust performance.
SD	**1968 Gasoline Engine Warranty Maintenance Service** Service typical of gasoline engines in 1968 through 1970 models of passenger cars and some trucks operating under engine manufacturers' warranties in effect during those model years. Also may apply to certain 1971 and or later models, as specified (or recommended) in the owners' manuals. Oils designed for this service provide more protection against high and low temperature engine deposits, wear, rust, and corrosion in gasoline engines than oils which are satisfactory for API Engine Service Category SC and may be used when API Engine Service Category SC is recommended.	Oil meeting the 1968-1971 requirements of the automobile manufacturers. Intended primarily for use in passenger cars. Provides low temperature antisludge and antirust performance.
SE	**1972 Gasoline Engine Warranty Maintenance Service** Service typical of gasoline engines in passenger cars and some trucks beginning with 1972 and certain 1971 models operating under engine manufacturers' warranties. Oils designed for this service provide more protection against oil oxidation, high temperature engine deposits, rust, and corrosion in gasoline engines than oils which are satisfactory for API Engine Service Categories SD or SC and may be used when either of these categories are recommended.	Oil meeting the 1972-1979 requirements of the automobile manufacturers. Intended primarily for use in passenger cars. Provides high temperature antioxidation, low temperature antisludge, and antirust performance.

TABLE 12 DESIGNATION, IDENTIFICATION AND DESCRIPTIONS OF CATEGORIES (REF. 17)

TABLE 12
(Continued)

Letter Designation	API Engine Service Description	ASTM Engine Oil Description
SF	**1980 Gasoline Engine Warranty Maintenance Service** Service typical of gasoline engines in passenger cars and some trucks beginning with the 1980 model operating under engine manufacturers' recommended maintenance procedures. Oils developed for this service provide increased oxidation stability and improved anti-wear performance relative to oils which meet the minimum requirements for API Service Category SE. These oils also provide protection against engine deposits, rust, and corrosion. Oils meeting API Service Category SF may be used where API Service Categories SE, SD, or SC are recommended.	Oil meeting the 1980 warranty requirements of the automobile manufacturers. Intended primarily for use in gasoline engine passenger cars. Provides protection against sludge, varnish, rust, wear, and high-temperature thickening.
CA for Diesel Engine Service	**Light Duty Diesel Engine Service** Service typical of diesel engines operated in mild to moderate duty with high-quality fuels and occasionally has included gasoline engines in mild service. Oils designed for this service provide protection from bearing corrosion and from ring belt deposits in some naturally aspirated diesel engines when using fuels of such quality that they impose no unusual requirements for wear and deposit protection. They were widely used in the late 1940s and 1950s but should not be used in any engine unless specifically recommended by the equipment manufacturer.	Oil meeting the requirements of MIL-L-2104A. For use in gasoline and naturally aspirated diesel engines operated on low sulfur fuel. The MIL-L-2104A Specification was issued in 1954.
CB for Diesel Engine Service	**Moderate Duty Diesel Engine Service** Service typical of diesel engines operated in mild to moderate duty, but with lower quality fuels which necessitate more protection from wear and deposits. Occasionally has included gasoline engines in mild service. Oils designed for this service were introduced in 1949. Such oils provide necessary protection from bearing corrosion and from high temperature deposits in normally aspirated diesel engines with higher sulfur fuels.	Oil for use in gasoline and naturally aspirated diesel engines. Includes MIL-L-2104A oils where the diesel engine test was run using high sulfur fuel.
CC for Diesel Engine Service	**Moderate Duty Diesel and Gasoline Engine Service** Service typical of lightly supercharged diesel engines operated in moderate to severe duty and has included certain heavy duty, gasoline engines. Oils designed for this service were introduced in 1961 and used in many trucks and in industrial and construction equipment and farm tractors. These oils provide protection from high temperature deposits in lightly supercharged diesels and also from rust, corrosion, and low temperature deposits in gasoline engines.	Oil meeting requirements of MIL-L-2104B. Provides low temperature antisludge, antirust, and lightly supercharged diesel engine performance. The MIL-L-2104B specification was issued in 1964.
CD for Diesel Engine Service	**Severe Duty Diesel Engine Service** Service typical of supercharged diesel engines in high speed, high output duty requiring highly effective control of wear and deposits. Oils designed for this service were introduced in 1955, and provide protection from bearing corrosion and from high temperature deposits in supercharged diesel engines when using fuels of a wide quality range.	Oil meeting Caterpillar Tractor Co. certification requirements for Superior Lubricants (Series 3) for Caterpillar diesel engines. Provides moderately supercharged diesel engine performance. The certification of Series 3 oil was established by Caterpillar Tractor Co. in 1955. The related MIL-L-45199 specification was issued in 1958.

particular tests and the nature of the criteria of performance associated with the tests can be found in the SAE Handbook. The tests consist of several engine tests to simulate different types of automotive operating conditions. There are different tests and pass/fail criteria for spark ignited Otto engines and compression ignited diesel engines. The performance criteria change frequently and are determined by joint committee efforts among SAE, API, and ASTM lubricant activities.

There are also performance specifications in other areas of application, the most notable of which is probably the military specifications utilized by various military agencies.

10. CONCLUSIONS

The brief overview of lubricant properties and behavior presented in this chapter serves as an introduction to a rather complex field very dependent on both science and art. It should give the reader a feel for the complexity of the field and the state-of-the-art. It also provides an introduction to the broader literature that exists in the field.

11. REFERENCES

1. Neale, M.J., Ed., "Tribology Handbook," Butterworth, London (1973).

2. O'Connor, J.J., and Boyd, J., Eds., "Standard Handbook of Lubrication Engineering," McGraw-Hill, New York (1968).

3. Gunderson, R.C., and Hart, A.W., Eds., "Synthetic Lubricants," Reinhold Publishing Company, New York (1962).

4. Bowden, F.P., and Tabor, D., "The Friction and Lubrication of Solids," Oxford University Press, New York (1954) and, "Friction and Lubrication of Solids, II," Oxford University Press (1964).

5. Fein, R.S., "AWN-A Proposed Quantitative Measure of Wear Protection," Lubrication Engineering 31 (1975) 581 - 582.

6. Rowe, C.N., "Lubricated Wear," Chapter in Reference (8).

7. Sakurai, T., and Sato, K., "Study of Corrosivity and Correleation Between Chemical Reactivity and Load-Carrying Capacity of Oils Containing Extreme Pressure Agents," ASLE Trans. 9 (1966) 76 - 85.

8. Peterson, M.B., and Winer, W.O., Eds., "Wear Control Handbook," ASME, New York (1981).

9. ASTM Book of Standards, American Society for Testing and Materials, Philadelphia, Pennsylvania (1981).

10. ASME Pressure Viscosity Report, I and II (1953).

11. Bair, S., and Winer, W.O., "Some Observations in High Pressure Rheology of Lubricants," ASME Paper No. 81-LUB-17 (to be published in Trans. ASME, Journal of Lubrication Technology, 1982).

12. Jones, W.R., Johnson, R.L., Sanborn, D.M., and Winer, W.O., "Viscosity-Pressure Measurements of Several Lubricants to $5.5 \times 10^8 N/m^2$ ($8 \times 10^4 psi$) and 149C (300F)," Trans. ASLE, 18, No. 4, (1975) 249 - 262.

13. Alsaad, M., Biar, S., Sanborn, D.M., and Winer, W.O., "Glass Transitions in Lubricants: Its Relation to Elastohydrodynamic Lubrication (EHD)," Trans. ASME, Journal of Lubrication Technology, 100 (1978) 404 - 417.

14. Bair, S., and Winer, W.O., "Some Observations on the Relationship between Lubricant Mechanical and Dielectric Transitions under Pressure," Trans. ASME, Journal of Lubrication Technology, 102 (1980) 229 - 235.

15. Bair, S. and Winer, W.O., "Shear Strength Measurements of Lubricants at High Pressure," Trans. ASME, Journal of Lubrication Technology, 101 (1979)) 251 - 257.

16. Bair, S. and Winer, W.O., "A Rheological Model for Elastohydrodynamic Contacts based on Primary Laboratory Data," Trans. ASME, Journal of Lubrication Technology 101 (1979) 258 - 265.

17. SAE Handbook, 1981, Society of Automotive Engineers (1981).

12. ACKNOWLEDGEMENTS

Figure Number	Source (Chapter Reference)/Citation
1, 2	Reproduced from the Tribology Handbook, edited by M.J. Neale (Butterworths 1973) with the permission of Dr. A.R. Lansdown.
5	Ref. 4, with permission of D. Tabor, F.R.S.
6, 8	With permission of R.S. Fein.
9	Ref. 7, with permission of the American Society of Lubrication Engineers, publishers of Lubrication Engineering and ASLE Transactions.
10	With permission of R.S. Fein.
15	Ref. 1, Tribology Handbook, with permission of the editor, M.J. Neale and publisher, Butterworths.
17, 19, 20(A)(B)	Courtesy of Texaco's Magazine, Lubrication.
33	Ref. 17, Reprinted with permission © 1981 Society of Automotive Engineers, Inc.

Tables

2	Ref. 1, Reprinted by permission of the Council of the Institution of Mechanical Engineers.
4, 5	Courtesy of Texaco's Magazine, Lubrication.
9, 10	Ref. 17, Reprinted with permission © 1981 Society of Automotive Engineers, Inc.
11	With permission of National Lubricating Grease Institute
12	Ref. 17, Reprinted with permission © 1981 Society of Automotive Engineers, Inc.

CONTAMINATION IN FLUID SYSTEMS

Dr. E.C. Fitch
Oklahoma State University

1. INTRODUCTION

Contaminant is the scourge of all fluid dependent systems - including lubrication, hydraulic, pneumatic, liquid/gaseous fuel and coolant types. Operational success of most modern machine systems depends on its control. Synonymous to wear control for fluid systems, contamination control is central to most modern technological systems. By utilizing this fundamental factor, the user can achieve efficiency, reliability, and service longevity. Productivity goals and customer loyalty can only be gained and maintained by achieving contamination control of the integrated fluid systems.

Contaminant, as universally defined today, is any foreign or unwanted energy or substance that can have deleterious effects on system operation, service life, or reliability. The presence of contaminant in the fluid of a system has a catalytic effect upon component performance deteriorating processes (such as tribological and fluid-to-surface type wear mechanisms).

Although the scope of the subject is too broad to present in any depth in this discussion, the practice of contamination control can be viewed which will help fluid systems engineers gain a working perspective of the subject. Details have invariably obscured the critical relationships between the controlling factors and hampered progress. This chapter will attempt to provide a practical introduction to the subject of fluid contamination. Material will be presented under the following headings: theory, types, analyses, sources, consequences, and control/prevention.

2. THEORY

Contaminants in fluid-dependent systems result from change that takes place during manufacture of the components and the system, change that occurs within the system during operation, and change that occurs in the system when it is exposed to the environment. The resulting contaminant must be accurately characterized and assessed before the user can appraise the seriousness of its manifestation.

The level of contaminant can vary from point to point in the system, from time to time, and unfortunately by the method of sampling. The contamination level is always highest just upstream of the contaminant separator and, of course, immediately downstream of the major ingression sources.

The ingression duty cycle corresponds to the severity of the actual work and respective work environment of the system and is the most important factor influencing contamination level on a time basis. Flow through the contaminant separator (or the residence time of the fluid in the system) also determines the magnitude of the contamination level. The only remaining factor influencing the contamination level of the fluid is the actual performance characteristics of the contaminant separator, i.e., the better it is, the lower the level.

A mathematical model can be written to express the contamination level of a circulating type system, as illustrated in Figure 1. This model would apply to all fluid system types, with the possible exception of a "single pass" fuel system. The theoretical relationship would be an accounting of the material ingressed per minute (R), the material removed by a filter having a filtration ratio of Beta, and the material still present in the system as reflected downstream of the filter (N_d).

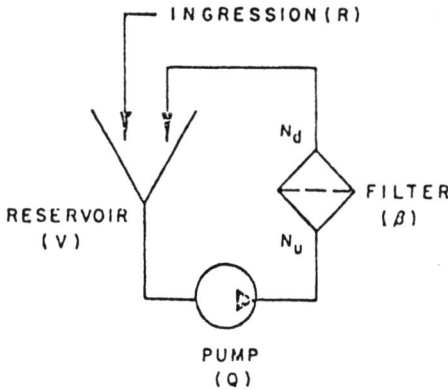

FIGURE 1 BASIC CIRCUIT

During the transient period, the ratio of the volume of the reservoir (V) to the flow rate (Q) of the pump can influence, along with Beta of the filter, the time constant of the contamination level curve of the system. However, since this transient period is generally very short in most modern systems, only the steady-state part of the differential equation is of real interest. For all practical purposes, the steady-state value of the downstream contamination level of the fluid can be described by the following equation;

$$\bar{N}_d = \frac{R}{(\beta - 1) Q} \qquad (1)$$

The filtration ratio, Beta, is a function of both upstream and downstream particle concentration per unit volume of fluid as defined by

$$\beta = \frac{N_u}{N_d} \qquad (2)$$

at the steady-state condition when the standard contaminant (AC Fine Test Dust) is the ingressed material. All particle-related terms in Eq. (1) are cumulative (i.e., includes all particles above a given size). For example, if the reference size is 10 micrometers, the units of "R" would be the number of particles 10 micrometers and larger which ingress the system per minute; while the filtration ratio would be identified as Beta Ten.

Cumulative efficiency of a filter is defined as

$$E_c = \frac{N_u - N_d}{N_u} \tag{3}$$

Using Eq. (2), the cumulative effeciency can be written as a function of only Beta, or

$$E_c = \frac{\beta - 1}{\beta} \tag{4}$$

This relationship is reflected in the efficiency graph shown in Figure 2.

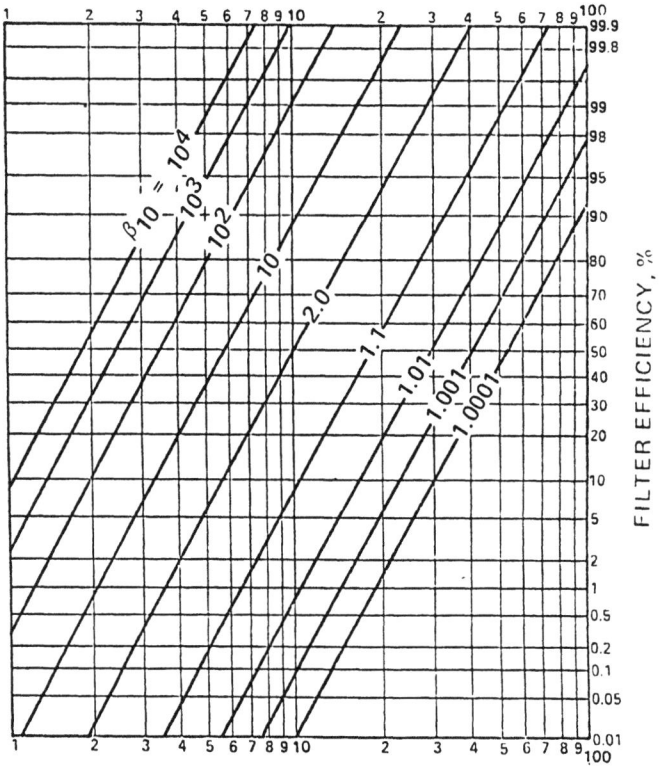

FIGURE 2 BETA VERSUS EFFICIENCY

Note from Eq. (1) that the contamination level of the system can be reduced an order of magnitude by simply changing either R, Q, or Beta appropriately. Control possibilities are evident and should be recognized.

Contaminant service life of a component or system depends on the balance existing between the two influencing factors - contamination level of the fluid and contaminant tolerance of the component or system in question. The contaminant tolerance level exhibited by a fluid component depends upon its contaminant sensitivity and operating conditions. The contaminant sensitivity of a fluid component, an inherent characteristic, depends upon three design aspects - basic mechanism, fabrication materials, and unit loading. Operating conditions which may be imposed on the fluid component include not only fluid properties but also the severity of the pressure, temperature and speed duty cycle.

All fluid components are sensitive in some degree to particulate contaminant entrained in the fluid. The term, "contaminant sensitivity," refers to degradation in performance occurring when a component is exposed to a specific contamination environment. For example, for a pump, rationalize that some finite fluid delivery volume is lost for every particle passing through its pumping chamber. This lost delivery volume (due to exposure to a given particle size) can be expressed as

$$Q' = \frac{\text{volume lost}}{\text{particle}} \times \frac{\text{particle exposure}}{\text{time}} \qquad (5)$$

Of course, the rate of particle exposure is simply the product of the particle concentration per unit volume and the flow rate.

Contaminant wear results from particle interaction with component surfaces. Contaminant wear can be reflected in two different ways; by actual wear debris generated from the critical surfaces and degradation in a responsive performance parameter of the component resulting from surface deterioration and clearance changes. The block diagram shown in Figure 3 relates these concepts for a hydraulic pump. Operating conditions and the exposed contaminant level cause the destruction of critical surfaces. The results are wear debris and clearance changes. Wear debris can be measured by a Ferrograph and reported in terms of the D54 density. The clearance change can be measured in terms of flow readings and reported in terms of flow degradation. The block diagram shows why an interrelation exists between flow degradation of a pump and its Ferrographic reading, a most important conclusion to the study of fluid contamination control.

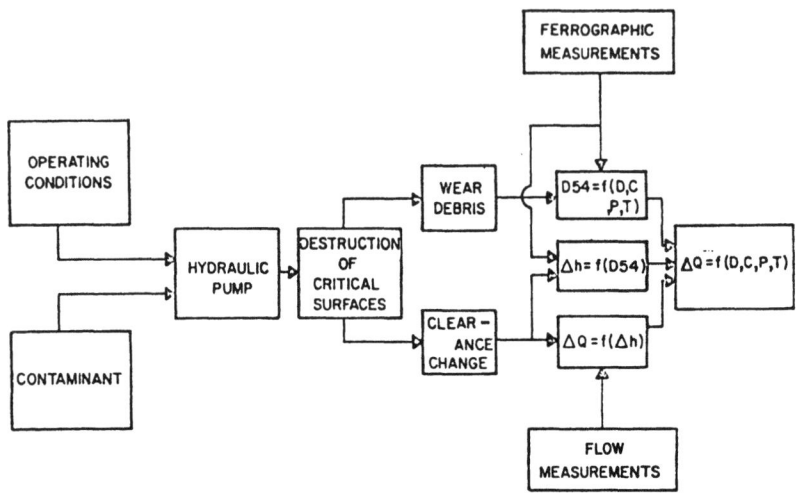

FIGURE 3   PERFORMANCE VERSUS WEAR

Research and application in fluid contamination control have been hampered by the complexity of the interacting parameters. Only recently have researchers gained sufficient insight in its practice, permitting recognition and understanding of the controlling parameters of the system and adequate description of the influence of each parameter on the system as a whole.

The Contamination Control Balance, illustrated in Figure 4, attempts to convey the nature of the interacting parameters in a pictorial form. The Balance should provide necessary insight for implementation of the theory and prove valuable in helping the reader understand and explain many principles on which contamination control is based.

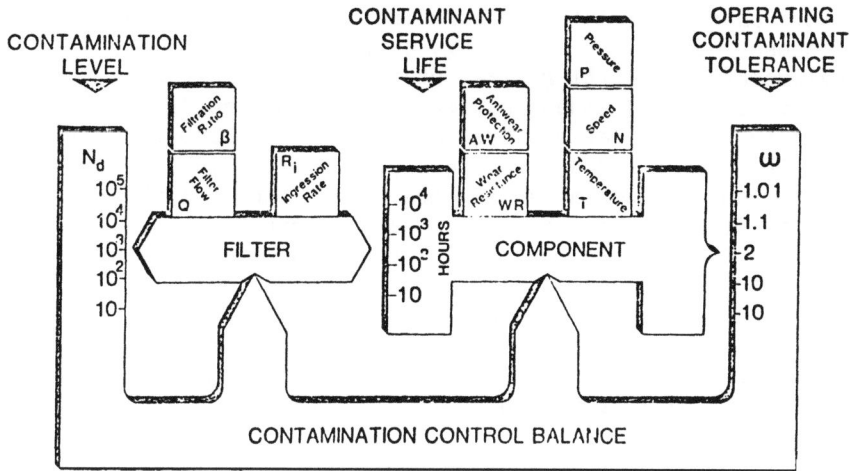

FIGURE 4   CONTAMINATION CONTROL BALANCE

3. CONTAMINANT TYPES

Contaminant, unwanted material or energy in the system, can exist as a gas, liquid, or solid. A gas can either be in the dissolved state or entrained in the liquid as bubbles or slugs. A liquid, depending on its volume and compatibility with the host fluid, can be free, dissolve, or emulsified. Since the term, liquid, encompasses all deformable, cohesive matter, precipitates (such as gums and biological debris) are included in this category. Solid matter found in the fluid is better known as particulates, or just particles. Material type contaminants of interest in fluid systems include particulate, aeration, tramp liquid (water), chemical and microbial.

Unwanted energy in fluid systems can deteriorate both the fluid and exposed system components and be the prime factor for failure, poor reliability and short service life. Forms of energy which can be highly detrimental to fluid-dependent systems are radiation, static electricity, corrosion, magnetism, and thermal energy.

Particulate contamination is characterized by specific descriptive, identifying terms which can be classified as follows:

- inherent physical and chemical properties
- environmental behavior
- geometrical shape
- size distribution
- concentration or population density.

Actual characterization factors include particle density, hardness, compaction, settling, dispersion, transport, agglomeration, limiting size, state, shape, size, size distribution, and concentration. Particle size distributions found in fluid-dependent systems are not normal but skewed in the direction of greater numbers of small particles. When plotted on the cumulative log-log^2 graph, these system distributions often exhibit a linear characteristic. If the distribution plot is not linear, problems usually exist.

Aeration is due to dissolved or suspended air in the system fluid. As long as air remains dissolved, no problems occur; but, in low pressure areas, it can become suspended air and the system will exhibit many undesirable characteristics such as the following:

- lower bulk modulus
- loss of power
- gaseous cavitation
- noisy operation
- accelerated fluid oxidation
- loss of lubricity
- higher temperatures
- loss of stiffness.

Water, a serious contaminant in most mineral-base fluids, is always present in operating systems to some degree, either in its free or dissolved form. Free water may be present as a precipitant or emulsified with the host fluid. Water is a major cause of rust and corrosion and the breakdown of additive packages in the fluid. When both water and dirt are present in the fluid, a synergistic effect results. Also, water in a system at subfreezing temperatures can transform into ice crystals that jam valves, clog filters, and cause controls to become erratic, the same problems as caused by sand particles and metal chips.

Chemical contaminants include solvents, fluid-breakdown residue, and surface-active agents. Literature abounds with horror stories of the contamination of complex fluid systems by chemicals. Incompatible chemicals can hydrolyze in the system, forming acids which in turn attack internal metallic surfaces. Surface-active chemicals, called surfactants, produce a variety of unfavorable conditions in fluid systems, particularly fuel systems.

Microbial contaminant is an ever present threat, particularly when water exists in the system. Once active growth has commenced in the aqueous phase of the fluid, it is generally self-sustaining due to the fact that water is one of the end products of an infestation. Microbial contamination results in the following:

- short fluid life
- degraded surface finish
- short filter life
- rapid corrosion
- obnoxious smells and fluid discolorations.

Radiation contamination is radiant energy in, on, or around the fluid system. It can change the state of a material, induce a chemical reaction, change a physical characteristic, or alter an electrical or magnetic characteristic of a material. In a fluid system, the fluid itself is most susceptible to damage by radiation. Changes have been noted in viscosity, acidity, volatility, foaming, coking, flash point, autogeneous ignition temperatures, and oxidation stability. Elastomers, plastics, and resins (which make up many of the seals, packings, and similar components of fluid systems) are also severely vulnerable to radiation contamination.

Static electricity or electrokinetic contamination in fluid systems result from fluid moving past surfaces within the system. At the solid-fluid interface, ions become detached, causing a streaming current, and the potential difference along the direction of flow is referred to as the streaming potential. This resulting charge depends on the flow velocity, properties of the fluid, and the surface area of the solid-liquid interface. This contaminant causes serious erosion in valves, and the charge generation has been responsible for many industrial fires.

Corrosion contamination includes electrochemical reactions that occur within the confines of a closed fluid system. This reaction, an oxidation and reduction process, can destroy critical surfaces and generate gross amounts of particulate contamination (rust). Types of corrosion identified include galvanic, pitting, crevice, intergranular, erosion, stress/cracking, and fretting.

Magnetism, a contaminant in a fluid dependent system when unshielded magnetic fields exist near control actuators, can force ferromagnetic particles in the fluid to move and collect in critical areas. Such capturing of particles can have serious consequences when they bridge, clog, or jam critical orifices, clearances, or invade load-bearing wedge-shaped openings.

Thermal energy (either excess or lack of it) can be just as devastating a contaminant to a fluid system as any other unwanted or foreign substance or energy. Temperatures above design conditions can cause a chain reaction leading to the total destruction of the system, early seal failure and fluid breakdown. The effect of subfreezing temperatures on the system can be highly detrimental to the proper operation and functioning of the system. Low temperatures result in high input power; pump cavitation; seals loose their memory, become distorted and crystallize; valves respond slowly; filters bypass or rupture, etc.

Regardless of its form, if the degree of manifestation of contaminant is intolerable to the components of the system, it must be reduced in level and controlled, limiting its ingression or removing it from the system.

4. ANALYSIS

The credibility of contaminant analysis depends just as much on having a representative sample of the contaminant as on the accuracy of the evaluation technique. To be representative of the contaminant in the system, the sample must be undefiled by exposed surfaces or by the environment with which it may come in contact prior to being appraised. Furthermore, for a sample to be representative, it must contain a full contaminant spectrum from a nondiscriminatory part of the system. Finally, the contamination assessment method has a direct influence on the potential value of the results.

Sample container cleanliness is the first prerequisite for achieving valid contaminant analysis. Furthermore, such control must be exercised over sampling appendages and any other exposed wetted surfaces within the analyzer process. Cleanliness control must also be maintained over the fluids used to flush, rinse, or dilute the sample. When particle evaluation methods do not restrict particle entry or consider the effect of tramp particles, the cleanliness of the environment surrounding the analysis station can be suspect.

Sampling method is the second prerequisite needed to ensure the validity of contaminant analysis of a fluid. Validity concerns the propriety of the sample with respect to the contaminant population it represents. Using an inappropriate sampling technique can compromise sample validity, as can withdrawing the fluid from a biased contaminant-level location or obtaining the sample during a nonrepresentative period of system activity. Selection of the sampling method and its proper application are critical aspects of the overall analysis effort. Obviously, if a sample is removed from a cyclonic or particle size stratified zone, the specimen is not representative. Finally, contamination profiles are dynamic in nature and continually change (depending on the environment, wear of exclusion devices, and the characteristics of the filter). Hence, the sampling period, with respect to the machine or to its system work and environmental cycle, is of paramount importance in obtaining a representative fluid sample.

Particle dispersion is a major requirement if the analysis method is to discriminate between discrete particles in the fluid. Of course, dilution helps minimize the interaction between particles and reduce flocculation and agglomeration; but, dispersing agents may be needed to stabilize the suspension. Poorly dispersed particles often represent the biggest single problem in size analysis of particulate contamination. When flocs form, particle size values are usually too large and the size distribution is too broad.

Particle settling is often a serious problem - particularly when particle density is great and fluid viscosity low. Particles must be maintained in suspension so that the full size distribution is presented representatively to the sensing zone of the analyzer, or a biased analysis results. The use of magnetic stirrers (for nonferrous particles) is an effective solution but inappropriate for most wear-type system samples. The settling rate of discrete particles in a given fluid can be assessed by Stokes Law or by conducting repeated analyses and establishing the rate at which the number of particles of a given size declines. Compensation for significant particle settling must be made or fluid viscosity increased (through dilution with a more viscous fluid or by refrigeration to hold heavier particles in suspension). The cumulative particle-size distribution curve steepens as the settling of particles continues unabated.

Aeration occurs when air bubbles are dispersed in a fluid. Particle counting instruments usually count them as if they are particulate matter. In fact, air entrained in the sample during agitation prior to analysis can totally mask the true size distribution of the particles. The size of the air bubbles depends on the surface tension of the fluid in which the air is dispersed. Normally most of the particles are small, producing an upward inflection of the cumulative particle size distribution curve at the small particle sizes. Researchers have developed effective techniques to give fast deaeration with minimal difficulty.

Undissolved liquid (e.g. water) in a sample can totally mask the true size distribution of the particles, just like air. Sometimes such a liquid is inadvertently added with dilutent, or it may have been contained originally in the sample of fluid. In either case, agitation of the sample just prior to anlysis effectively disperses the small droplets throughout the sample and results in an erroneous distribution of the particulate matter in the fluid. Undissolved and finely dispersed liquid droplets in a fluid cause an unrealistically large particle population and produce an effective flattening of the cumulative distribution curve.

Once a representative contaminant sample has been obtained and properly prepared (deaerated, dehydrated, dispersed and diluted); its assessment, reporting, and interpretation are all subjects of concern. Factors which deserve consideration are the following:

- effective analysis methods
- analysis method calibration
- cleanliness level reporting
- analysis interpretation.

Many methods are available for analyzing distribution of particle sizes in a fluid. Each method utilizes some particular property or combination of properties to distinguish one size particle from another. In some methods, the size dimension is measured directly; whereas, in others, the dimension is derived from measured physical behavior of the particles. Furthermore, some methods detect and measure each particle; whereas, in other methods, particles are evaluated as a unit.

Analysis methods can be classified as optical, electrical, geometric, gravitational, cylconic, magnetic, particle concentration, or ultrasonic. Optical methods include imaging (microscopes) and light extinction methods (such as employed by HIAC/ROYCO). Electrical types measure the change in fluid resistance when a particle passes through an aperature. Geometric methods include the ever popular sieving method and the modern version, the microsieve. Gravitational methods cover both sedimentation and elutration techniques. Inertial type separators perform cyclonic classification. Ferrography is the basic magnetic type analysis method. Particle concentration methods include gravimetric, silting index, turbidity, optical density, and the simple patch test. Ultrasonics is making a comeback, based on the work at Brown University.

Automatic particle counters are the most popular means of assessing the contamination level of fluids in industry today. These instruments would not be popular, however, if standards did not exist with which to calibrate them. For many years, users of particle counters were totally dependent on the manufacturer of the counter for calibration and achieving a commonality between laboratories. This dependency ended in the fluid power industry with the adoption of a calibration procedure that utilizes AC Fine Test Dust as the calibration medium.

A reproducibility survey was conducted prior to the adoption of the procedure by ISO and has become a standard (ISO 4402). This round-robin indicated that laboratories which had previously calibrated their counters using ACFTD procedure showed excellent reproducibility; whereas, those which depended upon intuition and nonstandard calibration techniques had unsatisfactory performance. Typical results of this survey are shown in Figure 5. The international survey demonstrated that, after each counter was calibrated per the ACFTD procedure, particles could be counted with an average standard deviation of $2/3\sqrt{N}$ (where N is the average number of particle counts). That is, if 10,000 particles are being counted, then the deviation would be 66.7 particles.

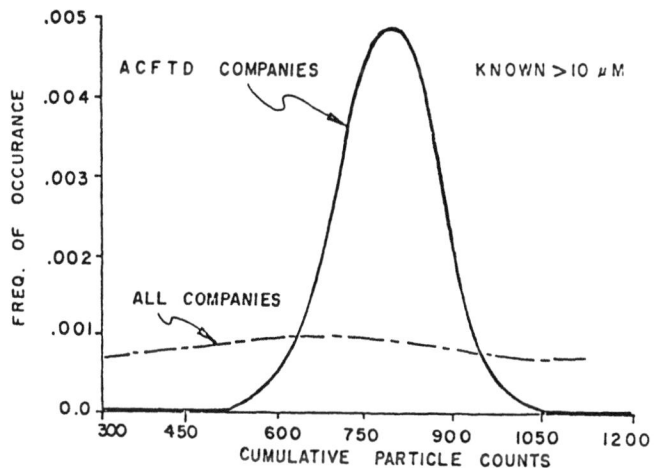

FIGURE 5   RESULTS OF INTERNATIONAL SURVEY

Categorizing particle counts to define cleanliness classes was instituted in the early 1960's by several organizations. Of these early attempts, only NAS 1638 is still in use within the aerospace field. All the early cleanliness codes had one common flaw - they were based on a fixed particle size distribution. Interestingly enough, this fixed distribution, selected as typical of field system ditributions of that period, corresponds roughly to that of AC Fine Test Dust. Undoubtedly, particle distributions from those early systems displayed the results of poor filter bypass control, which tended to flatten the distributions like AC Fine Test Dust.

The ISO Solid Contaminant Code (ISO 4406), approved in the late 1970's, represents a significant milestone in the area of fluid-contamination control. It provides a simple, unmistakable, meaningful, and consistent means of communication between suppliers and users. It applies to all types of fluid systems, and a theoretically infinite number of slopes is available to describe the contamination level of a fluid.

The ISO Code is assigned on the basis of the number of particles per unit volume greater than 5 and 15 micrometers in size. These two sizes were selected because of feeling that the concentration at the smaller size would give an accurate assessment of the silting condition of the fluid, while the population of the particles greater than 15 micrometers would reflect the prevalence of wear catalysts. Thus, particle size distribution by the ISO coding system, is described by a 5 micrometer range number and a 15 micrometer range number (with the two numbers separated by a solidus). When particle size distribution is plotted on the graph shown in Figure 6, the ISO Code can be read by noting the corresponding Range Numbers on the graph directly where the curve crosses the 5 and 15 micrometer particle size.

6. SOURCES

As far as functioning, service life, and reliability of a fluid system are concerned, the specific source of contaminant makes little difference. The fact that it exists, is entrained in the fluid, and gains exposure to critical surfaces is of fundamental importance. The term "ingressed contaminant" is reserved for contaminant from "wherever", the magnitude of which must be controlled.

Forms and types of contaminants present in the fluid encompass the complete range of materials and fluids used in

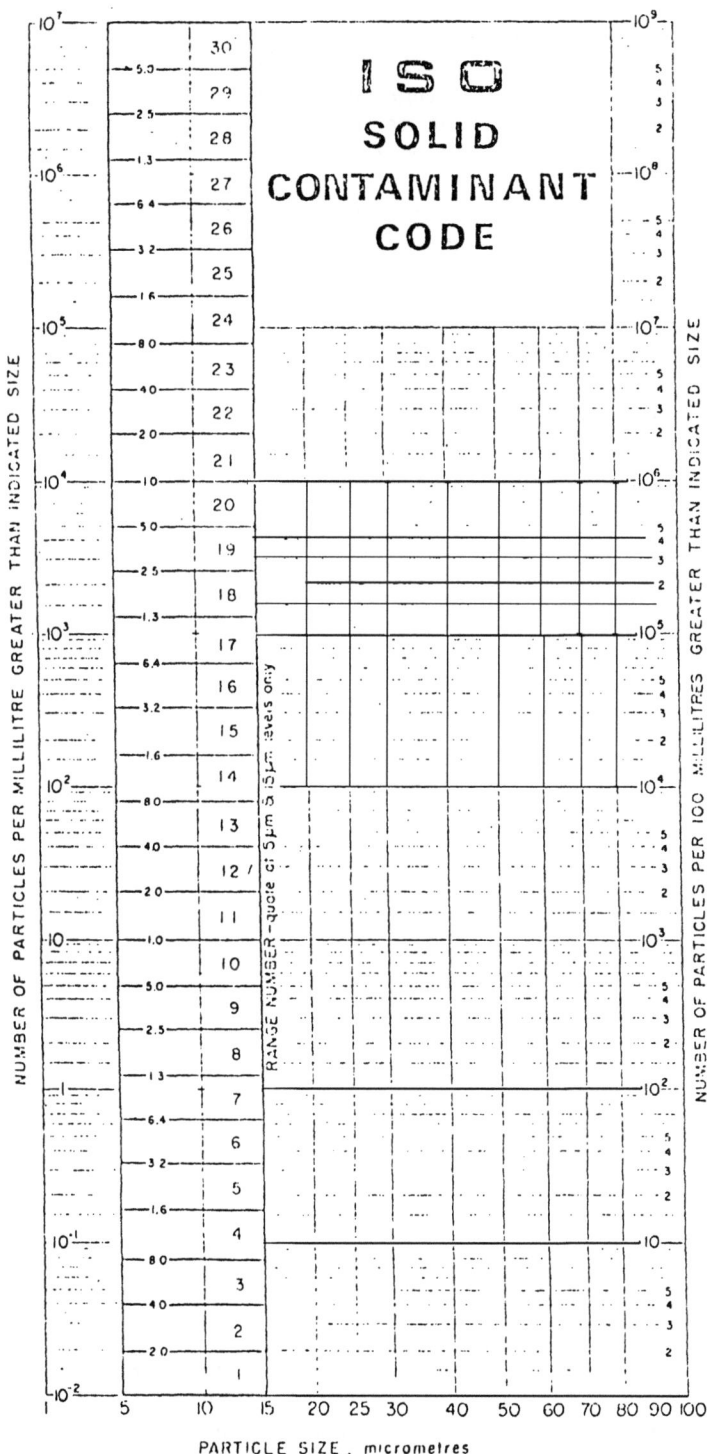

FIGURE 6 THE ISO SOLID CONTAMINANT CODE

producing and operating the system, as well as contaminant characteristics of the ambient environment in which the system has operated. Even the ingression mechanism affects the type system contaminants, such mechanisms as the wear modes, chemical reactions, and ingestion paths.

Specific origin of system contaminant is difficult to identify because sources of particulates are as numerous as the material constituents involved in the system. The dominant contaminant in a system is often far more important than the spectrum. Dominant contaminants are generally good indicators of system status; prevalent wear modes, internal state condition, severity of duty cycle, and degree of environmental hostility.

The sources of system contaminants can be classified as follows:

- Implanted contaminant
- Generated contaminant
- Ingested contaminant
- Escaped contaminant
- Induced contaminant
- Biological contaminant.

Implanted contaminant originates from manufacturing, handling, packaging, dispatching, and assembling process. This type of contaminant is the most critical to the system during new system break-in and commissioning exercises. Components such as lines and reservoirs with surface soil can be hazardous, particularly when the system is first placed in operation. But, if surfaces are rusty and corroded, they can cause damage throughout the life of the system, these are true perpetual ingression sources. The way to minimize implanted contaminant is to ensure that component surfaces are clean and protected prior to assembly. Once a system has been assembled, it should be flushed and checked by the Roll-off Cleanliness procedure (T2.9.8M-1979) recently approved by the National Fluid Power Association (NFPA). Filling the system with new "dirty oil" is another more subtle way of implanting contaminant in the system.

Generated contaminant constantly contributes ingressed contaminant. This contaminant represents the spoils of system inactivity, operation, and deployment. The magnitude of generated contaminant is normally controlled by respecting the severity limits of "rated" operation, environment, and fluid cleanliness. Barring the influence of human failings, only when the classical wearout mode occurs should the generated type of ingressed contaminant be alarming. Pump tests conducted by the Milwaukee School of Engineering and reported in SAE Paper 690866 showed that the average contaminant generated per hour is as shown in Figure 7. The Ferrograph offers an excellent method for monitoring generated contaminant from a fluid system, as illustrated in Figure 8 for a pump break-in test.

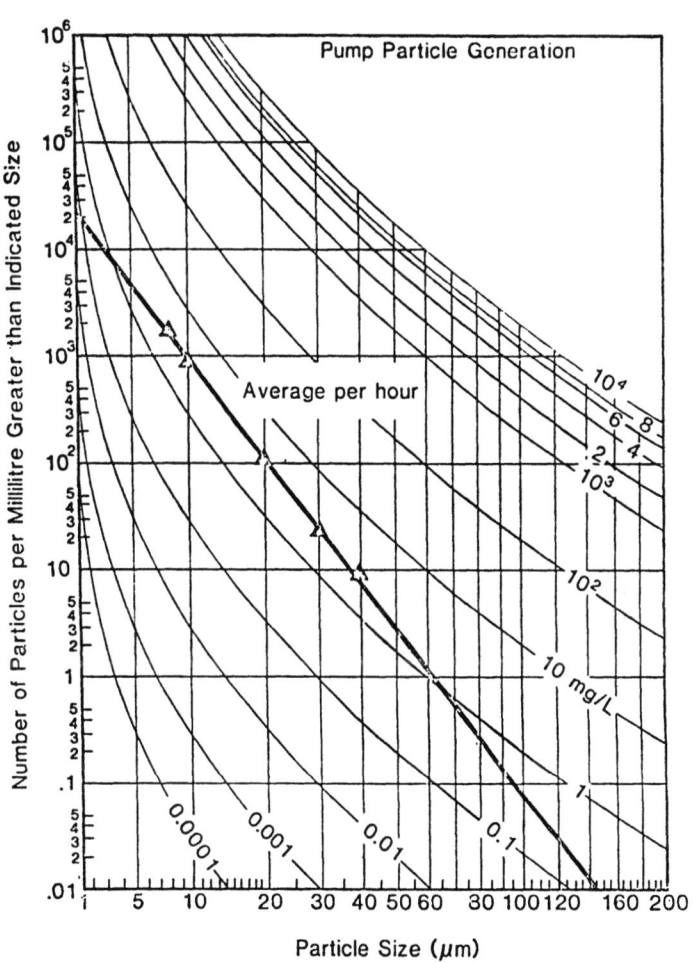

FIGURE 7   PUMP CONTAMINANT GENERATION RATE

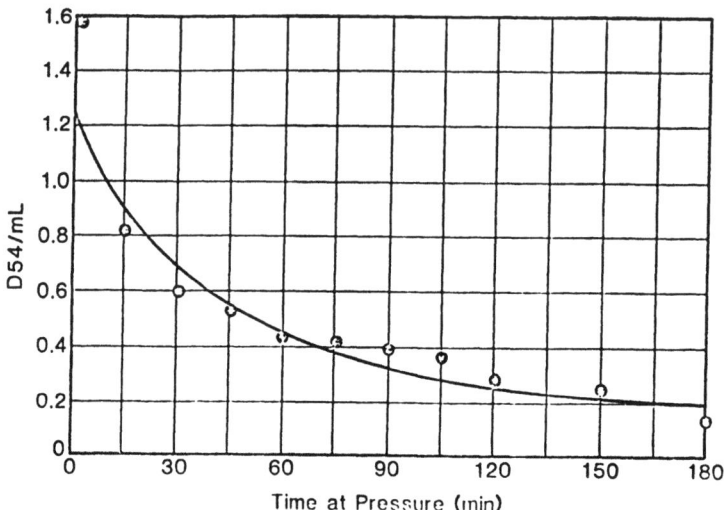

FIGURE 8   WEAR VERSUS PUMP BREAK-IN TIME

Ingested contaminant should be of utmost concern when the system must operate in hostile, dirty environments. Absolute exclusion of contaminant from the ambience is an idealistic thought. Field system studies have shown that ingested contaminant from wiper seals and reservoir breathers can totally dominate the contaminant ingression picture. When conditions warrant, this type of contaminant deserves almost "vigilante" attention.

Escaped contaminant is particles that have migrated through or around the filter. This type of contaminant must be identified quickly because the contamination level of the fluid is almost totally dependent upon the filter, not only for capturing particles but also for retaining them until they can be externally disposed. This contaminant results from a sneak flow path past the filter (an unseated relief or bypass valve or a cut in the filter medium) as well as from a flow surge that allows the viscous drag forces on the captured particles to exceed surface held forces of the fibers of the filter. Escaped contaminant is one of the major causes of early system failure and must be controlled in every conceivable way.

Induced contaminant is defilement from careless, improper, malicious, or hurried acts of system intrusions. This contaminant is often an operator's or user's nightmare. It exists because someone did the wrong thing - a farm worker adding water to the hydraulic system in order to finish a so-called perishable activity or mine personnel adding coal dust to the hydraulic reservoir to raise the level of the fluid to energize a float switch. Strict maintenance practices and operating policies are the only solution to this problem, rules that must be established by the one who must pay the consequences (the cost of equipment overhaul and the loss of production during downtime).

Microbial contaminant can only be arrested by the addition of a biocide to the fluid or by removal of one of the key physiological or nutritional requirements needed for microbes to flourish, that is, water, energy, or nutrients. Ingression of microbes may not be stopped, but the conditions for their survival and growth may be eliminated in the system. Unless arrested, microbial growth (regardless of the strain) results in a slime and usually an increase in the viscosity of the host fluid.

6. CONSEQUENCES

Every component has a limiting value for the contamination level which must be respected if the system goals for safety, performance, reliability, and service life are to be achieved. The consequences or penalty for not maintaining the contamination level below its limiting value is the loss of acceptable performace, a perceptible if not catastrophic failure. A perceptible failure includes loss of performance caused by abrasive, erosive, corrosive, electrostatic, and adhesive wear; wear types of which operator awareness generally exists. Catastrophic failures include such dreaded events as ruptures, collapses, lockups, obstructions, seizing, clogging, jamming, silting, and obliterations.

Degradation in performance can be related in most instances to the amount of contaminant wear (wear debris generation) occurring in the component. This relationship is a fundamental aspect of the contaminant sensitivity of components.

Sensitivity of a component to contaminant can be assessed by exposing it to ever-increasing sizes of contaminant, while measuring the influence of each size on the designated performance parameter. The performance parameter for a pump is flow; for a valve, it might be control pressure; for a hydraulic cylinder, the position hold capability; and, for a hydraulic pressure seal, static or dynamic leakage.

The test circuit needed to evaluate the contaminant sensitivity of a fluid component is schematically illustrated in Figure 9. With the test component operated at specified conditions, contaminants of various size ranges are exposed to the component for 30 minutes and then filtered. After each exposure period, the performance parameter is measured and recorded. The performance value after exposure is divided by the original performance value to obtain a performance degradation ratio. The performance degradation for both flow and pressure for a pressure-compensated pump is shown in Figure 10. In this example, the compensated pressure increases, which could lead to a dangerous safety situation. Flow degradation, as contaminant size ranges are exposed to the pump, continues to degrade in a normal fashion throughout pump life.

Another approach to contaminant sensitivity testing of fluid components is to monitor the amount of wear debris generated after each size exposure of contaminant. This can be done today using Ferrography. Standard contaminant sensitivity tests require test contaminant concentrations of 300 mg/l to yield measurable performance degradation values. Using Ferrography, only 10 - 20 mg/l concentrations have proved sufficiently sensitive and discriminatory on gear pumps to show even small changes in wear rates due to exposure to the standard size ranges of test contaminant. Debris generation rate has been correlated quite effectively with performance degradation in the case of gear pumps - Figure 11 illustrates both methods for two different pumps.

The assumptions that water in fluid systems simply vaporize, dissolve or emulsify and do no harm are entirely erroneous. The fact is that water can be extremely detrimental to the system long before it has a chance to "vaporize out," and the total elimination of dissolved water in oil is highly unlikely by the simple act of heating. Water-induced effects need investigation as much as any other contaminant of the system if high performance and reliability are to be achieved.

In most systems, there is very little chance of avoiding the entrance of moisture into the system. The U.S. Army studied water absorption tendencies of hydraulic fluids and found that fluids exposed for 30 days to an 80 percent relative humidity experienced an increase in water content as much as 804 percent.

The presence of water can grossly disrupt the "balance" within the chemical system of oil and interfere with the normal performance of additives through processes that currently are not fully understood. However, it is known that water reacts with zinc dialkyl dithiophosphate (ZDDP), an antiwear additive, and totally destroys its effectiveness. In fact, most investigators agree that ZDDP decomposes in the presence of water to form free sulfur or hydrogen sulfide. Thus, highly corrosive hydro-sulfurous acids can form that can destroy critical mating surfaces within the component when water is available.

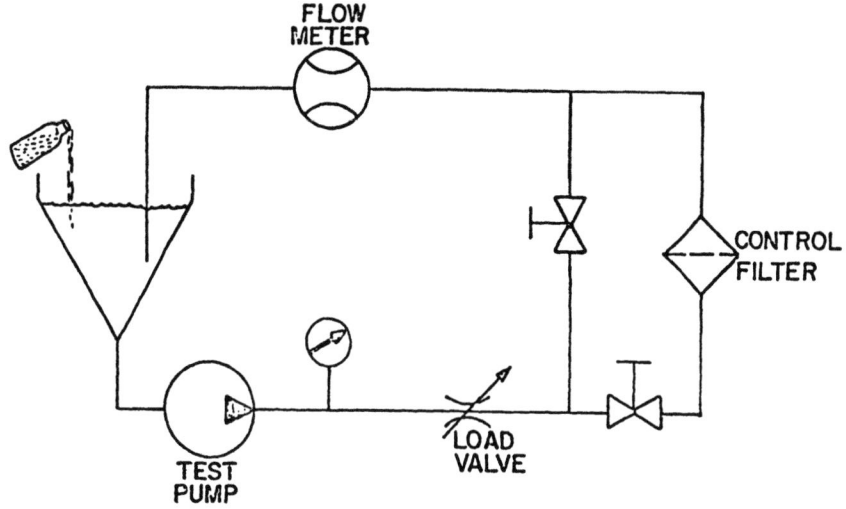

FIGURE 9 SCHEMATIC OF PUMP CONTAMINANT SENSITIVITY TEST CIRCUIT

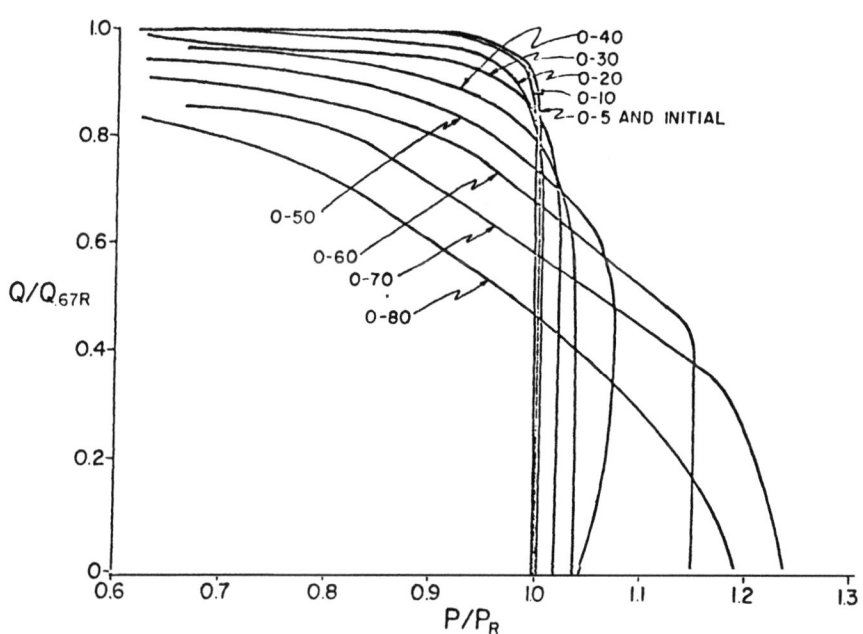

FIGURE 10 PRESSURE COMPENSATED PUMP CONTAMINANT TEST RESULTS

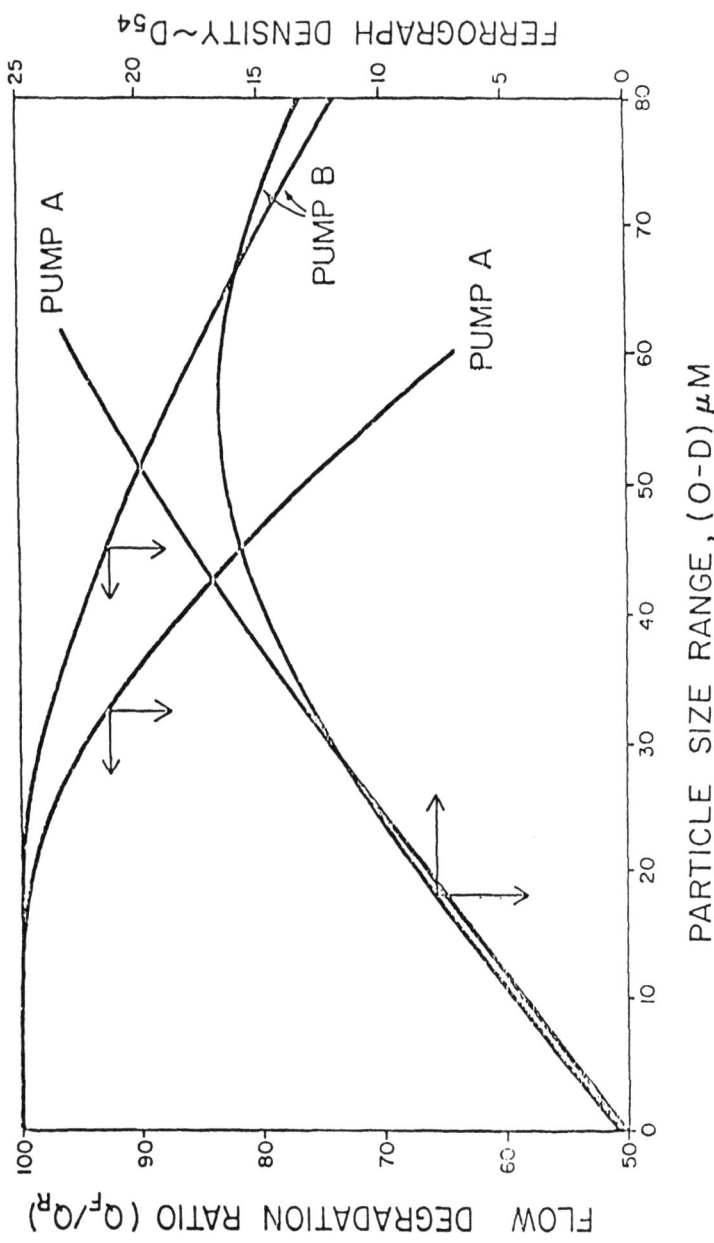

FIGURE 11 CONTAMINANT SENSITIVITY VERSUS PUMP WEAR DEBRIS

7. CONTROL/PREVENTION

In a fluid system, control over contamination is achieved and contaminant related failure is prevented simply by maintaining the contamination level of the fluid below the contaminant tolerance level of the components. If this balance is not possible, three basic options exist as follows:

- Increase the separation performance of the filter
- Decrease the contaminant ingression rate of the system
- Improve the contaminant tolerance of the components.

The separation performance of a filter depends upon three factors, namely;

- fluid residence time
- structural integrity of filter
- particle capture/retention capability of filter.

Fluid resistance time is equal to the circulating volume of the system divided by the flow rate through the filter. The lower the residence time, the higher the separation performance and the lower the contamination level. For any given filter, the lower the residence time, the shorter the time constant of the system - i.e., the faster the system responds in "cleaning-up" or reducing the fluid contamination level of the system. The more times that fluid is processed through the filter, the more opportunities that exist for the filter to remove the contaminating particles.

Structural integrity of a filter assembly is one of the most important factors in contamination control. If structural deficiencies exist in an assembly, all other "good" and desirable features that the assembly might possess can be totally overshadowed. Such deficiencies can result in unfiltered flow paths large enough to negate completely the intended function of the filter in the system.

To verify the absence of such deficiencies in the filter, five ISO approved standards exist to help assess the structural integrity of the filter elements:

- Fabrication Integrity (ISO 2942) - reveals defects in the element
- Collapse/Burst (ISO 2941) - shows ressitance of structure to differential pressure
- Material Compatibility (ISO 2943) - assesses deterioration tendency of element in hot system fluid
- End Load (ISO 3723) - shows axial compression resistance of element after Material Compatibility Test
- Flow Fatigue (ISO 3724) - reveals cyclic flow endurance of filter medium.

Other structural deficiencies in filters are common throughout the industry, and ISO procedures are not as yet available for their identification and assessment. These deficiencies include the following:

- Element to housing seal - sealing effectiveness under static and dynamic flow and mechanical vibrating conditions
- Bypass valve leakage - versus temperature, pressure differential, and mechanical vibration
- Silting susceptibility of bypass valve - under both open and closed conditions
- Cold soak bypass response - for filter element rupture prevention under cold startup conditions.

Filter media offer two types of capture/retention sites for particles entrained in the influent:

- Structural trap - where particles are held by mechanical means, like in a screen - called absorption
- Surface adhesion - where particles are held by surface forces, like electrostatic forces of van der Wall's attractive forces, called adsorption

Particles reach the capture site by action of one or more transport mechanisms, as illustrated in Figure 12.

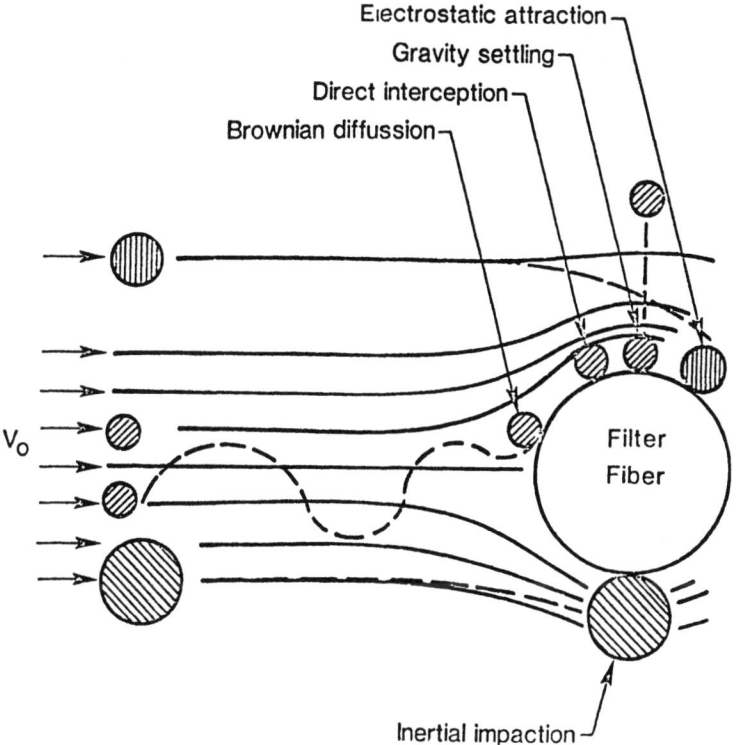

FIGURE 12 PARTICLE CAPTURE MECHANISMS

Evidence is mounting that particles retained by adsorption are mobile when perturbated flow or vibrating conditions occur. Hence, all particles below the sieve (pore) size of the media that are basically trapped by adsorption can become re-entrained by almost any dynamic action imposed on the filter, hydraulic or mechanical. For this reason, the trend toward silt control (small pore size) media is not only for the removal of such particles but to prevent "clouding" the fluid with silt size particles every time shock conditions occur. Clouding results in a highly visible concentration of sub-sieve size particles that were able to overcome the surface adhesion. This effect represents a dangerous situation to any fluid system.

The overall particle separation (absorption and adsorption) performance of a filter assembly can be assessed most accurately by conducting the ISO Multipass Filtration Test. This test (ISO 4572) is a static flow test so a discrepancy may exist between the rated performance (Beta Ten filtration ratio) and actual performance. An unsteady flow version of the standard multipass filter test has been developed and is ready for industrial perusal and promotion as a standard.

The ISO Multipass Filtration Test is restricted to the rating of filters which have Beta Ten values below 75 (because sample container cleanliness can start having a noticeable effect above the 75 value). Hence, a multipass test version for ultra-fine filters is being advanced and made ready for round-robin testing.

This test requires upstream and downstream in-line particle counting. Since AC Fine Test Dust will be used, the filtration ratio derived from the test can still be called "Beta." But, the rating size will probably be three instead of ten, and the upstream gravimetric level will probably be 3 mg/l rather than 10 mg/l. Instead of a Beta Ten rating system as ISO approved for "fine" filters, the "ultra-fine" filters will be rated by Beta Three values, if repeatability and reproducibility are both verified by the intra-lab testing.

The reader should realize that a Beta rating for a filter is tied to AC Fine Test Dust as the ingressed contaminant. If the particle size distribution of the ingressed contaminant in the field is not the same as that of AC Fine Test Dust, then a discrepancy in field performance prediction can exist. Fortunately, for those of us making predictions and unfortunately for those living with poor exclusion devices, field ingression distributions and the AC Fine Test Dust distribution are quite close, and the use of Beta in contamination control equations has worked very well up till now.

The future is not going to be as kind in this regard as the past has been. Contamination control managers are getting smart, controlling ingression sources much better, and altering the slope of field ingression. This means that, in order to make accurate performance predictions in the future, the Beta ratings (an accurate, well defined standard rating system which should continue) must be transformed mathematically into a performance oriented term which is independent of the ingressed contaminant.

The development of a unique size distribution rating for filter media has been the object of an ongoing research effort at the Fluid Power Research Center for several years. By ignoring surface forces as a particle retention mechanism, a unique rating can be derived. The accuracy of the new filtration model depends on the amount of adsorption which actually occurs in the system. Surge flow allows the model to predict particle separation performance quite accurately for any particle size distribution being ingressed. Work is continuing and an exchange of ideas between laboratories could be mutually advantageous.

One of the most perplexing and often depressing problems which commonly confronts machine system personnel is filter bypassing. Difficulty arises in pin-pointing the exact path of the sneak flow. It could result from a misalignment of the element in the housing, an unseated seal, a weak spring forcing the element against the sealing surface, a jostling action between the element and housing that repeatedly unseats the seal caused by the machine travelling over rough terrain, flow surges causing hydrostatic forces sufficiently great to drive the element off the seat, and "on-and-on." The effect of bypass is serious and, more often than not, will negate the otherwise good effects of the filter. Bypassing can be identified on the ISO Multipass Filtration Test and from fluid samples taken both upstream and downstream of the filter.

From what we have seen, it is important to locate filters where they are not subject to flow surges and severe mechanical vibrations, particularly low frequency, jerky movements. Hence, locations downstream from valves, motors, cylinders or anything else that contributes to perturbed flow should be avoided. Look for or create an ideal location where steady flow exists. Debris-catching filters downstream of pumps and "last chance" type filter screens ahead of critical components (servo-valves) make good engineering, maintenance, and economic sense.

All other factors remaining constant, the contamination level of the fluid is proportional to the contamination level of the environment. This is due to the fact that exclusion elements such as reservoir breathers and wiper seals for cylinder rods are proportional devices - that is, they let a certain proportion of each size particle pass through.

Environmental contaminant adhering to an exposed cylinder rod can be pumped into the confines of the cylinder and system when the rod retracts if the particles are not properly scraped off and rejected by effective wiper seal action. An effective wiper seal hugs a cylinder rod for an acceptable service life interval. Once a seal relaxes and allows environmental contaminant to ingress, the seal lip begins to wear, and ingestion continues at a an accelerating rate until the seal is worthless.

Wiper seal service life is a function of cycle distance traveled, environmental contamination level, wiper-seal sensitivity or ingressivity rating, and maximum level of acceptable system contaminant ingression. The SAE (J-1195) Wiper Seal Test establishes wiper seal ingressivity.

Fluid component contaminant tolerance can be improved when

consideration is given to the type wear processes involved and contaminant properties contributing to the wear process. Contaminant entrained in the system fluid essentially goes wherever the fluid goes. If the fluid is used as a bearing lubricant, entrained contaminant has the opportunity to abrade or fatigue relative moving surfaces. Similarly, fluid exuding through annular leakage paths can become lodged and destroy the critical surfaces involved.

Another factor involved in the wear process is unit loading of mating parts. When particulate matter harder than the attendant surfaces separates them from each other, the severity of the abrasion depends on the normal force on the surface, the lateral movement of the surface, and the shear strength of the contaminant particles. Reduction in unit surface loading is an effective way to increase contaminant tolerances.

Whenever system debris lodges in valve clearances, the problem of stiction arises. Valves with high spool-positioning force levels are more tolerant of contaminant than those having low force capability. In fact, if valve spool forces are large enough, valve hysteresis reaches a limiting value above which additional contamination has little effect. Thus, valves with the smallest contact area between the spool land and its mating sleeve are least susceptible to contamination. By undercutting, or shortening valve lands not needed for metering and leakage control, more friction effects of contaminants are reduced.

Metering orifices are never intended to be used to shear particles. Therefore, when they become damaged, the valve operates erratically. Leakage of fluid resulting from such damage results in unequal pressures being exerted on the control valve and promotes actuator drifting. Particles intercepted by an orifice and collected within the flow passage result in a phenomenon called "silting," which again causes unequal pressures and actuator drift.

Particular attention should always be given to avoiding dead zones, where contamination can settle and cause direct interference damage and/or possibly become dislodged by flow surges or vibrations and slug critical downstream components. Contaminant tolerance of a valve can be improved by the following:

- Reducing spool/bore eccentricity
- Increasing the hardness of the spool/bore
- Reducing the asperity height of the mating surfaces.

8. SUMMARY

Contamination control is still not a science; but for those who have been following the progress of knowledge generation in this field, we are getting closer. Much research needs to be done. Hopefully, this discussion will spur some action in other parts of the world. Contamination control is a world problem, and we should all participate in formulating the science.

9. REFERENCE

Fitch, E.C., "An Encyclopedia of Fluid Contamination Control", Hemisphere Publishing Co., Washington, D.C., 1978, Revised 1980.

TRIBOLOGICAL FAILURES AND MECHANICAL DESIGN

M. B. Peterson
Wear Sciences Corp.

A. J. Koury
Naval Air Systems Command

1. INTRODUCTION

Studies show that approximately 40 - 60 % of operating costs are maintenance related. This is illustrated in Figure 1 which shows the per flight hour costs of operating an A6E aircraft. It can be seen that almost 40% of the direct costs are maintenance related. A further analysis of the costs indicate that about 70% are unscheduled maintenance at the squadron level due to component failures or removal of components in anticipation of failure. These costs are only the direct cost of failures and do not include such items as stocking parts, purchase, down time, and loss of equipment. When these are added in, it can be seen that maintenance due to failures is expensive and serves no useful purpose if the failures can be avoided. In order to reduce these costs, a variety of different approaches to maintenance have been tried. These are listed in Table 1.

Table 1

ALTERNATIVE MAINTENANCE POLICIES TO REDUCE FAILURES

Preventive Maintenance	Condition Monitoring
Analytical Maintenance	Contract Maintenance
Failure-Free Design	Leasing
Modularization	Failure Warranties
On Condition Maintenance	

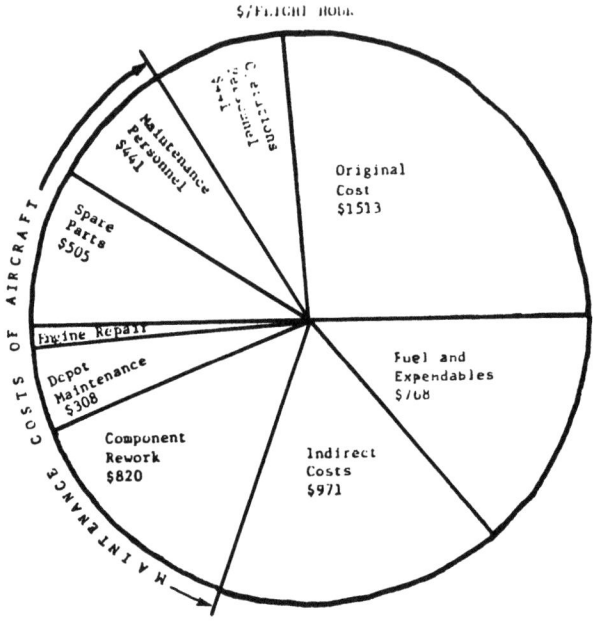

FIGURE 1 COST OF OWNERSHIP OF ONE A6E AIRCRAFT

Before the 1950's the preventive maintenance concept was promoted almost exclusively, the idea being that prevention of failures and equipment deterioration was always preferred to breakdowns and inadequate performance. However, as labor costs rose, and as technology produced better equipment, management began to question the need for certain maintenance actions. Furthermore, there was the impression that constant tinkering was in itself a significant cause for equipment malfunctions. The search for alternative maintenance policies has lead to a variety of approaches. These are described in the following paragraphs.

On-condition maintenance refers to performing maintenance only when required as dictated by machinery behavior. The difficulty with this approach was that a few catastrophic failures could eliminate any savings that resulted from elimination of preventive maintenance. Thus, attention was directed to methods of prediction of incipient failure. The expansion of this concept became "condition monitoring." By a variety of techniques (instrumentation, trend analysis, life projections, etc.) the condition of the equipment and its components was known and maintenance was performed only when the equipment deteriorated beyond prescribed limits.

During the same period of time there was a renewed interest in growth in contract maintenance. Essentially, the manufacturer of a given piece of equipment will perform maintenance on that equipment for a fixed fee. Such an approach fixes the costs of the user and provides certain benefits to the supplier such as service information, source of repeat sales, and for good equipment, an additional source of profit from a sale. The ultimate in contract maintenance is the lease where ownership is retained by the leasor who is responsible for the equipment. This approach does not reduce failures but rather stabilizes the cost.

Analytical maintenance refers to a maintenance program where service performance is recorded and the design modified to eliminate field failures. Such a program has as its basis a computerized maintenance actions system. By reviewing all maintenance actions and categorizing them, specific design, inspection, or repair problems become obvious. The essence of such a program is that the owner essentially accepts responsibility for the design. One of the results of the Analytical Maintenance Program has been the realization that much of the maintenance costs are not due to repair or replacement but rather due to removal and the need for general disassembly to reach the faulty component. As a result there has been increased use of modularization. Here, failures are accepted as inevitable and the technology was devoted to more rapid change. In some equipment like radar sets, it was estimated that this approach reduced costs by 50%. One problem with replaceable modules is that there is frequently an increase in "no defect" removals; any suggestion of a problem and the module is changed.

The understanding of service problems along with the growth in predicting reliability and life cycle costs has lead to an extension of the warranty idea. In this concept, the manufacturer guarantees a certain operating time between and/or a certain number of maintenance man-hours. The article could be as small as a valve or as large as a ship or aircraft. The advantage of this approach is that it forces the manufacturer to consider the durability as well as the performance of his design. To give such warranties it is, of course, necessary to be able to predict and control failures.

The conclusion which can be drawn from this chain of events is that there is great need to reduce service failures, and we are rapidly approaching an era where such failures will not be tolerated. Owners of equipment want maintenance-free, failure-free equipment and will exert their influence in the market place. As a result of this trend, component technologists must understand failures and learn how they can be prevented.

2. NATURE OF FAILURE PROCESS

If failures are to be prevented, it is necessary to understand the failure processes and how they occur in service. Some understanding may be gained from reference to Table 2. The designer starts with certain inputs: the requirements (function, cost, life); the conditions (operational and environmental); his past experience with similar components; and a knowledge of failure processes. From this he prepares several alternative designs to meet the requirements. Once a compatible design is chosen it is manufactured, assembled, put into service, maintained, and eventually removed from services. During this total life cycle, three different kinds of failures occur: inadequate design, poor quality control, or "conditional" changes.

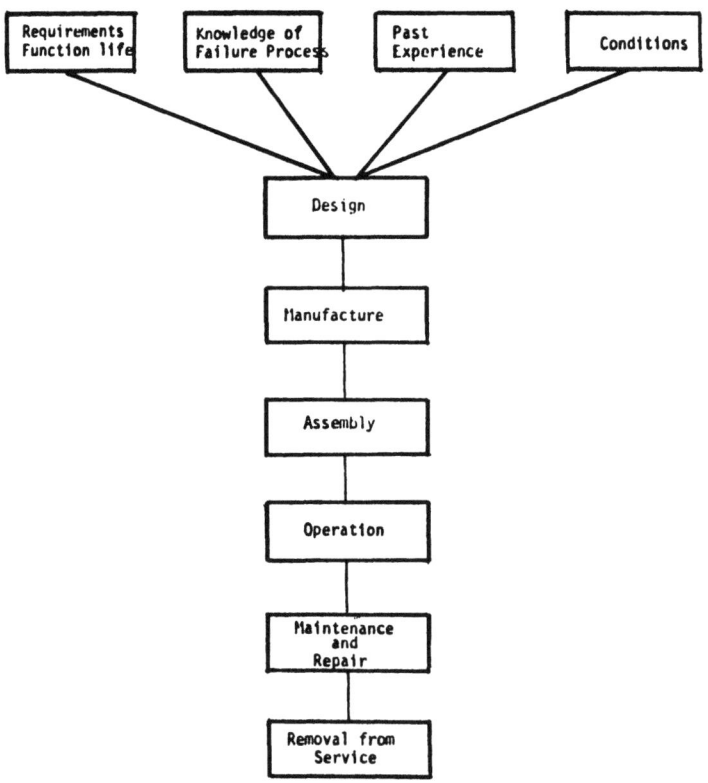

TABLE 2   EQUIPMENT DESIGN AND OPERATION

Inadequate designs can be the result of mistakes in the design, however, they are usually the result of inadequate inputs; conditions are not exactly known, techniques are not available for life prediction, certain failure processes are not adequately understood and most important, specific resistance to failure of a given design cannot be estimated.

Once a design is selected which meets the requirements, is able to be manufacturedand is compatible with other components, it must be fabricated and assembled. This process can introduce a whole series of quality control problems. There are variations in materials, design tolerances, and assembly procedures which, in essence, change the design.

In operation, certain dissipative processes are initiated which will eventually determine the life of the component or part. At this stage of the life cycle, failures usually occur because the conditions have changed from those anticipated in the design. Some of the most significant kinds of changes are listed in Table 3. These changes, of course, may be due to operational mistakes or unanticipated changes like contamination.

Table 3

CONDITIONS RESPONSIBLE FOR FAILURES

Misalignment and Positioning	Contamination
Excessive Loads or Temperatures	Vibration
Material Degradation	Shock Loads
Inadequate Lubrication	Damage

This general view of the failure process is significant in that each different kind of failure can be prevented in a different manner as shown in Table 4. Design failures can be avoided by improved procedures or better information. In particular, the designer needs better information on the actual conditions encountered in service, how this affects component failure, and how designs can be modified to reduce these kinds of failures. If quality control is the problem, then tighter inspections and specifications are needed. If operation or environmental changes occur, then this situation must somehow be incorporated into the design; either the original design or by a design modification.

Table 4

APPROACHES TO FAILURE PREVENTION

Kind of Failure	Failure Prevention
Inadequate Designs	
Mistakes	Improved Design Procedures
Inadequate Inputs	Improved information on condition or failure prevention techniques.
Quality Control	Better inspections and specifications
Conditional Changes	
Mistakes	"Fool Proof" Designs
Unanticipated Conditions	Survey to define and modify the design to accept the condition.

    If one is trying to reduce failures he could proceed on a broad front using all the listed approaches. However, it may be more effective to first determine the kinds of failure which most predominate in a given application and then to approach failure prevention in the most appropriate manner.

    In order to determine which approach to apply to naval aircraft, a special investigation was conducted in order to identify the parts which fail most frequently and the respective type of malfunction[1]. These data from the Navy 3M system are shown in Table 5 for the A6 aircraft and Table 6 for the H46 aircraft. Each component on these lists was reviewed to determine the nature of its failure process. The results of this investigation are shown in Table 7. It can be seen that most of the aircraft failures are caused by changing conditions which are given secondary consideration in the design process. In other words, they are caused by one of the conditions listed in Table 3. These failure modes were, of course, known to the designer but he did not possess sufficient information for adequate control.

TABLE 5   A6 AIRCRAFT STRUCTURE   330/A/C/   1 YEAR MANHOURS

Work Unit Code	Identification	Total Hours	Total Incidences	Corroded	Broken	Cracked	Leaking	Stuck	Worn	Loose	Improper Lubrication
1121500	Wing Skin	12,778	1752	11,429	331	170	534	0	127	8	0
1113000	Misc Access Doors/Panels	11,122	2295	5,723	3,206	1,057	143	41	299	316	10
1113100	Forward Engine Access Door	9,491	1659	4,846	3,505	533	8	237	147	91	10
1113800	Access Panel	7,775	1646	4,080	2,071	762	72	6	290	162	18
1432100	Flaperon Actuator	6,724	139	2	19	2	5,622	60	10	7	4
148140C	Slat Assembly	4,895	533	3,959	595	346	7	54	183	4	3
1121200	Wing Outer Panel	4,591	296	2,880	159	783	591	0	103	1	0
1121100	Wing Inner Panel	4,558	378	3,991	27	87	434	2	0	0	3
1411200	Flaperon Skin	4,124	442	3,170	45	37	0	40	3	0	0
1118300	Boarding Ladder Assembly	3,972	879	2,331	774	501	6	27	283	19	6
1442100	Rudder Actuator	3,919	99	9	92	0	2,789	42	79	23	0
1452100	Stabilizer Actuator	3,496	83	16	62	0	2,272	1	20	0	0
1131200	Vert Stabilizer Skin	3,275	495	3,024	110	55	0	0	12	14	0
1116100	Ext Equipment Platform	2,494	538	1,127	872	180	15	43	123	79	4
1121400	Wing Leading Edge	2,578	407	1,778	79	60	70	2	244	21	0
1113500	Aft Engine Access Door	2,457	420	1,975	317	74	0	18	35	22	1
1113600	Tail Pipe Access Door	2,383	369	1,819	368	60	12	0	106	7	0
1311100	MLG Shock Strut	2,385	313	1,114	36	465	492	122	10	2	116
1311000	MLG Mechanical Components	2,365	291	1,144	217	77	47	162	99	12	578
1321100	NLG Shock Strut	2,559	167	1,464	9	3	897	4	16	4	147
1451100	Hor Stabilizer Cable	2,453	46	4	1,079	0	170	0	349	6	0

TABLE 6   H46 AIRCRAFT   300 A/C   1 YEAR MANHOURS

Work Unit Code	Identification	Total Hours	Total Incidences	Corroded	Broken	Cracked	Leaking	Stuck	Worn	Loose	Improper Lubrication
1341800	Brake Assembly	4833	621	111	0	10	2260	0	2205	0	1
1521000	Fwd Rotary Wing Head	3253	214	333	171	0	2685	0	5	13	49
1321100	Alg Shock Strut	2258	200	76	194	62	1523	89	122	105	0
1426700	Upper Boost Dual Act Cyl	2211	269	16	80	19	1547	47	213	119	0
1522000	Aft Rotary Wing Head	2008	181	253	102	1	1630	0	13	7	1
1421700	Cyl Stick Dual Boost Act	1790	205	5	10	132	1174	91	114	24	0
133110C	Wheel & Tire Assembly	1589	253	60	217	27	1523	0	174	0	5
1155000	Aft LH Clamshell	1468	207	49	389	231	0	6	558	113	0
2611000	Forward Transmission	1429	39	55	187	158	556	7	169	6	0
1155A00	Fwd LH Clamshell	1252	176	103	398	277	0	1	427	51	0
1155800	Aft RH Clamshell	1189	288	55	588	172	0	0	249	115	0
1311100	MLG Shock Strut	1062	158	85	58	46	1051	1	113	75	0
1156300	Aft Fuselage Step Assembly	1010	199	27	678	268	27	0	12	3	0
142880	Dual Boost Actuator	1000	149	27	8	11	595	37	24	77	0

The same sort of analysis has been conducted for surface ships[2]. For corrective maintenance actions, 12,250 items were reviewed from 20 ships which had operated on the average of 3.7 years at a rate of about 2000 hours per year.

Table 7

COMMON AIRCRAFT FAILURES

Component	Failure Process
Wing Skin	Paint Damage due to Debris
Access Doors	Cracks and Paint Damage due to use of Door as Work Platform.
Panels	Structural Flexing Breaks Seal Between Panel and Fastener Causing Corrosion

Table 7
(cont.)

Component	Failure Process
Actuators and Struts	Dirt catches in seal -- Abrades rod causing leaking
Cables	Inadequate lubrication of cables
Wing Head	Seal leakage due to wear and distortion
Control Bearings	Corrosion due to water and salt in bearings

For illustrative purposes, the data for the propulsion system are presented. An analysis of these systems for the offending components is given in Table 8. It is interesting to note that of the 20 systems and hundreds of components, only a few dominate the maintenance expensives of the ship.

Table 8

PROPULSION SYSTEM REPAIR
17 SHIPS

	Reported Incidences
Air Start Motor (6)	25
Heads, Valves (96)	513
Injection Fuel Pumps	718
Injectors (96)	859
Fuel Booster Pump (6)	134
Salt Water Pump (6)	287
Governor (6)	66
Piston, Rings, Liners (96)	7
Turbocharger (6)	43

Table 8 (cont.)

	Reported Incidences
Fresh Water Pump (6)	17
Assessory Drive Gears (6)	7
Holset Coupling (6)	25
Lube Oil Pump (6)	2
Pedestal Bearing	15
Fawick Clutch	12
Reduction Gear (3)	16
Reduction Gear LO Pump (3)	15
Reduction Gear SW Pump (3)	9
Valves	8
Exhaust Leaks	37
Emergency Trip Wire	27

To be of assistance, the underlying cause of the malfunction must be stated, and costs and frequency attributed to those causes. This was accomplished by reviewing each maintenance item and attributing a cause to each, where possible. These resulting costs are listed in Table 9. The cost of the maintenance item was determined as the cost of material plus labor at a rate of $6.57/hour. This was the direct labor charge exclusive of overhead. If overhead is included, a figure about three times that amount should be used.

Table 9

COSTS OF CORRECTIVE MAINTENANCE

Wear	$1,420,513
Contamination giving corrosion	2,373,797
Leaks	505,590
Vibration	579,756
Corrosion	973,820
Broken	481,922
Contamination giving wear	3,674,622
Misalignment	282,482
Design faults giving wear	32,930
Vibration giving wear	33,549
Contamination control	565,939
Calibration	88,802
TOTAL	$11,013,722

In this table the phrase "contamination giving corrosion" means that system contamination leads to corrosion.

Preventative maintenance costs were obtained by reviewing each item in the planned maintenance schedule and again attributing the costs to independent categories. Here it is not a repair but usually a measurement to determine the need for repair. For example, a measurement of a piston ring to determine if it had to be replaced would be attributed to wear. The categories turn out to be somewhat different than that for corrective maintenance as shown in Table 10.

Table 10

PREVENTIVE MAINTENANCE COSTS

Wear	$2,125,420
Leaks	25,440
Vibration	71,800
Corrosion	214,900
Broken	4,080
Misalignment	40,840
Contamination control	1,511,380
Calibration	64,020
Check component operation	551,132
Lubricate	193,340
Record engine data	742,880
TOTAL	$5,545,232

The shipyard repair data was only available for 10 ships over a period of two years. This information had to be prorated to 20 ships over the 3.7 years in order to be compared with the previous data. These data are shown in Table 11.

Table 11

SHIPYARD REPAIR COSTS

Wear	$ 54,575
Leaks	4,440
Vibration	3,700
Broken	259,000

Table 11
(cont.)

Contamination giving wear	293,410
Contamination	11,100
Design fault giving wear	1,827,060
TOTAL	$2,453,285

A summary of all costs is given in Table 12. From Table 12 it can be seen that a large amount of the total maintenance costs can be attributed to contamination, wear, and corrosion as shown in Table 13.

Table 12

SUMMARY OF COSTS

Wear	$3,600,508
Contamination giving corrosion	2,373,787
Leaks	535,470
Vibration	705,286
Corrosion	1,265,479
Broken	745,102
Contamination giving wear	3,968,032
Misalignment	323,322
Contamination control	2,088,419
Calibration	152,822
Design wear	1,859,990
Vibration giving wear	33,549
Testing component operation	551,132
Lubricate	193,340
Record Engine data	742,880

Table 13

COSTS PER OPERATING HOUR

	Cost	Cost/Ship Hour
Contamination	$8,430,238	$56.96
Wear	5,494,047	37.12
Corrosion	1,265,479	8.55
Broken	745,102	5.03
Vibration	705,286	4.77
Fuel Cost	----	75.00

The conclusions here are very much the same as that for the aircraft:

- A few components are responsible for the bulk of the failures.
- Most failures are caused by conditions which exist in these components which were not anticipated in design.
- Components are needed which are more resistant to conditions such as contamination, vibration, misalignmnet, etc.

Although the literature is not extensive these general conclusions also apply to other applications[3, 4, 5, 6].

3. FAILURE OF TRIBOLOGICAL COMPONENTS

3.1 Classification

If one attempts to classify failures, certain difficulties are encountered, as illustrated in Table 14 which lists causes of sliding bearing failures. The difficulty is that the list contains a variety of unequal elements which do not lend themselves to a rational approach to failure prevention. Some are causes, others are the result of the failure process. Some may cause failure in certain bearings but not in others. Wear is a process characteristic of all bearings and may or may not lead to a dimension change of significant magnitude to interfere with the function of the part. In order to better understand failure processes, it is sometimes convenient to break the failure process into the categories shown in Table 15. There are sources, conditions, dissipative processes, characteristic component failure modes, observable results, and reasons for replacement and repair. These categories are illustrated using the example of the bushing. The failure process begins with severe misalignment due to structural deformation of a base plate and eventually leads to seizure. It should be noted that all categories need not be different. The observed result is an inoperable bearing and that is the reason for replacement. However, under the same circumstances, siezure or welding may not have occurred, only a large increase in friction. The observable result may have been "noise" and the reason for replacement might be "anticipation of failure" if high friction was tolerable.

Table 14

CAUSES OF BEARING FAILURE

Oil passages plugged	Misalignment
System low on oil	Shock leads
Poor surface finish	Poor materials
Lubricant Degradation	Corrosion
Lubricant Deterioration	Fatigue
Excessive temperature	Wear
Differential thermal expansion	Improper crush
Excessive sliding	Scoring
Poor thermal conductivity	Dirt
Improper dimensions	Improper viscosity

Dimensional instability

Table 15

FAILURE PROCESS CATEGORIES

Category	Bushing Example
Source	Misaligned Shaft due to Structural Distortion
Condition	Restricted Area of Contact High Pressure/High Temperature
Dissipative Process	Thermal Softening
Characteristic Failure Mode	Seizure (welding)
Observable Result	Inoperable
Reason for Removal	Inoperable

The important point here is not the uniqueness of the categories but approaches to prevention. Causes may be important in legal situations, but they are too varied to be of much use in failure prevention. In the bushing example cited, the original cause may have been the settling of the earth but this is of little help in prevention. For prevention of failures, it is important to know the conditions and dissipative processes which can lead to failure and the characteristic failure modes of the various components. If these are known then these conditions may be avoided or the components may be modified to prevent the characteristic failure modes.

A large amount of literature was reviewed on the tribological components in order to understand the failure processes and to suggest means of failure prevention. A collation of data leads to the results shown in Table 16, which now refers only to the tribological components. For each component a given condition will aggravate one of the dissipative processes which will lead to one of the failure modes (DIRT ——> CUTTING ——> WORN), (WATER ——> CORROSION ——> SURFACE DAMAGE), etc.

Table 16

TRIBOLOGICAL FAILURES

Conditions Leading to Failure	Dissipative Processes	Component Failure Modes
Dirt/Contamination	Stress Cycling	Inadequate Friction
Inadequate Lube Supply	Heating & Heat Cycling	Instability
Shock Loads	Plastic Flow	Distortion/Deformation

Table 16
(cont.)

Conditions Leading to Failure	Dissipative Processes	Component Failure Modes
Water in Lubricant	Fatigue (Crack propagation)	Fractured/ Broken
High Temperatures	Adhesion and Transfer	Surface Damage
Inadequate Film Thickness	Cutting/Tearing	Worn
Overload/Speed	Material Diffusion	Seizure
Vibration	Corrosion	
Improper fit		
Inadequate Materials		
Inadequate Lubricant		
Misalignment		
Improper Mounting		

The different characteristic failure modes are related to specific components in Table 17. The terms used in a given category are customary names which are used to describe a given kind of failure. For example, "fade" is used to describe inadequate friction characteristics of brake materials. Thus it can be seen that in the tribology area there are seven major types of failure which occur. The real question is whether these can be appropriately linked to specific conditions and whether appropriate prevention techniques can be identified. This is given more detailed consideration in the following sections where specific components are considered.

Component	Inadequate Friction	Instabilities	Distortion Deformation	Fracture	Surface Damage	Seizure	Worn
Bushings	Excessive Torque	Stick Slip	Plastic Flow	-	-	Galling Sliding Seizure	Shaft or Bushing Dimension Chnge
Seals	High Friction	Whirl	Creep Coining Distortion	-	Tearing Deposits	-	Seal Wear or Shaft Grooving
Rolling Contact Bearings			Ball Indentation	Cage or Race Fracture	Spalling Pitting	Smearing	Cage Wear Roller Wear Race Wear Land Wear
Brakes	Fade	Squeal	-	-	Heat Check	-	Wear
Gears	-	-	Surface Rippling	Tooth Breakage Case Crush		Scoring	Tooth Wear
Splines	-	-	-	-	-	Galling	Tooth Wear
Cables	-	-	-	Strand Breakage	-	-	-
Fits	Over 0 Torque	-	-	-	-	Galling	Fretting
Fasteners		-	Cold Flow	FRACTURE	-	-	WEAR
O-Rings							WEAR

TABLE 17   COMPONENT FAILURE MODES

3.2 Dry Bushings[6 - 15]

Dry bushings refer to a shaft and a journal where fluid lubricants are not used. There are several kinds of bushings, the most common being polymers, solid lubricant filled polymers, and static metal bushings. Bushings are widely used in small devices (pumps, fans, valves, etc.) where speeds and loads are low and a large number of cycles are not accumulated during the lifetime of the part.

A failure chart for the polymers and filled polymers is given in Table 18. The most common failure modes are distortion, seizure, and wear. Since these materials are much softer than the shaft material they are very sensitive to load and temperature effects. Although Table 18 appears to be quite complicated, it merely says that the common failure processes are:

- Bearing deformation due to high ambient temperatures or shock loads.
- Seizure due to inadequate clearances in design.
- Wear due to contamination, operation at high load and speeds, or at high pressures (low contact areas); wear due to high temperature operation, or use of a poor shaft finish.

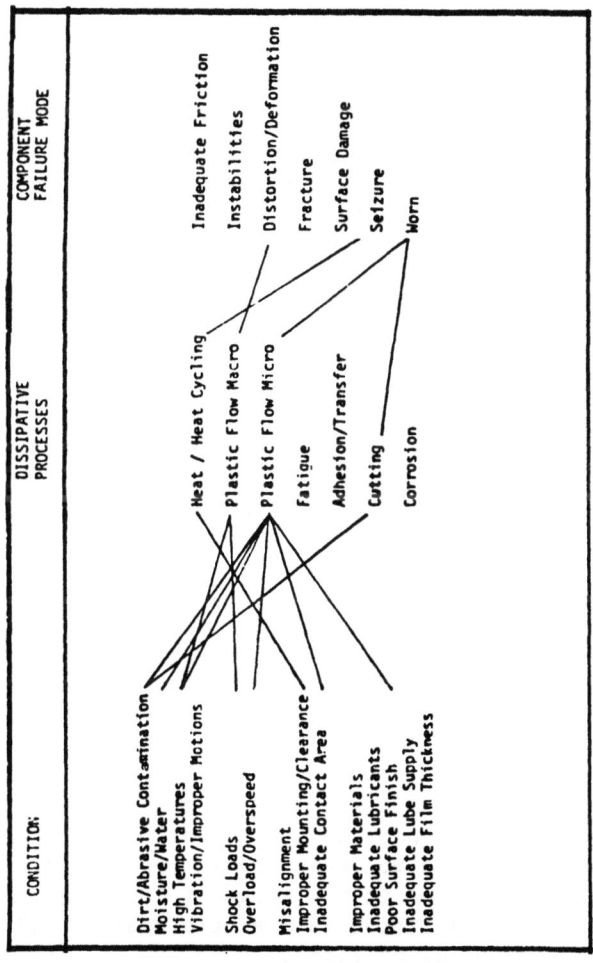

TABLE 18   DRY BUSHING FAILURE MODES (POLYMERS & FILLED POLYMERS)

The most common approach to failure prevention is to change materials. A wide variety of materials are available which operate under a variety of conditions and stress levels[16]. Even with a given material like teflon, a variety of fillers are available for different stress and temperature levels. Usually, it is not the design loads and ambient temperatures that are at fault. Rather, it is the shock loads not anticipated in the design and the unexpected temperatures of operation which create a problem. Materials are available for temperatures to 600°F and pressures to 1000 psi (depending upon the desired life).

Seizure is strictly a design problem. Due to unequal thermal expansion, the metal shaft may grow at a faster rate than the bearing material so at a certain temperature the clearance is reduced to zero and seizure may occur. Prevention of this kind of failure is simple; clearances should be adjusted to allow for thermal expansion.

Excessive wear can be attributed to a variety of conditions. Wear is known to increase rapidly at a given interface temperature[17]. The interface temperature is a function of operating conditions (load, speed, contact area) and the bearing ambient temperature. To prevent excessive wear, the interface temperature must be lowered. This, of course, can be accomplished by reducing the load or speed and/or increasing the contact area. However, these adjustments are often not possible, in which case the materials must be changed. Almost all manufacturers provide data on the operating limits (pressure and velocity) for their respective materials. Although these data are obtained under limited conditions, they do provide a good guideline for selecting bushing materials. Excessive wear based upon this condition is a design fault. If changing materials is difficult, cooling may be appropriate. The results presented in Reference 18, show that considerable convective cooling will result with small amounts of air flow across a bearing. This is not difficult to accomplish since it usually means redirecting the natural air flow about the moving shaft. Other means of cooling also exist such as hollow shafts and materials with better thermal conductivities.

Since bushing materials are soft, they are sensitive to shaft roughness. Designers usually tend to use rough surfaces since it is less expensive, however, in order to minimize wear, finishes should be in the range of 10 to 20 RMS. Improving the surface finish is a simple design modification and should be investigated first.

Wear due to contamination, either abrasive particles or water, is very frequent. Elimination of the contamination (by sealing) is difficult since the purpose of using dry bushings is design simplification. However, some approaches are available based upon the literature which was accumulated. Grooves cut in the surface of the bearing collect the abrasive particles and reduce the wear[19]. Reducing the clearance or extending the bushing length prevents abrasives from entering the bearing[20]. Putting a teflon coating on the surface of a hinge pin also acted as a seal[21]. This can often be accomplished relatively easily.

Water or other fluid contaminants are known to increase wear. The low wear of many plastic and filled plastic materials is contingent upon the transfer of a thin film to the shaft surface. This transfer film reduces the effective surface roughness and thus reduces wear. If water or other contaminants prevents this film from forming then wear will be higher in magnitude, being a function of the surface roughness. Wear can also be reduced by coating the shaft with the bushing material prior to operation.

Thus, the literature provides a variety of techniques to reduce failures depending upon the nature of the failure process. These are summarized in Table 19.

Table 19

DRY BUSHINGS
APPROACHES TO FAILURE PREVENTION

Component Failure Mode	Failure Prevention
Distortion/ Deformation	Change to higher Temperature or higher strength materials.
Seizure	Check bushing clearance Increase clearance

Table 19
(cont.)

Component Failure Mode	Failure Prevention
Wear	Change to higher temperature material Cool Surfaces Improve surface roughness Use 10 - 20 RMS Coat shaft Reduce Clearance Increase Bearing Length

3.3 Fluid Lubricated Bearings/Bushings[22-27]

Fluid lubricated journal bearings are of two varieties: boundary lubricated and fluid film lubricated. Although there is no clear distinction, the fluid film bearings are designed to operate with a given lubricant film thickness while boundary lubricated bearings may or may not, depending upon the operating conditions and the materials. Boundary lubricated bearings are usually made from bronzes, copper alloys, steels, cast iron, plastics, or ceramics, although any material may be used.

The failure chart for boundary bearings/bushings is shown in Table 20. Basically, the same failure processses apply as for dry bushings except that this bearing type depends upon the presence of a lubricant at the interface. If this lubricant is not present, failure is almost inevitable. The lubricant provides the protection against adhesive processes, thus compromises have been made in that property of the materials. Consequently, any condition which tends to remove the lubricant will tend to increase the probability of failure. These conditions include: high temperatures, excess loads and speeds, severe misalignment, inadequate contact area, and insufficient or interrupted lubricant supply. This increased adhesion can either lead to seizure or to inadequate friction, depending upon the requirements of the bearing.

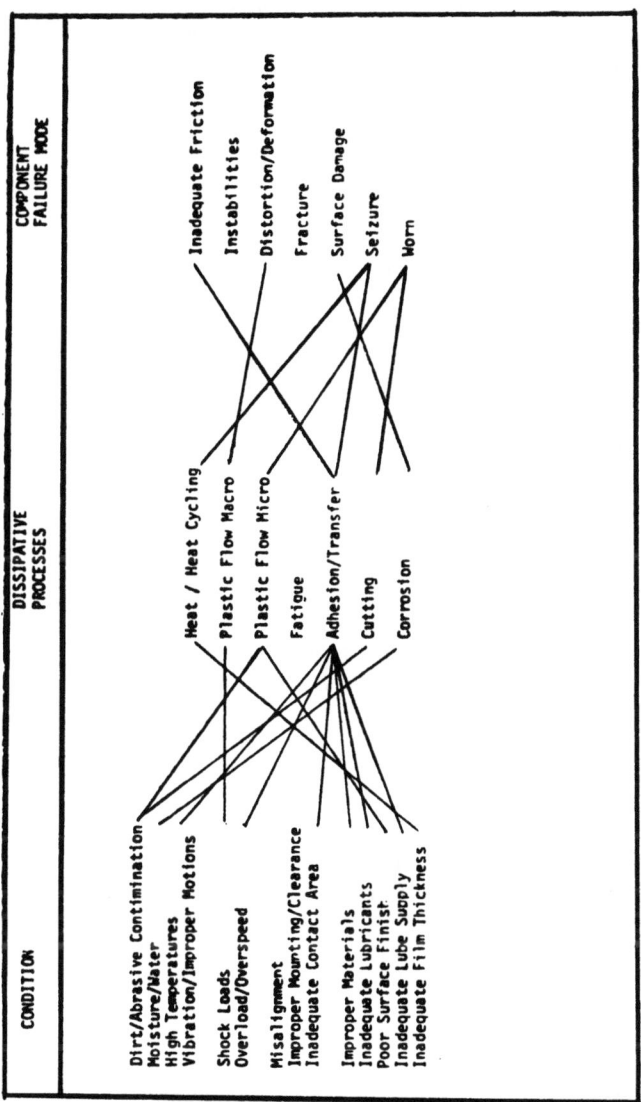

TABLE 20 FLUID LUBRICATED BUSHINGS (BOUNDARY LUBRICATED)

Failure prevention techniques can concentrate on removing the condition or reducing adhesion as shown in Table 21. A detailed investigation of bearing failures shows that most of the problems arise from contamination which increases the wear and reduces the lubricant flow to the bearing. A second factor is neglect. Many of the lubricant points are almost inaccessible and thus do not receive relubrication. The problem in this case is not adhesion but lubricant supply and distribution, debris removal, and sealing. The minimization of this factor will take a special bearing design with grooving and an internal reservoir. Simple designs of such bearings are available.

Table 21

APPROACHES TO PREVENTING SEIZURES

Reducing Adhesion

Use Nonsoluble Materials
Increase Shaft or Bearing Material Hardness
Add Low Adhesion Coatings
Use Lubricant with Improved Additives
Use Less Ductile Materials

Eliminate Condition

Increased Cooling or Heat Removal
Grooving for Better Lubricant Distribution
Use Positive Lubricant Supply System
Increase Contact Area

3.4 Seals[23-45]

Seals may be divided into two main categories: static seals such as gaskets and O-rings, and dynamic seals which have a sliding interface. Dynamic seals may be face seals, circumferential seals, or lip seals. Lip seals are given primary consideration because of their preponderance in aircraft structures and because an anlysis indicates that they are one of the main causes of tribological maintenance. The technology of lip seals has advanced rapidly in the past 20 years. There is now a much better understanding of their mechanism of sealing and the factors which control their life. This new understanding has produced new designs and design guides[46, 47]. This new technology provides a variety of means to prevent and correct service failures. In addition, a variety of new elastomeric materials have been adapted to seal usage which greatly expands their range of operating usefulness. A seal has become a precise engineered component, a fact not yet recognized by maintenance and manufacturing personnel. As a result, seals are frequently damaged before they are used. They should be treated with the same respect as is given rolling element bearings. Such care would reduce failures dramatically. The seal interface consists of a thin lubricant film approximately .0001" in thickness. This film thickness must be maintained if a satisfactory life is to be expected. If the film becomes too thick, leakage will occur. If it becomes too thin, the interface will be inadequately lubricated and surface damage and increased wear will occur. This, too, will result in increased leakage. The factors which control the film thickness are: load, velocity, temperature, alignment, shaft roughness, and the supply of lubricant. Factors which affect its change are: misalignments, wear or surface damage, and changes in the material properties due to environmental effects. These factors are illustrated in the failure chart shown in Table 22. Leakage is not, of course, the only technical reason why a seal is replaced. Some removals result from noise, vibration, and convenience. Seals are often replaced whether they are defective or not, in order to assure a failure-free duty cycle.

Leakage is caused by a variety of conditions as illustrated in Table 22. Many of these conditions (high ambient temperatures, overloads or overspeeds, inadequate lubricants, poor lubricant supply and inadequate film thickness) lead to plastic deformation damage to the seal interface which results in leakage. Many of these same conditions lead to overheating at the interface which causes the seal to distort, again causing leakage.

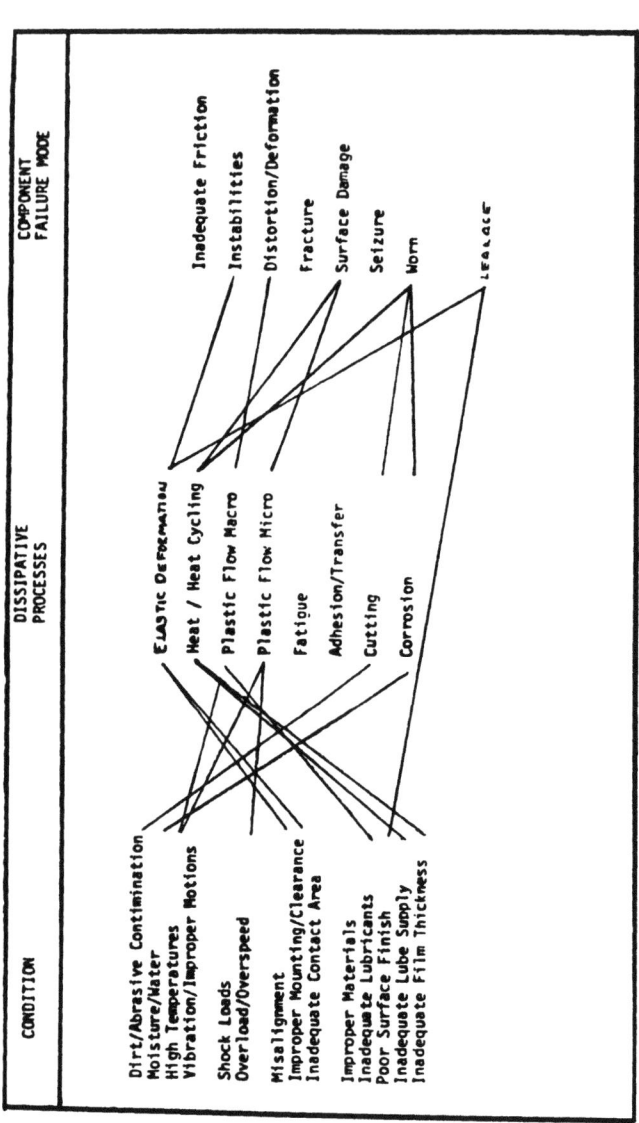

TABLE 22 SEAL FAILURE MODES

A second major area of failure is leakage caused by seal elastic deformations which exceed the material compliance and present a direct leakage path. This most frequently happens at high speeds when the seal is unable to follow shaft movements. Misalignments may be caused by shaft "out of roundness," nonparallel center lines of shaft and seal, or eccentricity.

Another major failure area is caused by contamination, either abrasive dirt particles, moisture, or oxidized or decomposed lubricant. These conditions lead to increased seal wear. Certain conditions (misalignment, finish) can in themselves produce leakage if they become excessive.

Primary failure modes can then be classified as wear and surface damage, distortions of the required seal geometry, and dynamic instabilities which produce leakage. Based upon this, some approaches to failure prevention are shown in Table 23.

Table 23

APPROACHES TO PREVENTION
OF SEAL FAILURES

Wear and Surface Damage	Add wear sleeves or coatings of more wear or corrosion resistant materials.
	Improve seal/shaft alignments.
	Improve lubricant supply to interface cooling.
	Hydrodynamic lip seals.
	Use fluoroelastometer seal.
Seal Deformation	Improve seal/shaft alignments.
Instabilities	Improve seal/shaft alignments.
	Improve lubricant distribution.
	Clean interface.

3.5 Rolling Contact Bearings[48-69]

A properly applied, well lubricated, properly installed rolling contact bearing operating in a contamination free environment, can have a life of greater than 10 years in continuous operation. Yet, in service it is one of the most frequently replaced parts. This is usually not a fault of the design but of environmental factors such as dirt, corrosion, lack of adequate lubrication, and excess temperatures. Although these can be avoided in the design, they are often overlooked or unspecified. Designs are based upon static or dynamic load capacity. The static capacity is the bearing load necessary to yield a given indentation in the race. The dynamic load capacity is the operating life at a given load based upon bearing fatigue. This life is determined by running a large number of bearings to failure at a given load and then extrapolating through a range of loads. It is found that these loads are conservative for other failure modes such as wear or fracture. Unfortunately, most bearings do not operate under the ideal conditions of laboratory tests and much shorter lives are found in the service environment.

The failure modes for rolling contact bearings are listed in Table 24. Most of the common failure modes are exhibited by rolling contact bearings. In contamination free bearings, failure is most common in bearings which have appreciable sliding or with those which have very thin lubricant films. Thin films are characteristic of low speed, high load bearings.

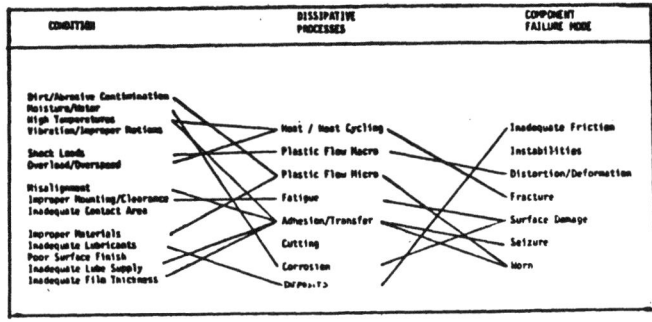

TABLE 24   FAILURE MODES - ROLLING CONTACT BEARINGS

Inadequate friction is a special case for instrument bearings which require constant torque. The bearings removed from service would be considered satisfactory for most other applications. The friction variations are usually due to small amounts of deposits in the contact area or small variations in concentricity. Bearing deformations are usually the result of shock loads which cause indentations. Fracture of the race can also result from shock loads or from thermal expansion of the bearing or the housing due to high temperatures. Actually, fracture of the cage is more common than fracture of the race and usually takes place after considerable cage wear. Cage fractures result from centrifugal stresses, misalignment, or accelerating forces on the cage due to uneven ball loads. There are very few ways to prevent these stress failures except to over design the bearing, that is, to use a much larger bearing than the application requires. Such failures are often isolated cases attributable to quality control in the bearing. If they are the fault of the machine, then the condition should be eliminated.

Seizure failures are most often due to insufficient lubrication. They are usually referred to as "smearing," a term which describes the surface appearance of the bearing. "Smearing" failures can also be the result of high temperatures in the contact area. At high temperatures the lubricating capabilities of the lubricant are destroyed and the effect is the same as having insufficient lubricant. In the failure process, high friction first occurs; there is a transition from fluid film friction ($f < .01$) to boundary friction ($f <= .10$) and then to dry friction ($f => .40$). This increase in friction brings about a forty fold temperature rise in a small contact area, causing the failure area to enlarge. This process continues until a definite event occurs which reduces the friction or the surfaces are damaged. This damage can result in seizure with certain materials and wear with others. The cage is often more sensitive to this type of failure and is the first to be influenced. Coatings are often used to prevent seizure failures. In most cases they provide auxiliary lubrication to compensate for the depleted fluid. In other cases, the coatings will extend the operating temperature range before lubricant failures occur. This is particularly true for soft metal platings (silver, gold, etc.) which provide improved heat conduction and flow, reduced surface roughness, and thinner effective lubricant films. Coatings could also be used for another purpose. If the coating melts it could provide the friction reducing event required to reverse the failure process. This approach has not been used in practice but is worthy of consideration.

Surface damage is the most common bearing failure mode. The specific kinds of surface damage are spalling due to fatigue, and pitting due to localized surface fatigue, corrosion, or debris indentation. If the bearing is not removed because of the pitting, continued operation will eventually lead to its removal because of either spalling or excessive wear. A survey of the literature indicates that a large number of studies have been directed at the understanding of the variables influencing fatigue damage.

Changing these variables appropriately will increase the fatigue life as shown in Table 25. Several important prevention techniques can be noted. For example, more frequent lubrication insures a cleaner lubricant and increased fatigue life. Longer fatigue life can also be obtained by use of improved materials. Precise correlations with specific material properties are difficult because different materials are processed differently. However, small increases in material hardness are known to greatly increase fatigue life. Certain materials like the tool steels (M50) are usually substituted for increased life, however, the increased life may be more related to their processing than to the material. Recent evidence has shown that all surface distress can be drastically reduced if the effective film thickness is increased. Variables which increase film thickness are reduced surface roughness, increased viscosity, reduced temperatures, lower loads, increased speeds, and larger contact areas.

Table 25

FACTORS INFLUENCING FATIGUE LIFE

Variable	Effect on Fatigue Life
Load	Varies $\frac{1}{L^3}$
Velocity	Depends upon velocity effect on temperature, film thickness and centrifugal loading
Temperature	Reduce life by decreasing viscosity
Water in Lubricant	Reduces fatigue life
Lubricant film thickness	Thicker film increases life
Viscosity	Higher viscosity increase life
Abrasives in lubricant	Reduces fatigue life
Reactive lubricants or or additives	Reduces fatigue life
Solid lubricant additives	Increases fatigue life

Table 25
(cont.)

Variable	Effect on Fatigue Life
Materials	Specific for material and processes
Hardness	Higher hardness increases life
Roughness	Increased roughness decreases life
Surface damage	Reduces fatigue life
Stress Concentrations	Reduces fatigue life
Residual compressive stresses	Increases life

Wear is a common failure mode only in those bearings which experience appreciable sliding. This wear is usually associated with the cage since it rubs against the ball and the race land. Improved materials and appropriate application of solid lubricants can eliminate many cage wear problems[70, 71]. Coatings are also used. Race wear is almost always due to dirt or corrosive materials in the lubricant. If the conditions are not appropriate to accelerate fatigue damage, then excessive wear will result.

4. SUMMARY OF RESULTS

In the past 10 years there has been considerable reevaluation of maintenance policies with emphasis on reduced maintenance while still requiring high levels of reliability, availability, and durability. To achieve this, several different approaches to maintenance are being considered. Condition monitoring avoids preventive maintenance and many inspections and performs maintenance only after some indication is given that it is required. This "indication" may be a sensor which monitors a given condition, service data which indicates that a particular maintenance task is required, or some analytical life prediction scheme. Each of these techniques are currently in use to some degree with instrumentated receiving the greatest emphasis. Contract maintenance, more characteristic of commmercial markets, transfers maintenance responsibilities to another who presumably can achieve economies of scale and specialize in a particular technology or system. Very often the maintenance is performed by the equipment manufacturer. This is a particularly useful arrangement since it provides an immediate source of information for the manufacturer and provides economic incentives to reduce maintenance. Adequate engineering is also available to make the required design changes. Failure warranties are similar to contract maintenance except the actual maintenance is performed by the equipment user. However, the manufacturer is responsible for maintenance costs. In these cases, the manufacturer and the user agree on what is to be warranted, how failures are defined, and the cost.

Whatever the approach, reduced failures are desirable since the user ultimately pays the service costs. The critical element of a maintenance policy is to completely understand the nature of the service problems encountered. Once such information is available the most cost effective appraoch to meet maintenance objectives can be devised. This may be the introduction of new technology (maintenanace technology) or the use of more conventional approaches. With condition monitoring it is only necessary to monitor these critical problem areas rather than all possible failure modes. The literature suggests that condition monitoring has been successful in these instances. Where it has been unsuccessful is where it has been too broadly based. An understanding of the failure modes also allows a better understanding of where warranties are required and where they will be cost effective. Thus it can be concluded that new approaches to failure control should be investigated.

This chapter classifies failures using a variety of categories. The source of failures were classified as inadequate designs, poor quality control, or changes in operating conditions. Most failures were found to result from unexpected changes in operating conditions; for example, a bearing designed for 200°F actually operates at 300°F. The changing conditions most responsible for failure were factors such as contamination, misalignments, vibration, and inadequate lubrication supply techniques. Alternative component designs are required which will remove these conditions or provide tribological components which are more tolerant to these conditions. A review of the literature indicates that sufficient technology is available to provide failure-free components, however, data are not available to predict the life benefits so gained. Such data are needed to justify the costs of change. For example, is a higher cost, longer life bearing more cost effective than the more frequent replacement of a lower cost one.

5. REFERENCES

1. Peterson, M.B., "Aircraft Lubrication and Lubrication Systems," Final Report USN Contract N00014-78-C-0785 (1980)

2. Peterson, M.V., Koury, A.J., Devine, M.J., and Minuti, D., "Costs and Causes of Maintenance in a Ship's Diesel Propulsion System," ASME Paper 77 WA/LuB2 p.4 (1977).

3. Baker, R. and Hollingsworth, D.J., "A Computerized Methodology for the Identification of Aircraft Equipment Items for Reliability Improvement," NTIS AD-A059 p. 566 (1978).

4. Mickelson, S.D., McCoy, K.J., Glick, G.L. and Allen, C.W., "Extending the Life of Household Appliances," ASME Paper No. 76-DE=2 p.8 (1976).

5. Bucsek, G.F., "Analysis of Gas Turbine Failure Modes," NTIS AD-B003 229 p. 76 (1974).

6. Lancaster, J.K., "Breakdown and Surface Fatigue of Carbons During Repeated Sliding," Wear Vol 6 No 6, p. 467 (1963).

7. Lewis, R.B., "Predicting Wear of Sliding Plastic Bearing Surfaces," Mech. Eng. Vol. 86 No. 10 p. 32 (1964).

8. O'Rourke, J.J., "Fundamentals of Friction, PV, and Wear of Fluorocarbon Resins," Modern Plastics Vol 42, No. 9 (1965).

9. Pratt, G.C., "Plastic-Based Bearings," Chapter 8 in *Lubrication and Lubricants* by E.R. Braithwaite, Elseveir, London (1967).

10. Pinchbeck, P.H., "A Review of Plastic Bearings," Wear Vol 5 No. 1, p. 85 (1962).

11. Steign, R.P., "Friction and Wear of Plastics," Metals Engineering Quarterly, Vol 7, No. 2, p. 8 (1967).

12. Harris, B., "Little Known Facts Affecting Teflon Fabric Bearing Life," SAE Paper No. 800676, p. 9 (1980).

13. Docksell, S., Huffman, J.L., "U.S. Army Helicopter Rod End Bearing Reliability and Maintainability Investigation," NTIS AD-768 843, p. 123 (1973).

14. Barnes, W.C., "Military and Commercial Aircraft Bearing Field Experience," NTIS AD-861 738 p. 473 (1969).

15. Williams, F.J., VanWyk, J.W., and Lipp, L.C., "Static, Dynamic, and Fatigue Load Influence on Solid Lubricant Compact Bearings," Lubrication Eng., Vol 30 N2, p. 76 (1974).

16. Carson, R.W., "A Special Review of Self Lubricating Bearings," Product Eng. vol 35, p. 79 (1964).

17. Crease, A.B., "Design Data for the Wear Performance of Rubbing Bearing Surfaces," Tribology, Vol 5 No 1, p. 15 (1973).

18. Murray, S.F., Peterson, M.B., and Kennedy, F., "Wear of Cast Bronze Bearings," Annual Report INCRA Project #210 (1975).

19. Glaeser, W., "Bushings," <u>Wear Control Handbook</u>, M.B. Peterson and W.O. Winer, ASME 1980.

20. Garkunov, D.N., "Investigation of the Wear of Mating Parts with Reciprocating Rotational Motion," <u>Friction and Wear in Machinery</u>, (ASME) Vol. 14 p. 81 (1962).

21. Peterson, M.B., Gabel, M.K., Devine, M.J., and Minuti, D.V., "Wear Control for Naval Aircraft Components," Final Report USN Contract No. N62269-72-C-0764 (1976).

22. Rafique, S.O., "Failures of Plain Bearings and Their Causes," Proc. Instn. Mech. Engrs., 178, Pt, 3N, p. 180 (1964).

23. Murray, S.F. and Peterson, M.B., "Evaluations of Molded Phenolics as Oil-Lubricated Bearing Materials," Paper No. 63-WA-301, Presented at Winter Ann. Mtg., ASME, Phil., Pa.(Nov. 1963).

24. Karpe, S.A., "Study of Turbine System Bearing Failures Generally Classified as the Machining Type," NTIS AD-454 595, (1964).

25. Harbage, A., "Investigation of Rubber Stern Tube and Strut Bearings for Contrarotating Service," NTIS AS-648 973, p. 32 (1967).

26. Craig, Jr., W.D., "Effects of Surface Treatment, Material and Proportions of Friction and Seizure of Plain Journal Bearings," NTIS AD-810 186L, p. 34 (1967).

27. Williams, F.J., and Ascani, Jr., L., "Analytical Investigation of Wing Pivot Bearings for Variable Sweep Wing Aircraft," NTIS AD-476 968, p. 117 (1965).

28. King, A.L., "Bibliography on Fluid Sealing," British Hydromechanics Research Assn, Essex, England (1962).

29. Stair, W.K., "Bibliography on Dynamic Shaft Seals," University of Tennessee ME-5-62-3, (1962).

30. Ewbank, W.J., "Dynamic Seals - A Review of the Recent Literature," ASME Paper No. 67-WA/LUB-24.

31. Findlay, J.A., "Cavitation in Mechanical Face Seal," Trans. of ASME, J. of Lubr. Tech. 90, 2, (1968).

32. Hamaker, J., "New Materials in Mechanical Face Seals," Nat'l Conference on Fluid Power, Proc. 19, p. 58 (1965).

33. Wasil, T.J., and McCleary, G.P., "Sealing Corrosive Materials," Lubr. Eng. 23, 6, p. 234 (1967).

34. Orcutt, F.K., Smalley, A.J., "Investigation of the Operation and Failure of Mechanical Shaft Seals," MTI 68TR44, Prepared for ONR under Contract No. N00014-67-C-0500.

35. "Damping Vibrations Reduces Seal Failure," Product Engineer, p. 80 (1966).

36. Symons, J.D., "Shaft Geometry A Major Factor in Oil Seal Performance," J. of Lubr. Tech., Trans. of ASME 90, 2, (1968).

37. Symons, J.D., "Engineering Facts About Lip Seals," Trans. of the Soc. of Autmotive Eng., p. 614 (1963).

38. Dega, R.L., "Zero Leakage-Results of an Advanced Lip Seal Technology," Trans. of ASME, J. of Lub. Tech. 90, 2, p. 463 (1968).

39. Heyn, W.O., "Shaft Surface Finish on Important Part of the Sealing System," Trans. of ASME, J. of Lub. Tech. 90, 2, p. 375 (1968).

40. Heffner, F.E., "A General Method for Correlating Labyrinth Seal Leak Rate Data," J. of Basic Engr. p. 265 (1960).

41. Shepler, P.R., "Split Ring Seals," Machine Design, Seals Reference Issue, (Mar. 1967).

42. Ludwig, L.P., and Greiner, H.F., "Design Considerations in Mechanical Face Seals for Improved Performance I. Basic Considerations," ASME Paper No. 77-WA/LUB3, p. 10 (1977).

43. McNally, L., "Increased Pump Seal Life," Hydrocarbon Processing, V58, No. 1, p. 107 (1979).

44. McKibbin, A.H., and Parks, A.J., "Aircraft Gas Turbine Face Seals. Problems and Promises," Fourth Int. Conf. on Fluid Sealing, ASLE SP-2, p. 28 (1969).

45. Harrison, E.S., "Mechanical Seal Design and Lubrication for Corrosive Applications," ASLE Paper 76-AM-6B-1 p. 4 (1976).

46. McGrew, J.M., "Handbook for Seals in Naval Aircraft," Contract N62269-74-C-0379, ARP Project Report (1976).

47. Hayden, T.S., and Keller, Jr., C.H., "Design Guide for Helicopter Transmission Seals," NASA CR 120997.

48. Kaufman, H.N., and Walp, H.P., "Interpreting Service Damage in Rolling Type Bearings - A Manual on Ball and Roller Bearing Design," Amer. Soc. of Lubr. Eng., (1953).

49. Bisson, E.E., and Anderson, W.J., "Adavanced Bearing Technology," Natl. Aeronautics and Space Admin., Wash., D.C., NASA SP-38 (1964).

50. Simpson, F.F., "Failure of Rolling Contact Bearings," Proc. Instn. Mech. Engrs., 169, Pt., 3D, p. 248 (1964).

51. Wren, J.J., and Moyer, C.A., "Modes of Fatigue Failures in Rolling Element Bearings," Proc. Instn. Mech. Engrs., 179, Pt. 3D, p. 236 (1964).

52. Edigaryan, F.S., "Basic Types of Wear in Anti-Friction Bearings," Russian Engr. Jour., XLVI, 4, (1966).

53. Tallian, T.E., "Special Research Report on Rolling Contact Failure Control Through Lubrication," SKF Report AL66Q028, (Sep. 1966).

54. Cheng, H.S., and Orcutt, F.K., "Summary Report on Elastohydrodynamic Lubrication and Failure of Rolling Contacts," NTIS AD-481 826, (1966).

55. Smalley, A.J., et. al., "Review of Failure Mechanisms in Highly Loaded Rolling and Sliding Contacts," NTIS AD-657 337 (1967).

56. Eschmann, P., "Rolling Bearing Wear Life," ASME Paper 67-WA/Lub-2, 67.

57. Halling, J., and Brothers, B.G., "Wear Due to the Microslip Between a Rolling Body and its Track," ASME Paper 64-Lub-30, (1964).

58. Lundberg, G., and Palmgren, A., "Dynamic Capacity of Rolling Bearings," Acta Polytech, Mec. Eng. Ser., 1, 3, (1947).

59. Dawson, P.H., "Effect of Metallic Contact on the Pitting of Lubricated Rolling Surfaces," J. Mech. Eng. Sci., 4, 1, (1962).

60. Gentile, A.J., and Martin, A.D., "The Effects of Prior Metallurgically Induced Compressive Residual Stress on the Metallurgical and Endurance Properties of Overload Tested Ball Bearings," Paper No. 65-WA/CF-7, ASME (1965).

61. Zaretsky, E.V., Parker, R.J., and Anderson, W.J., "Component Hardness Differences and Their Effect on Bearing Fatigue," Jour. of Lub. Tech., Trans. ASME 89, t, (1967).

62. Rounds, F.G., "Some Effects of Additives on Rolling Contact Fatigue," ASLE Trans. 10, 3, p. 243 (1967).

63. Reichenbach, G.L., and Syninta, W.D., "An Electron Microscope Study of Rolling Contact Fatigue," ASLE Trans 8, p. 217 (1965).

64. Rounds, F.G., "Rounds of Base Oil Viscosity and Type on Ball Bearing Fatigue," ASLE Trans. 5, p. 172 (1962).

65. "Bearing Lubrication Under Severe Conditions," NTIS AD-408 646, (1963).

66. Daugherty, T.L., and Rosenfeld, M.S., "Progress Toward Long Life Bearings," NTIS AD-662-193 (1967).

67. Denhard, W.F., Freeman, A.P., Singer, H.E., "Failure Analysis of Critical Ball Bearings," NTIS AD-470 397 (1965).

68. Moyar, G.J., and Morrow, A.V., "Surface Failure of Bearings and Other Rolling Elements," Univ. of Illinois Engr. Experiment Station, Bulletin 469 (1964).

69. Littman, W.E., "The Mechanism of Contact Fatigue," NASA Symposium Interdisciplinary Approach to the Lubrication of Concentrated Contacts," p. 8.1 (1969).

70. Johnson, R.L., Swikert, M.A., and Bisson, E.E., "Investigation of Wear and Friction Properties under Sliding Conditions of Some Materials Suitable for Cages of Rolling Contact Bearings," NACA Report 1062 (1952).

71. Devine, M.J., Lamson, E.R., and Bowen, J.H., "Inorganic Solid Film Lubricants," Journal of Chemical and Engineering Data, Vol. 6, No. 1 (1961).

TRIBO-TESTING

M. Godet

D. Berthe
G. Dalmay
L. Flamand
A. Floquet
N. Gadallah
D. Play

Institute National des Sciences Appliquees

1. INTRODUCTION

Tribo-testing is a critical facet of tribological technology. A key element in the machinery wear integrity optimization process is the existence of viable wear test techniques and approaches. This tribo-testing element will be discussed in this chapter with respect to extrapolation and simulation.

2. EXTRAPOLATION IN TRIBOLOGY

The difficulties encountered in attempting to extrapolate friction and wear data, obtained on laboratory rigs, to industrial problems is discussed under this section in the light of the Three Body model. Third body (or intermediate film) rheology along with explicit transverse and longitudinal boundary conditions are shown to be necessary before extrapolation can be expected. Various domains of tribology (thick film lubrication, solid lubricants, dry bearings, etc...) are explored to see how much of the information required for extrapolation is available in each domain. Simulative testing must be undertaken when extrapolation requirements are not met.

2.1 Generalities

Three types of experimental programs are performed today in tribology. They differ from each other both in purpose and method. They can be grouped under the headings: friction mechanisms, material testing, and simulation.

Research in friction mechanism is most often carried out on fairly simple tribometers. Experiments are performed under carefully controlled conditions and working surfaces can be examined with sophisticated equipment. The purpose here is to analyze the basic mechanisms of friction, independently of any specific application. This type of work is reported in References 1 - 3.

This section will not be concerned with experiments run to improve the basic understanding of friction but with those run for practical purposes as outlined later. Let us therefore attempt to define briefly what is meant by material testing in tribology and by simulation.

2.1.2 Material Testing in Tribology

The purpose of material testing in general is the determination of intrinsic material constants or limits which characterize material behavior over a large range of conditions. Once determined, these constants or limits can serve to predict the response of the material in a given environment. One classical example is the determination of Young's modulus, E, Poisson's ratio, ν, and the proportional limit, σ_o, in solid mechanics. The intention in material testing is to go from the particular, which is the test environment, to the general, which is the field of application, Table 1. Tests are few and standardized. The key word is "extrapolation."

Materials testing is not necessarily viewed in this fashion in tribology. It is now clear that there is no such thing as an intrinsic friction property of a given or even of a pair of materials[5]. Extrapolation of values of friction or wear, from one test condition to another, is condemed and even material ranking is known to be dangerous. Tests are unlimited in number and are rarely standardized[6,7,8]. Indeed there are at least as many tests as there are applications, and existing tests serve to qualify a material for a given usage (i.e. for specific loads, P, velocities, U, temperatures, T, and configurations, G) but do not yield parameters which can be related to the "intrinsic friction property" mentioned above.

The only true satisfactory work which can be listed under the heading of material testing in tribology is the study of

Types of tests	Solid mechanics	Tribology
	1 (traction)	∞
Results	E, ν, σ_0	f vs P,U,T,G ; w vs P,U,T,G
Consequence	extrapolation	no extrapolation

TABLE 1 MATERIAL TESTING

lubricant behavior under different pressures and shear rates[9, 10, 11]. Constitutive laws which apply over a fairly wide range of conditions are developed and performances can be predicted[12 - 16]. The approach here is very similar to the one referred to earlier in solid mechanics. It can be transposed to tribology, as the material which governs the friction process (i.e. the lubricant) is identified, it can be isolated and tested separately.

The situation is different in "dry friction". As, in most instances, the material which governs friction is neither identified nor isolated and therefore cannot be tested[17]. This situation will be considered later in this discussion.

2.1.3 Simulation

A detailed discussion of simulation will be presented later in this chapter[18]. Simulation is a global process, directed to the determination "in vitro" of the performance of a given mechanical component, isolated from a complex engineering environment[19]. This process is carried out independently of both basic friction phenomena and material evaluation. It is intended to solve a particular technological problem[20 - 23]. The procedure is therefore not standard and has to be redefined for each application. Many examples of material assessment through simulation are found in the literature[24]. By definition, no extrapolation is expected from simulation. The simulation can be compared to a black box, entry conditions are fed to it and results are transposed to the application with a minimum amount of interpretation. The difficulty resides in carefully designing the black box.

2.2 Third Bodies

Before getting into the details of material testing in tribology, it is necessary to try and set the pattern which exists and see what is known today and what should be known later to extrapolate laboratory results to full scale industrial applications as is done in structural design for instance.

Let us therefore recall a few well established facts around which we will attempt to construct our reasoning.

- Solid bodies adhere to each other in the absence of surface films [25]. Surface fimls are found where repeated sliding between solids is encountered. These films can be formed artificially, generated "in situ" by chemical action or wear, and entrained kinematically in the contact[26]. These films will be grouped under the heading of third bodies[27]. Oil and oxide films form effective third bodies.

- A contact is therefore made up of two first bodies (i.e. machine elements such as gears) and an intermediate film or third body.

- Third bodies can be defined in a general or "material sense" as a zone situated between the two first bodies enclosed by frontiers close to the surface which mark a change in composition or structure from the bulk material, Figure 1.

- Third bodies can also be defined in a more restrictive or "kinematic sense" (when the first bodies undergo relative motion) as the thickness across which the difference in velocity is accommodated.

The first definition is preferred by material scientists, while the second is more useful when the problem is analyzed in terms of continuum mechanics.

- Third bodies can be continuous or discontinuous, thus generating respectively "full" or "empty" contacts, Figure 2[17]. Continuous third bodies are found for instance in classical EHD, Figure 3. Different third body patterns are obtained with empty contacts. Solid lubricants exhibit essentially longitudinal patterns, Figure 4. Elastomers exhibit lateral patterns, Figure 5[30]. Further random patterns can also be expected to exist.

 - Third body origin is diverse. The contact can be supplied in third body by either tangential or normal feed[31]. In normal feed, the third body is formed "in situ" through wear. Resulting wear particles are compacted during motion and separate the first bodies thus serving as load carrying agents. This situation is well illustrated during the first pass of experiments run with chalk, plastics, carbon brushes, etc[32]. Normal feed implies wear, with the notable exceptions of hydrostatic and porous bearings.

In tangential feed, the third body is drawn through the contact by the moving bodies causing separation of these bodies and thus acting as load-carrying agents. This situation is found with both liquid and solid lubricants and also when wear debris originally formed through normal feed is recirculated by the first bodies. Tangential feed prevents wear. Table 2 presents a few representative cases.

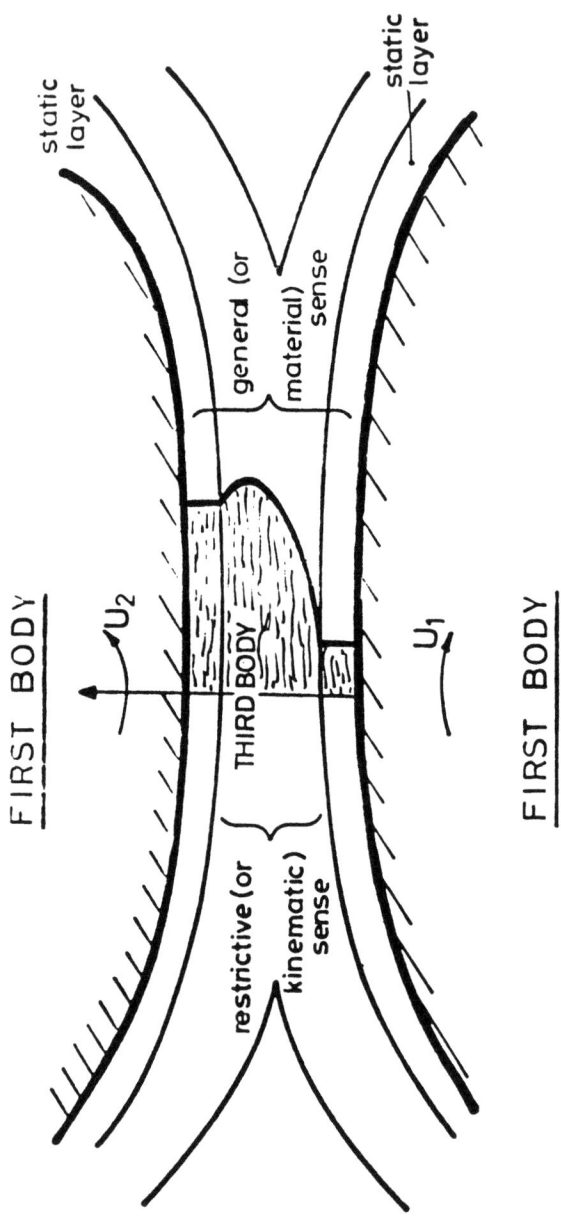

FIGURE 1 THIRD BODY DEFINITIONS

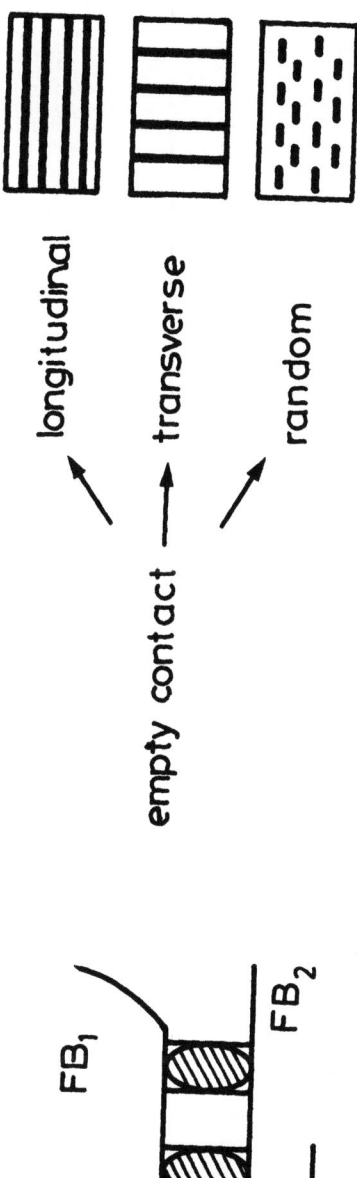

FIGURE 2 EMPTY CONTACTS

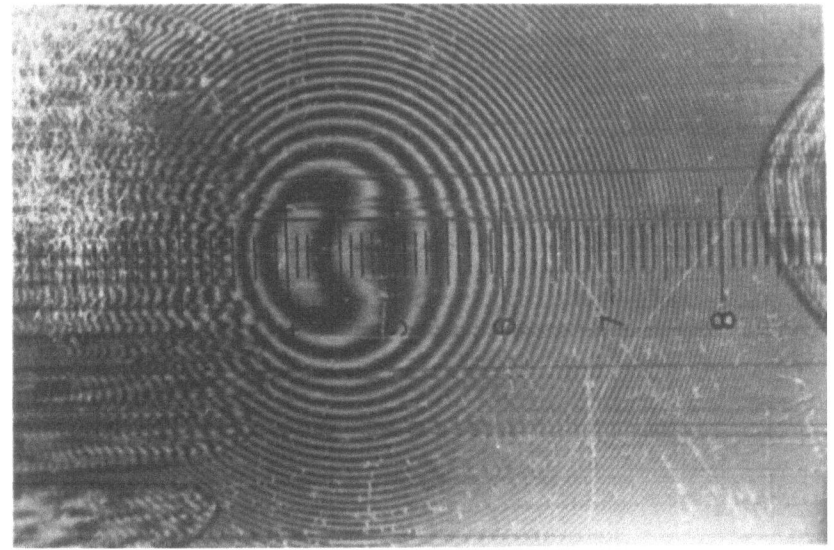

FIGURE 3   FULL CONTACT (EHD REF. 28)

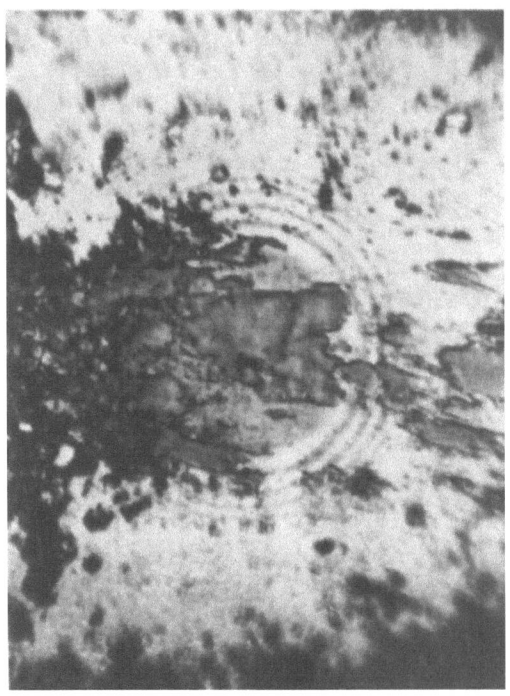

FIGURE 4   EMPTY CONTACT-LONGITUDINAL (REF. 29)

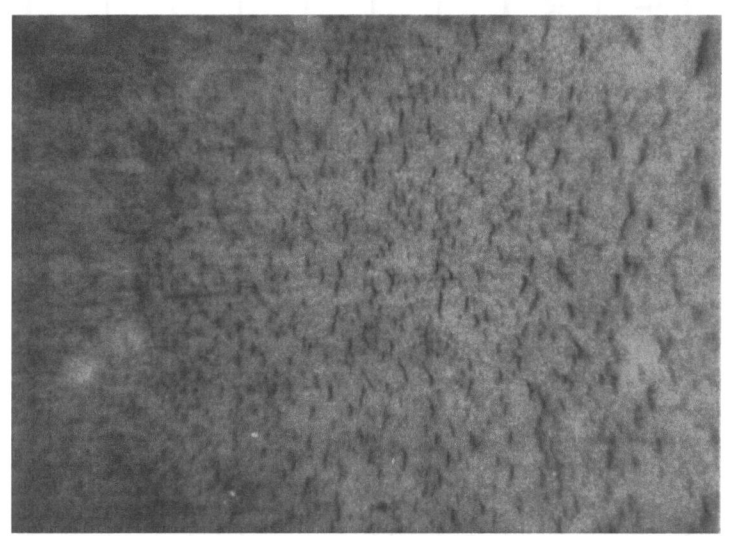

FIGURE 5 EMPTY CONTACT - TRANSVERSE (REF. 17)

cases studied	feed		
	pure normal	pure tangential	mixed
I hydrodynamic lubrication		O	
II thick film: solid lubricant		O	
III chalk	O	①	O
IV boundary lubrication		O	O
V rubber	O		O
VI plastics	O	①	O
VII pencils	O		

① : first pass

TABLE 2 NORMAL AND TANGENTIAL FEED

In both hydrodynamic and solid lubrication, feed is exclusively tangential. Visualization shows that when chalk is rubbed against glass, feed is first normal and wear particles are formed, which are deposited on the track and recirculated in the contact during the following passes thus initiating tangential feed[32]. Similar behavior was noted with some plastics. Boundary films are reaction products formed on the surface by active components of the lubricant[33]. Feed is therefore mixed as the third body is partially formed from first body material (normal feed) and is also drawn kinematically (tangential feed) into the contact. Rubber rolls are first formed through normal feed and depending on configuration, are recirculated tangentially. Pencils which are rarely called upon to write over a "used" track, present an interesting, if not very common case of pure normal feed[34-35].

In conclusion, the above remarks suggest that third bodies:

1) separate partially or totally first bodies,
2) are fed either normally or tangentially to the contact,
3) are transported along the contact,
4) are load-carrying agents which permit relative motion.

Accordingly, from an industrial point of view, a tribology problem is solved when for a given set of dynamic, thermal, and geometric conditions, a material combination is found which is capable of:

a) sustaining heavy loads (of the order of tons) transmitted by the first bodies,

b) accommodating differences in velocity (of the order of meters per second) between the first bodies.

That combination must therefore:

a) withstand large normal pressures ($1 < p_{max} < 5 \times 10^3$ MPa) i.e. possess good volume properties, or more specifically, a <u>high shear resistance,</u>

b) accept high shear rates ($10^3 < u/h < 10^7$ sec^{-1}) i.e. posssess good surface properties or more specifically a <u>low shear resistance.</u>

Hence right from the start, it is clear that no one homogeneous material can satisfy both requirements simultaneously and that more complex combinations have to be sought. The simplest combination is the "Three Body Model" which has been described above. Clearly first bodies withstand the high loads while third bodies accommodate the high shears and transfer the loads.

2.3 Material Testing in Tribology

2.3.1 Idealization of the Problem

If it is true that a minimum working combination in tribology is characterized by the three body model, two classical material tests are required if both first bodies are identical and three are necessary if they are different. Further the conditions at the interface between first and third bodies must be specified. This argument sets the pattern of what follows. To simplify this discussion, it will be assumed that:

1) first bodies are identified and characterized and that no further testing is required,
2) contacts are full, and thick films are considered,
3) surfaces are smooth,
4) stationary conditions prevail.

Three important factors are thus left out of the discussion:

1) roughness effects on load-carrying capacity,
2) short and long term transients; the first being concerned with changes in dynamic or thermal conditions, the second being concerned with contact modifications during tests generally grouped under the heading of running-in[36].
3) empty contacts.

Finally, wear will not be discussed but it will be understood that, all other things being equal, a decrease in load-carrying capacity causes an increase in wear. Even within these important limitations, extrapolation from material tests results in tribology, other than in lubrication, will be shown to be practically impossible. This point will be developed later.

2.3.2 Levels in Extrapolation

Let us therefore try to define what, in this context, is meant by extrapolation. Three arbitrary levels of extrapolation with three different levels of efficiency can be defined, Table 3.

In the **first level**, one standardized and repeatable test yields "coefficients" and "limits" (such as Young's modulus, heat conductivity etc..) which are introduced in a "theory" that presents "absolute" results for a given engineering application. This procedure has been applied, at least in principle, for close to a century in classical strength of materials problems. The number of problems that can be solved in unlimited. From an engineering point of view, Level I efficiency is remarkable, and material testing can only reach that efficiency when all governing parameters or coefficients, within known limits, are identified and clearly <u>expressed</u> in terms of <u>mechanics</u>. Hence, theory predominates. Level I is used in tribology to solve straight medium pressure hydrodynamic problems.

In the second level, the mathematical theory is incomplete or more often non-existant because the governing parameters are either not identified physically or not expressed in terms of mechanics because of mathematical difficulties. However, definite trends can be drawn either from material and surface science studies or from experimental programs which can guide the engineer in the choice of materials for the conditions considered. The number of practical problems that can be solved largely justifies the approach. The answers furnished in Level II are comparative or relative, and the number of practical problems that can be solved is large. Level II extrapolation efficiency is adaquate and this approach is commonly used in metallurgy, corrosion problems, etc.

In the third level, little is known of either the mechanics or the pyhsics of the problem. Neither theory nor established trends exist. One test gives one point for a particular condition and nothing can be inferred concerning the system performance if that condition is varied only slightly. The answer funished in Level III is discrete or punctual. Extrapolation efficiency is nil as there is no basis for extrapolation. The description of Level III can appear to be somewhat extreme, however it is fairly close to reality when one considers, as noted in Table 3, the number of tests necessary to come forth with a workable industrial solution. The expense of this approach is of course formidable.

2.4 First Level Extrapolation in Tribology

2.4.1 First Level Requirements

An essay on testing in tribology should attempt to identify the level of extrapolation reached in various branches of this discipline and see what can be done, if anything, to move up to the level above. The most efficient way is to list the requirements imposed by Level I extrapolation and see how close representative branches come to it.

In order to reach Level I efficiency in tribology, it is necessary to predict load-carrying capacity or third body thickness and friction in a system working under given operating

EXTRA-POLATION LEVEL	TESTS		UTILISATION			type of information	number of cases treated / efficiency		examples
	number	results	formal	phenom.	empiric.				
I	1	coeff. limits	x			absolute	∞	10	strength of materials
II	10	trends		x		relative	>>1	7	metallurgy
III	1000	points			x	discrete	1	1	fretting wear

TABLE 3   LEVELS IN EXTRAPOLATION

conditions. These predictions require a theoretical analysis. Let us therefore attempt to identify the information needed by theory in order to solve the elementary three body case presented earlier (i.e. full contacts and smooth surfaces). The approach used in thick film lubrication which, as noted earlier, is the only case solved in a satisfactory manner will serve as guide. By contrast, lubrication will be compared to dry friction. It proves easier to start from the solution and work backwards thus meeting each aspect required to get that solution to work, and concentrating on third bodies as (section 2.3) all information and theories concerning first bodies are assumed to be available. This is usually the case in practice.

Load-carrying capacity and friction are forces. They result from the vectorial sum of normal and tangential stresses acting at contact. Independently of the origin of these stresses, both components are expressed in terms of mechanics. It is necessary therefore, to identify the information needed to develop a third body mechanics theory. Third body mechanics, in turn, is nothing more than thin film mechanics of continua (TFMC). TFMC equations, both dynamic and thermal, are formed around constitutive equations and require boundary conditions for their solution as they are partial differential equations. Constitutive equations can be written when the behavior (or rheology) of the third body considered is known. Kinematic boundary conditions, Figure 6, are divided in transverse and longitudinal conditions. The transverse conditions express the interaction between first and third bodies in terms of either displacement or stress. The longitudinal conditions take in the factors which govern the extent of the third body. As such they govern "entry" and "exit" conditions. Thermal boundary conditions specify heat fluxes or temperatures in both the transverse and longitudinal directions.

2.4.2 Third Body Rheology

As mentioned in Section 2.1, a considerable amount of work has been done recently in order to determine the rheological properties of lubricants under different pressures and strain rates. Results have been applied with success, in thick film lubrication. Constitutive equations have been developed.

In dry friction, however, the situation is radically different as the third body is rarely identified. An intermediate situation exists in polymer work where third bodies, both transfer films and polymer surfaces films, have been detected and their composition determined through surface analysis[37]. In these cases, third bodies are found not to be necessarily homogeneous as composition gradients and structure variations are noted across their thickness. These variations

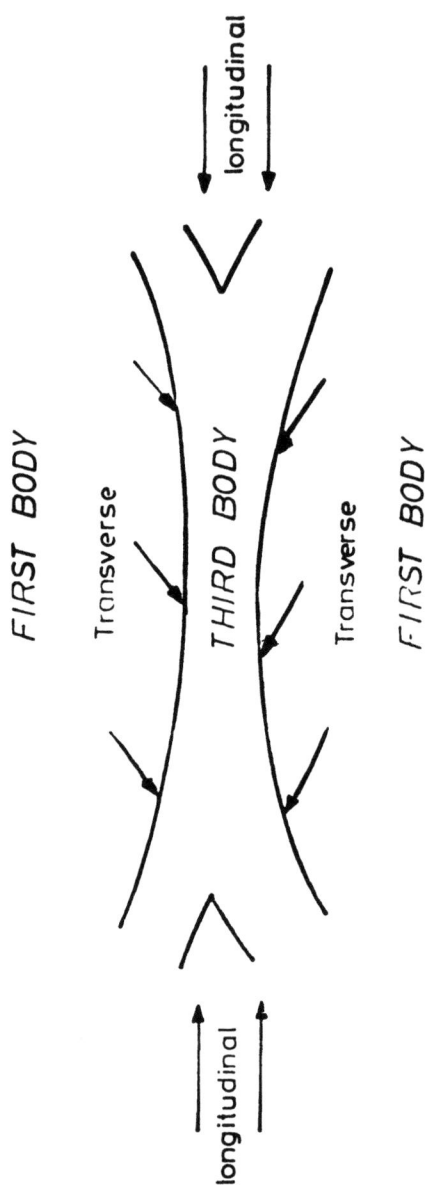

FIGURE 6 BOUNDARY CONDITIONS

underline the difference noted earlier between third bodies defined in the general sense and the strict sense. However, in a purely formal basis, this composition gradient would not constitute an insurmountable obstacle for analysis if the corresponding variations in rheology were known. The real difficulties lie elsewhere, and result from the fact that:

- The third body approach has not been completely worked through and all of its consequences analyzed and acted upon. This obstacle is one of attitude.

- Very little third body data are available particularly in cases where normal feed prevails. This obstacle results from difficulties encountered in the development of microrheometers capable of analyzing minute volumes of material.

Note, however, that as far as Level I extrapolation is concerned, precise information concerning third body composition is not needed. The only information demanded is that required to write the constitutive equations.

2.4.3 Boundary Conditions

In thick film lubrication, transverse boundary conditions are clearly expressed in mechanical terms by the "no slip at the wall" condition irregardless of the shearing stress. This condition seems to be quite satisfactory in most instances even though:

- It has been questioned recently under particular running conditions[39].

- The physics of the problem are not completely worked out[40].

From an engineering point of view, the second point is interesting as it suggests that a simple global mechanical expression of complex liquid solid interface phenomena is enough to reach Level I efficiency in a great many applications. However, concerning both entry and exit conditions, flow continuity conditions are in most instances sufficient to express longitudinal boundary conditions[42]. Thermal boundary conditions are often untractable mathematically, but basic concepts are clearly outlined for most cases and will not be discussed further here[43 - 46].

In "dry friction", the problem is not even approached in these terms and progess in the expression of boundary conditions appears remote as the respective "scientific partners" do not talk the same language. Statement of the type "beyond temperature, T_o, material A is seen to deposit transfer films when rubbed against material B under such and such condition," which conclude the work done by physicists are not yet exploitable in terms of mechanics to reach Level I efficiency[48]. Such statements are of course of a great help in Level II work. Clearly some effort must be expended, if third body analysis is to progress, and thus insure that the underlying physical concepts of material transfer incorporate stress and displacement notions[26].

Little is known about longitudinal conditions in "dry friction".

2.5 Extrapolation Levels in Different Areas of Tribology

The preceeding paragraphs attempted to identify the information required in thin film mechanics studies to reach Level I extrapolation. It appears useful now to see how close typical tribology systems come to that ideal, Table 4. As stated earlier, in order to calculate load-carrying capacity and friction in full smooth contacts, it is necessary to:

1) determine the values of third body rheological parameters (i.e. coefficients and limits),
2) express both transverse and longitudinal boundary conditions in both kinematic and thermal cases,
3) have available the corresponding thin film mechanics of continuum theory and the necessary programs to apply that theory to the particular problem under study.

In order to give a full account of the situation, and understand how far along in third body analysis the particular subject has advanced, it is necessary to specify whether or not the third body has been identified both in composition and as a load-carrying capacity agent. Further, an attempt will be made to predict whether efficiency or extrapolation level changes can be expected shortly. Finally, priority areas of work and application are noted in each instance. Table 4 lists seven cases chosen arbitrarily. The first three are taken from the thick film lubrication domain and obviously belong to the Level I class of solution. This does not mean that all aspects are either known or expressed in usable terms from an applied mechanics point of view, but only that most problems which belong to these cases can be solved theoretically:

- Some basic problems concerning feed and recirculation exist in bearings which are representative of low pressure (L.P.) hydrodynamic problems[49].

E.P.	steels 1000°C	plastics	sol. lub.	hydrodynamics				CONTACT		
reaction prod.	debris	debris	ox M...	oils				T. B.		
				VLP	HP	LP		SPEC.		
F	E	E	E	?	F	F	F:Full ; E:Empty	TYPE		
N.T	N.T	N.T	T	T	T	T	N: Normal / T: Tangentiel	FEED		
P	N	P	Y	Y	Y	Y	Y:Yes N:No	ID.		THIRD BODY
P	N	N	Y	Y	Y	Y	P: Partly	Rh.		
U	U	U	U	K	K	K	K: Known	Coef.		
U	U	P	U	P	P	K	U:Unkn. P:Partl.	Lim.		
N	N	N	N	N	N	N	E: Expressed / N:Not exp.	PH	TRANS	Boundary conditions
N	N	N	N	T	T	T	T: Transcribed / N:Not trans.	M		
N	N	N	N	P	P	P	E: Expressed / N:Not exp. P:Part.	PH	LONG.	
N	N	N	N	P	T	T	T: Transcribed / N:Not tran. P:Partl.	M		
N	N	N	N	Ex	Ex	Ex	Ex: Existant / P. Partial N: No	FORM.		SOLUTIONS
Y	P	Y	N				Y:Yes N:No P:Part.	PHEN.		
Y	Y	Y	N					EMPI.		
III	III	III	II	I	I	I	LEVEL	1981	EXTRA.	
M	L	M	M	S	S	S	S: Short M:Med / L: Long term	Prog.		
id. 3rd B. rheol. theory	id. 3rd B	debris rheo. B.C. theory	rheo theory	lim. B.C.	BC (entry) lim. (HP) ph.of fluids	B.C.* recirculation	priority		FUTURE WORK	
gears, cams etc	oven steel works	dry bearings	slow bearing fretting	seals	ball bearings gears	bearings	examples		APPLICATIONS	

*B.C. Boundary Conditions

TABLE 4   EXTRAPOLATION LEVELS IN DIFFERENT AREAS OF TRIBOLOGY

- Longitudinal boundary conditions are not completely understood in ball bearings which are representative of high pressure (H.P.) lubrication[50].

- The conditions which define cavitation zones in seals representative of very low pressure (VLP) lubrication, are not always clearly defined[51]. Note that a significant advance in third body high pressure, high strain rate rheology has been accomplished in the last ten years as a result of the work undertaken on traction of different oils.

The classification of the last four cases in Level II or III is somewhat arbitrary. Indeed trends are found in the four areas, but tests must be run to validate each solution. There is a significant difference however, between the four when expressed in terms of closeness to Level I extrapolation. The substantial difference between Case IV (solid lubrication) and the other cases is that:

1) the third body is identified,
2) it exists in sufficiently large quantities to insure that satisfactory rheological determinations can be conducted.

However, many problems remain concerning:

1) contact fullness,

2) boundary conditions,

3) corresponding TFMC theory.

Contact fullness and longitudinal boundary conditions, particularly entry conditions, are related as feed is controlled by conditions at inlet. Transverse boundary conditions must also be looked into with great detail as indications here are scarce. Solid lubrication however, appears to be a likely area for progress. Attempts should be made in this case, even if in very limited applications, to reach Level I efficiency. The situation is clearly much less promising in plastic wear studies. This skepticism as mentioned earlier, results from the complexity of the third body composition[37]. The whole process of load-carrying capacity in high temperature refractory steel friction, has proven to be of such complexity that there is very little hope of reaching any form of theoretical development in the near future[52]. However, visualization studies have given new lines of approach which can help breach the gap between Level III and Level II extrapolation. The very same ideas can be developed concerning E.P. studies.

Table 4 is fairly eloquent as areas of ignorance are shaded while areas of competence are left plain. One sees at a glance, the distance that separates the well coded disciplines such as lubrication from those such as fretting, which from a mechanical point of view, currently offer little hope.

2.6 Conclusion

In conclusion, this section has attempted to show that the notion of testing in tribology is as varied as the discipline itself. Tribology does indeed include elements of all three levels of extrapolation which command these radically different approaches in testing.

Level I requires material tests in the most classical sense. Material constants are determined and included in well established computer programs. Experiments are conducted occassionally to "verify" a result. "Indicative" tests can also be run to choose the best method among different levels of complexity in theoretical approaches. Thick film lubrication is the only example of Level I extrapolation in tribology. Progress in lubrication will come from the application of known methods to more amd more complex technological situations and from transfer of new mathematical methods from older and therefore more developed sciences such as fluid mechanics.

A significant effort is needed in micro-rheometry and in the understanding of both transverse and longitudinal boundary conditions. Further, developments of thin film mechanics of continuum solutions with varied constitutive equations and different longitudinal and transverse boundary conditions should be encouraged. Application of these solutions to specific cases should be followed with extreme care and with a well developed critical sense in order to detect unfounded claims and unrealistic mathematical developments. Precise answers to industrial problems come from simulation in Level II and III cases.

3. SIMULATIVE TESTING

Straight material testing has been shown to be not applicable to tribology as it does not yield "intrinsic" parameters which can later be used to extrapolate test results to real applications. The design engineer must thus rely on simulative testing.

The purpose of this section of the chapter is to present summaries of three cases of simulative testing which have been worked on in the laboratory over the last few years. Detailed accounts of the methods and the relevant bibliographies are found in the original published papers. The first case considers rib/roller contacts in tapered roller bearings; the second, gear damage simulation; and the third, simulation of helicopter oscillating dry bearings.

In simple terms, it can be said that engineers call upon data tables, extrapolation techniques, and of course experience to produce new designs. It was suggested in the first part of this chapter that three levels of extrapolation are found in engineering, depending upon the subject considered. Examples of each level are found in tribology. In Level I, all governing parameters are identified and taken into account in fully developed theories mostly through efficient computer programs. Straight fluid bearing design is classified in Level I.

The distinction between Level II and Level III is not as clear and easy to express. In tribology, Level II would include:

a) Cases are included in which the lubricant (intermediate film or third body) is identified and available in quantities sufficient for analysis, but whose rheology and also feed (longitudinal boundary conditions) and adhesion (transverse boundary conditions) aspects are unknown. Solid lubricants can be classified in Level II.

b) Cases are included in which the lubricant is identified and characterized rheologically but present particular feed conditions. Special bearing configurations belong to Level II.

c) Cases are included in which the lubricant is identified and characterized rheologically with classical feed on longitudinal boundary conditions, but for which thick films cannot be formed. Here, Level I techniques can predict partial separation but cannot foretell the damage which results from that condition. Partial elastohydrodynamic cases including highly loaded gears, belong under Level II.

Each case lacks some information which prevents it from being listed in the Level I table. However, trends exist which can be used in an engineering sense as guides to help solve the problem.

Level III would include all cases in which the third body is not identified. Dry bearings and most of the dry friction applications are recorded under that heading.

In the last few years through "Systems Analysis", a much needed and determined effort to tidy up simulative testing has been conducted[53]. This approach which is presented elsewhere, will not be discussed here. Its aims, however, are similar to those pursued in this treatment. The emphasis in this discussion is in the careful identification of the governing parameters and their simulation.

In order to avoid being side-tracted, the logic behind simulative testing, and its relation to testing in tribology is as follows;

1) All mechanical rubbing components are made out of at least three elements; two first and one third body.
2) Satisfactory material testing requires characterization of all three elements along with the definition of conditions at both first and third body interfaces. Only under these conditions can Level I efficiency be reached.
3) When these conditions cannot be satisfied, the engineer who seeks an answer to a specific industrial problem, as opposed to the scientist who attempts to understand a phenomenon, has to turn to simulation.
4) Consequently, test devices can be, or rather should be, radically different in fundamental and industrial laboratories.

3.1 Simulation of Rib-Roller End Contact in Tapered Roller Bearings

This discussion is part of a collaborative program between the Société Nouvelle de Roulement (SNR) in Annecy (FRANCE) and the Laboratoire de Mécaniques des Contacts of INSA. Detailed information is presented in References 54 and 55.

Figure 7 shows a section of a tapered roller bearing in which three contacts are found:

- rolling elements and cage (not shown)
- cone 3 and roller 2 line contact
- rib 1 and roller 2 point contact

The problem under consideration is the last contact. Without entering the details of the problems outlined elsewhere, the rib and roller end contact presents a certain number of original points which must be understood;

a) The hydrodynamic domain is very limited with the pressure build up entry abscissa, Figure 7, being very small which corresponds to starved conditions.

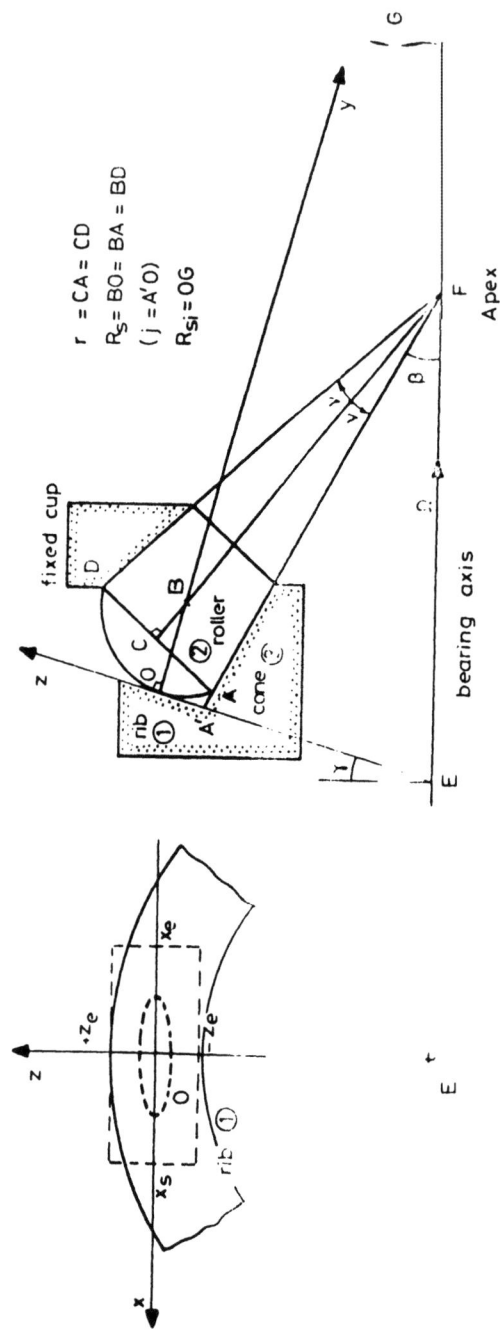

FIGURE 7 RIB ROLLER END CONTACT GEOMETRY

b) The large axis of the contact ellipse is parallel to the rolling direction, thus lateral or transverse flow is very important.
c) The kinematics of the problem are complex as rolling, sliding and spinning coexist.
d) Loads are comparatively small, thus the lubrication regime is either piezo viscous or lightly loaded elastohydrodynamic.

Classical theoretical methods can be used for the piezoviscous situation however, existing point contact elastohydrodynamic solutions cannot be applied because of the peculiar geometry. Consequently, under this case the isoviscous and piezoviscous problem will be solved but experimental simulation will be instituted for the entire range of conditions.

Respective simulation will be divided in two parts:

- Simulation and investigation will be initiated concerning the particular contact condition of only one contact on the INSA interferometric apparatus.

- Simulation will also be initiated concerning the entire bearing on the SNR special device in which rib and cone contacts are isolated.

In the original study, optimal geometry, film thickness and friction were studied. The optimal geometry could be determined by choosing the spherical roller end radius which gave the largest film thickness for given set of conditions.

3.1.1 The Two Simulators

Figure 8, a and b, show a schematic view and an actual photograph of the INSA one contact simulator. The apparatus is made of a flat glass disc and a steel toric specimen. The disc is driven by a variable speed motor which is fixed on a hydrostatic bearing possessing one degree of freedom in order to measure the lateral friction force. The toric specimen is fastened to a shaft connected to a second variable speed motor which itself is held by a cylindrical hydrostatic bearing. This bearing allows both axial and angular displacements. The axial displacement is used to measure the traction force in the rolling direction and the angular displacement is used to apply and measure the load.

The maximum height variation at contact location on the disc and on the toric specimen, is less than 0.5 µm. Surface roughnesses are respectively 0.005 µm and 0.01 µm CLA. The lower

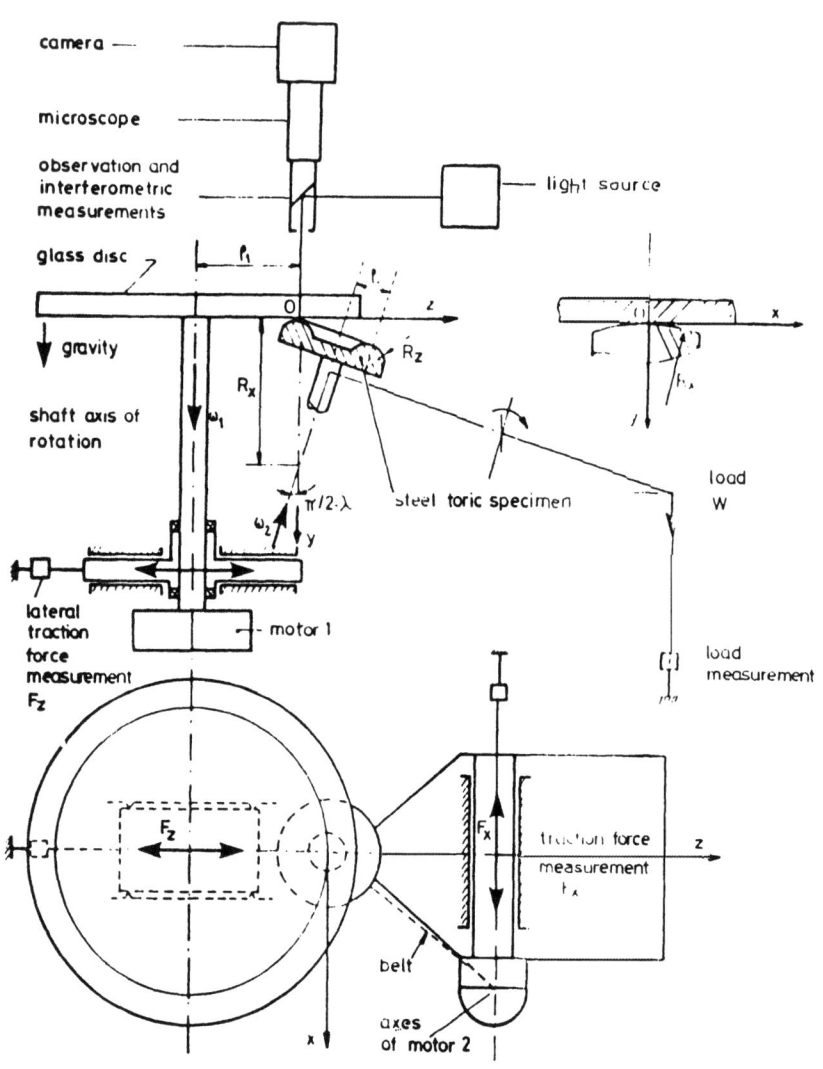

FIGURE 8(A)   SCHEMATIC VIEW OF THE APPARATUS

FIGURE 8(B)　ONE CONTACT SIMULATOR

face of the disc is coated with a chromium layer and the upper face with an antireflecting coating. A classical optical system composed of a microscope, light source and camera creates an interferometric pattern.

Simultaneous measurements of load, angular speeds of both specimens, inlet oil temperature, traction force on the toric specimen in the rolling direction, lateral traction force on the disc, and film thickness are performed for this peculiar contact under hydrodynamic and elastohydrodynamic conditions.

A test rig was built by SNR to study the friction of the rib-roller end contact in order to determine an optimum geometric configuration as well as to study influence of the surface roughness on friction of the SNR tapered roller bearing 32314 BA, used primarily in truck axle applications.

Figure 9 shows a schematic view of the SNR apparatus in which rib and cone are separated in order to measure individually rib and cone friction. Hydrostatic pads are used to minimize friction generated between the bearing rib and cone and intermediate parts which are necessary for load application. These pads allow friction torque measurements. Axial load is applied on the main shaft by a hydraulic jack and is measured with a strain gauge transducer. Load is transmitted through the rollers to the cone and rib which are supported by the hydrostatic pads. Friction torques are also measured with strain gage transducers. Load and speed and their variations with time are regulated electronically. The rib-roller end contact temperature is measured by a thermocouple embedded 1mm below the surface of the rib. Inlet oil temperature is also monitored.

3.1.2 Specific Conditions To Be Simulated in the Rib-Roller End Contact

The mechanical parameters required to define the running conditions in the axis Oxyz at the point of contact O are:

- the two principal radii of curvature of the rib (R_{1x} and $R_{1z} = \infty$) and of the roller end which is spherical ($R_{2x} = R_{2z} = R_s$).

- the three linear and angular velocity components of the rib ($U_{1x}(0)$, $U_{1y}(0)$, $U_{1z}(0)$, and Ω_{1x}, Ω_{1y}, Ω_{1z}) and those of the roller end ($U_{2x}(0)$, $U_{2y}(0)$, $U_{2z}(0)$ and Ω_{2x}, Ω_{2y}, Ω_{2z}),

    - the extent of the contact region,
    - the surface roughness of surfaces 1 and 2,

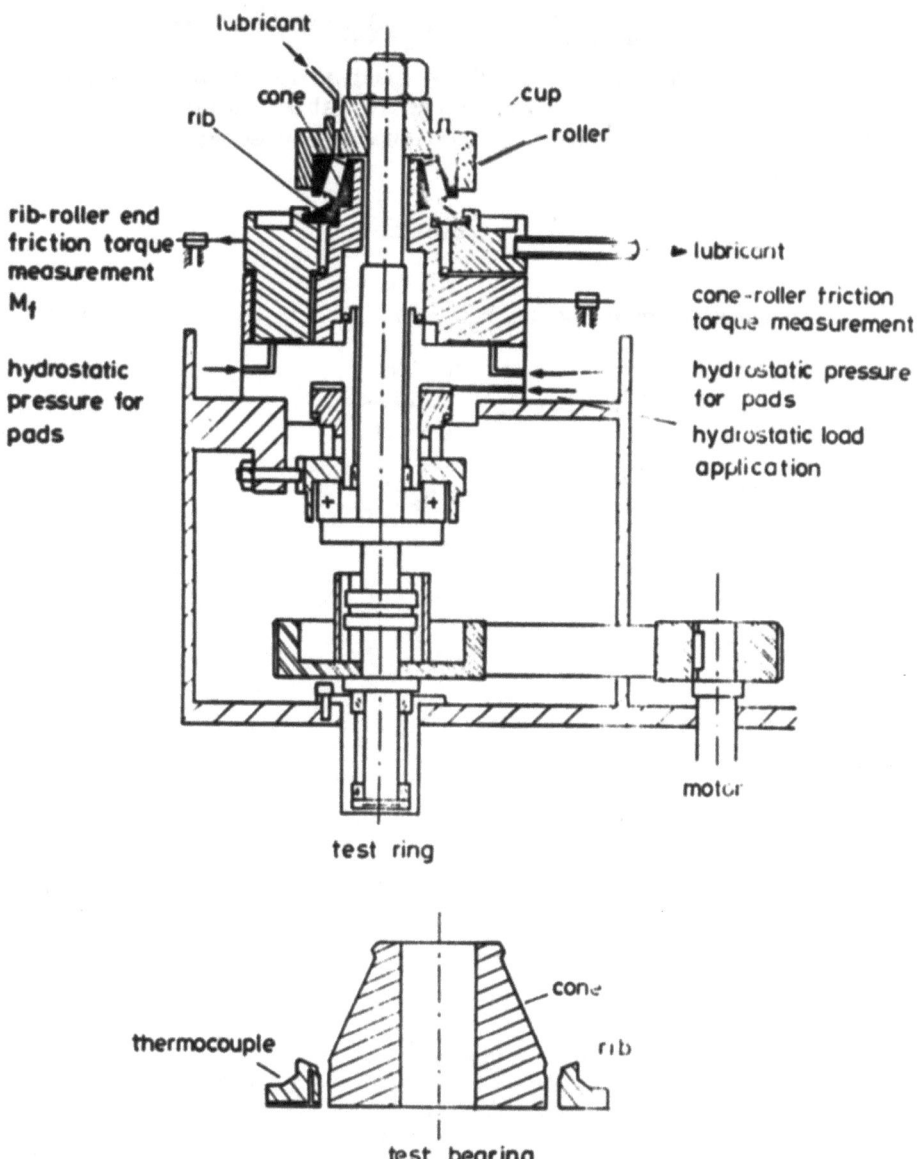

FIGURE 9 S.N.R. APPARATUS

- the applied load along the y axis,
- lubricant and material characteristics.

For the SNR bearing n° 32314 studied here, the kinematic and geometric parameters are given in Table 5 where Ω is the angular speed of the bearing in radians per second.

Table 5

Interference radius $R_{si} = OG$		0.21040 m
Equivalent radius $R_x = \dfrac{R_s \; OG}{OG - R_s}$		0.72874 m
Equivalent radius $R_z = R_s$		0.163262 m
	$U_{1x}(0)$	0.03124 Ω m/s
	$U_{1y}(0)$	0
	$U_{1z}(0)$	0
Rib (1)		
	Ω_{1x}	0
	Ω_{1y}	0.5614 Ω rd/s
	Ω_{1z}	0.1485 Ω rd/s
	$U_{2x}(0)$	0.01917 Ω m/s
	$U_{2y}(0)$	0
	$U_{2z}(0)$	0
Roller end (2)		
	Ω_{2x}	0
	Ω_{2y}	-2.8821 Ω rd/s
	Ω_{2z}	0.11743 Ω rd/s

As mentioned earlier the rib-roller end contact is an unusual lubrication problem. More specifically:

- The contact is lightly loaded (0 to 10^3 N) with the maximum Hertz pressure usually lower than 0.3 GPa.

- The velocities which govern the performance of this epicyclic contact are characterized by the rolling speed $U_{1x}(0) + U_{2x}(0)$, the sliding speed $U_{2x}(0) - U_{1x}(0)$ and the spinnning of both bodies Ω_{1x} and Ω_{2y}. The value of the rolling speed is similar to the one of the roller-cone contact but the sliding speed and spinning are independent.

- The contact geometry in which R_x is large ($R_x \sim$ 1 to 10 m) has a radius ratio, $R = R_z/R_x$, less than unity ($\sim 0.001 <$ R $< \sim 0.5$).

- The domain size in which the pressure can build up is very small thus the contact is severely starved (xe $\sim$ 2.4 mm and $z_e \cong 0.66$ mm).

- Thermal effects must be taken into account as sliding speed and spinning are important.

3.1.3 Results

This discussion will not list all the results obtained in this study. The discussion will, however, present the method used to simulate the contact chosen and show points of agreement between the three approaches.

3.1.3.1 Theory

An optimal roller end spherical radius was determined by hydrodynamic theory. This was achieved by simultaneously changing γ and R_s to ensure that the contact point, "o", is situated half way along the rib and choosing the situation which yields the largest film thickness. For Ω equal to 1000 rpm, the film thickness is on the order if 1μm for an oil of viscosity 0.03 Pl. The prime parameter is R_s, and the optimum values for load and traction forces are found for

$$\frac{R_s}{R_{si}} \sim 0.85 \text{ to } 0.90,$$

where R_{si} is the interference radius value of the spherical roller end. Thermal effects were shown to be much less important than geometric and domain size effects.

3.1.3.2. Experimental Results (INSA rig)

Tests were run for the particular geometric and kinematic conditions that prevail in this contact. More specifically; film thicknesses and traction forces were obtained at ambient temperature with low viscosity mineral oils under the following running conditions;

- The contact is formed by a toric steel specimen whose principal radii of curvature are Rx ≃ 1.374 m and Rz ≃ 0.663 m and a flat glass disc.

- The applied loads varied from 10 to 320 N giving maximum hertz pressures up to ~ 0.1 GPa.

- Rolling, sliding and spinning speeds are those found in the rib-roller end contacts.

Good agreement between the previous hydrodynamic theory and these experiments is obtained in the hydrodynamic regime for both load and traction forces. Departure from fully flooded elastohydrodynamic theory is due to domain size and starvation effects. These results show that hydrodynamic and the elastohydrodynamic lubrication conditions can be maintained at the rib-roller end contact for high loads and high speeds.

3.1.3.3 Experimental Results for the SNR Rig

Test rig speed and speed variation can be programmed between 0 and 1000 rpm. Oil inlet temperature is regulated at 63° C. Axial loads, ω_a, vary by steps up to 10^5 N and generate maximum Hertz pressures between 20 and 300 M Pa. Representative values for contact ellipse dimensions are a ≃ 2.48 mm and c ≃ 0.78 mm for R_s/R_{si} ≃ 0.85 and a contact load ω ≃ 10^3 N. Rib and roller end surface approximate roughnesses measured along a direction perpendicular to the grinding direction are respectively 0.18 μm CLA and 0.08 μm CLA. Hence, the composite surface roughness is ε ≃ 0.20 μm CLA.

Sixteen different geometries suggested by the previous theoretical analysis, were tested. For a given geometry γ and R_s are specified. The SNR criterion through friction torque measurement, is based on the minimal speed, Ω, which gives thick film operating conditions. The optimum contact geometry configuration which can be deduced is

$$\frac{R_s}{R_{si}} \approx 0.88$$

which is very close to the theoreticl hydrodynamic prediction given in the preceeding paragraph.

A typical friction coefficient versus speed curve is given in Figure 10 for a given load.

An example of the good correlation between the SNR rig and the following is presented in Figure 10.

1) the INSA experiment in the EHD zone.
2) theory in the hydrodynamic regime.

3.1.4 Discussion

The hydrodynamic approach performed for rigid contact surfaces is useful in the understanding of the behavior of this lightly loaded contact and clearly shows that the geometry, the domain size and starvation are the main factors which govern the respective lubrication. Obviously, the surface deformation and the surface roughness along with thermal effects are also important factors which must be taken into account.

The INSA simulator gives a confirmation as to the importance of the geometry, the kinematic conditions, and the domain size in the rib-roller end lubrication process. These factors remain important when the experiments are extended to the elastohydrodynamic regime.

Good theoretical correlation was obtained with the SNR experimental results for a tapered roller bearing operating in the hydrodynamic or elastohydrodynamic regime. Clearly, low speed conditions which might bring about glass/metal contact cannot be simulated in the INSA rig. This constitutes one of the limitations of the system. However, the complexity of the INSA simulator is justified due to the fact that the inlet and kinematic conditions along with the lateral flow conditions could only be simulated in this manner.

3.2 Gear Simulation

The following nomenclature is utilized in this section:

E_1, E_2 = Young's modulus W = normal load

E' = reduced modulus

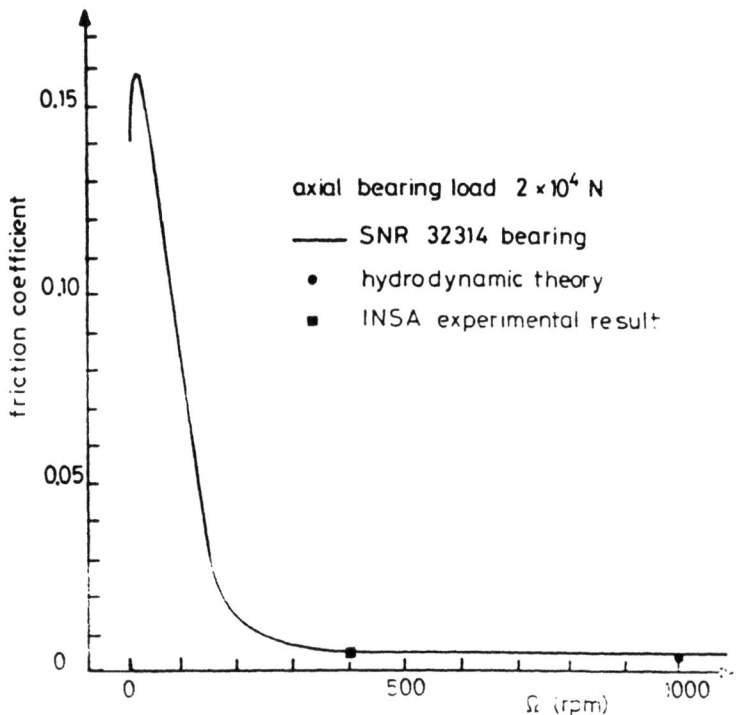

FIGURE 10 FRICTION COEFFICIENT VERSUS ROTATING SPEED FOR A GIVEN LOAD. COMPARISON OF THE INSA AND SNR RESULTS.

H_v	= Vickers' hardness	α	= piezoviscosity coefficient
h	= EHD film thickness (Dowson and Higginson)	β	= thermosviscosity coefficient
		$\varepsilon_1, \varepsilon_2$	= asperities heights
L	= track width	λ	= slide/roll ratio
p	= pressure	μ	= viscosity
R	= reduced radius of curvature	ν_1, ν_2	= Poisson's ratio
R_1, R_2	= radius of curvature	$\dot{\rho}$	= density
T	= temperature	$\sigma = \sigma_1^2 + \sigma_2^2$	= composite roughness
U_1, U_2	= rolling speeds	σ_1, σ_2	= RMS roughness

This case study is discussed in detail in Reference 56 and 57.

The purpose of this effort is to compare the results furnished by a gear machine and a disc machine. Both are run under closely controlled conditions (defined below) in order to establish whether, under what appears to be the most favorable conditons, correlation between gear and disc surface distress is possible. The damage considered is surface durability commonly known as "pitting." Gear tests were run on the modified FZG machine of the Societe Nationale Industrielle Aerospatiale of Suresnes (France). Disc tests were performed on the variable center disc machine of the Institut National des Sciences Appliquees de Lyon (France).

Figure 11, a and b, give respectively a schematic view, and an actual photograph of the disc machine. Table 6 lists the machine characteristics.

3.2.1 Simulation

Loads, speeds, oil film thickness (h), surface roughness (T), roughness ratio (σ/h), gear materials, surface treatments, and lubricants are known to affect surface durability. They are

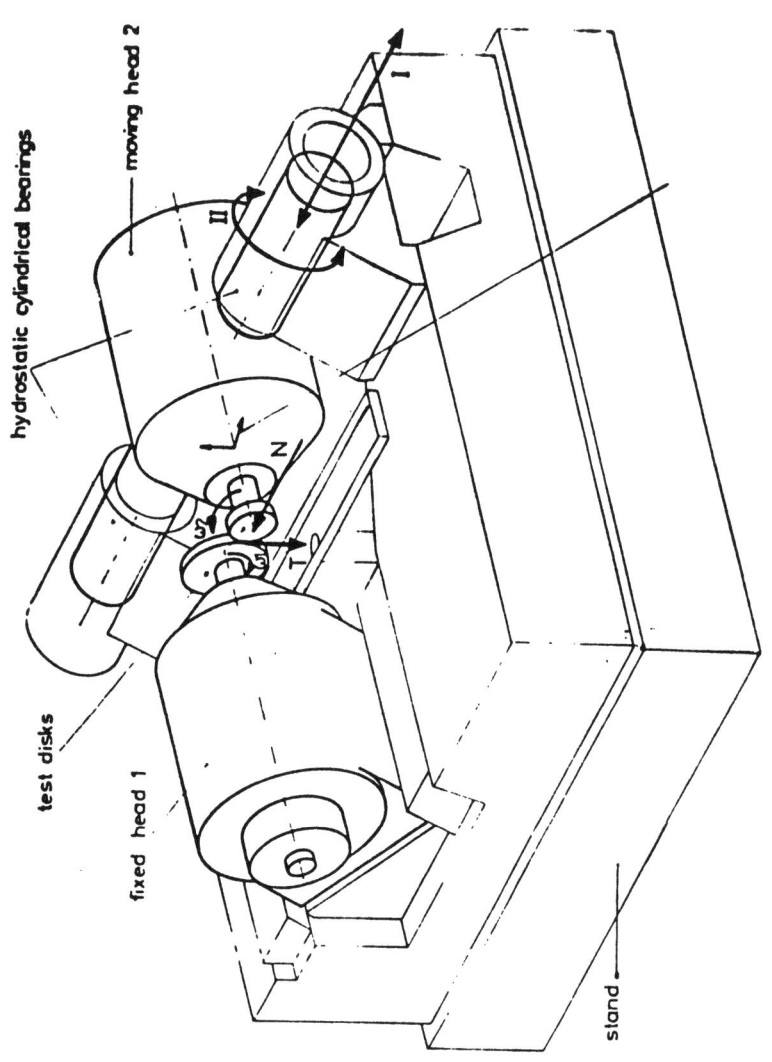

FIGURE 11(A) DISC MACHINE

FIGURE 11(B) GENERAL VIEW OF DISC MACHINE

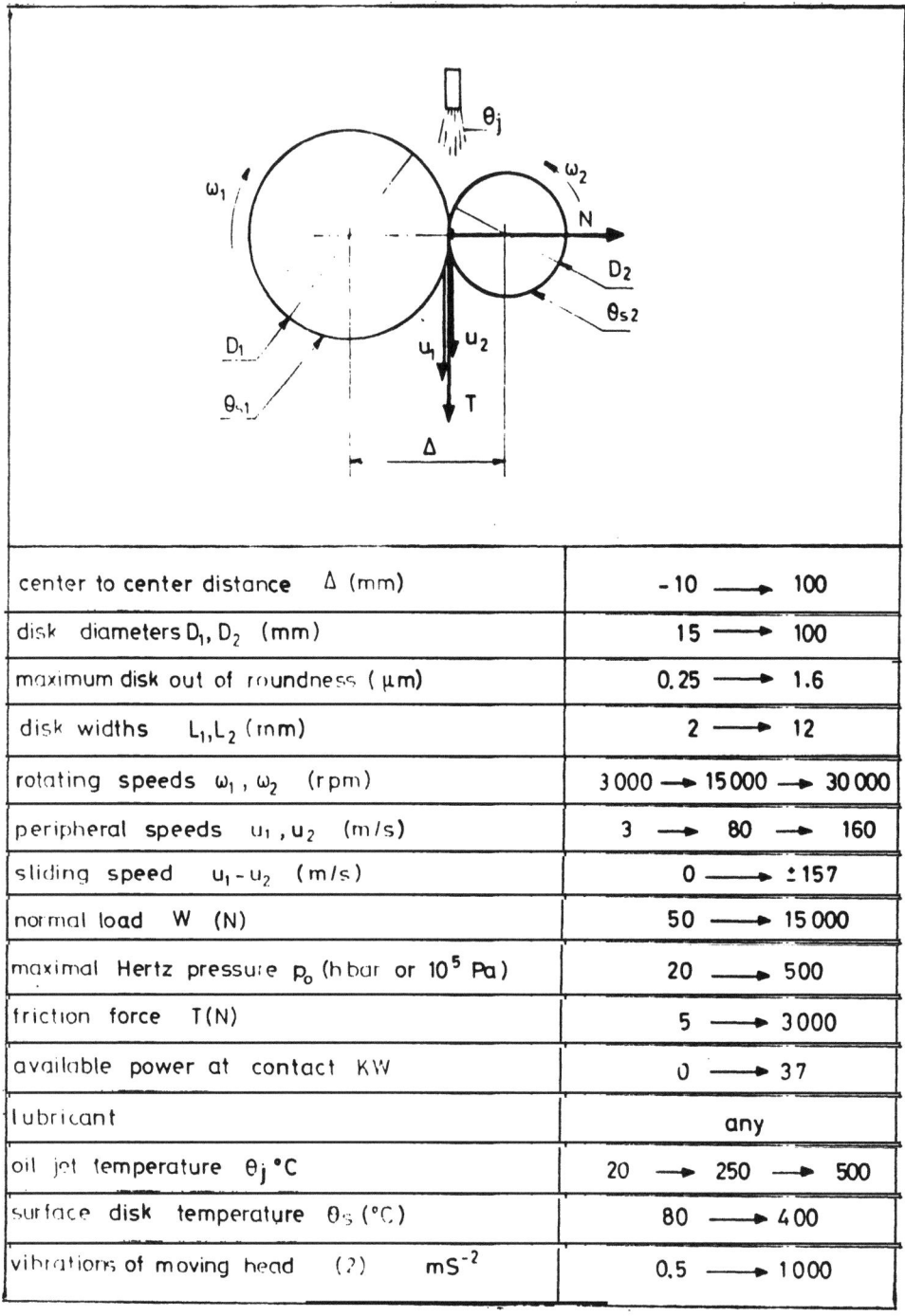

center to center distance Δ (mm)	$-10 \longrightarrow 100$
disk diameters D_1, D_2 (mm)	$15 \longrightarrow 100$
maximum disk out of roundness (μm)	$0.25 \longrightarrow 1.6$
disk widths L_1, L_2 (mm)	$2 \longrightarrow 12$
rotating speeds ω_1, ω_2 (rpm)	$3000 \longrightarrow 15000 \longrightarrow 30000$
peripheral speeds u_1, u_2 (m/s)	$3 \longrightarrow 80 \longrightarrow 160$
sliding speed $u_1 - u_2$ (m/s)	$0 \longrightarrow \pm 157$
normal load W (N)	$50 \longrightarrow 15000$
maximal Hertz pressure p_o (hbar or 10^5 Pa)	$20 \longrightarrow 500$
friction force T (N)	$5 \longrightarrow 3000$
available power at contact KW	$0 \longrightarrow 37$
lubricant	any
oil jet temperature θ_j °C	$20 \longrightarrow 250 \longrightarrow 500$
surface disk temperature θ_s (°C)	$80 \longrightarrow 400$
vibrations of moving head (?) mS^{-2}	$0.5 \longrightarrow 1000$

TABLE 6 DISC MACHINE CHARACTERISTICS

not independent of each other and have to be considered singularly or grouped in the simulation.

In this discussion, simulation is based on the faithful reproduction, in a disc machine, of the Contact Mechanical Definition (CMD) prevalent at a particular point along the gear profile. CMD includes the mechanical parameters which are taken into account in the recent elastohydrodynamic theories of rough surfaces (EHDR) and the material and environmental parameters. The mechanical parameters of CMD are listed in Table 7a. Material and environmental aspects cannot obviously be defined as clearly, and identical conditions must therefore be created. Consequently both discs and gears were taken from the same blank and treated in the same oven. Both types of specimens were ground with approximately the same finish expressed in terms of both CLA and R_t. The lubricants tested are taken from the same batch. Both gears and discs were run in the air.

Simulation of gears by discs performed in this effort is, however, incapable of reproducing faithfully:

- surface roughness direction,

- hydrodynamic transients introduced by the variations in CMD which occur along the gear profile,

- the thermal effects,

- any eventual interaction between two neighboring but different points along the profile.

Surface roughness direction is different in the gears and discs tested. Gear surfaces are generated on a Maag machine which yields a cross-striated pattern, while the discs are finished on a cylindrical grinder which give longitudinal roughness. This difference is known to produce differences in scuffing. The effect on pitting is less clear as the film thickness differences introduced by roughness is small at the low slide/roll ratios at which pitting is produced.

Film thickness is known to be modified by transients, however the change noted under the EHD conditions found in gears is negligible.

Oil viscosity at the contact entry which is known to control film thickness, cannot be measured in gears. Oil entry temperature can only be approximated with trailing thermocouples on the discs. Simulation of thermal effects is therefore very difficult. No significant discrepancies in the simulation results are however, expected from these effects, as all tests

CONTACT MECHANICAL DEFINITION (CMD)	
geometry	$R, \ 1/R = 1/R_1 + 1/R_2$
	L
roughness	$f(\varepsilon_1) \ ; \ f(\varepsilon_2)$
kinematics	$U_1 \ , \ U_2$
normal load	W
material and lubricant characteristics	μ_0
	α
	$E', \ \dfrac{1}{E'} = \dfrac{1}{2}(\dfrac{1-\nu_1^2}{E_1} + \dfrac{1-\nu_2^2}{E_2})$

TABLE 7(A)   MECHANICAL PARAMETERS OF CMD

were run with fairly fluid oils (7.1 10^{-3} Pa.s) at a relatively high temperature (80° C). At this high temperture the viscosity temperture gradient is very low, therefore any variation in film thickness can be expected to be small. Interaction effects on simulation results will be shown to be of importance and will be discussed later.

Note that correct simulation of a given point along the gear profile can only be achieved on a fairly sophisticated machine with a variable center to center distance and independently controlled drives on each disc.

3.2.2 Test Procedure

Gear characteristics and corresponding EHDR parameters at the profile point chosen for simulation are listed in Table 7b. The load program for both machines is given in Figure 12. The test procedures are slightly different as the load is applied at rest in the gear machine and at speed in the disc tester. A temperature controlled oil jet (80°C) lubricates the discs at contact entry while the gears are splash lubricated. The gear rig bath temperature is also controlled at 80° C. Lubricant filtration is not present in either the gear or the disc rig. Track observations are performed during the tests in both machines according to the schedule given in Figure 12. Oil changes are also indicated in this figure. Disc and gear roughness are measured before and after each completed test. Gear profiles are recorded at the end of the test. Friction force is constantly monitored on the disc machine. Vibration level is also monitored in both cases and tests are stopped when the vibration reaches a given level, usually resulting from a classical spall or by any form of surface damage which gives a signal of equal magnitude. Tooth breakage obviously causes the gear test to stop.

3.2.3 Surface Damage

Various forms of surface damage were obtained during these tests as follows:

1) micropits: craters roughly 20 μm deep and equal in diameter. They come either isolated or grouped, Figure 13. According to theory they should occur above a roughness ratio of 1.25.

2) sponges and flats: an area which can extend over several mm^2 and in which a high concentration of micropits is observed. Average depth of sponges can be slightly higher than that of micropits. The sponge areas can exhibit flats, Figure 14.

GENERAL GEAR PARAMETERS				CONTACT MECHANICAL DEFINITION RETAINED		
CHARACTERISTICS	PINION 1	GEAR 2				
external diameter (mm)	88.5	105.5	geometry	R_i		$2R_1 = 31.4$ mm $2R_2 = 38.5$ mm
pitch line diameter (mm)	81	99		L_i		$L_1 = 4$ mm $L_2 = 10$ mm
number of teeth	27	33	roughness	Ra_1 Ra_2		$0.5 \rightarrow 1 \mu m$
modulus (mm)	3			Rt_1 Rt_2		$3.6 \rightarrow 8 \mu m$
operating pressure angle	22°27'		kinematics	U_1		6.627 m/s
addendum modification factor	0.33	0.20		U_2		6.047 m/s
base circle diameter (mm)	70.144	93.2		λ		0.045
speed (r p m)	3666*	3000	linear normal load N/mm	W		$455 \rightarrow 832$
linear normal load (N/mm)	445 $\rightarrow$ 632		material and lubricant characteristics	E		2.3×10^{11} Pa
				μ		synthetic oil 7.5 cSt (80°C)
face width (mm)	15			α		$\simeq 1.3 \times 10^{-8}$ Pa^{-1}

TABLE 7(B) GEAR CHARACTERISTICS & CORRESPONDING EHDR AT THE PROFILE POINT

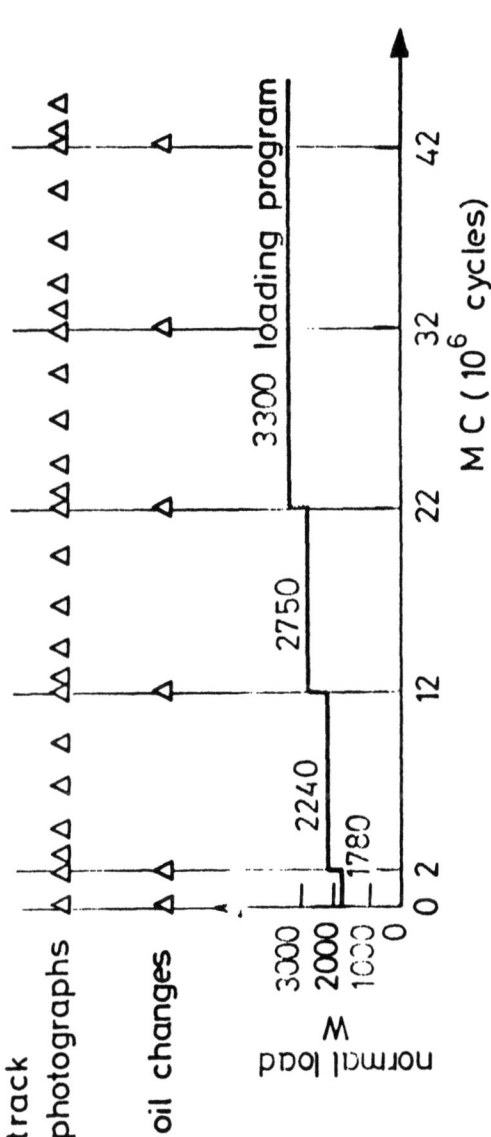

FIGURE 12 TEST PROCEDURE

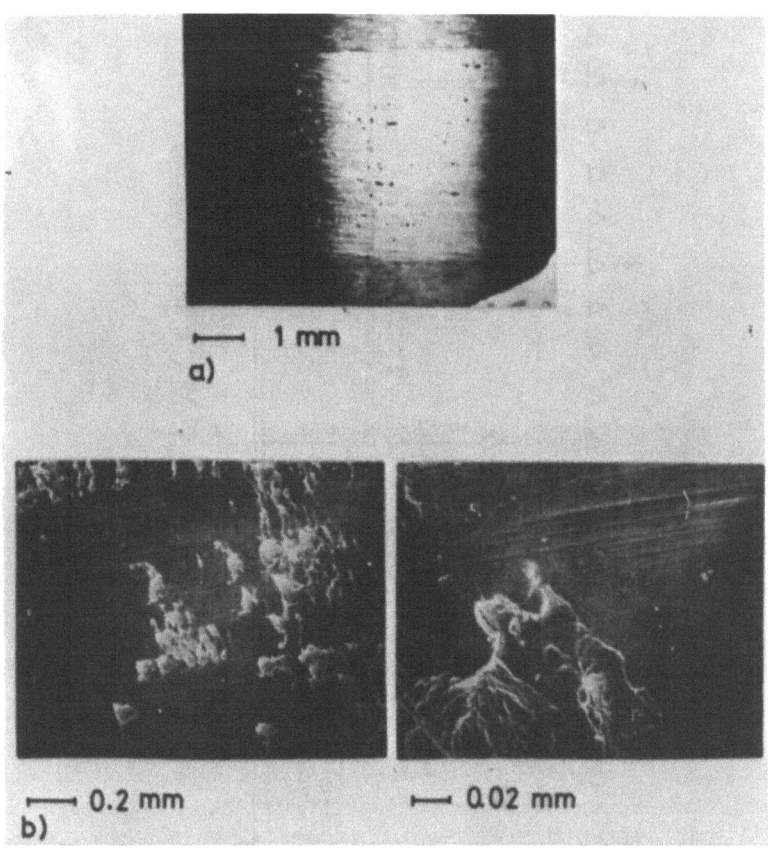

FIGURE 13   MICROPITS (A) OPTICAL, (B) S.E.M.

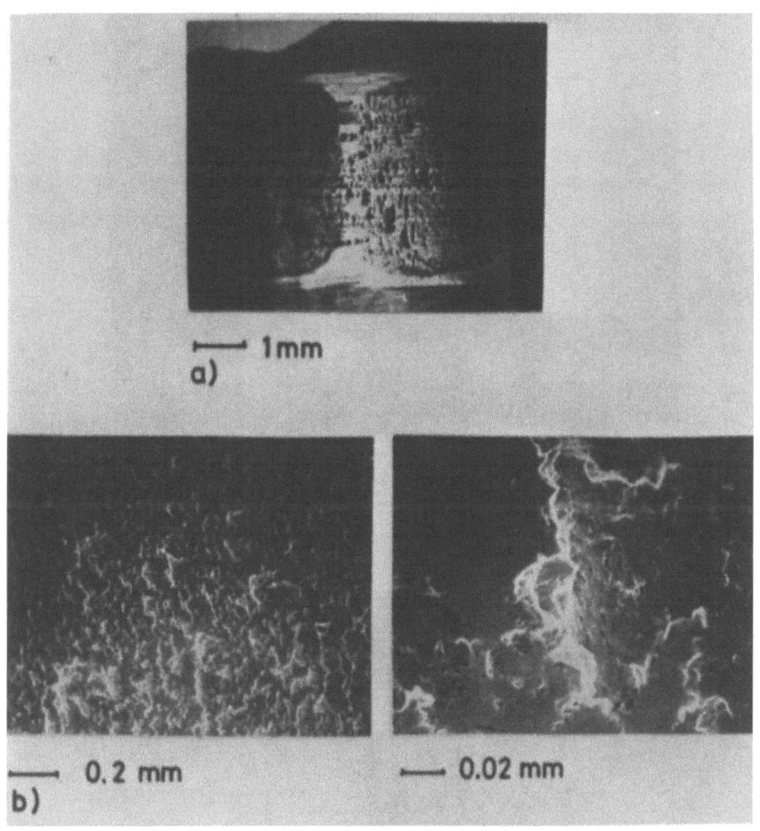

FIGURE 14   SPONGE AND FLATS

3) macropits: a shallow zone in which micropits have joined to form a void of several mm^2 in area. Macropits must not be mistaken for spalls which are 200 µm deep, Figure 15.

4) surface wear: a fairly evenly worn track from which the sponge zones were swept away. This condition is enhanced by sliding, Figure 16.

5) spalls: subsurface initiated fatigue pits abundantly described in the literature, Figure 17.

Clearly, macropits and sponges only characterize different densities of micropits but their individual appearance is sufficiently different to be noted. Running conditions are not affected by micropits and sponges. Flats, however, increase the

⊢——⊣ 1 mm

FIGURE 15 MICROPITS AND MACROPITS

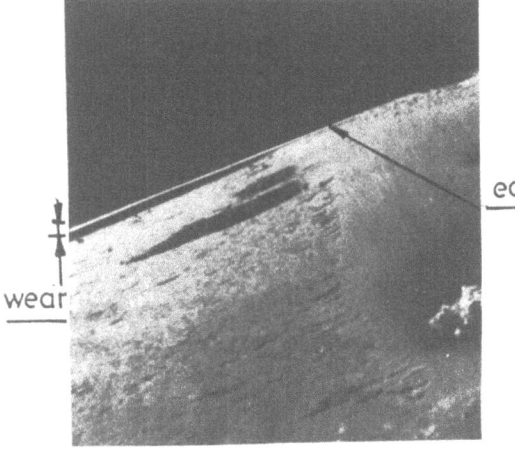

FIGURE 16 SLOW DISC SURFACE WEAR

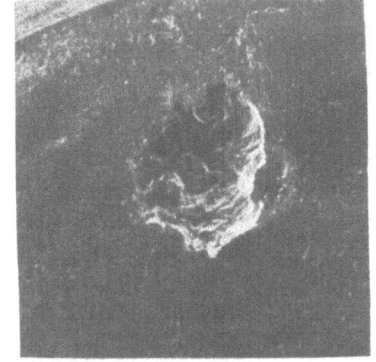

⊢—⊣ 0.2 mm

FIGURE 17 SPALL SURROUNDED BY MICROPITS

vibration level. Macropits grow in time and only become dangerous when they reach spall size which increase the vibration level significantly. In general terms, it appears that if material is removed by one cause or another, over an area equivalent in size to the hertzian band, the test cannot be pursued.

Spalls do not systematically develop in densely pitted areas. One spall however, was found below a macropit. In accordance with results found in the literature, the slow track suffers greater damage than the fast one and unless marked otherwise, results presented below deal with the slow track.

3.2.4 Results

Correlation studies were performed for 3 gear material combinations, Table 8, and one synthetic lubricant of the diester type. A minimum and maximum of respectively 3 to 5 runs were made for each test on both maachines. The final number of runs was a function of data scatter. Life reported is the arithmetic average of values measured in each test. Maximum dispersion with respect to the average value is approximately $\pm 60\%$, which is relatively low for fatigue tests. The number of runs is limited by time as each, including all measurements and controls, takes an average of two weeks. Test results are presented in Table 9.

In Table 9 the columns list the different forms of damage encountered in both mechanisms. The table also shows the roughness ratio, σ/h, at the beginning and at the end of the tests for both gear and disc. Correlation observed between discs and gears will not be judged strictly on life before final damage but on the existence and progression of similar forms of surface distress which might lead to different failures in the two systems.

Table 9 clearly shows that the correlation on micropit formation is very good between discs and gears for all three materials. Micropits appear between one or two megacycles (MC). They grow in number and reach a high density on material combination II. Their number is limited in material combination I and II. No macropits were observed on material I and III for both gears and dics, but extensive formation is observed with II in both mechanisms. Correlation therfore is also satisfactory for macropits.

Spalls were generated early on gears and discs with material combination I. The final damage was thus the same in both cases. Disc life was however, on the average 6 times longer than gear life which is better than the differences noted elsewhere in the literature. Correlation is adequate in this case.

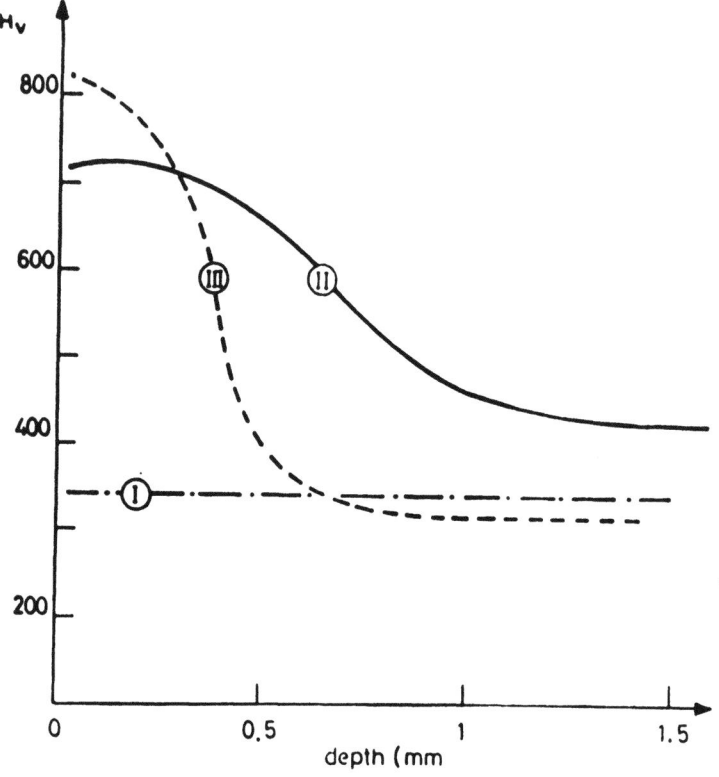

material combination	composition (%)					thermal treatment	grinding after treatment
	C	Ni	Cr	Mo	V		
I	.35	3.80	1.7	.30		through hardened	yes
II	.16	3.25				case hardened	yes
III	.32		3	1	.2	nitrided	no

lubricant (diester)
– viscosity variations : $\mu = \mu_o \exp\left[\alpha p + \beta\left(\frac{1}{T} - \frac{1}{T_o}\right)\right]$
– thermoviscosity coef. : $\beta = 4230\,°K$
– piezoviscosity coef. : $\alpha = 1.32\ 10^{-8}\ Pa^{-1}$
– viscosity at $T_o = 313\,°K$: $2.3\ 10^{-2}\ Pa.s$
– density : $\rho = 0.94$

TABLE 8 MATERIALS AND LUBRICANT

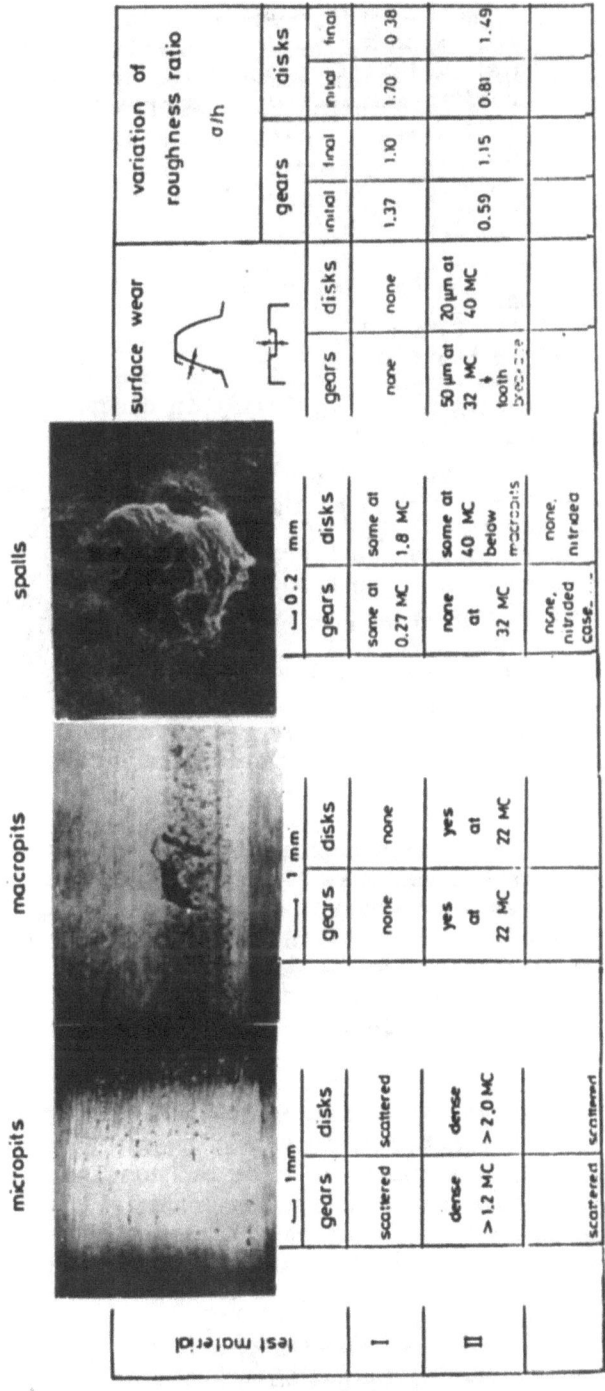

test material	micropits		macropits		spalls		surface wear		variation of roughness ratio σ/h			
									gears		disks	
	gears	disks	gears	disks	gears	disks	gears	disks	initial	final	initial	final
I	scattered	scattered	none	none	some at 0.27 MC	some at 1.8 MC	none	none	1.37	1.10	1.70	0.38
II	dense >1.2 MC	dense >2.0 MC	yes at 22 MC	yes at 22 MC	none at 32 MC	some at 40 MC below macropits	50 μm at 32 MC ↓ tooth breakage	20 μm at 40 MC	0.59	1.15	0.81	1.49
	scattered	scattered			none, nitrided case	none, nitrided						

TABLE 9   RESULTS

The situation found with material combinations II and III is more complex. With material combination II in gears, macropits were noted at 22 MC and accompanied by a relatively high wear rate. This situation led to tooth breakage, occuring at 32 MC. In discs, macropits were observed also at 22 MC with spalls forming below the macropits at 40 MC. Wear was significant but less than in gears. Clearly the micropit density, which increased in discs subjected to low slide/roll ratios, were swept away in gears. This situation is the result of the sliding which occurred in the neighboring zone and which, because of proximity, affected the zone simulated. If true, this hypothesis highlights the possibile interaction noted earlier, which can exist between points along the gear profile that see different CMD. Clearly this interaction does not exist in discs. To substantiate this conclusion, a heavily micropitted disc of material combination II, was tested under a higher slide/roll ratio (0.08). The test disc pits were rapidly worn away. However, final disc surface finish was not as good as that found in gears which is not surprising if one considers the initial roughness in that particular test. Two hypotheses can be advanced to explain the absence of spalls in the gears of material combination II. The first is that high wear retards spall formation. The second is that tooth rupture may have occurred before spall intitiation. The correlation which is clearly unsatisfactory if final damage is the only criteria considered, becomes more acceptable if the causes which lead to the two different forms of final damage are compared.

With materials combination III, case crushing was noted in both discs and gears at 20 and 60 MC respectively. Early disc failure was attributed to edge effects. Disc tests were stopped because of high vibration, while gear tests were terminated by tooth breakage. The final damage appeared to be different in both mechanisms but the cause was the same. Here again correlation is acceptable if, as above, the causes which lead to these two different forms of final damage are compared.

Finally, the variation in surface roughness during the tests follow the same trend in both disc and gear for a given material. Running-in was noted for material combination I and III. Surface deterioration was measured for material combination II. This leads to an increase of roughness ratio for material II and to a decrease for materials I and III for both disc and gear testing.

3.2.5 Conclusion

Simulation of gears with discs of different material combinations was performed on a high performance disc machine. The point simulated was chosen from previous experiments on

gears where damage was observed to the the most important.
Results show the following:

- The first forms of damage, i.e. micropits observed during the first few megacycles, are identical for the three materials in both gears and discs.

- The progression from micropits to macropits also followed the same trends in both mechanisms.

- Final damage in the two mechanisms however, is not always the same. Indeed if spalls are noted in both gears and discs for material I, the situation is different for material II and III where gear tooth rupture was noted. However, gear tooth rupture observed with materials III followed case-crushing which was also noted on discs and the correlation can therefore be considered to be acceptable. The tooth rupture observed with material II was due to high wear, which was not noted on the corresponding discs although they were severly pitted.

All results therefore clearly show that as long as gear geometry at the point simulated is not affected by wear, good agreement is found between discs and gears. Leaving roughness variation effects aside, the tests show that if the gear profiles are modified by wear, the gear CMD is altered while if disc wear occurs, the disc CMD is unchanged. This results from the fact that the variation in radii of curvature due to wear is negligible. Hence as long as the CMD of both mechanisms is identical, surface damage is the same. As soon as the CMD of gear differs from that of discs, mechanism damage is necessarily different and the correlation fails.

Concerning interaction, sliding causes densely pitted surfaces to wear uniformly. Interaction is imposed by the necessity of maintaining profile continuity during the wearing process. A given surface element cannot wear independently of its neighbor and a given point on the gear surface does not necessarily show the damage which corresponds intrinsically to that of the CMD calculated at that point. It can show however, a form of damage which is modified due to the interaction of neighboring points having different CMD and thus different forms of damage.

This study, which was conduced from 1973 to 1975, has been followed by quite considerable work on different materials and for different running conditions. The new work confirms what was observed originally and allows the following points to be highlighted:

a) Micropit density is very sensitive to material heat treatment. Materials with identical surface hardness but which have been produced in two separate batches can yield different micropit densities. These densities in turn, due to durface/volume interaction, can significantly modify life.

b) Slight changes in disc ratio do not appear to significantly alter surface damage when:

   1) the slide/roll ratio is low
   2) the hertz pressure is simulated independently of load.

c) Attention must be paid to synchronization effects. In the type of disc machine used here speed is controlled very accurately. However, a point of one of the discs does not come into contact repeatedly with a corresponding point of the other disc. With gears, by definition, synchronization is guaranteed. Synchronization effects have been noted at various times by different authors.

3.3 Simulation of Helicopter Oscillating Dry Bearings

Design of dry bearings is known to present problems in particular applications as guide lines are often missing in practice. Dry bearings are commonly used in aircraft, in transport, and in general also in industrial machinery. Classical criteria of the PV type approach (Pressure x Velocity = Constant) have been proven to be unsatisfactory. Significant improvements have been brought following the introduction of dimensional considerations. However, these improvements are unable to optimize high performance bearings. This section will deal briefly with these design problems. Further information can be found in References 58 and 59.

3.3.1 Governing Parameters in Dry Bearing Simulation

It is well extablished that in all contact conditions which perform satisfactorily, a thin film or third body separates both machine elements. The third-body composition varies from one type of contact to another and is rarely homogenous. In contacts formed by a moving steel ring, Figure 18, or shaft and a fixed bearing made out of friction material such as a plastic liner, a thin transfer film or third-body one (TB_1) covers the steel ring (first-body on FB_1) and under some conditions a layer of amorphous material known as third-body two (TB_2) is packed on the surface of the liner (first-body two, FB_2). As third-body formation, elimination and endurance govern both friction and

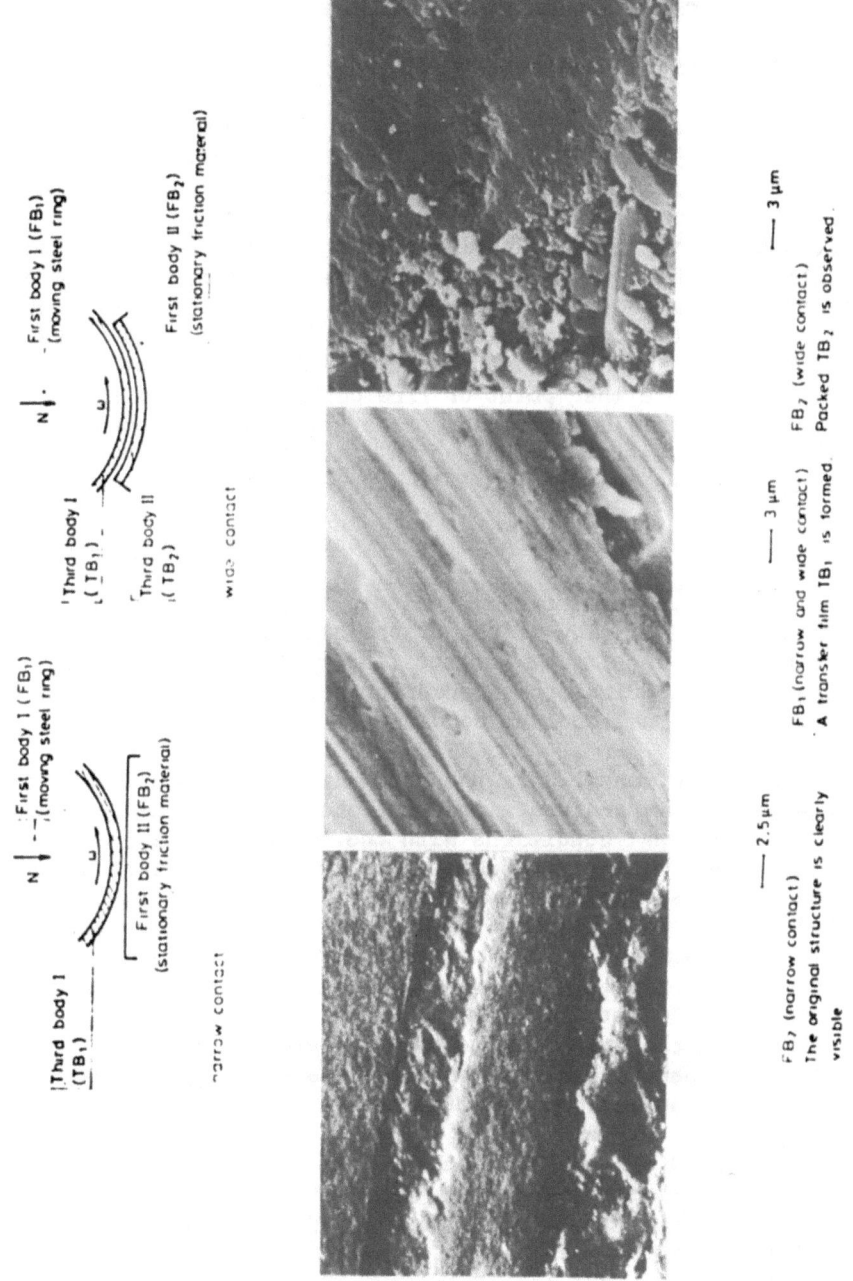

FIGURE 18 THIRD BODIES OBSERVED IN NARROW & WIDE CONTACTS

wear, the parameters which control these phenomena must be identified before a simulation can be attempted.

3.3.1.1 Contact Geometry

The effect of contact geometry on TB formation is significant. As an illustration, Figure 18 shows SEM pictures of both TB_1 and TB_2 formed in narrow (or Hertzian) and in wide (or distributed) contacts when a steel ring rubs against a carbon-reinforced resin under identical average pressures and linear velocities. TB_1 is formed in both cases, although TB_2 was only observed in the wide contact. Clearly, narrow and wide contacts are not satisfactory definitions but the observations noted are sufficient to indicate that bearing simulation cannot be performed with either point or line contact machines and that distributed loads have to be considered. Further, contact microgeometry cannot be ignored in simulation as wear or TB elimination has been shown to depend on surface roughness.

3.3.1.2 Contact Temperatures -- Effects of Boundary Conditions

Contact temperatures are believed to govern friction and wear in plastics. The effect of the heat path or the actual design of the machine on contact temperatures for a given energy input is first considered for a given energy input and for given external cooling conditions.

The experimental data presented in Figure 19(a) shows the variation of temperature with frictional energy measured under the same external conditions with the same test specimens but for two different heat paths. The thermocouple used is situated close to the surface of the stationary rubbing specimen. Curves I and II are characteristic of heat paths I and II respectively, which were imposed by two different cooling conditions. Further evidence is given in Figure 19(b) in which both the variation of friction and temperature with time are recorded. Conditions are identical in all four tests with only the heat paths varied as indicated. Heat path I led to liner destruction although satisfactory operation was obtained in the other three cases.

Figure 19 clearly shows that relatively minor differences in heat paths, determine the performance of a given frictional material for the same P.V. value. The differences achieved artificially in Figure 19, exist naturally between two industrial designs which use the same friction material. They must therefore be taken into account in any simulation attempt. Quite obviously, at the design stage, it is impossible to determine experimentally the heat path of a new mechanism. Therefore, this determination has to be done theoretically.

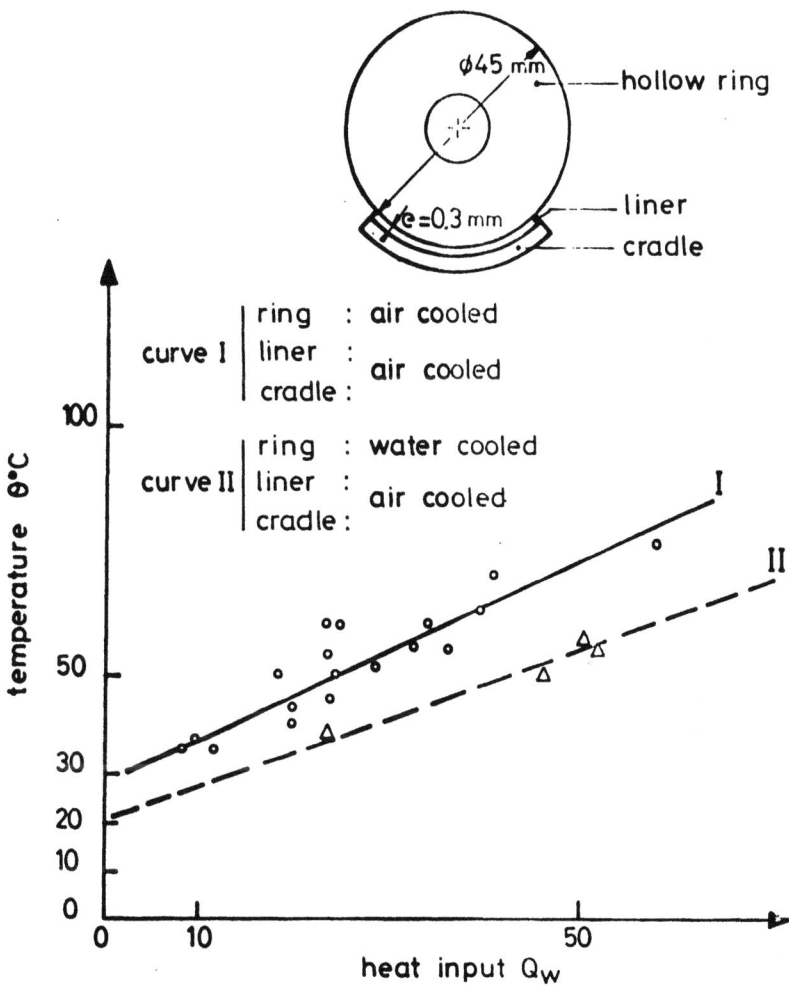

FIGURE 19(A)   TWO CHARACTERISTIC HEAT PATHS OF THE SAME MACHINE

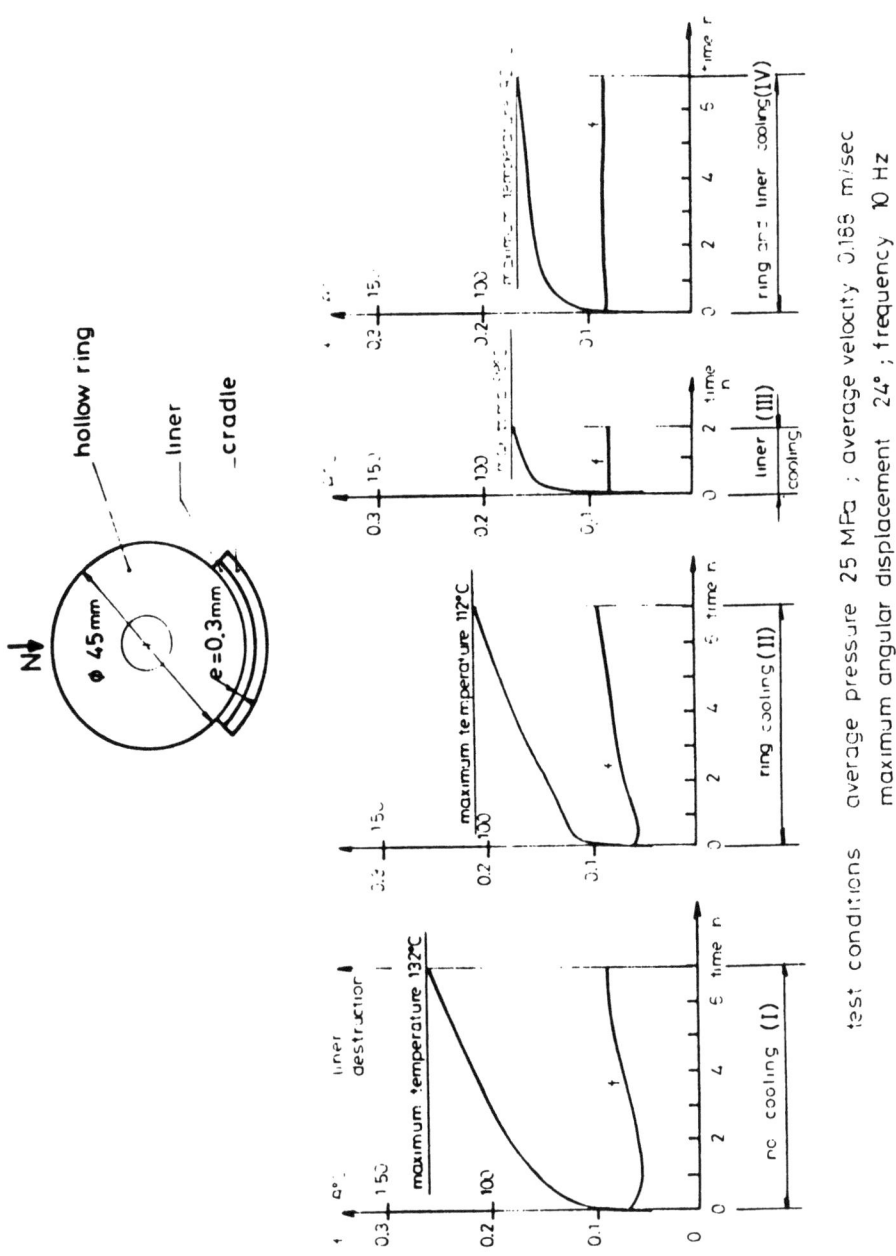

FIGURE 19(B) EFFECT OF FOUR DIFFERENT COOLING CONDITIONS ON TEMPERATURE AND FRICTION OF A STEEL PLASTIC LINER COMBINATION OPERATING UNDER OSCILLATORY MOTION

3.3.1.3 Contact Temperatures - Effect of Kinematics

Figure 20 shows the effect of the type of motion on contact temperature. The analytical techniques used to obtain these results are presented elsewhere and will not be discussed here. It is sufficient for our purpose to note that, for the same energy input, the temperature calculated for a steel ring rubbing against a plastic sector to be 126° C if the ring motion is limited to small angular displacements and only 80° C if this ring rotates continuously at 25 rev min^{-1}. Thus, the heat path of the machine clearly depends on the nature of the movement of its parts which therefore has to be taken into account in the simulation.

3.3.1.4 Minimum Sliding Distances

It was stated earlier that third bodies governed friction and wear. It was also shown that the nature of the motion determines the contact temperatures which inturn influence TB formation. However, independently of contact temperature, there is no evidence concerning the effect of the nature of the motion on TB formation. Third bodies are created during sliding but the minimum sliding distance necessary to generate them for a given material combination is not known.

3.3.1.5 Load, Pressures and Velocities

All test machines have fixed specimen geometries and information concerning the variation of both friction and wear with load is available. However, no data on the effect of pressure or pressure distribution for a given load are given, as this requires changes in specimen geometries. Thus, in simulation attempts, assumptions have to be made to relate the load and pressure conditions which exist in the real system to those created in the simulator.

Instantaneous contact velocities can be calculated accurately in all applications. In reciprocating systems the velocity variation can be complex and is therefore practically impossible to reproduce exactly on a simulator. Hence, equivalent velocity conditions must be defined for the simulator.

3.3.1.6 Surface Treatment and Environment

Figure 21 shows the difference observed in both friction, temperature, and wear when a plain and a phosphated steel ring rub against a carbonfiber reinforced resin. Clearly, third-body stability is enhanced by surface treatment and wear is considerably decreased. Environmental effects on the friction and wear of plastics are also well known and it is therfore

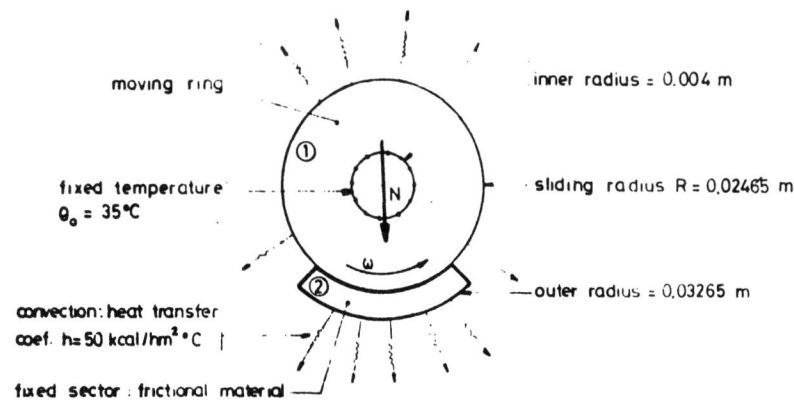
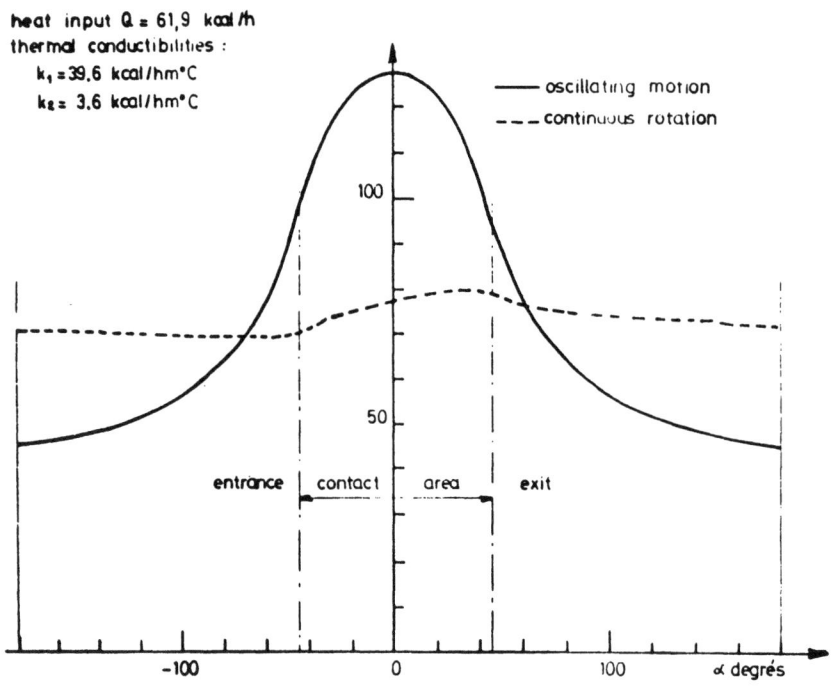

FIGURE 20 VARIATION OF CONTACT TEMPERATURE WITH POSITION FOR OSCILLATORY MOTION AND CONTINUOUS ROTATION

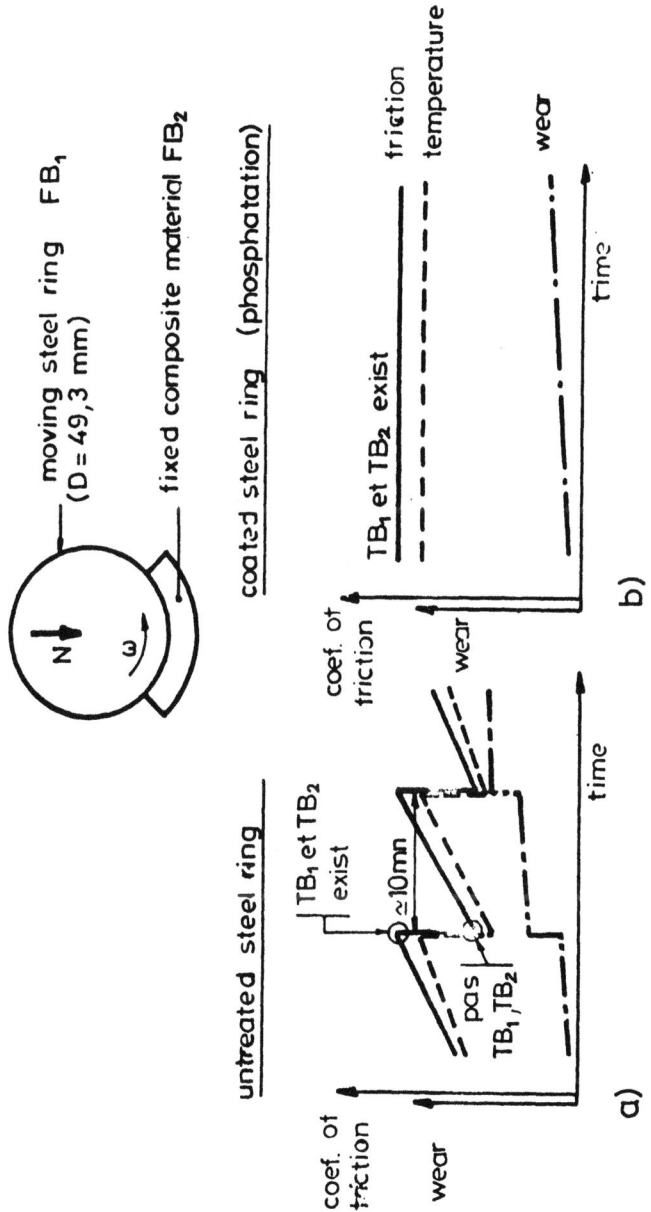

FIGURE 21 EFFECT OF SURFACE TREATMENT ON THIRD-BODY STABILITY AND THUS OF FRICTION TEMPERATURE AND WEAR

necessary for simulation to take into account both surface treatment and environment.

3.3.1.7 Wear Stimulation

Wear controls bearing life. However, the problem of wear simulation is very difficult and extrapolation of data from simulator results to the real application is still uncertain. This uncertainty arises from the fact that the basic phenomena are unquantified. A first step, however, is to consider that, for a given material combination, wear rates differ significantly with third-body stability, Figure 21. If the parameters which control TB stability are identical in both the simulator and the real bearing, TB behavior will be identical in both cases. Thus, a low wear condition on the simulator will lead to a low wear situation in the bearing even if the scale factors cannot be defined. The limiting values for the low wear rates imposed for the simulator studies were based on earlier experience obtained on materials tested both on the simulator and on the full-scale rig during a short preliminary investigation. A wear rate of 0.8×10^{-3} μm m^{-1}, determined on the simulator, was chosen.

3.3.2 Simulation Parameters

It is clear that contact geometry, contact temperature, contact kinematics, pressures, load, and velocities must be considered individually if adequate simulation is to be achieved. Environmental and surface treatment requirements must also be met. The full-scale simulation of the dry bearing described in Table 10 leads to the construction of a test machine that is well beyond the scope of this effort and defeats the very purpose of simulation which, as much as possible, must be sophisticated but which must be simple and economic to run.

It was therefore decided to build a simulator capable of reproducing:

a) the contact microgeometry and macrogeometry;

b) the maximum and average contact pressures and their variation in time due to cyclic loading;

c) the type of motion and the average velocity;

d) the contact temperature (this must be met independently of a, b, and c);

e) the environment.

Bearing dimensions and operating conditions	
Dimensions :	
bearing shaft diameter	0.125 m
bearing width	0.019 m
Load and pressures :	
cyclic loading	$N = N_o (1 + \sin \omega t)$
frequency $f = \omega/2\pi$	4.4 Hz
maximum amplitude 2No	$0 < 2N_o < 22\,000$ Newtons
maximum pressure p_{max}	$0 < p_{max} < 12$ MPa
average pressure	$0 < p_a < 9.3$ MPa
Displacement :	
cyclic motion	$\alpha = \alpha_1 \sin \omega t$
frequency $f = \omega/2\pi$	4.4 Hz
maximum angular displacement	$0 < 2\alpha_4 < 20°$
Material and environment :	
shaft material	440 C steel (untreated)
surface roughness	0.4 µm CLA
friction material	to be defined
atmosphere	air
Limits :	
maximum coefficient of friction for the highest load 0.15 maximum wear rate 0.8×10^{-3} µm/m	

TABLE 10 BEARING DIMENSIONS AND OPERATING CONDITIONS

3.3.2.1 The Simulator, Figure 22

The contact used for the simulator was that of a ring rubbing against a 90^0 sector. Motion can be continuous or oscillating and load is either constant or cyclic. The heat path can be varied by different cooling techniques. The simulator characteristics are given in Table 11 along with the contact parameters found in dry bearings. The conditions that can be simulated are cross-hatched.

The test conditions for the bearing described in Table 10 and established from simulation principles are given in Table 12. Four different woven friction liner materials which were retained after an exhaustive preliminary study formed the list of test materials. The simulator tests were then performed. As an example, the results obtained with the optimal test liner are presented and the performed calculations are described below.

Table 13 gives the results obtained on the simulator with a Dacron weave liner filled with Teflon. Five cyclic loads and four angular displacements were tested with the load and displacement frequency being 10 Hz.

3.3.2.2 Experimental Procedure

The test specimen ring is cleaned ultrasonically in a dimethylethyl ketone bath. The sector is wiped with a clean cloth and the load applied after the ring is set in motion. Each test lasts 30 hours. Friction and temperature are recorded continuously and wear is measured at predetermined times. The wear rate is calculated after running-in and is expressed in terms of wear thickness in micrometers per length of travel in meters.

3.3.3 Simulation

The different steps in the simulation procedures required in the design of high performance bearings are outlined in Table 14.

(1) The bearing running conditions together with the limiting requirements of the friction force, wear rate and maximum contact temperatures are clearly defined. A thermal model for the bearing and its housing is developed.
(2) The load and speed to be applied to the simulation are determined from simulation principle.
(3) A short list of promising liner materials is established after either a literature survey or a preliminary gross ranking campaign.

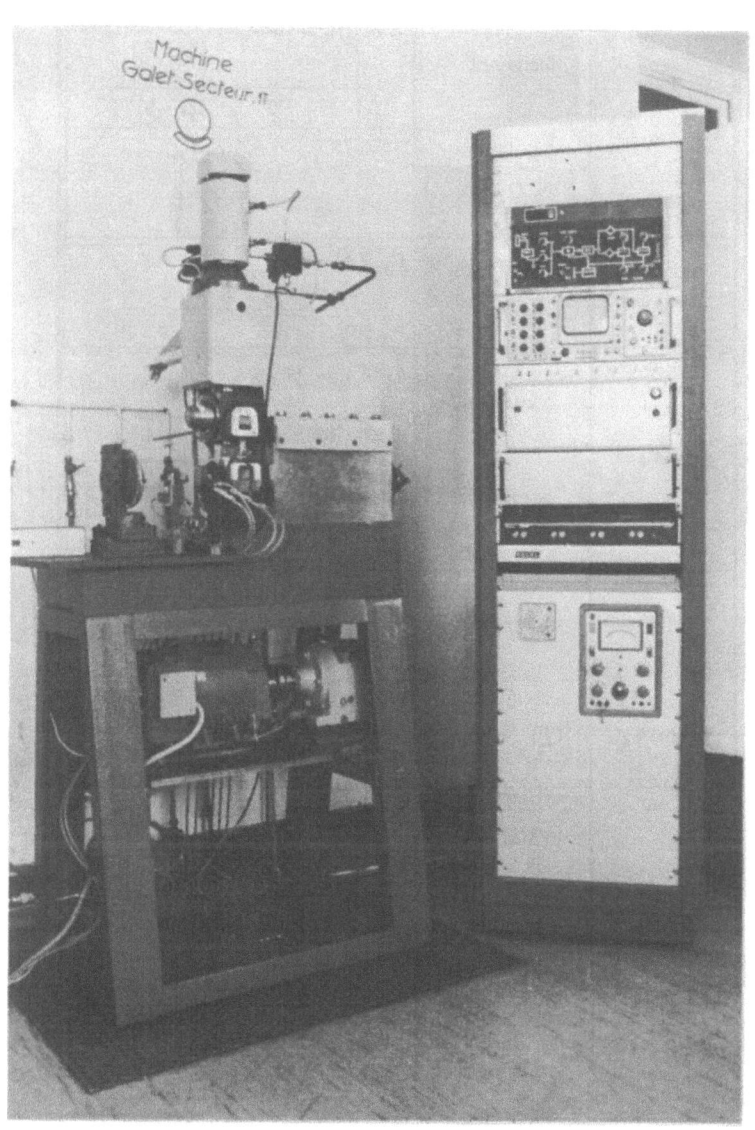

FIGURE 22   GENERAL VIEW OF THE APPARATUS

PARAMETERS		RANGE OF VARIATIONS
load	constant	N or $2N_0$ in Newtons; 0.1, 1, 10, 10^2, 10^3, 10^4
	pulsed $N=N_0(1+\sin\omega t)$	ω rd/s; 0.1, 1, 10
contact pressure	constant / pulsed	$P_{av} \times 10^7$ Pa; 0.01, 0.1, 1, 5, 10
movement	continuous	V m/s; 0.001, 0.01, 0.1, 1, 10
	oscillation $\alpha = \alpha_1 \sin(\omega t + \varphi)$	$\alpha°$: 0, 90, 180, 360; $\varphi°$: -180, 0, 180
geometry	diameter D	D: m; 10^{-3}, 10^{-2}, 10^{-1}, 1
	L/D ratio	L/D; 0.2, 0.6, 1
	sector arc	$\zeta°$; 45, 180, 360
contact temperature	constant / cyclic	T°C; -200, 0, 200
environment	air or various gases	oxydant, air, reductor; RH%: 0, 50, 100

TABLE 11   CONTACT PARAMETERS IN PLASTIC BEARINGS

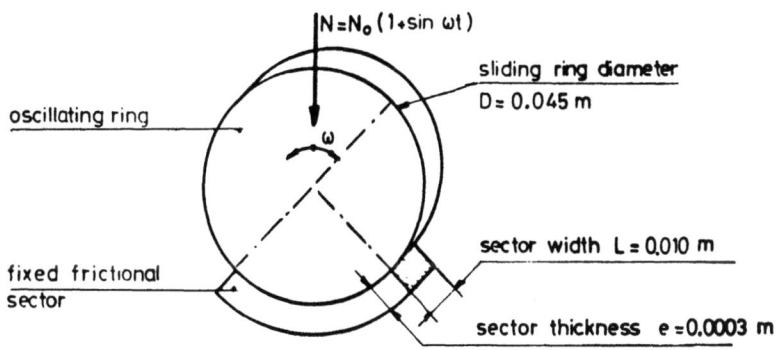

TEST CONDITIONS						
cyclic loading at 10 Hz	N_o (N)	660	1000	1330	1760	1930
	p_{moy} for $2N_o \times 10^7$ Pa	0.41	0.62	0.82	1.09	1.2
maximum angular displacement α at 10 Hz	α (°)	6	12		18	24
	V_{moy} (m/s)	0.050	0.090		0.140	0.190
material for ring	440 C steel ± 60 HRc, 0.4 µm CLA					
thermal conditions	air cooled shaft air cooled support					

TABLE 12   SIMULATION TEST CONDITIONS

			maximum angular displacement at 10 Hz				
			6°	12°	18°	24°	
			$V_{moy}=0.05$ m/s	$V_{moy}=0.09$ m/s	$V_{moy}=0.14$ m/s	$V_{moy}=0.19$ m/s	
cyclic load $N = N_0(1+\sin\omega t)$ at 10 Hz	660 (1+sin ωt)	f	0.28	0.32	0.25	0.19	0.33
		W*	8.7	19.9	23.3	23.5	40.9
		θ °C	35	43	60	54	70
		u̇**	1.18	0.74	0.41	0.665	0.79
	1000 (1+sinωt)	f	0.21	0.25	0.21	0.18	
		W*	9.9	23.5	29.6	33.8	
		θ °C	37	45	60	60	
		u̇**	1.9	0.74	0.4	0.6	
	1330 (1+sinωt)	f	0.185	0.16	0.13	0.13	
		W*	11.6	20	24.4	32.51	
		θ °C	35	40	50	56	
		u̇**	3.55	1.6	0.4	0.4	
	1750 (1+sinωt)	f		0.11	0.115		
		W*		18.1	28.4		
		θ °C		50	52		
		u̇**		0.6	0.59		
	1930 (1+sinωt)	f				0.1	0.15
		W*				36.3	58.05
		θ °C				55	76
		u̇**				0.82	0.76

material : teflon filled dacron weave cloth
*heat input (watt) - **rate of wear (x 10³ µm/m)

TABLE 13 EXPERIMENTAL RESULTS

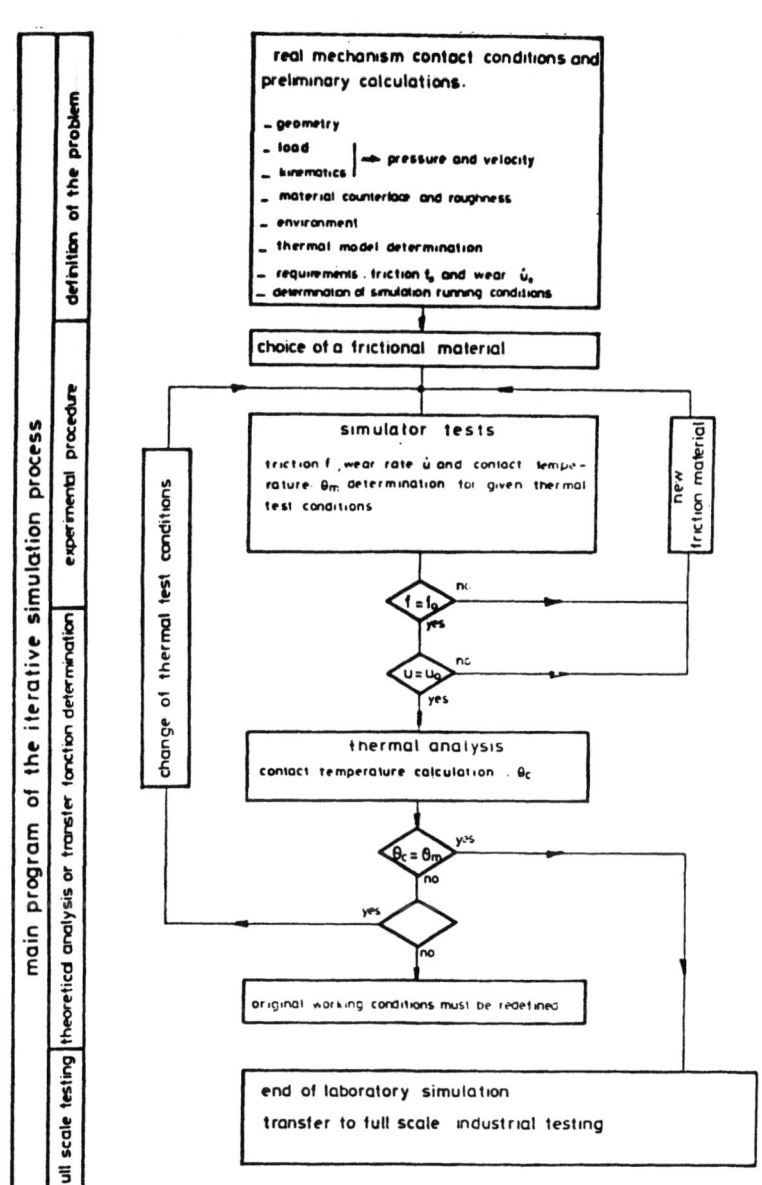

TABLE 14 SIMULATION FLOW CHART

(4) Simulator tests are run and the coefficient of friction, f, the wear rate, u and the contact temperature, θ_m, are recorded or estimated. The performances noted are compared with the requirements listed in step 1. If the requirements are not satisfied, a new material is chosen and step 4 is started again.

(5) If the requirements are met, the contact temperature, θ_c, in the real application is calculated for a representative value of the coefficient of friction measured in step 4. The values of θ_c and θ_m are then compared.

(6) If the calculated and the measured temperatures θ_c and θ_m differ significantly, the thermal condition in the simulator are changed and step 4 is started again with the new conditions. This process is repeated until both values agree. Experience shows that industrial applications have larger thermal masses than laboratory benches and as a result θ_m is usually larger than θ_c. Convergence however, can be obtained by cooling the test bench.

(7) If θ_m and θ_c cannot be made equal, the original working conditions must be redefined. If θ_m equals θ_c, the laboratory simulation is terminated and full scale industrial testing can begin.

3.3.4 Thermal Analysis of the Real Mechanism

Step 5 of the simulation flow chart, Table 14, requires that contact temperatures be calculated in the actual mechanism. The two-dimensional thermal model studies by finite elements, Figure 23, includes three elements: the inner shaft, the liner, and the outer bearing. In most applications, the liner is glued to the outer bearing and this disposition is introduced in the model for the calculation performed for simulation purposes. As such, the maximum temperature of 82° was calculated.

Returning to the simulation analysis, Table 8 shows that a maximum temperature of 76° was measured one millimeter below the contact during the tests. The value calculated in the real application at that distance is 75°. As the agreement between the theoretical and experimental values outlined in the simulator flow chart Table 14 is satisfactory, the laboratory simulation is therefore ended.

3.3.5 Full Scale Tests

Full scale tests were conducted under this regime. As expected, no detailed observations concerning friction and wear rate values were available. However, good agreement between

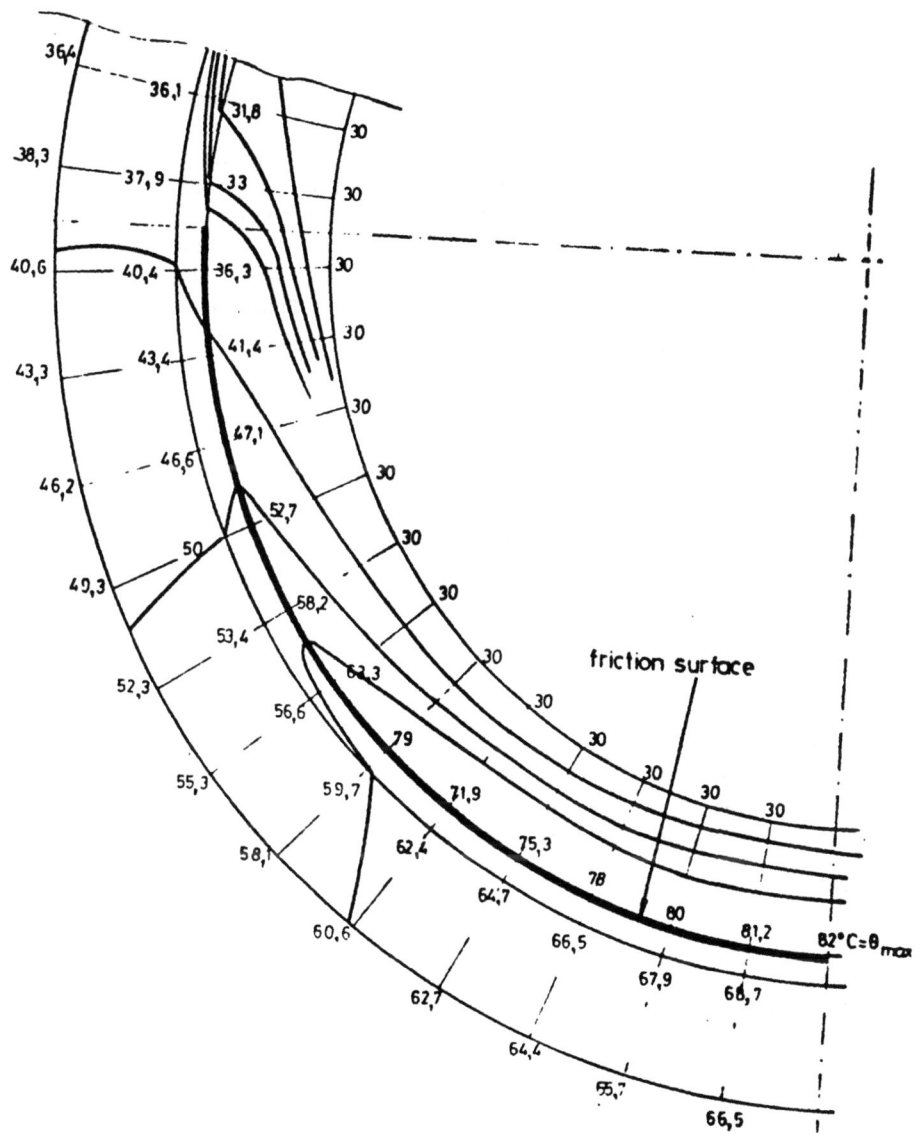

FIGURE 23 TEMPERATURE DISTRIBUTION WITH LINER GLUED ON OUTER BEARING (CONFIGURATION I)

measured and calculated temperatures was observed and the performance qualitatively confirms simulation test results.

3.3.6 Conclusions

Simulation based on the reproduction of contact pressures and temperatures, average velocity, type of motion, and contact geometry appears to give valuable results. As such, it constitutes a first step in the actual design of high performance bearings whose life performance could not be predicted earlier with any measure of certainty. The method suggested here is still cumbersome and can be improved. It is hoped that, as general calculation programs aand specific test data are accumulated, many steps in the iterative procedure outlined can be eliminated for each application. Further, computer aided design techniques are being tried in order to enhance the design process.

3.4 General Conclusion

Simulative testing has been shown to be capable of producing useful answers to industrial tribological questions when intrinsic coefficients, boundary conditions, and theories are not available. As such, it represents a very important link between fundamentals and application. The examples considered here have been chosen among the more complicated and sophisticated simulation attempts. This was the case because the information sought tends towards absolute performance prediction rather than comparative behavior. Clearly accurate absolute prediction can only be expected when Level I requirements are met. However, the case studies presented here can give an indication as to how well the governing parameters have been identified over a given range of mechanical conditions and thus serve to help isolate the factors which must be introduced eventually in a more general theory. To be efficient, from an industrial point of view, simulative testing must follow closely the advances in basic sciences in order to capitalize on their contribution. As such, tribotesting is a dynamic subject when progresses along with basic understanding.

4.0 Bibliography

1. Bowden, F.P., and Tabor, D., "Friction and Lubrication of Solids," Vol. I and II. Clarendon Press 1950 and 1964.

2. Buckley, D., "Surface Effects in Adhesion, Friction, Wear and Lubrication," Elsevier 1981.

3. Mugis, D., and Barquins, M., "Fracture Mechanics and the Adherence of Viscoelastic Bodies," J. Applied Physics, Vol. II (1978) 1989 - 2024.

4. Dowson, D., Godet, M., and Taylor, C.M., "The Wear of Nonmetallic Materials," MEP London 1976.

5. Godet, M., and Play, D., "Mechanical Aspects of Dry Friction and Wear Testing," Ref. 4, p. 77 - 86.

6. Barwell, F.T., "Bearing Systems," Oxford University Press 1979 (Chap. XII).

7. Benzing, R., Goldblatt, I, Hopkins, V., Jamison, W., Mecklenburg, D., and Peterson, M., "Friction and Wear Devices," 2nd Ed. ASLE 1976.

8. Pieuchot, A., Blouet, J., Gras, R., Alfred, R., and Courtel, R., "Methodolgie et calssement des essais de frottement et de leurs resultats," Materiaux Electricite (GAMI) Oct. 1969.

9. Paul, G., "Time Dependent Viscosity Following a Pressure Rise Measured on an Impact Viscometer," ASLE Trans. Vol. 19, no. 1, pp. 17 - 22, (1975).

10. Johnson, K.L., and Robert, A.D., "Observation of Viscoelastic Behavior of an EHD Lubricant Film," Proc. Roy. Soc. London A.337, pp. 217 - 242 (1974).

11. Alsaad, M., Bair, S., Sanborn, P.M., and Winer, O., "Glass Transitions in Lubricants, Its Relation to EHD Lubrication," ASME Trans. Vol. 100, pp. 404 - 417 (July 1978).

12. Bair, S., and Winer, W.O., "A Rheological Model for EHD Contacts Based on Primary Laboratory Data," ASME-JOLT, Vol. 101, July 79, pp. 258.

13. Tevaarwerk, J., and Johnson, K.L., "A Simple Constitutive Equation for EHD Oil films," Wear, Vol. 35, pp. 315 - 316 (1975).

14. Gupta, P.K., Flamand, L., Berthe, D., and Godet, M., "On the Traction Behavior of Several Lubricants," ASME-JOLT, Jan. 1981, pp. 55 - 64.

15. Dowson, D., Taylor, C.M., Godet, M., and Berthe, D., "EHD and Related Subjects," MEP London 1979 (7 papers from page 144 to 214).

16. Houpert, L., Flamand, L., and Berthe, D., "Rheological and Thermal Effects in Lubricated EHD Contacts," To be published in AMSE-JOLT 1981.

17. Godet, M., Play, D., and Berthe, D., "An Attempt to Provide a Unified Treatment of Tribology Load-Carrying Capacity, Transport and Continuum Mechanics," ASME-JOLT April 1980, Vol. 102, pp. 153 - 163.

18. Godet, M., Play, D., Floquet, A., Flamand, L., Berthe, D., and Dalmaz, G., "Simulation in Tribology," to be published.

19. Czichos, H., Tribology," Elsevier 1978 7-3, p. 264.

20. Play, D., "Simulating Contact Conditions in Dry Bearings," Tribology International Oct. 1978, pp. 295 - 301.

21. Play, D., "Testing Alloys for Use in Heat Treatment Furnaces," Tribology International June 1978, pp. 193 - 196.

22. Flamand, L., Berthe, D., and Godet, M., "Simulation of Hertzian Contacts Found in Spur Gears with a High Performance Disc Machine," ASME-JMD, Vol. 103, Jan. 1981.

23. Dalmaz, G., Teissier, J.F., and Dudragne, G., "Friction Improvement in Cycloidal Motion Contacts: Rib Roller and Contact in Tapered Roller Bearings," (Ref. 47 to be published).

24. Connell, R.A., Summers, G.G., Shephard, J.P., and Shone, E.B., "The Use of Filled PTFE as a Ball Valve Seat Material," Ref. 4, p. 235 - 238.

25. Ref. 1 (Vol. I, Chap. VII). Ref. 2 (Chap. IX).

26. Godet, M., and Play, D., "Introduction to Tribology," Colloques internationaux du CNRS (France) no. 233 "Polymeres et Lubrification." p. 361 - 376.

27. Play, D., "Portance et transport des troisiemes corps en frottement sec," These Dr. es-Sciences, INSA-UCB (Lyon-France) Oct. 1979.

28. Dalmaz, G., "L'hydrodynamique du contact spher-plan," These UCB Lyon, Mai 1971.

29. Slinely, H.E., "Dynamics of Solid Lubrication as Observed by Optical Microscopy," ASLE Trans. Vol. 21, no. 2, 1977, pp. 109 - 117.

30. Aharoni, S.M., "Wear of Polymers by Roll Formation," Wear 25 (1973) p. 309 - 327.

31. Play, D., and Godet, M., "Self Protection of High Wear Materials," ASLE Trans. Vol. 22. no. 1, pp. 56 - 64 (1979).

32. Play, D., and Godet, M., "Visualization of Chalk Wear," Ref. 4, p. 221 - 229.

33. Georges, J.M., and Mathia, T., "Considerations sur les mecanismes de la lubrification limite," Journal de Mechanique Appliquee, Vol. 2, no. 2, (1978) p. 231 - 266.

34. El Sanabary, Ahmed Fouad, "Effect des proprietes rheologiques de polymer sur l'usure," These Dr. Ing. INSA-UCB Lyon, Dec. 1980.

35. El Sanabary, A.F., Play, D., and Godet, M., "Volume Effects in Polymer Transfer," Submitted for publication ASME-JOLT.

36. Berthe, D., Flamand, L., Foucher, D., and Godet, M., "Micropitting in Hertzian Contacts," ASEM-JOLT Oct. 1980, Vol. 102, p. 478 - 4890.

37. Lancaster, J.K., Play, D., Godet, M., Verall, A.P., and Waghorne, R., "Third Body Formation and the Wear of PTFE Fiber Based Dry Bearings," ASME-JOLT, Vol. 102, no. 2, April 1980, pp. 236 - 246.

38. Dowson, D.D., Taylor, C.M., Godet, M., and Berthe, D., "Thermal Effects in Tribology," MEP London 1980.

39. Rozeanu, L., and Snarsky, L., "Second Order Thermal Effects in Lubrication," Ref. 38 p. 95 - 100.

40. Sieberg, A., "Fluid Mechanical Effects of Polymeric Surfaces Phases," Colloques internationaux du CNRS no. 233 Polymères et Lubrification, p. 82 - 86.

41. Dowson, D., Godet, M., and Taylor, C.M., "Cavitation and Related Phenomena in Lubrication," MEP London 1975.

42. Elrod, H.G., and Adams, M.L., "A Computer Program for Cavitation and Starvation Problems," Ref. 41, p. 37 - 43.

43. Floquet, A., Play, D., and Godet, M., "Contribution a j'etude thermique du frottement sec dans les paliers," Journal de Mechanique Appliquee, Vol. 2, no. 4, 1978, p. 499 - 539.

44. Floquet, A., Play, D., and Godet, M., "Surface Temperatures in Distributed Contacts. Application to Bearing Design," ASME-JOLT, Vol. 99, no. 2, p. 277 - 283, 1977.

45. Pinkus, O., and Wilcox, D.F., "Thermal Effects in Fluid Film Bearings," Ref. 38, p. 3 - 24.

46. Boncampain, R., and Frene, J., "Thermohydrodynamic Analysis of a Finite Journal Bearing; Static and Dynamic Characteristics," Ref. 38 p. 33 - 42.

47. Dowson, D., Taylor, C.M., Godet, M., and Berthe, D., "Friction and Traction," Westbury House, IPC Science and Technology 1981 (to be published 1981).

48. Brendle, M., and Colin, G., "The Frictional and Transfer Behavior of Compacted Solid Lubricants on Smooth Metallic Surfaces," Ref. 47 (to be published).

49. Barwell, F.T., and Tingard, S., "The Thermal Equilibrium of Plain Journal Bearings," Ref. 38 p. 24 - 32.

50. Chiu, Y.P., "An Analysis and Prediction of Lubricant Film Starvation in Rolling Contact Systems," ASLE Trans., Vol. 17, p. 22 - 35, 1974.

51. Haardt, R., "Flow Considerations Around the Cavitation Area in Radical Face Seals," Ref. 41, p. 221 - 227.

52. Play, D., and Godet, M., "Relation Between CR-NI Steels and Debris Transport at High Temperatures (950°C)," ASME-JOLT, Vol. 102, no. 2, p. 247 - 253, April 1980.

53. Czichos, H., "Tribology," Elsevier Scientific Publishing Co. 1978.

54. Dalmaz, G., Tessier, J.F., and Dudragne, G., "Friction Improvement in Cycloidal Motion Contacts: Rib-Roller End Contact in Tapered Roller Bearings," I FRICTION and TRACTION by D.D. Dowson et al., Westbury House, IPC Science and Technology Press 1981.

55. Gadallah, N., "Effects de la geometrie et de la Cinematique l'epaisseur de film et la force de frottement dans un contact poncutel lubrifie", These de Docteur-Ingenieur LYON 1981.

56. Flamand, L., Berthe, D., and Godet, M., "Simulation of the Contacts Found in Spur Gear with a High Performance Disc Machine," ASME-JMD. Vol. 103 Jan. 1981, p. 204 - 209.

57. Flamand, L., Berthe, D., "A Brief Discussion of Different Forms of Wear Observed in Hertzian Contacts at Low Slide/Roll Ratios," in "Effects of Surface Roughness in Lubrication," by Dowson et al MET Ltd., London 1978.

58. Play, D., and Godet, M., "Design of High Performance Dry Bearings," Wear 41 (1977) 25 - 44.

59. Play, D., "Simulating Contact Conditions in Dry Bearings," Tribology Int. Oct. 1978 p. 295 - 301.

5. ACKNOWLEDGEMENTS

Permission granted: ASME, Flamand, L., Berthe, D., and Godet, M., "Simulation of the Contacts Found in Spur Gear with a High Performance Disc Machine,", ASME-JMD, Vol. 103, January, 1981, p.204-209.

MONITORING

D. Scott
Teeside Polytechnic

1. INTRODUCTION

Many advances in engineering have been made by turning failure into success. Thus, the history of engineering in the past century and a half is in part a story of failure in service followed by improved materials and design incorporating greater reliability and less maintenance as a result of successful failure investigation[1]. However, in the present and foreseeable future world economic situation, prevention of failure in service is more beneficial than any lessons which can be learned from the failure investigation.

Most modern machines are complex and expensive with their vital parts totally enclosed. The practice of withdrawing such machines from service at periodic intervals for examination and maintenance to avoid failure, involves expensive dismantling. Means have thus been developed to assess machinery condition while in operation and allow predictive maintenance to be carried out when most convenient before the breakdown point is reached[2].

Condition monitoring is thus concerned with extracting information from machines to indicate their condition[3, 4]. This can be done by various techniques and the resulting information used for planning machine operation and maintenance in order to improve reliability, safety and economy of operation.

This discussion outlines the philosophy of condition monitoring, describes the various techniques available, reviews their application, compares their effectiveness and discusses economical aspects of their use.

2. APPROACHES

Machines can be run until they fail and then repaired. This is feasible if failure does not involve personnel risk and production losses and if spare machines are readily available and repair is relatively inexpensive. However, in many instances breakdown maintenance can be expensive in terms of lost output and machine destruction. In some situations it may be dangerous.

A better method is to stop machines at regular intervals based on past experience, for planned preventative maintenance in order to reduce the chance of unplanned stoppages through breakdown. A compromise however, is required between too frequent maintenance which reduces productivity and increases expense and too long an interval which involves the risk of some unacceptable failures in service.

Economically, a more acceptable method is to carry out preventative or on-condition maintenance at irregular intervals determined by the actual condition of the machine. Thus the main function of condition monitoring is to provide knowledge of machine condition and its rate of change which is essential for the successful operation of on-condition maintenance. The knowledge may be obtained by selecting a suitable parameter for measuring deterioration and recording its value continuously or at suitable intervals.

Assessing the trend of this measurement can provide a useful lead time in warning of incipient machine failure. The overall level of the condition monitoring activity appropriate to a particular plant or industry is usually decided by the potential economic savings.

Monitoring a complete plant necessitates a large number of measurements of numerous parameters and as such the cost of such monitoring systems is high. Complete condition monitoring is only applicable to very few major plants of strategic national importance. Only key machines in a plant may be monitored but this may also involve fairly expensive equipment and facilities. This method is generally applicable to most larger industrial establishments. Generally, monitoring a few critical components of key machines based on past failure experience may be adequate to avoid expensive unplanned maintenance and loss of production. This method is widely applicable in industry and can be readily used on existing equipment to show economic advantages.

3. TECHNIQUES

Although there are numerous techniques and a large amount of instrumentation available there are only four basic methods of machinery condition monitoring. These are visual, performance, vibration, and wear debris monitoring. There are two levels of assessment. The first determines that some wear or deterioration is occurring in a system or machine and the second determines the specific component of the system or machine which is wearing or deteriorating.

In visual monitoring, machine components are inspected for the direct observation of wear and change using microscopes, stroboscopes, thermography, dye penetrants, and X-rays. In performance monitoring, changes in machine performance such as output and operating parameters are measured to establish how well the machines are performing their intended duties. Parameters such as operating temperature and power losses are important. Vibration, noise, and shock measuring techniques serve to indicate wear, misalignment, and changes in component tolerances. A considerable selection of transducers, recorders, and signal analysis procedures are available. Wear debris monitoring involves examination of machine lubricants for the presence of entrained wear debris, contaminants, and lubricant degradation products. Wear debris can be detected directly, assessed by collection methods, or assessed by lubricant analysis using spectroscopy or Ferrography.

Monitoring systems may be classified into three categories; manual, semiautomatic, and fully automated. In a manual system, the data are collected manually, processed manually, and displayed manually by reports or graphs. Manual systems are simple to implement and are relatively inexpensive to design and operate. For a small operation seeking limited performance monitoring objectives, a manual system can be a cost effective solution. One objection to a manual system is the added work load it imposes on the observer who normally has a primary responsibility other than monitoring which can result in missing or inaccurate readings. Some degradation in the quality of the data can occur due to inaccuarcies inherent in the processing method. Tedious smoothing or statistical analysis techniques which can improve data analysis, are usually avoided. Manual data systems must deal with essentially steady state phenomena since a high probability exists that transient or rapidly varying data will not be observed or will be missed due to human limitations or to the simple instrumentation which may not be capable or accurately following such data.

Semiautomated systems combine manual and automated capabilities. Usually the manual portion is data collection,

transcription and conversion to a form compatible with the
automated portion of the system which handles data storage,
processing, and output or display of the processed information.
There is usually a time-lag inherent in semiautomated systems
between data collection and data output by a central location.
Monitoring is therefore aimed at trending performance degradation
and/or identifying anomalous behavior. A semiautomated system is
easy to design and operate but exhibits a problem, in common with
the manual system, of the inaccuracy inherent in manual reading
and recording of instruments. A benefit of computer processing
in a semiautomated system is the ease with which the data can be
organized for comparative purposes.

A full automated system from data acquisition through data
display is capable of gathering and processing high amounts of
data, handling signals of all types, maintaining a high level of
accuracy, and of processing and displaying data in real time.
The computer can be used to perform selective recording using
software programs to accomplish data compression. Automated
monitoring represent the ultimate in capability but the cost of
such systems is high and the analysis software requires
considerable development.

4. HARDWARE

Machinery condition monitoring requires hardware that falls
into four categories; transducers, electronic equipment, display
or output devices, and interface hardware[5].

Transducers perform the task of measuring machine parameters
and converting the information to an electrical signal
proportional to the value measured. They are thus a basic
building block in any machinery condition monitoring system.
Electronic hardware conditions transducer output signals to a
common format, processes the signals according to a predetermined
decision sequence, and selects information for recording or
output. Display/output devices provide information in the form
of flags, lights, instruments, recordings, or print outs for
machine safety, diagnostics, or performance trending. Interface
hardware consists of equipment for troubleshooting or for remote
data processing on a more extensive basis for trending or
storage.

Transducers used for monitoring, convert measured parameters
such a speed, pressure, and temperature into proportional
electrical signals. Selection of the parameters necessary to
accomplish the established objectives of the condition monitoring
system determines the basic transducer types required.
Transducer selection is based on several variables. Transducer
accuracy is critical to the accuracy of the condition monitoring

system, as system accuracy will always be less than transducer accuracy. Thus, system requirements determine the transducer accuracy required. Systems reliability is only as reliable as transducer reliability, so that selection of transducers of the required reliability is of prime importance. The environment in which the transducer is required to operate also determines the required design specification. Transducer cost is a direct function of the required environmental, accuracy, and reliability specifications.

4.1 Temperature Measurement

The operating temperature of a machine is a useful indication of the effectiveness of its operation. For example, the exhaust gas temperature (EGT) of a gas turbine engine is an important control parameter.

There are several methods generally used in measuring temperatures in a condition monitoring system. These include thermocouples, thermal resistance temperature sensors, bi-metallic switches, and radiation pyrometers. The most commonly used sensor is the thermocouple which operates on the basis of the electro motive force (EMF) developed in a circuit of two dissimilar metals. The EMF generated is proportional to the temperature measured. Thermocouples are standard equipment, are reasonably linear within a given range, are stable, and are capable of measuring wide temperature ranges. Thermocouples may be connected in parallel so that if one fails, the average temperature can still be measured with little loss of accuracy. The disadvantages of thermocouples are a small signal level in the order of millivolts and the necessity for cold junction compensation. Thermocouples are usually supplied in configurations with distinct temperature ranges for each configuration, Table 1.

Table 1

Temperature Ranges for Some Typical Thermocouples

Thermocouple Material	Temperature Range
Copper – Constantan	$-300°$ to $700°F$
Iron – Constantan	$0°$ to $1400°F$
Chromel – Alumel	$0°$ to $2300°F$
Platinum – Rhodium	$0°$ to $2700°F$

In the thermal resistance sensor, the resistance element produces a change in electrical resistance with respect to temperature which is measured with some form of Wheatstone bridge. Resistance temperature sensors may be either metallic or semiconductors. Some typical temperature ranges of thermal resistance sensors are given in Table 2.

Table 2

Temperature Ranges for Some Typical Thermal Resistance Sensors

Resistor Material	Temperature Range
Platinum	$-450°$ to $1200°F$
Copper	$-325°$ to $300°F$
Nickel	$-100°$ to $300°F$
Thermistors	$-100°$ to $500°F$

Metallic resistance sensors are linear, have a high signal level, and are very stable. However, maximum temperature range is the limiting factor. A semiconductior resistance sensor or thermistor has advantages of a high signal level and low cost. The disadvantages are a narrow measurement range, nonlinear output, and lower accuracy than a metallic resistance sensor. Thermistors also suffer some of the inherent problems of semiconductors such as difficulty in control of operating characteristics in batch manufacture. They are not used where a high degree of accuracy is essential.

Bi-metallic switches may be used where a discrete signal is required. They have the advantage of accurately switching within a very limited temperature range, require no signal conditioning and are usually inexpensive. They are used mainly for specific applications.

Radiation pyrometers utilize the intensity of emitted radiation as a measure of temperature. They consist of an optical focusing sensor head for collecting the radiant from a selected area typically .1 inch diameter, steel sheathed flexible fiber optical cables, a silicon detector and necessary signal conditioning. The advantages of radiation pyrometers include high accuracy of temperature measurement, rapid response to temperature changes, and a high temperature range. The disadvantages are susceptibility to lens coking in certain environments and complicated signal conditioning requirements.

They are potentially attractive for gas turbine blade temperature measurement[6,7].

Other methods of temperature indication range from simple, inexpensive temperature indicating labels and paints, low melting point metals, hardness changes in selected alloys to thermography, or thermal imaging[8,9,10]. The latter technique converts the infrared radiation picked up by a scan of the surface with an infrared camera into a proportional electrical signal.

The signal is amplified and displayed on a cathode ray tube or recorded on photographic film.

4.2 Pressure Measurement

Performance monitoring usually requires accurate pressure measurement so that pressure transducers are common to many condition monitoring systems. Pressure transducers may be passive or active. A passive pressure transducer, such as a metal foil or semiconductor strain gauge requires excitation for output whereas an active pressure sensor generates an output voltage without excitation.

In a pressure transducer, the strain gauge is connected into a Wheatstone bridge circuit and the strain on the strain gauge produces a resistance proportional to the pressure displacing it. The most commonly used pressure transducer, the metal foil strain gauge, provides very accurate measurement and has low sensitivity to thermal effects, shock, and vibration. It is capable of either static or dynamic measurement with A.C. or D.C. excitation and has continuous resolution. The only disadvantage is a relatively high cost when designed to meet aircraft specifications. A compromise is often required between high resolution, durability, and capital cost. Semiconductor strain gauge pressure transducers possess high sensitivity and are smaller and cheaper than metal foil strain gauges but their use is limited owing to their poor reproducibility, reliability, and accuracy due to temperature effects.

The only active pressure transducer used, piezoelectric, is based on the principle that asymmetrical crystalline materials produce an electrical potential on the application of strain or stress. The most widely used crystals are quartz, tourmaline, rochelle salt, and barium titanate. Piezoelectric crystals are most widely used in accelerometers. When used in pressure transducers, they can only be used for dynamic measurement as current is generated only under dynamic loading of the crystal. They are mainly used where accuracy in dynamic response is only required. They are usable to high temperatures and are very rugged. Their static response is poor and requires a change amplifier to render the signal usable.

4.3 Vibration Measurement

For vibration measurement it is important to consider not only the vibration pick ups but the monitoring system as a whole. The overall vibration system objectives must be finalized prior to transducer selection. Software capabilities should indicate the final system requirements. Capability may range up to a full diagnostic program able to fault isolate to component or plant level. This capability dictates the complexity of the hardware required to monitor vibration.

In any system, other than a gross overall displacement check, some frequency filtering of transducer outputs is required in the form of band pass filters in the signal conditioning or software capability to provide such filtering. Overall objectives dictate filtering requirements. It may be desirable to concentrate on one or two octave bands in the frequency spectrum associated with machine rotational speeds or to filter many bands to diagnose the complete operational spectrum of machine vibrations. Filtering also eliminates bands not associated with the machine frequency spectrum. The transducers used in vibration monitoring are velocity pick ups and accelerometers, depending upon system requirements.

Although the degree of accuracy available is relatively high the inherent design of velocity pick ups presents disadvantages. Vibration is measured by the relative motion of a coil with respect to a magnetic field and as it is a mechanical device it is subject to wear and reliability problems. Velocity pick ups are unsuitable for the higher frequency ranges (4000 + Hz) due to a relatively low natural frequency. It is usually necessary to integrate the output to reduce the effects of signal noise and provide the displacement. This limits the use of velocity pick ups to an overall gross system check rather than a more sophisticated analysis.

Accelerometers are constructed of a stack of piezoelectric crystals in compression with a mass mounted on the stack. An output is generated when the accelerometer is subjected to vibration, but the low level output requires substantial amplification to obtain a usable signal. Advantages of the accelerometer include no moving parts, high reliability, and a high natural frequency. Integration of the accelerometer output provides a velocity measurement for more sophisticated diagnostic work. A disadvantage is the low level output which necessitates shielded cabling to reduce interference from extraneous noise on the vibration signal. Higher output accelerometers are being developed to increase signal to noise ratio and to reduce the effects of noise. Some very sophisticated condition monitoring systems have been developed based on gas turbine engine parameter interrelationships[11].

4.4 Position Transducers

A potentiometer is a variable resistance transducer used to measure displacement. In aircraft it may be used to measure fuel control settings, engine and airframe displacements. Potentiometric transducers which have a large electrical output and are available for any desired displacement range, are inexpensive and require simple signal conditioning. A major disadvantage is that mechanical contact renders the potentiometer subject to wear and vibratory stresses which curtail the useful life.

A synchro is basically a variable transformer which provides an electrical output proportional to the angular displacement of its shaft. An alternating current excitation is provided to the synchro and the phase relation changes as the shaft is rotated. An advantage of a synchro over a potentiometer is that brushless construction eliminates mechanical contact; however, signal conditioning requirements are more complicated than for a potentiometer.

A linear variable differential transformer (LVDT) used to measure linear displacements is an inductance transducer which produces an electrical output proportional to the displacement of a moveable magnetic core. A primary coil and two secondary coils surround the core. These secondary coils symmetrically spaced on either side of the primary coil are connected in a series-opposing circuit and motion of the magnetic core varies the mutual conductance of each secondary coil to the primary coil which determines the voltage induced in the secondary coils. If the core is central, the voltage induced is identical and $180°$ out of phase so there is no net output. If the core moves off center, one secondary coil has a greater voltage induced. The output of an LVDT is linear within its range of displacement and there is no physical contact between the coil and core so that there is no deterioration by wear. An LVDT is not affected by mechanical overload and has excellent dynamic response due to a low core mass and the absence of friction but it requires complex signal conditioning.

4.5 Lubricant Monitoring

Lubricant monitoring provides a means of detecting abnormal conditions of oil wetted components. Magnetic chip detectors, screens and oil filters with pop out buttons to indicate an excessive oil pressure drop are inexpensive methods. Chip detectors and screens are checked periodically to determine particulate concentration. Some magnetic chip detectors contain an electrical circuit which is completed when a metal particle makes contact with the magnet to provide a signal that there is an abnormal lubricant condition. The disadvantage of these methods is that they do not provide continuous monitoring. They must be properly located in the lubrication system and may not provide a warning in sufficient time to allow remedial action to prevent secondary damage.

An on-line lubricant system monitoring method has been developed based on the principle of light scattering for particulate debris detection and light attenuation for chemical or thermal lubricant degradation. As the lubricant passes through the transducer it causes a rotor to turn. The rotor contains fluid passages and optical references which are alternatively inserted in a light measuring system as the rotor revolves. The optical paths utilize sealed fiber optics to conduct light into and out of the fluid and to change its direction. One photo sensor is mounted radially to view the light beam at $90°$ to provide the scattering output. The attenuation sensor views the axial component of transmitted light. The output of each sensor is a series of pulses alternating between reference and signal. The output requires extensive signal conditioning which is expensive.

5. WEAR DEBRIS ANALYSIS

As the history of a wear process is recorded in the debris produced, an attractive method of monitoring the condition of machinery is by the careful analysis of wear debris and contaminants in the lubricant used[12]. By measuring the quantity and observing the nature of the wear debris, it is possible to obtain an indication of the condition of the various machine components which are lubricant washed. As the failure of load-carrying lubricated surfaces is usually a slow progressive process, wear debris analysis allows advanced warning of surface deterioration towards failure.

Wear debris monitoring may be carried out by direct detection using inductive or capacitance detectors and filters which detect the presence of conducting debris. Monitoring may also be carried out by debris collection using removable magnetic plugs, filters, and centrifuges. However, wear debris analysis is usually applied to lubricant samples. Elemental analysis of lubricant samples may be carried out using spectrographic oil analysis procedures (SOAP). Wear particle analysis is carried out by particle counters and Ferrography. Great care is necessary with oil sampling. A sample of lubricant should be taken, if possible, from the machine or system while it is in operation or in less than two minutes after stopping in order to ensure that most particles are still in suspension.

5.1 Spectrographic Oil Analysis Procedure

SOAP may be carried out using either atomic absorption or atomic emission spectroscopy[13,14]. The atomic absorption spectrophotometer which costs less than other spectrometric methods operates on the principle that atoms absorb only light of their own specific wavelength. The lubricant sample suitably diluted is vaporized in a flame and each element required, determined separately using a source lamp which emits light of wavelength characteristic of the element. Light is absorbed on passage through the flame and comparison with a reference beam allows a direct measurement of element concentration. Standards of known element concentration are used for instrument calibration.

The atomic emission spectrometer measures the characteristic wavelength of light emitted when elements are excited by an electrical discharge. Usually a small quantity of lubricant picked up on the periphery of a rotating graphite disc is vaporized in an electric discharge to promote emission of spectra by metallic elements present. In a direct reading instrument, photomultipliers aligned to spectral lines of interest are used to measure element concentrations simultaneously. A print out of element concentration in parts per million by weight reduces test time. Complex calculations are necessary and spectral interference may require compensation. However, the method is very useful for large numbers of specimens handled by a central control unit. A big disadvantage of SOAP is that it is blind to large particles which can be most dangerous in precision machinery[18].

X-ray fluoresence by exposure of a medium to a source of irradiation can detect the same elements as atomic absorption plus phosphorous, sulphur and chlorine. Although expensive it has considerable potential as an on-line continuous elemental concentration detector for very critical systems.

5.2 Particle Counting

Particle counting can be carried out using atomic counters such as the HIAC, Coulter, and Quantimet. The HIAC and Coulter counters provide a detailed particle size distribution in a fluid by counting particles individually and then automatically totalling. The HIAC counter works on a light blockage system and the Coulter counter by electric pulses caused by the passage of particles through an electrolyte.

The Coulter counter is more involved than the HIAC as the latter was specifically designed to analyze hydraulic fluids whereas the former was developed for blood cell counts[16].

Quantitative study of particles collected on filters or ferrograms can be carried out with the Quantimet image analyzing computer which can be used to discriminate between metal, oxides and polymeric material by light contrast relative to background. Numerical outputs include numerical concentration, particle size, ratio of length to width, and statistical distribution of particle size. Simple statistical parameters provide a clear indication of wear transitions occurring in multi-component systems[17].

A new area of particle detection in condition monitoring form the gas stream of a gas turbine is by electrostatic discharge[18, 19]. The electrostatic discharge is due to metal debris oxidation (burning) and metal rubbing or pitting. Metal rub produces a positive charge and metal burning (over temperature) a negative charge. Monitoring was first accomplished with a metal rod in the gas turbine tail pipe, but it has been found that a combination of an electrostatic wire grid and metal ring provided a much better indication of turbine engine distress due to monitoring the entire exhaust instead of just a portion by a rod.

A short review of methods of examination of debris and lubricant contaminants has been published[20].

5.3 Ferrography

Ferrography is a method of recovering particles from a fluid and depositing them on a substrate according to size and magnetic susceptibility for analysis[21, 22, 23]. The direct reading (DR) Ferrograph, a simple instrument for use at plant or depot level, is used to determine the amount and size distribution of wear particles in a sample of lubricant from which significant data can be derived. Use of a simple equation provides a single figure for the severity of wear or other index and experience can determine ranges to allow a code of monitoring practice to be established. Full Ferrographic analysis using the bichromatic microscope, electron microscopy, heating techniques, and Quantimet can be used to supplement the information[22, 24, 25]. Trend analysis has proved to be potentially attractive[26].

Particles generated by different wear mechanisms have characteristics which can be identified with the different wear mechanisms[21, 24]. Rubbing or adhesive wear particles found in the lubricant of most machines have the form of platelets and are indicative of normal permissable wear, Figure 1. Cutting or

abrasive wear particles take the form of miniature spirals and
loops, Figure 2. An accumulation of such particles is indicative
of a serious abrasive wear process. Particles consisting of
compounds can result from an oxidizing or corrosive environment,
Figure 3. Steel spherical particles, Figure 4, are a
characteristic feature associated fatigue crack propagation in
rolling contacts[27]. Specific regimes of wear have been
classified by the nature of the particles produced by surfaces in
sliding contact[28]. Different mechanisms such as rolling
bearings, gears, and sliding bearings produce distinctive
particles. An atlas of such particles is available[29]. Atlases
of characteristic nonmetallic and nonferrous particles are in
preparation.

The condition monitoring of nonlubricant washed,
inaccessible components by the extraction of particles from
exhaust or gas streams has been investigated[30]. Lubricants,
lubricant additives, lubricant contaminants, and friction polymer
and their influence on machinery conditions can be monitored[21].
Solvents are available to fluidize grease to prepare ferrograms.
An on-line Ferrograph can be incorporated in a machine or system
to allow continuous condition monitoring. Wear debris in a
lubricated systems if isolated by a high gradient magnetic field
and quantitative measurement of the debris is achieved by a
surface effect capacative sensor.

Ferrography has been successfully applied to the conditions
monitoring of natural and prosthetic joints. Particles retrieved
from the synovial fluid or saline washings of joints by
centrifuging are treated with a magnetizing solution containing
erbium ions and the resulting suspension used to make
ferrograms[31]. Scanning electron microscopy, in conjunction with
X-ray energy analysis, is a powerful tool for particle
identification[32]. The ability to monitor the condition of
prosthetic joints ferrographically should aid the development of
implant materials and the design of artificial joints.
Ferrographic synovial fluid analysis should augment the
understanding or the etiology and pathogenesis of degenerative
arthritis and provide a method for the diagnosis, documentation
and prognostication and treatment of the disease.

Monitoring of the content of nonmagnetic or nonmetallic
materials in fluids can be effected by the use of special fluids
which contain both lubricant solvating solvents and water
miscible solvents to dissolve the lubricant and allow magnetic
salts to become absorbed on the nonmetallic or nonmagnetic wear
particles so that they can be precipitated by the Ferrograph.

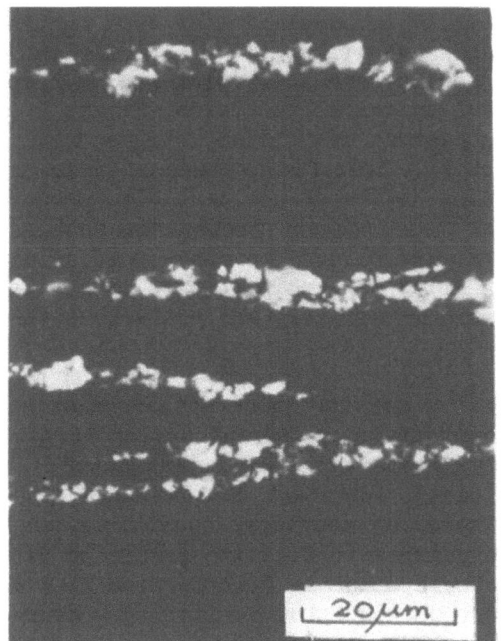

FIGURE 1   RUBBING WEAR PARTICLES

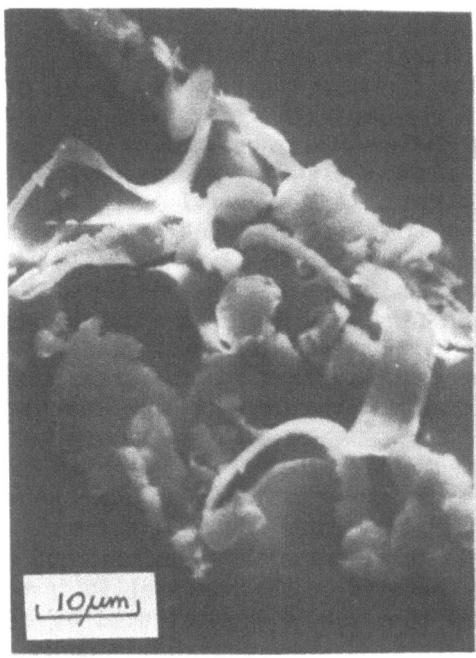

FIGURE 2   CUTTING WEAR PARTICLES

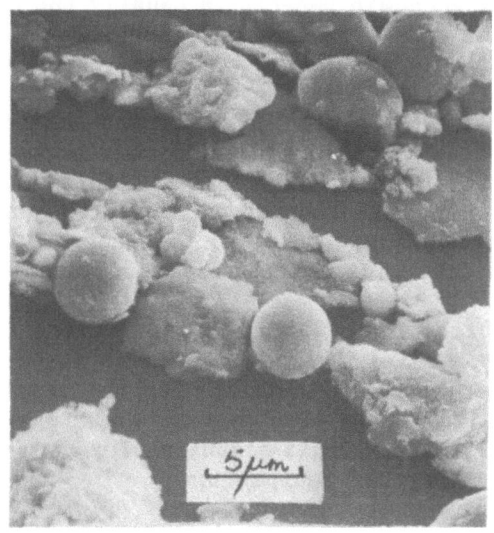

FIGURE 3 SPHERICAL PARTICLES

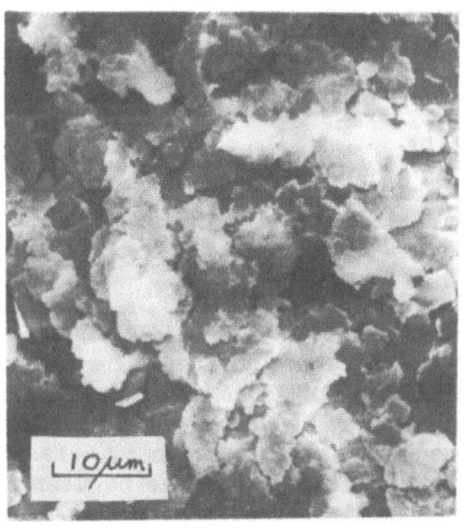

FIGURE 4 OXIDIZED PARTICLES

6. THE SIGNIFICANCE OF MONITORING

Overall economic benefit is usually the most important consideration affecting the decision to apply machinery condition monitoring and in particular the benefits arising from the reduction of output losses and maintenance costs. Energy and fuel conservation are also an important consideration as a machine such as a gas turbine will consume considerably less fuel to perform its required duty than a machine which has suffered deterioration. Prevention of failure also conserves material required for new components and the energy required to extract, refine, and fabricate the materials of construction.

The application of condition monitoring is essential where a safety risk is likely to arise, such as in plant handling dangerous materials and machines for the transportation of people. Monitoring is also desirable where accurate and advanced planning of maintenance is essential for normally inaccessible continuously operated equipment.

Condition monitoring enables the early detection of faults while damage is still slight in plants of recent design which may have development problems. It can provide useful information to guide design improvements[33]. Where operators cannot be expected to detect faults in expensive or sophisticated vital equipment, condition monitoring enables incipient faults to be detected to allow remedial measures before expensive failure or expensive consequential damage occurs. Condition monitoring may be significant when a plant manufacturer or lubricant supplier can offer a service to users of equipment thus reducing the cost to each and allowing a useful feedback to prove product design and development and the formation of improved lubricants and additive packages.

Machinery condition monitoring may also be affected economically by the use of equipment already available for quality control, process control, or servicing requirements.

7. EFFECTIVENESS OF MONITORING

Benefits which can be derived from the successful monitoring of the condition of plant and machinery include increased availability resulting in increased output from the capital invested and reduced maintenance costs. Machine running time can be increased by maximizing the time between overhauls by safely switching from regular periodic maintenance to on-condition maintenance. Overhaul time can also be reduced as the nature of the problem or the specific module in trouble for complex modular constructed machine is known so that spares required and essential personnel are available. If damage consequential to initial damage is avoided, considerable savings in time and capital can be effected. The storage of spares and the expensive tying up of capital can be eliminated if advanced warning of requirements is known to enable procurement of the spares in time.

Machinery condition monitoring can allow more efficient plant operation by controlling output and operating life to the best compromise. Recorded experience of the operation of most machinery aids the specification of improved design of future machines.

8. THE ECONOMICS OF MONITORING

If safety is the primary factor, then machinery condition monitoring is essential no matter what the cost. However, machinery conditions should generally only be considered where savings are considerably in excess of the cost of monitoring. Condition monitoring is most cost effective with capital intensive plants utilizing continuous operation of high cost machinery. It is not likely to be effective for lots of small cheap machines where ample spare capacity is readily available.

The main source of financial savings which can be made from machinery condition monitoring in an industrial establishment, are a reduction in the loss of production due to breakdown and the cost of maintaining plant and machinery. As a rough initial guide, industries are likely to achieve savings of the order of 1.2% of their added value output calculated from total sales revenue minus the cost of raw materials and energy brought in[3]. With expensive machines, which are expensive to maintain, it is essential to delay overhaul as long as possible but to avoid expensive consequential damage from simple intial wear.

To illustrate potential cost savings by effective machinery condition monitoring and the safe switch from periodic routine to on-condition maintenance, the cost of overhauling a modern large jet engine is approximately 12½% of its initial capital cost of £½M. It has been reported that a simple bearing failure in a fully integrated steel mill can lead to a total shutdown which at full output rate may cost up to £300 per minute[34, 35]. A similar bearing failure in a modern generator set could involve Central Electricity Generating Board in a loss of up to £20 per minute till the set was again operational. This being the difference in the cost of generating electricity with a large modern generator and smaller stand by equipment. A similar bearing failure in the U.S.A. has been quoted to cost $28,000 per day[36]. It has been reported that the total cost of wear for a U.S. Naval aircraft amounted to $243 per flight hour and of this total $140 was due to unscheduled maintenance[37]. Although SOAP has often been

criticized, the U.S. Defense Department spends £40M on oil
analysis to predict only certain types of failure in one power
system, the aircraft gas turbine, but this expenditure effects
savings of twice this amount in terms of direct repair costs[36].
A high safety factor is also achieved.

Owing to environmental problems, it has not been possible to
replace some older paper mills and so the aging plants are being
worked continuously at full or overload to maintain the required
output. In one mill failure of a roller bearing on a main roll
results in total shut down involving a loss in overall profit of
£100 per hour. As it can take up to 24 hours to effect repairs
and experience has shown that in other mills, of the 75 vital
bearings, some 25 failures are experienced per year. Prevention
of even one failure by monitoring would be cost effective, but an
efficient monitoring system may be expected to prevent most of
this failure and unscheduled maintenance.

9. SELECTION OF A MONITORING SYSTEM

The selection of an appropriate method for monitoring the
condition of a machine must be based on a consideration of which
of its many components are most likely to fail and in what way.
Possible methods for monitoring these components can then be
considered. It may be possible to choose a single method by
which all likely failures of components in a machine may be
detected with acceptable efficiency. In many instances however,
it is necessary to use many techniques as specific information
from each technique allows comprehensive data to be accumulated
from which a suitable lead time to failure can be established
before any sequential damage has occurred. For instance in
Germany, concern had been expressed regarding the use of a single
method of aircraft engine condition monitoring particularly in
the light of experience[38]. Deterioration of vital components had
occurred and sometimes not been detected so that expensive
failures in service had resulted. Equally expensive had been
engine removal on the basis of SOAP results when no deterioration
had been found. As in many cases, it is difficult to make a
correct decision regarding engine removal on the basis of SOAP
alone. Ferrography has been used to advantage to supplement
SOAP. Experience with the use of both techniques has indicated
that ferrographic analysis can detect the incidence and build up
of wear debris too large to be detected by SOAP and still too
small to be detected conveniently by MDP. The use of three
techniques cover the usual size range of wear debris, as shown in
Figure 5, to prevent failure and false alarms by SOAP and to
prevent secondary expensive damage by particles large enough to
be picked up by MDP[39].

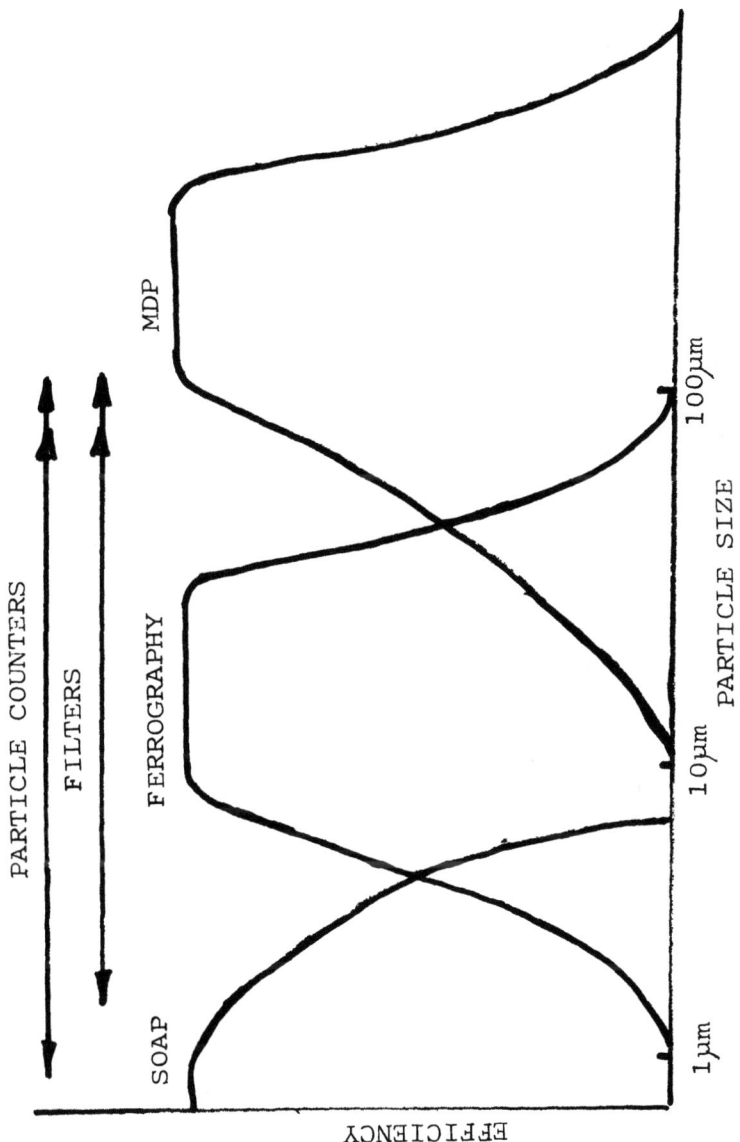

FIGURE 5 PARTICLE SIZE RANGE EFFICIENCY OF DIFFERENT
TECHNIQUES OF WEAR DEBRIS ANALYSIS

The monitoring of nonlubricated vital components, such as turbine blades and discs of a gas turbine, has proven to be a difficult problem. Although Boroscope examination is currently used, it is difficult to carry out effectively. Over the past few years there has been a gradual acceptance of vibration analysis, although this has proved to be neither the simplest nor the most effective method to use. There is the problem of interpreting the data caused in part by the fact that the vibration signal tells as much about the general dynamics of the machine as its damaged state. Modern software, to effectively separate the information, requires extensive investigation and development and thus is expensive.

10. CONCLUSIONS

For machinery condition monitoring, the simplest method providing the accuracy of prediction required should always be selected. The simplest method is usually the cheapest to purchase and the easiest to install and operate. The more sophisticated the equipment used, the less reliable is it likely to be and generally the more prone to misuse and misinterpretation. Sophisticated equipment also requires specialist operators and maintenance.

The time taken to interpret the result of monitoring after measurement has been completed, must be less than the failure propagation rate. Where possible, the monitoring should be controlled by the person responsible for making the maintenance decision and thus monitoring on the flight deck of an aircraft, on board a ship, or in the control room of an industrial plant is preferable to control by a remote central laboratory. Continuous monitoring is preferable to periodic measurement and alarm signals can be beneficial even with fully automated systems.

The best method of monitoring is usually derived from long experience arising from the monitoring of the particular machine. An optimum economic service life may be determined for many expensive to operate machines. For example, with the present escalating trend in fuel price, keeping a jet engine in service too long may involve extra expensive fuel consumption with a worn engine together with eventual expensive overhaul due to extensive replacements caused by overall general long term wear. Withdrawal at the optimum service life may result in a less expensive overhaul with a return to improved fuel consumption. The optimum period will probably shorten with increase in fuel cost compared with increase in maintenance costs.

11. REFERENCES

1. Scott, D, and Smith, A.I., "Improvement of Design and Materials by Failure Analysis and the Prognostic Approach to Reliability," Inst. Mech. Engrs. Conf. Pub. 22 (1973), London.

2. Scott, D., and Westcott, V.C., "Predictive Maintenance by Ferrography," Wear, 44 (1977) 173 - 182.

3. Neale, M.J., "A Guide to the Condition Monitoring of Machinery," H.M.S.O. London (1977).

4. Collacott, R.A., "Mechanical Fault Diagnosis and Condition Monitoring," Chapman & Hall, London (1977).

5. Scott, D., "Hardware and Instrumentation - State-Of-The-Art," In "Performance Monitoring and AIDS Seminar/Workshop," A.D.I. Transportation New York (1981).

6. Barber, R., "A Radiation Pyrometer Designed for the In Flight Measurement of Turbine Blade Temperature," S.A.E. Paper 690432 (1969).

7. Curwen, K.C., "Turbine Blade Pyrometer System in the Control of the Concorde Engine," Kollsman Instruments Ltd, Southampton (1975).

8. Wilson, R.W., "The Diagnosis of Engineering Failures," South African Mech. Eng. November. 11. (1972).

9. Belcher, P.R., and Wilson, R.W., "Templungs," The Engineer 221 (1966) 305.

10. Rogers, L.M., "The Application of Thermography to Plant Condition Monitoring," British Steel Corporation Report TB/TH/71 (1971) Sheffield.

11. Urban, L.A., "Gas Turbine Engine Parameters Interrelationships," 1969. Hamilton Standard (United Technologies). U.S.A.

12. Scott, D., "Particle Tribology," Proc. Inst. Mech. Engrs. London 189 (1975) 623 - 633.

13. Barrett, G.M., "Spectrographic Analysis of Crankcase Lubricating Oils as a Guide to Preventive Maintenance of Locomotive Diesel Engines," Proc. Inst. Loco. Engrs. January (1961).

14. Davies, A.E., "Principles and Practice of Aircraft Powerplant Maintenance," Trans: Inst: Marine Engrs. 84 (1972) 441 - 447.

15.. Seifert, W.W., and Westcott, V.C., "Investigation of Iron Content of Lubricating Oils by Ferrograph and Emission Spectrometer," Wear 73 (1973) 239 - 249.

16. Kubitschek, H.E., "Electronic Measurement of Particle Size," Research, 13 (4) (1960) 128 - 135.

17. Odi-Owei, S., Prince, A.L., and Roylance, B.J., "An Assessment of Quantimet as an Aid in the Analysis of Wear Debris in Ferrography," Wear 40 (1976) 237 - 283.

18. Couch, R.P., Rossback, D.R., and Burgess, R.W., "Sensing Incipient Jet Engine Failure with Electrostatic Probes," Symposium for Airbreathing Propulsion, Monterey CA. (1972) AF Flight Dynamics Laboratory, Wright Patterson AFB. Ohio.

19. Burgess, R.W., "An Investigation of the Detection of Charged Metal Particles in a Jet Engine Exhaust by a Cylindrical Electrostatic Probe," AFIT-EN, Wright Patterson AFB Ohio (1972) Thesis AD 745540.

20. Scott, D., "Examination of Debris and Lubricant Contaminants," Proc. Inst. Mech. Engrs. London 187 (3G) (1968) 83 - 86.

21. Bowen, E.R., Scott, D., Seifert, W.W., and Westcott, V.C., "Ferrography," Tribology Int., 9 (3) (1976) 109 - 115.

22. Scott, D., Seifert, W.W., and Westcott, V.C., "The Particles of Wear," Scientific Amer. 230 (5) (1974) 88 - 97.

23. Scott, D., and Westcott, V.C., "Ferrography," Proc. Eurotrib. 77 Band 1 paper 70 (1977) 1 - 6.

24. Scott, D., "Debris Examination in a Prognostic Approach to Failure Prevention," Wear 44 (1975) 15 - 22.

25. Scott, D., and Mills, G.H., "Debris Examination in the SEM - A Prognostic Approach to Failure Prevention," Scanning Electron Microscopy Pt. IV (1974) 883 - 888 I.I.T. Chicago.

26. Scott, D., and McCullagh, P.J., "Condition Monitoring of Gas Turbines - An Exploratory Investigation of Trend Analysis," Wear 49 (1978) 373 - 389.

27. Scott, D., and Mills, G.H., "Spherical Debris - Its Occurrence, Formation and Significance in Rolling Contact," Wear, 24 (1973) 235 - 242.

28. Reda, A.A., Bowen, E.R., and Westcott, V.C., "Characteristics of Particles Generated at the Interface Between Sliding Steel Surfaces," Wear 34 (1975) 261 - 273.

29. Bowen, E.R., and Westcott, V.C., "Wear Particle Atlas," Foxboro Analytical, Burlington, U.S.A. (1976).

30. Scott, D., and Mills, G.H., "An Exploratory Investigation of the Application of Ferrography to the Monitoring of Machinery from the Gas Stream," Wear 48 (1978) 201 - 208.

31. Mears, D.C., Hanley, E.N., Rutkowski, R., and Westcott, V.C., "Ferrography Analysis of Wear Particles in Arithroplastic Joints," J. Biomed. Mater. Res. 12 (1978) 867 - 875.

32. Scott, D., Russell, A., and Westcott, V.C., "Recent Developments in Ferrographic Particle Analysis and Their Application in Tribology," Proc. Eurotrib. 81, Warsaw, Poland. (1981) in Press.

33. Scott, D., and Westcott, V.C., "Ferrography - An Advanced Design Aid for the 80's," Wear 34 (1975) 251 - 256.

34. Braithwaite, E.R., "$Mo S_2$ Second Thoughts," Industrial Lubrication 21, (8) (1969) 241 - 247.

35. Scott, D., "Introduction to Tribology," In Fundamentals of Tribology. Ed. N.P. Suh and N. Saka, M.I.T. Press (1978) 1 - 16 Cambridge, Mass. U.S.A.

36. Ling, F.F., "Socio-economic Impacts of Tribology," Proc. Tribology Workshop (1974) National Science Foundation, U.S.A. 32 - 64.

37. Devine, M.J., (Ed.) Proc. Workshop on Wear Control to Allow Product Durability, (1977) Naval Air Development Center, Warminster, P.A. U.S.A.

38. Hoffman, W., "Some Experience with Ferrography in Monitoring the Condition of Aircraft Engines in Germany," Wear 65 (1981) 201 - 208.

39. Jones, M.H., Private correspondence - To be published.

TRIBOLOGY: THE MULTIDISCIPLINARY APPROACH

B. R. Reason
Cranfield Institute of Technology

1. THE ROLE OF THE DISCIPLINES

1.1 The Importance of Tribological History

As with many other aspects of technology, the rate of expansion of research, innovation, analysis, synthesis, and product development in Tribology has probably followed some form of exponential curve if such expansion is to be quantified, however recondite the connection, in terms of the rate of production of technical papers in the field.

On this basis, it is highly improbable that any single individual could be conversant either with the scope of the whole spectrum in depth or offer a prognosis on the detailed ramifications resulting from the dissemination of such information on current and future tribological evolution.

Bearing this in mind, however, it is clearly important that the individual tribologist does not extend this proposition to the point where he will eschew any attempt to view the subject as a unified entity, merely because of the extent and variety of the available material. Indeed, it is the opinion of the author that there is a growing tendency in technology in general towards the cultivation of the individual and specific tree, unfortunately at the expense of the wood, and that this philosophy, especially in the field of Tribology, could have consequences which might be litte short of disasterous if the results of the manifest efforts in the field of research are not widely implemented at the grass roots of industrial utilization.

In the present world climate of limited natural resources and rapidly shrinking energy reserves, it is the prime requisite, if not the moral duty, of the industrial designer to employ as far as possible, both the cropus of information which has been placed at his disposal and the specific resources available to him in his particular milieu.

The question, however, remains as to how such a goal is to be accomplished, bearing in mind not only the multiplicity of the applications of the subject, but also the interdisciplinary nature of the complete tribological spectrum in the context of its current utilization.

The human mind is, however, a flexible system, highly developed in a role of adaptation to ever-changing situations. Fortuitously, also, it relies on its experience and especially on precedential information in framing both its present problems and establishing its future solutions. In this, like the development of language itself, it relies heavily on the chronological experience of other minds, for in the knowledge of what has gone before lie the seeds of what will ultimately follow.

It is to this hierarchy of past experience, therefore, that we must initially turn to establish the foundation of our study of the interaction of the various disciplines, in order to place them both in historical time sequence and to present specific developments in what, with the benefit of hindsight, we may justifiably call a 'temporal perspective', remembering always that we stand as a judiciary with this perrogative of time denied, both to ourselves in our current problems and to our tribological antecedents when they were coping with theirs.

Indeed, the author considers it of paramount importance if any fundamental understanding of the interdisciplinary nature of the tribological spectrum is to be achieved, that the seeker after such insight turns to the process of tribology history itself.

By this process, he will firmly establish a broad foundation on which any conceptualized integration of the various disciplines within the tribological framework must necessarily rest. For this reason, the historical aspect will be treated in some depth.

1.2 Development of Tribology From 1850 to 1925

Historically, the 'Age of Steam Power' is considered by many text-books to begin around the year 1850 and to continue unabated until the end of the century[1]. Undoubtedly, the period saw a great upsurge of interest in the training and education of engineers in all the industrialized countries of the world. Technical education flourished and academic chairs were established in many branches of engineering, particularly in the U.S.A., U.K., and the European continent and Russia. In the light of this, it is to a certain extent understandable why prior to 1850, lubrication, per se, had received little or no attention - the demand for bearings capable of coping with heavy duty while producing low frictional loss, simply did not exist.

As steam power became more efficient and bearing loads and speed escalated, problems inevitably arose with bearing failures, either through rapidly diminished life or of a more catastrophic nature.

The coming of the railways introduced bearings for locomotives and rolling stock in large quantities with the concomitant problem of their efficient lubrication. Indeed, it was the 'father of the railway', George Stephenson, who became the first president of the Institution of Mechanical Engineers when it was first formed in England in 1847.

It was not, therefore, entirely fortuitous that around the 1850's several of its members were publishing papers on the efficient lubrication of railway axle boxes. Mineral oil was already being produced for use in the Lancashire cotton mills and its use as a rival to the more conventional vegetable and animal greases in lubrication was a subject of some contention, the basic problem then, as it is today, being its efficient containment in the vicinity of the bearing.

1.3 Journal Bearing Friction - The Early Workers

Experimental work with loaded half-bearings with and without lubricants had been published by Hirn in 1854, who noted that the coefficient of friction was directly proportional to speed at constant temperature and was also directly related to the lubricant's viscosity[2]. Although this work revealed the essential prerequisites of efficient lubrication, it received little or no recognition since it contravened the established laws of dry friction proposed by Coulomb and confirmed by Morin[3,4]. Hirn's work, which included studies on air as a lubricant in addition to a range of animal, vegetable, and mineral oils, was later to recieve due recognition from Robert Thurston, the first President of the American Society of Mechanical Engineers[5].

Thurston, himself a university graduate, was essentially desirous of conserving energy through efficient use of materials and lubricants and sought diligently to convince his contemporaries of this necessity. He published much experimental data on friction coefficients for a wide spectrum of lubricants and recognized the importance of mechanical test machines for performance evaluation of bearings. Arising from his work in this area he established that, for a lubricated bearing, a point of minimum friction coefficient existed within a range of increasing load.

1.4 Petrov's Analysis and 'Mediate Friction'

It was a Russian, Petrov, however, who, motivated both by the inevitable problems with railway axle boxes and the desire to find a market for Caucasian mineral oil, (which had been first produced in 1876) undertook an extended program of friction measurements on these units. Using oils of differing viscosities, he was the first to establish the relationship between working oil viscosity and bearing power loss. More important however, was his discovery of 'mediate friction' i.e., that power loss arose from an intermediary, the lubricant, which physically separated the bearing surface, thereby resolving the hitherto contentious problem of surface contact friction into one of simple viscous drag. Employing the established viscous friction law of Newton in a simple analytical model he produced, in a paper of (1833), basic equations for viscous power losses[6, 7]. Through further extensive experimental work and an analysis of Hirn's results, Petrov established the validity of his concept of 'mediate friction' and laid the foundation for the classical hydrodynamic analysis that was to follow shortly.

1.5 Beauchamp Tower - 'The Gentleman' and the 'Wooden Plug'

Meanwhile, in England, the Institution of Mechanical Engineers had not been idle. A Research Committee had been formed to encourage studies into prevalent engineering problems and at a Council Meeting in 1879, one of the subjects felt worthy of investigation 'should time and money be found to be sufficient' was 'Friction between solid bodies at high velocities'.

The need for this, it would appear, arose from a conglomeration of contradictory evidence in various kinetic friction experiments and a 'gentleman' was sought who would be in 'a position to take the subject in hand'. Happily both the 'time' and 'money' and the 'gentleman' finally appeared for, on April 20th, a Mr. Beauchamp Tower was duly appointed. His test machine, a simulated axle box bearing, was quickly assembled and commissioned and experiments were commenced.

Tower produced his 'First Report' to the Institution in 1883 and its findings, though he may not have appreciated it at the time, were to have momentous results in the history of Tribology. Initially, Tower found that erratic friction results arose with the common methods of lubrication prevailing at the time and resorted to an oil bath system which gave better consistency; thus establishing the importance of adequate lubrication[8].

Towards the end of the initial experiments the bearing, having seized, was stripped. While the brass was out, a ½" diameter hole was drilled into the half brass for a lubricator. On running the machine on reassembly, oil flowed out of this hole and to prevent its egress the hole was blocked with a wooden plug. On restarting the machine, the plug was observed to be slowly forced out by the oil, indicating a 'considerable pressure'. A pressure gauge, being screwed in, registered a pressure over twice that of the average on the bearing cross-section and in Tower's words 'showed conclusively that the brass was actually floating on a film of oil, subject to a pressure due to the load'.

This discovery, that a pressure sufficient to separate the surface, was generated within the lubricant marked a fundamental experimental breakthrough, confirming Petrov's concept of mediate friction and initiating a hydrodynamic analysis which was to form the bedrock of fluid bearing design.

Within two years, in 1885, Tower's second paper, devoted to mapping the complete pressure envelope over the area of the bearing surface had been produced. The accuracy of his measurements can be judged from the fact that the integrated pressure gave a load of 7,988 lb f, the applied load being 8,008 lb f[9].

1.6 Reynold's Equation

It has often been said that good experimentation produced good theory and Tower's work was no exception. In 1886, Professor Osborne Reynolds submitted his classical paper 'On the Theory of Lubrication and its application to Mr. Beauchamp Tower's Experiments' to the Royal Society[10]. Indeed, there is evidence that he was already at work on this in 1884, for he delivered a paper 'On the Function of Journals' in that year to the British Association for the Advancement of Science in Montreal. It was also clear at this meeting that both Lord

Rayleigh and Professor Stokes had discussed Tower's work in relation to the concept of the hydrodynamic taper wedge from remarks made by Lord Rayleigh during his presidential address. However, it was to be Reynolds who launced the Hydrodynamic Theory which is the cornerstone of fluid-film lubrication as we know it today.

The basic differential equation formulated by Reynolds

$$\frac{\partial}{\partial x}\left[\frac{h^3}{\eta}\frac{\partial p}{\partial x}\right] + \frac{\partial}{\partial z}\left[\frac{h^3}{\eta}\frac{\partial p}{\partial z}\right] = 6 \ (u_0 - u_1) \frac{\partial h}{\partial x} + 6h \frac{\partial}{\partial x} (u_0 + u_1) + 12V$$

expresses differential functions of pressure in two coordinates in terms of the relative movements of the bearing surfaces and requires a sequential process of integration to produce equations characterizing the bearing's performance. The conceptual intuition behind its formulation implies a highly erudite mind and an outstanding capacity for logical deduction and imagination.

From this basic differential equation, Reynold's derived analytical expressions for pressure distribution and load-carrying capacity per unit width of bearing (i.e., neglecting the effect of side leakage), but found difficulty in expressing these for bearings operating at high duty because of the lack of convergence in certain trigonometrical expressions in a series solution.

1.7 The Sommerfeld Solution

It was, in fact, a nuclear physicist, Arnold Sommerfeld, whose analytical solution in 1904 to the problem posed by the Reynold's equation, paved the way to the establishment of a concrete foundation for future fluid bearing analysis and design[11]. Neglecting the effect of side leakage reduces the Reynold's equation to the form:

$$\frac{dp}{dx} = 6 \ \eta_U \left[\frac{h - h_0}{h^3}\right]$$

which expresses the slope of the pressure curve in the direction of motion without restriction on the geometry of the fluid film. The equation could therefore be applied to bearing systems other than circular, the problem of determining the pressure distribution being one of overcoming the erected integrals resulting from the substitution of the film thickness 'h' in terms of the (x, θ) coordinate, Figure 1.

In overcoming three erected integrals for a circular bearing (with 'h' expressed in terms of the variable coordinate 'θ) Sommerfeld employed the standard substitution δ = tan (θ/2) to solve the first integral on the assumption that the oil film was continuous around the bearing. Having obtained a solution to this first integral, Sommerfeld used reduction formulae to obtain solutions to the remaining two.

It is of interest in passing to discover that the so-called 'Sommerfeld Transformation' expression:

$$\cos \psi = \frac{\varepsilon + \cos\theta}{1 + \varepsilon\cos\theta}$$ (where 'ε' is the eccentricity ratio, Figure 1)

widely attributed to Sommerfeld was probably never actually used by him at all. Professor Duncan Dowson, in his magnificent book 'A History of Tribology', a most comprehensive and exhaustive piece of historical scholarship, writes: "Perhaps the most amazing finding which emerges from a careful study of this classical paper is that the mathematical transformation attributed to Sommerfeld was apparently never used by him. He simply used the standard substitution δ = tan (θ/2) and reduction formulae to solve the integrals which yielded expressions for pressure, load-carrying capacity, and viscous traction in his solution of the full 360° journal bearing problem[12].

Professor Dowson goes on to say that, although Sommerfeld used a function which can be rearranged to form a similar expression to the 'Transformation', it was probably Boswall in 1928, who first wrote it in its now familiar form[13].

1.8 Slider Bearing Studies - The Michell Analysis

As already indicated, both Reynolds and Sommerfeld obtained solutions to the full fluid bearing by neglecting side leakage effects. It was left to an Australian, Michell, in 1905, to obtain a solution in which side leakage effects were included, in this case for the problem of the slider bearing[15].

Expressing the fluid pressure as a series of terms, whose solution involved Bessel functions with specified boundary conditions, Michell employed numerical procedures to obtain

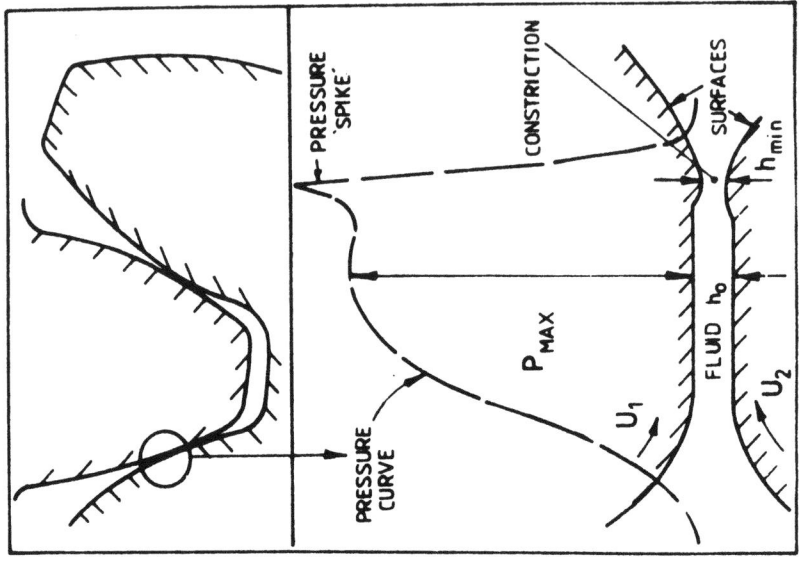

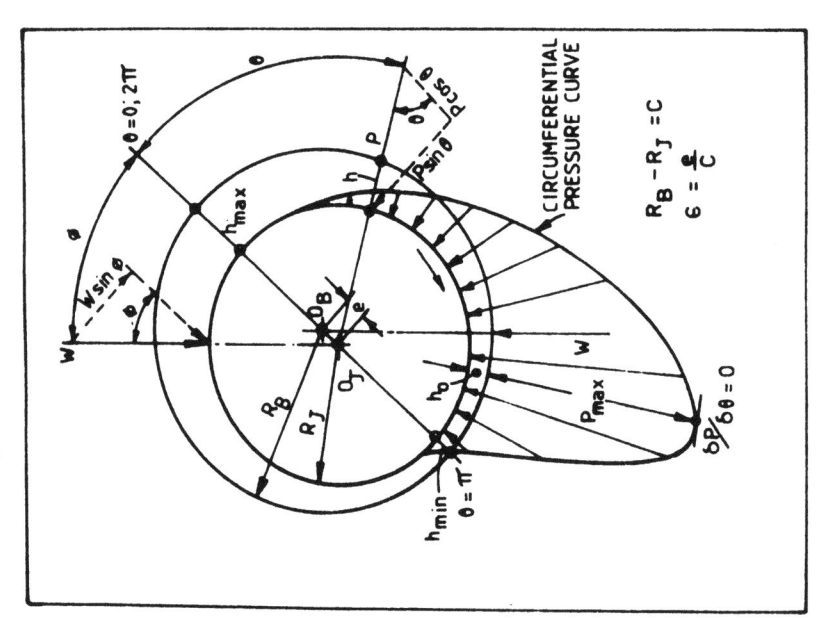

FIGURE 1 FLUID FILM PRESSURE GRADIENT

the first analytical three-dimensional solution of the
Reynold's equation for planar surfaces, producing pressure
envelopes for four rectangular pad geometries.

Here was a mathematical exposition of the highest order,
the first unrestricted solution of the Reynold's equation,
worthy of the great master himself. It showed, as may be seen
from Figure 2, the dramatic effect of the fluid side leakage
on the axial pressure gradient directly corroborating the
experimental findings of Tower, some twenty years earlier.

It is noteworthy that, in addition to being a brilliant
analyst, Michell was a practicing engineer, his interest in
thrust bearings arising from problems with these units in
centrifugal pumps and water turbines in Australian
hydroelectric plant. This culminated in the invention and
patenting of the tilting pad thrust bearing, Figure 2, in
1905, the year of his analytical publication.

1.9 Viscometry

In this period towards the end of the century,
experimentalists and inventors in the lubricant field had been
particularly active. Reynolds, himself, when presenting his
classical paper in 1886, included empirical temperature
viscosity relationships for olive oil based on his own
findings and for water and air based on work by Poiseuille in
1846, and Maxwell in 1860[16, 17].

The viscometer itself had been introduced by Slotte in
1881 and in 1884 Engler's viscometer had been officially
adopted by a section of the German railways for the comparison
of lubricants[18, 19]. In England, in the same year, Redwood
had produced his own instrument describing it in a paper in
1886[20]. During this meeting, he produced an instrument
developed for the same purpose by a Mr. G. M. Saybolt from the
Standard Oil Company U.S.A. All the instruments measured oil
viscosity as an efflux in seconds at a standard temperature
and expressed their results, either as a time ratio with a
standard fluid, or in absolute time. These instruments were,
with development, related to each other by empirical equations
and became the foundation for standard industrial viscometric
systems.

We are today apt to treat viscosity with a certain sang-
froid as a basic concept easily cognized when considering
viscous energy dissipation. It must be appreciated, however,
that, prior to about 1880, viscosity was an exceedingly
nebulous quality, although Dolfus as far back as 1831, had
demonstrated an efflux instrument which he called 'viscometre'

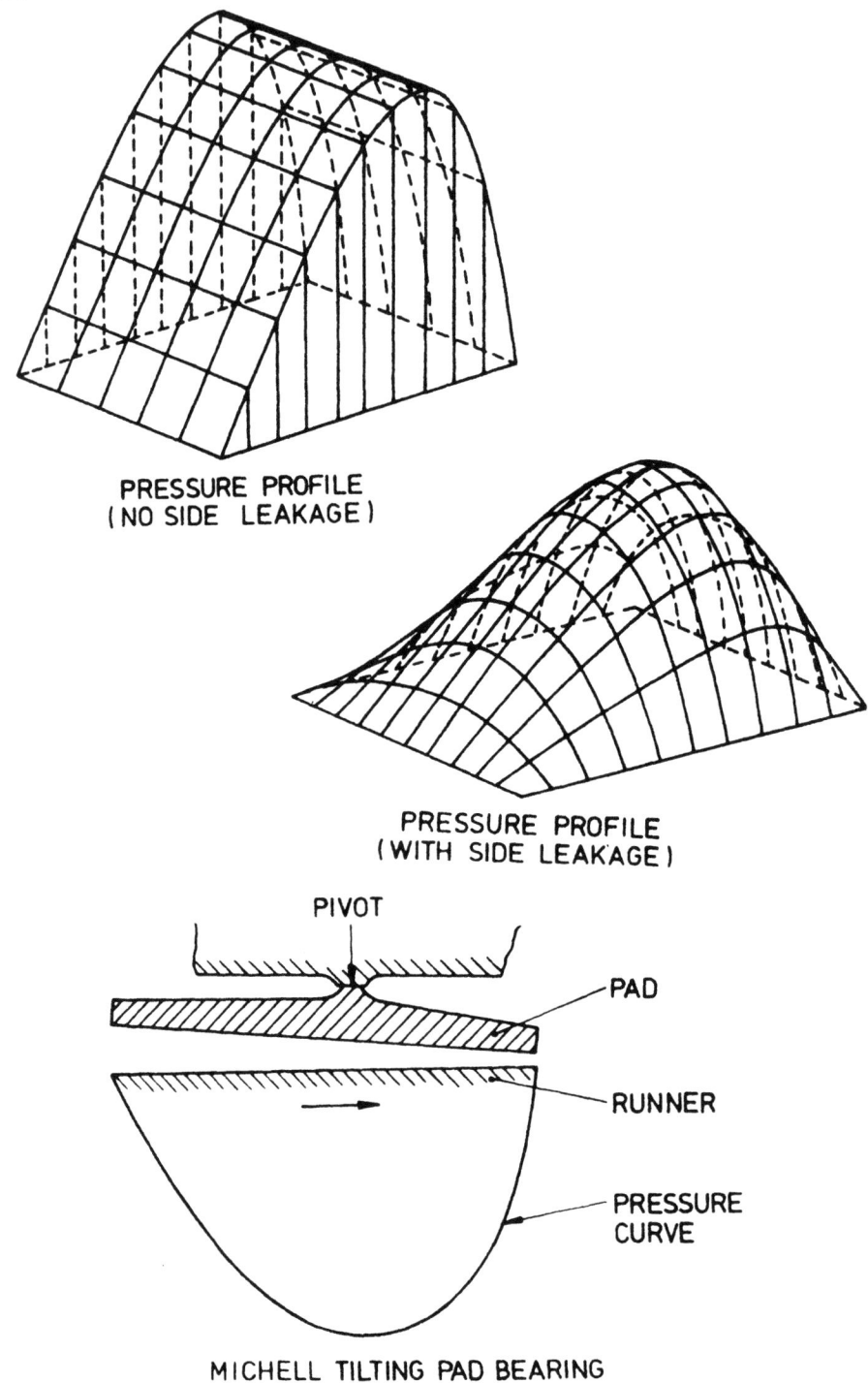

FIGURE 2 FLUID SIDE LEAKAGE PRESSURE GRADIENT EFFECTS

deriving, on the basis of the liquid's efflux time, an 'index of liquidity'[21]. 'Fluidity' of liquids had certainly been appreciated but the more obvious properties of density or specific gravity were generally considered by engineers to be a lubricant's cardinal property in bearing applications prior to the last quarter Century*. The escalation of bearing duty concomitant with increasing engine power in this period, however, rapidly exposed the limitations of the established animal and vegetable lubricants and this, together with their relatively high cost, inititiated their eventual demise to be supplemented by the, by then, ubiquitous mineral oil. The genre of this age of lubrication investigation is reflected in books by Thurston in 1885 and by Archbutt and Deeley in 1900 and provides most illuminating reading[5, 22].

1.10 Hydrodynamics - The Aftermath

Bearing testing, per se, had not stagnated since Beauchamp Tower's monumental breakthrough in 1883. Indeed, in some respects his results may well have had a catalytic reaction on other experimentalists. In 1886, Goodman read a student paper to the Institution of Civil Engineers in which he reported the first direct measurement of fluid film thickness between bearing surfaces using a micrometer device and a simple electrical circuit, thus confirming Petrov's concepts three years earlier and lending experimental credence to Reynold's paper which was published shortly after it in the same years[23, 10].

In the year 1892, a most intriguing phenomenon was observed by Albert Kingsbury in the United States. Kingsbury was investigating screw thread friction and in order to apply loads to his test machine, employed a 6" diameter piston within a cylinder. With the cylinder in the vertical position, he found that the piston could be rapidly spun without frictional resistance from the cylinder wall. The experiment was repeated with the cylinder in a horizontal position with the same result. From these simple experiments, Kingsbury postulated that the piston was rotating on a film of air and building an air journal bearing test rig, measured circumferential and axial pressures at specific points on the bearing surface. Kingsbury was unaware of Reynold's work when he first demonstrated the test rig in Washington in 1896, but was shown the paper at a later demonstration before the Bureau of Steam Engineering at the Navy Department. Subsequent to reading Reynold's paper, he published his paper 'Experiments with an air lubricated bearing' in 1897, a classical paper, in

* Maxwell in 1860 had already discussed viscosity and 'viscosity coefficient'[17].

many ways years before its time[24]. Reynold's paper also encouraged Kingsbury to employ the new hydrodynamic concepts to the hitherto troublesome problem of thrust bearing lubrication and in 1898, instigated the design and construction of a pivoted-pad thrust bearing, the bearing pads pivoting on spherical seatings. The bearing was first used by the Westinghouse Co. in 1904, by whom Kingsbury was employed; certain teething troubles arose however, and it was replaced by a ball bearing. By this time Michell had completed his analysis and obtained an English patent for his bearing in 1905. Parallel development of both bearings continuing from this period onward, Kingsbury being awarded a patent in the U.S.A. in 1910. Unlike Leibniz and Newton, however, and the wrangle over the development of the infinitesimal calculus relations between Kinsbury and Michell were amicable, each respecting the other's independent development[25].

1.11 Rolling Contact Bearings

In Germany Friherr Drais with his invention of the 'Draisine' in 1818, inititated the velocipede, later to be known as the bicycle. This technological development spread throughout Europe and in the process attention naturally became focused, both objectively and subjectively, on the problem of its tendency to resist locomotion.

Frenetic activity arose in an 'effort to reduce effort' following the first patent in 1962, granted to A. L. Thirion for a ball bearing device for such a machine. This culminated towards the end of the nineteenth century, not only in a new industry, but in a new field of study in metallurgy, materials science, and manufacturing techniques arising from the high contact stress engendered in ball contacts; case hardening, developed at the turn of the century, together with better precision in manufacture, providing extended utilization.

1.12 Contact Mechanics

As a corollary to these developments, the science of contact mechanics became greatly extended, yet, perversely, one of its foremost contributors, Heinrich Hertz, became involved in the subject through an interest in optics and the contact between glass lenses, rather than steel balls.

In 1881, Hertz presented his now classical paper on the subject to the Physical Society of Berlin the expression derived enabling deformations and stresses to be calculated for generalized elastic surfaces in counter-conformal contact[26]. This work, as we shall see, was destined to play a very significant role in tribological studies some sixty-five years later.

Experimental support for Hertzian contact theory was provided around the turn of the century by Professor Richard Stribeck, working in Berlin.

Stribeck both considered the mechanical properties of bearing steels arising from heat treatment and tested the Hertzian predictions by mechanical testing ball assemblies in a variety of contact situations. He established, for a radial ball bearing of 'n' balls under load, that a simple expression 4.37 W/n (Stribeck's formula) expresses the greatest load taken by any ball but suggested the number be approximated to '5' to allow a factor of safety[27].

1.13 Stribeck & Gümbel

Stribeck also studied friction in hydrodynamic bearings in addition to rolling contact bearings and in 1902 confirmed, with a series of carefully controlled experiments, the point of minimum friction previously reported by Thurston, today recognized as the transition point between incipient asperity contact and a full fluid film[28]. In doing so he ended the controversy which had existed during the fifty years, prior to this time, regarding the mechanism of friction in journal bearings. Many engineers today speak of the characteristic form of the journal bearing friction curve as the 'Stribeck Curve', on the basis of this early pioneering work.

Stribeck's results were analyzed in 1914 by a compatriot of his Dr. Ludwig Gumbel[29]. Gumbel demonstrated that the family of curves produced by Stribeck could be unified into one single curve, a dimensionless parameter

$$\frac{\eta \omega}{p}$$

being plotted against the friction coefficient. This grouping was later to be known as the 'Gumbel number'. Gumbel also considered the problem of cavitation in oil films and suggested possible boundary conditions.

1.14 Dimensionless Groups - The 'Hersey Number'

Although dimensionless groups had been utilized sporadically by workers prior to 1914, it was the Pi Theorems of Buckingham of the National Bureau of Standards, U.S.A., who placed such approaches on a firm foundation[30]. The technique proved attractive to engineers and Mayo D. Hersey (1914) was to become the first worker to apply such an approach to journal bearing analysis showing that hydrodynamic friction could be expressed as a function of

$$\frac{\eta n}{p}$$

a restatement of the Gümbel number in terms of 'n' the speed of rotation[31]. The term appears today in the upper case form

$$\frac{ZN}{P}$$

where 'Z' is the abbreviation for the German word 'Zahigkeit' (viscosity).

1.15 Fluid Analysis - The Consolidation

Starting with Reynolds and extending through Sommerfeld and Michell, fluid analysis was continued for gaseous fluid by Harrison in 1913[32]. Although unaware of Sommerfeld's solution, he proceeded to solve the Reynold's equation for a full journal bearing, drawing attention in the process to discrepancy between the journal and bush reactive torque arising from the journal's eccentricity, Figure 1. He extended this work to compressible fluid flow between parallel plates and on the basis of mass flow continuity and neglecting side leakage, numerically integrated the Reynold's equation. He then compared results for incompressible and compressible flows against those obtained experimentally by Kingsbury; excellent agreement being obtained. Between 1913 and 1919, Harrison treated inclined surfaces under varying load and speed and in 1919 produced a paper on dynamically loaded journal bearings which stands today as a pioneering work years before the practical problems of bearing dynamics had presented themselves to the designer[33]. In the same paper, he also produced an analytical study of surface profile effects in pivot bearings.

1.16 The Problem of Counter-Conformal Contacts

The success attained by analysis on conformal surface contact problems led to studies on the more complex problems of counter-conformal surfaces.

In 1916, in the journal 'Engineering', Martin produced an analysis of the contact conditions between two lubricated discs loaded together and rotating in nominal line contact, recognizing that such a configuration closely modelled the conjunction conditions between straight spur gears[34]. Treating the system as isoelastic and the lubricant as isoviscous he obtained an analytical expression for the minimum film thickness in the contact area, i.e.

$$h_m = 2.44748 \cdot \left[\frac{\eta URL}{W}\right]$$

Although this expression yielded a magnitude of film thickness considerably less than the measured surface finish of gears at that time, the implication of the approach is of considerable significance since it represented an adherence to concepts of classical hydrodynamic lubrication initiated by the Reynold's paper of 1886. However, around this time, a feeling was growing that there were other factors which influenced the lubrication between sliding surfaces in close proximity.

1.17 Boundary Lubrication

Young, as far back as 1805, had considered the equilibrium of a liquid droplet on the surface of a solid in terms of the interfacial surface tensions and Tomlinson in 1867 and 1875 had discussed the influence of surface contamination upon gas release at interfacial contacts between solids and liquids in terms of adhesive forces[35, 36, 37]. Lord Rayleigh in 1918, only two years after Martin's analysis, in a paper entitled 'On the lubricating and other properties of thin oily films' referred to his earlier experiments on surface tension and wrote: "We see that the phenomena here in question probably lie outside the usual theory of lubrication, where the layer of lubricant is assumed to be at least many molecules thick[30]."

Already Langmuir in 1917 was involved with work on thin surface films, principally on the idea of oriented monomolecular layers as were Hawkins, Davis, and Clark in the U.S.A., in a paper of the same year[39, 40].

In 1919, Hardy and Hardy spoke of the nature of an 'adsorbed' layer of lubricant onto a surface and considered that the lubrication depended wholly on the chemical composition of the fluid[41]. In the following year, a report from the Department of Scientific and Industrial Research (D.S.I.R) recognized three distinct stages of lubrication governed by three laws: 1. Dry friction. 2. 'Greasy' friction. 3. Viscous friction; while Hardy, again in 1920 used the term 'boundary conditions'[42], [43].

However, it was both Hardy and Doubleday in joint papers to the Royal Society in 1922 who laid the foundation of the school of boundary lubrication and molecular chain theory, emphasising the idea of multilayer orientated films, rather than a mono-layer and stating what, in retrospect, was to prove to be a portentous conclusion: " ... Boundary Lubrication differs so greatly from complete lubrication (i.e., full fluid) as to suggest that there is a *discontinuity between the two states.*" (Author's italics)[44], [45].

Thus, Hardy and Doubleday papers not only provide a starting point from which a whole new world of surface studies evolved, but provided the inititation of a whisper which, however unintentional, was later to grow into a murmur of dissent not so much against the concepts of classical hydrodynamic theory in its rightful domain, since this had already been proven beyond reasonable doubt, but in that the newly crowned 'prince' governed exclusively in all tribological realms, specifically in certain conformal contact situations but much more particularly in the realm of the counter-conformal contact, as exemplified by lubricated gear teeth. Here, if anywhere, it seemed, the apparent failure of Martin's analysis to predict realistic oil film thickness by classical hydromechanics was a domain where a new pretender to that particular throne might reasonably raise his standard in revolt. In reality, nearly thirty years were to elapse before, to coin a metaphor, oil, and not the fatty acid molecule, was to be poured on the troubled waters of the counter-conformal conflict in the shape of an analytical approach worthy of the pragmatism of Reynolds. Proposed by Grubin and Vinagrodov, Moscow 1949, it combined classical hydrodynamic theory with the elasticity of the contacting materials and the effect of pressure on the lubricant's viscosity and predicted, at last, finite fluid film thicknesses albeit of some one or two orders less than those that their compatriot, Petrov, had first brought to the notice of the world some eighty-three years previously; eighty-three years, in truth, from a 'Russian thick film' to a 'Russian thin film'[46].

1.18 The 'Great Schism'

Such a point in the early 1920's therefore, might justifiably be referred to by the tribological historian, as the point of the 'Great Schism', where an arbitrary division of the lubrication spectrum arose. The first camp, the area of full fluid lubrication, frequented by the industrial bearing user and designer in a search for greater efficiency and reliability in the face of ever mounting bearing duty, progeny of the classical hydrodynamic tradition, secure in the knowledge of their analytical lineage and the genealogy of their illustrious antecedents the experimentalists; the second, the new embryonic school of the surface contact investigators whose banner, born initially by the boundary lubrication workers, would eventually rally together a new generation of studies in the physics and chemistry of thin films and dry surfaces, contact mechanics (including friction and later wear processes), lubricant and material testing procedures, metallurgy and surface topography, together with a rapid development in surface techniques and lubricant and additive formulations.

1.19 The Concept of 'Oiliness'

By 1925, the schism might be said to be fully developed, though certain experimentalists had a 'foot in both camps'. Kingsbury, himself, as far back as 1903, had concluded that there was a friction reducing property in a lubricant under boundary friction which was separate and distinct from viscosity; he called this 'oiliness' and his was among the first attempts to describe how certain lubricants appeared to develop resistance to film rupture under these extreme conditions. The definition of 'oiliness' adopted by the Society of Automotive Engineers (S.A.E.) is indicative of the uncertainty of the phenomenon; for them oiliness was a term signifying, to quote: "differences in friction greater than can be accounted for on the basis of viscosity when comparing different lubricants under identical test conditions" i.e., that lubricants can possess boundary friction capability entirely independent of their viscosity. Water, for example, had appreciable viscosity, but practically no 'oiliness' in the concept of those times.

1.20 Hydrodynamic Limits - The Work of Stanton

Other workers, however, had mixed feelings about precisely when and where boundary lubrication conditions could be said to apply. A Lubricants and Lubrication Inquiry Committee had been set up in England by the D.I.S.R., in 1920; the general concensus was that, in gear teeth at least, the high localized stresses precluded hydrodynamic action and that boundary lubrication would pertain. One member of the Committee, Dr. T. E. Stanton was prepared to swim against the tide of enthusiasm of the 'molecular chain school' and realizing the impracticability of measuring pressures between rotating gear teeth simulated such conditions by employing a conformal bearing with large radial clearance[47]. Using a pressure gauge, (like his famous predecessor, Beauchamp Tower), he first measured pressures around a small load-carrying arc, recording peak pressures of some 3½ ton/in^2 and obtained the theoretical pressure profile for the contact configuration from hydrodynamic theory. The degree of correlation was outstanding and did much to reinforce confidence that hydrodynamic action could still be manifest under such severe conditions. This early experiment undoubtedly provided inspiration for the subsequent work on elastohydrodynamic action in gear teeth that was to commence immediately after the Second World War. Ironically, Stanton himself believed that boundary lubrication applied to piston rings after carrying out experiments in 1925; his sliding velocities were, however, much slower than later workers' who suggested that hydrodynamic films operated for most of the stroke, except at the extremities of travel, where boundary films were manifest[48].

1.21 Outcome

It is not proposed within the confines of this discussion, to continue looking at the chronology of tribological progression further, since it has been illustrated how early liaison between the disciplines developed towards the end of the nineteenth century and how the Great Schism evolved towards the first quarter of the twentieth, which tended to dissolve it. Clearly, factors such as the rise to preeminence of the internal combustion engine brought a certain further separatism in compartmented lubricant speicifications and testing procedures, together with a continuance of the two separate viscosity concepts, the efflux time specification, principally employed by the oil industry, and the absolute or dynamic viscosity concepts used exclusively by hydrodynamicists and bearing designers.

1.22 Factors Assisting Early Liaison

What is important, however, is to distill, on the basis of the chronology that has been discussed, the factors which up to 1925, either singly or in combination, resulted in the close liaison between workers in the field, since some of these early formative movements towards a multidisciplinary approach might gainfully be considered in the context of the present day problems of compartmented specialization. These factors may be roughly generalized as follows:

    (1) The realization by experimentalists, theoreticians, industry, learned societies and government laboratories that common lubrication problems existed with the development of steam power and the

willingness of each section to cooperate with the other in attempts at common solutions. This might be termed the 'Problem Recognition and Cooperative Factors'.

(2) The formation by the learned societies of special committees and groups, both to foster cooperation between a cross-section of workers in the relevant areas by serving jointly on various panels and to disseminate information by meetings and the publication of papers. In certain cases, as already seen, such groups were prepared to support relevant research by direct funding. This might be termed the 'Support and Dissemination Factor'.

(3) The emphasis on the interrelationship between theory and practice as exemplified by the dual roles undertaken by academician/engineers such as Hirn, Petrov, Thurston, Reynolds, Stribeck, Goodman and Michell, to name a few. This flexibility of the individual, both to be prepared to assimilate practical experience from personal contact with experimental apparatus and, at the time, to apply high-level theoretical knowledge to the solutions of practical engineering problems, would appear to be a hallmark of this period when technical innovation and analysis flowered in the tribological field as never before. This might be termed then, the 'Individual Duality Factor".

(4) The establishment of centers for the training and education of specialists in disciplines specifically relevant to the problems arising with the coming of the new technology and the orientation of such studies to the practicalities of solutions in the real world of engineering and industry. Such a step represented a partial break with tradition and, for the time being at least, the destruction of the academic 'White Tower' syndrome, which had bedevilled certain educational institutions. This may be called "The Orientated Training Factor".

2. THE INTERLINKING OF THE DISCIPLINES

Following the end of the Second World War, a rapid interlinking between the disciplines began to take place in the field of Tribology. This is attributable to five primary causes, three technical, one economic, and one conservational. Chronologically, these may be categorized as follows:

(1) The rapid post-war development of chemical additives for lubricants and the increased application of synthetic fluids, solid lubricants, and plastic materials.

(2) The successful development of the elastohydrodynamic theory for counter-conformal contacts and its experimental vindication.

(3) The increasing sophistication of techniques for surface examination and analysis, parametric measurement and data recording and processing.

(4) The economic implications of an overall multi-disciplinary approach to tribology following the findings of the Jost Report in 1966.

(5) The current emphasis on material and energy resource conservation arising as an aftermath of the world oil crisis.

These factors will be reviewed in order to see their influence on the interaction of the disciplines.

2.1 Additives, Synthetic and Solid Lubricants and Plastics

As previously noted, boundary lubrication studies began in earnest in the early 1920's and by the mid-thirties large quantities of vegetable oil and animal fats were being used by the petroleum industry. Such compounds are rich in fatty acids with their active polar group, a prime prerequisite in forming the metallic soap films on sliding surfaces with dramatically reduced friction coefficients. Since, for the most part, the constituent compounds of mineral oil are nonpolar, they make poor boundary lubricants. Thus, the inclusion of fatty acids in lubricating oil and greases, gave them an added lubricating dimension.

Heavy sliding conditions induce high surface temperatures and when these rise above about $150^\circ C$ metallic soap films decompose. This problem arose with the advent of the so-called 'Extreme Pressure' (E.P.) lubricant, basically compounds of sulphur, phosphorous, and chlorine. Strictly, these should be termed 'Extreme Temperature' (E.T.) lubricants, since they react chemically with the metal surfaces only at higher temperatures to produce low coefficient layers by a process of controlled corrosion. A wide range of other ingredients was also in use by 1937 in this type of application, among these being lead soaps.

Prior to the war then, attention had been focused on reducing friction and wear by additives. After the war this was turned to improvements in the performance of the lubricant itself.

The rapid change of the viscosity of mineral oils with temperature was occupying the motor industry at this time. 5W, 10W and 20W oils were initiated around 1950 to assist starting in sub-zero temperatures, but this led to high consumption in normal running. By 1952, the first multi-grade oils were being introduced; high molecular weight polymetric materials such as polymethacrylate and polyisobutylene being added to the lubricant base stock to inhibit the natural drop in viscosity with increasing temperature. In addition to these V.I. (Viscosity Index) improvers, came a wide range of other additives such as antioxidants, corrosion inhibitors, detergent/dispersant additives, pour point depressants and anti-foam additives, these being developed and extensively employed for use in both oils and greases. A wide range of literature is available on the subject of additives, excellent reviews being presented in References 49 and 50.

The invention of the gas turbine towards the end of the war, resulted in an intense period of activity in the field of synthetic lubricants in its aftermath. Zorn, in 1939, had experimented with castor oil, but storage deterioration presented problems [51].

The first successful materials were the organic diesters, reaction products of organic acids and alcohols, a promising compound being di (2 ethyl-hexyl) sebacate, a derivative of sebacic acid and originally used as a rubber plasticizer. With the rise of engine power, bearing temperatures rose causing some problems in gear lubrication in the turbines. This was overcome by further research, complex esters (derivatives of polyethylene glycol, sebacic acid and alcohols) being used to thicken the diesters. Inevitably, however, with bulk lubricant temperatures in excess of 150°C, thermal breakdown at a position known as the 'β' hydrogen, (Figure 3), in this material took place and an entirely new material, trimethylopropane-tripelargonate (a derivative of pelargonic acid) was developed, known, because of the absence of the 'β' hydrogen in its structure, as a 'hindered ester', (Figure 3). The bulk capability of this material was in excess of 200°C. Ester type lubricants currently serve as the 'bread and butter' lubricant for the aviation industry.

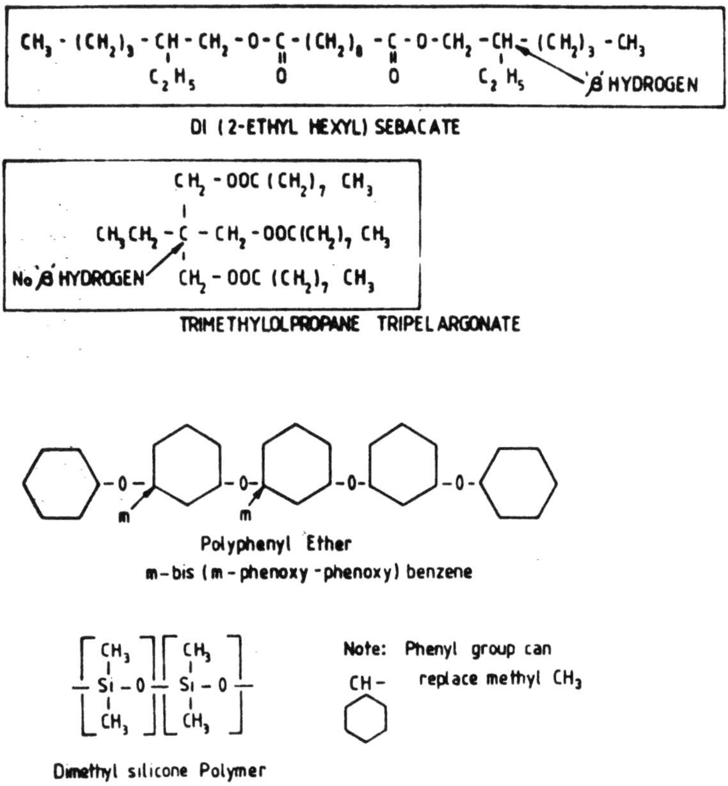
β HYDROGEN
DI (2-ETHYL HEXYL) SEBACATE
No β HYDROGEN
TRIMETHYLOLPROPANE TRIPELARGONATE
m
m
Polyphenyl Ether
m-bis (m-phenoxy-phenoxy) benzene
Note: Phenyl group can
replace methyl CH3
Dimethyl silicone Polymer

FIGURE 3 LUBRICANT CHEMISTRY EVOLUTION

A third generation arose with the coming of the supersonic 'Concorde' with bearings operating in the region of 260°C, together with gear lubrication problems at high temperatures. The economics of developing entirely new lubricants proved daunting and the pelargonates were 'stretched' by employing new additives to inhibit thermal degradation of the base fluid and the older additives themselves. A new concept in mechanical testing of the lubricants and additives was employed at this time in which test rigs simulated actual working conditions in the turbine or complete turbine components were, themselves, used as mechanical test rigs. This direct mechanical testing enabled a wide range of lubricants and additives to be examined directly on a 'trial and error' basis which represented a breakaway from the traditional 'benchtop' measurements of lubricant properties for specification purposes and enhanced the importance of mechanical testing in overall lubricant performance evaluation.

Military requirements, principally for supersonic fighter aircraft, pushed thermal demands on fluids to the extremes of the temperature spectrum in the late 1960's and new materials were produced, principally polymers of aromatic phenols such as polyphenyl ether, (Figure 3), or aliphatic polymers containing inorganic silica in place of carbon, (Figure 3). The latter had some of the best V.I.s available but were poor boundary lubricants especially on ferrous surfaces and did not readily accept additives. The former, while not having this drawback, were exceedingly viscous in cold conditions and had to be 'diluted' with ester type oils. Both, however, had very good high temperature capability; in the region of 300 - 370°C.

Other synthetic fluids were also developed for applications in specific lubrication situations, such as silicate esters for low temperature lubrication and corrosion prevention (compounded with other synthetics), or as base stock for high temperature hydraulic fluids. Phosphate esters were used as fire resistant hydraulic fluids in mining, aircraft, and ships. Polyglycol ethers were employed in cutting and machining operations and in certain boundary lubrication conditions or, in its water soluble form, for brake fluid and specialized hydraulic oils. Halogenated compounds (for use in high flammability process situations) and Silanes (for high temperature hydraulic oil and grease base stock) began to find use. Thus the work of the organic and inorganic chemist, combined with the study of surface contact mechanics and mechanical testing, followed from the evolution of additives and synthetic fluids.

Rheology was added to the spectrum in the study of the behavior of grease, particularly in the combination of synthetic oils with clay and other bases to produce high temperature greases of the 'Bentonite' type.

Solid lubricants, in the form of graphite and molybdenum disulphide, were employed in tribological applications during the nineteenth century, but research on these and other materials escalated in the period 1950-1965, particularly in the U.S.A., with the challenge of the space age (environments of extreme temperatures in rocket engines and hard vacuum in space.) A wide range of loose powders, metals, oxides, molybdates, and tungstates were investigated in addition to layer lattice salts. Mixtures of graphite with soft oxides and salts in a variety of environments were researched, as were coatings of oxides of lead and silica in 'duplex' structure ceramics and ceramic-bonded calcium fluoride.

Thus, disciplines of powder metallurgy, inorganic chemistry, surface failure mechanics, friction materials science, ceramic composite material science, and surface physics were combining with surface examination and analysis techniques to spread the multidisciplinary umbrella. A comprehensive treatment of non-conventional lubricants is given in reference 52.

Nonmetallic bearing materials were already being discussed in 1937 with regard to rubber and graphite, but the largest application by far, has been in the use of polymetric materials[53]. These divide into two classes of plastic; thermoplastic and thermosetting. In the former category, nylon found initial use but in terms of minimum friction no commercial plastic surpasses polytetrafluoroethylene (P.T.F.E.), formulated early in the 1950's, followed by polyamides (1959) and polyacetal, polypropylene and polyethylene in the later years.

Thermosetting materials such as phenolic resins were introduced in combination with fabrics for so-called synthetic resin bearings and were used in German rolling mills as early as 1928, water lubricated bearings of this type being also used in similar applications in the U.S.A. in the 1930's[54].

The importance of plastic bearings is their ability to operate without a lubricant and in the food, textile, and processing industries they are extensively used for this reason. In the domestic market, they find application in lightly loaded mechanisms and their general inertness makes them suitable in environments where metallic corrosion is a problem and in medical applications. Additionally, solid or synthetic lubricants may be incorporated in the plastic to enhance its tribological performance. P.T.F.E. is also finding increasing usage, itself, as a 'filler' in powdered metal bearings, in addition to the normal fluid and solid lubricants used in these units. Thus plastics and powder metallurgy must rightly be included in our multidisciplinary spectrum.

2.2 Elasto-Hydrodynamics

It might be justifiably said that no single area of twentieth-century tribological study has yielded such profound success, both in the theoretical attack on the problem and in the vindication of the theory by experimental results, than in the area of elasto-hydrodynamic lubrication (E.H.L.).

In certain respects, it recalls the heyday of Petrov, Beauchamp Tower, Reynolds, and Sommerfeld in welding together theory and practice into a unified whole. It differs in one important respect, however, in that in the last quarter of the twentieth century, experimentalism had begotten theory, while some fifty years after the converse was to hold; elasto-hydrodynamic theory spawning the experiments. Two things, however, were certain; first, that through all the vicissitudes of the 'counter-conformal conflict' classical hydromechanics had emerged triumphant, and second, as a corollary to this, the 'Great Schism' of twenty-five years earlier had, at last, been healed. No doubt the chairman of the 'Committee on Tribology', Peter Jost, had such thoughts in mind when in a jocular misquote of Karl Marx, he addressed a Tribology Conference with the appeal: "Tribologists of the world unite! You have nothing to lose but your molecular chains![55]"

How then had such a transition from the days of Martin's theory come about? In 1936 and 1938, Peppler had considered elastic deformation effects between gear teeth, but concluded, erroneously, that fluid pressure could never exceed the required Hertzian contact pressure[56,57]. Gatcombe in 1945, analysed pressure viscosity effects in the oil film but although this increased the thickness predicted by Martin, it did not satisfy the criterion of a fluid film above the height of asperity roughness[58]. Hersey and Lowdenslager in 1950 produced similar results to this[59]. Blok in 1952 indicated that, for an exponential pressure viscosity relationship, a limiting film thickness giving infinite fluid pressure arose if the surfaces were considered iso-elastic; for this condition the predicted load was some two to three times Martin's load; while earlier, in 1950, Poritsky had suggested a Hertzian type pressure distribution with a parallel film, modified by deformations at inlet and exit to permit finite gradients of fluid pressure, (Figure 1)[60,61].

Prior to most of this work, however, almost at the end of the war, a Russian, Ertel, produced a rationale of the problem in two intuitive steps of masterly insight. First, that with an exponential law for pressure/viscosity effects in the two dimensional Reynold's equation, a practically parallel film results, i.e., the pressure in the Hertzian zone is the same whether a lubricant is there or not and second, outside this zone, the fluid pressure in not high enough to change the shape of the conjunction from that of the dry contact. Using these concepts together with the Hertzian equations for the contact geometry, Grubin in 1949, produced an analysis, primarily of the inlet conditions, and by fitting an algebraic expression to a family of iterated integral curves, obtained a single formula for the mean dimensionless film thickness in the contact zone;[46]

$$H_o = \frac{1.95 \, (GU)^{8/11}}{W^{1/11}}$$

It is seen that essentially the film thickness is hardly affected by load, the fluid behaving almost like a solid under the high pressure. This equation at once predicted film thicknesses which were orders of magnitude greater than Martin's theory and established the probability of full fluid films in the contact area.

In a qualitative discussion of the pressure distribution, Grubin postulated the existence of a 'pressure spike' arising from a sudden convergence of the surfaces in the exit region, (Figure 1).

Characteristics of the elasto-hydrodynamic contact predicted by Grubin were confirmed by Petruesevich in 1951 for a limited range of numerical solutions, but, specifically, his solutions produced both the contraction in film thickness at exit and the pressure spike, sometimes referred to as the 'Grubin/Petruesevich' spike [62].

An extended numerical solution of the problem for a wide range of operating conditions was obtained by Dowson and Higginson in 1959, using iteration[63]. A detailed account of this work, together with a comprehensive review of the chronology of elasto-hydrodynamic lubrication may be found in the book by these authors, Reference 64.

For minimim film thickness calculation i.e., at the exit contraction, the Dowson-Higginson formula is given by:

$$H_m = \frac{h_m}{R} = \frac{2.65\, U^{0.7} G^{0.54}}{W^{0.13}}$$

For frictional and other calculations involving the major area of contact, the Grubin formula is applicable. A comprehensive study of isothermal solutions of the problem was given by Dowson and Whitaker in 1964[65].

Thermal effects were considered by Sternlicht, Lewis, and Flyn (1961) and Cheng and Sternlicht (1964)[66,67]. Rheological effects were studied by Bell in 1961, Crook (1963), and Dyson (1970). Christensen had considered surface roughness effects (1962) in normal approach contacts while Dawson (1961) studied disc/pitting[68-72].

A series of elegant papers had been published by Crook (1958-1963) in which he measured film thicknesses in the conjunction by electrical capacitance, a similar technique being used by Dyson (1966) with excellent correlation being obtained[73,74]. Similar work was done by Sibley (1960) using X-ray collimation techniques[75]. Optical methods were used for this purpose by Kirk (1962) and Cameron who, using interferometry techniques, studied a large range of contact situations[76,77]. This method is given in some detail by Ford et. al. in Reference 78.

One of the most dramatic measurements was the pressure profile in the conjunction by Kannel (1964), using Maganin' (a material whose resistance varies with pressure) being deposited as a thin strip on one contact surface[79]. Cheng and Orcut (1966) and Hamilton and Moore (1971) extended this work with certain pressure spikes being obtained[80,81]. Temperature measurements have been made by many of the above workers using trailing or buried thermocouples and strip transducers. Infrared techniques have been used by Hamilton and Moore and Prof. Winer and his colleagues in recent years[82].

Thus, a wide range of disciplines: Hydrodynamics, Elasticity, Fatigue, Thermodynamics and Heat Transfer, Fluid Physics, Rheology, Surface Topography, Electronics, Transducers and Instrumentation, Heat Radiation, Optics and Surface Contact Mechanics had been collectively interwoven through the study of elasto-hydrodynamics in a concerted multidisciplinary approach.

2.3 Surface Studies, Measurement, Monitoring and Processing

The sheer scale and scope of the explosion of methods and devices for surface examination and analysis, parametric measurements, data recording and processing, coupled with the advent of digital and analogue computers in the fields of theoretical studies, data storage and process control, places any extensive treatment in depth beyond the bounds of the present discourse. A broad outline only, of this realm of multi-disciplinary interaction, within our Tribological spectrum, must therfore, necessarily suffice.

In essence, this section may, quite arbitrarily, be broken into the following groups, bearing in mind their tribological relevance:

(1) Friction and Wear, Surface Examination, and Measurement
(2) Mechanics and Properties of Materials and Lubricants
(3) Measuring Techniques and Parametric Measuring Devices
(4) Mechanical Testing, Bench Testing, and Field Evaluation
(5) Condition Monitoring Applications.

It is emphasized that this is entirely subjective, since there is obviously no lack of overlap between the areas chosen.

2.3.1 Friction, Wear and Surface Examination Analysis

From a cursory examination of the subjects of friction and wear, it might appear paradoxical that while these topics are clearly associated, studies on the former began as far back as 1699, with the historical work of Guillaume Amontons, while studies on the latter topic have only really accelerated in the period following the Second World War[83]. Two factors which may help to explain this dichotomy are as follows. First, where power loss in tribological situations could be mitigated, this would, in general, be opted for. Mechanical integrity (or the lack of it) manifests itself much more dramatically than wear integrity and since frictional power loss in machines was of prime importance (especially when input power was limited) studies would be necessarily directed to overcoming the all too obvious outcomes of these problems. Second, a greater appreciation has been manifest, both during and after the war, of the importance of long-term reliability and life of machine components and therefore attention has been directed to the importance of surface studies.

To further as well as implement these studies, a range of new techniques and equipment has evolved which broadens, even further, the scope of multidisciplinary tribology.

During the thirties, the stylus profilometry methods were invented which, after the war, became linked to the digital computer enabling three dimensional topographical models of surfaces to be produced; an example being the work of Edmonds described in Reference 84.

Optical methods of surface examination had, of course, been used by metallurgists for many years, but, with the marketing of the 'Quantimet' image analyzer in 1963, surface contours could be examined and analyzed in still greater detail. Optical interferometry itself had also been used to aid optical surface examination but by far the greatest advance in surface examination and particle morpholgy came with the invention of the electron microscope (Scott and Scott 1957) and the scanning electron microscope (Salt 1970). A further breakthrough came with the electron probe analyzer (Cay and Quinn 1972) in which a collimated electron beam causes the excited surface atoms to produce characteristic X-rays, which may be detected by a spectrometer giving, essentially, the chemical constituents of the material. For surface examination, low energy electron diffraction (L.E.E.D.) and Auger analysis (Cheng 1971 and Buckley and Pepper 1972) have been used. Wear studies by Jones (1976) illustrate the utility of the method[85].

The generation of hydrogen peroxide (H_2O_2) at newly machined surfaces was explained by Kramer (1950) on the basis of the so-called exo-electron emission from the oxide films forming on the surfaces and this phenomenon has been used by March and Rabinowicz (1976) for incipient fatigue investigations using a rolling four-ball machine[86, 87].

Electrical conductance and contact resistivity techniques have been used by numerous workers for the investigation of surface asperity contact mechanics and the influence of surface oxide and other gaseous and solid contaminants on surface interaction[88, 89]. Such work has, also, much importance in the field of electrical engineering, such as switch and brush contact problems, in addition to mechanical engineering applications. Since friction and wear are always manifest when surfaces physically interact in rubbing contact (welding and tearing being considered to be incurred at the asperity peaks) a great deal of attention has been directed, both experimentally and analytically, to the mechanism of such interactions. These subjects have included surface metallurgy, surface physics, surface topography, surface chemistry and chemical kinetics, surface and asperity heat transfer, contact mechanics, material elasticity and plasticity, gas chromotography and chemical spectroscopy, in addition to a wide variety of surface treatments, metallurgical, mechanical and chemical. Clearly, in this work, the emergence of new techniques and equipment for surface examination will, as already indicated by precedent, have a vital role to play in future tribological development, among these being the likely evolution of laser and holographic techniques both in static and dynamic situations.

2.3.2 Mechanics and Properties of Materials and Lubricants

With the inclusion of stress and elasticity into tribological analysis, the importance of techniques to measure the behavior of materials subject to plastic and elastic deformation in a range of relatively severe environments exemplified by temperature, pressure, high overload, transient shock, vibration, and general excess duty have become increasingly important.

In the field of direct and indirect strain measurement alone, standard techniques such as strain gauges, mechanical and optical methods, photoelasticity, photostress, and brittle lacquers have been reinforced with the developement of solid-state equipment. Metrology, itself has advanced with increasing use of electronic comparators, air gauges and a variety of displacement sensors. Lasers are among the new tools finding increasing use in this area, while X-ray/crystallography techniques have been employed in the study of crystal lattice movement, interfacial slip and residual stress measurements[90].

Fluid physics have been used in the study of lubricant behavior under high pressure, particularly from the viscometric standpoint, while analytical work has been carried out on the rheology of liquid and semifluid lubricants with reference to their visco-plastic behavior and relaxation time[91]. Studies of lubrication in semi-turbulent and full turbulence regimes have been extensive while the effect of extremely high shear rates (of the order of 10^6 sec^{-1} and above) has received attention as the bearing peripheral surface speeds have increased[92, 93].

With the extended use of hydraulic equipment, the behavior of fluids under cavitation conditions is a subject of considerable study together with the propensity of lubricants to entrain and release gases[94]. Lubricant and additive behavior at very fine levels of filtration (<0.5 microns) has received attention, as have situations in face where ultra-flat surface (<0.1 micron) waviness has been achieved[95].

2.3.3 Measuring Techniques and Parametric Measuring Devices

Probably no other field of technology has had such an impact on tribological activity as the revolution that has taken place in the field of electronics during the post-war years. For purposes of exposition, the subject may be divided into the area of data measurement and the area of data recording and processing. In the former, the development and increasing sophistication of instrumentation transducers for fluid pressure and film thickness measurement, micro thermocouples and thermistors, load cells, linear and angular displacement transducers, accelerometers, flow measurement devices and particle counters has enabled the experimental tribologist to quantify, with a hitherto unknown accuracy and diversification, the manifest interactions of tribological phenomena over a multifarious spectrum of activity.

In certain areas, micro-miniaturism had added a new dimension to what was previously considered possible, a particular example being the foil strain gauge, enabling these devices to be incorporated into 'custom built' transducer units for particular applications. The author and his colleagues have been active in this field for several years, a particular outcome of this work being the discovery of the phenomenon of fluid 'tensile stress' in journal bearings[96]. Thus, invention and the development of new manufacturing techniques have ever increasing ramifications in the progression of tribological discovery.

Likewise, in the field of data recording and processing, the invention of the transistor and a variety of solid state devices opened up an entirely new domain of data recording and processing almost overnight, particularly in the field of signal amplification. Operational amplifiers, initially used in analogue computers, have been employed in conjunction with tribological transducers for signal conditioning, giving compact and highly stable systems[97,98]. Transient signal recording was enhanced with storage oscilloscopes and other devices such as ultraviolet recorders and digital storage recorders, high speed chart and tape recorders and with the coming of the microprocessor, automated testing data storage and processing, coupled with robotic control, opened up an entirely new field of tribological possibilities.

In the field of theoretical tribology the coming of the digital computer has enabled problems, otherwise intractable to classical analysis, to be circumvented or refined and finite difference techniques, which lend themselves admirably to computer processes, have been extensively exploited[99]. The so-called 'mini' and micro computer adds further possibilities of diversification. The author has found computer graphics to be an exceedingly useful tool in examining pressure envelopes in hydrodynamic journal bearings, where groove configuration effects may be directly portrayed, Figure 4.

2.3.4 Mechanical Testing, Bench Testing and Field Evaluation

This area is so vast that nothing but a cursory glance can be made within the confines of the present exposition. Bench testing, and to a lesser extent, mechanical and other rig testing, grew from the need to standardize lubricants and materials between organizations at both national and international levels. The oil companies and bearing manufacturers themselves, from the basic necessity of product uniformity, initiated standardized procedures; the former with, primarily bench tests, the latter (particularly the manufacturers of rolling contact bearings) with product batch-testing in mechanical test rigs. A great deal of tribological information of a multidisciplinary nature arose from this source, while the formation of test bodies such as the American Society of Automotive Manufacturers (A.S.A.M.) and the Institute of Petroleum (I.P.) helped to disseminate this information and standardize test procedures. Government institutions and the armed forces in various countries, also laid down close specifications and quality control for bearings, gears, materials, and lubricants.

Notwithstanding the value of laboratory testing, it was quickly realized that simulation is no substitute for actuality, and a growing trend became manifest, particularly after the Second World War, for either testing products in a true working environment (as has been outlined in the mechanical rig-testing of the 'Concorde' lubricants) or through cooperation between a product manufacturer and a product user in extensive 'field' trials. This had both the advantage of operations in a real environment under a wide spectrum of working conditions and of providing information, on a statistical basis, of product performance. Through such organized data acquisition, information is compiled for both diagnostic and prognostic purposes, enabling, in many cases, early remedial action to be initiated.

2.3.5 Condition Monitoring Applications

In situations where high mechanical and wear integrity are demanded, or plant economics mitigate against complete shutdown or service withdrawal, condition monitoring may be

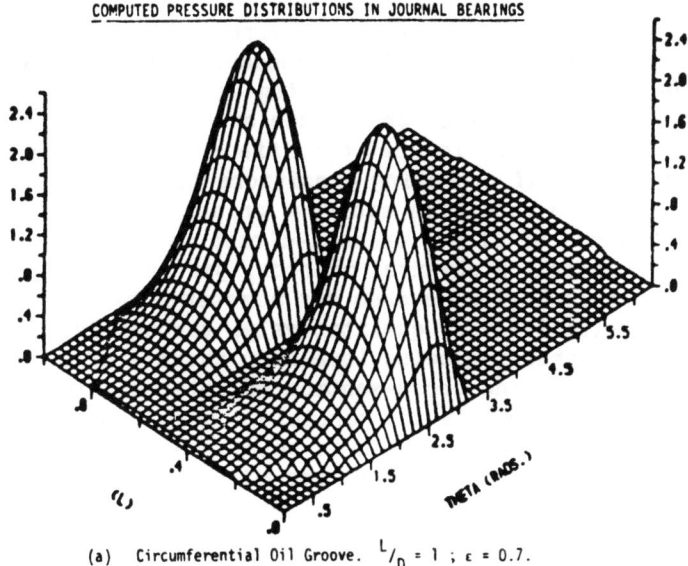

(a) Circumferential Oil Groove. $L/D = 1$; $\varepsilon = 0.7$.

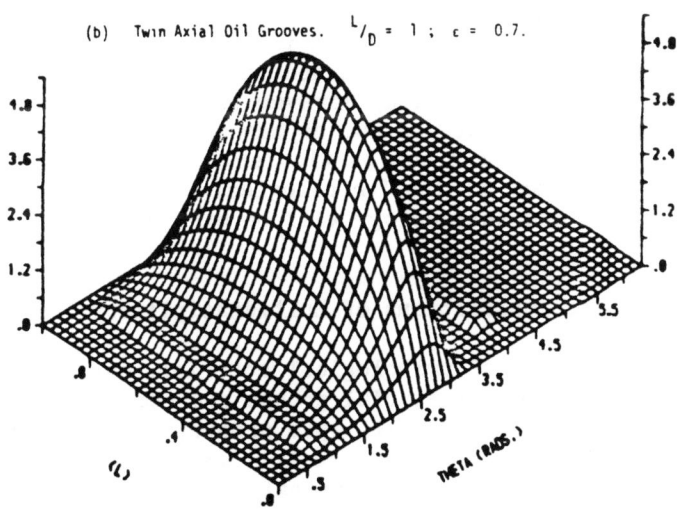

(b) Twin Axial Oil Grooves. $L/D = 1$; $\varepsilon = 0.7$.

FIGURE 4   COMPUTED PRESSURE DISTRIBUTIONS IN JOURNAL BEARINGS

employed in which 'in-situ' information is extracted from the machine, or its components, on a temporal 'in-service' basis. In the tribological realm, this procedure can be of extreme importance, particularly in assessing wear integrity situations. A variety of techniques exist to establish if wear is taking place; in what specific areas (if this is established) and if operating conditions fluctuate, in what situation is the condition manifest or accentuated. Apart from techniques of direct or semi-direct observation such as optical, chemical or physical methods, performance changes within the tribological system can sometimes be assessed, either quantitatively or qualitatively, through increased power losses, changing noise levels or thermal effects. This alone involves a wide cross-section of inter-disciplinary techniques. Wear debris and particle contamination is receiving particular attention in this sphere and it would appear that automated equipment and micro-processor control are areas where future development may well be extended as the overall economics of machine reliability and efficiency are more fully appreciated.

This third section on the interlinking of the disciplines therefore, has indicated the vast extension of tribological technology and activity under the multidisciplinary umbrella particularly in the post-war years, and the impetus that advances in technology and invention have initiated.

The exponential growth curve mentioned in the opening section of the discussion is thus seen to be no misnomer in describing the rate of change which is taking place. Clearly, the implications are that some methodology with regard to categorization must be applied, however loosely, if the modern day tribologist is not to end up, as has already been intimated, in tending his own particular 'tree' to the exclusion of the rest of the 'tribological wood'. This characterization and methodology will be examined later in the discussion.

2.4 The Economic Implications of Multidisciplinary Tribology

In the chronolgy of tribological coteries, certain meetings or formations of special groups are remembered for their significance. For overall epic proportions the great 1937 'Discussion on Lubrication and Lubricants' organized in London by the Institution of Mechanical Engineers must rank as an historical landmark. Some 600 delegates registered, 136 papers were given, and an exhibition of lubricants, bearings, materials, test machines, and research equipment, held at the Science Museum, was visited by some 18,300 persons in two weeks.

In post-war years, this impetus was continued by the 1957 'Conference on Lubrication and Wear' organized again by the Institution with over 1,000 delegates from 21 different countries participating; 104 papers being given. In a prolific decade of meetings from 1963 to 1972, the Institution promoted a vast spread of tribological papers, engendering a growing tide of multidisciplinary participation in research and development work.

The economic aspect, particularly in maintenance and replacement cost reduction and savings commensurate with increased life of machinery, was beginning to make itself apparent generally and studies in this area in particular were already underway in certain of the Eastern-Bloc countries, notably East Germany.

In 1964, the British Minister of Education and Science, Lord Bowden of Chesterfield, requested H. Peter Jost to '... consult with persons and bodies to establish the present position of lubrication, education and research in this country, and to give an opinion on industry's needs thereof', terms of reference whose parlance echo, with a certain faint nostalgia, the palmy days of 'The Mechanicals' and Beauchamp Tower!

On this basis a Lubrication Engineering Working Group of fourteen members under the chairmanship of Peter Jost was inaugurated, holding ten meetings and eight investigatory sessions to hear expert evidence.

Views were sought from wide and diverse sources in the United Kingdom and abroad, some 400 colleges and universities, industrial companies, corporated, and government research establishments and laboratories, government organizations and private individuals being consulted. From this mass of information, the Group formed the opinion that, at a conservative estimate, a potential saving to British Industry of some £515 million pounds per annum could be achieved arising from improvements in education and research at all levels, from the industrial shop floor to the university post-graduate researcher.

Early assessment of the entire gamut of the investigation led to the irrevocable conclusion that the word 'lubrication' held such subjective connotations that an entirely new generic term was required. This was not merely an exercise in semantics, but a genuine effort to produce a single work encompassing the entire range of the interlinking multidisciplinary fields of activity. With commendable scholastic zeal the English Dictionary Department of the

Oxford University Press was approached and, with their help, a single word 'Tribology' was chosen. The word, derived from the Greek 'tribos' (rubbing), was considered by some, at the time, to be the converse of what good lubrication practice was trying to acheive. Nevertheless, it was after due discussion, adopted - not only by the Group, but, with a dignified and wholly appropriate passage of time, by the concensus of industrial opinion. Its definition was as follows:

> 'Tribology is the Science and Technology of interacting surfaces in relative motion and the practices related thereto.'

The Report of the Working Group entitled 'Lubrication (Tribology) Education and Research - A Report on the Present Position and Industries' Needs (H.M.S.O. 1966), provided within its innocuous blue covers, the seeds of an economic bombshell.

The Jost Report, as it subsequently came to be called, made certain recommendations. In essence, these were as follows:

Educational

General and specialized courses at all levels from the shop floor upwards should be initiated, including the enlargment of undergraduate and postgraduate training; this would include specialized M.Sc. courses.

Research

Institutes specializing in Tribology should be established for basic and applied research, postgraduate instruction and industrial liaison. A two-way bridge with industry should be formed; for example through contract research and industrial research training.

General

An information center in Tribology should be established and a design and practice handbook published. The Institution of Mechanical Engineers was invited to extend the membership and activities of its 'Lubrication and Wear Group'.

In the light of the economic implications of the Jost Report, it is not surprising that, unlike so many good working committee reports before, it did not simply acquire dust on some ministerial shelf. Governments may be accused of lethargy in matters concerning the overall public 'good' but when pecuniary considerations of significant magnitude arise their 'response time' is little short of miraculous. So it proved to be, but, leaving aside the overtones of a £515 million carrot, the Working Group were to be congratulated on framing their report in so cogent a manner that any 'stick-waving' amounted to little more than a gesture to ritual. By 1972, most of the recommendations of the 1966 Report had been completely implemented.

The rest of Europe was not slow in following this initial impetus. In 1973, an International Tribology Council was formed with Peter Jost as its first President and by 1974 some twenty-one countries were represented on the Council. Thus Tribology flourished throughout Europe, its multidisciplinary nature pervading all quarters and at all levels in a spreading and interliking network. The long-term effects of this cannot, even at this time of writing, be fully assessed but the economic implications are manifestly obvious. Temporarily this genesis could not have been more opportune in the light of the world energy crisis that was, so swiftly, to follow.

2.5 The World Oil Crisis and the Conservation of Energy

Although to the casual observer the face of the petroleum oil industry may appear outwardly sanguine, the passage of time has seen violent changes reflected in its complexion, Simpson, Reference 100.

Perhaps the very nature of its parturition and dynamic evolution has given it a natural predilection for drama on the grand scale, but no single event since the war years has so thoroughly shattered the economic quiescence of the West than the onset of the world oil crisis. Almost overnight, it seemed the attitude of nations to the energy hoards stored in their economic coffers changed from that of inebriate spend-thrifts to that of Draconian misers.

In the halcyon days before the 'OPEC' spectre sat at the feast, concepts such as 'Energy Conservation', 'Total Energy', and 'Total Technology' bore little or no credence. Suddenly, however, such strange terms became vibrant with meaning as this newly acquired truth penetrated into the entrenched bastions of governmental and industrial hierarchy, those august bodies establishing, with a hitherto uncharacteristic swiftness, that, at least as far as their fossil energy reserves were concerned, they were living on 'borrowed time'.

Yet this, in fact, was no newly created credo, no nascent economic theory of energy relativity. The semantic trappings might be different, the body of the argument carried, for certain tribologists at least, a decidedly familiar ring. Petrov, in the long paper of 1883 on 'Friction in Machines and the Effect of the Lubricant' produced formulae for estimating power loss and showed how a proper selection of lubricants dramatically reduced wastage of energy. He was awarded the Lomonosoff Prize of the Imperial Russian Academy of Sciences for his pains, and there the matter rested. Robert H. Thurston, impressed by the need for greater economy in the use of power, published in 1885 an extended edition of an earlier book entitled 'Friction and Lost Work in Machinery and Millwork', which went largely unnoticed by his fellow engineers in America. His, if anyone's, was a lone voice crying in the wilderness of these exuberant years of early oil exploitation.

Mayo Hershey, the doyen of American lubrication during the years of its meteoric expansion, had written in papers of 1933 and 1949 on the importance of the study of power loss in any engineering education[101, 102]. Through his now classical paper of 1936 'The Oil-Shed Fallacy, Attacking the Problems of Lubrication by Rational Methods', his inner feelings on the matter became abundantly clear[103]. Though the presentation is in a jocular vein and makes both humorous and delightful reading, the underlying current of concern and sincerity cannot pass unnoticed.

If any sort of conviction was lacking, however, a paper given by a European authority, Dr. Georg Vogelpohl, (1951) to the Third World Petroluem Congress at The Hague, must surely have shattered all illusions. In a study of world energy production and dissipation, Vogelpohl estimated that one third to one half of the world's energy production is consumed in friction. Even to the converted, this makes startling reading. The opening pages of Prof. Dudley Fuller's book: 'Theory and Practice of Lubrication for Engineers' contains further material on the energy question and is worthy of scrutiny[104].

In conclusion, it is evident that any future prognosis must wait for the outcome of current diagnosis. What is abundantly clear, however, is the role that multidisciplinary tribology must play if any sort of satisfactory outcome is to be acheived. To quote Mayo Hersey on the necessity for a broad base of scientific and technical education in the field of lubrication: "... it is a kind of pyramid, the attainable height depending on the extent of the base." It is the individual blocks that make the multidisciplinary base of our pyramid, however, and we must now turn to these in order to see how they interact and how then might be concisely assembled.

3. THE INTERACTION OF THE DISCIPLINES

3.1 The Surface Concept

We have, through our study of some of the chronology of Tribology and the several factors that, either acting singly or in combination, have influenced the interlinking of the disciplines, now reached a foundation point where it is possible to describe the interaction of the disciplines within specific boundaries. With Tribology reduced to an overall concept of proximate surface interaction, we may conveniently use this orthodoxy to illustrate the interaction of the disciplines themselves, at least as an initial starting point. Such illustration has already been used by Barwell and may, with certain modifications, be extended here to encompass a wider spectrum[105].

3.2 Function of a Lubricant

The author, in lectures to his students, has referred to the concept of 'Function of a Lubricant' which may, broadly, be defined by four categories.

(1) To eliminate or reduce wear between surfaces in relative motion.
(2) To reduce friction between such surfaces.
(3) To afford protection of the surfaces in hostile environments.
(4) To effect transfer of heat from the surfaces.

With regard to (1), we have already seen that we may eliminate wear entirely if surface asperity contact or particle contamination are precluded. Alas, with (2), however, we can never entirely eliminate friction, since every fluid film in viscous shear will dissipate energy.

Hostile environments may follow broad classes, examples being excessive load and speed, temperature extremes, chemical and abrasive environments, general contamination, radiation and space environments.

The importance of heat transfer is clerly self-evident; the lubricant and surfaces playing, in actuality, dual roles both being, in themselves, sources of generating heat (by viscous friction and/or asperity contact), while simultaneously acting as agents for its dissipation. If the former outweighs the latter, then a 'thermal spiral' is initiated and the surfaces seize.

3.3 The 'Surface Entity'

As a preliminary step let us postulate a hypothetical pair of surfaces and direct our thinking to the phenomenon of their interaction. Our hypothetical working surfaces may be characterized by the sequential order of their production and utilization e.g.,

(1) Production of Bulk Material
(2) Surface Heat Treatment (if any)
(3) Final Finishing Operations
(4) Application of Lubricant to Surface
(5) Bearing Working Environment and Duty.

Considering these (in the above) we will examine each as follows:

(1) The bulk material properties (mechanical strength, ductility, etc) will depend on the mechanical forming and heat treatment. Bulk intercrystalline stress will be manifest.
(2) Surface heat treatment may produce different metallurgical structures with, possibly, complex residual stresses in the surface giving intercrystalline changes in the component's outer skin.
(3) Final finishing operations dramatically affect tribological characteristics. Essentially, both the overall surface profile and surface topography will be determined by the methods employed at this stage. Additionally, a distorted skin microstructure will result with further lattice crystalline stress and, possibly, work hardening. Chemical amalgams also form, particularly with additive-loaded cooling fluids, in the elevated temperature environment between the metal removal agent and the component. Oxide films will quickly arise on the freshly cut area, permeating into the microstructure, together with contaminants such as water vapor, gases, and other chemicals all adding to the original amalgam. This type of surface can, in fact, react electronically itself, giving the possibility of inciting the lubricant's oxidation at the common interface by catalytic action. Thus, the component's surface skin will have entirely differnt frictional, hardness and strength properties from that of the original 'parent' metal before a lubricant is even used.
(4) Application of the lubricant to the surfaces immediately initiates further chemical reactions. Surface adsorbed layers form from reactions with

additives, such as metal passifiers and corrosion inhibitors, while fatty acids produce long-chain polar compounded layers over the whole surface, giving further chemical complexity.

(5) Bearing working environment clearly modifies further our tribological picture. At higher temperature the initial surface chemistry is modified. E.P. additives, if present, will generate a controlled corrosion; asperity contacts modify overall surface profile and topography and, in their act of contact, generate heat locally with concomitant mechanical and chemical changes to the working surfaces. High loads produce elastic distortion and internal stresses which may, as in the case of pitting, adversely affect the whole mechanical integrity of the surface itself. High speeds increase viscous shear stress, reduce additive stability and produce further heat which both lowers the lubricant viscosity and causes thermal distortions. Debris, particles, chemical and gas entrainment mechanically and chemically affect the working surfaces and inhibit the action of the lubricants and additives. Embedded particles produce friction and wear of bearing surfaces, while cavitation may initiate surface damage in babbit or other soft materials.

Through this labyrinth of acting and reacting phenomena, therefore, we may now begin directly to establish the interaction of the disciplines themselves in relation to their applicability to the various phenomena. Figure 5 illustrates a schematic of our hypothetical surfaces for a hydrodynamic or elasto-hydrodynamic condition and gives immediately the required correlation of the disciplines in one entity. This we will designate, for further reference, 'The Surface Entity'.

4. MULTIDISCIPLINARY THINKING

4.1 The Importance of a 'Rational'

In the light of the phenomenist approach we have employed in the schematic of Figure 5 to illustrate the interaction of the disciplines through the interaction of the phenomena, it would clearly be of value to the tribologist anxious to have an eye on the 'tribological wood', if an overall rational, or 'tribological philosophy' could be established, not only in uniting the disciplines within a single entity but to establish a basic conceptualization of that entity in both its generalized form and function. Further, this philosophy could

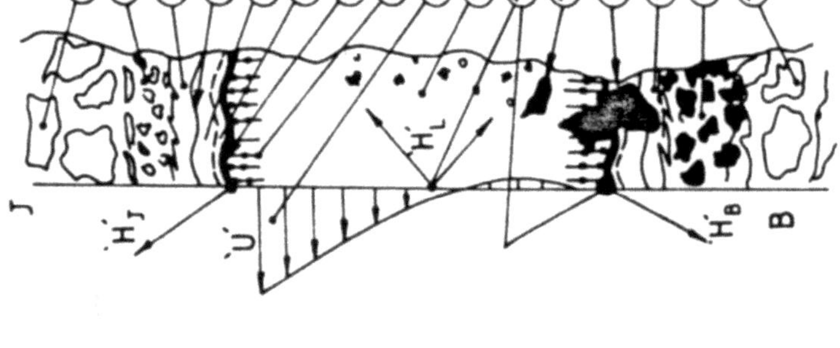

No	DESCRIPTION	DISCIPLINE
1	Bearing Material.	Metallurgy, Mechanics of Materials.
2	Heat Treatment Structure.	Metallurgy, Production Heat Treatment.
3	Distorted Microstructure from Working.	Crystalogaphy, Electron Physics.
4	General Envelope of Surface.	Metrology, Production Engineering.
5	Surface Asperity Demarcation Zone.	Surface Topography, Metrology.
6	Oxides, Adsorbed Gas and Contaminants.	Inorganic Chemistry, Surface Physics.
7	E P and other Reactive Additive Layer.	Organic and Surface Chemistry.
8	Fatty Acid Polar Chain Layer.	Boundary Lubrication, Organic Chemistry.
9	Fluid Viscous Shear Generating Heat.	Hydrodynamics, Viscometry.
10	Generated Fluid Pressure.	Hydrodynamics, Instrumentation.
11	Heat Transfer to Lubricant and Surfaces.	Heat Transfer, Temperature Measurement.
12	Particles and Gas in Entrained Fluid.	Microscopy, Monitoring, Fluid Physics.
13	Embedded Particle Contamination.	Contact Mechanics, Friction and Wear.
14	Mechanical and Thermal Distortion.	Elasticity, Plasticity, Thermal Physics.
15	Soft Babbit (Duplex Structure).	Metallurgy, Materials Science, Microscopy.
16	Rigid Backing Material.	Mechanics of Materials, Materials Testing.

FIGURE 5 INTERACTION OF DISCIPLINES IN THE ELASTO/HYDRODYNAMIC REGIME

of the Hersey number, until we reach the point of minimum
friction coefficient sometimes referred to as 'Leloup's Point'
or the 'unlatching point' by workers such as Gümbel[106, 28].
To the left of this point, asperity contact (i.e., Regime B)
will be initiated, a progressively larger portion of the load
being carried by this form of contact as the value of the
Hersey number decreases, see Figure 6.

This situation is modelled by Figure 7 where we have a
condition of mixed friction, part hydrodynamic, arising from
the shearing of the pressurized lubricant trapped between the
asperities, and what might loosely be described as 'boundary
lubrication' at the interfaces of the sliding asperities.
This situation is exceedingly complex since, insofar as the
asperities are concerned, their deformation may be either
elastic or plastic and their lubrication effected either by
metallic soap films, E.P. additive action (if the temperature
is high enough) or both, combined with 'chemical amalgam
lubrication' from the initial surface films.

What is known, a priori, is that at these moving asperity
conjunctions intense heat flashes are manifest, surfficiently
hot to locally weld asperities in a transitory manner before
they are torn apart, thus creating further crystal lattice
deformation in the surface skin of the material.

4.3 The Interaction of Friction, Wear and Temperature

In experimental work on the above phenomenon, the author
employed a unique design of a Four-Ball machine capable of
continuously monitoring wear, friction, and contact temperature
as the test proceeded[107]. Figure 8 (a trace of results using
an 'E.P.' loaded oil) shows an initially rapid rate of wear
and temperature rise with time, the former quickly becoming
essentially constant while the latter reduces to a uniform
level as the formation of the wear scar decreases the contact
pressure. Heat soak into the bulk material follows, until the
bulk temperature reaches a level for the E.P. additive to
react chemically with the surface; the formation of the E.P.
film causing an immediate drop in friction and therefore
contact temperature. As the latter is lowered, the E.P.
reaction ceases, the protective film is worn away and the
cycle is repeated in a series of 'saw tooth' profiles, Figure
3. Thermocouples placed in the 'skin' of the materials
(approx. 5×10^{-3} ins. below the contact) showed,
additionally, violent temperature oscillations while bulk
temperature response, itself, was relatively gradual, Figure
9. This tends to support the Flash Temperature concept
proposed by Blok (1937), in that there are two components of
temperature, a 'skin' and a 'bulk', in this type of

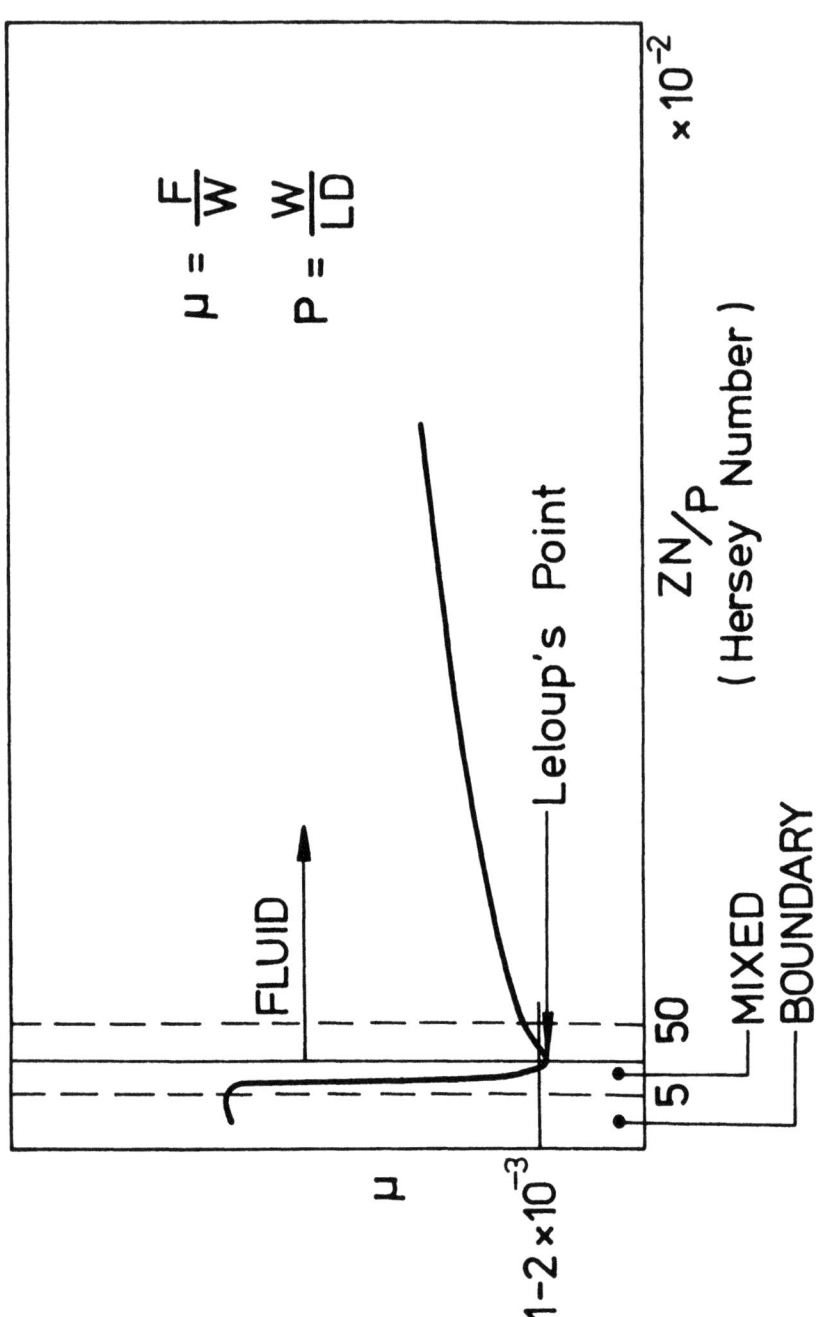

FIGURE 6 STRIBECK CURVE

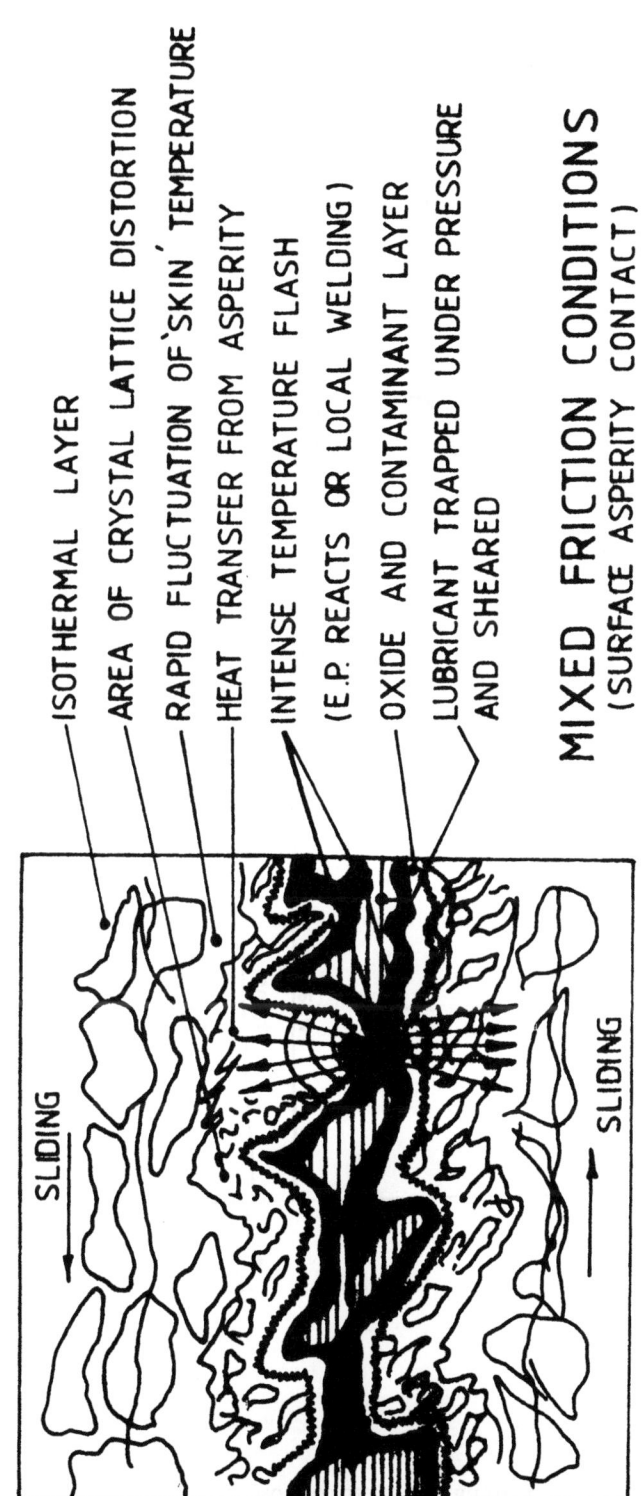

FIGURE 7 MIXED FRICTION CONDITIONS (SURFACE ASPERITY CONTACT)

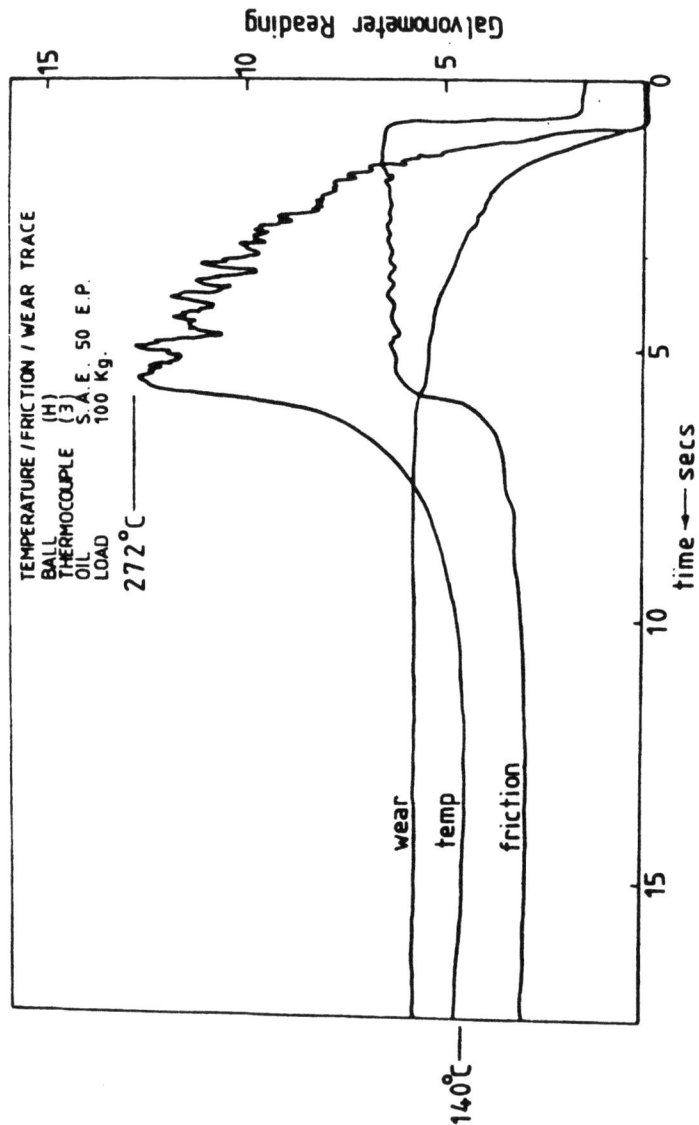

FIGURE 8 TYPICAL TEMPERATURE/FRICTION/WEAR TRACE FOR EXTENDED TEST RUN WITH AN E.P. LOADED OIL

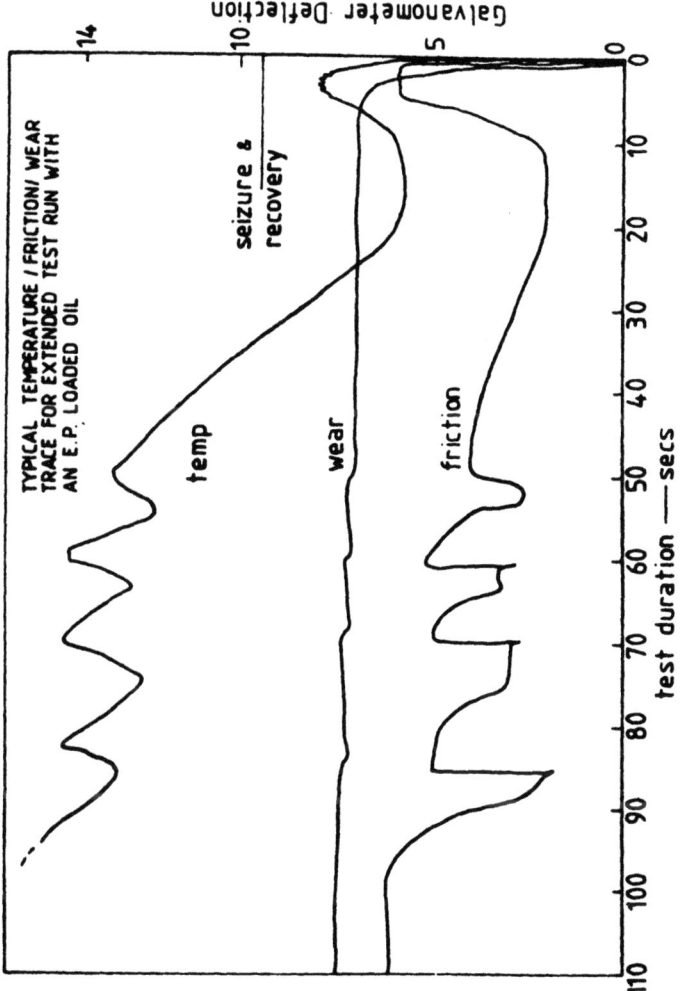

FIGURE 9   TEMPERATURE/FRICTION/WEAR TRACE

then be extended to embrace the question of the creation of that entity, and its subsequent development and utilization in an overall holistic methodology (including the maintenance of its efficient operation).

To do this, however, we must proceed in a series of logistic steps producing, from an initial fundamentalism, a progressional scheme of conceptual development.

4.2 The 'Micro' and the 'Macro' Models

As intimated above we must begin any logistical process from some generalized approach. Let us start, then, by dividing our surface contact into two geometrical categories:

(1) Conformal Surfaces
(2) Counterconformal Surfaces

which will be considered to be capable of operating in two distinct regimes:

(A) Hydrodynamic or Elasto-hydrodynamic
(B) Surface Asperity Contact or Mixed Friction Regime

<u>Regime A</u> - Using Figure 1 we may, therefore, initially define the confines of our exoteric boundaries (those which lend themselves to normal processes of synthesis) for either conformal or counterconformal conditions which we will term the 'Macro Model'. From the relationship between the configurations and idealized form of its components, we may establish, through a knowledge of geometry, trigonometry, kinematics, mechanics and dynamics, the outer function or 'External Manifestations' (E.M.) of the system.

Returning to Figure 5, our esoteric multidisciplinary 'Surface Entity' which we may term the 'Micro Model' of the system [containing its own inner, interlinked functions or 'Internal Manifestations (I.M.)] we proceed to place this, conceptually, at any relevant point within the conjoining surfaces of our 'Macro Model'. Thus, we immediately establish a unified whole, representing an extension of our concepts into one entity, possessing within itself both 'Internal and External Manifestations'.

<u>Regime B</u> - The next step on the logistical path is to return to our 'Surface Entity' or 'Micro Model' of Figure 5 and imagine a gradual, but progressive, diminution of the fluid film between the surfaces. A corollary to this is to consider a translation of the working condition along the abscissa of the 'Stribeck' curve leading to a reduction in the magnitude

contact[108]. Thus, the parametric (and therefore the multidisciplinary) nature of these interactions may be inferred from a study of such results, at least on a qualitative basis.

Returning to our 'Micro Model' we can, therefore, for a general liquid-lubricated contact, examine its 'Internal Manifestation' in the various regimes of lubrication (specified by translation through Hersey numbers) while simultaneously inducting it into our 'Macro Model'.

For visco-elastic lubricants the 'hydrodynamics' of the 'Surface Entity' model (as conceived) will not be altered, since heat will be generated whatever the lubricant. In such cases, as with solid lubricated or dry contact conditions, the friction coefficient itself may be used, instead of the Hersey number, to conceptualize parametric changes.

4.4 The Tribological Entity (T.E.)

The entity we have thus created represents an apperceptive paradigm or conceptual tribological model, by means of which we may interpret, through its 'External Manifestations' (its E.M.'s), the interaction with its outer environment over a multiplicity of situations and, as a result of this prescribed role, its 'Internal Manifestations' (its I.M.'s) arising from the reactive and interlinking phenomena between the proximate surfaces, characterized and assessed through the relevant disciplines and their own multiple interaction.

This created entity the author will term the 'Tribological Entity', later to be expanded into a larger overall complex embracing, as has already been intimated, not only the entity, per se, but its production development and utilization. It would be convenient at this stage to first categorize the disciplines specifically, on the basis both of the discussion so far and in the light of the ultimate object being considered.

4.5 The Categorization of the Disciplines

It is to be emphasized at the outset that any categorization of disciplines assembled with the object of utilizing such an aggregate in the process of tribological thought (in the multi-disciplinary sense) must be, by its very nature, entirely subjective. The author, while being aware of this, makes no apology for the schematic to be presented but makes this point initially, not to avoid objections but at least to mitigage extraneous digression. Methodology, in any field, is subjective in nature, any proponent being influenced both by his individual psyche and the conditioning process of his personal milieu; the author being, certainly, no exception to this universal verity. With this small caveat in mind, we will categorize the disciplines into eight primary 'cells' or 'blocks' with one 'Service Discipline'. These will be termed 'Central Disciplines of Tribology' and are defined as follows:

1. Discipline of Mechanical Engineering
2. Discipline of Materials Science
3. Discipline of Production Engineering
4. Discipline of Tribo-Engineering
5. Discipline of Chemistry
6. Discipline of Chemical Engineering
7. Discipline of Physics
8. Discipline of Electrical and Electronic Engineering together with:
9. Discipline of Mathematics (Service Discipline)

Each discipline may be delineated in terms of fields or subjects relevant to tribologial applications. In the context of the ground that has already been covered in the exposition so far, and in the light of the material to be presented, we may conveniently place this list in the form of an appendix. This also has the advantage of compactness in the interest of a clearer cognitive presentation, see Appendix 1.

A second subclass of disciplines is also to be recognized in which peripheral tribological activity has some relevance and which, in turn, motivates special tribological thought. These may be termed 'Peripheral Disciplines of Tribology' as follows;

1. Applied Medicine and Bio-Engineering
2. Management and Product Entity Planning
3. Energy Conservation
4. Ecology
5. Geo-Mechanics

4.6 The 'Hersey Pyramid'

Having now erected a model of the initial concept of a 'Tribological Entity' and defined the interdisciplinary cells or blocks in schematic form, our next task is to begin to integrate them into the 'Hersey Pyramid' referred to earlier in the text. At the risk of laboring the metaphor, we may describe the 'blueprint' of the structure as the <u>concept</u> of the 'Tribological Entity' and the disciplines as the building blocks.

Our next task is to produce a method jointly utilizing these to produce a working actuality, with all its manifest ramifications. This will be treated as an 'Application Methodology' requiring a basic elucidation.

5. APPLICATION METHODOLOGY

5.1 Concept of 'Extended Systems'

Before describing the details of the methodological approach, it is necessary to extend our concept of the 'Tribological Entity' itself.

We must recognize that, in applying an overall multi-disciplinary approach to the problem of 'interacting surfaces in relative motion' and to take the 'Jostian' definition of Tribology to its definitve conclusion, 'the practices related thereto', it is necessary to extend our concept of an appropriate system beyond that of the narrow constraint of 'bearing' or 'gear' which our initial conformal and counterconformal illustration may have, however inadvertently, implied.

A workpiece, turning against a cutting tool and surrounded by an environment of cutting fluid, is clearly not a bearing, although a pair of proximate surfaces covered with articular cartilage and operating within a closed environment of synovial fluid might be considered as such; the likely generic description adopted however, certainly by the medical profession, would be 'animal joint'! The importance of the coefficient of friction between the surfaces of the objects in which we encase our feet and the contacting surfaces which support our weight is not readily obvious to the majority of our fellow men unless environmental conditions, such as a polished floor or a children's winter sled, bring these matters to their abrupt attention.

From the foregoing, it is seen that the number of various systems falling under the aegis of our definition of 'Tribological Entity' is likely to be exceedingly large.

5.2 'T.E.' 'E.M.' and 'I.M.'

In an initial appraisal of an approach to a methodology, we will, for brevity, use 'T.E.' for 'Tribological Entity' and 'E.M' and 'I.M.' for its 'External and Internal Manifestations'. We see that the 'T.E.' exists to perform its 'E.M.', but in order to do so, it must be capable of maintaining the integrity of its 'I.M.' within the confines envisaged for it by its creators. This gives us, then, an extended concept of the 'T.E.' as a self-maintaining entity within certain constraints, among which we may place both temporal constraint and service or maintenance constraints in addition to the normal design and production ones.

In the case of the cutting tool, for example, the 'T.E.'s 'E.M.' is to produce workpieces to whatever criteria have been specified; maximum throughput, minimum cost or predetermined tolerance, relative to its own specified integrity. To accomplish this, the 'I.M.' may involve maximum life of the 'T.E.' through, say, minimum friction and wear, together with minimum service and replacement costs for both tool and lubricant. All these considerations are inextricably bound with design parameters (such as tool geometry, material and lubricant choice and machine speed and feed) and must, in our present paradigm, be characterized by our 'Surface Entity' concept and background procedural planning (service and replacement costs, design adaptability, procurement feasibility) together with background factors such as product and market research and process optimization.

From this simple example we begin to see the relevance of certain subsections of the disciplines we have listed in our schematic of Appendix 1.

5.3 The Overall Tribological Entity ('O.T.E.')

In characterizing the 'T.E.' we have initially assumed a singularly specific system. There is no reason, however, why this concept should not be extended to embrace a series of tribological systems in terms of an 'Overall Tribological Entity' (O.T.E.) characterized by an 'Overall External Manifestation' (O.E.M.)

The 'O.E.M.' of a gearbox for example is, basically, to produce speed changes between its input and output power transmitters and/or to change their direction of operation. This system may contain complementary 'T.E.'s such as bearings, gears, fluid seals and slideways, together with external peripheral equipment serving the main unit, with their own subsets of 'T.E.'s e.g., pumps, valves, centrifuges and pressure actuators. The tribologist must therefore consider such a system as an integrated whole, conceptually linked with actual tribological problems arising as a result of the 'O.E.M.'.

Nor should the concept be restricted to a single machine complex. In optimizing the efficient lubrication of a steel mill, for example, or in setting up a tribological research laboratory, the modern tribologist must examine a further gamut of factors, not all technical; examples being overall economics, health and safety constraints, ecological factors and homo-ramifications.

5.4 The Function Factor

An application methodology which has seen extensive use at the Cranfield Institute of Technology over a wide spectrum ranging from specific industrial consultancy to the planning and commissioning of a complete tribological research laboratory, integrates the concept of the 'O.T.E.' to a number of interrogative factors termed 'Function Factors', which the author has found, may be gainfully employed where some form of 'O.T.E.' has, in the broadest sense, to be 'created and sustained'. The factors, which may be used either singly or in combination, are categorized as follows:

1. The Design Function Factor (D.F.F.)
2. The Production Function Factor (P.F.F.)
3. The Service Function Factor (S.F.F.)
4. The Market Function Factor (M.F.F.)
5. The Homo Function Factor (H.F.F.)

Each factor may be delineated in a series of subclasses. For brevity, as with the subclasses of the disciplines, we will place these for reference in appendix form, Appendix 2. Their elucidation and utilization will be illustrated in the following sections.

6. A MULTIDISCIPLINARY APPROACH IN DESIGN, DEVELOPMENT, AND MANUFACTURING OF AN O.T.E. AND ITS SUSTAINMENT

6.1 Methodological Background

The methodology to be enumerated in attempting to fulfil the assignment inferred in the title to this section, is not hypothetical. Its basis is one of simple pragmatism which, through dint of application and a process of evolution over a number of years, has assumed, rightly or wrongly, the trappings of a methodology. Its foundation rests on three 'Definitions' and three 'Enumerations' i.e.,

<u>Definition</u>

Step 1. The Definition of the 'Requirement'
Step 2. The Definition of the 'Problem'
Step 3. The Definition of the 'Constraints'

Enumeration

Step 1. The Function Factors
Step 2. The Disciplines Involved
Step 3. The Overall Analysis, Decisions and Results

6.2 The Use of Case Studies

The initial two specifications, although appearing hardly worthy of scrutiny, are considered by the writer to be of cardinal importance. To illustrate this point two case studies will be given, covering each.

Case 1. A requirement arose, during the manufacture of micro-transducers for a tribological research rig, for lengths of micro-bore (5×10^{-3} ins. dia) tubing to be cut. The author and his colleagues went to considerable pains to overcome the problem but without success, the severed end always becoming deformed in spite of a variety of different approaches. In examining a sample of the tube, forwarded by the manufacturers, the author suddenly realized that both ends were cleanly cut. A telephone call elicited the technique and the problem was immediately solved, although the manufacturers agreed they had, themselves, spent time and effort on the solution. This illustrates an erroneous definition of the requirement: 'We must solve this problem' instead of 'We must get this problem solved'. Time and effort may often be better conserved by lateral thinking, rather than longitudinal.

Case 2. The second example had much more serious ramifications involving the definition of a problem. A manufacturer of industrial machines placed a problem involving gas sealing between moving surfaces with a tribological organization. An analysis was made involving computer modelling of the hydrodynamic contact between the seal and the sliding surface. Recommendations were proposed and implemented, but the problem remained. At a later date the author happened to be involved, in a consultative capacity, on the problem and requested actual components from the machine instead of the drawings that had previously been furnished. Examination of the surfaces revealed extensive wear, the problem being, in actuality, one of surface contact mechanics and not hydrodynamics.

We see, from these two simple case studies, the importance of establishing, as quickly as possible, both the initial requirements and the nature of the problem. Clearly the whole concatination of procedural effort is totally negated if either of the initial definitions proves to be erroneous.

Proceeding then on the assumption that proper definitions have been forthcoming, we apply Step 3. to our problem, the 'Definition of Constraints'.

6.3 Constraints

Constraints may be both complex and multifarious, but it has been found useful to attempt to establish a distinction between what may be termed 'Primary' and 'Secondary' constraints. The classification of these will in general depend on what is being attempted and will differ from one project to another.

To avoid obfuscation, we will list the constraints in twelve broadly definable areas and leave the reader to extend or particularize them to his own subjective satisfaction.

6.3.1 Constraint Classifications

(1) Feasibility, (2) Environment, (3) Performance and/or Development, (4) Complexity, (5) Economic, (6) Spacial, (7) Weight, (8) Number, (9) Manufacturing, (10) Maintenance/Service, (11) Market, (12) Homo (e.g., Skill, Knowledge, Effect).

6.4 The Methodology Process

Having listed our constraints we must introduce them into our methodology, together with the 'Function Factors'. In general, we consider *all* constraints but in particular, we reduce them, in most applications, to a specific few; in fact, the fewer the better since, as we shall gather from the case studies to follow, initial constraints often engender the formation of new ones.

The methodology is essentially an iterative process in that, having stated the terms of reference of the problem and specified 'Primary' and 'Secondary' constraints, we form an 'initial decision'. This being done we scan the 'Function Factors' interrogatively and note down their relevance to our requirements. It is sometimes convenient at this stage to include the Step 2. Enumeration; 'The Discipline Involved' as we proceed or, alternatively, leave this until later. We may conveniently list the steps as follows:

1. State Terms of References through Definitions
2. Specify Primary and Secondary Constraints
3. Make Initial Decisions
4. Initiate 1st Function Factor Scan (Appendix 2)
5. Produce Intermediate Decisions
6. Specify New Constraints (if any)
7. Initiate 2nd Function Factor Scan
8. Repeat to Final Decisions
9. Intiate Actions as Necessary
10. Examine Results and Produce Conclusions.

The process is conveniently illustrated by two typical case studies, one for an 'O.T.E.' and one for a 'T.E.'. (Note: Conditions considered especially important carry underlining; thus <u>A10</u>).

7. CASE STUDIES

7.1 Case Study No.1, Type 'O.T.E.'

<u>Subject</u>: 'To Plan and Commission an Advanced-Level Tribological Research Laboratory'

<u>Constraints</u>

Primary   -   (6)* Spacial (Given Floor Area)
                 (8)   Number (Maximum Rig Variety)
                 (3)   Performance (High Quality Research)

Secondary - (5)   Economic (Budget Limit)
                 (3)   Development (Research Variation)
                 (10) Service and Maintenance (Minimum Cost)

<u>Intial Decision</u>

1. Small Size of O.T.E.'s (For Maximum Number)
2. High Integrity (Reliability, Low Maintenance Cost)
3. High Precision (Reliability and Quality of Results)

<u>1st Function Factor Scan</u>

D.F.F.   (A1)** Mechanical Integrity
             (A4)   Rig Adaptability
             (A7)   Quick Dismantling
             (A8)   Extended Life
             (A10)   Good Development Potential
             (A12)   <u>Minimum Costs</u>

P.F.F.   (B6)   Internal/External Choice
            (B8)   Good Quality Control
            (B10) External Limitations (Small Rig
                   Availability, Small Transducers, High Costs).

* See Constraint Classification PSI.
** See Appendix 2

S.S.F.   (C1)   Signal Conditioning, Recording Equipment
            (C3)   'In Service' Rig Development
            (C7)   No Field Data
            (C9)   Special Monitoring Instruments
            (C11) Mainly Student Skill
            (C12) Minimum Service and Maintenance Costs

M.F.F.   (D1)   Micro-Miniature Transducers Not Available
            (D4)   Manufactured Transducer and Signal
                   Conditioning Equipment Expensive.

Intermediate Decisions

1. (A1 - A12) Design Own Test Rigs
2. (B6 - B11) Build Own Test Rigs. Manufacture Rig Components Internally. Buy Ancillaries Externally.
3. (B8, B10, C9, C12, D1, D4) Design, Build and Develop own Micro-Miniature Transducers.

Decision Constraints

Primary   1. **Feasibility** (Transducer Design, New Concepts)
            2. **Manufacture** (Physical Size of Transducers, High Accuracy of Rig Components)

Secondary   1. Available Student/Technician Skill
            2. Complexity ('Simple' Transducer Designs Required)
            3. Economic (Signal Conditioning Equipment Costs)

2nd Function Factor Scan

Subject: 'Micro Transducers and Instrumentation'

            (A5)   Assess Design of Micro Transducers and Signal
                       Conditioning Systems
            (B1) (B9) Assess Overall Manufacturing Costs of
                       Transducer Designs
            (B1) (B10) Assess Total Cost of Manufactured Signal
                       Conditioning Systems
            (C11) (D1) (D9) Initiate Small Student Projects on
                       Transducers

      (C1) (C9)   Assess Internal Instrumentation Potential (Manufacture of Equipment)
      (B6) (B10) (C6)   Research 'Market Capability'/Cost Situations
      (B6)   Good Probability of Student Interest and Enthusiasm
      → Motivation. Possibility of Students Training other Students
      → All Round Homo/Economic Benefit.

Final Decision   Proceed as Intermediate Decisions, **But** Build Signal Conditioning Equipment Internally.

Results   Test Rigs Built. Transducers Designed and Built. Students Acquire High Degree of Instrumentation Knowledge and Training. Good Student Research Facilities at Relatively Low Costs. Rapid Expansion of Laboratory. Diversification of Research. Economical Contract Research Capability. Unification of Multidisciplinary Methodological Approach.

Disciplines Utilized or Involved

(I) 1, 4, 6 (II) 1 (III) 1, 3, 4 (IV) 5, 6, 8, 9, 10, 11 (VII) 3, 12 (VIII) 1, 2, 3, 5.

7.2 Case Study No. 2. Type 'T.E.'

Subject: 'To Investigate the Performance of Porous Bush Journal Bearings in Domestic Fan Heater Units in the Light of Premature Bearing Failures'

      (Work undertaken for U.K. Government - 'British Gas')

Constraints

Primary   (3) Performance (Wear Integrity)
           (5) Economic (Speicifed Contract Budget)
           (3) Development (Units Already in Mass Production - i.e. Minimum Modification)

Secondary (9) Manufacturing (Fixed Procedure)
            (2) Environment (High Temperature: $100^\circ C$ plus)

Initial Decision

      (1) Investigate Bearing Operating Performance
      (2) Look at Possibility of Environment Simulation
      (3) Research Manufacturer's - Fan Heater Units

1st Function Factor Scan

D.F.F.	(A1)	Unknown Mechanical Integrity
	(A2)	Unknown Excess Duty
	(A4)	Limited Design/Manufacture Flexibility
	(A5)	Basic Design Simple
	(A7)	Limited Dismantling Efficacy
	(A8)	Seeking Approx. 15,000 Hrs. Service Life
	(A9)	Likely Long-Term Employment
	(A11)	Geometry and Material Fixed
	(A12)	Essentially Minimum Cost
P.F.F.	(B1)	Mass Produced
	(B2)	Little Flexibility
	(B5)	Lubricant Options, Lubricant Impregnation Flexibility
	(B7)	None
	(B8)	Little Information - Need to Investigate
	(B9)	Low Costs Imperative
	(B10)	Units and Bearings Freely Available
S.F.F.	(C1)	Felt Lubricant Reservoirs
	(C2)	None
	(C3)	None
	(C4)	Fan Units Interchangeable in Service. (Difficulty in Changing Bearings)
	(C5)	None (Complete Fan Unit Replaced)
	(C6)	Unsatisfactory - Requires Definite Improvement
	(C7)	Minimal
	(C9)	Parameters to be Decided
M.F.F.	(D1)	Improved Reliability
	(D2)	Large Fixed Demand
	(D3)	Possible Development in Other Areas
	(D4)	Limited - Since Units Cheap
	(D5)	Relatively Low Unit/Volume Cost
	(D6)	Negligible
H.F.F.	(F1)	High Utilitarian Appeal
	(E6)	User Complaints (Service Life, Noise During Failure) <u>Customer Satisfaction Important</u>

Intermediate Decisions

1. Build Test Rig for Mechanical Evaluation
2. Batch Test Statistically (12 Units/Test)
3. Accelerate Test Time by Overload
4. Test Complete Unit 'In-Situ'
5. Use Extended Tests (Full 24 Hours if possible)
6. Monitor Bearing Temperature and Friction Continuously

7. Attempt to Simulate Environment
8. Restrict Transmitted Vibrations from Test Rig to Test Units.

2nd Function Factor Scan - Subject: 'Test Rig'

D.F.F.
- (A1) Long Term Wear/Mechanical Integrity
- (A2) Testing Overload Capability
- (A4) Bearing Interchangeability in Fan Unit
- (A5) Basically Simple Design - Continuous Running Capability
- (A6) Ease of Fan Unit Changes
- (A7) Minimum Dismantling Time
- (A10) Acquired Results Considered for Other Applications
- (A11) No Extraneous Forces of Vibrations to be Transmitted; Accurate Monitoring on a Continuous Basis
- (A12) Overall Costs within Budget

P.F.F.
- (B2) Minimum Internal Commitment
- (B6) Manufacture Components Externally
- (B8) High Quality Control on Manufacturing (Unit Interchangeability)
- (B9) Rig Costs to be an Absolute Minimum (Component Quotations Required)
- (B10) None - Manufacturer Willing to Supply Bearings and Fan Units, Gratis.

S.F.F.
- (C1) Data Logger and Tape Punch for Monitoring
- (C3) Rig Adaptability in Service
- (C5) Bearing Interchangeability
- (C9) Continuous Monitoring of Bearing Temperature and Torque - Fixed Load and Speed
- (C10) Engage a Full-Time Research Officer.

H.F.F.
- (E2) Minimum Manual Attention
- (E6) Specialist Research Officer

Final Decisions on Rig

1. Multi-Unit Test Rig - Extended Tests under Fixed Overload
2. Operate Loaded Units on Air Bearings (No Vibration or Extraneous Friction)
3. Proceed with Mechanical Testing Only, Initially. (On Basis of Cost and Complexity)
4. Fully Automate Parametric Monitoring. Computer Plot and Process Data.

Result from Rig Tests

Rig showed wear integrity to vary between batches of bearings. **Not necessarily load dependent.** Higher friction and temperature noted in certain cases with **lighter** loads, but **correlated with load** when lubricant impregnation carried out in laboratory.

Final Decision

Quality control of lubricant impregnation during manufacture suspect. Likely to be prime cause of the reported premature failures.

Results

New research initiated on impregnation techniques and lubricant and additive optimization. New numerical solution of porous bush bearing lubrication produced and published[109].

Further research investigations in this field currently proceeding, including Mechanical Modelling. Extensive multi-disciplinary interaction. Methodology qualified.

Disciplines Utilized or Involved, (See Appendix 1);

```
   (I)    1, 2, 3, 5, 6, 7, 8
  (II)    1, 2, 4, 5, 7, 9
 (III)    1, 3, 4
  (IV)    1, 2, 3, 4, 5, 7, 9, 10, 11
   (V)    1, 2, 4, 5, 6
 (VII)    2, 3, 8, 12
(VIII)    1, 2, 3, 4
  (IX)    1, 2, 3, 5, 6
```

8. CONCLUSION

On the simple premise that human nature remains relatively immutable throughout the processes of time, although allowing that expressionistic or behavioristic patterns are undoubtedly conditioned by events in any particular milieu, we can by an investigatory examination of historical epochs in which progress, innovation, and a general flowering of human cooperation and endeavor arose, elicit an overall appreciation of the various motivating forces contributing to such a renaissance. Conversely, by a like process we can observe, in general terms, factors contributary to the antithesis of such a desirable homogeneity. These are the facts that history, itself, makes plain.

Technology is considered by the author to be no exception to this rule and he has therefore, throughout this expose, stressed the importance of obtaining insight into the recognition of our current situation from a study of the history of tribological development. Segregative attitudes chiefly arise when the overall commonality of problems becomes lost or obscured.

We are all, consciously or unconsciously engaged, insofar as the tribological field is concerned, in multidisciplinary action of some sort and it is, perhaps, through the creation of some basic paradigm that our appreciation of this fact is heightened. This the author has attempted to do and, using the methodology and case studies outlined, has indicated the possibility of generating an overall holistic approach to tribology. It is hoped that this may, with due allowance, assist tribologists, (through multidisciplinary insight rather than direct revolution or schism) towards that most desirable process, to recapitulate the 'Jostian' metaphor, of finally casting off their 'molecular chains'.

9. REFERENCES

1. Burstall, A.F., "A History of Mechanical Engineering," (Faber and Faber, London) Ch. VIII 286 - 365.

2. Hirn, G., "Sur les principaux phenomenes qui presentant les frottements mediats," Bull. Soc. Ind., Mulhouse, 26, (1854), 188 - 277.

3. Coulomb, C.A., "Theorie des machines simples, en ayant egard au frottement de leurs parties et a la roideur des cordages," Mem. Math. Phys. X. Paris, (1785), 161 - 342.

4. Morin, A.J., "Nouvelles experiences faites a Metz en 1833 sur le frottement sur la transmission du mouvement par le choc sur la resistance des milieux imparfaits a la penetration des projectiles, et sur le frottement pendant le choc," Mem. Savans Etrang. (Paris), vi, 641 - 785; Ann. Min. X (1836), 27 - 56.

5. Thurston, R.H., "A Treatise on Friction and Lost Work in Machinery and Millwork," (Wiley, New York) 7th Edn., (1903).

6. Newton, L., Mechanical Principles (London 1688), Cajoris revision of Motte's translation, University of California Press, 1946.

7. Petrov, N., "Friction in Machines and the Effect of the Lubricant," (a) in Russian: Eng. Jnl., St. Petersburg, (1833), 71 - 140, 288 - 279, 377 - 436, 636 - 564. (b) German translation by L. Wurzel, Hamburg, L. Voss, (1877), 187.

8. Tower, B., "First Report on Friction Experiments (Friction of Lubricated Bearings)," Proc. Inst. Mech. Engrs., Nov. 1883, 632 - 659. (See also: 'Adjourned Discussion', Proc. Inst. Mech. Engrs. Jan. 1884, 29 - 35).

9. Tower, B., "Second Report on Friction Experiments (Experiments on the Oil Pressure in a Bearing)," Proc. Inst. Mech. Engrs., Jan. 1885, 58 - 70.

10. Reynolds, O., "On the Theory of Lubrication and Its Application to Mr. Beauchamp Towers's Experiments, Including an Experimental Determination of the Viscosity of Olive Oil," Phil. Trans. R.Soc. 177, 1886, 157 - 234.

11. Sommerfeld, A., "Zur hydrodynamischen Theorie der Schmiermittelreibung," Z. Math. Phys., 50, (1904), 97 - 155.

12. Dowson, D., "History of Tribology," (Longmans, London), 1979.

13. Boswall, R.O., "The Theory of Film Lubrication," (Longmans, London).

14. Michell, A.G.M., "The Lubrication of Plane Surfaces," Z.Math. Phys., 52 Pt. 2 (1905), 123 - 137.

15. Michell, A.G.M., "Improvements in Thrust and Like Bearings," British Patent No. 875.

16. Poiseuille, J.L.M., "Memoires savante etrangers," Vol. 9, 1846, 433.

17. Maxwell, C., "Illustrations of the Dynamical Theory of Gases," (186), Phil. Mag. Series 44, 14, 19.

18. Slotte, K.F., Wied, Ann. 14, 1881, 13.

19. Engler, C., "Ein Apparat zur Bestimmung der sogenannten Viskositat der Schmierole," Chem. Z.9. 1885, 189 - 90.

20. Redwood, B., "On Viscosimetry or Viscometry," (1866), J. Soc. Chem. Ind. 5, 121 - 9.

21. Dolfuss, D., See: Forbes, R.J. 'Petroleum' in "A History of Technolgy," Vol. 5, "The Late Nineteenth Century" ca. 1850 - 1900, 102 - 23, Singer, C., Holmyard, E.J., Hall, A.R. and Williams, T.I. (eds.) (Clarendon Press, Oxford).

22. Archbutt, E., and Deeley, R.M., "Lubrication and Lubricants; a Treatise on the Theory and Practice of Lubrication and on the Nature, Properties and Testing of Lubricants," (1900), Charles Griffin, London.

23. Goodman, J., "Recent Researches in Friction - Part II," Proc. Instn. Civ. Engrs., ixxxv, Session 1885-6, Pt. III, 3-36.

24. Kingsbury, A., "Experiments with an Air-Lubricated Journal," (1897) J. Am. Soc. Nav. Engrs. 9, 267-92.

25. More, L.T., Isaac Newton, (1934), New York p. 565 et seq.

26. Hertz, H., "On the Contact of Elastic Solids," (1881), J. reine und angew. Math, 92, 156-71.

27. Stribeck, R., "Kugellager für beliebige Belastungen," (1901), Z. Ver. Dt. Ing., 45, No. 3, 73-125.

28. Stribeck, R., "Die Wesentlichen Eigenschaften der Gleit - und Rollenlager," (1902), Z. Ver. Dt. Ing., 46, No. 38, 1341-8; 1432-8, No. 39, 1463-70.

29. Gumbel, L., "Das Problem der Lagerreibung," (1914), Mbl. berl. Bez. ver. dt. Ing. (VDI), 1 Apr. and No. 5, May, 6 June, 87-104 and 109-120 (also July 1916).

30. Buckingham, E., "On Physically Similar Systems; Illustrations of the Use of Dimensionally Similar Equations," (1914), Phys. Rev. 4, 347-76 (Oct.).

31. Hersey, M.D., "The Laws of Lubrication of Horizontal Journal Bearings," (1914) J.Wash. Acad. Sci. 4, 542-52.

32. Harrison, W.J., "The Hydrodynamic Theory of Lubrication with Special Reference to Air as a Lubricant," (1913), Trans. Camb. Phil. Soc. xxii (1912-25), 6-54.

33. Harrison, W.J., "The Hydrodynamic Lubrication of a Cylindrical Bearing under a Variable Load, and of a Pivot Bearing," (1919), Trans. Camb. Phil. Soc. xxii (1912-23), 373-88.

34. Martin, H.M., "Lubrication of Gear Teeth," (1916) 'Engineering', (London), 102, 199.

35. Young, T., "An Essay on the Cohesion of Fluids," (1805), Phil. Trans, R. Soc. Pt. I, 65-87.

36. Tomlinson, C., "On the So-Called 'Inactive' Conditions of Solids," (1867), Phil. Mag. xlix 305.

37. Tomlinson, C., "On the Action of Solids in Liberating Gas from Solution," (1875), Phil. Mag. xlix 302-7.

38. Lord Rayleigh, "On the Lubricating and Other Properties of Thin Oily Films," (1918), Phil. Mag. J. Sci. sixth series, 35, No. 206 (Feb.) 157-63.

39. Langmuir, I., "The Constitution and Fundamental Properties of Solids and Liquids: II Liquids," (1917), J. Am. Chem. Soc., 39, 1848-1906.

40. Hawkins, W.D., Davies, E.C.H., and Clark, G.L., "The Orientation of Molecules in the Surface of Liquids, the Energy Relations at the Surfaces; Solubility, Adsorption, Emulsification, Molecular Association and the Effects of Acids and Bases on Interfacial Tension," (1917), J. Am. Chem. Soc. 39, Apr., 541-96.

41. Hardy, W.B., and Hardy, J.K., "Note on Static Friction and on the Lubricating Properties of Certain Chemical Substances," (1919), Phil. Mag. S6. 38. 32-40.

42. D.S.I.R. Committee, Review of Existing Knowledge of Lubrication. Department of Industrial and Scientific Research (1920).

43. Hardy, W.B., "Problem of Lubrication," Address to the Royal Institution of Great Britain, (1920); (See: 'Collected Scientific Papers of Sir William Bate Hardy', Cambridge, U.P. (1936), 639-44).

44. Hardy, W.B., and Doubleday, I., "Boundary Lubrication - The Temperature Coefficient," (1922a), Proc. R. Soc., A101, 487-92.

45. Hardy, W.B., and Doubleday, I., "Boundary Lubrication - The Paraffin Series," (1922b), Proc. R. Soc., A100, 550-74.

46. Grubin, A.N., and Vinogradova, E.I., "Investigation of the Contact of Machine Components," (1949), Kh.F. Ketova (ed.), Central Scientific Research Institute for Technology and Mechanical Engineering (Moscow), Book No. 30, (D.S.I.R. translation No. 337).

47. Stanton, T.E., "On the Characteristics of Cylindrical Journal Lubrication at High Values of Eccentricity," (1925), Proc. R. Soc. A. cii. 241-55.

48. Stanton, T.E., "The Friction of Pistons and Piston Rings," (1925), 'The Engineer', 139, 70, 72.

49. Ford, J.F., "Lubricating Oil Additives - A Chemist's Eye View," (1968). J. Inst. Pet., 54, 535, Jly. 1968, 198-210.

50. "Interdisciplinary Approach to Liquid Lubricant Technology," (1972), Symposium Lewis Research Centre, Cleveland U.S.A., Jan. 11-13 (NASA S.P. 318).

51. Zorn, H., "Esters as Lubricants," (1947), U.S. Army Air Force P. 433-475, Translation No. F-YS-957 RE.

52. Bisson, E.E., and Anderson, W.J., "Advanced Bearing Technology," (1964), (National Aeronautics and Space Administration) (NASA), Washington, DC.

53. Fogg, A., and Hunwicks, S.A., "Some Experiments with Water-Lubricated Rubber Bearings," (1937), Instn. Mech. Engrs., "Proceedings of the General Discussion on Lubrication and Lubricants," 1, 101-6.

54. Arens, J., "Bakelised Bearings for Rolling Mills," (1936), 'Engineering', No. 141, p. 593.

55. "Committee on Tribology" (1966) - See Section 2.4 of the present work.

56. Peppler, W., "Untersuchungen Uber Druckubertragung bei belasteten und geschmierten umlaufenden achsparallelen Zylindern," (1936), Maschinenelemente-Tagung, Aachen 1935, 42, V.D.I., Verlag Berlin 1936.

57. Peppler, W., "Druckubertragung an geschmierten, zylindrcischen Gleit und Walzflachen,", V.D.I., Forschungsheft 391.

58. Gatcombe, E.K., "Lubrication Characteristics of Involute Spur-Gears," A Theoretical Investigation (1945), Trans. Amer. Soc. Mech. Engrs., 67 177.

59. Hersey, M.D., and Lowdenslager, D.B., "Film Thickness Between Gear Teeth," (1950), Trans. Amer. Soc. Mech. Engrs., 72, 1035.

60. Blok, H., Discussion. Gear Lubrication Symposium, Part II, The Lubrication of Gears (1952), J. Inst. Petrol. 38, 673.

61. Poritsky, H., "Stresses and Deflections of Cylindrical Bodies in Contact with Application to Contact of Gears and of Locomotive Wheels," (1950), J. Apple, Mech., Trans. Amer. Soc. Mech. Engrs., 72, 191.

62. Petrusevich, A.I., "Fundamental Conclusions from the Contact-Hydrodynamic Theory of Lubrication," (1951), Izo. Akad. Nauk. SSSR. (OTN) 2, 209.

63. Dowson, D., and Higginson, G.R., "A Numerical Solution to the Elasto-hydrodynamic Problem," (1959), J. Mech. Engng. Sci. 1, No. 1, 6-15.

64. Ibid. Elasto-hydrodynamic Lubrication (1966), Pergamon Press - Oxford.

65. Dowson, D., and Whitaker, A.V., "The Isothermal Lubrication of Cylinders," (1964), A.S.L.E./A.S.M.E. International Lubrication Conference, Washington DC, 13-16 Oct. 1964, A.S.L.E. Preprint No. 64 - LC - 22.

66. Sternlicht, V., Lewis, P., and Flynn, "Theory of Lubrication and Failure of Rolling Contacts," (1961), Trans. Amer. Soc. Mech. Engrs., J. of Basic Eng. 83, Series D, 2, 213.

67. Cheng, H.S., and Sternlicht, B., "A Numerical Solution for the Pressure, Temperature and Film Thickness Between Two Infinitely Long Lubricated Rolling and Sliding Cylinders Under Heavy Loads," (1964). A.S.M.E. Paper No. 64, Lub. II, A.S.M.E./A.S.L.E. International Lubrication Conference, Washington DC, 13-16 Oct., 1964.

68. Bell, I.F., "Elasto-hydrodynamic Effects in Lubrication," (1961), M.Sc. Thesis, University of Manchester.

69. Crook, A.W., "The Lubrication of Rollers, IV, Measurements of Friction and Effective Viscosity," (1963), Phil. Trans. A 225, 281.

70. Dyson, A., "Flow Properties of Mineral Oils in Elasto-hydrodynamic Lubrication," (1967), Phil. Trans. R. Soc., Lond., No. 1093 A528, 529-64.

71. Christensen, H., "The Oil Film in a Closing Gap," (1962), Proc. Roy. Soc. A266 312.

72. Dawson, P.H., "The Pitting of Lubricated Gear Teeth and Rollers," (1961), "Power Transmission," 30, No. 351, 208.

73. Crook, A.S., "The Lubrication of Rollers," (1958) (I) Phil. Trans. A250, 387. (II) Film Thickness with Relation to Viscosity and Speed, Phil. Trans. A254, 223. (III) A Theoretical Discussion of Friction and the Temperatures in the Oil Film, Phil.

74. Dyson, A., Naylor, H., and Wilson, A.R., "The Measurement of Oil Film Thickness in Elasto-hydrodynamic Contacts," (1966). Proc. Instn. Mech. Engrs. 180, (3B), 119-34.

75. Sibley, L.B., Bell, J., Orcutt, F.K., and Allen, C.H., "A Study of the Influence of Lubricant Properties on the Performance of Aircraft Gas Turbine Rolling Contact Bearings," (1960), WADD Technical Report, 60-189.

76. Kirk, M.T., "Hydrodynamic Lubrication of Perspex," (1962), Nature, 194, 965.

77. Cameron, A., and Gohar, R., "Theoretical and Experimental Studies of the Oil Film in Lubricated Point Contact," (1966), Proc. R. Soc. A291, 520-36.

78. Foord, C.A., Wedeven, L.D., Westlake, F.J., and Cameron, A., "Optical Elasto-hydrodynamics," (1970), Proc. Instn. Mech. Engrs., 184, Pt. I, 487-503.

79. Kannel, J.W., Bell, J.C., and Allen, C.M., "Methods for Determining Pressure Distribution in Lubricated Rolling Contact," (1964). A.S.L.E. Paper No. 64 LC-24, A.S.M.E. International Lubrication Conference (Washington, DC) 13-16 Oct., 1964.

80. Cheng, H.S., and Orcutt, F.K., "A Correlation Between the Theoretical and Experimental Results of the Elasto-hydrodynamic Lubrication," (1966). Proc. Instn. Mech. Engrs., 180, Pt. 3B, 158-168.

81. Hamilton, D.B., and Moore, S.L., "Deformation and Pressure in an Elasto-hydrodynamic Contact," (1971), Proc. R. Soc., London, A322, 313-330.

82. Ausherman, V.K., Nagaraj, H.S., Sanborn, D.M., and Winer, W.O., "Infrared Temperature Mapping in Elasto-hydrodynamic Lubrication," (1976), Trans. Am. Soc. Mech. Engrs. J. Lubr. Technol. F98 No. 2, 236-43.

83. Amontans, G., "De la Resistance causee dans les machines," (1699), "Memoires de l'Academie Royale A. (Chez Gerard Kuyper, Amsterdam, 1706), 257-82.

84. Edmonds, M.J., "The Effect of Surface Configurations on Thermal Energy Transfer Across Pressed Contacts," Ph.D. Thesis, Cranfield Institute of Technology, Sept. 1978.

85. Jones, M.H., "Element Concentration Analyses of Films Generated on a Phosphor Bronze Pin Worn Against Steel Under Conditions of Boundary Lubrication," (1976), Trans. A.S.L.E. (Paper No. 76-LC-2B-3), 21, No. 2.

86. Kramer, J., "Der Metallische Zustand," (1950), Vandenboeck und Ruprecht, Gottingen.

87. March, P.A., and Rabinowicz, E., "Exo-electron Emission for the Study of Surface Fatigue Wear," (1976), Trans. A.S.L.E. 20, 315-20.

88. Bowden, F.P., and Tabor, D., "The Friction and Lubrication of Solids," (1970), (O.U.P., London 1950, 377).

89. Llewellyn Jones, F., "The Physics of Electrical Contacts," (1957) (Clarendon Press, Oxford).

90. Culity, B.D., "Elements of X-Ray Diffraction," (Addison-Wesley, Reading Mass) (1956), 431.

91. Hutton, J.F., Ref. 50, 187-261.

92. A.S.M.E. "Film Lubrication: Turbulence and Related Phenomena," (1974). Trans. Am. Soc. Mech. Engrs.; J. Lubri. Technol., 96, F, No. 1, Jan.

93. Cheng, H.S., Ref. 50, 271-75, 282-86, 306-7.

94. "Cavitation and Related Phenomena in Lubrication," (1975) 1st Leeds-Lyon Symposium on Tribology (Mech. Eng. Pub., London).

95. Fern, A.G., and Nau, B.S., Seals (Oxford University Press) Engineering Design Guides.

96. Dyer, D., and Reason, B.R., "A Study of Tensile Stresses in a Journal Bearing Oil Film," (1976). J.Mech. Eng. Sci. 18, 46-52.

97. Reason, B.R., and Schwarz, V.A., "A Thermal Prediction Technique for Extending In-service Life of Roller Bearing Assemblies," (1981), 33rd Meeting, Mechanical Failures Prevention Group, Apr. 21-23 N.B.S., Gaithersburg, U.S.A.

98. McFarlane, C., and Reason, B.R., "Experimental Studies in the Operating Performance of a Hybrid Air Journal Bearing with Particular Reference to Pressure Profile Measurement," (1981), 8th International Gas Bearing Symposium, Leicester, U.K.

99. Reason, B.R., and Dyer, D., "A Numerical Solution for the Hydrodynamic Lubrication of Finite Porous Journal Bearings," (1973), Proc. Instn. Mech. Eng. 187, 7/73, 71-8.

100. Simpson, A., "The Seven Sisters: The Great Oil Companies and the World They Made," (1975). (Hodder and Stoughton, London).

101. Hersey, M.D., "Notes on the History of Lubrication," (1933) Pt. I, J. Am. Soc. Nav. Engrs., xiv, No. 4, 1933.

102. Hersey, M.D., and Hopkins, R.F., "Observations on Educating the Engineer," (1949) Internationaler Kongress fur Ingenieur-Ausbildung, Ed. Roether Verlag, Darmstadt.

103. Hersey, M.D., "The Oil-shed Fallacy. Attacking the Problems of Lubrication by Rational Methods," (1936), Technology Review 38, 181-2, 192, 194, 195, 198.

104. Fuller, D.D., "Theory and Practice of Lubrication for Engineers," (1956), (Wiley, New York).

105. Barwell, F.T., "Bearing Systems - Principles and Practice," (1979), (Oxford University Press).

106. LeLoup, L., "Report on Investigations of Thrust of Bearings," (1949), Rev. universelle mines, Belgium, 9th Series, Vol. 5, 1949, 258-272.

107. Reason, B.R., "The 'Cranfield' Four-ball Machine: A New Development in Lubricant Testing," (1975), International Tribology Symposium "Tribology for the Eighties," Sept. 1975, Paisley, Scotland, U.K.

108. Blok, H., "Seizure-Delay" Method For Determining the Seizure Protection of E.P. Lubricants. (1939), S.A.E. J. Trans.), 1939, 44, No. 5, May, 193.

109. Reason, B.F., and Siew,, A.H., "A Numerical Solution to the Coupled Problem of the Hydrodynamic Porous Journal Bearing," (1981), International Conference on Numerical Methods for Coupled Problems, University of Swansea, Wales, U.K., Sept. 1981.

10. NOTATION

C	= bearing radial clearance
D	= diameter of shaft/bearing
e	= bearing eccentricity
F	= frictional force
G	= materials parameter
h	= local fluid film thickness
h_o	= film thickness at maximum pressure
h_m	= minimum film thickness
H	= $hm/_R$, dimensionless film thickness
L	= bearing length
N	= rotational speed
p	= local fluid pressure
P	= W/LD, specific pressure
R_1, R_2	= local radii of curvature, $\frac{1}{R} = \frac{1}{R_1} + \frac{1}{R_2}$
U, U_o, U_1	= surface velocities
V	= surface velocity
W	= load on bearing

x	=	coordinate in direction of U
y	=	coordinate through fluid film
z	=	coordinate in direction of V
ε	=	e/c, bearing eccentricity ratio
θ	=	angular coordinate
η	=	fluid viscosity
μ	=	$\frac{F}{W}$, coefficient of friction

11. APPENDICES

11.1 Principal Interdisciplinary Fields of Tribology

11.2 Schematic for Methodology in the Design, Production, and Marketing of an Overall Entity

Appendix 1

PRINCIPAL INTERDISCIPLINARY FIELDS OF TRIBOLOGY

I. DISCIPLINE OF MECHANICAL ENGINEERING	II. DISCIPLINE OF MATERIALS SCIENCE
1. Machine and Element Design	1. Material Integrity Science
2. Hydrodynamics and Fluid Mechanics	2. Metallurgical Engineering
3. Heat Transfer and Thermodynamics	3. Polymer, Ceramic, Mineral and Composite Materials
4. Mechanics of Materials	4. Friction Material Science
5. Metrology and Surface Topography	5. Surface Failure Mechanics
6. Mechanical Testing and Condition Evaluation	6. Contamination and Corrosion Science
7. Environmental and Field Evaluation Methodology	7. Hard and Soft Material Science
8. Mechanical Modelling Techniques	8. X-Ray Crystallography and Electron Microscopy
9. Vibration Technology	9. Analysis Techniques: 'Ferrography', Traces, Surface Topology, 'Quantimet' etc.

III. DISCIPLINE OF PRODUCTION ENGINEERING	IV. DISCIPLINE OF TRIBO-ENGINEERING
1. Material Removal Technology: Cutting, Grinding, Lapping, Stamping etc.	1. Lubrication Function and Regime Specification
2. Material Displacement Technology: Forming, Forging, Drawing, Rolling, Extruding, Spinning, Rivetting.	2. Lubricant and Material Specification Methodology
3. Material Shaping Technology: Casting, Moulding, Fabrication, Powder Metallurgy.	3. Lubricant and Cutting Fluid Applications Technology
4. Special Process Technology: Electro-Chemical, Ultra-Sonic and Spark Machining Technology, Surface Plating, Friction Welding, Surface Treatment and Coating Technology, Heat Treatment, Chemical Treatment.	4. Tribological Technology in Hostile Environments
	5. Tribological Systems Technology
	6. Tribological Field Appraisal Methodology
	7. Tribological Malfunction and Diagnosis Assessment
	8. Special Applications Tribology
	9. Application Constraint Specification
5. Tribological Aspects of Production Methodology.	10. Ancillary Element/Systems Technology
	11. Categorised Field Applications

Appendix 1
(Continued)

PRINCIPAL INTERDISCIPLINARY FIELDS OF TRIBOLOGY

V. DISCIPLINE OF CHEMISTRY
SUB-SECTIONS OF DISCIPLINE

1. Liquid Lubricant (Petroleum & Synthetic) Technology
2. Additive Formulation and Development Technology
3. Solid and Semi-Solid Lubricant Technology
4. Lubricant Analysis, Evaluation and Testing Methodology.
5. Surface Contact Chemistry (including Surface Kinetics).
6. Lubricant and Surface Contamination Chemistry
7. Polymer and Special Tribological Materials Chemistry.
8. Surface Process Development Chemistry
9. Toxicity, Health and Ecological Monitoring Methodology.
10. Chemical Analysis Techniques: Spectography, Surface Chemical Analysis, Gas Chromotography, Specific Tests etc

VI. DISCIPLINE OF CHEMICAL ENGINEERING
SUB-SECTIONS OF DISCIPLINE

1. Petroleum Lubricant Production Technology
2. Synthetic Lubricant and Additive Production Technology
3. Solid and Semi-Solid Lubricant Production Technology
4. Polymetric and Special Tribological Material Production Technology
5. Process Fluid Production Technology
6. Lubricant Handling, Dispersing and Storage Methodology
7. Lubricant and Materials Marketing Management

VII. DISCIPLINE OF PHYSICS
SUB-SECTIONS OF DISCIPLINE

1. Rheology
2. Surface Contact Physics
3. Optics and Electro-Optics
4. X-Ray Defraction and Crystallography
5. Irradiation Technology
6. Acoustic Emission and Condition Monitoring
7. Laser and Holographic Technology
8. Contamination, Filtration and Particle Monitoring
9. Vacuum Physics
10. Extreme Temperature Physics
11. Radiation and Atomic Physics
12. Transducer and Instrumentation Physics.

VIII. DISCIPLINE OF ELECTRICAL AND ELECTRONIC ENGINEERING
SUB-SECTIONS OF DISCIPLINE

1. Electrical Instrumentation and Transducer Technology
2. Signal Transmission and Data Acquisition Technology
3. Signal Conditioning and Display Methodology
4. Electrical Contact Science
5. Electronic and Solid State Engineering
6. Magnetic and M.H.D. Bearing Technology
7. Electrical Process Technology
8. Electrical Discharge Science
9. Electrical Analogue Simulation

Appendix 1
(Continued)

PRINCIPAL INTERDISCIPLINARY FIELDS OF TRIBOLOGY

IX. DISCIPLINE OF MATHEMATICS (SERVICE DISCIPLINE)

1. Classical Analytical Methods
2. Finite Difference and Finite Element Methods
3. Analogue and Digital Computer Methods
4. Computer Aided Design Techniques
5. Statistical and Probability Methods
6. Mathematical Modelling Techniques
7. Mathematical Comparology

I. DISCIPLINE OF MECHANICAL ENGINEERING
II. DISCIPLINE OF MATERIALS SCIENCE
III. DISCIPLINE OF PRODUCTION ENGINEERING
IV. DISCIPLINE OF TRIBO-ENGINEERING
V. DISCIPLINE OF CHEMISTRY
VI. DISCIPLINE OF CHEMICAL ENGINEERING
VII. DISCIPLINE OF PHYSICS
VIII. DISCIPLINE OF ELECTRICAL AND ELECTRONIC ENGINEERING

Appendix 2

SCHEMATIC FOR METHODOLOGY IN THE DESIGN, PRODUCTION AND MARKETING OF AN OVERALL ENTITY

A. DESIGN FUNCTION FACTOR (D.F.F.)
1. Mechanical/Wear Integrity
2. Excess Duty Potential
3. Energy Conservation Efficacy
4. Design/Manufacture Flexibility
5. Design Complexity Assessment
6. Specific Component Accessibility
7. Component Dismantling Efficacy
8. Component/Entity Life Requirement
9. Component/Entity Obsolescence Norm
10. Entity Development Latency
11. Specific Design Constraints
12. Influencing Economic Factors

B. PRODUCTION FUNCTION FACTOR (P.F.F.)
1. Production Potential
2. Manufacturing Commitment
3. Material Commitment/Availability
4. Machine Availability/Adaptability
5. Process Availability/Adaptability
6. Manufacturing Options
7. Material Options
8. Quality Control Capability
9. Production Economic Constraints
10. External Component Availability/Cost
11. Entity/Component Production Time
12. External Processes/Delivery Constraints

C. SERVICE FUNCTION FACTOR (S.F.F.)
1. Peripheral Service Requirement
2. Component Maintenance Requirement
3. In-Service Adaptability
4. Service Interchangeability
5. Unit Replacement Capability
6. Statistical Field Integrity Norm
7. Field Service Data Availability
8. Field Induction Methodology
9. Entity/Component Monitoring Requirement
10. Service Environmental Constraints
11. Service Skill Constraints
12. Service Economic Status

D. MARKET FUNCTION FACTOR (M.F.F.)
1. Market Research Specification
2. Overall Market Requirements
3. Latent Market Potential
4. Market Option Status
5. Entity Marketing Cost (Unit and Volume)
6. Entity Operating Cost Evaluation

E. HOMO-FUNCTION FACTOR (H.F.F.)
1. Utilitarian and Psychological Appeal
2. Ergonomic Efficacy
3. Health and Safety Constraints
4. Ecological Acceptability
5. Dispensing and Handling Constraints
6. Homo-Liability Ramifications

APPENDICES

Appendix A

PRESSURIZED BEARINGS

D. Koshal
Liverpool Polytechnic

W. B. Rowe
Brighton Polytechnic

1. INTRODUCTION

Introducing the subject of pressurized bearings it is important to distinguish between hydrostatic bearings more generally known as externally-pressurized, and pressure-fed hydrodynamic bearings which are alternatively termed self-acting bearings. The two types of journal bearings are illustrated in Figure 1. The term hydrostatic usually refers to externally pressurized liquid bearings rather than gas bearings. Hybrid bearings combine hydrostatic and hydrodynamic features. The following discussion is concerned with both liquid and gas bearings.

1.1. Hydrodynamic Bearings - Advantages and Disadvantages

The hydrodynamic bearing is supplied with liquid under pressure at a point near the maximum film thickness. Rotation of the journal draws the lubricant into the heavily loaded region and separates the two surfaces. Although such bearings may be fed under pressure, the point of entry is normally opposite the applied load or near the point of maximum film thickness. Hence the pressure does not act to support the load, in fact, quite the opposite occurs. The reason for pressure-feeding is to ensure that a sufficient volume of oil or other fluid enters the bearing clearance. This is necessary to ensure sufficient oil to maintain a full film throughout the loaded portion of the bearing and also to prevent overheating. Since the load support mechanism depends

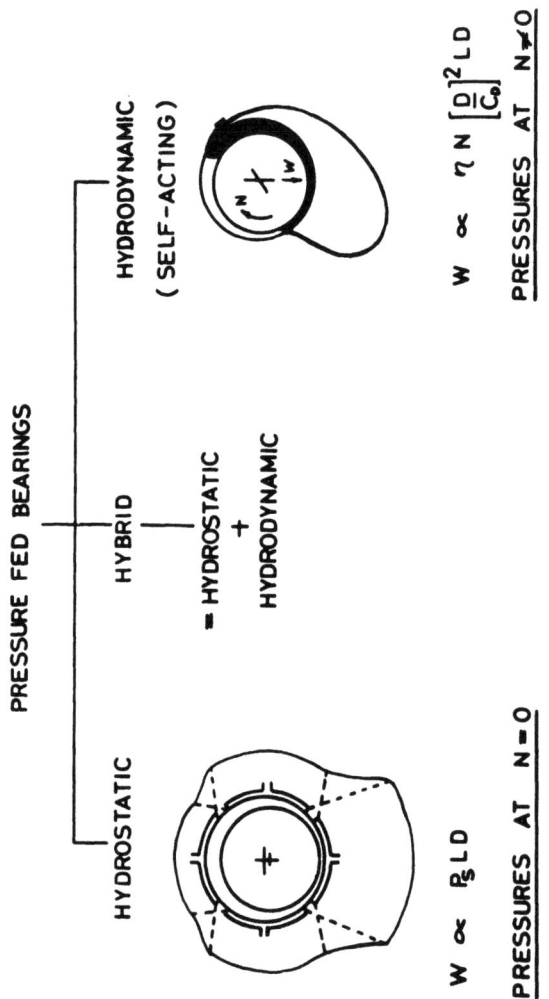

FIGURE 1 TWO TYPES OF JOURNAL BEARINGS

entirely on movement of the bearing surfaces towards the thin film region, it follows that there is no fluid-film load support at zero speed. Starting and stopping hydrodynamic bearings under load will inevitably lead to wear. The bearings illustrated in Figure 2 are all examples of self-acting or hydrodynamic bearings which may or may not be pressure-fed. The attraction of most hydrodynamic bearings of which there are a wide diversity of types, is the compactness and simplicity of the system for a high load-carrying capacity at appropriate speeds. Possible problems at high speeds which may be overcome with hydrostatic bearings, are whirl instability and high running temperatures. Gas bearings have been reported at extremely high speeds although the operating film pressures are much lower.

1.2 Hydrostatic Bearing - Applications

The hydrostatic bearing provides fluid-film load support by a completely different mechanism. As shown in Figure 1, the fluid is supplied to a number of entry ports around the bearing from a source at a supply pressure P_s, which must be high enough to support the load. The pressures in the bearing film are controlled so that the highest pressures act to oppose the applied loads. As a result of the pressurized supply and the associated control devices, the fluid-film support does not depend on speed and such bearings may operate at any speed down to zero with full fluid-film separation of the surfaces. The ability to operate at zero speed and high speeds with any load capacity determined by the supply pressure is one of the principal advantages of hydrostatic bearings. In addition, high stiffness may be achieved and in the case of liquid lubricated bearings, the stiffness can be designed independently of the load. This allows the designer to determine the bearing performance to suit the requirements of the machine. Other features of hydrostatic bearings are low starting torque, zero friction at zero speed, absence of start-up wear, high accuracy of location and smoothness of movement, good dynamic stability and cool operation with suitable design. The disadvantages are the requirements for more expensive and bulky hydraulic equipment, control restrictors, and effective filtration to prevent blockages in the supply. Some examples of liquid hydrostatic bearings are shown in Figure 3 and will be discussed in more detail in connection with the design of bearing systems.

Hydrostatic bearings have been employed very successfully for many years in a number of low-speed machines which require high load support and low friction in order to achieve high precision in positioning. In the USA several radio telescopes have been supported on hydrostatic bearings. A notable

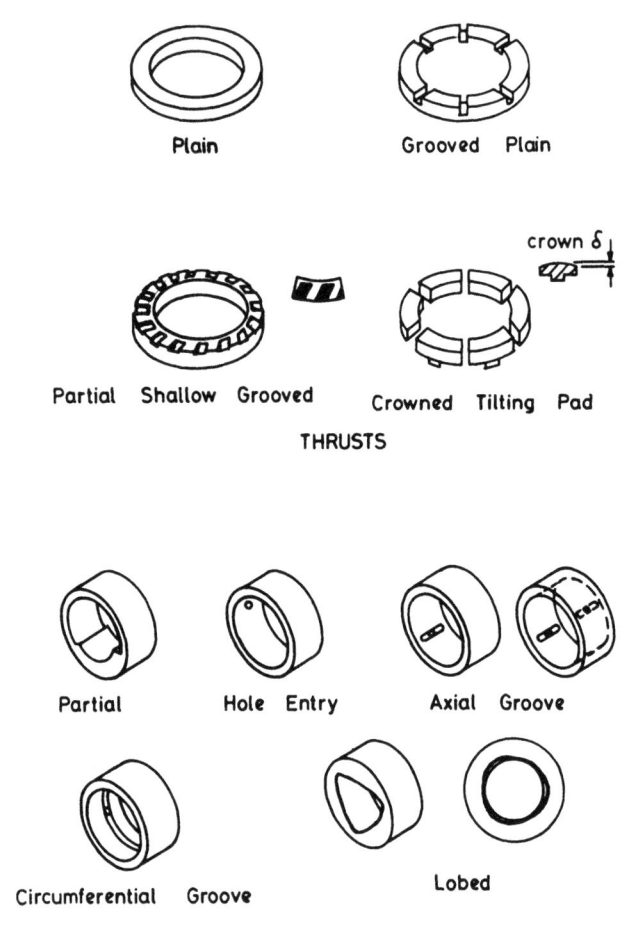

FIGURE 2   SOME EXAMPLES OF HYDRODYNAMIC BEARINGS

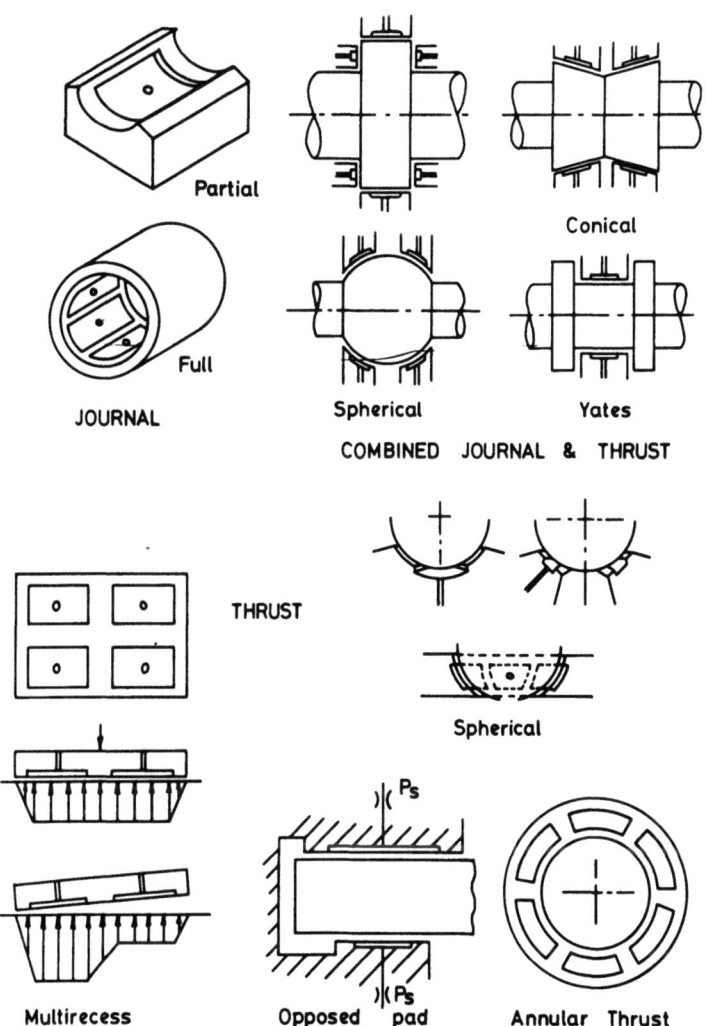

FIGURE 3   EXAMPLES OF HYDROSTATIC BEARING CONFIGURATIONS

example includes the 63 m (210 ft.) diameter, Goldstone radio antenna. Hydrostatic bearings have also been widely applied in machine tool slideways and spindles for low wear, high positioning accuracy, and high bearing stiffness. It is important that the bearings of machine tools are not subject to wear which reduces the resistance to chatter in metal-cutting operations and makes it difficult to maintain machining tolerances and production rate. Such features are particularly important for automatically controlled machine tools. Hydrostatic spindle bearings have been employed in a range of metal-cutting, grinding, and measuring machines at low and high speeds.

Other applications include experimental apparatus such as support bearings for bearing test-rig journals and dynamometers, as well as bearings and seals in hydraulic motors where a ready source of pressurized oil is available. A feature of hydrostatic bearings which could be important in some applications is a strong vibration damping action which has been applied for noise damping.

1.3 Externally Pressurized Gas Bearing Applications

Externally pressurized bearings may also be supplied with air or other convenient gas. A similar range of thrust and journal bearings may be designed, although gas bearings are usually designed as plain bearings for stability. Two typical gas bearing configurations are shown in Figure 4. Gas bearings offer very low friction and are often operated from the workshop compressed air supply. Careful filtration is essential.

Externally pressurized gas bearings have many of the advantages of liquid lubricated hydrostatic bearings including the possibility of constructing extremely high precision movements. Exceptionally low friction even at high speeds is a distinctive feature of E.P. gas bearings which tend to be completely cool running. A feature which may be of importance is the freedom from contamination when employing a clean inert gas. Slot-entry bearings have the additional advantage that they are commercially available from the Horstmann Gauge Company.

E.P. gas bearings are more often operated with lower pressures than liquid hydrostatic bearings for convenience and safety. This means that lower loads can be supported for the same size of bearing. E.P. gas bearing applications include high-speed dentists drills, turbine flow-meters, gyroscope bearings, grinding machine spindles, and roundness measuring machines. In each case, the virtual elimination of friction and wear is of overriding importance for the achievement of accuracy of positioning, reliability, or very high speeds.

1.4 Hybrid Bearing Applications

When a hydrostatic bearing operates at speed there is extra load support from the fluid-film due to hydrodynamic effects. The bearing is then said to operate in a hybrid manner.

A similar effect can be achieved with externally pressurized gas bearings, although the application of hybrid gas bearings has not received the same degree of consideration as hybrid liquid bearings.

Hybrid bearings may be designed to optimize both the hydrostatic and the hydrodynamic performance to achieve high load-carrying ability economically. Hybrid bearings, when designed appropriately, perform as superior hydrodynamic bearings at speed with the attractive features of hydrostatic bearings at low and high speeds. These features include good load capacity and stiffness independent of speed, low start-up torque and absence of wear, high accuracy of location and smoothness of motion, and good dynamic stability and cool operation when appropriately designed. The additional features are high overload capacity at high eccentricity, the ability to employ higher viscosity oils at high speeds, and the tolerance of wider variations in manufacturing clearance. The hybrid bearing is also superior to both axial groove and circumferential groove hydrodynamic bearings where heavy dynamic loading is applied in widely varying directions. The disadvantage of hybrid bearings is the same as for hydrostatic bearings; it is necessary to provide auxiliary hydraulic equipment. However, it is possible that a lower pressure system will suffice in view of the high overload capability. Hybrid bearings avoid recesses for maximum hydrodynamic effect and hence, in appearance and construction have more in common with externally pressurized gas bearings as shown in Figure 4.

There are obvious applications where plain hybrid bearings would have performance advantages. These applications are in high-speed machines where hydrodynamic bearings tend to suffer from whirl instability as in generator sets and turbines. Machine tools for intermittent cutting operations, or where shock loads and heavy overloads may occur on occasions, are also suitable applications.

A special class of hybrid bearing is the jacking bearing, in which the bearing is jacked hydrostatically under pressure to separate the bearing surfaces and hence avoid wear and high

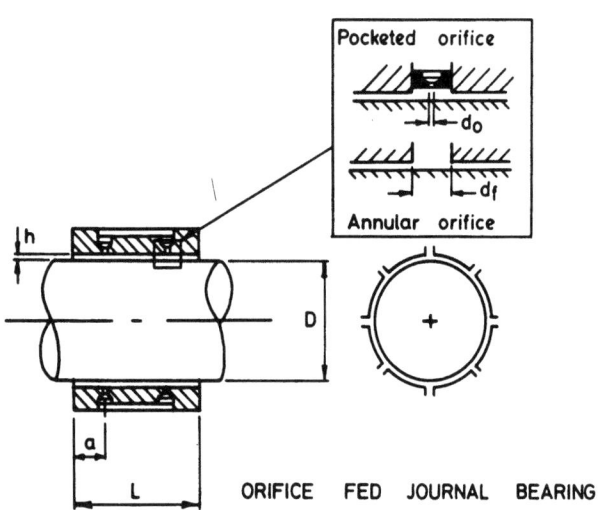
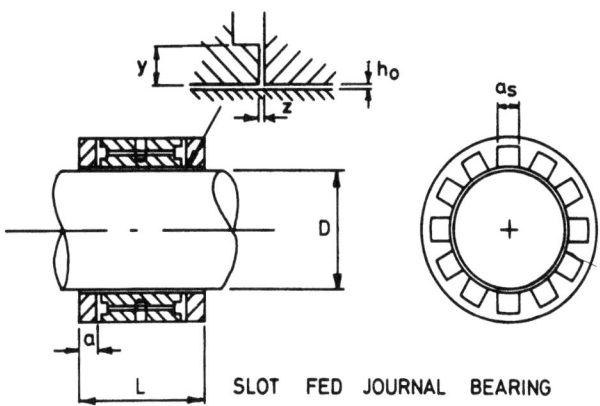

FIGURE 4 TYPICAL GAS BEARING AND PLAIN HYBRID BEARING CONFIGURATIONS

friction under starting and stopping conditions. At speed the supply pressure may be reduced or even switched off to allow the bearing to operate purely hydrodynamically.

2. EXTERNALLY PRESSURIZED BEARING SYSTEMS

A hydrostatic bearing essentially involves two resistances in series. The fluid flows through an orifice restrictor or other control resistance and then through the resistance formed by the bearing clearance.

The basic principle of a hydrostatic bearing is illustrated in Figure 5. A fixed displacement pump supplies oil to a capillary restrictor at a constant supply pressure, P_s, which is controlled by a relief valve. The pressure is reduced through the capillary and arrives at the entry to the bearing at recess pressure, P_r. The pressure further reduces as the oil flows through the restriction caused by the bearing gap, h. If the bearing gap decreases by a distance, δ, the restriction to flow through the bearing increases and the recess pressure increases. The relationship between bearing clearance and recess pressure will depend on such features as the nature of the control device, in this case a capillary, and parameters such as the supply pressure and the oil viscosity. An important aspect of hydrostatic bearing design is careful attention to filtration. The main filter is often placed in the return line from the relief valve to the tank. This filter operates to extract particles whenever the pump is switched on. A blocked orifice or capillary restrictor can easily occur and bearing failure may result. A line filter of the type used in carburetor fuel lines is a simple and useful additional protection for the restrictor.

Figure 6 indicates the principal types of externally pressurized bearings. In each case, the bearing may be designed for liquid or gas operation although the detail design will require to be quite different.

Liquid and gas hydrostatic bearing configurations mainly fall into three categories; journal bearings which allow shaft rotation, thrust bearings which allow sliding on a flat plane, and combined journal and thrust bearings which allow shaft rotation with axial constraint. Some examples of these three groups of liquid lubricated bearings are shown in Figure 3.

A common oil-lubricated journal bearing configuration is the cylindrical bearing containing 4, 5, or 6 recesses. Each recess must be controlled by its own restrictor. Recesses are not an essential feature of hydrostatic journal bearings and

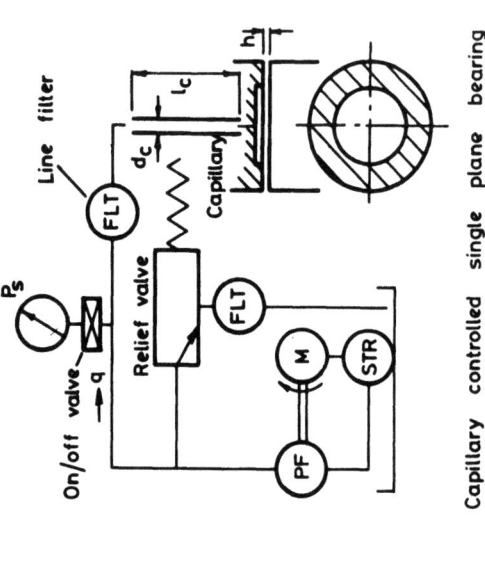
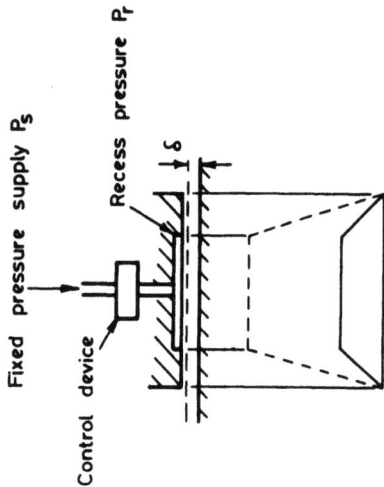

FIGURE 5 BASIC HYDROSTATIC BEARING SYSTEM AND TYPICAL PRESSURE DISTRIBUTION

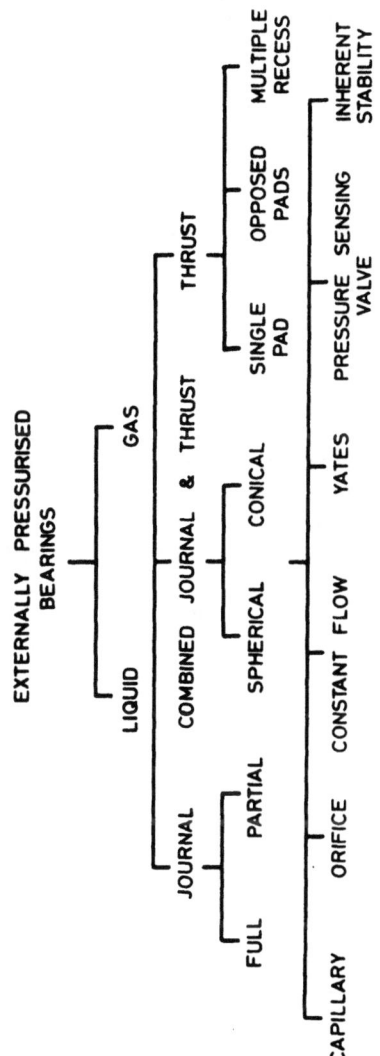

FIGURE 6   EXTERNALLY PRESSURIZED BEARINGS

the alternative plain bearing configurations may offer advantages for manufacturing simplicity and reliability.

For very large oil-lubricated journal bearings and high speed journals, the designer may well experience problems in specifying reasonable tolerances for manufacture. There may also be problems of high power consumption and flow-rate. Both problems may be reduced by employing partial journal bearings. Usually a pedestal will contain two partial bearings angularly disposed to cradle the journal.

Most journal bearings require axial constraint and a common method is to employ a thrust flange supported between two annular recessed pads. Thrust bearings for journals tend to require relatively high flow-rates due to the problems of achieving sufficiently wide bearing lands in the space available. This is a problem which may be partially overcome by designing a caliper arrangement instead of the usual $360°$ thrust pads. An economical alternative is the conical configuration which can withstand radial and axial loads. The spherical arrangement allows for misalignment in addition to combined radial and axial loads. In this case, it is necessary to decide whether this advantage is worth the extra manufacturing complexity. The Yates bearing achieves radial and axial loads in a simple and economical way. The oil which leaks from a conventional journal bearing has to escape through the thrust bearings. It has been found that such a system allows quite substantial axial loads to be supported in addition to normal radial loading.

The simplest example of a flat sliding bearing is the circular or rectangular single-recess pad. However, a single plane thrust pad must always be held down by a positive force. Where the load reverses in direction, as in some machine tool slideway systems, opposed pads would be employed. The thrust pad with a single recess has virtually no resistance to tilt. This may be an advantage in a spherical bearing which is designed to allow free rotation in any direction, but for most machines pads must be arranged in a suitable pattern to ensure alignment of the bearing surfaces. The rectangular multi-recess bearing pad for a linear slide and the multi-recess annular thrust pad for rotary movement, may both be employed where tilt resistance is required. Figures 7 and 8 show an example of a recent application of an annular thrust pad to support the analyzing magnet of the Nuclear Structure Facility at the SRC Daresbury Laboratory. This bearing was designed in collaboration between Liverpool Polytechnic and Daresbury.

FIGURE 7 THE ANALYZING MAGNET AT THE BASE OF THE NUCLEAR
STRUCTURE FACILITY AT THE SRC DARESBURY LABORATORY
SUPPORTED ON A HYDROSTATIC BEARINGS

FIGURE 8   ANNULAR THRUST PAD BEARING FOR DARESBURY LABORATORY

The magnet bends the beam of ions through 90° towards the experimental equipment. The diameter of the bearing is 1.8 m (72 in.); it carries a load of 53 tonnes (53 tons) and its surface is flat to an accuracy of 5 microns (0.0002 in.). The operating requirements involved maintaining the vertical center-line of the apparatus within 0.1 mm (0.004 in.) radius at a height of 6.3 m (21 ft.) above the bearing face, while the magnet may be rotated to direct the beam into any one of three experimental areas. This was successfully achieved with a supply pressure P_s, of 1.1 MN/m^2 (162 lbf/in^2) and a pressure ratio $\beta = 0.5$. Removable capillary tubes were used for bearing control for reasons of simplicity and ease of cleaning.

The machining and construction of thrust pads is usually a simple matter. For journal bearings, several methods have been employed for producing recesses. These include milling or grinding, which is difficult for small and medium size engineering bearings of less than 250 mm (10 in.) diameter. Other methods include electrical discharge machining and fabrication. The latter method is the simplest in the absence of EDM facilities. Examples of the fabrication of journal bearings is illustrated by Figures 9 and 10. Figure 9 shows the parts of a recessed journal bearing and Figure 10 the parts of a slot-fed journal bearing suitable for hydrostatic or hybrid operation with either liquid or gas lubrication.

Figure 6 also indicates a range of control principles which may be employed in hydrostatic bearing design. Some method of control is essential, as already explained, and a separate control device must normally be employed for each bearing recess. External devices may include capillary or slot restrictors, orifices, constant flow valves or pumps, and pressure-sensing valves. Some bearing configurations have inherent control due to a shallow recess or the particular combination of journal and thrust bearings as in the Yates bearing. The stability of the bearing film clearance and the stiffness of the bearing film are both dependent on the pressure/flow-rate characteristics of the control device. Some pressure/flow-rate characteristics are illustrated in Figure 11. It will be seen that flow-rate through a capillary is directly proportional to the pressure difference along the capillary and hence the flow-rate is inversely proportional to recess pressure. This gives the lowest oil film stiffness as indicated by line 1 on the film thickness versus load ratio diagram. The orifice, line 2, and the constant flow supply, line 3, both yield stiffer bearings and a pressure sensing valve, line 4 may be tuned to achieve a region of infinite or even negative static oil film stiffness. In practice, negative stiffness is not recommended as it can lead to limit-cycle

FIGURE 9 A FOUR-RECESS HYDROSTATIC JOURNAL BEARING

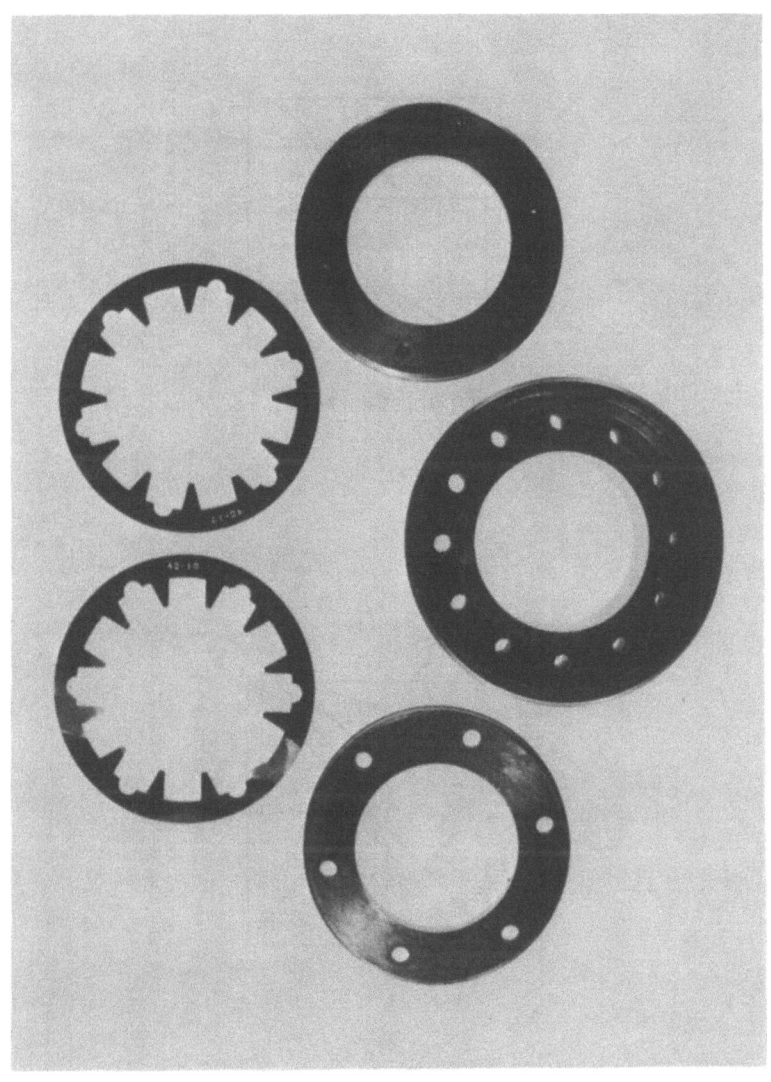

FIGURE 10 A SLOT-ENTRY JOURNAL BEARING FOR GAS OR LIQUID OPERATION

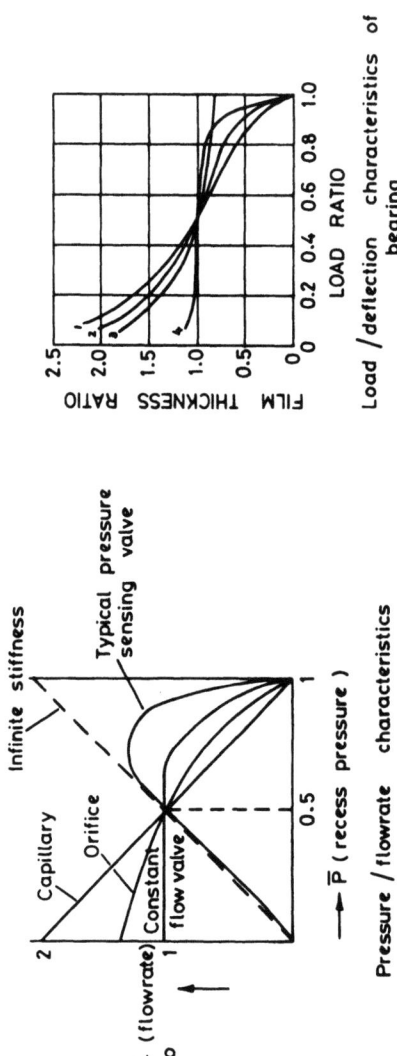

FIGURE 11 EFFECT OF TYPE OF CONTROL DEVICE

oscillation of the bearing system. It is also found that with
bearings which are more highly tuned for high stiffness, it is
necessary to maintain a more strict control on other
parameters such as oil temperature and bearing clearance. For
this reason, capillary restrictors are recommended for most
purposes. A simple form of pressure sensing valve, applied to
a journal bearing, is illustrated in Figure 12. In this
example the control device is a double diaphragm valve. When
a recess pressure increases in reaction to an applied load on
the bearing, it causes the diaphragm to deflect. This reduces
the restriction in the supply line and increases the flow-
rate. The flexibility of this device may also be an advantage
for preventing the build up of silt in the restrictors.
Figure 12 shows this principle applied to a 6-recess bearing
employed in a high removal - rate centerless grinding machine.
Three such valves are necessary to control the six recesses.
The novelty of the system illustrated is the use of the
bearings for grinding force measurement and also for dynamic
wheel balancing. Force measurements could also have been
achieved with capillary control although the extremely high
bearing stiffness was considered an advantage for achieving
accuracy.

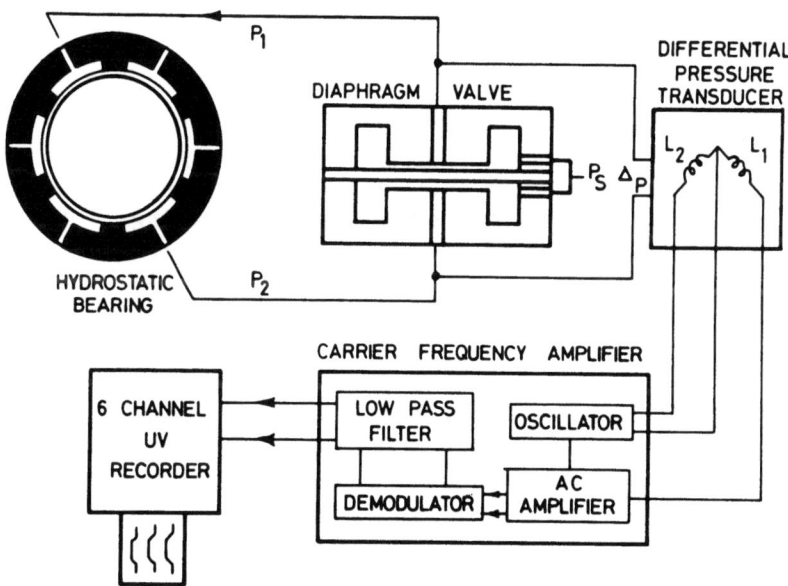

FIGURE 12 FORCE MEASUREMENT AND BALANCING WITH HYDROSTATIC
          BEARINGS

As previously mentioned, the detail design and
construction of externally pressurized gas bearings tends to
be quite different from the design of liquid lubricated
hydrostatic bearings. Whereas many liquid-fed bearings

involve substantial recesses as shown in the previous diagram, such recesses are likely to prove unacceptable for a gas bearing. Gas bearings tend to be prone to a vibrational instability known as pneumatic hammer if large gas volumes are stored between the restrictor and entry to the bearing clearance. For this reason orifice and lost restrictors are more commonly employed than capillary restrictors or other valves which may be less stable. Considerable research has been carried out on porous pad restrictor inserts which potentially yield high bearing stiffness and good stability if the problems of manufacture and controlling the porosity are overcome.

Externally pressurized gas bearings also tend to differ from liquid hydrostatic bearings in the materials employed. It is essential that all materials should be corrosion resistant and hence stainless steel and brass are used extensively. Greater precautions are also necessary to allow for a possible gas or power supply failure since the bearings operate dry. In the case of gas-supply failure, drive motors must be isolated and stopped quickly. A pressurized gas reservoir offers some protection by allowing time for braking and run-down of the motor.

3. DESIGN OF EXTERNALLY PRESSURIZED BEARINGS

The bibliography in Section 5 of this discussion will direct the reader to sources of design data. The following notes are intended to give an indication of load-carrying capacity of plane pads and cylindrical journal bearings. For brevity, the following guides are necessarily over-simplified. However, greater accuracy and sophistication can be achieved by reference to the published data.

3.1 Design Procedures for Hydrostatic Journal Bearings

The geometry of a typical bearing is shown in Figure 13. The land width, 'a', and the recess pressure, P_r, are the main parameters which govern the flow-rate from a recess. The total flow-rate from the bearing is given by:

$$q = \frac{P_r \pi D h_o^3}{6 a \eta}$$

At higher speeds more flow will be required to maintain a cool bearing. As a guide, it is usually recommended to design the bearing for minimum total power dissipation where the total power is equal to pumping power plus friction power.

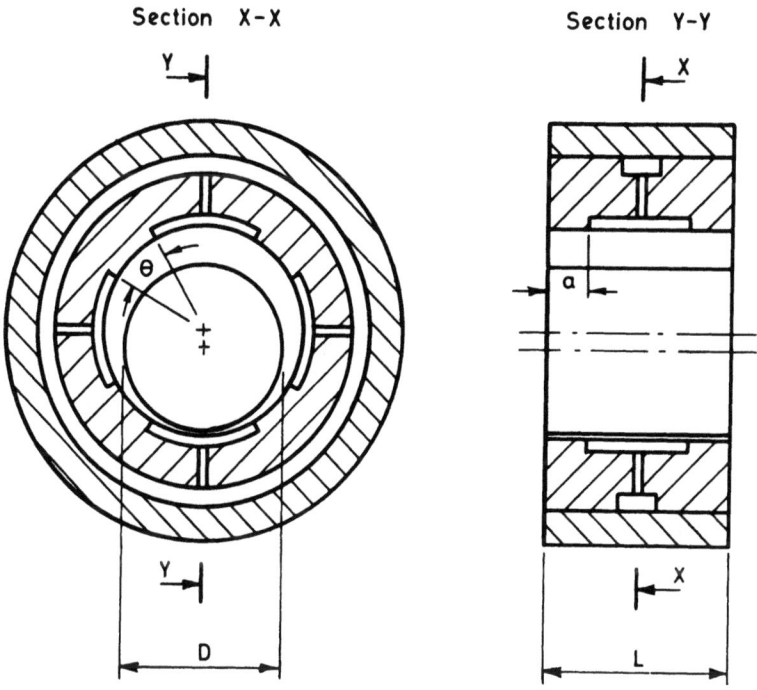

FIGURE 13   THE GEOMETRY OF A CAPILLARY – CONTROLLED FOUR-RECESS HYDROSTATIC JOURNAL BEARING

Design for minimum power is assumed in the flow-chart, Figure 14, and forms a basis for the selection of bearing parameters For minimum power, the ratio of friction power to pumping power should lie in the range

$$1 \leq K \leq 3$$

An approximate load capacity is given by

$$W = \tfrac{1}{4} P_s LD \text{ where } P_r = \tfrac{1}{2} P_s$$

Stiffness

$$\lambda = \frac{\tfrac{1}{2} \times P_s LD}{h_o}$$

Temperature rise (°F) = $0.015 \times P_s$ (lbf/in^2)

Temperature rise (°C) = $4 \times 10^{-6} \times P_s$ (N/m^2)

For the widest tolerances on bearing clearance, the minimum clearance condition should not be less than 2/3rds the maximum clearance condition. Figures 15 and 16 are design charts which give some indication of the selection of clearance and other design parameters. The recess pressure at the maximum clearance (at K = 1), is adjusted to make $P_r = 0.4\ P_s$. This gives $P_r = 0.7\ P_s$ and K = 3 at the minimum clearance. At the minimum clearance condition the temperature rise may be doubled.

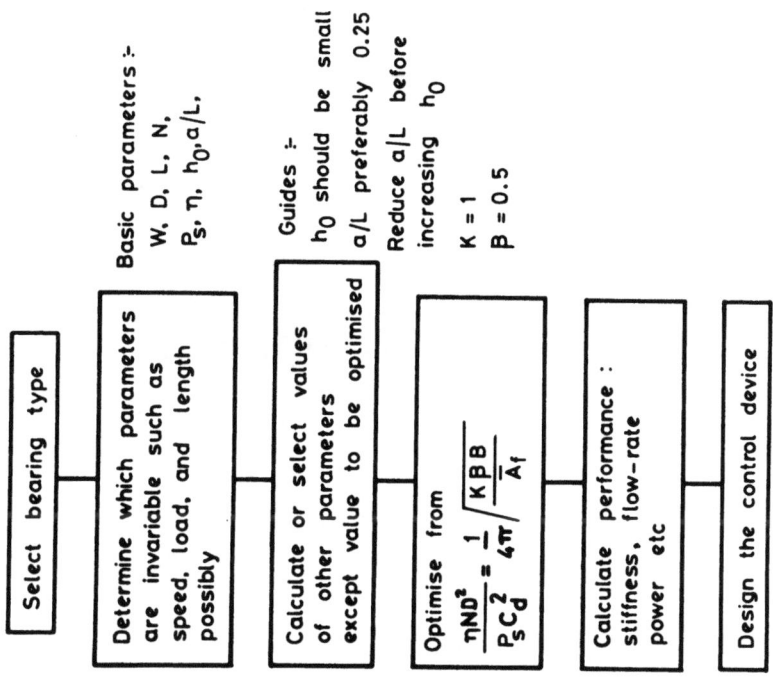

FIGURE 14 FLOW CHART ILLUSTRATING THE BASIS OF DESIGN PROCEDURES FOR JOURNAL BEARINGS

3.2 Design of Hybrid Journal Bearings (Non-recessed)

Figures 17, 18, and 19 indicate the type of bearing loads achievable with hybrid bearings. To obtain greater load capacity from hybrid bearings, the speed may be increased or the equivalent effect may be obtained by setting K = 12 in the flow chart for hydrostatic bearings. Figure 17 compares a plain hybrid bearing, at a relatively low-speed (K = 1), with a capillary controlled recessed journal bearing operating at the same speed corresponding to the minimum power condition. The hybrid bearing is also compared with an axial groove hydrodynamic bearing. The hybrid bearing supports the highest loads up to an eccentricity ratio of 0.8. Figures 18 and 19 give an indication of load capacity at higher speeds corresponding to higher values of K.

3.3 Hydrostatic Plane Pads (Recessed)

Design procedures follow similar considerations to those for journal bearings, so that the following approximate guides apply:

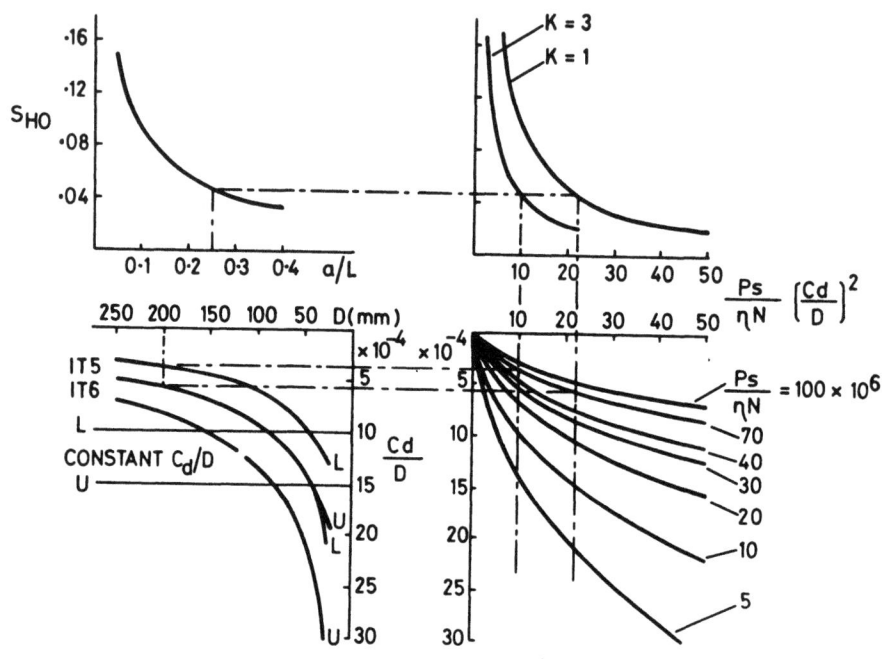

FIGURE 15 RECESSED JOURNAL BEARINGS - TOLERANCES

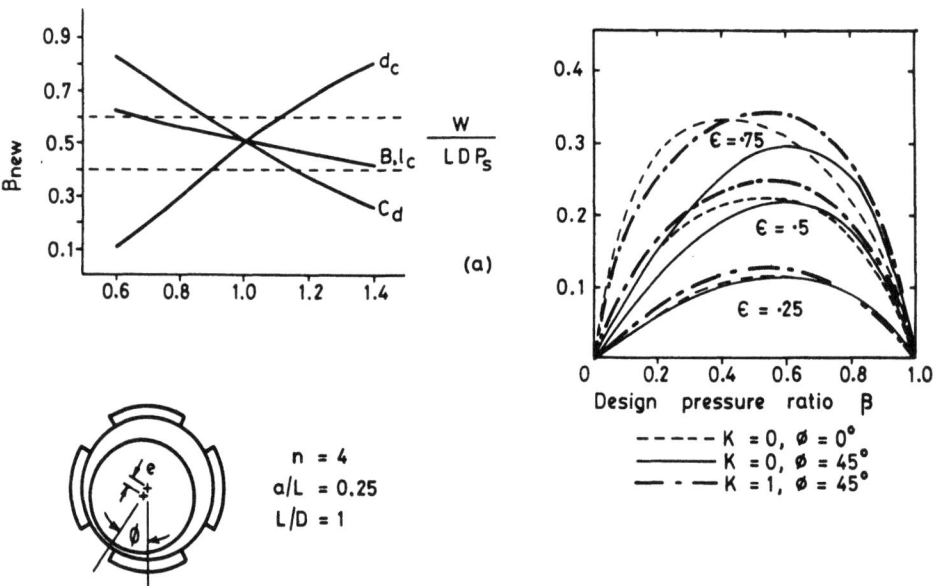

FIGURE 16 PERFORMANCE OF FOUR RECESS HYDROSTATIC JOURNAL BEARINGS

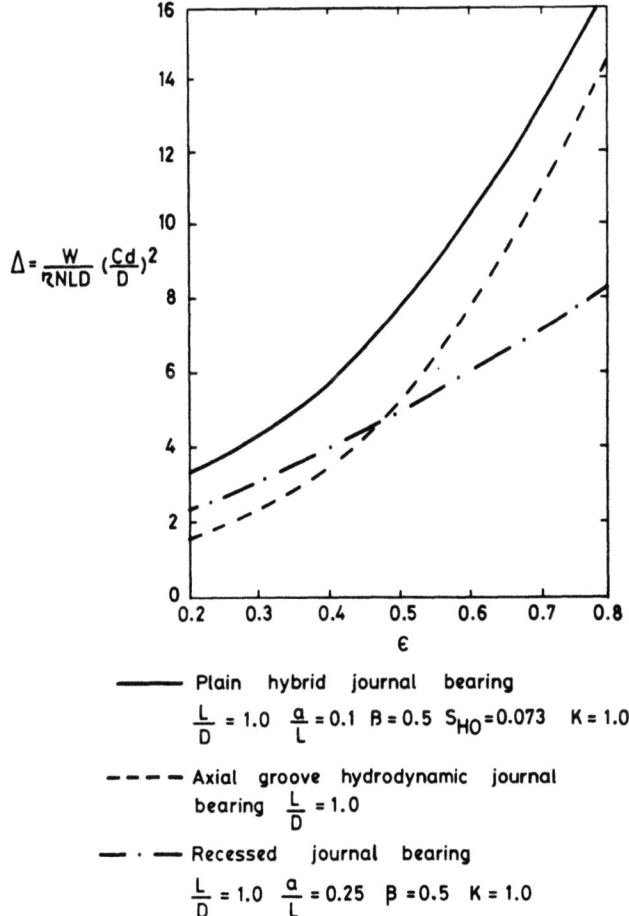

FIGURE 17 COMPARISON OF SLOT HYBRID JOURNAL BEARING WITH AXIAL GROOVE HYDRODYNAMIC JOURNAL BEARING

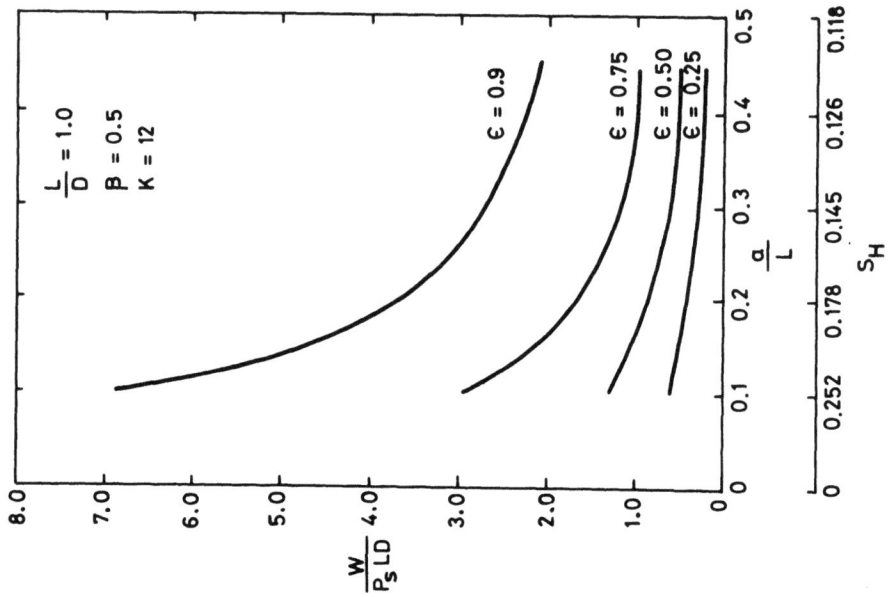

FIGURE 18   SLOT HYBRID BEARING

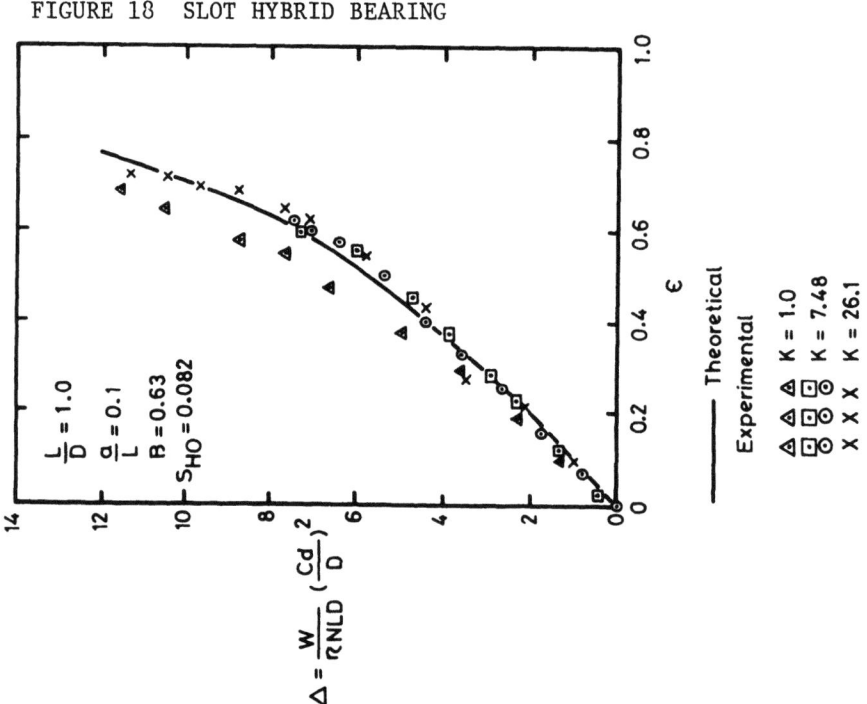

FIGURE 19   VARIATION OF INVERSE SOMMERFELD NUMBER WITH ECCENTRICITY RATIO

Recess pressure

$$P_r = \tfrac{1}{2} P_s$$

Load

$$W = P_r \times \text{Effective area}$$

Effective area = ½ x (total area + recess area)

Stiffness

$$\lambda = \frac{1\tfrac{1}{2} \times W}{h_o} \qquad \ldots\ldots \text{(Capillary)}$$

$$\lambda = \frac{2 \times W}{h_o} \qquad \ldots\ldots \text{(Orifice)}$$

Flow rate

$$q = \frac{P_r h_o^3}{12\, a\eta} \times \text{(mid-land perimeter of bearing)}$$

3.4 Design of Externally Pressurized Gas Bearings - Plane Pads

The load capacity of plane pad gas bearings is subject to greater errors in calculation than liquid bearings. The load capacity depends on a number of factors including the pad shapes, the inlet configuration, and the type of restriction. Experimental results have been given by Wunsch of the National Engineering Laboratory for square pads and these results may also be applied as an approximate guide for circular pads. The design employed by Wunsch involved an orifice with a small inlet pocket as in the journal bearing, Figure 4. The size of the pocket was found to be important. Increasing the pocket size increases the static stiffness and the load carried by a particular size of pad. However, the dynamic stiffness deteriorates and the likelihood of pneumatic hammer is increased. For this reason, it is important to avoid deep pockets. Orifices are typically inserted in plugs of 3.175 mm (⅛ in.) diameter at a depth of 0.075 mm (0.003 in.) to 0.225

mm (0.009 in.) to form a small pocket. A small pocket is a considerable advantage for load and stiffness. The following results were obtained for square pads

$$W_{max} = 0.2 \, P_g^{1.1} \, a^{0.8} \quad \ldots\ldots \text{(Pocket orifice)}$$

$$W_{max} = 0.13 \, P_g^{1.1} \, a^{0.8} \quad \ldots\ldots \text{(Inherent orifice)}$$

where P_g = gauge pressure

a = pad area

A stable air-bearing at the natural frequency tends to act like a simple mass-spring system. It was found by Scholes and Wunsch that the natural frequency could be calculated approximately from the static stiffness and the mass.

$$\omega_n = \frac{\lambda_{st}}{m}$$

On thrust bearings, the natural frequency can be increased by a double-sided system of opposed pads.

Air flow depends strongly on clearance and hence involves the question of the minimum clearance which can be maintained taking account of the necessity for tolerances and structural deformation under load.

An example of typical flow requirements for square pads, as described above, varying in size from 1600 mm² (1 in.²) to 6400 mm² (4 in.²) supplied at pressures up to 0.54 MN/m² (80 lbf/in²) guage may be found in the NEL Report by Nimmo. It was found that these bearings consumed up to 0.28 1/s (0.6 ft.³/min) of free air. In each case, the orifice diameter was 0.5 mm (0.020 in.).

3.5 Design of Externally Pressurized Gas Bearings - Journals

The basic configuration is illustrated in Figure 4, for a double-entry bearing i.e. 2 rows of restrictors at $\frac{a}{L} = 0.25$. An alternative for small $\frac{L}{D}$ ratio bearings is the single-entry configuration i.e. 1 row of restrictors at $\frac{a}{L} = 0.5$. The following is an approximate quide.

n = 8 Number of pocketed orifices per row (orifice entry)

n = 12 Number of inlet slots (slot-entry)

K_{go} = 0.4 Gauge pressure ratio

$\frac{L}{D}$ = 1.0 Length/diameter ratio of bearing

W = ¼ × $(P_o - P_a) D^2$ ..... Single entry

W = 0.4 × $(P_o - P_a) D^2$ ..... Double entry

Typical flow-rate =

0.12 l/s (0.25 ft.^{3}/min) free air with C_d = 0.025 mm (0.001 in.) at 0.54 MN/m^2 (80 lbf/in.^{2}) gauge supply pressure.

Further information for double-entry bearings is given in Figure 20.

4. NOTATION

a, l	Bearing land width
a_s	Width of slot restriction
C_d	Diametrical clearance
d_c	Capillary diameter
e	Eccentricity
h	Film thickness
h_o	Design value of film thickness
l_c	Capillary length
n	Number of recesses, orifices, slots etc.
P_r	Recess pressure
q	Bearing flow-rate
y	Length of flow-path through slot restriction
z	Film thickness in slot restriction
D	Bearing diameter
K	Design parameter = Ratio of Friction Power to Pumping Power

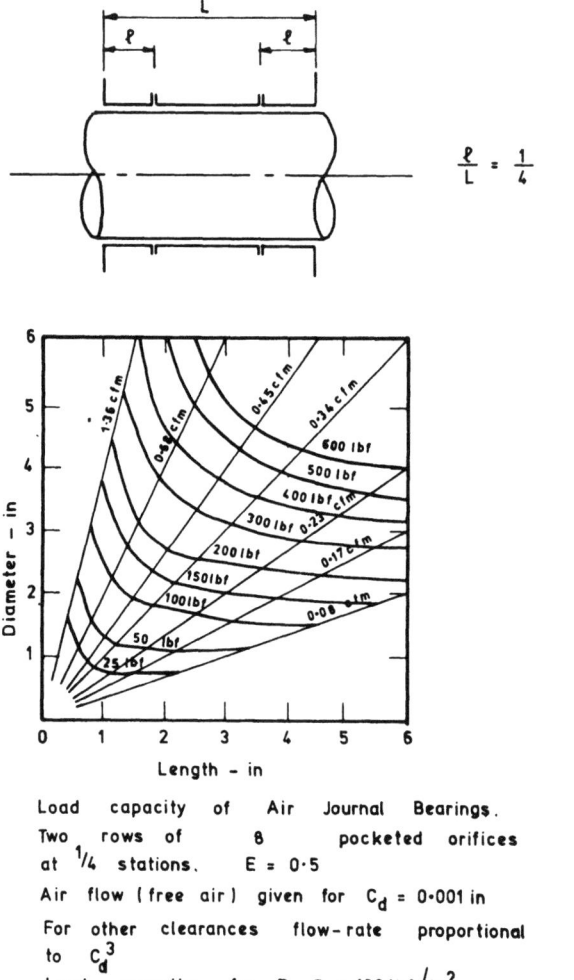

FIGURE 20   DOUBLE ENTRY BEARINGS

L	Bearing length
N	Rotational speed in rev/s
P_g	Gauge supply pressure
P_s, P_o	Supply pressure
S_h	Design parameter = $\frac{\eta N}{P_s} (\frac{D}{c_d})^2$
S_{ho}	Value of S_h when K = 1
W	Bearing load supported by fluid film
β	Pressure ratio P_r/P_s
Δ	Inverse Sommerfeld Number = $\frac{W}{\eta NLD} (\frac{c_d}{D})^2$
ε	Eccentricity ratio e/h_o
η	Dynamic viscosity
λ	Bearing film stiffness

5. A DESIGN BIBLIOGRAPHY

5.1 Hydrostatic Bearings

1. "Hydrostatic Lubrication" by D.D. Fuller, Machine Design, U.S.A
   - Pt. 1. "Oil Pad Bearings" June 1947
   - Pt. 2. "Oil Lifts" July 1947
   - Pt. 3. "Step Bearings" August 1947
   - Pt. 4. "Oil Cushions" September 1947

2. "Design of Hydrostatic Bearings" by H.C. Rippel, Machine Design, U.S.A.
   - Pt. 1. "Basic Concepts" pp. 108 - 117, 1.8.63
   - Pt. 2. "Controlling Flow" pp. 122 - 126, 15.8.63
   - Pt. 3. "Influence of Restrictors" pp. 132 - 138, 29.8.63
   - Pt. 4. "Bearing Friction" pp. 170 - 172, 12.9.63
   - Pt. 5. "Bearing Temperature" pp. 182 - 190, 26.9.63
   - Pt. 6. "Practical Flat Pad Design" pp. 201 - 208, 10.10.63
   - Pt. 7. "Conical and Spherical Pads" pp 185 - 192, 24.10.63
   - Pt. 8. "Cylindrical Pads" pp 189 - 194, 7.11.63
   - Pt. 9. "Journal Bearings" pp. 199 - 206, 21.11.63
   - Pt. 10. "Multi-recess Bearings" pp 158 - 162, 5.12.63

3. "Hydrostatic Bearing Design" J.P. O'Donoghue and W.B. Rowe, Tribology International Vol. 2, No. 1, pp. 25 - 71, Feb. 1969.

4. "Hydrostatic Bearings for Machine Tools" F.M. Stansfield, Machinery Publishing Company, 1970.

5. "Design Procedures for Hydrostatic Bearings" W.B. Rowe and J.P. O'Donoghue, Machinery Publishing Company, 1970.

6. "The Tribology Handbook - Section A9 Hydrostatic Bearings" Edited by M.J. Neale, Butterworth 1973.

7. "Externally Pressurized Bearings - Pt. 1. Journal Bearing Selection" K.J. Stout and W.B. Rowe, Tribology International pp. 98 - 106 Vol. 7, No. 3, June 1974.

8. "Externally Pressurized Bearings - Pt. 3. Design of Hydrostatic Bearings Including Tolerancing Procedures", K.J. Stout and W.B. Rowe, Tribology International, pp. 195 - 212, Vol. 7, No. 5, October 1974.

5.2 Externally Pressurized Gas Bearings

1. "Effect of Bearing Area and Supply Pressure on Flat Air Bearings Under Steady Loading" H.L. Wunsch and W.M. Nimmo, N.E.L. Report No. 38, 1962.

2. "Gas Lubricated Bearings" Edited by N.S. Grassam and J.W. Powell, Butterworth, 1964.

3. "Air Flow Data for Flat Air Bearings" W.M. Nimmo, N.E.L. Report No. 174, 1965.

4. "The Design of Air-Bearing Slideways" H.L. Wunsch, N.E.L. Report No. 201, October 1965.

5. "Design of Aerostatic Bearings" J.W. Powell, Machinery Publishing Company, 1970.

6. "The Tribology Handbook - Section A10 - Externally Pressurized Gas Bearings" Edited by M.J. Neale, Butterworth, 1973.

7. "Externally Pressurized Bearings - Pt. 1. Journal Bearing Selection" K.J. Stout and W.B. Rowe, Tribology International, pp. 98 - 106 Vol. 7, No. 3, June 1974.

8. "Externally Pressurized Bearings Pt. 2. Design of Gas Bearings Including a Tolerancing Procedure" K.J. Stout and W.B. Rowe, Tribology International pp. 169 - 180 Vol. 7, No. 4, August 1974.

5.3 Hybrid Bearings

1. "Fluid-Film Journal Bearings Operating in a Hybrid Mode. Part 1 - Theoretical Analysis and Design and Part II - Experimental Investigation", D. Koshal and W.B. Rowe, Trans. A.S.M.E., 1980.

Appendix B

TRIBOLOGICAL INVESTIGATIONS OF THE CONTACT MECHANICS IN A
ROTARY POSITIVE DISPLACEMENT MACHINE

A. Kumar
B. Reason
Cranfield Institute of Technology

1. INTRODUCTION

This is an initial report of a tribological study undertaken on the sliding contact mechanics of a rotary positive displacement machine. The overall objective of this study was to enhance the operating life efficiency of the machine.

The innovative mechanical design relies on the sliding contact between a tribological pair of elements, namely, a polymeric and a cast iron element.

The overall operating efficiency of the machine is susceptible to any wear taking place in the sliding contact zone, which alters the running clearances in the machine. Any tribological solution to minimize this aspect would, therefore, enhance the overall performance of the machine.

2. CONTACT CONFIGURATION

The physical configuration of the sliding contact between the two rotating elements may be imagined as a cylinder (cast iron) meshing orthogonally with two planer discs (polymer). Specifically, the assembly represents a cast iron worm gear meshing with polymeric pinions.

The cast iron rotor, driven at synchronous speed, drives the pinions at a fixed speed ratio. However, the linear velocity at any point on the meshing surface varies throughout the contact path.

Essentially, the two rotating elements form a noncontacting seal system to contain the working fluid, 'FREON', for the short period that the two are in contact. Thus, the efficiency of the seal is governed by the wear between the elements.

3. OUTLINE OF THE PROBLEM

The sealing action was designed to operate hydrodynamically with a finite fluid film of the transported fluid. The fluid film was not single phase, gas or liquid, but was a two-phase mixture thus negating a 'true' hydrodynamic action.

In practice, however, asperity contact between the two rotating surfaces is sufficient to cause the initial condition of a nominal straight line contact of the polymer pinion to change to a finite area contact. The initial wear is very evident at the early stages of the machine operation. The problem can thus be stated as follows:

(1) To establish the actual operating conditions in the contact zone.

(2) On the outcome of (1) to modify the material selection or specification to maximize operating life and efficiency.

(3) To provide, if possible, some form of boundary lubrication.

4. DESIGN APPROACH

To understand the complex contact condition, it was necessary to obtain 'in situ' data from the machine operating at its actual running condition. Past experience of the wear in complex sliding contact indicated the limitation of an analytical approach[1]. Any analysis (be it a theoretical study or via a separate model simulation) was unlikely to produce meaningful results or the required tribological optimization.

At its most basic level, a simple planer tribological pair of surfaces of cast iron and polymer operating in conditions of mixed friction and using 'FREON' as a lubricant is not readily amenable to analysis. In the machine itself, both the geometry and the kinematics of the contact zone were complex. Influencing factors include the variation of the surface sliding speed and the phase change of the working fluid. Another factor was the behavior of such a tribological

pair at the relatively low temperature ($-10°C$) since a
literature search had produced limited information on such
conditions.

Clearly, since a multiplicity of interacting variables
existed in the contact zone, with varying degrees of
tribological relevance, it was necessary to decide on the most
significant. Another primary consideration was whether any
signal output could be measured and transmitted from the
transitory surface contact during the actual operation of the
machine.

The difficulty in physically achieving this, lies in four
distinct areas:

(1) The general inaccessibility of the contact within
the machine.

(2) The importance of keeping any machine modification
to an absolute minimum both from the cost and time
aspects.

(3) The necessity of transmitting the signals from a
rotating component to the monitoring devices.

(4) The spacial restrictions within the contact area
proper, necessitating the use of miniature
transducers.

5. INSTRUMENTATION

5.1 Choice of Instrumentation

Although several specific variables were available for
measurement in the sliding contact, temperature and fluid
pressure measurement were considered to be of prime importance.
The temperature would indicate the severity of the surface
friction; the pressure would indicate the variation of the
pressure within the transported fluid. The pressure would
also indicate the efficiency of the contact geometry in
producing the quasi-hydrodynamic film.

Spacially, the installation of the micro-thermocouples
presented less of a design problem than that of placing micro-
pressure transducers in the contact zone. Temperature, as a
variable, was therefore selected for a preliminary
measurement. The choice of one initial parameter also reduced
the overall cost. The problem resolved itself, therefore,
into the development of a system for monitoring the transducer

signal from the rotating pinion, together with the sealing arrangement and the machine modifications.

5.2 Instrumentation System

The complete instrumentation circuit is shown in Figure 1, and illustrates the signal transmission, display, and storage. Basically, the signal from the temperature sensors was passed from the rotating component via high quality silver/silver graphite slip rings which were air cooled to minimize thermal drift. Maximum peak-to-peak noise in these units is 50 µV/mA at 10,000 r.p.m. Any 'Seebeck' effects were minimized by mounting rotating junctions on a junction disc integral with the pinion shaft and rotating in ambient air.

Signals from the stationary side of the slip rings were fed, via a junction box, to a cathode ray oscilloscope for display and photography. The signal was also fed to a 'Fluke 2200 B' sixty channel data logger for signal conditioning and data storage. The makers specify an accuracy of $\pm 0.1°C$. An optical pick-up was used, in conjunction with a disc, to provide a time datum for the oscilloscope traces.

5.3 Temperature Sensors

Thermocouples were manufactured in the laboratory from copper/constantan wire 0.127 mm (0.005 ins.) diameter. The basic design requirements for the temperature sensors were as follows:

(1) Small tip mass to delineate rapid temperature transients.

(2) High mechanical integrity and fatigue strength at the sliding contact.

(3) Good output characteristics to minimize signal/noise ratio.

5.4 Method of Installation of the Sensors

The contact edge sensors were installed by milling five grooves (four in the leading edge and one in the trailing edge) in one of the teeth of the polymer pinions, the grooves being 1.0 mm wide x 8 mm long x 0.5 mm deep. Each sensor was cemented in place using an epoxy resin adhesive and the excess smoothed to the profile of the pinion, upon hardening. Tapping holes, drilled in the pinion, led the sensor wires out through the middle of the shaft to the slip rings.

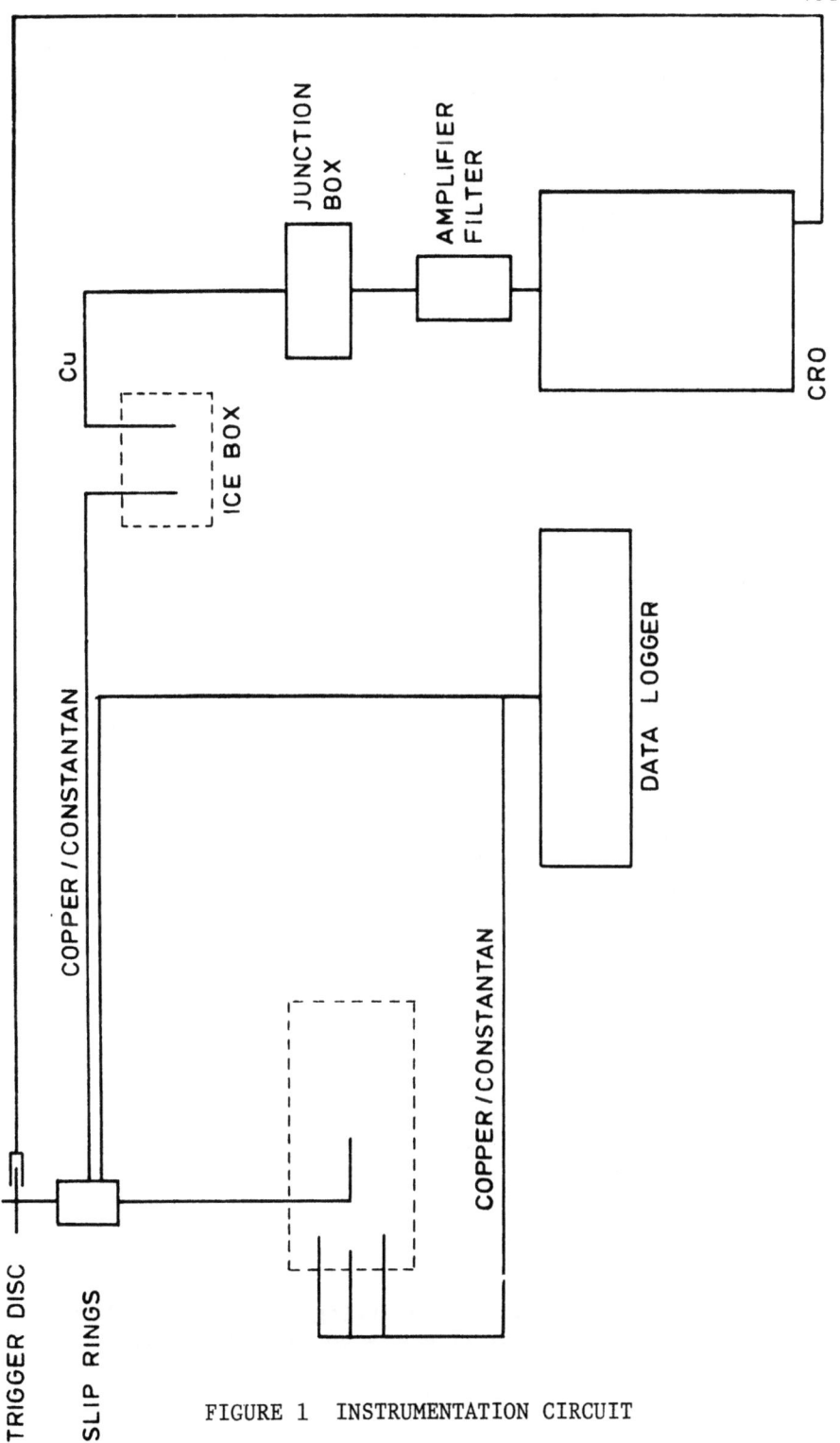

FIGURE 1 INSTRUMENTATION CIRCUIT

Additionally, a reference sensor was placed in the pinion to datum the mean temperature of the 'FREON'.

5.5 Testing Stages

From Reference 1 it was already known that, for ferrous surfaces in sliding contact, a two-zone temperature distribution was measured. This consists of a surface skin effect and a bulk material effect. The former gives temperature transients within a skin some 0.5 mm (0.020 ins.) thick while the latter shows a gradual change of temperature with time. There was no evidence to suggest that this effect would be manifest in polymeric material.

Testing was, therefore, initiated in two stages. Initially, the edge sensors were located 1.5 mm (0.060 ins.) below the edge for the first tests. A second pinion had its sensors mounted flush with the contact surface; the thermocouple beads being honed back to the contact edge. This arrangement would, in effect, sense the temperature level of the contact surface proper.

6. RESULTS

Tests were run for over 200 hours during which time the output signals from the sensors were monitored and photographed. Plate 1 shows a typical surface thermocouple trace at the initial and final periods of the tests. The trigger, in each case, is the lower trace. From these, the temperature transients during each cycle are seen to occur during the 10 ms of engagement.

The test run with the sensors placed below the contact line showed no transients at all, a mean constant temperature of $-15°C$ being recorded throughout the test.

It must again be emphasized that this work reports merely the preliminary findings. Further refinements have been achieved and will be presented at a later date. However, the essential nature of the contact mechanics, as far as the temperature effect is concerned, is considered to have been established.

7. DISCUSSION

The most noticeable facts emerging from a study of the photographic traces (Plate 1) may be stated as follows:

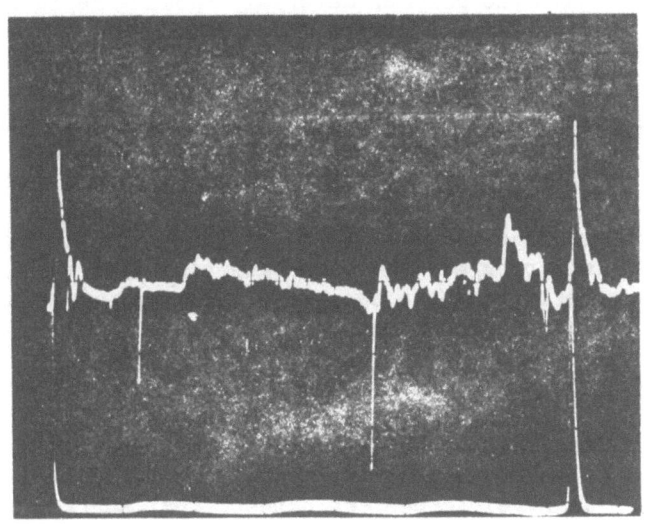

PLATE 1A. PROFILE AFTER 1 HOUR.

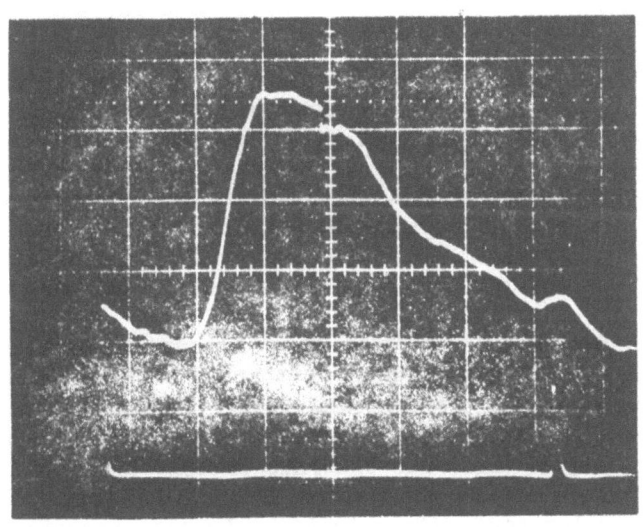

PLATE 1B. PROFILE AFTER 200 HOURS.

PLATE 1 TEMPERATURE PROFILES

(1) As with ferrous surfaces, there is a 'skin' effect with the polymer material, showing rapid temperature fluctuations during the initial running period, Plate 1. A.

(2) As in Reference 1, there is a distinct change in both the magnitude of the temperature and the general form of the temperature profile with time. Plate 1. B. shows that after 200 hours running the surface temperature fluctuations in the skin have disappeared, the temperature profile merely reflecting the changes of temperature of the transported fluid.

(3) The 'bulk temperature', as with the ferrous surfaces, does not reflect the rapid skin transients experienced during the initial running-in, and remains largely constant throughout the tests. This reinforces the concept that surface transient effects convey heat to the bulk material in a gradual manner.

Figure 2, taken from Reference 1, shows curves of temperature plotted against distance below the contact surface for a variety of test duration times. From this, two basic points emerge:

(1) The subsurface temperature gradient is markedly affected by the duration of the running time; the longer the time, the less the gradient.

(2) In the surface skin proper and in the bulk material some distance below the skin, the temperature gradient is linear, while in the transition zone between the two, the effect in nonlinear.

Essentially, the 'heat sink' mechanism mentioned earlier may be seen to apply, heat constantly being generated within the surface skin transferring itself to the thermal reservoir of the bulk material, the temperature gradient between the two being reduced (and thus the rate of heat transfer) as the mean temperature of the thermal reservoir is raised. This results in a smaller concomitant increase in heat transfer if the friction (and therefore the heat generated) remains constant.

This mechanism appears to apply for ferrous materials but the same could be true for a polymeric material with a lower heat dissipation property.

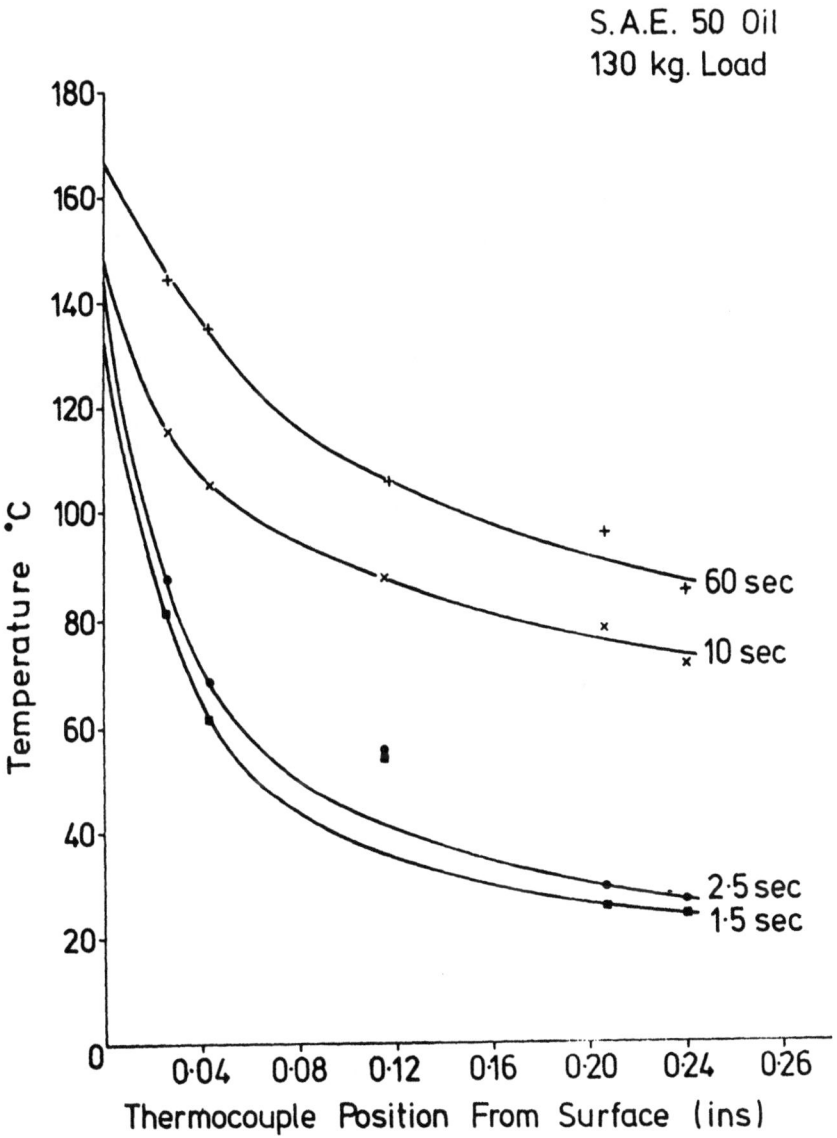

FIGURE 2   EXTRAPOLATED TEMPERATURE TO SURFACE

Considering Plate 1 from this perspective, we may postulate a similar mechanism for the polymeric material contact. Initially, the pinion makes a nominal line contact with the ferrous surface of the rotor. Specific pressures are high and rapid temperature fluctuations are manifest in the contact region, frictional heat being concentrated on the asperities as they interact with one another in the regions of high local pressure. This produces localized temperature spikes. Clearly, the frictional heat intensity is a maximum as the surface asperity area is small, giving low heat transfer.

This heat flux is rapidly dissipated, not internally to the bulk material of the polymeric pinion, while it remains at a constant temperature, but rather to the two phase mixture in the contact zone. Some heat transfer to the rotor surface, (because of its better thermal conductivity) is also probable. However, as the asperities are rapidly worn away, the effective area of the tooth flank that 'sees' the rotor groove increases, producing better heat transfer to the polymer pinion. Concomitant with this, two effects are manifest; first, specific pressures decrease with increasing area, thus decreasing local contact frictional energy. Second, a higher proportion of the load is carried by the quasi-hydrodynamic fluid film in the contact zone, thus decreasing frictional energy generation yet further.

Plate 1. B. shows the run-in surface absorbing (and therefore retaining for a longer period) a higher proportion of the total heat now generated in the contact by the transportation of the 'FREON'. This gives an exponential type decay, the heat transfer being largely governed by the heat transfer coefficient of the polymeric material.

This effect is modelled in Figure 3 which indicates the direct effect of the changed contact area on the proportion of heat transferred to the pinion material in terms of the change in the temperature profile seen in Plate 1, A and B. Initial measurement of the amount of wear at the contact edge of the pinion showed an average wear scar of some 0.5 mm (0.020 ins.) width (after 200 hours of running) from the nominal line contact at the start of the tests.

8. CONCLUSIONS

It has become clear from this preliminary study, of the striking overall similarity in the surface mechanics of the polymeric material and cast iron to the situation of two hardened ferrous surfaces[1]. The similarity of the two zone temperature distribution in these two tribological systems is

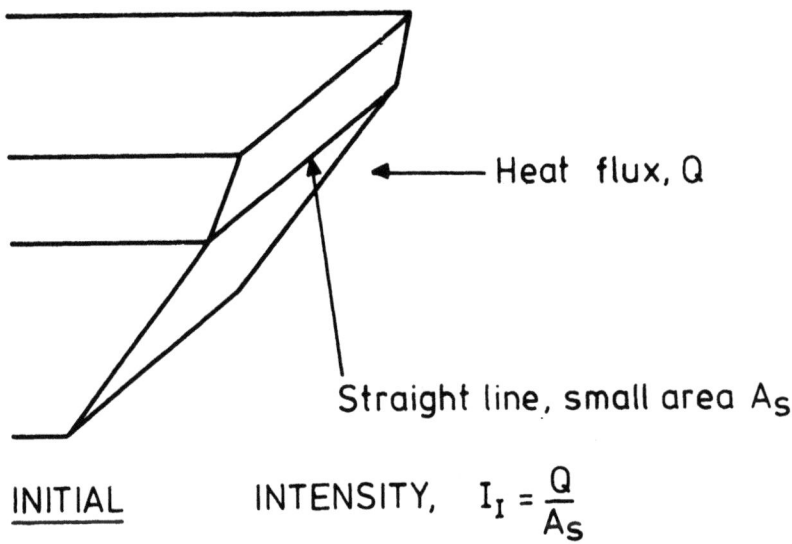
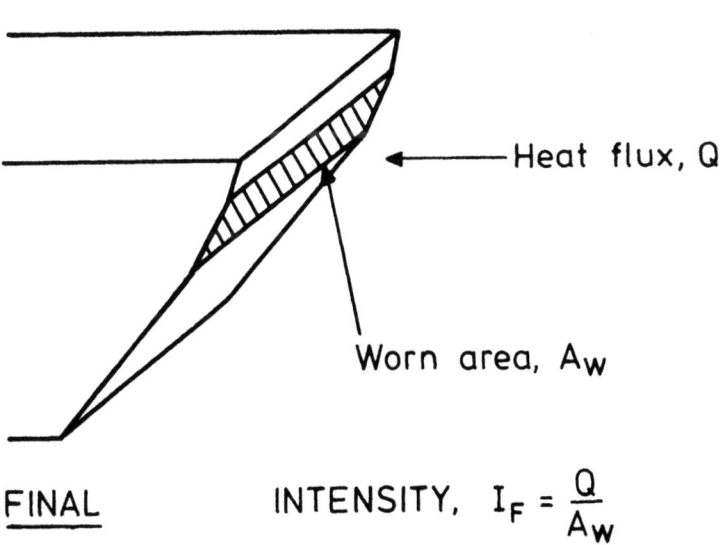

FIGURE 3 IDEALIZED CONTACT CONFIGURATIONS

the most surprising aspect of this study. This suggests that the surface asperity wear, and its related 'skin heat transfer' effect, may be manifest in a wide variety of contact conditions.

The major experimental challenge of monitoring, displaying and processing the transducer signal from the contact zone with the machine operating under actual operation has been achieved. Further experimental results are encouraging and it is hoped, with enhancement of the data acquisition system, a better understanding of the contact mechanics will be obtained.

The project illustrates the importance of a multi-disciplinary tribological approach towards improving the operating life and efficiency of an industrial machine. These improvements will enhance cost effectiveness, for this type of rotary positive displacement machine.

9. REFERENCES

1. Reason, B.R., "The 'Cranfield' Four-Ball Machine: A New Development in Lubricant Testing," (1975), International Tribology Symposium "Tribology for the Eighties", Sept. 1975, Paisley, Scotland, U.K.

Appendix C

THREE DIMENSIONAL TOPOGRAPHICAL DESCRIPTIONS OF SOLID SURFACES

B. Snaith
Cranfield Institute of Technology

M. J. Edmonds
Brighton Polytechnic

S. D. Probert
Cranfield Institute of Technology

1. INTRODUCTION

The discussion describes a surface measuring system developed for accurate three-dimensional surface characterization and the quantification of surface degradation due to wear or any other deformation process. The system is based on a stylus profile-tracing instrument and incorporates an automatically controlled parallel-profile tracing technique, with micro-computer data handling and processing. The advantages of such a technique for three-dimensional assessment of surfaces, as against the more usual two-dimensional assessments, are highlighted in the typical quantitative and qualitative surface representations obtained from the system.

2. NOMENCLATURE

m Mean absolute profile slope over sampling length, radians

R_q Root-mean-square roughness over sampling length, μm

R_k Kurtosis of the ordinate distribution density

R_s Skewness of the ordinate distribution density

R_z Ten-point height, μm

R_{max} Maximum peak-to-valley height within a sampling length, μm

t_p Bearing ratio at a depth p (expressed as a percentage of R_{max}) below the highest peak, %

λ_a Average wavelength, μm

3. SURFACE ASSESSMENT

The need for accurate surface descriptions is well appreciated in the field of tribological research. Consequently, during the last two decades the expenditure of considerable effort has resulted in important progress with respect to (i) the choice of parameters which truly and uniquely characterize a surface, as well as (ii) the invention, development, and commercial production of reliable, highly sensitive equipment for measuring these parameters. Problems concerning element (i) occur because of the large number of parameters available, some of which are interrelated and may not uniquely specify a surface. Also in making the choice, careful consideration must be given to the particular application[1]. Most previous and present surface analysis techniques are based on the digitized interpretation of recorded profiles, obtained from stylus instruments performing single profile traces. The value of the technique has been greatly enhanced by the application of parallel-profile tracing[2-5]. More truly representative surface descriptors, emerge from this technique as a result of multiple parallel tracings, and averaging the recorded data. Added to this is the ability to compute topographic rather than only single profile data. Surface topographical descriptions in the form of isometric and contour maps, may also be produced.

This discussion describes a three-dimensional surface measuring system employing parallel-profile tracing and presents examples of its output, which are relevant to the examination of surfaces before and after they have been subjected to friction and wear.

4. MEASURING SYSTEM

The system developed at Cranfield Institute of Technology has been based on a Talysurf 4 stylus profile-tracing instrument. It incorporates an automatically controlled, three-dimensional relocation stage, and a micro-computer data handling and processing facility.

The relocation stage, Figure 1, consists of two moving tables: one which traverses parallel to the locked stylus arm and the other indexing perpendicular to the arm thereby enabling parallel tracing to be performed. The former is driven through a pulley arrangement and reduction gearbox by means of a D.C. motor. Slow speeds of traverse (i.e. 1mm/sec or 0.25mm/sec) can be chosen according to the degree of surface roughness likely to be encountered and fast return speed is used in the reverse direction, i.e. when the stylus has been raised off the surface. The straightness accuracy of the slide-way is to within 1µm over the complete 150mm range of traverse. An optical encoder, coupled into the reduction gearbox, enables accurate positioning of the traversing table. Calibration results using an interferometer yield a basic encoder step length of 1.532µm $\pm$ 0.004µm over the full range of traverse. The sampling interval has a minimal value of 3.06µm (corresponding to every second encoder pulse) or can be any multiple of this up to 61.2µm. The indexing table, mounted on the lower traversing table, is driven by a stepper motor. This gives a minimal step-interval between the parallel traces of 2.501µm $\pm$ 0.034µm over a maximum 25mm traverse.

Lifting of the stylus arm, in order to raise the stylus off the surface to permit the fast return, is achieved by a motor-driven lifting arm connected to a light stirrup placed around the stylus carrying arm. No restrictions of the normal operation of the stylus instrument, i.e. single traces and meter cut-off options, have been incorporated into the design. Disconnection of the lifting stirrup and unlocking of the stylus arm enables the instrument to be used in its normal mode. A kinematic relocation facility, using the three-ball technique, is incorporated on top of the traversing table. Relocation of this table and the indexing table to within $\pm$1µm is possible by means of the positional display units on the associated electronic control console. This console consists of several modules, each interlinked to provide a fully automatic operation cycle. Pre-settings of the required sampling/parallel step intervals and number of traces, together with the traverse position at which sampling will commence and end, are made on the console. A general view of the system is shown in Figure 2.

Ouput from the stylus head is fed directly to a North Star Horizon micro-computer through an 8-bit ADC input channel and recorded onto floppy discs. Data handling is software-controlled; the unit possessing a capacity of 16×10^3

FIGURE 1   THREE-DIMENSIONAL RELOCATION STAGE

FIGURE 2 VIEW OF THE MEASURING SYSTEM

readings per stylus trace and up to 250 ensemble averages. At this stage some standard surface parameters may be evaluated on the micro-computer which provides an output to a line printer. For detailed statistical analyses and qualitative representation of the surface in the form of contour maps, data are transferred via a direct link to a GEC 4070 interactive computer, Figure 3.

5. RESULTS AND DISCUSSION

5.1 Surface Parameters

The effects of the measuring procedure on certain defined parameters for a surface are shown in Table 1. In particular, a quantitative illustration of the effect of performing parallel tracing and averaging the profile data may be observed. For turned and ground surfaces, some of the standard surface parameters computed on the micro-computer at various ensemble averages are given. R.M.S. roughness, ten-point height and mean surface slope for both surfaces show most distinct changes on increasing from single profile analyses to averaging from two parallel traces. Similarly the average wavelength parameter, which is particularly sensitive to the wear of the surface, significantly changes according to how many traces are analyzed. The skewness and kurtosis parameters, which have been used to assess the running-in of contacting surfaces show different effects as the number of profiles increases[6]. For both surfaces the skewness values show only minor changes, the ground surface exhibits a slightly negative skew and the turned surface a small degree of positive skew, see Figure 4. This does not occur for the recorded kurtosis parameter in that, for the ground surface at one profile trace, the implied distribution is platykurtic changing to strongly leptokurtic on increasing the number of traces taken. Consideration must be given (when assessing such parameters) to the problems of large scatter due to random sampling[7]. The power spectrum, Figure 5, becomes more evenly distributed as the number of profiles taken increases. It has been suggested that the power spectral density function derived by Fourier analysis, as presented here, is an insensitive indicator of wear and that the Walsh power spectrum may be more appropriate when attempting to quantitatively describe wear[8].

The commonly-used bearing area curves for both specimens, Figure 6, provide an interesting comparison between topographic and profile measurements. A total of 100 parallel profile traces, each trace containing 128 samples, makes up the array used for the computation of topographical curves. As would be expected, the single profile assessment taken along

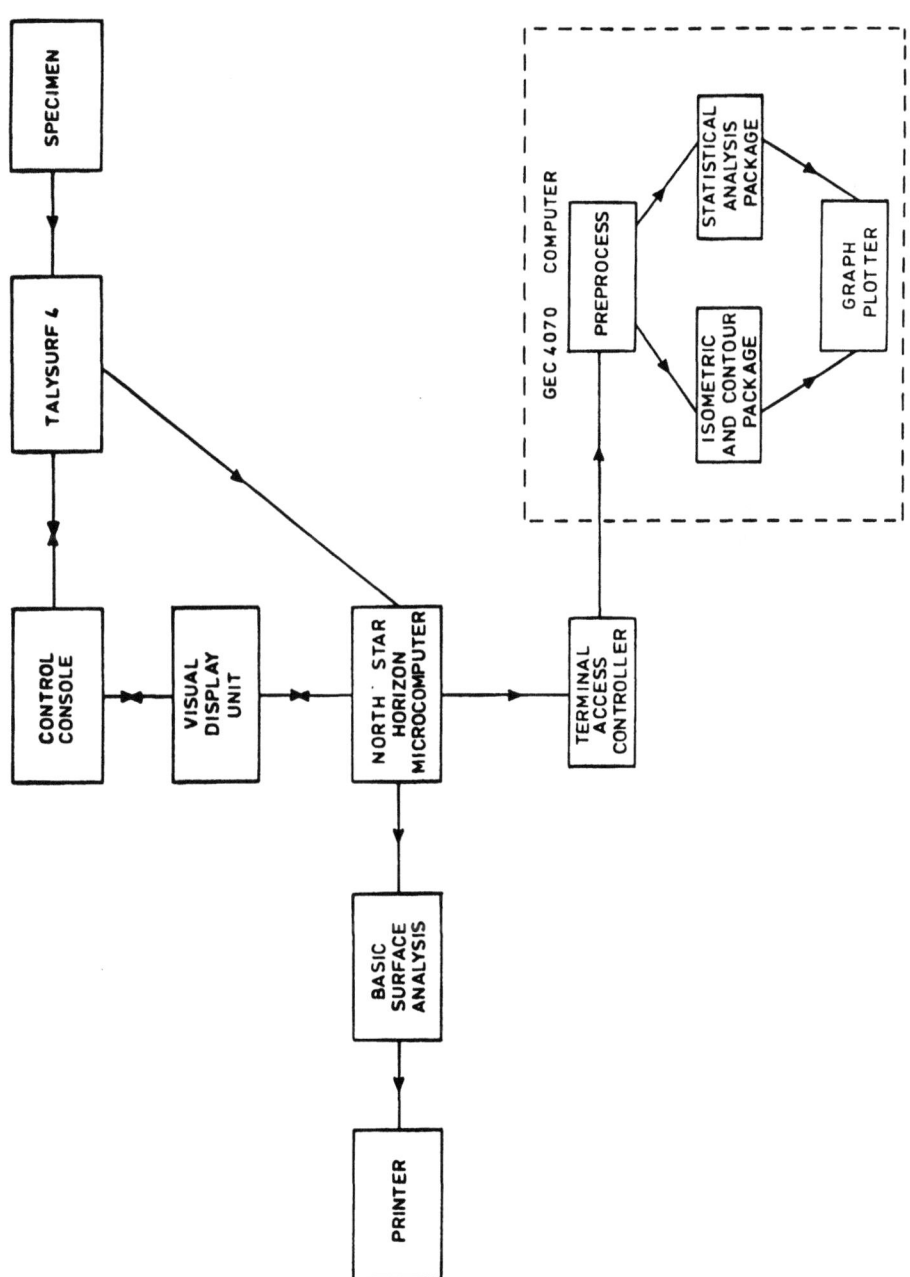

FIGURE 3   DATA HANDLING AND PROCESSING SEQUENCE OF OPERATIONS

Surface Examined	Parameters	R.M.S. Roughness R_q	Ten Point Height R_z	Mean Surface Slope m	Average Wavelength λ_q	Skewness R_s	Kurtosis R_k
Ground-Surface (along lay) parallel step interval of 12.5 μm	1 profile trace	0.286	1.68	0.042	42.78	-0.37	-5.30
	2 profile traces	0.848	4.58	0.181	29.44	-0.39	6.50
	5 " "	0.937	4.13	0.190	30.99	-0.45	7.72
	10 " "	0.829	4.62	0.144	36.17	-0.53	19.4
Turned-Surface (along lay) parallel step interval of 25 μm	1 profile trace	16.05	62.52	0.335	301.0	0.19	-7.06
	2 profile traces	17.54	85.01	0.990	111.3	0.33	-3.90
	5 " "	17.44	71.76	0.950	115.4	0.31	-4.72
	10 " "	17.89	79.79	0.995	112.9	0.32	-0.18

ALL ANALYSES: 2048 SAMPLES AT A 3.06 μm SAMPLING INTERVAL

TABLE 1   SURFACE TEXTURE PARAMETERS AS EVALUATED AT VARIOUS ENSEMBLE AVERAGES

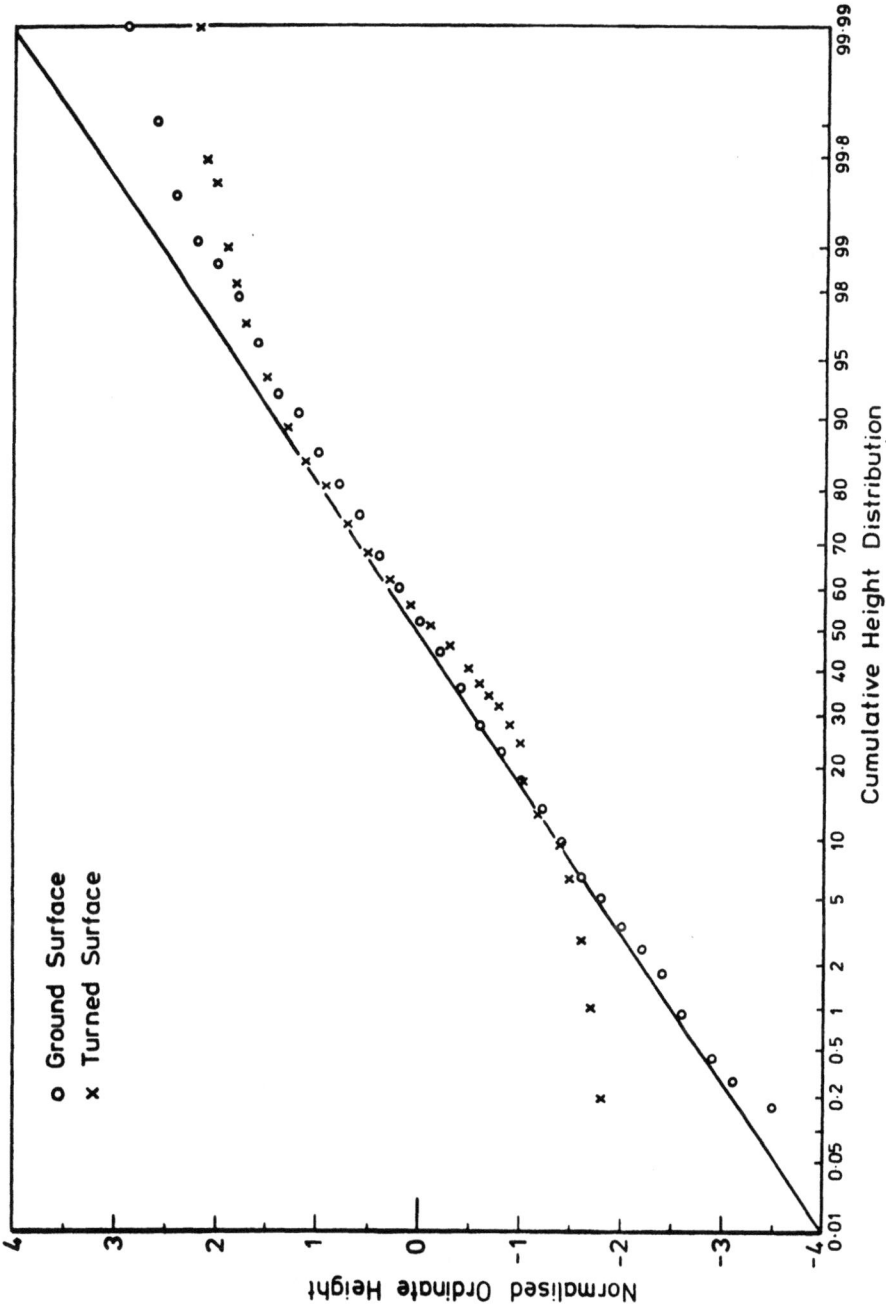

FIGURE 4   CUMULATIVE ORDINATE HEIGHT DISTRIBUTIONS

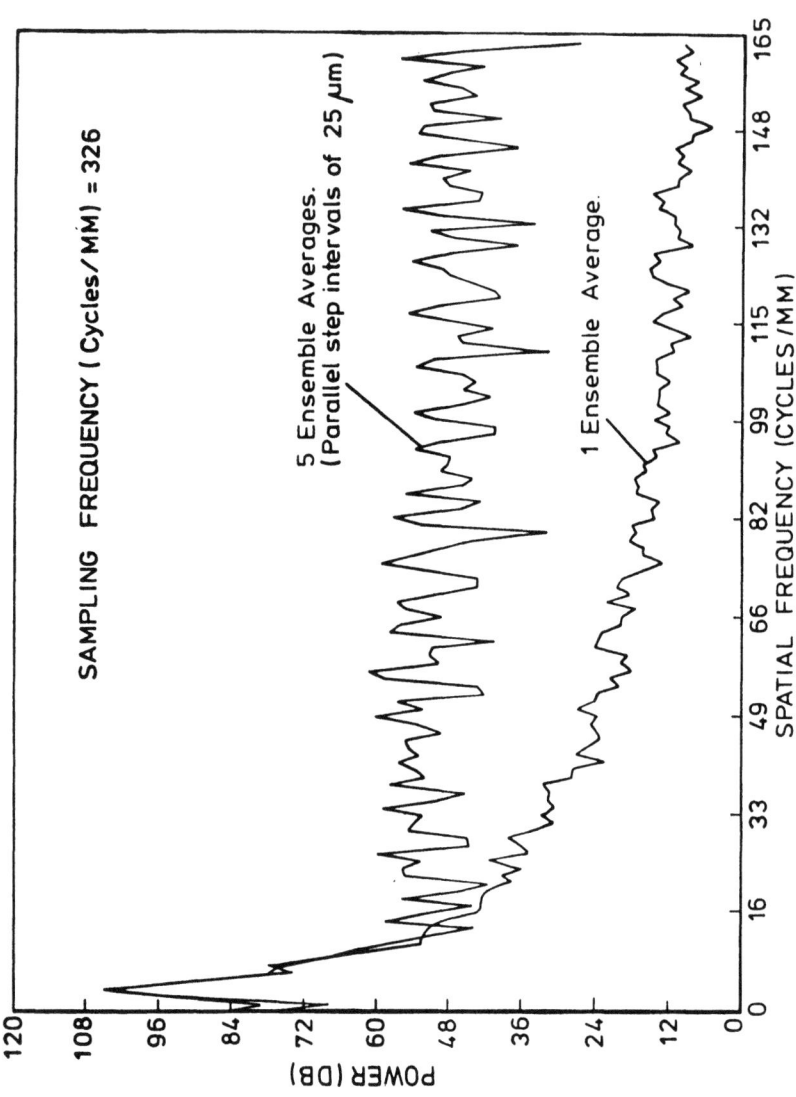

FIGURE 5 POWER SPECTRUM FOR A TURNED-SURFACE FOR TWO DIFFERENT ENSEMBLE AVERAGES

the lay of the periodic turned surface gives a far better
description of bearing area than the assessment from a
profile, taken at random, across the surface lay, Figure 6a.
For the more random ground specimen, topographic, single
profile and ten profile assessments are collated and compared
in Figure 6b.

5.2 Surface Maps

An isometric view of a turned surface is depicted in
Figure 7. This covers 1.57 x 1.25mm^2 of nominal area, made up
from 100 parallel traces at a step interval of 12.5 μm, each
trace containing 128 samples taken at 12.24μm intervals. The
associated surface height distribution map, Figure 8, contains
ten contour levels each at an interval of 8.6μm. Smoother
surfaces may be described using the experimental measuring
system described; Figure 9 illustrating a lapped surface of
0.021μm r.m.s. roughness. The assessment time required for
surface tracing, data transfer, and plotting over such an area
is typically about 90 minutes.

Such plots can assist in quantifying surface degradation
due to wear or any other deformation process. The relocation
facility enables the accurate assessment of the same area of
surface before and after damage has been inflicted. Inverted
maps (i.e. with the valleys represented as peaks) may be
produced to assist in the examination of local deformations
within valleys[9].

6. CONCLUSION

The experimental facility described provides accurate
measurements and pictorial representations of surfaces in
three-dimensions. Comparison of single profile assessments
and averaged multiple profiles clearly highlights the
advantage of this three-dimensional profilometric technique
for the characterization of surfaces. Such systems enable
useful quantitative assessments of surface wear to be obtained
as well as to examine the relationships between profile and
surface statistics.

To reduce computational time, the future development of
the micro-computer software is now envisaged. Also, a
completely software-controlled operation by the micro-computer
is planned.

7. ACKNOWLEDGEMENT

The authors are indebted to the Science Research Council
for support of this project.

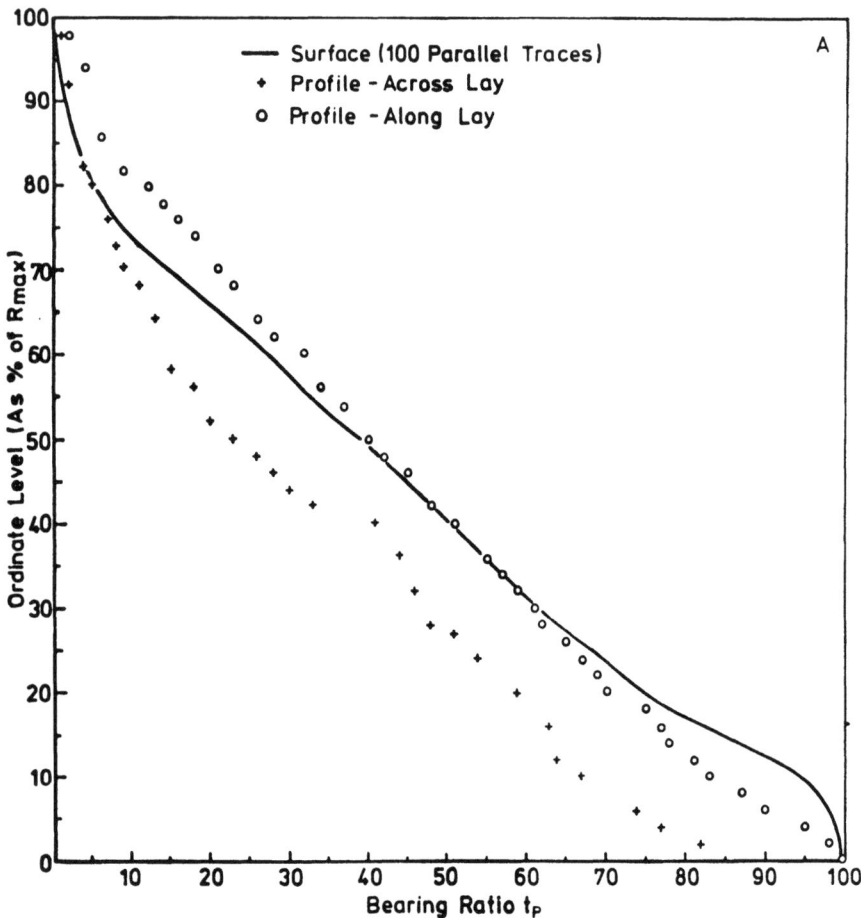

FIGURE 6(A) TOPOGRAPHICAL AND PROFILE BEARING AREA CURVES - TURNED-SURFACE BEARING CURVES

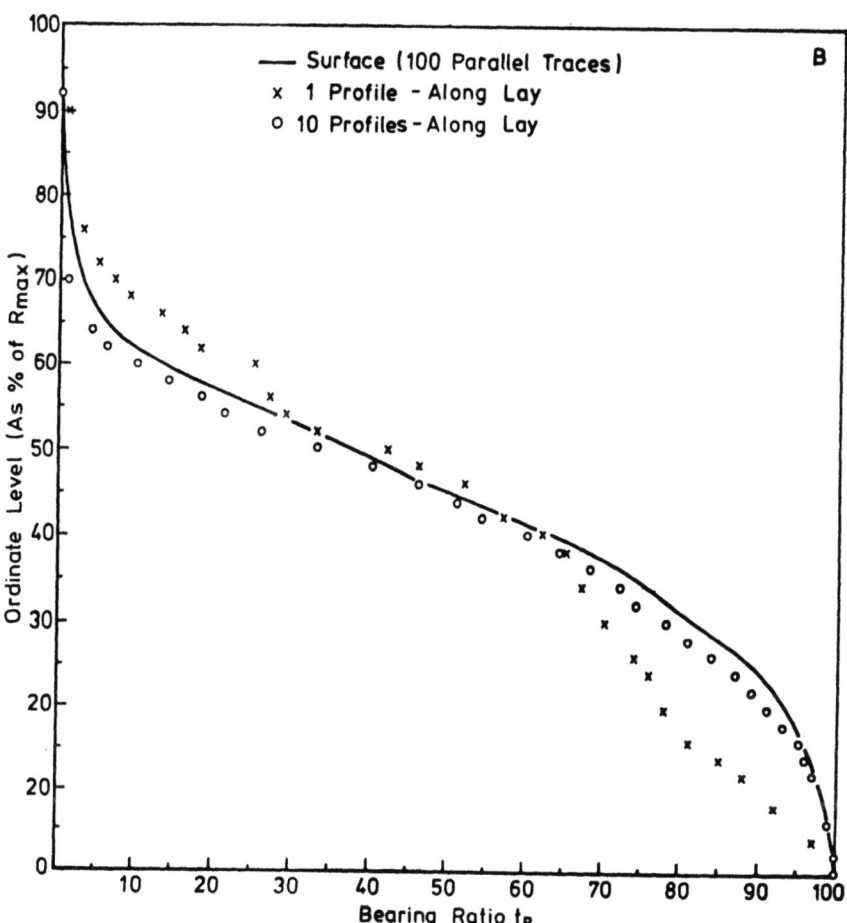

FIGURE 6(B) TOPOGRAPHICAL AND PROFILE BEARING AREA CURVES - GROUND-SURFACE BEARING CURVES

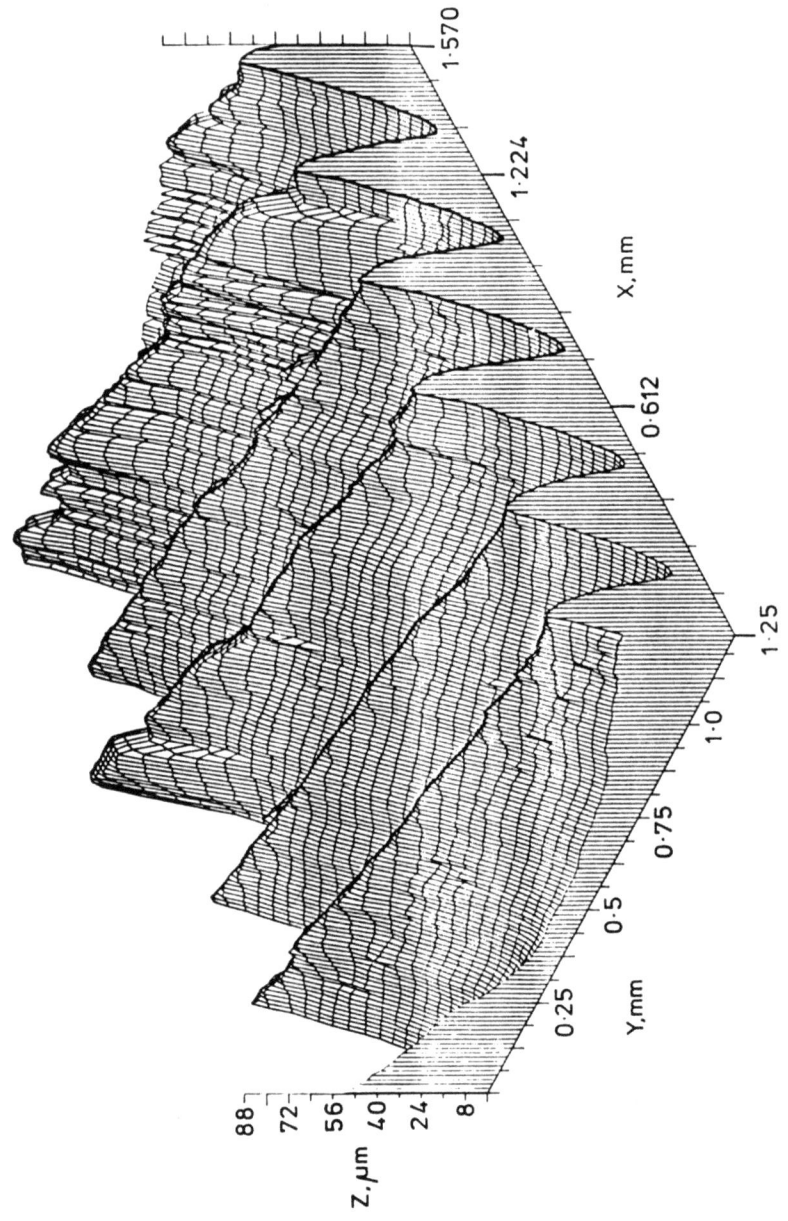

FIGURE 7   ISOMETRIC VIEW OF A TURNED-SURFACE

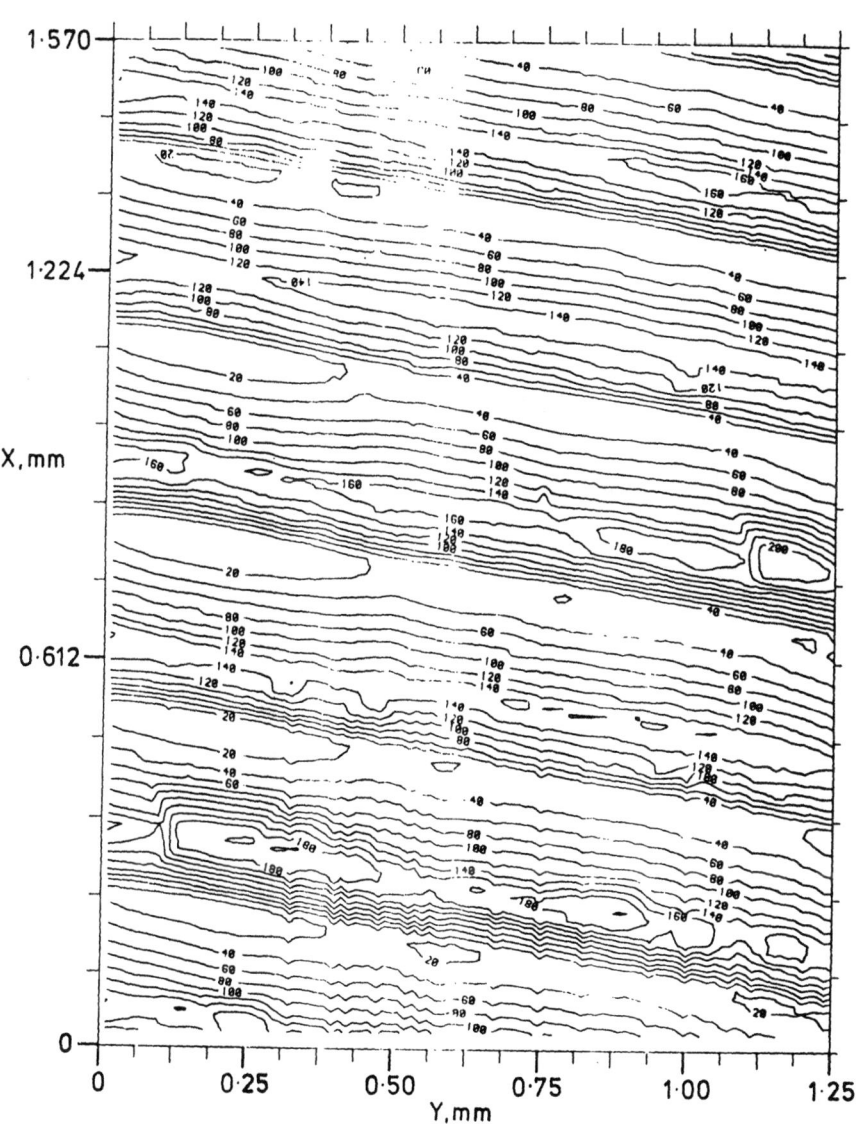

FIGURE 8   CONTOUR PLOT OF THE TURNED-SURFACE DEPICTED
IN FIGURE 7

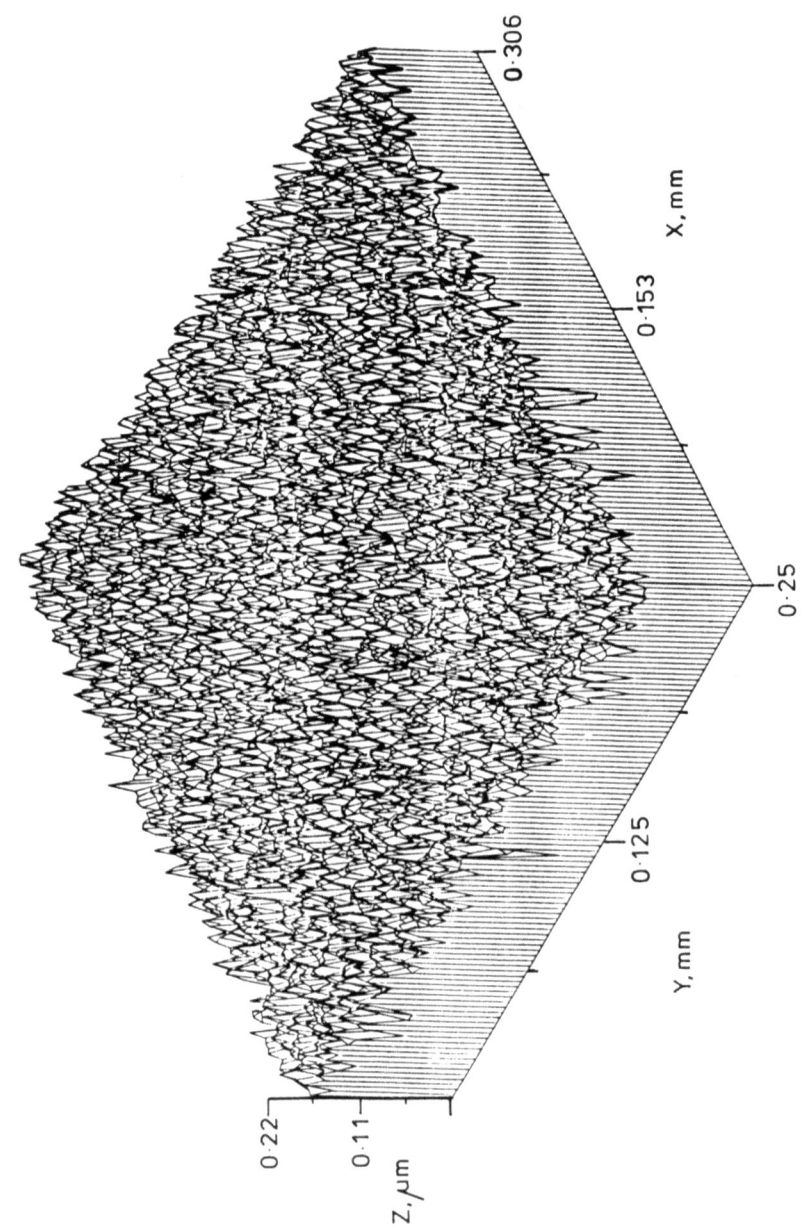

FIGURE 9   ISOMETRIC VIEW OF A LAPPED SURFACE

8. REFERENCES

1. Thomas, T.R., "Characterization of Surface Roughness," Prec. Eng., Vol. 3, No. 2, 1981, 97 - 104.

2. Williamson, J.B.P., "The Microtopography of Solid Surfaces," Proc. I. Mech. Eng. 182 (3K) 1967/68, 21 - 30.

3. Grieve, D.J., Kaliszer, H., and Rowe, G.W., "A Normal Wear Process Examined by Measurements of Surface Topography," Ann. CIRP, Vol. 18, 1970, 585 - 592.

4. Sayles, R.S. and Thomas, T.R., "Mapping a Small Area of Surface," J. of Physics E, Vol. 9, 1976, 855 - 861.

5. Williams, A., and Irdus, N., "Detection and Measurement of Damage to Surfaces after Static Contact Loading," Proc. Int. Conf. on Metrology and Properties of Engineering Surfaces, Leicester Poly., U.K., April 1979. Edited by K.J. Stout and T.R. Thomas, Elsevier, Sequoia, S.A., Lausanne, 281 - 291.

6. King, T.G., Watson, W., and Stout, K.J., "Modelling of Micro-Geometry of Lubricated Wear," Proc. 4th Leeds-Lyon Symp., MEP, London, 1978, 333 - 443.

7. Thomas, T.R., and Charlton, G., "Variation of Roughness Parameters on some Typical Manufactures Surfaces," Prec. Eng., Vol. 3, No. 2, 1981, 91 - 96.

8. Smith, E.H., and Walmsley, W.M., "Walsh Functions and Their Use in the Assessment of Surface Texture," Proc. Int. Conf. on Metrology and Properties of Engineering Surfaces. Leicester Poly, U.K. April 1979. Edited by K.J. Stout and T.R. Thomas, Elsevier, Sequoia, S.A. Lausanne.

9. Snaith, B., Edmonds, M.J., and Probert, S.D., "Use of a Profilometer for Surface Mapping," Prec. Eng., Vol. 3, No. 2, 1981, 87 - 90.

If you have any concerns about our products,
you can contact us on
ProductSafety@springernature.com

In case Publisher is established outside the EU,
the EU authorized representative is:
**Springer Nature Customer Service Center GmbH
Europaplatz 3, 69115 Heidelberg, Germany**

Printed by Libri Plureos GmbH
in Hamburg, Germany

TRIBOLOGICAL TECHNOLOGY – VOLUME I

NATO ADVANCED STUDY INSTITUTES SERIES

Proceedings of the Advanced Study Institute Programme, which aims at the dissemination of advanced knowledge and the formation of contacts among scientists from different countries.

The series is published by an international board of publishers in conjunction with NATO Scientific Affairs Division

A	Life Sciences	Plenum Publishing Corporation
B	Physics	London and New York
C	Mathematical and Physical Sciences	D. Reidel Publishing Company Dordrecht and Boston
D	Behavioural and Social Sciences	Martinus Nijhoff Publishers The Hague, Boston and London
E	Applied Sciences	

Series E: Applied Sciences — No. 56

TRIBOLOGICAL TECHNOLOGY VOLUME I

Proceedings of the NATO Advanced Study Institute on Tribological Technology, Maratea, Italy, September 13 – 26, 1981

edited by

Peter B. Senholzi
Mechanical Technology Incorporated
1656 Homewood Landing Road
Annapolis, Maryland, U.S.A.

1982

Springer-Science+Business Media, B.V.

Distributors:

for the United States and Canada
Kluwer Boston, Inc.
190 Old Derby Street
Hingham, MA 02043
USA

for all other countries
Kluwer Academic Publishers Group
Distribution Center
P.O.Box 322
3300 AH Dordrecht
The Netherlands

ISBN 978-94-011-9809-7   ISBN 978-94-011-9807-3 (eBook)
DOI 10.1007/978-94-011-9807-3

Copyright © 1982 by **Springer Science+Business Media Dordrecht**
Originally published by Martinus Nijhoff Publishers in 1982
Softcover reprint of the hardcover 1st edition 1982

All rights reserved. No part of this publication may be reproduced, stored in a retrieval system, or transmitted, in any form or by any means, mechanical, photocopying, recording, or otherwise, without the prior written permission of the publishers, Springer-Science+Business Media, B.V.

CONTENTS OF VOLUME I

Acknowledgements   VI

Chapter I
Peter B. Senholzi: Introduction   1

Chapter II
Horst Czichos: Scope of Tribology   7

Chapter III
Nam P. Suh: Surface Interactions   37

Chapter IV
A.W.J. de Gee: Materials in Tribotechnical Applications   209

Chapter V
M.J. Edmonds: Surfaces   287

ACKNOWLEDGEMENTS

The Tribological Technology ASI Program was cosponsored by NATO and the U.S. Office of Naval Research. Dr. M. di Lullo has served as the NATO sponsor and Messrs. H. Martin and K. Ellingsworth have served as Office of Naval Research sponsors. Credit is given to these individuals for their generous assistance and guidance that was provided during the organization and conductance of the ASI Program.

A special acknowledgement is extended to the respective ASI technical contributors who have taken time from their busy schedules in order to ensure the successful conductance of the ASI Program and compilation of the text (especially Brian Reason and Horst Czichos who served on the ASI Organizing Committee). My staff members, Nanette Brown and Alan Maciejewski, are also to be acknowledged for their assistance in the areas of program organization, material compilation, and text editing. Appreciation for the extensive manual typing responsibilities is extended to Susan G. Rohn. Special recognition and thanks is extended to my dedicated assistant, Phyllis Pittman, who expended considerable time and energy in the successful completion of all facets of the ASI Program. Last but not least, I would like to extend my gratitude to my wife, Patricia, who provided considerable support as well as exhibited extreme patience throughout the conductance of this Program.

INTRODUCTION

Peter B. Senholzi
Manager, Applications Engineering
Mechanical Technology Incorporated

In most discussions concerning tribology, authors felt compelled to introduce the topic by first defining the term. In this discussion, however, the term will not be defined until after the circumstances that spawned its existence have been summarized.

This discussion will revolve around two facets of mechanical system design, that of structural integrity and wear integrity. In the not too distant past, the evolutionary process of machine design had in most cases attained a stage of development where the structural integrity of a system was ensured. Simply stated, this means that the machine components would not "break." However, at this development phase, little emphasis was placed on component wear integrity. This meant that although machinery components would not break, they would be subject to high wear/wear-out rates. Frequent component replacements were used in order to compensate for this situation. A "throw away" philosophy was thus adopted. This philosophy, however, resulted in many undesirable secondary effects. Mechanical systems were characterized by expensive logistics, costly maintenance, limited availability, limited lives, and substantial energy consumption. This energy consumption resulted from primary effects of high equipment frictional levels (causing the high wear rate) and such secondary effects as spare parts production and transportation. As a result of these factors, the "throw-away" philosophy was a very vulnerable luxury.

During the last fifteen or so years, environmental factors with respect to mechanical system operation have changed drastically. Respective operating organizations have been faced with such factors as limited natural resources (material and energy), limited economic resources, spiraling manpower costs, and resulting spiraling capital equipment costs. These factors are compounded by a substantial inflation rate. This "new" operation environment comes into direct conflict with the existing "throw away" philosophy.

In order to resolve this conflict, the machine design process was forced to address the facet of wear integrity. However, the resolution is not as simple as it appears. Machine wear and friction are a function of such assorted variables as lubrication, additives, materials, surfaces, contamination, design, etc. No existing discipline was fully and effectively qualified to address the total of these variables and, hence, the problem of component wear integrity. An unresolvable situation was thus created under the existing technological structure.

One of the first to verbalize this situation were the British in the Jost Report published in 1966, which summarized the findings of the now famous Jost Committee. Established in 1964 by the UK Minister of State for Education and Science, the Committee was directed to define the current position of lubrication education and research in the UK and to give an opinion on the needs of industry in this field.

The resulting Jost Report outlined this wear integrity problem and recommended a multidisciplinary approach to its solution. The document labeled the proposed approach "tribology," and defined it as "the science and technology of interacting surfaces in relative motion and of the practices related thereto." This term included the subjects of friction, lubrication, and wear, e.g.

- The physics, chemistry, mechanics, and metallurgy of interacting surfaces in relative motion including the phenomena of friction and of wear.

- Fluid film lubrication, e.g., hydrostatic, hydrodynamic, aerostatic, and aerodynamic.

- Lubrication other than fluid film, e.g., boundary and solid lubrication.

- Lubrication in special conditions, e.g., during metal deformation and cutting processes.

- The properties and operational behavior of bearing materials.

- The engineering of bearings and bearing surfaces (e.g., plain and rolling bearings, piston rings, machine slides, gear teeth, etc.) including their design, manufacture, and operation.

- The engineering of bearing environments.

- The properties and operational behavior of fluid, semi-fluid, gaseous, and solid lubricants and of allied materials.

- The quality control and inspection of lubricants.

- The handling, dispensing, and application of lubricants.

- The management and organization of lubrication.

The report went on to provide justification in support of its recommendations by quantifying the potential cost savings accrued through the application of the multidisciplinary tribology approach. It was probably this quantification effort that was the most startling aspect of the Jost Report. Cost savings for British industry at that time (1966), were estimated to be approximately $500 million per year.

Although this quantification effort has raised much discussion as to the legitimacy of several included factors, the mere magnitude emphasizes the potential of tribology. Based on the findings of the Jost Report and motivated by the projected savings figure, Britain has since initiated a substantial tribological technology implementation program. The remainder of Europe, to a slightly lesser extent, has followed suit.

The United States is several years behind Europe in the adoption of tribological technology. This lag results in part, from the fact that U.S. organizations were not confronted with the conflict between the "throw away" philosophy and contradictory operating environmental trends until the early 1970's, several years after the British encounter. The difference in encounter dates probably results from economic factors, an explanation of which is beyond the scope of this discussion.

The relatively new nature of tribological technology, coupled with its multidisciplinary foundation and its varying adoption schedule, renders it a fertile area for academic treatment. It is this scientific basis that formed the lecture content of the NATO Advanced Study Institute held in Maratea, Italy, in September of 1981.

The Tribological Technology ASI Lecture Program was designed to develop multidisciplinary tribological technology from a fundamental basis through an applications viewpoint. This development approach consists of eleven discreet lectures. These lectures form the foundation of this text and are organized as follows:

Scope of Tribology

The first lecture, chapter two, provides a description of the scope of tribology. Such aspects as background, economics, justification, terminology, applications, and implications are covered.

Surface Interactions

The third chapter provides a foundation for the Tribological Technology. It is comprised of fundamental discussions concerned with interaction theory and interaction types.

Surface Interaction Elements

Chapters four through eight provide in-depth discussions of individual interaction elements. Respective element discussions include:

A. Materials
B. Surfaces
C. Lubrication
D. Lubricants
E. Contamination

Tribological Failures and Mechanical Design

The ninth chapter provides discussions of tribological failure definition, failure types, consequences, and failure control and prevention measures.

Tribo-Testing

Chapter ten covers testing approaches available for interaction element research, development, or quality assurance. This testing segment includes mechanism, component, and full system approaches.

Monitoring

Chapter eleven, "Monitoring," presents discussions concerned with the tracking of the wear process. Primary emphasis under this chapter involves equipment diagnostic approaches, diagnostic equipment, philosophy, and effectiveness.

Tribological Multidisciplinary Approach

The final chapter discusses a total tribological approach from equipment design through manufacture and operation. The previous chapters are integrated under this discussion.

A schematic of this lecture program is illustrated in Figure 1. This approach and content provides a logical, comprehensive treatment of tribological technology.

The above respective lectures have been compiled into the present volume. Each author is an expert of international standing and was subjected to minimal constraints with respect to the formulation of individual written chapters. As a result, individual author styles and emphasis have not been suppressed in this volume.

The attached appendix contains several technical papers addressing relevant tribological research efforts. These papers were solicited and presented as part of the ASI program.

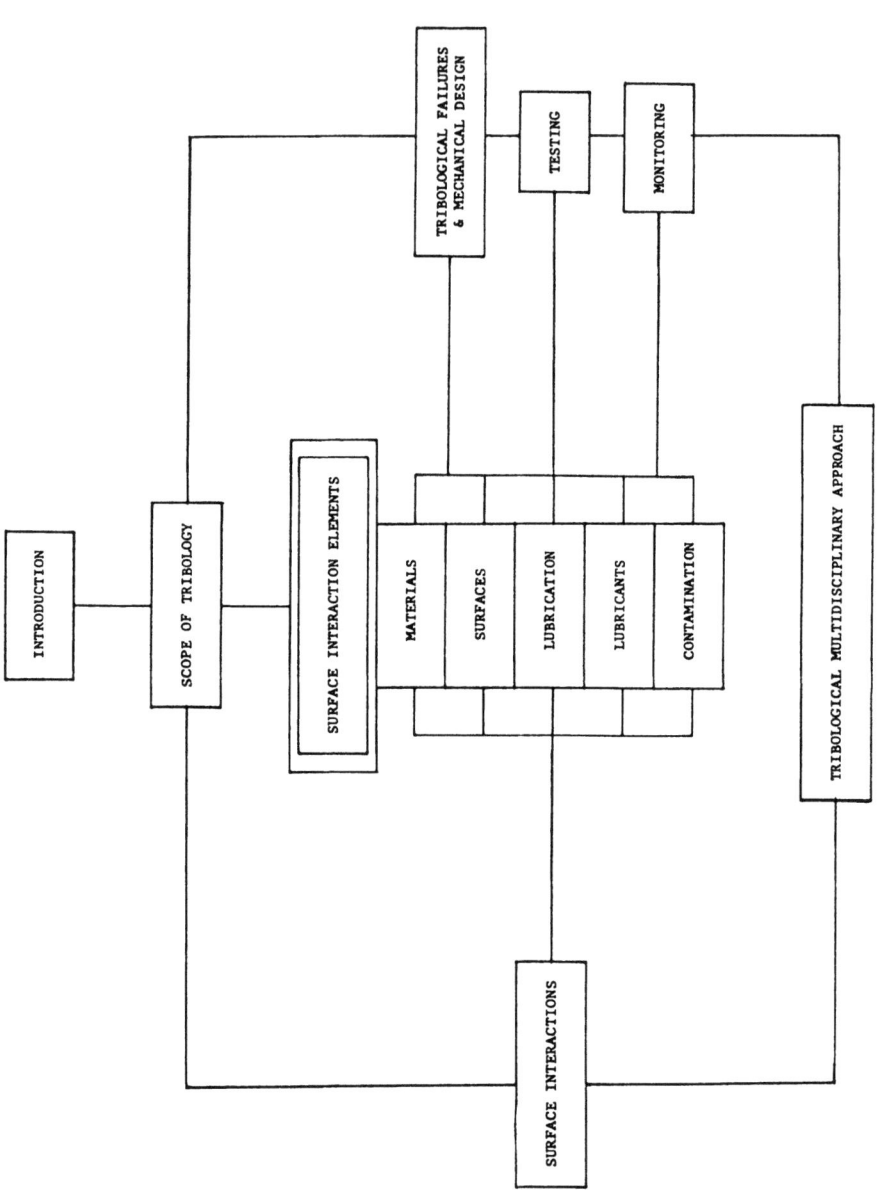

FIGURE 1   TRIBOLOGICAL TECHNOLOGY ASI PROGRAM APPROACH

SCOPE OF TRIBOLOGY

Horst Czichos
Berlin-Dahlem, West Germany

1. INTRODUCTION

The term "TRIBOLOGY" was coined by a British committee in 1966 from the word "tribos", which means "rubbing" in classic Greek. Tribology is defined as: "The science and technology of interacting surfaces in relative motion and of related subjects and practices." Since its definition tribology has been widely recognized as a new general concept embracing all aspects of the transmission and dissipation of energy and materials in mechanical equipment including the various topics of friction, wear, lubrication, and related fields of science and technology. This discussion provides a general description of tribology, its scope, applications, and implications. The material is presented under the following headings: background, economic impact, terminology, fields of application, tribological parameters, and tribological system effectiveness.

2. BACKGROUND

The term "TRIBOLOGY" was coined by a British committee in 1966 from the word "tribos", which means "rubbing" in classic Greek[19]. Since its definition 15 years ago, the term tribology has been widely recognized as a new general concept embracing all aspects of the transmission and dissipation of energy and materials in mechanical equipment including the various topics of friction, wear, lubrication, and related fields of science and technology. As friction is responsible for a major loss of useful mechanical energy, and wear is a major reason for replacing equipment; a better understanding and utilization of the principles of tribology is particularly important for the conservation of energy and materials in engineering design.

The historical development of tribology is described comprehensively in the famous book by Dowson "History of Tribology," for which the reader is referred to[8]. In chronological terms, the development of the studies of the main areas of tribology - friction, lubrication, and wear - fall into two periods of time with the end of the Second World War marking the demarcation. In terms of subject-matter there seems little doubt that, if in the previous 75 years the main interest was on "lubrication" and in the time before on "friction", the present main interest seems to be on "wear." This development is connected with the increasing awareness that the complex friction and wear problems in contemporary mechanical engineering can only be treated successfully by a unifying approach which takes into account systematically all important influencing parameters and processes[7].

After the Second World War, a general trend in mechanical engineering towards higher loads, higher velocities, and higher operating temperatures was observed in connection with attempts to reduce the weights of the moving parts in machinery. Since that time there has been a steady increase of interest in the problems of friction, lubrication, and wear in connection with the increasing worldwide attempts on the conservation of energy and materials. Tokens of this increasing interest were, among other things, the appearance of the first international journal entitled WEAR (since 1957) entirely devoted to the science and technology of friction, wear, and lubrication. A further new journal, the TRANSACTIONS of ASLE, has been published (since 1958) by the American Society of Lubrication Engineers. In West Germany a research programme ("Schwerpunkt-Programm") had been launched in 1961 sponsored by the Deutsche Forschungsgemeinschaft (DFG). About the same time in England, a Working Group was set up to investigate the present state of lubrication education and research and to give an opinion on the needs of industry thereof. In 1966 this Working Group published a report (the so-called "Jost Report"), in which for the first time the term "tribology" appeared[19]. The origin of tribology was described in the report "The Introduction of A New Technology" (1973) as follows[20]:

In the early 1960's there was a steep increase in the reported failure of plant equipment and machinery due to wear and associated causes. At the same time, increased technology, increased capital intensity of plant equipment and the use of more continuous processes made breakdowns of such plant equipment and machinery more costly, competitively more serious, and therefore even less desirable than before. This trend was recognized by specialists involved in the subjects of wear, friction, and lubrication. The situation seemed to call for more and better education in the subjects and for more and better coordinated research.

While trying to establish the reasons for the wide neglect of the subject in the past, despite its technological and economic importance, three principal reasons for this neglect may be mentioned:

  (i)   The interdisciplinary nature of the subject which included the disciplines of mechanical engineering, physics, metallurgy, and chemistry.
  (ii) The fact that only with the advances in technological development in production methods of recent years had attention been focused on the importance and interdependence of the constituents of this interdisciplinary subject.
  (iii) The term "lubrication," used in its narrower sense, had not only prevented many people from fully appreciating the economic significance of the subject matter, but it was also a misnomer for the description of the sphere of "transference of force from one moving surface to another" (whether the purpose of the transfer of such forces was associated with high friction, e.g., on brakes, clutches, conveyors or alternatively with low friction, e.g., with bearings, slides, etc.)

After consultation with the Editors of the Supplement of the Oxford English Dictionary, the term "Tribology" (Triboscience or Tribotechnology) was recommended for describing the subject matter. Tribology is defined as:

> "The science and technology of interacting surfaces in relative motion and of related subjects and practices."

3. IMPACT OF TRIBOLOGY

In considering the impact of tribology, three aspects may be emphasized: the scientific, the multidisciplinary, and the economic aspect of tribology.

From a scientific point of view, tribology is to be considered as being the discipline which tries to explain the most dominant irreversible processes in nature and technology. It is well known that all macroscopic processes in nature are irreversible. Science in its "pure" theories has largely omitted this irreversibility since the laws of "ideal" processes were much easier to develop. In this connection, an interesting remark of the mathematician John V. Neumann should be quoted concerning an important branch of tribology, namely lubrication[7]. John V. Neumann pointed out that during most of the development of hydromechanics until about 1900, the scientific interest was mainly concentrated on the development of "pure" theories of fluid mechanics by solving beautiful mathematic problems neglecting terms of internal fluid friction. These approximations, however, had almost nothing to do with real fluids. He characterized the theorist making such analyses as a man who studied the flow of "dry" water. In contrast, tribology is attempting to investigate the irreversible processes of mechanics in detail and to explain the complex effects of friction-and-wear-induced energy and materials dissipation.

A second important aspect is the multidisciplinary nature of tribology. Since tribology is defined as "science and technology of interacting surfaces in relative motion", it includes not only the work of physicists, chemists, and material scientists interested in the surface properties of materials, but also the work of engineers who use "interacting surfaces" for the transmission of motion, forces, work etc., in various types of machinery. Therefore, tribology is connected with several branches of science and technology, such as physics, chemistry, materials science, mechanical engineering, and lubrication engineering. Consequently, the attempts for the solution of tribological problems require the combined effort of these branches.

The importance of the economic aspects of tribology has often been emphasized in the literature in recent years[23]. The 1966 Jost Report, in which the term tribology was coined, estimated that by the application of mainly known principles of science and technology of tribology, the economy of Great Britain could save in the region of £515 million p.a. (at 1965 values), as follows, see Figure 1[19].

		£ million
(a)	Reduction in energy consumption through lower friction	28
(b)	Reduction in manpower through better lubrication	10
(c)	Savings in lubricant costs	10
(d)	Savings in maintenance and replacement costs	230
(e)	Savings of losses consequential upon breakdown	115

(f) Savings in capital expenditure
due to higher utilization ratios
and greater mechanical efficiency   22
(g) Savings in investments through
increased life of machinery   100

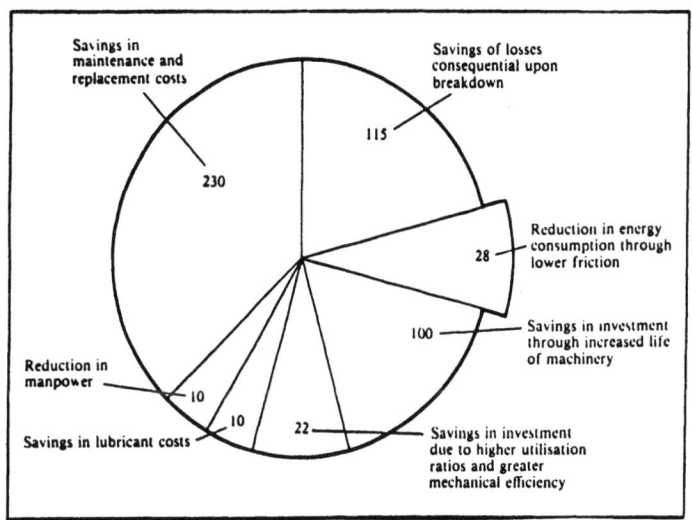

FIGURE 1   ESTIMATED SAVINGS IN TRIBOLOGY IN THE UK
(JOST REPORT, 1966)

In West Germany a 280 page report on Tribology was published in 1976 by the Federal Ministry of Research and Technology (BMFT) as a basis of a pertinent research programme[21]. In this report, it was estimated that the economic losses due to friction and wear in West Germany amount to more than 10 billion DM p.a. (at 1975 values) which is equivalent to 1% of the GNP.

Also, in 1976 a plan for "energy conservation through tribology" was published by the American Society of Mechanical Engineers (ASME) supported by the US Department of Energy and the US Office of Naval Research[22]. The American investigation found that three major sectors, transportation, electric utilities, and industry, accounted for about 80%. Half of this energy was rejected in the course of its utilization. In the transportation and utility sectors, 75% and 65% respectively, resulted primarily from thermal cycle limitations but also from friction and sealing losses. In the industrial sector, waste of approximately 25% was caused primarily by mechanical losses, such as friction, wear, and leakage past seals. The total energy used, that rejected, the potential savings through tribology, and - in enlarged form - the areas in which such savings could be made are illustrated in Figure 2. Even if all the figures cited in this section are taken as rough estimates only; they clearly indicate the importance of tribology for the conservation of energy and materials.

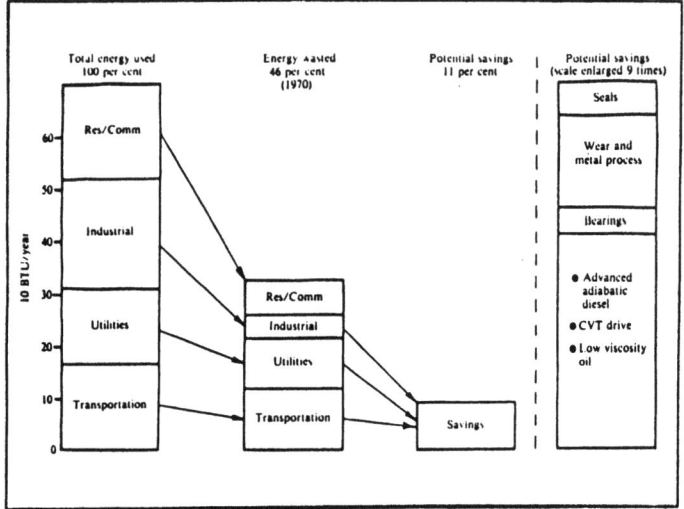

FIGURE 2   ESTIMATED LOSSES & SAVINGS IN THE US (ASME REPORT, 1977)

4. TERMINOLOGY

Owing to the multidisciplinary nature of tribology, the terminology used in this field has many origins. Thus, misleading or ambiguous terms are sometimes used. In 1969 the International Research Group on Wear of Engineering materials, working under the auspices of OECD, has compiled a "Glossary on Terms and Definitions" for the field of tribology. This Glossary has been recognized as a key word index and is included in the new ASME Wear Control Handbook published in 1980[16]. It contains definitions of the following items:

(i)   500 general tribological terms,
(ii)   93 terms on bearing types, and
(iii)   58 terms on oils and greases.

In addition, the Glossary contains an eight-language index consisting of translations of the tribological terms into the following languages: English, French, German, Italian, Spanish, Japanese (Romaji), Japanese (Kanji), Arabic, and Portuguese.

With this Glossary on Terms and Definitions, the basis for a multilingual terminology for the field of tribology has been laid. Because it is not possible to repeat details of the Glossary within the frame of this discussion, only the definitions of the three most important tribological terms will be cited here from the Glossary[16].

Friction: The resisting force tangenial to the common boundary between two bodies when, under the action of an external force, one body moves or tends to move relative to the surface of the other.

Wear: The progressive loss of substance from the operating surface of a body occurring as a result of relative motion at the surface.

Lubrication: The reduction of frictional resistance and wear or other forms of surface deterioration between two load-bearing surfaces by the application of a lubricant.

Because of the great variety of types, modes, or processes of friction, lubrication and wear which determine the behavior of any tribological entity, it is necessary to have a sub-classification of these basic terms. For these sub-classifications different criteria are used such as:

- kinematics or type of motion (sliding, rolling, impact, etc.)
- type of materials (solid, fluid, metal, polymer, etc.)
- type of interfacial tribological process (hydrodynamics, adhesion, abrasion, etc.)

Whereas in most cases it is possible to describe the basis of the kinematics or the type of material, the type of friction and lubrication mode of a given tribological entity (e.g. sliding friction or rolling friction; solid lubrication or fluid lubrication; etc.) there are considerable difficulties in attempting a logically consistent classification of wear phenomena. In 1979 the German Institute for Standardization (DIN) published a standard on wear (DIN 50 320 "WEAR") in which a classification of the terms of wear is given based on an analogy to the (classical) field of the strength of materials[18]. In order to classify the strength, deformation, or failure of a material, the following specifications must be given:

(i) type of material
(ii) type of external stress (e.g. compression, tension, bending, etc.)
(iii) type of internal deformation or fracture mechanism (e.g. ductile fracture, brittle fracture)

In analogy, in order to classify the type of wear occurring within a tribological entity, the following specifications must be given:

(i) the materials involved in wear (solid, fluid, etc.)
(ii) type of kinematic tribological action (sliding, rolling, impact, etc.)
(iii) type of inerfacial wear mechanism (adhesion, abrasion, surface fatigue, etc.)

The resulting classification of DIN 50 320 for the field of wear is shown in Figure 3. In the classification scheme shown in Figure 3, the types of wear are named according to the kinematics. This is in analogy both to the classification used in the field of the strength of materials, as mentioned above, and to the common classification of friction like sliding friction, rolling friction, etc. During the occurrence of the different types of wear, in the interface of the wear couple, one or more different wear mechanisms may act. It is generally accepted that there are four main wear mechanisms which are described in DIN 50 320 as follows[18]:

- Adhesion: Formation and rupture of interfacial adhesive bonds (e.g. "cold welded junctions," "scuffing")

- Abrasion: Removal of materials by a scratching process (micro-cutting process)

- Surface Fatigue: Fatigue and crack formation in surface regions by tribological stress cycles resulting in separations of materials (e.g. "pitting")

- Tribo-chemical Reactions: Development of reaction products as a result of chemical reactions taking place between the wear couple and the interfacial medium.

In conclusion with the terms and definitions of the Glossary as contained in the Wear Control Handbook and the classification of wear of the Wear Standard DIN 50 320, a multilingual terminology for the field of tribology is available[16, 18]. It is strongly recommended that this terminology be used in order to avoid ambiguities and to improve understanding and communication in the field of tribology on an international scale.

System structure	Tribological action (symbols)		Type of wear	Effective mechanisms (individually or combined)			
				Adhesion	Abrasion	Surface fatigue	Tribo-chemical reactions
Solid — interfacial medium (full fluid film separation) — solid	sliding rolling impact		—			×	×
Solid — solid (with solid friction, boundary, lubrication, mixed lubrication)	sliding		sliding wear	×	×	×	×
	rolling		rolling wear	×	×	×	×
	impact		impact wear	×	×	×	×
	oscillation		fretting wear	×	×	×	×
Solid — solid and particles	sliding		sliding abrasion		×		
	sliding		sliding abrasion (three body abrasion)		×		
	rolling		rolling abrasion (three body abrasion)		×		
Solid — fluid with particles	flow		particle erosion (erosion wear)		×	×	×
Solid — gas with particles	flow		fluid erosion (erosion wear)		×	×	×
	impact		impact particle wear		×	×	×
Solid — fluid	flow oscillation		material cavitation, cavitation erosion			×	×
	impact		drop erosion			×	×

FIGURE 3   CLASSIFICATION OF WEAR PHENOMENA

5. FIELDS OF APPLICATION

The many technical aims and functions to be realized through "interacting surfaces in relative motion" lead to a great variety of tribo-technical components and tribo-engineering systems and processes. It can be shown that the technical functions realized through interacting material surfaces can be broadly classified in the following groups[7]:

   a) Transmission of MOTION
   b) Transmission of WORK (or POWER)
   c) Generation or reproduction of INFORMATION
   d) Transportation of MATERIALS
   e) Forming of MATERIALS

The various engineering systems in which friction and wear processes occur can be easily classified according to their function in considering the pertinent inputs and outputs. A broad classification is given in Table 1. Invariably, motion is a characteristic of any tribo-mechanical system. This motion may constitute a transfer of work, materials, or information. In some instances, the purpose of a system may be to change a rate of motion or to eliminate it altogether. It is also often desired to restrict motion, i.e., to reduce the number of degrees of freedom a machine element may possess. In other instances, materials are not merely moved but also changed in state or form. Mechanical devices which produce or transfer information are still common, but are being steadily replaced by devices in which there is little or no mechanical motion; for example, the replacement of mechanical clocks by digital electronic clocks.

In all the engineering systems compiled in Table 1, friction, wear, and lubrication processes of various kinds occur. In order to optimize the functional technical behavior of these systems and to minimize friction-and- wear-induced energy and material losses, tribological knowledge must be applied with respect to all the basic engineering activities, namely:

   - design
   - manufacture
   - operation
   - condition monitoring
   - repair
   - recycling

6. TRIBOLOGICAL PARAMETERS

Any proper analysis and solution of tribological problems require the consideration of numerous parameters and influencing factors. Through the application of the recent methods of systems analysis, it is possible to compile and to classify the main tribological parameters[7].

The procedure of a systematic analysis of tribological parameters will be explained on the basis of the simplified representation of a tribological system, Figure 4, taken from DIN 50 320[18].

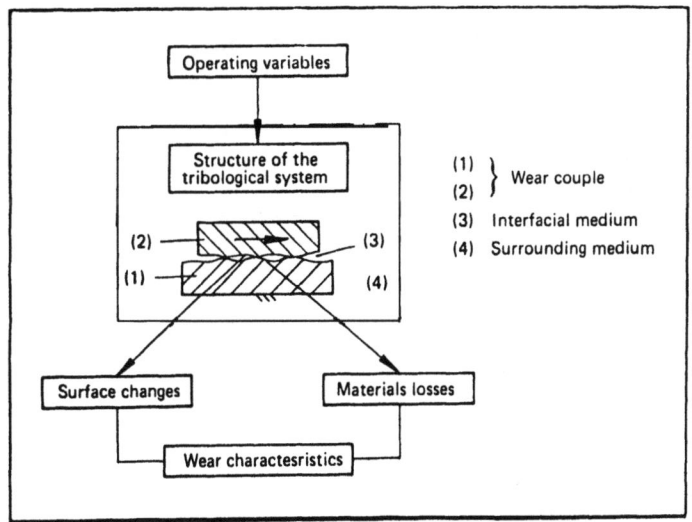

FIGURE 4   PARAMETER GROUPS OF A TRIBOLOGICAL SYSTEM

For the systematic analysis of the relevant tribological parameters, first the components and substances directly participating in the tribological process have to be conceptually separated from the other parts of the technical equipment considered. The components and substances directly participating in the tribological process are termed the "elements" or "components" of the tribological system. These elements, together with their tribologically relevant "properties" and the "interactions" between them, form the "structure of the tribological system." The operating quantities which act on the elements of the tribological system from outside form the "operating variables." The energy and material losses which occur through the action of the operating variables on the structure of the tribological system are to be described by tribological loss-characteristics.

It follows that for a systematic discussion of tribological parameters, the following four basic aspects must be considered:

I   Functional technical purpose of the tribo-system
II   Operating variables
III Structure of the tribo-systems, consisting of:
    a) materials components involved
    b) relevant properties of the components
    c) interactions between the components
IV Tribological loss-characteristics

6.1 Functional Technical Purpose of Tribological Systems

The characterization of the functional technical purpose of a tribological system is a first needed overall description. The technical aims realized through interacting moving surfaces in tribological entities may range from aerospace applications to biomechanical joints. A broad classification has already been given in Section 5 (see Table 1).

Inputs and outputs needed for technical function		Primary technical function of the system	Examples
Main inputs $\{X\}$	Main outputs $\{Y\}$		
Motion + Work	Motion	Guidance of motion Coupling of motion Annihilation of motion	Bearings Clutches Brakes
	Work	Power transmission (mech., hydr., pneum.)	Gears
	Information	Generation of information	Clocks; Cams and followers
		Reproduction of information	Data transducer (audio, video; tape or record)
Motion + Materials	Materials	Transportation	Wheel/rail Pipeline
		Forming of materials	Wiredrawing

TABLE 1 CLASSIFICATION OF TECHNICAL FUNCTIONS OF TRIBOLOGICAL SYSTEMS

The most general technical purpose of a tribological system is the guidance of motion through various types of "bearings." The other basic groups are the transmission of mechanical work, the transmission of information - for instance the control of machine functions with cams - and the forming of materials. It follows that basically the technical function of tribological systems is connected with the transmission or transformation of one or more of the basic quantities:

- motion
- work
- information
- materials.

In using these four basic quantities or related quantities, the technical function of the various tribological systems may be classified in terms of the input-output relations of these quantities.

6.2 Operating Variables

Having identified the functional technical purpose of a tribological system under discussion, it is then necessary to characterize and to compile its basic operating variables.

The most characteristic operating variable of a tribological system is the type of relative motion between tribo-element (1) and tribo-element (2). The basic types of motion are:

- sliding
- rolling
- spin
- impact.

It can be shown that every type of relative motion between system components can be expressed as a superposition of these four basic types of motion. In addition to the characterization of the type of motion, its dependence on time should be specified, for example:

- continuous
- oscillating
- reciprocating
- intermittent.

The other basic operating variables are the following quantities:

- load F_N
- velocity v
- temperature T
- distance of motion s
- operating duration t.

For some tribological systems, these physical operating variables are accompanied by material inputs, e.g., the flow rate of the lubricant. Some disturbing inputs may also be present, e.g., vibration and radiation. In addition to these parameters, it may often be necessary to also specify composite or derived quantities, e.g. contact pressure, temperature gradients, etc.

6.3 Structure of Tribological Systems

In addition to the description and classification of the various tribo-mechanical systems according to their external function and the compilation of pertinent operating variables, the question of the internal "structure" of these systems is to be discussed.

As described above, the structure of a tribological system is given by the systems elements (that is the material components of the system), their relevant properties and their inter-relations, described formally by the set $S = \{A, P, R\}$, see Figure 5.

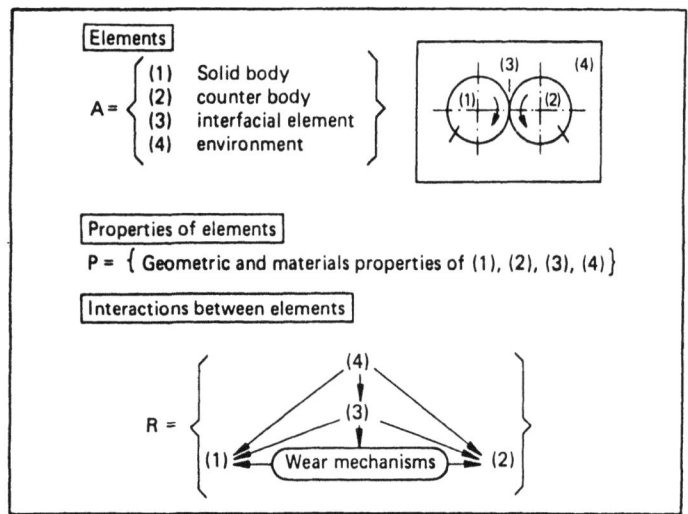

FIGURE 5   STRUCTURE OF A TRIBOLOGICAL SYSTEM $S = \{A, P, R\}$

6.3.1   Elements of the System, $A = \{a_i\}$

If the system's envelope is located as closely as possible around the "interacting surfaces in relative motion," it appears that in most tribological systems four different basic elements are involved in the friction and wear processes.

As illustrated in Figure 5 for a simple two-disc configuration, the components that form the pair of the "interacting surfaces in relative motion" are the moving elements, (1) and (2) respectively. The other two basic elements are the interfacial lubricant (3) (if any) and the environmental atmosphere (4). These main elements are linked to others or may be composed of sub-constituents. For example, element (3), the lubricant, may consist of a base oil and additives. In Table 2 the elementary elements or components (1), (2), (3), (4) of tribological systems are listed as examples from every group of the basic tribological systems compiled in Table 1.

Tribological system (or process)	Elements of the system			
	Tribo-element (1) (moving or stationary)	Tribo-element (2) (moving or stationary	Interfacial medium (3)	Surrounding medium (4)
Sliding bearing	shaft	bushing	lubricant	air
Band clutch	shaft	band	–	air
Disc brake	disc	pad	contaminant	air
Worm gear set	worm	gear	gear oil	air
Cam and follower	cam	follower	lubricant	air
Printing unit	print-head	paper	dye	air
Audio pick-up	record	sapphire tip	–	air
Electrical contact	ring	brush	spray	cover gas
Locomotion	wheel	rail	contaminant	air
Pipeline	fluid	pipeline	–	–
Wiredrawing	wire	die	borax	air
Hot extrusion	billet	die	glass	air
Turning	workpiece	cutting tool	cutting fluid	air

TABLE 2   COMPILATION OF TYPICAL TRIBO-ENGINEERING SYSTEMS

6.3.2 Properties of the Elements, $P = \{P(a_i)\}$

The behavior of any tribological system is influenced by many properties of the basic elements (1), (2), (3), (4) all described above. Owing to the great variety of tribo-mechanical systems and tribological processes, it is very difficult to provide a comprehensive general compilation of the tribologically relevant properties of the systems elements. The following properties of the elements have been found to be of primary concern for the tribological behavior of the system:

(i)   Properties of the Tribo-Elements (1) and (2) (moving elements)
The tribologically relevant properties of the elements (1) and (2) can be subdivided into "volume" properties and "surface" properties. At least the

following main properties of the elements (1) and (2) should be specified:
- Volume properties: geometry, chemical compositions, and metallurgical structure; materials data, including elastic modulus, hardness, density, and thermal conductivity.
- surface properties: surface roughness and surface composition.

(ii) Properties of the Lubricant (3)
The main relevant properties of the lubricant are the viscosity and its dependence upon temperature and pressure, together with the chemical composition of the lubricant.

(iii) Properties of the Environmental Atmosphere (4)
The main relevant properties of the atmosphere are its chemical composition and the amount and pressure of its components, especially water vapor.

To assist in the compilation of tribologically relevant parameters of the elements of tribo-mechanical systems, a data sheet has been developed and published as an appendix to the Wear Standard DIN 50 3[18].

6.3.3 Interactions Between the System's Elements, $R = \{R(a_i, a_j)\}$

The tribological interactions between the elements of a mechanical system, i.e. the contact, friction, lubrication, and wear processes, are of paramount interest in the description of any tribo-mechanical system. In Figure 6, basic tribological processes which are known today are expressed in the form of highly simplified schematic diagrams for systems of increasing complexity, i.e. increasing number of interacting elements.

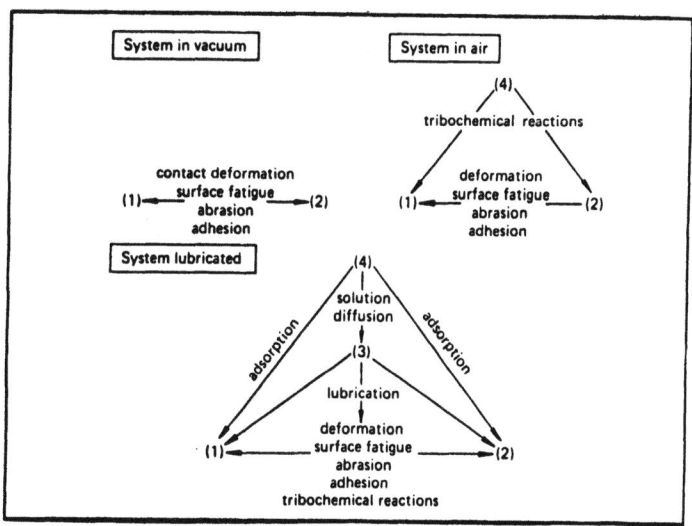

FIGURE 6   SCHEMATIC REPRESENTATION OF TRIBOLOGICAL INTERACTIONS BETWEEN SYSTEM COMPONENTS

In an ultrahigh vacuum, the simplest tribological system consists only of the two interacting partners (1) and (2). The main possible tribological interactions between moving and stationary elements are then covered by the terms contact deformation, surface fatigue, abrasion, and adhesion. In air, i.e. under dry friction condition, these processes are supplemented by interactions between the moving and the stationary partners (1) and (2) and the atmosphere (4). Through these interactions tribo-chemical reactions result. Finally, in a lubricated system, tribological processses between the elements are given by interactions between all four basic elements. In this case, the direct (contact) interactions between moving and stationary elements are prevented or influenced through the different mechanisms of lubrication. Depending on the thickness of the lubrication film, different lubrication regimes result.

Also here, interactions between (4) and (3) with (1) and (2) should be taken into account. For instance the diffusion of atmospheric oxygen into the lubricant (4)→(3), followed by oxidation processes between the lubricant and the moving and stationary partners (3)→(1), and (2), can distinctly influence the mechanisms of mixed and boundary lubrication.

6.4 Tribological Characteristics

The tribological characteristics, i.e. the characteristics that describe the dynamic changes of a tribo-mechanical system as a consequence of the action of friction and wear processes, may be divided into the following three groups:

- (a) Tribo-induced changes of system's structure
- (b) Tribo-induced energy losses
- (c) Tribo-induced material losses.

Depending on the tribological processes which occur within a tribo-mechanical system, the tribo-induced changes of a system's structure (a) concern:

- (i) the destruction of elements or the creation of new elements in a tribo-system, e.g. the degradation of a lubricant or, on the contrary, the creation of "frictional polymers,"
- (ii) changes in the properties of the elements, for instance, changes in contact topography and surface composition,
- (iii) changes in the interrelations between the elements, for instance, changes of the wear mechanisms under the action of the operating variables, or changes in the lubrication mode.

The other two main tribological characteristics (b) and (c), the friction-induced energy losses and the wear-induced material losses, may be expressed formally as:

friction losses = f (operating variables; system's structure)
wear losses = f (operating variables; system's structure)

Consequently, friction coefficient, f, and wear rate, w, of a tribological system may be expressed formally as:

$$\boxed{f = f(X;S)} \qquad \boxed{w = f(X;S)}$$

where X : Operating variables

$S = \{A, P, R\}$: System's structure

Although the parameter groups X and S are not independent variables since they are connected with each other through the tribological interrelations R, the above symbolic representation of friction and wear characteristics can be conveniently used as a starting point for the practical application of a systems methodology in order to influence or to mitigate friction- and wear-induced energy and material losses.

7. TRIBOLOGICAL SYSTEM EFFECTIVENESS

The friction, lubrication, and wear processes involved in any tribological system determine its technical functional behavior. These processes may change the properties of the components, as for example the surface roughness of the interacting material elements, and may thus influence the performance of the whole system. In this final section, first the influence of tribological processes on the dynamic behavior of tribological systems will be considered. Then the aspects of failure, safety, and reliability of tribological systems will be discussed.

7.1 Dynamics of Tribological Systems

The functional behavior of any tribo-mechanical system is connected, by definition, with relative motion of one or more components of the system. As described in Section 5, this motion may constitute a transfer of work, information, or material through the tribo-mechanical system. In any case, the motion and the dynamics of the whole system are influenced by the interfacial friction processes between the moving components. Clearly, the details of the influences of tribological processes on the dynamics of motion may be quite different for the various tribo-mechanical systems, for example, ball bearings, slideways, or metal machining systems. Considering the functional transmission of motion through the various types of tribo-mechanical systems, the influences of the tribological processes may lead to unwanted vibrations of the moving parts and to "stick-slip" motion. These disturbing influences on the functional behavior can be observed in various tribo-mechanical systems; from the squealing of brakes to the chattering of machine tools during cutting processes.

Many tribo-mechanical systems whose functional purpose is connected with the transmission of motion can be modelled in a simplified manner by the configuration shown in Figure 7. The model system consists of a body (1) of mass m_1, moving relatively to its counterpart (2) of mass m_2 fixed to the ground via a spring with a spring constant C_{s2} and a damper with a damper constant C_d. The body (1) is driven via the spring C_{s1} at constant velocity $v_o = s/t$.

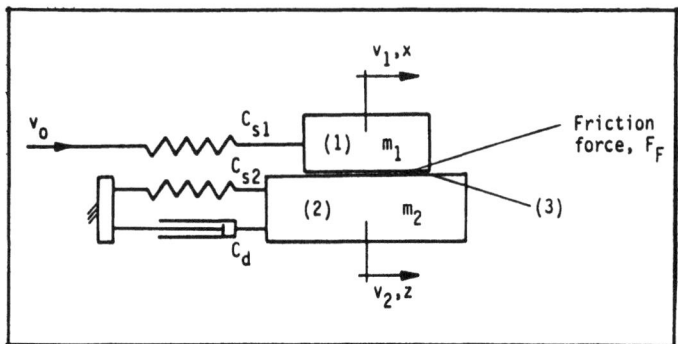

FIGURE 7   MODEL OF A TRIBOLOGICAL SYSTEM

The motion of body (1) of velocity v_1 and distance x relative to body (2) of velocity v_2 and distance z is influenced by the friction force F_F acting in the interface (3) between body (1) and body (2). From the following simple qualitative consideration it follows that the type of motion is determined by the value of the friction force at $v_{rel} = 0$ and the dependence of the friction force on the velocity $F_F = f(v)$. Let the initial state of the system, shown in Figure 7, be such that the springs C_{s1} and C_{s2} are uncompressed and m_1 and m_2 are at rest. When the motion of velocity v_o is introduced there will be no movememt of m_1 relative to m_2 ("stick phase") until the driving force on m_1 is high enough to overcome the (static) friction force between m_1 and m_2. If then the motion of m_1 relative to m_2 starts ("slip" phase), the springs decompress. Thus, the driving force is lowered by a certain amount. If now the driving force on m_1 falls below the (kinetic) friction force, a second "stick" phase may evolve.

This in turn leads to an increase of the driving force until the motion of the second slip phase starts, and so on.

Through appropriate modelling it may be possible to characterize the dynamic behavior of tribological systems by differential equations and to simulate it with the help of an analogue computer.

From the results of an analogue computer simulation, some general conclusions on the dynamic behavior of tribological systems may be drawn 7

Generally speaking for a lubricated sliding system, depending on the "operating point" within the Stribeck curve, a different dynamic behaviour of the tribo- mechanical system can be observed. Without going into numerical details, the following three different general patterns of the motion behavior of the modelled tribo- mechanical system of Figure 7 can be distinguished:

(i) For the condition of friction around the minimum of the Stribeck curve, the system is unstable and the motion following a disturbance is divergent, i.e., the system excites itself to vibrations, as shown in Figure 8.

(ii) For the conditions of friction on the left part of the Stribeck curve the typical stick-slip motion diagram results, as in Figure 9.

(iii) For the conditions of friction on the right part of Stribeck's curve the system is stable, i.e. vibrations introduced to the system are damped automatically. This behavior can be seen in Figure 10 for five different slopes to the right part of the Stribeck curve and five operating values of the friction coefficient f.

The results confirm the experimental observation that stick-slip effects are likely to occur if the slope of the velocity curve is negative or equal to zero, $\frac{df}{dv} \leq 0$, as in the left part of the Stribeck curve. Thus, stick-slip effects may occur only under conditions of solid friction, boundary, or mixed lubrication, but are unlikely to occur under conditions of hydrodynamic lubrication.

7.2 Failure and Reliability of Tribological Systems

In reviewing the causes of failure of mechanical systems, Collacot, in his book on "Mechanical Fault Diagnosis and Condition Monitoring" distinguishes between the following main aspects[6]:

(a) Service failures
(b) Fatigue
(c) Excessive deformation
(d) Wear
(e) Corrosion
(f) Blockaages
(g) Design, manufacturing and assembly causes of failure

This compilation of the main classes of failure already indicates that there are several nontribological causes which may lead to failure of mechanical equipment. This can also be seen from the data of Tables 3 and 4 in which failure characteristics of typical tribo-mechanical systems, namely rolling and sliding bearings and mechanical clutches, are listed. The data have been compiled by the insurance company ALLIANZ[7]. Table 3 contains the results of investigations of the causes of 1400 rolling-bearing failures and 530 sliding-bearing

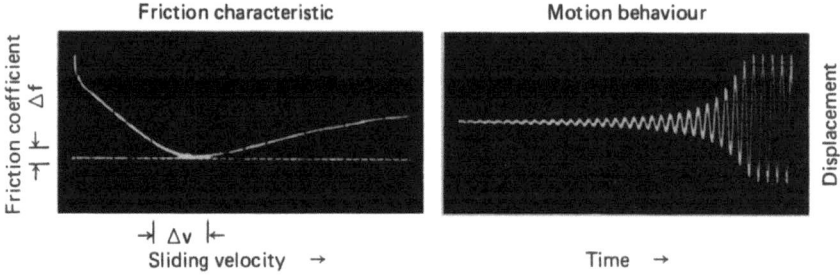

FIGURE 8 INSTABLE MOTION BEHAVIOR FOR CONDITIONS AT THE MINIMUM OF THE STRIBECK CURVE

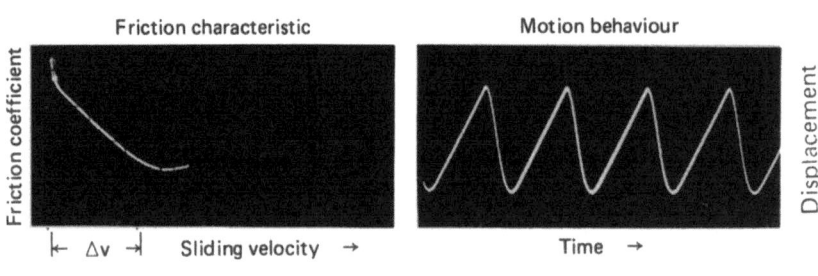

FIGURE 9 STICK-SLIP OSCILLATIONS FOR CONDITIONS ON THE LEFT PART OF THE STRIBECK CURVE

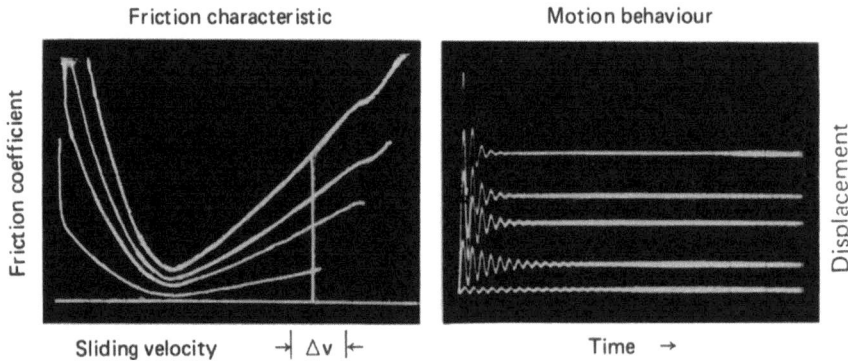

FIGURE 10 STABLE MOTION BEHAVIOR FOR CONDITIONS ON THE RIGHT PART OF THE STRIBECK CURVE

failures. It can be seen that for these tribo-mechanical systems, about 30% of the functional failures are due to wear processes. Table 4 contains the percentage of the main types of damage of mechanical clutches. Also in this case, the damage occurring at the load-transmitting surfaces has been found to constitute about 30% of the total failure causes. These examples show that, besides the tribo-induced causes of failure, various nontribological causes may lead to failure of mechanical equipment.

Causes of failure	Occurrence (%)	
	rolling bearings	sliding bearings
manufacturing faults	14.4	10.7
design and calculation faults	13.8	9.1
materials faults of components	1.9	3.6
service faults, maintenance faults, failure of monitoring equipment	37.4	39.1
wear	28.5	30.5
failure through external causes	4.0	7.0

TABLE 3   CAUSES OF FAILURE OF ROLLING AND SLIDING BEARINGS

Damage types	Occurrence (%)
breakages due to overstressing	60
scuffing, seizure	18
mechanical and corrosive surface damage	15
cracks	5
deflections, deformations	2

TABLE 4   TYPES OF DAMAGE OF MECHANICAL CLUTCHES

Next, the question may be discussed, how the tribological processes, i.e. the friction and wear processes, influence and disturb the behavior of a given tribological system. This is illustrated schematically in Figure 11.

In the upper part of Figure 11, a typical tribological system, namely a gear box, is shown schematically. The technical function of the system is to transform certain inputs, namely angular velocity and torque, into outputs which are used technically. The technical function can then be described formally as a transformation of the inputs into the outputs via a certain transfer function.

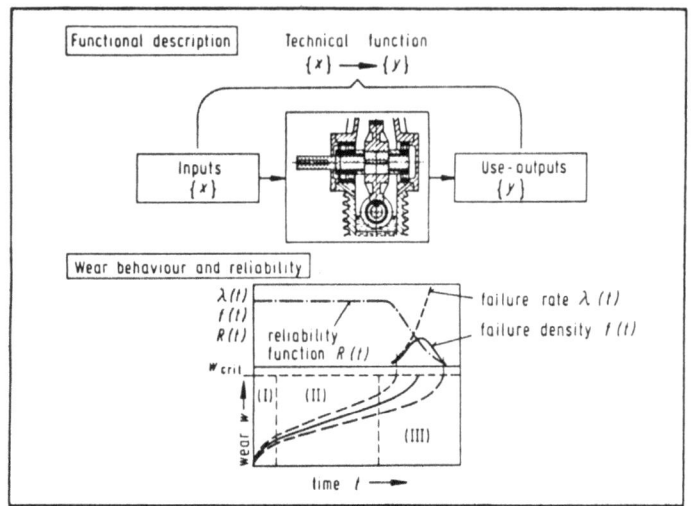

FIGURE 11 INFLUENCE OF WEAR-INDUCED CHANGES OF SYSTEMS STRUCTURE ON FUNCTION AND RELIABILITY

Now, the dynamic performance of the system is accompanied by perturbations on their function and structure. For example, through the action of wear processes, the properties of the moving components may be changed and a certain material loss-output may result. As illustrated in the lower part of Figure 11, for the loss-output of a tribo-mechanical system, three main different characteristics may be distinguished which are often observed experimentally:

(i)   Self accommodation (running in)
(ii) Steady state
(iii) Self acceleration

These three modes of changes in the systems structure may follow each other in time, as indicated in the graph in the lower part of Figure 11. If then the wear rate reaches a maximum admissible level, the systems structure has changed in such a way that the functional input-output relations of the system are disturbed severely. Repeated measurements show random variations in the data, as indicated by the dashed lines in the wear diagram of Figure 11. From sample functions of the wear process, a distribution, a failure rate, and a corresponding reliability function results.

In a quantitative way the reliability of a mechanical system may be characterized mathematically as follows:

Generally, the reliability of a mechanical system is expressed by a probabilistic function R(t) based on the following definitions:

$F(t)$: probability distribution function of the time of failure

$f(t) = \dfrac{dF(t)}{dt}$: density function

$\lambda(t) = \dfrac{F(t)}{1-F(t)}$: failure rate

($\lambda(t)dt$ is a conditional probability that the system will fail during the time t + dt under the condition that the system is safe until the time t)

$R(t) = 1-F(t)$: reliability function

MTBF : mean time to failure (measure of reliability for repairable equipment

In some cases, the failure rate (λt) of a component in a system can be estimated from the point of view of the physical behavior of the material used. Empirically and sometimes theoretically, the following probabilities are proposed:

(a) EXPONENTIAL DISTRIBUTION

$\lambda(t)$ = constant = c
$f(t) = C \cdot \exp(-Ct)$
$R(t) = \exp(-Ct)$

In this case the failure rate is constant. It means physically that any failure occurs accidentally without any accumulation of fatigue-like effects during its service time under certain stresses. Many kinds of electronic components follow this type of failure. Components in a machine break down in this mode when the failure is brittle fracture.

(b) RAYLEIGH DISTRIBUTION

$$\lambda(t) = Ct$$

$$f(t) = Ct \cdot \exp\left(\frac{-Ct^2}{2}\right)$$

In this case the failure rate increases with time. The constant, C, indicates the rate of deterioration of the component which depends upon the stress level applied to it.

(c) NORMAL DISTRIBUTION (TRUNCATED)

$$f(t) = (1/s\sqrt{2\pi}) \exp\left\{-\frac{1}{2}\left(\frac{t-\mu}{s}\right)^2\right\}$$

Many components of machines obey this distribution, especially if the failure occurs due to wear processes. The failure rate of this distribution cannot be expressed in a simple form.

(d) WEIBULL DISTRIBUTION

$$\lambda(t) = \frac{C}{t_o} t^{C-1}$$

$$f(t) = \frac{C}{t_o} t^{C-1} \exp(-t^C/t_o)$$

This is a distribution with two paramenters, t_o, the nominal life and the constant C. The distribution is found to represent failure of many kinds of mechanical systems, such as fatigue in ball bearings.

(e) GAMMA DISTRIBUTION

$$f(t) = C\frac{(Ct)^{x-1}}{\Gamma(x)}\exp(-Ct)$$

where $\Gamma(x)$ is a Gamma function. This is also a distribution with two parameters. Theoretically, the importance of this distribution is attributed to the fact that the equation is an x-fold convolution of the exponential function. It means physically that a component fails at x-th shock which occurs as a Poisson statistical process.

These distributions are representative ones which appear in the failure process of various components and systems. As a general overview, in Table 5, a compilation of the phenomena of deterioration and the mode of failure in connection with the underlying physical processes is given (the table is due to Yoshikawa)[7].

Physical process	Deterioration (Mode of failure)						Probability distribution
	Topological change		Geometrical		Physical property		
	Fixing	Separation	Micro	Macro	Bulk	Surface	
Fracture		●					Exponential
Yielding				●			Exponential
Fatigue		●	●		●		Weibull
Creep		●		●	●		Normal
Diffusion					●		Normal
Corrosion						●	Rayleigh
Erosion						●	Rayleigh
Rusting	●					●	Rayleigh
Wear			●			●	Normal
Adhesion	●						Gamma
Staining			●			●	Exponential

TABLE 5 PHENOMENA OF DETERIORATION AND MODE OF FAILURE

It can be seen that different failure modes and different elementary failure processes are associated with different types of failure distribution functions. It follows, in turn, that conclusions on the type of failure mechanism may be drawn from the experimental determination of failure distribution curves. For a great deal of tribological systems failing as a consequence of wear processes, the failure behavior is characterized by the normal distribution or the Weibull distribution. If for a given type of tribological system the failure mode and the type of failure distribution are known, this knowledge can be used to improve the reliability of the system.

To conclude the discussion on the failure and reliability of tribological systems, the dependence of the failure rate on the operating duration of a system should be considered. If the failure rate is plotted as a function of time, a curve known as the "bath-tub-curve" is found, as shown in Figure 12.

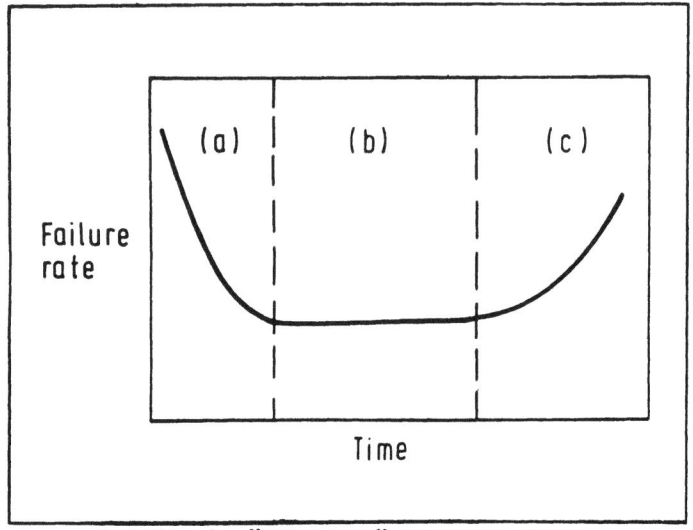

FIGURE 12   "BATH-TUB" FAILURE CURVE

In this curve three regimes can be distinguished:

(a)   early failures
(b)   random failures
(c)   wear-out failures.

None of the distribution curves discussed above have this bath-tub-shaped failure curve, but an approximation may be obtained by selecting an appropriate probability density function for each of the three regimes. Regime (a) describes the region of the "infant death" of the system. This regime is characterized by a decrease of the failure rate with time, for example in running-in. The regime (b) of constant failure rate is the region of normal running. Here, failure occurs as a consequence of statistically independent factors. Regime (c) is characterized by an increase of the failure rate with time. Here, failure may be due to aging effects. As described above, for a great deal of tribo-induced failures the failure rate increases with time. Thus, region (c) of the "bath-tub-curve" of Figure 12 appears to be relevant for the normal mode of wear-induced failure of mechanical systems.

8. REFERENCES

The literature in the field of tribology is extremely voluminous owing to the vast number of topics included. At present, 6000 - 8000 new titles are published per year. An international and comprehensive bibliography is provided by the DOCUMENTATION TRIBOLOGY. Information about this DOCUMENTATION can be obtained from the Bundesanstalt fur Materialprufung (BAM), Dokumentations-stelle Rheologie und Tribologie, Unter den Eichen 87, D-1000 Berlin 45, West Germany.

In the following, only a selection of books in the field of tribology are listed, which were published after 1970.

8.1 Recent Books on Tribology (Selection)

1. Barwell, F.T., "Bearing Systems: Principles and Practices," (Oxford: University Press, 1978).

2. Benzing, R., Goldblatt, I., Hopkins, V., Jamison, W., Mecklenburg K., Peterson, M., "Friction and Wear Devices," (Park Ridge: American Society of Mechanical Engineers, 1976).

3. Bowden, F.P. and Tabor, D., "Friction - An Introduction to Tribology," (London: Heinemann, 1973).

4. Buckley, D.H., "Surface Effects in Adhesion, Friction, Wear, and Lubrication," (Amsterdam: Elsevier, 1977).

5. Cameron, A., "Basic Lubrication Theory," (London: Longman, 1971).

6. Collacott, R.A., "Mechanical Fault Diagnosis and Condition Monitoring," (London: Chapman and Hall, 1977).

7. Czichos, H., "Tribology - A Systems Approach to the Science and Technology of Friction, Lubrication and Wear," (Amsterdam: Elsevier, 1978).

8. Dowson, D., "History of Tribology," (London: Longman, 1979).

9. Dumbleton, J.H., "Tribology of Natural and Artificial Joints," (Amsterdam: Elsevier, 1981).

10. Engel, P., "Impact Wear of Materials," (Amsterdam: Elsevier, 1978).

11. Habig, K.H., "Hardness and Wear of Materials (in German)," (Munchen: Hanser, 1980).

12. Halling, J. (Ed.), "Principles of Tribology, "(London: MacMillan, 1975).

13. Ling, F.F., "Surface Mechanics," (New York: Wiley, 1973).

14. Moore, D.F., "Principles and Applications of Tribology," (Oxford: Pergamon, 1975).

15. Neale, M.J. (Ed.), "Tribology Handbook," (London: Butterworths, 1973).

16. Peterson, M.B. and Winer, W.O. (Eds.), "Wear Control Handbook," (New York: ASME, 1980).

17. Suh, N.P. and Saka, N. (Eds.), "Fundamentals of Tribology," (Cambridge, Mass: MIT Press, 1980).

8.2 Terminology

18. Standard DIN 50 320. "Wear: Terms, Systematic Analysis of Wear Processes, Classification of Wear Phenomena," (Berlin: Beuth Verlag, Dec. 1979).

8.3 Tribology Reports

19. "Lubrication (Tribology) Education and Research (Jost Report)," Department of Education and Science, HMSO, London, 1966.

20. "The Introduction of a New Technology," Department of Trade and Industry, HMSO, 1973.

21. "Research Report T 76-38 Tribologie (Code BMFT-FBT 76-38)." Bundesministerium fur Forschung und Technologie (Federal Ministry for Research and Technology), West Germany 1976.

22. "Strategy for Energy Conservation Through Tribology," ASME, New York, Nov. 1977.

23. Jost, H.P. and Schofield, J., "Energy Saving Through Tribology: A Techno-Economic Study", Proc. Inst. Mech. Engrs, London, Vol. 195 No. 16 (1981) 151-173.

SURFACE INTERACTIONS

Nam P. Suh
Massachussetts Institute of Technology
Cambridge, MA 02139
U.S.A.

1. INTRODUCTION

1.1 Three Apsects of Tribological Problems

Tribology is concerned with the science of the interface between two surfaces in relative motion. The nature and the consequence of the interactions that take place at the interface control the friction and wear behavior of the materials involved. During the interactions, forces are transmitted; energy is consumed; physical and chemical natures of the materials are changed; and the geometry of the surface topography is altered. All these consequences of the surface interactions satisfy the laws of nature. Therefore, the essence of tribology is two-fold: understanding the nature of these interactions and dealing with the technological problems created when such interactions occur. The purpose of this series of lectures is to expound on the basic mechanisms that govern the interfacial behavior.

There are three fundamental aspects to the tribological science that must be understood to deal with the technological problems:

(i) The effect of environment on surface characteristics through the physicochemical interactions.
(ii) The interaction of the surfaces in contact which results in force generation and transmission between the surfaces.

(iii) The behavior of the material near the surface in response to the external force acting on the surface.

An understanding of all three aspects of tribology requires a multidisciplinary background which is difficult to acquire. As a consequence, most researchers and practitioners only deal with a limited aspect of tribology. Perhaps, this may be one of the reasons why progress in the tribology field has been slow.

To tribologists who are mainly concerned with boundary lubrication, the first two aspects of tribology (i.e., physicochemical interactions of the surface with the environment and the asperity interactions) are of primary interest. However, the third aspect of tribology (i.e., the macroscopic deformation of the surface in response to the external force) cannot be ignored since it affects the two aspects. The physical and chemical interactions of lubricants (and gaseous environment) with the sliding surface produce compounds which are actually present at the interface. These materials in turn affect the nature of force generation and transmission at the asperity contact, and thus, the frictional force. Unfortunately, the mechanisms of boundary lubrication are yet to be fully understood. To date, lubricants and additives to lubricants have been developed through much trial and error. Although careful chemical analyses of the layers formed on the surface have been done, the mechanisms by which these additives impart beneficial effects have not yet been established.

Those interested in studying the frictional behavior of materials must comprehend the details of the chemical and physical interactions at the interface. This requires understanding of all three aspects of tribology. The contribution of each aspect of tribology to friction changes as a function of sliding distance (or time) and the environment. Therefore, the coefficient of friction of the same pair of materials differs depending on the specific application.

Wear is primarily a consequence of the response of materials to given surface tractions. In order to understand wear, it is therefore necessary to deal with the mechanics of deformation of the surface layer and the mechanisms of force transmission at the surface. In a given tribological situation, wear particles can be generated by different mechanisms. This is often confusing to those who have not been initiated into the tribology field, since they expect a single mechanism to be responsible for all wear particle generation. What is important is to recognize the wear rate determining wear process in a given situation.

It should also be noted that the surface properties, both chemical and physical, are different from the bulk properties. The chemical composition of the surface layer may be different from the bulk; dislocations have different mobility and experience different forces; electronic configurations at the surface may be different from the bulk configurations; and even the atomic position at the surface differs from that of the bulk. Therefore, it is necessary to establish the fundamental understanding of the surface chemistry and physics in order to make future progress in tribology.

1.2 Phenomenological Observations Related to Friction and Early Theories

Friction exists between any two sliding surfaces in relative motion. This fact has been known to mankind from prehistoric days and has been used to man's advantage, albeit intuitively and empirically in most cases. Many of the macroscopic phenomenological observations made on frictional behavior still form the basis for much of the current engineering practice. In recent decades, the scientific efforts made have been attempting to provide rational explanations to these ancient observations through scientific investigations. Therefore, it is most appropriate to discuss the phenomenological aspect of friction as part of these introductory remarks. Frictional behavior is affected by the following factors:

a) Kinematics of the surfaces in contact, i.e., the direction and the magnitude of the relative motion between the surfaces in contact.
b) Externally applied load and/or displacements.
c) Environmental conditions such as temperature and lubricants.
d) Surface topography.
e) Material Properties.

The above list of important factors that control friction clearly indicates that the coefficient of friction (i.e, the ratio of tangential force to the normal load) is not a simple material property.

Most metals behave in such a manner that the coefficient of friction under normal sliding conditions is, to a first approximation, independent of normal load and sliding speed. However, as the normal load is increased to a very high value, such as found in metal cutting, the coefficient of friction often decreases with the normal load and the sliding speed. In the intermediate normal load and speed ranges, the frictional coefficient may reach a peak value as shown in Figure 1.1.

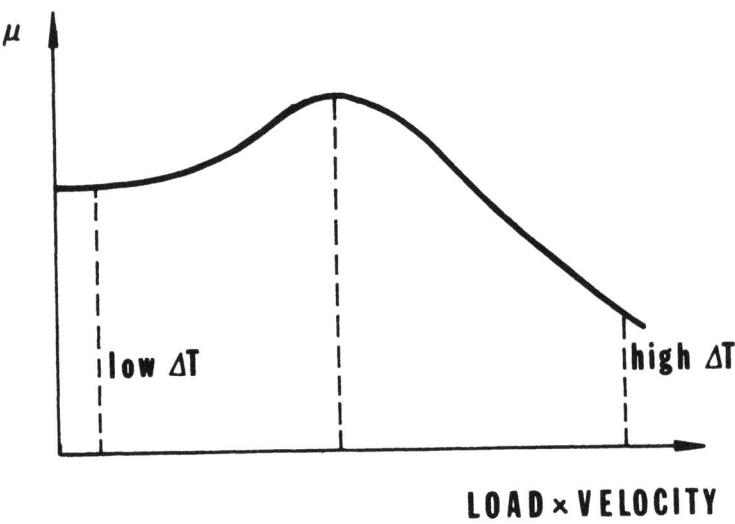

FIGURE 1.1 QUALITATIVE REPRESENTATION OF THE COEFFICIENT OF FRICTION OF METALS VERSUS LOAD TIME SPEEDS. ΔT IS THE INTERFACIAL TEMPERATURE RISE

Polymeric materials behave differently from metals in that their coefficient of friction is a function of the normal load and sliding velocity. As the load is increased, most polymers, be it a thermoplastic or a thermoset, exhibit lower friction coefficients, even at a low sliding speed as shown in Figure 1.2. As the sliding speed is increased, the coefficient of friction increases. At a very high speed, the coefficient of friction decreases with an increase in sliding speed, as shown in Figure 1.3.

In the past these observations were explained in terms of the adhesion model. In essence, this theory assumes that the surface consist of asperities and the interface consist of asperity contacts, as shown in Figure 1.4. The real area of contact is much smaller than the apparent area of contact in most cases, except in such applications as in metal cutting where the real area of contact approaches the apparent area of contact. When a relative motion is imparted to the interface by applying a tangential force, each asperity contact welds together and shears to accomodate the relative motion. Then,

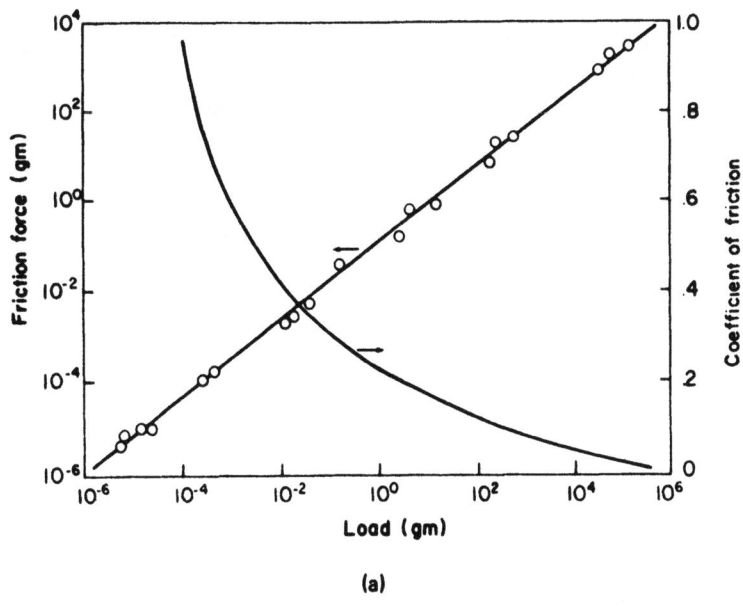

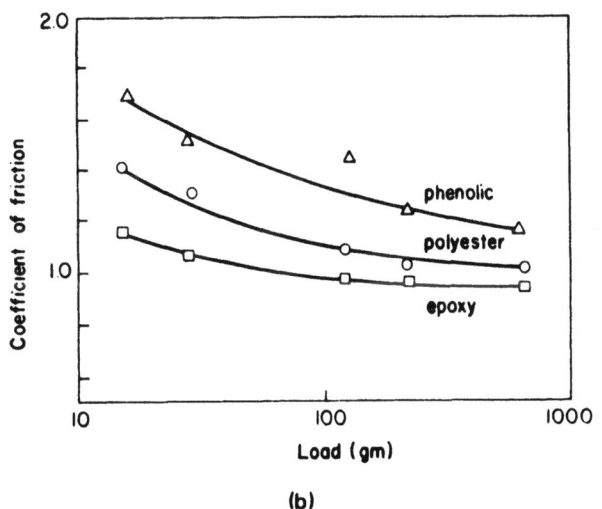

FIGURE 1.2 COEFFICIENT OF FRICTION OF THERMOPLASTICS AND THERMOSETTING PLASTICS

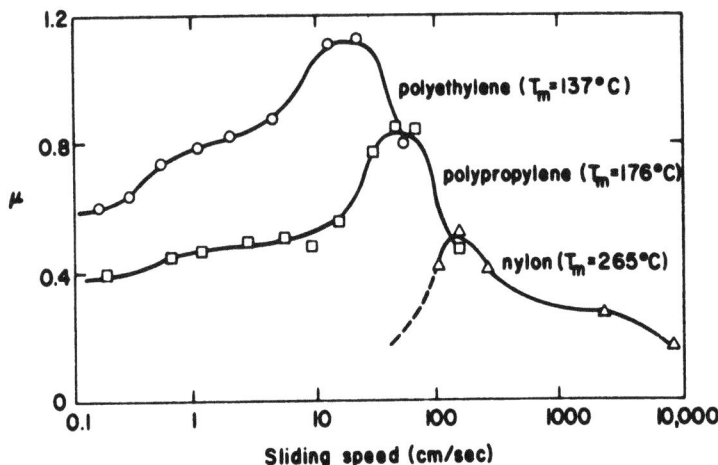

FIGURE 1.3   COEFFICIENT OF FRICTION OF THERMOPLASTICS AS A FUNCTION OF SLIDING SPEED

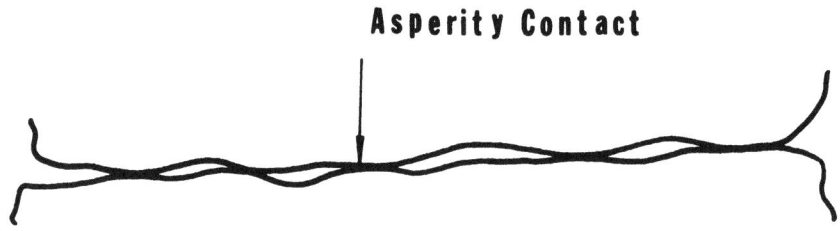

FIGURE 1.4   ASPERITY CONTACT

to a very rough first approximation, the frictional force is given by

$$F = A_r \tau \tag{1.1}$$

where τ is the shear stress at each junction and A_r is the real area of contact. The real area of contact was related to the hardness of the material based on the assumption that the real area of contact must be large enough to support the given normal load. That is,

$$A_r = \frac{L}{H} \tag{1.2}$$

where H is the indentation hardness of the metal. τ is the critical shear strength of the material which must be overcome to satisfy kinematics of the sliding motion. A great deal of discussion has taken place as to how τ is related to H. Many believers of the adhesion theory assumed that τ must be greater than the critical shear strength of the bulk material, due to the work-hardening of the surface layer during sliding. This type of argument was necessitated by the low predicted values of the friction coefficient. If we assume that τ is equal to the critical shear strength of the bulk k, then

$$F = A_r \tau = \frac{L}{H}\tau = \frac{L}{6k}k = \frac{L}{6}$$

$$\mu = \frac{F}{L} = \frac{1}{6} \tag{1.3}$$

which is much smaller than the typical observed values under steady state sliding conditions. In order to improve the correlation between the experimental results and the adhesion theory, a number of theories have been advanced.

One of these theories is due to Rabinowicz, who argued because of the surface energy of adhesion, the actual area of contact is much larger than that given by Eq. (1.2)[1]. If the overall surface energy change is denoted by W_{ab} (i.e., the surface energy of adhesion), then

$$W_{ab} = \gamma_a + \gamma_b - \gamma_{ab} \tag{1.4}$$

where γ_a and γ_b are the surface energies of the two contacting surfaces and γ_{ab} is the interface energy. The sum W_{ab} is always positive, i.e., the overall energy is decreased by bonding. Idealizing the indentation of asperities as an indentation by a conical indenter of material b penetrating into a half-space of material a, as shown in Figure 1.5, the work done by the normal load L during an infinitesmal indentation dx may be equated to the difference in the work done in deforming the material plastically and the surface energy change, i.e.,

$$L \, dx = \pi r^2 \, H \, dx - (2\pi r) \, W_{ab} \frac{dx}{\sin\theta} \quad (1.5)$$

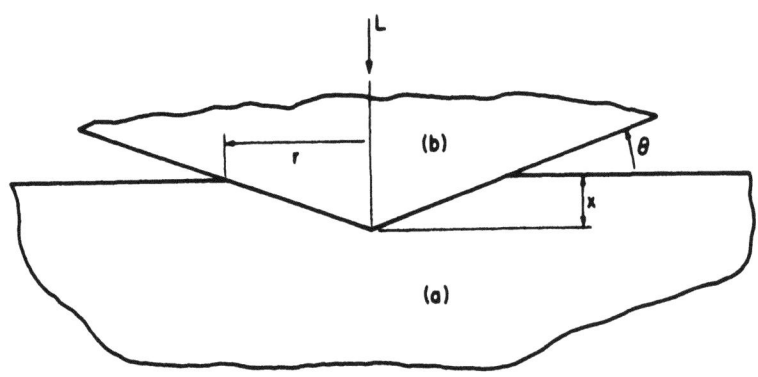

FIGURE 1.5 INDENTATION OF CONICAL INDENTER TO REPRESENT THE INDENTATION OF ASPERITIES

Eq. (1.5) may be rewritten for the area in contact as

$$\pi r^2 = \frac{L}{H} + \frac{2\pi r}{\sin\theta} \frac{W_{ab}}{H} \quad (1.6)$$

Eq. (1.6) states that when the interfacial energy change is included in considerations the projected area (πr^2) is larger than that given by Eq. (1.2) by an amount ($2\pi r/\sin\theta$) (W_{ab}/H). Substituting Eq. (1.6) into Eq. (1.3) the expression for the coefficient of friction gives

$$\mu = \frac{F}{L} = \frac{k}{H} \frac{1}{1 - 2 W_{ab}/rH \sin\theta} = \frac{k}{H} (1 + K \frac{W_{ab}}{H}) \quad (1.7)$$

where k is a geometric factor. According to Eq. (1.7), the coefficient of friction is high when the ratio of the surface energy of adhesion W_{ab} to hardness H is large and when roughness angle θ is small. In support of this theory

Rabinowicz presented a correlation to μ vs. W_{ab}/H as shown in Figure 1.6.

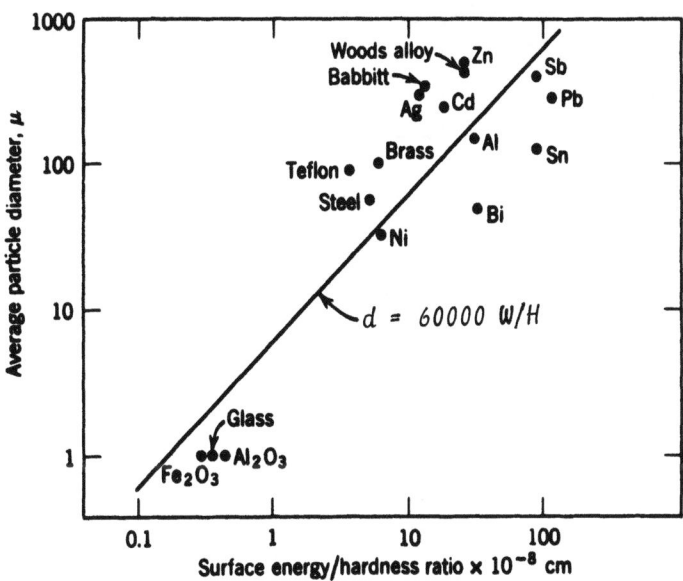

FIGURE 1.6 PLOT OF AVERAGE WEAR PARTICLE DIAMETER AGAINST THE W/H RATIO FOR METALLIC AND NONMETALLIC MATERIALS

In spite of the correlation, some basic questions raise issues on the importance of surface energy considerations in determining friction. The most obvious difficulty is the relatively small magnitude of the surface energy change in comparison to the total work done. An order of magnitude analysis of Eq. (1.6) shows that the first term of RHS of Eq. (1.6) is much larger than the second term of RHS. Another difficulty is that most surfaces are so highly contaminated by absorbants and impurity atoms in the metal that the validity of experimentally measured W_{ab} for various surfaces is in doubt[2]. It will be shown in a later section that the correlation shown in Figure 1.6 can also be explained in terms of the mechanical properties of the surface.

Another adhesion model which also stresses the real area of contact is advanced by A. P. Green[3,4]. Green analyzed the deformation of the surface asperity contact using the slip line field for a rigid-perfectly plastic material. The plasticity analysis of the junction shown in Figure 1.7 showed that the coefficient of friction can be larger than 0.17 as shown in Figure 1.8. For a typical surface, the angles δ and θ are less than 10°, yielding μ of approximately 0.6. Based on the analysis and experiments with plasticine, Green

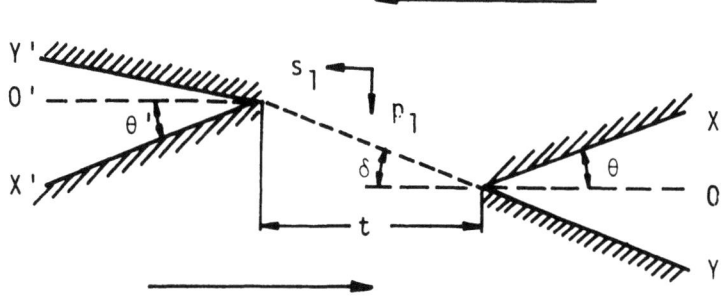

FIGURE 1.7 STRONG JUNCTION DURING STEADY SLIDING

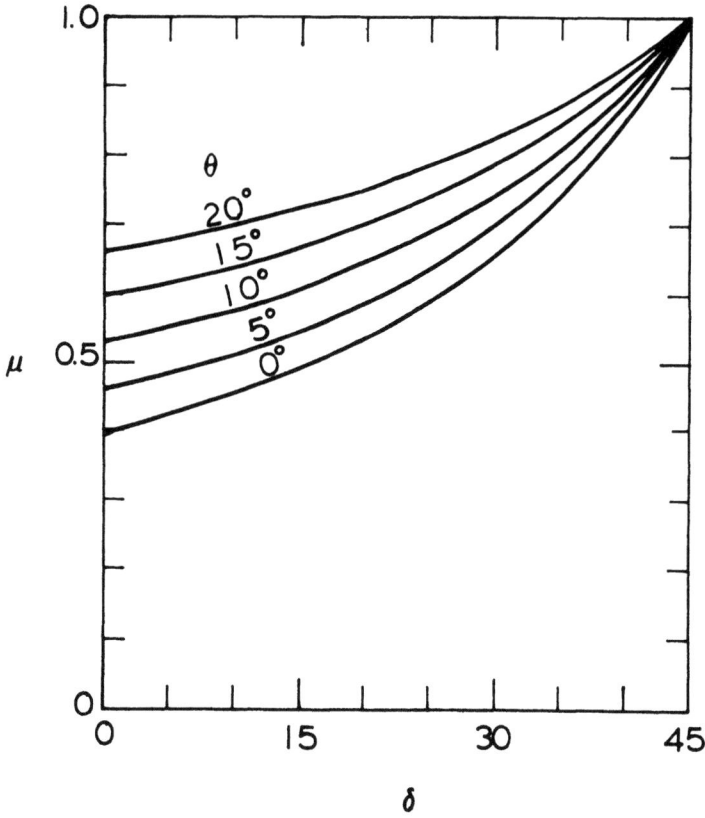

FIGURE 1.8 THEORETICAL RESULTS OF ARCHARD FOR STRONG JUNCTIONS FOR μ VS. SLOPE OF ASPERITIES

concluded that some of the strongly adhering junctions may also support a tensile stress during the deformation process. Since the total compressive normal load must be supported by the asperity junctions in compression, the equilibrium consideration requires the real area of contact under the steady state sliding situation be greater than when all the junctions are under compression. Therefore, he reasoned that the coefficient of friction can be extremely large. This line of thought cannot, however, explain large coefficients of friction observed, even when adhesion is absent.

One of the early attempts to explain friction was to relate it to surface roughness, because surface is not generally smooth, consisting of asperities (i.e., short range perturbations from the mean) and waviness (i.e., long range pertubations from the mean). The roughness theory assumed that the frictional force is equal to the force required to climb up the asperity of slope θ. Then, the coefficient of friction is given by:

$$\mu = \tan \theta \qquad (1.8)$$

It is, however, clear that asperities undergo deformation due to the sliding action rather than simply sliding over each other. Moreover, asperities cannot continue to climb up asperities throughout the sliding action.

Another school of thought attributes the frictional force to a combined effect of adhesion, plowing, and roughness. Shaw and Macks expressed the frictional force to be the sum of the adhesion component, given by Eqs. (1.1) and (1.2), and the plowing component, which can be expressed as

$$\mu = \frac{\tau}{H} + \tan \theta + P \qquad (1.9)$$

where P accounts for the plowing force[5]. Kragelskii also believes that plowing contributes to the frictional force as well as adhesion[6]. He expressed the coefficient of friction as

$$\mu = \frac{\tau_o}{P_r} + \beta + K a_h \ h/r \qquad (1.10)$$

where τ_o is the shear strength of the surface due to molecular bonds when the normal pressure is equal to zero; P_r is the contact pressure which, for a surface that has been run-in, depends on elastic modulus and the surface roughness parameter; a_h is the coefficient of hysteresis losses in friction; β is the molecular bond strengthening coefficient; h is the height of an asperity; r is the asperity radius; and K is the friction parameter. In many sliding situations the ideas embodied in developing Eqs. (1.9) and (1.10) are closer to the real picture than the adhesion theory of friction. However, as shall be shown later, the frictional force is not a constant as implied by these equations; Eqs. (1.9) and (1.10) may not satisfy the equilibrium condition for a vertical component of forces; and contrary to Eq. (1.10), even after run-in, there can be plastic deformation due to adhesion which will affect the real area of contact. A new theory of friction which overcomes these difficulties is presented in a later section.

The friction behavior is greatly affected by the interfacial temperature and the environment. The interfacial temperature and environment govern the degree of adhesion, the nature of chemical reaction at the interface, and the hardness of the surfaces. In normal sliding situations at low speeds, the interfacial temperature rise is small, relative to its melting points. In high speed cutting of metals the interfacial temperature rise can be quite significant. As a consequence, the frictional behavior can be very different in two situations. At low temperatures, mechanical behavior of solids dominate the frictional behavior; whereas at high temperatures (i.e., when the homologous temperature is greater than 1/2), chemical interactions and mutual solubility may govern the frictional behavior.

1.3 Phenomenological Observations Related to Wear and Earlier Theories

Wear of materials occurs by many different mechanisms depending on the materials, the environmental and operating conditions, and the geometry of the wearing bodies. These wear mechanisms can be classified into two groups as shown in Table 1.1; those primarily dominated by the mechanical behavior of solids and those primarily dominated by chemical behavior of materials. In many wear situations there are many mechanisms operating simultaneously, but there is usually the rate determining process which must be identified to deal with the wear problem. What determines the dominant wear behavior are mechanical properties, chemical stability of materials, temperature, and operating conditions. More commonly known phenomenological aspects of the wear behavior of metals and polymers under sliding conditions will be described in this section.

When two materials slide against each other to a very rough approximation, the wear volume W is linearly proportional to the distance slide, S, and normal load, L,

A) Wear processes which are primarily dominated by the mechanical behavior of materials under a given loading condition

Type	Typical Characteristics & Definations	Observed in
Sliding Wear (Delamination Wear)	Plastic deformation, crack nuclaeation and propagation in the subsurface	Sliders, bearings gears, cams where surfaces undergo relative motoin.
Fretting Wear	The early stages of freetting wear is the same as sliding wear but depends on relative amplitude. The entrapped wear particles can have signifacant amplitude is important	Press fit parts with a small relative sliding motion.
Abrasive Wear	Hard particles or hard surface asperities plowing and cuuting the surface in relative motion	Sliding surface, earth removing equipment.
Erosive Wear (Solid particle Impingement)	Due to solid particle impingement, large sub-surface defermation, crack nucleantion and propagation	Turbines, pipes for coal slurries, helicopter blades.
Fatigue Wear	Fatigue crack propagation takes place, normally perpendicular to the surface, without gross plastic deformation under cyclic loading conditions	Ball bearings, roller bearings, glassy solid sliders.

B) Wear processes which are primarily controlled by chemical process and thermally activated processes

Type	Typical Characteristics & Definations	Observed in
Solution Wear	Formation of new compounds of a lower free energy of formations; high temperature; no gross plastic deformation, atomic level wear process.	Carbide tools in cutting steek at high speeds.
Diffusive Wear	Diffusion of elements across the interface.	HSS tool in cutting steel at high speeds.
Oxidative Wear	Formation of weak, mechanically incompatible oxide layer	Sliding Surface in highly oxidative environment (not common)
Corrosive Wear	Corrosion of grain boundaries and formation of pits	Lubricated and corrosive atmosphere

TABLE 1.1 CLASSIFICATION OF WEAR

but inversely proportional to the hardness of the material. This may be expressed as

$$W \alpha \frac{LS}{3H} \qquad (1.11)$$

Eq. (11) is normally written as

$$W = K \frac{LS}{3H} \qquad (1.12)$$

where K is a dimensionless proportionality constant commonly known as the wear coefficient. The factor 3 is a result of Archard's model for adhesion theory of wear[7]. This will be discussed later. There are many exceptions to the above statement. For example, a soft, commercially pure copper can be much more resistant to wear than AISI 1045 steel which is much harder. In normal sliding situations the wear coefficient for most metals is in the range of 10^{-4} to 10^{-3}.

In abrasive wear, the surface of a softer metal is plowed by wear particles or hard asperities. Abrasive wear follows the relationship given by Eq. (1.12) reasonably well, i.e., the harder the material the less is the wear rate. The wear coefficient of typical abrasive wear is of the order of 10^{-2} to 10^{-1}.

The wear particles generated by the abrasive mechanism resemble metal chips generated by cutting action.

In fretting wear the interface undergoes a small oscillatory motion, which results in wear of materials. The wear coefficient in this case depends on the amplitude of oscillation, as shown in Figure 1.9, when the relative displacement at the interface is less than a critical value. At large amplitudes of oscillation, the fretting wear coefficient approaches those of unidirectional sliding wear.

At high sliding speeds and loads, such as in metal cutting, the wear rate depends sensitively on the chemical nature of the material. This is shown in Figure 1.10 for various carbides cutting steel. In this case, the wear rate correlates with temperature, but does not increase linearly with normal load and the sliding distance. The hardness of the tool material has little bearing on the wear life of these tools. On the other hand, an alumina tool will not wear due to chemical instability but rather by mechanical deformation of the surface layer[8,9].

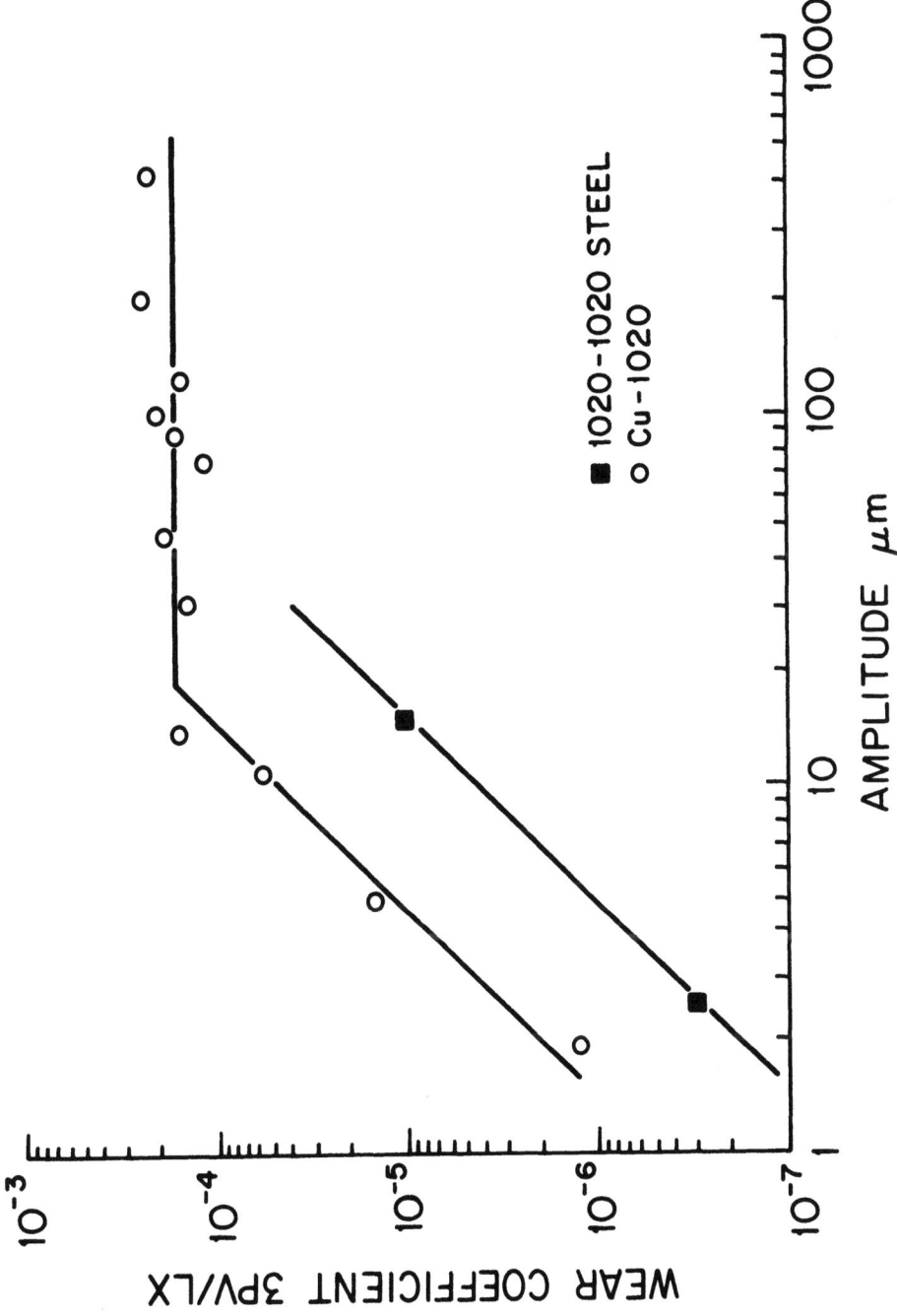

FIGURE 1.9 FRETTING WEAR COEFFICIENT AS A FUNCTION OF THE DISPLACEMENT AMPLITUDE

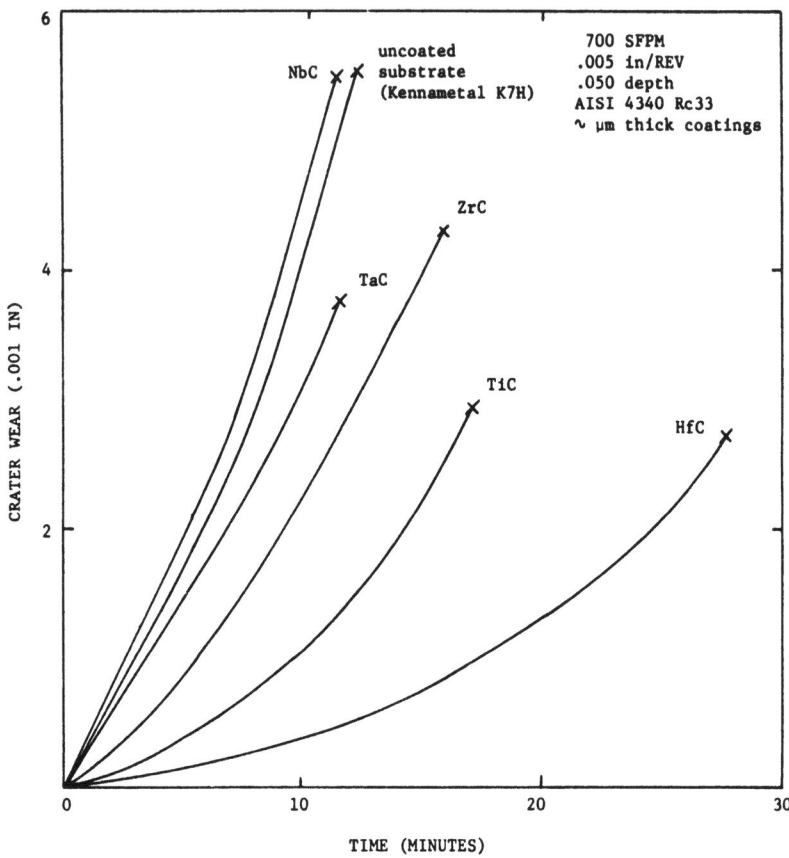

FIGURE 1.10   WEAR RATE OF VARIOUS CARBIDES

All these wear phenomena, except abrasive wear, were explained in terms of the adhesion theory of wear until the advent of the delamination theory of wear in 1972 for sliding wear and solution wear theory in 1979[10, 9]. The adhesion theory states that the wear of materials is due to the welding of asperity junctions which create a hemispheric wear particle when the weaker material near the welded junction fractures. Archard developed a mathematical model to describe this process, which yielded Eq. (1.12)[7]. The difficulty with this model was that there were too many exceptions and that this theory could not form the basis for improvement and development of wear resistant materials. It also violates the conservation law for energy because the actual work done is several orders of magnitude larger than the fracture energy.

The fact that this adhesion is not a complete description of the sliding wear behavior may be appreciated by examining the physical significance of the wear coefficient. The wear coefficient is a dimensionless quantity defined by Eq. (1.12) which may be rewritten as:

$$K = \frac{3WH}{LS} \qquad (1.13)$$

Since L/H is the real area of contact and since the cross-sectional area of the plastically deformed subsurface zone under the asperity contact A_p is of the order of A_r, Shaw showed that Eq (1.13) may be rewritten as[11]:

$$K = \frac{3WH}{LS} = \frac{W}{A_p S} = \frac{\text{Worn Volume}}{\text{Volume of the plastically deformed zone}} \qquad (1.14)$$

Therefore, the wear coefficient, K, for <u>sliding wear</u> may be interpreted as a dimensionless quantity that represents the ratio of the worn volume to the volume of the plastically deformed zone. Since K is the order of 10^{-4} to 10^{-3}, the volume of the material removed by wear is a very small fraction of the material undergoing plastic deformation below the asperity contact. Therefore, it is clear that sliding wear cannot be properly understood without comprehending the plastic deformation process at the subsurface. This process is the primary mode of energy dissipation during sliding wear.

The abrasive wear was modeled in the past as a cutting process. This assumes that an abrasve particle leaves a wear track of the same cross-sectional shape. For example, if the abrasive grain can be idealized as a cone, as shown in Figure 1.11, the wear volume, removed after the abrasive traverses a distance S, is the area shown by the shaded area. However, this type of over simplified model misrepresents the true picture. In order to illustrate this point, we may again examine the physical significance of wear coefficient but this time for abrasive wear, and compare its predictions with experimental results. It can be shown that when wear particles in the form of chips are generated without any plastic deformation, the specific energy μ (i.e., the work done to remove a unit volume of material by a cutting mechanism) is equal to the hardness of the material for the idealized conical model shown in Figure 1.11. Then, Eq. (1.12) may be rewritten as:

$$K = \frac{3WH}{\mu LS} \simeq \frac{Wu}{\mu LS} \qquad (1.15)$$

assuming $3\mu \simeq 1$. It is then seen that the dimensionless quantity K is simply the ratio of the work done to generate wear particles, in the form of cut chips, to the total external work done. Therefore, when the entire work done is

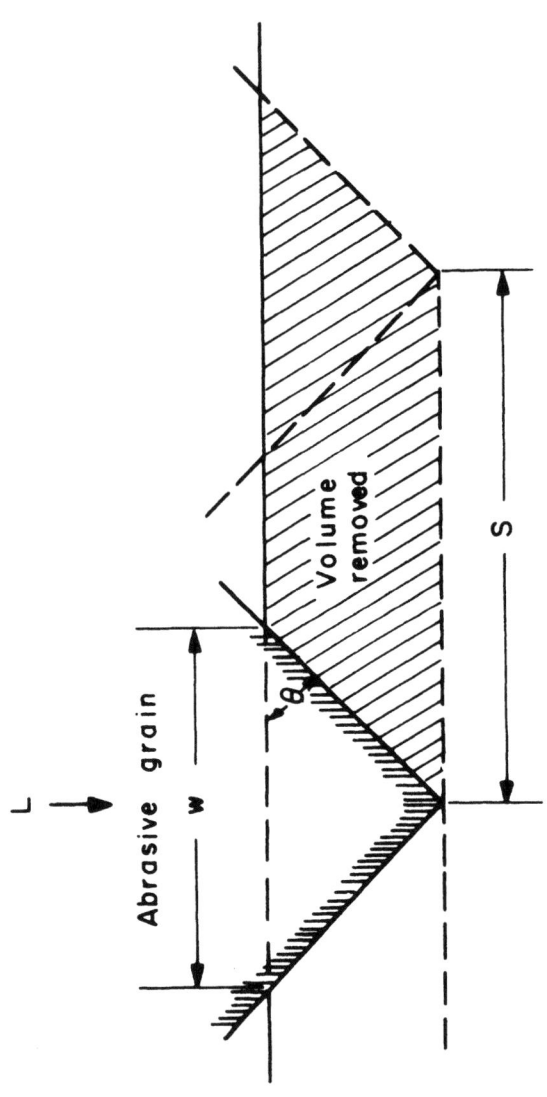

FIGURE 1.11 ABRASIVE WEAR MODEL

consumed to cut the surface, as the classical theories assumed, the wear coefficient should be nearly equal to unity. However, the experimentally determined maximum coefficients are one or two orders of magnitude less than unity.

The fact that a wear theory based on the cutting mechanism predicts too large a wear coefficient can also be seen from the results of the cutting test. Rabinowicz's theory, which is one of the more commonly cited works, states that the volume of material cut is equal to the volume displaced by an abraxise grain, i.e.,[1]

$$K = \frac{3 \tan \theta}{\pi} \qquad (1.16)$$

The same model for friction by plowing yields

$$\mu = \frac{\tan \theta}{\pi} \qquad (1.17)$$

Comparing Eqs. (1.16) and (1.17) the friction coefficient is related to the wear coefficient as

$$K = 3\mu \qquad (1.18)$$

When a diamond stylus with a cutting angle of $\theta = 35°$ was used to cut AISI 1095 steel, the experimentally determined values of K and μ were 0.23 and 0.6, respectively[12]. The theoretically predicted value for K was 0.67.

The discrepancy between the simple cutting model and the actual abrasive wear process is primarily caused by the deformation of the subsurface layer. A large plastic deformation occurs at and below the surface during abrasion. Plowing also forms ridges which have to be removed subsequently. For this reason, the wear rate depends on the ductility of the material; less ductile materials have generally higher wear coefficients than more ductile materials. Two examples are commercially pure nickel and copper, see Figure 1.12.

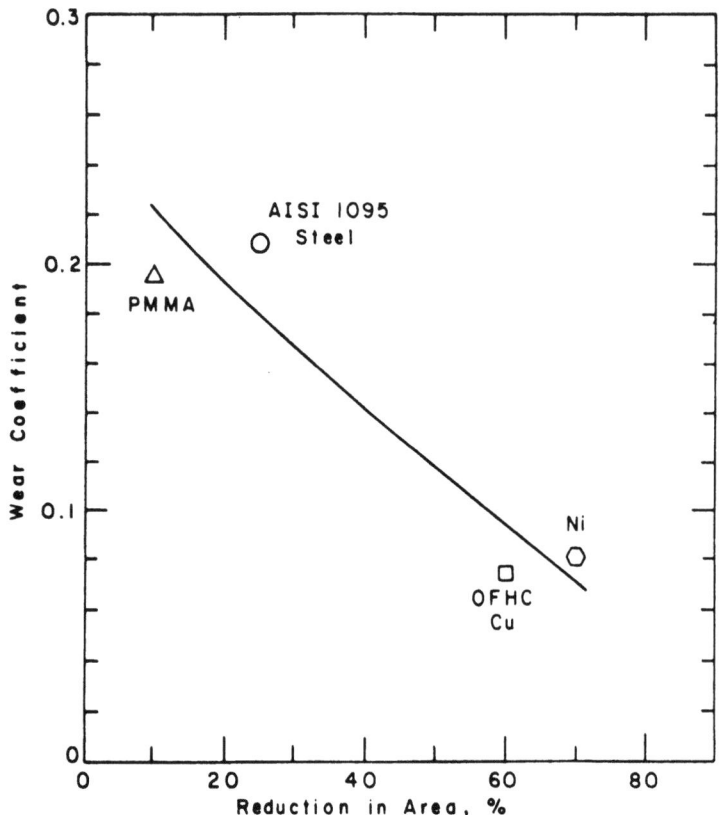

FIGURE 1.12 ABRASIVE WEAR RATE VERSUS DUCTIBILITY

The wear coefficient can be misleading because of its over-dependence on hardness. As shown in Table 1.2, the wear rate does not depend on the prior cold work, whereas the wear coefficient is very different between the annealed and cold-worked nickel. The large difference in numerical values of the wear coefficient, although the wear rate was the same, is due to the different values of the hardness used in computation. The cold-worked metal has the same wear rate as the annealed metal because the amount of cold work done during the wear process is much larger than the cold work done during metal processing and because the prior cold work might have induced "damages" which accelerates the wear process vis-a-vis annealed metals.

CONDITION	HARDNESS (kg/mm^2)	WEAR RATE (m^3/m)	WEAR COEFFICIENT	HARDNESS AFTER TEST
Annealed	88.5	8.07 x 10	0.053	240
Fully Cold Worked	242.0	8.66 x 10	0.157	242

* Material = Ni
  Applied Load = 4 kg
  Abrasive Size = 60 grit

TABLE 1.2 EFFECT OF COLD WORK ON ABRASION

A very useful piece of information to keep in mind in considering wear problems is that by knowing the wear coefficient, we can begin to speculate as to the cause of wear. With the caution that there are dangers in using numerical values indiscriminately, the following "ball park" figures are given for typical wear processes:

During sliding wear (be delamination or sometimes referred to as adhesive wear) 10^{-4} to 10^{-3}

Abrasive wear 10^{-2} to 1

Fretting wear (at large displacement amplitudes, > 100 μm) 10^{-4} to 10^{-2}

A rather extensive collection of wear data is available, but when special material combinations are to be used or when unique operating conditions exist, the most prudent thing to do is to perform actual tests in addition to using the wear coefficients available in handbooks[13].

1.4 Comments on Surface Topography

A great deal of work has been done to characterize the surface finish mathematically (see, for example, Whitehouse, Ref. 14). However, the relationships between the surface topography and the functional requirements for friction and wear are not yet fully understood. Based on the available information, it appears that the original surface finish is not very important in low speed dry sliding applications; whereas it is of importance when lubricants are used. In the former case, the surface geometry is drastically altered by sliding actions; whereas in the latter case, not only does the initial surface geometry change rapidly, but also the pressure and temperature in the lubricant and the metal surface depend sensitively on the surface topography. In this series of lectures, the effect of surface topography will be neglected.

1.5 Introduction to the Lecture Notes

There are two kinds of surface interactions: chemical and physical. Therefore, we will first examine the chemical and physical state of solid surface, followed by discussions of the chemical and the physical behavior of solid surfaces in Section 2. Then, in Section 3 the generation and transmission of forces at the interface will be examined, in order to understand the genesis of friction. Section 4 will deal with the response of materials when external forces are applied to the metallic and polymeric surface by the asperities and wear particles. Based on the basic analytical results of Section 4, delamination wear will be covered in Section 5. Finally a few concluding remarks will be made in Section 6.

1.6 References - Section 1

1. Rabinowicz, E., "Friction and Wear of Materials", Wiley, New York, 1965.

2. Buckley, D.H., "Definition and Effect of Chemical Properties of Surfaces in Friction, Wear and Lubrication", Fundamentals of Tribology (Ed. N.P. Suh and N. Saka), MIT Press, 1980.

3. Green, A.P., "Friction Between Unlubricated Metals: A Theoretical Analysis of the Junction Model", Proceedings of the Royal Society of London, A228 (1955) 191-204.

4. Green, A.P., "The Plastic Yeilding of Metal Junctions Due to Combined Shear and Pressure", Journal of the Mechanics and Physics of Solids, 2(1955) 197-211.

5. Shaw, M.C. and E.F. Macks, "Analysis and Lubrication of Bearings", McGraw-Hill, New York, 1949.

6. Kragelskii, I.V., "Friction Interaction of Solids", Soviet Journal of Friction and Wear, 1(1980) 7-20 (Translated by Allerton Press, Inc.)

7. Archard, J.F., "Contact and Rubbing of Flat Surfaces", Journal of Applied Physics, 24(1953) 981-988.

8. Suh, N.P., "New Theories of Wear and Their Implications for Tool Materials", Wear, 62(1980) 1-20.

9. Kramer, B.M. and N.P. Suh, "Tool Wear by Solution: Quantitative Understanding", Journal of Engineering for Industry, Trans. A.S.M.E., 102 (1980) 303-

10. Suh, N.P., "The Delamination Theory of Wear", Wear, 25 (1973) 111-124.

11. Shaw, M.C., "Dimensional Analysis for Wear Systems," Wear, 43 (1977) 263-266.

12. Sin, H.-C., N. Saka, and N.P. Suh, "Abrasive Wear Mechanisms and the Grit Size Effect", Wear, 55(1979) 163-190.

13. Rabinowicz, E., "Wear Coefficients - Metals", Wear Control Handbook, A.S.M.E., 1980.

14. Whitehouse, D.J., "The Effects of Surface Topography On Wear", Fundamentals of Tribology (Ed. N.P. Suh and N. Saka), MIT Press, 1980.

2. CHEMICAL AND PHYSICAL STATE OF THE SOLID SURFACE

2.1 Brief Introduction to Metals, Polymers and Ceramics

Metals are characterized by their metallic bonding and crystal structures. Most metals have body centered cubic (b.c.c.), face centered cubic (f.c.c.), and hexagonal close packed (h.c.p.) structures. Some have other structures such as tetragonal. The mechanical and chemical behavior and properties depend on crystallographic orientation. Plastic deformation occurs on closely packed planes along the closely packed directions. These crystallographic planes and directions of plastic deformation are called slip planes and slip directions, respectively. Similarly, the surface energy depends on the crystallographic plane; the surface energy of the close packed planes is typically the lowest.

All metals have defects in the form of vacancies and dislocations. Dislocations, which are line defects of atomic arrangement, lower the stress required to cause plastic deformation because of their mobility under stress. Workhardening is a result of many dislocations interacting with each other when the dislocation density increases with plastic deformation. A well annealed solid typically has a dislocation density of 10^6 cm^{-2}. As the number of dislocations increases, they form dislocation cells which can be as small as a few hundred angstroms in diameter. Dislocation behavior is of paramount importance in understanding the plastic deformation of crystalline solids. Although the dislocation behavior near the surface is extremely important in understanding all aspects of tribology, little is known about their structure and density of dislocations very near the surface (i.e., less than a hundred angstroms).

Most metals used in engineering applications are polycrystalline, made up of many grains of all orientations. Therefore, polycrystalline metals have isotropic properties on a macroscopic scale. Grain boundaries are interfaces between two grains of different orientation. Most grain boundaries are regions of random atomic arrangement of finite thickness and therefore, have higher energy than the bulk. Solutes, therefore, segregate to grain boundaries. In many situations polycrystalline solids with small grains have better mechanical properties; the yield strength and the toughness increases with decrease in the grain size.

Many metals used in engineering applications are alloys (both substitutional and interstitial) rather than pure elements. The alloying elements are added for grain refinement and to modify the chemical properites. Many metals used in tribological applications have hard ceramic phases which are introduced to increase hardness.

Polymers are covalently bonded solids. They are long chain molecules which consist mostly of hydrocarbons and other nonmetallic elements. Thermoplastics are linear long chain polymers, most of which either melt or soften as the temperature is increased, except "teflon" (polytetrafluroethylene) which decomposes before melting. Certain thermoplastics, such as polyethylene and polypropylene, are partially crystalline (i.e., crystalline regions are separated by amorphous regions). They have melting points as well as second order transition temperatures. They can undergo large plastic deformations. Glassy thermoplastics, such as polystyrene and polymethylmethacrylate (PMMA), are amorphous and brittle. Thermosetting plastics are those with 3-D network of covalent bonds and therefore amorphous and rigid at all temperatures, undergoing decomposition rather than melting at high temperatures. All these polymers are more corrosion resistant than most metals and chemically inert under normal operating conditions.

Typical polymers have a broad molecular weight distribution. As a consequence, when these polymers are solidified the molecular weight distribution near the surface or an interface is substantially different from those of the bulk. Therefore, in certain crystalline thermoplastics the surface layer has different mechanical properties from those of the bulk. The existence of such a layer is attributed to the preferential nucleation of high molecular species during the crystallation process at certain nucleation sites[1]. According to this reasoning, low molecular weight species of the polymer are rejected to the molten plastic-air interface during the crystallation process giving a weak surface layer. At a molten polymer-metal interface, a region of high cohesive strength is produced in the plastic if the metal surface provides nucleation sites. This is caused by rejection of the low molecular weight species from the interface into the bulk. In the absence of nucleation sites, a weak plastic layer results at the plastic-metal interface.

The first tribological application of polymers was about 50 years ago when phenolformaldelyde resin reinforced with fibers and fabrics was used to make bearings. Since then, particularly during the 1960's, a large number of new polymers were introduced for engineering uses, Table 2.1. Polymeric materials are increasingly becoming important in tribology, since in many applications it is not possible to establish hydrodynamic lubrication at all times due to the low sliding speeds (or oscillatory motions) or due to frequent stop-and-start situations. In some cases, such as in food and textile applications, the presence of lubricant is not acceptable due to the risk of contamination. In general, low cost, light weight, quiet operation, ease of fabrication, and resistance to corrosion have been the major reasons for use of polymeric bearing materials. In order to compensate for such properties as low thermal stability and low mechanical strength and to improve friction and wear properties, various fillers and fibers have been added to polymer matrix.

1839	Vulcanization of natural rubber
1907	Phenol-formaldehyde resin
1926	Alkyd resins
1931	Neoprene synthetic rubber
1937	Styrene-butadiene, acrylonitrile-butadiene rubbers
1938	Nylon 66
1941	Polyethylene
1942	Unsaturated polyesters for laminates
1943-45	Silicones, flurocarbon resins, polyurethanes, styreme-butadiene rubber
1947	Epoxy resins
1956	Linear polyethylene, acetals (polyoxymethylene)
1957	Polypropylene, polycarbonate
1959	cis-polyisoprene and cis-polybutadiene rubber
1960	Ethylene-propylene rubber
1962	Phenoxy resins, polyimide resins
1965	Polyphenylene oxide, polysulfones
af. 1965	Kenlar fibers

TABLE 2.1   INTRODUCTION OF PLASTICS USED IN TRIBOLOGY

Ceramic is a material which is a combination of one or more metals with a nonmetallic element, ussally oxygen, carbon, or nitrogen. The atoms are held together primarily by ionic bonding, with some covalent bonding. Ceramics, especially oxides and nitrides, have very low free energy of formation and therefore are very stable chemically. They do not readily react with other materials and have high melting points. Therefore, ceramics are used in high temperature tribology applications. Ceramics tend to be brittle and thus find limited use in applications that require toughness.

2.2 General Characteristics of a Solid Surface and Tribology.

A solid surface is created when a solid fractures or when a liquid solidifies. It is an asymmetric boundary between two regions: one region where the interatomic forces are greater than the thermal energy of each atom so that the atoms are closely packed, and the other region where there is gas with no "near neighbor" interatomic interactions. Because of the asymmetry, the atoms of the outer surface layer experience different forces than those in the bulk. Consequently, the electronic arrangement near the surface is considerably different from that of the bulk. In the case of many covalently bonded solids and a few metals, even the atomic arrangment reconstructs to new equilibrium configurations[2]. An example of such reconstruction is shown in Figure 2.1. One would not expect such a reconstruction of the surface of a perfect single crystal, when the surface is a close-packed plane, because there is no atomic arrangement with a lower free energy. However, even a small amount of adsorbed gas atoms seems to affect the structure of the surface by dislodging the surface atoms and forming a periodic array of substrate atoms and chemisorbed gas atoms, see Figure 2.1[3]. There must be a certain degree of reconstruction on most polycrystalline solid surfaces because of the random orientation of grains, except when the surface deforms due to the sliding action and exposes close packed slip planes.

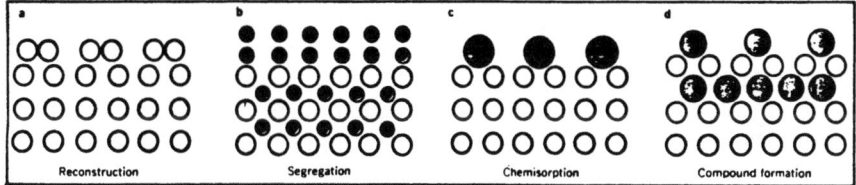

FIGURE 2.1 ATOMIC ARRANGEMENT AT THE SURFACE DUE TO RECONSTRUCTION OF THE SURFACE ATOMS. SEGREGATION OF SOLUTE ATOMS, CHEMISORPTION, AND DUE TO COMPOUND FORMATION BY MIXING OF ADSORBED ATOMS AND THE SUBSTRATE ATOMS (FROM ESTRUP, REF. 2)

In order to create a surface from the bulk, work must be done to break the atomic bonds. Consequently, the surface is at a state of higher free energy than the bulk; the surface energy is equal to the external work done in breaking the chemical bonds. The overall free energy of the surface and the surrounding is lowered when physisorption, chemisorption, and chemical reactions take place at the surface, see Figure 2.1. In physisorption, weak van der Waals-like forces act between an adsorbate and the substrate. Chemisorption bonds involve the mixing of the adsorbate and substrate wave functions, involving more reactive adsorbates such as H, O, CO and NO[4, 5]. Chemical reactions involve the formation of primary chemical bonds between the adsorbate and the substrate, resulting in a three-dimensional structure of a new compound. In tribology, chemisorption and chemical reactions at the surface are used to advantage in lowering frictional forces and minimizing the wear rates.

At the surface there is also an electrostatic potential difference associated with a charge double layer[6, 7]. Because of the asymmetry of repelling forces acting on the electrons of the outer-most shell of the surface atoms, one would expect to find excess electrons just outside the "original" surface and electron deficiency just inside the surface. Quantum mechanical calculation shows that there is a finite probability of finding an electron outside the metal surface in a vacuum. This difference in electronic density about the surface creates electrostatic potential at the surface which is more negative outside than inside the surface. This is referred to as the electrical double layer. These double layers are present at all interfaces, including the solid-electrolyte interface. The electric field created by this double layer can only penetrate about 1 angstrom into the metal surface due to the high mobility of conduction electrons, but can penetrate in excess of 1 µm in the case of insulators.

It is not clear how the reconstruction of the atomic structure and the changes in the electronic state at the surface affect adhesion and tribological behavior. However, it may be reasonable to assume that the directionally bonded solids (i.e., covalently bonded and ionic solids) do not readily adhere to other solids during sliding. Even in the case of identical metals sliding against each other, the reconstructed surfaces cannot instantaneously reestablish the original chemical bond at asperity contacts although the free energy is lowered by going back to the bulk state. This is because there is a finite activation energy barrier that the atoms and electrons must overcome in order to return to the bulk state. Therefore, welding of moving asperity junctions may not be common occurrences for some materials when the temperature at the interface is low.

The change in chemical composition due to segregation can have a significant effect on the tribological characteristics of a surface. It has been shown experimentally that a freshly exposed surface of a solid solution establishes a new composition at the surface. Buckley showed that the surface of 1% Al-Cu solid solution has a 6.5 times higher concentration of aluminum atoms at the surface than in the bulk, see Table 2.2, increasing the adhesion of these solid solutions to gold five-fold over pure copper[8]. Surface enrichment by segregation was also observed in other systems: nickel in iron, silver in palladium, gold in copper, copper in nickel, silver in gold, tin in copper, aluminum in iron, and platinum in osmium. The segregation may be attributed to the need to lower the strain energy created by the difference in atomic size, see Table 2.3.

Alloy	Ratio of the Surface Conc. to Bulk Concentration
Cu - 1 a/o Al	6.5
Cu - 5 a/o Al	4.5
Cu - 10 a/o Al	3.1
Cu - 1 a/o Sn	15.0 ± 2
Fe - 10 a/o Al	8.0

TABLE 2.2 MAXIMUM COVERAGE OF MINOR CONSTITUENT ON ALLOY SURFACES (FROM BUCKLEY, REF. 8)

Aluminum	(f.c.c.)	1.431
Titanium	(h.c.p.)	1.458
Vanadium	(b.c.c.)	1.316
Chromium	(b.c.c.)	1.249
Manganese	(cubic comp.)	1.12
Iron	(b.c.c.)	1.241
Cobalt	(h.c.p.)	1.248
Nickel	(f.c.c.)	1.245
Copper	(f.c.c.)	1.278
Zinc	(h.c.p.)	1.332
Silver	(f.c.c.)	1.444
Cadmium	(h.c.p.)	1.489
Tin	(Bc tetragonal)	1.509

TABLE 2.3   ATOMIC RADIUS (IN ANGSTOMS)

Most real surfaces are ordinarily covered with adsorbates and oxides, which minimizes the metal-to-metal contact at asperity contacts. In this sense adsorbates and oxides act as lubricants. Even heating metals in a vacuum at high temperatures cannot get rid of all adsorbates. One has to resort to argon ion bombardment and the like to remove all the extraneous elements[8]. The important consequence of a coherent oxide layer which is stongly bonded to a substrate is that it alters the dilocation behavior near the surface, making the surface harder than the bulk upon plastic deformation. It should be noted that not all oxide layers formed on a metal surface are stable.

Even in the absence of adsorbates, the surface may be contaminated by migration of a few ppm of interstitial impurities to the surface. A few ppm of carbon in iron was shown to change the surface chemistry[8]. Similar segregation has been seen in such systems as oxygen in platinum, phosphorous in iron, sulfur and carbon in nickel, sulfur in molybdenum, and sodium in lithium. In view of these contamination problems, one has to be extremely careful in using the surface energy data in tribological applications.

An interesting correlation has been made between the cohesive energy density (CED) of crystalline solids and adsorption of albumin protein molecules[9]. The work was done to check the following hypothesis made on protein adsorption: (1) the negatively charged albumin molecules accumulate on a solid surface due to the surface charges created as a consequence of proton diffusion into the solid; and (2) the solid with high CED, which have higher interatomic potential, should make it more difficult for protons to diffuse into the solid. The results show that diamond, which has the highest CED, adsorbs the least amount of proteins and pure metals such as gold, which have a very low CED, adsorbs the most. The adsorption of proteins is plotted as a function of CED of the substrate in Table 2.4.

MATERIAL	COHESIVE ENERGY DENSITY (CAL/CM^3)	SURFACE CONCENTRATION (μG/CM^2)
Diamond	49,700	0.1
Aluminum Oxide	34,200	0.1
Boron Nitride	31,000	0.9
Calcium Fluoride +	25,000	1.0, 3.4, 5.2
Magnesium Oxide +	15,800	3.5, 4.2, 10.0
Glass	15,000	1.5
Platinum	14,800	18.5
Stainless Steel	13,900	1.7
Gold	8,600	66.7
Aluminum	7,800	3.8
Silver	6,670	147.6

+ Adsorption dependent on plane of cleavage

TABLE 2.4 PROTEIN ADSORPTION OF VARIOUS MATERIALS AS A FUNCTION OF THEIR COHESIVE ENERGY DENSITIES

2.3 Chemical Interaction of Metal Surfaces with Lubricants

Metal surfaces are often lubricated with gases, liquids, and solids. Liquid lubricants with additives are the most commonly used means of lowering friction and wear. Lubricants are used to minimize the metal-to-metal contacts either by forming a new chemical compound on the surface or physically separating the surfaces, and thus lower the friction force. The mechanism of lubrication is in part dictated by the nature of interactions between the lubricant and the metal surface. The interactions can be in many different forms: chemical reaction which form new compounds on the surface; a physical interaction such as physisorption which attracts long chain molecules in the liquid to the surface due to van der Waal type electrostatic forces; or in the form of chemisorption involving strong chemical bonds between the liquid components and the metal surface.

Clean metal surfaces react with gases and long chemical molecules dissolved in liquids (for a review of the subject, see Godfrey, Reference 10). This phenomenon has been used to advantage in developing lubricants that will form a stable compound on the surface by inducing chemical reactions which minimize adhesion at the asperity contacts. Oxygen is an important element in boundary lubricants. Klaus, et al., have shown, using microanalytical techniques, that dissolved oxygen reacts with super refined mineral oil and the metal surface to form organometallics which reduce friction and wear[11].

Sulfur compounds are commonly used in boundary lubricants which react with the metal surface to form metal sulfides upon decomposition of these compounds under sliding conditions. These compounds are typically iron-sulfide and form on scuffed surface plateaus[12]. When oxygen is also present, a mixture of iron sufide and iron oxide form[13, 14]. In a gaseous environment with oxygen, iron sulfide changes into iron oxide, indicating that iron oxide is more stable than iron sulfides[15].

Phosphorous compounds such as tricresyl phosphate in mineral oil have been used in many lubricants. It is believed that they form iron phosphates. Similarly, chlorides also form a stable surface layer.

The most commonly used additive is zinc organo dithiophosphates. When this additive was added to a paraffin white mineral oil and a light hydraulic oil, brown and blue films were found to form on highly deformed steel surfaces[16]. Auger analysis and secondary ion mass spectroscopy show that the blue film is 50 μm thick and is primarily a metastable protoxide of iron, $Fe_{1-x}O$[17]. The brown film contained P, S, and Zn, and was composed of minerals such as sulfides, sulphates, phosphates, and thiophosphates. Georges, et. al., claim that the brown film is the reaction between zinc dithiophosphates and iron and contains some polymers of type $[Zn(PSO_2)]_n$.

This adsorption of long chained molecules on the surface also affect the mechanical properties of metals and other solids. Rebinder effect is the most well known phenomenon which describes the adsorption - induced reduction of the surface hardness[18]. The implication of such chemo-mechanical effects as the Rebinder effect on the tribological behavior of materials is not clearly known. A review article on surface effects in crystal plasticity by Latanision gives a good account of various surface phenomena under non-tribological loading conditions[6].

2.4 Mechanical Properties of Solid Surfaces

It was stated earlier that the chemical and physical properties of metallic and polymeric surfaces are different from the bulk. One of the physical properties tribologists are interested in is the flow strength of the materials near the surface. The question that has been investigated the most is whether the surface layer is softer or harder than the bulk when the specimen is subjected to uniform uniaxial deformation. This has been rather a controversial subject, since the experimental results support both views. According to Kramer the surface is harder, whereas Fourie holds the opposite view [19, 20]. Similar controversy also exists for highly worn surface layers[21, 22]. In order to understand the controversy, it is necessary to deal with dislocation dynamics near the surface, including generation and multiplication of dislocations and variation in dislocation density near the surface.

Kramer found that the original yield stress and the workhardening behavior of an aluminum monocrystal could be recovered, as shown in Figure 2.2, by chemically removing a 1 mm layer from the surface[19]. This experimental result was taken to mean that the Stage I workhardening is confined to the near-surface layer. Fourie, on the other hand, found the flow stress distribution of plastically deformed copper monocrystal to decrease near the surface, indicating that there was less workhardening near the surface[20]. Fourie's technique was first to deform a large specimen, slice them into thin sections ranging in thickness from 0.065 to 0.6 mm, and then reload them. The results are shown in Figure 2.3.

Many explanations have been advanced to deal with the soft-hard controversy. All these explanations are based on dislocation theory. One of the explanations for Kramer's results is that dislocations are generated near the surface layer by a Frank-Read type multiplication mechanism which can readily move into the interior of the solid, but cannot egress out of the surface because the surface (or a contaminated surface layer) acts as a barrier to dislocation motion and therefore the dislocations accumulate at the surface faster than the interior[6, 23]. Fourie's experimental results have been attributed to the fact that near the surface dislocations of opposite signs are not equally available, which are needed to form dislocation dipoles. Since dislocation dipoles are responsible for some of the workhardening in Stage II of

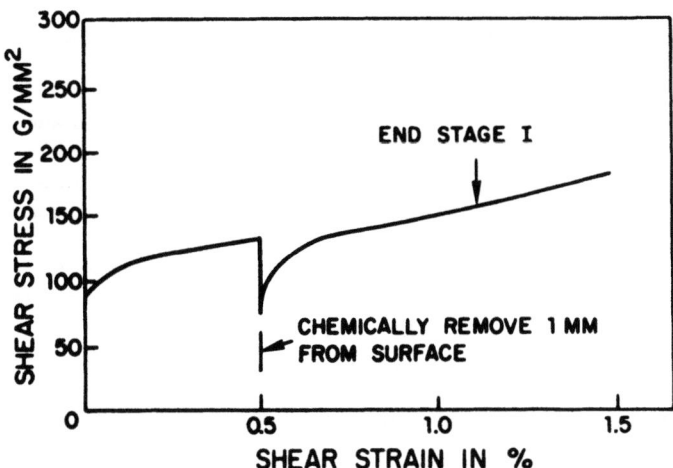

FIGURE 2.2 KRAMER'S EXPERIMENTAL RESULTS SHOWING THAT AN ALUMINUM MONOCRYSTAL RECOVERS THE ORIGINAL YIELD STRESS UPON RELOADING AFTER REMOVING A 1mm SURFACE LAYER ELECTROCHEMICALLY (FROM KRAMER AND DEWER, REF. 19B)

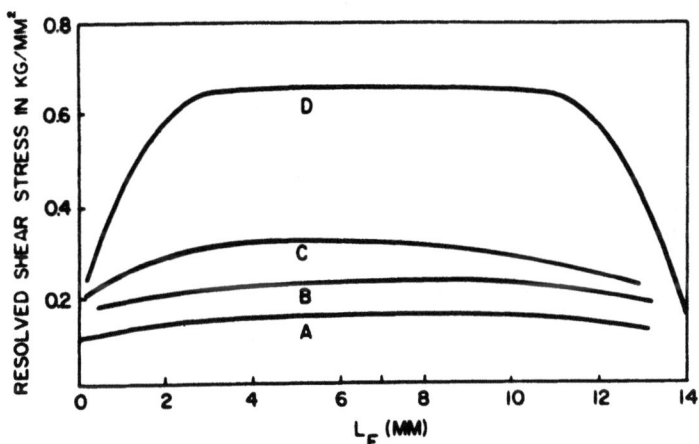

FIGURE 2.3 FOURIE'S EXPERIMENTAL RESULTS (20), WHICH SHOW THE FLOW STRESS DISTRIBUTION IN COPPER MONOCRYSTALS PLOTTED AS A FUNCTION OF THE LENGTH OF THE GLIDE PATH OF EDGE DISLOCATIONS, L_E, WHICH HAS THE VALUE 0 AND 14mm AT THE ORIGINAL SURFACES. CURVE A IS FOR AN AS-GROWN CRYSTAL: CURVES B,C, AND D ARE FOR PRESTRAINS OF 0.02, 0.029, AND 0.053, RESPECTIVELY (AFTER FOURIE, REF. 20)

f.c.c. metals, the surface layer does not workharden as much as the interior.

Dislocations very near and parallel to the surface experience image forces due to their proximity to the surface. When there is no continuous, coherent oxide layer adhering to the metal surface, the image force attracts the dislocations to the surface. When the image force is greater than the resisting drag force, commonly referred to as the dislocation friction stress, these dislocations are attracted to the surface and disappear, if the surface does not act as a barrier. Therefore, it is likely that the dislocations generated very near the surface during sliding may not be able to accumulate very near the surface, whereas the dislocations below this surface zone may be able to accumulate causing workhardening. When a hard, continuous oxide layer adheres to the metal surface, the boundary condition at the oxide-metal interface demands that a dislocation existing in the metal experiences a force which repels it from the surface. This would also be the case if a hard slider were moving over a soft metal surface. The magnitude of the image shear stress acting on a dislocation parallel to the surface is given by:

$$\tau_i = \frac{Gb}{4\pi (1-v) h} \qquad (2.1)$$

where G is the shear modulus, b the magnitude of the Burgers vector, v Poissons' ratio, and h the distance from the surface. This force is opposed by the dislocation friction stress σ_f and the change in the surface energy as the dislocations emerge from the surface. Assuming that the dislocation friction stress dominates, Eq. (2.1) may be written for the maximum depth in which the dislocations will not be stable as:

$$h = \frac{Gb}{4\pi (1-v)) \sigma_f} \qquad (2.2)$$

For silicon iron h is about 0.1 μm, whereas for pure copper it is about 10 to 20 μm.

Even if the lower dislocation density zone can exist very near the surface, it is expected that in polycrystalline metals only the dislocations in the outermost grains would be able to egress to the surface since the dislocations in the subsurface grains will encounter grain boundaries which are barriers to dislocation motion. Since the outermost grain of the worn surface undergoes a very large deformation, the resulting thickness of the grain can be extremely small. When it is less than the effective depth of image force given by

Eq. (2.2), the entire outer grain may be much softer than the subsurface grains. In no case can the soft layer be greater than the thickness of the outermost grain.

In order to estimate the thickness of the outermost grain, consider the deformation of a spherical grain of diameter D into an ellipsoid under the influence of shear stress. The thickness of a grain perpendicular to the surface c can be related to the equivalent strain $\bar{\varepsilon}$ of the surface layer as[24];

$$c = \frac{D}{\sqrt{3}\,\bar{\varepsilon}} \qquad (2.3)$$

The maximum equivalent strain for f.c.c. metals is very large, being of the order of 100, while it is less for AISI 1020 steel, being of the order of 20[25]. The diameter of a typical f.c.c. metal may be about 50 μm. Then, the thickness of the deformed grain c is about 0.3 μm. This is less than the effective depth of the image force for pure copper. This means that the dislocation accumulation in the outermost layer of a clean surface may be less than those in the subsurface grains and that the outermost layer of 0.3 μm may remain soft. In this case, the metal is the hardest at subsurface, i.e., in the vicinity of 0.3 μm below the surface. According to Eq. (2.3) and experimental results, pure f.c.c. metals with larger grains should have a thicker softer surface layer than two phase metals, such as AISI 1020 steel, since the maximum equivalent strain is less for the latter[29].

The experimental results for hardness variation near the surface are contradictory. Savitskii, Kirk and Swanson presented experimental results which show that the hardness is the maximum at subsurface and decreases toward the surface. These results are shown in Figure 2.4[21, 22]. In general however, there are more papers presented in the literature which claim that the hardness is the greatest at the surface[26]. None of the work published so far considered the grain size effects and the maximum strain of the surface layer. This controversy cannot be resolved until precise experiments are done usting a reliable hardness be resovled until precise experiments are done using a reliable hardness measuring technique.

The significance of the hardness question is that if the outermost surface layer is the softest, the adhesion of the flat asperity contact at the surface cannot account for the observed friction coefficients in metals, since shearing will always occur through this soft layer at low stress. Then

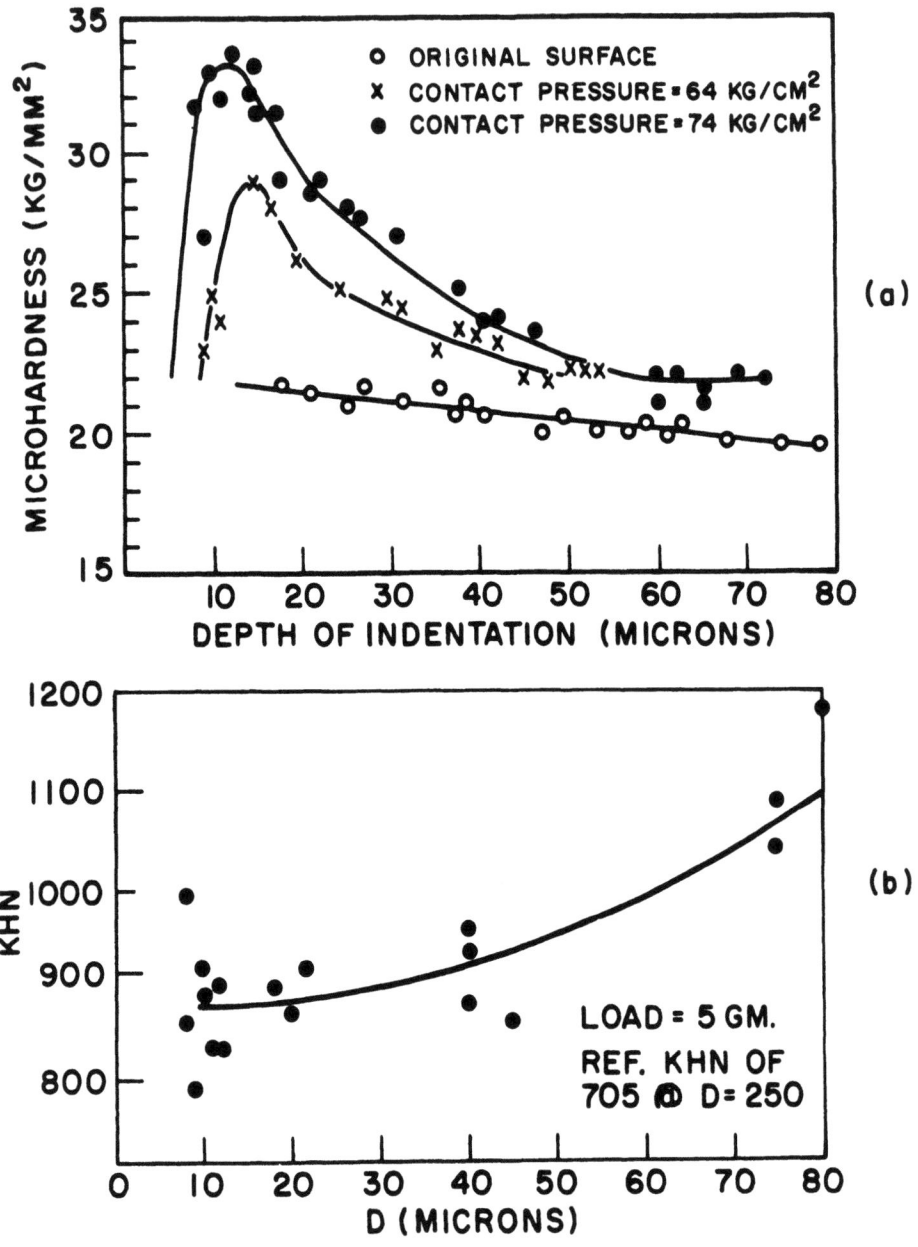

FIGURE 2.4 MEASUREMENTS OF THE SUBSURFACE HARDNESS VARIATION:
(A) INDENTATION HARDNESS OF THE WORN SURFACE OF ALUMINUM SPECIMENS UNDER VARYING INDENTATION LOAD (FROM REF. 21);
(B) INDENTATION HARDNESS OF COPPER SPECIMEN WEAR TESTED UNDER NORMAL LAOD OF 0.682kg (FROM REF. 22)

there must be other mechanisms that determine the frictional force and eventually the wear behavior.

2.5 Thermodynamic Analysis of an Interface

In earlier sections it was stated that the atomic structure of the surface of most metals (and probably other solids as well) is highly heterogeneous. The causes for the heterogeneity are the following:

a) Reconstruction of covalent solids and some metals,
b) Adsorbates due to physisorption and chemisorption,
c)) Oxide layers due to chemical reaction,
d) Segregation of solutes at the surface,
e) Reconstruction of the surface when adsorbates are present on the surface of metals, mixing the substrate atoms and adsorbate atoms.

The fact that these layers on the surface do not simply sit on the substrate but mix with the substrate atoms has been determined experimentally and through quantum mechanical arguments[27]. It can also be supported through thermodynamic arguments. In this section a thermodynamic analysis will be presented, following the treatment originally given by Cahn[28].

For the purpose of analysis, consider a thick oxide layer formed on a metal substrate. Then, the interface between the oxide layer and the substrate is the phase boundary between two different materials. The question we would like to answer is, "how thick is the interface?" If we assume that the interface has a finite thickness, the system consisting of the interface is locally a binary solution. If the mole fraction of oxygen, c, is a nonuniform property in the binary solution, the free energy per molecule, f, in the region of nonuniform composition will depend on both, the local composition and on the composition of the immediate environment. Assuming that the composition gradient is small compared with the reciprocal of the intermolecular distance, the concentration, c, and its derivatives may be taken as independent variables that can describe the local composition and the composition of the immediate environment. When f is a continuous function of these variables, it can be expanded in a Taylor series about f_o (the free energy per molecule of a solution of uniform composition c) as:

$$f(c, \nabla c, \nabla^2 c, \ldots) = f_o(c) + \sum_i L_i \left(\frac{\partial c}{\partial x_i}\right)_o + \sum_{ij} K_{ij} \quad (1)$$

$$\left(\frac{\partial^2 c}{\partial x_i \partial x_j}\right) + \frac{1}{2} \sum_{ij} K_{ij}^{(2)} \left(\frac{\partial c}{\partial x_i}\right)\left(\frac{\partial c}{\partial x_j}\right) + \ldots , \qquad (2.4)$$

where

$$L_i = \left[\frac{\partial f}{\partial\left(\frac{\partial c}{\partial x_i}\right)}\right]_o$$

$$K_{ij}^{(1)} = \left[\frac{\partial f}{\partial\left(\frac{\partial^2 c}{\partial x_i \partial x_j}\right)}\right]_o$$

$$K_{ij}^{(2)} = \left[\frac{\partial^2 f}{\partial\left(\frac{\partial c}{\partial x_i}\right)\partial\left(\frac{\partial c}{\partial x_j}\right)}\right]_o$$

i, j represents the successive substitution of spatial coordinates x, y and z, and the subscript o represents the value of the parameter of the uniform composition. In general, $K_{ij}^{(1)}$ and $K_{ij}^{(2)}$ are tensors reflecting the crystal symmetry and L_i's are components of a polarization vector in a polar crystal. For a cubic crystal or an isotropic medium the free energy must be invariant due to the symmetry operators of reflection ($X_i \rightarrow X_j$) and of rotation about a fourfold axis ($X_i \rightarrow X_j$). Therefore,

$$L_i = 0$$

$$K_{ij}^{(1)} = K_1 = \frac{\partial f}{\partial \nabla^2 c}\bigg|_o \qquad \text{if } i = j$$

$$K_{ij}^{(1)} = 0 \qquad \text{if } i = j \qquad (2.5)$$

$$K_{ij}^{(2)} = K_2 = \frac{\partial^2 f}{\partial(|\nabla c|)^2} \qquad \text{if } i = j$$

$$K_{ij}^{(2)} = 0 \qquad \text{if } i = j$$

Hence for a cubic lattice

$$f(c, \nabla c, \nabla^2 c, \ldots) = f_o(c) + K_1 \nabla^2 c + K_2 (\nabla c)^2 + \ldots \quad (2.6)$$

Integrating over a volume V of the solution, we obtain for the total free energy F of this volume:

$$F = N_v \int f \, dv = N_v \int v \{ f_o(c) + K_1 \nabla^2 c + K_2 (\nabla c)^2 + \ldots \} \, dv \quad (2.7)$$

where N_v is the number of molecules per unit volume. By applying the divergence theorem,

$$\int_V (K_1 \nabla^2 c) \, dv = -\int_V \left(\frac{dK_1}{dc}\right)(\nabla c)^2 dv + \int_s (K_1 \nabla_c \cdot n) ds \quad (2.8)$$

If we choose the boundary so that

$$\int_s (K_1 \nabla c \cdot n) \, ds = 0 \quad (2.9)$$

then

$$F = N_v \int \left[f_o(c) - \left(\frac{dK_1}{dc}\right)(\nabla c)^2 + K_2 (\nabla c)^2 + \right] dv \quad (2.10)$$

$$= N_v \int \left[f_o(c) + K (\nabla c)^2 + \ldots \right] dv$$

where

$$K = \frac{dK_1}{dc} + K_2 = -\left[\frac{\partial}{\partial c}\left(\frac{\partial f}{\partial \nabla^2_c}\right)\right]_o + \frac{\partial^2 f}{\partial(|\nabla c|)^2}$$

Eq. (2.10) states that the free energy is a sum of two contributions: one being the free energy that this volume would have in a homogenous solution and the other "gradient energy" which is a function of the local composition.

At a flat oxide/metal interface (two coexisting isotropic phases c_α and c_β), energy of nonequilibrium material of composition intermediate between c_α and c_β **may** be represented as a continuous function $f_o(c)$ of the form shown in Figure 2.5.

For one-dimensional composition change across the interface, and neglecting terms in derivatives higher than the second, we obtain for the total free energy F of the system as

$$F = AN_v - \int_{-\infty}^{\infty} \left[f_o(c) + K \left(\frac{dc}{dx}\right)^2 \right] dx \quad (2.11)$$

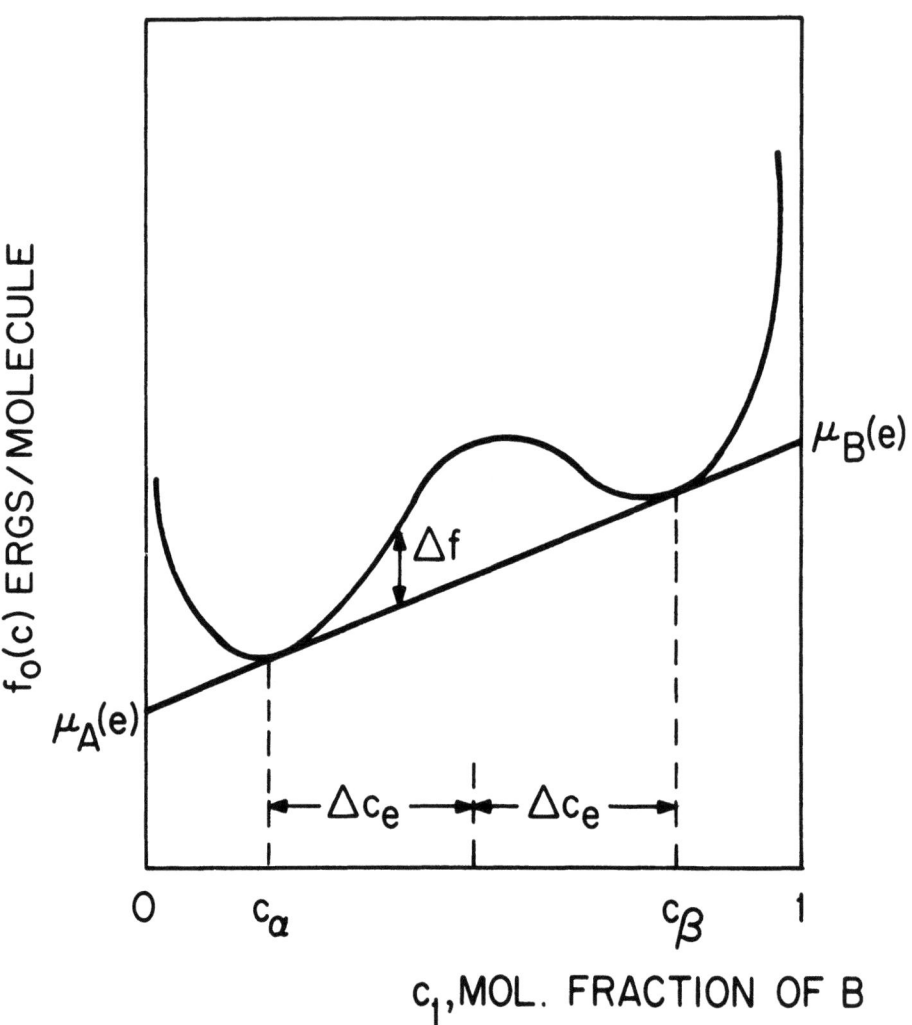

FIGURE 2.5 - FREE ENERGY OF A BINARY SYSTEM AS A FUNCTION OF COMPOSISTION (FROM CAHN, REF. 28).

The specific interfacial free energy, σ, is by definition the difference per unit area of interface between the actual free energy of the system and that which it would have if the properties of the phases are continuous throughout. Hence

$$\sigma = N_V \int_{-\infty}^{\infty} \left[f_o(c) + K \left(\frac{dc}{dx}\right)^2 - c\mu_B(e) - (1-c)\mu_A(e) \right] dx \quad (2.12)$$

$$= N_V \int_{-\infty}^{\infty} \left[\Delta f(c) + K \left(\frac{dc}{dx}\right)^2 \right] dx$$

where

$$\Delta f(c) = f_o - \left[c\mu_B(e) + (1-c)\mu_A(e) \right]$$

$$= c\left[\mu_B(c) - \mu_B(e)\right] + (1-c)\left[\mu_A(c) - \mu_A(e)\right]$$

In order to lower σ, the interfacial energy, highly diffused layer (i.e., small $K(dc/dx)^2$) is desirable, but it increases the $\Delta f(c)$ term by increasing the volume of the nonequilibrium composition, nonhomogeneous layer. Therefore, an optimum exists where $\{\Delta f(c) + K\left(\frac{dc}{dx}\right)^2\}$ term assumes a minimum value, i.e., the chemical potentials be constant through the system.

The foregoing thermodynamic treatment of the interface indicates that, in order to minimize the free energy, the atoms of adsorbates and oxide layers form a mixed layer of a finite thickness rather than simply sitting on the substrate. Therefore, most surfaces with contaminants are likely to be heterogeneous and thus may impede the egress of dislocations from the surface. Dislocation may still escape from the surface however, if they can move to the surface before adsorbates and oxides can form on the clean surface following the deformation of the surface layer.

2.6 Concluding Remarks

In this section the chemical and physical states of the outermost layer of the surface were examined. It was shown that the geometric arrangement, the chemical state, and the physical properties are highly heterogeneous. The substrates are normally covered with adsorbates, oxides and other chemical reaction products, and segregated solutes. These contaminants are likely to prevent adhesion in many sliding situations.

It was also shown that the hardness of the surface layer is different from that of the substrate, notwithstanding the controversy regarding the hardness/softness of the surface layer. The outermost grain of a polycrystalline metal can have a significantly different flow stress than the substrate if dislocation egress is not impeded by surface contaminants and heterogeneous structures. Further research must be done to clarify the mechanical properties of the outermost layer.

Finally, the state of the contaminant/substrate interface is examined from a thermodynamic point of view. It is shown that the atoms of contaminants are mixed with the atoms of the substrate in order to lower the free energy of the system. Because they are mixed, it is difficult to remove these contaminants by such techniques as heating. Consequently, it is safe to assume that the surface of most engineering materials is always contaminated. The high reactivity of solid surfaces can be used to minimize friction and wear. Additives are added to lubricants to form stable compounds on the surface, which lowers the frictional force by preventing plowing in addition to adhesion. This is discussed further in Section 3.

2.7 References - Section 2

1. Schonhorn, H., "Adhesion to Low Energy Polymers", Adhesion, Gordon and Breach Science Publishers, NY, 1969.

2. Estrup, P.J., "The Geometry of Surface Layers", Physics Today, April 1975, 33-41.

3. Somorjai, "Principles of Surface Chemistry", Prentice Hall, Englewood Cliffe, 1972.

4. Schrieffer, J.R., and P. Soven, "Theory of Electronic Structure", Physics Today, April 1975, 24-30.

5. Duke, C.B., "Electronic Structure of Clean Metal Interfaces", General Electric Report 68-C-33, Sept. 1968.

6. Latanision, R.M., "Surface Effects in Crystal Plasticity: General Review," Surface Effects in Crystal Plasticity, Noordhoff-Leyden, 1977.

7. Herring, C., in **Metal Interfaces**, ASM, 1, 1952.

8. Buckley, D.H., "Definition and Effect of Chemical Properties of Surfaces in Friction, Wear and Lubrication", Fundamentals of Tribology (Ed. N.P. Suh and N. Saka), MIT Press, 1980.

9. Suh, N.P. and L.T. Nguyen, "Correlation of Protein Adsorption with Cohesive Energy Density," to be published, 1981.

10. Godfrey, D., "Review of Usefulness of New Surface Analysis Instruments in Understanding Boundary Lubrication", Fundamentals of Tribology (Ed. N.P. Suh and N. Saka), MIT Press, 1980

11. Klaus, E.E., E.J. Tewksbury, and A.C. Bose, <u>Proc. Japan Soc. of Lubrication Engineers - American Soc. of Lubrication Engrs.</u>, Int. Lub. Cong., Tokyo, June 9-11, 1975 (Ed. T. Sakurai, Elsevier Pub. Co., NY, 1976, p. 39).

12. Coy, R.C. and T.F.J. Quinn, in Tribology Convention 1972, London, Sept. 27-28, 1972; Institution of Mechanical Engineers, London, 1973, p. 62.

13. Godfrey, D., "Chemical Changes in Steel Surfaces During Extreme Pressure Lubrication", Trans. American Soc. of Lubrication Engineers, 5 (1962) 57-66.

14. Tomaru, M., S. Hironaka, and T. Sakurai, "Effects of Oxygen on the Load-Carrying Action of Some Additives," Wear, 41 (1977) 117-140.

15. Buckley,D.H., "Oxygen and Sulfur Interactions With a Clean Iron Surface and the Effect of Rubbing Contact on These Interactions", Trans. Am. Soc. of Lub. Engineers. 17, (1974) 206-212.

16. Jahanmir, S., "Wear of AISI 4340 Steel Under Boundary Lubrication", Wear of Materials, ASME, 1981.

17. Georges, J.M., G. Meille, M. Jacquet, B. Lamy, and T. Mathia, "A Study of the Durability of Boundary Films", Wear, 42 (1977) 217-228.

18. Rebinder, P.A., Reports to the VI Congress of Physicists, Moscow, 29, 1928.

19. Kramer, I.R., "The Effect of Surface Removal on the Plastic Flow Characteristics of Metals", Trans. AIME, 227 (1963) 1003-1010; Kramer, I.R., and L.J. Demer, Trans. AIME, 221 (1961) 780.

20. Fourie, J.T., "The Flow Stress Gradient Between the Surface and the Center of Deformed Copper Single Crystals", Phil. Mag., 17 (1968) 735-756.

21. Savitskii, K.B., cited in I.V. Kragelskii, Friction and Wear, Butterworth, Washington, 1965.

22. Kirk, J.A. and T.D. Swanson, "Subsurface Effects During Sliding Wear", Wear, 35 (1975) 63-67.

23. Nabarro, F.R.N., "Surface Effects in Crystal Plasticity - Overview from the Crystal Plasticity Standpoint", Surface Effects in Crystal Plasticity (Eds. R.M. Latanision and J.T. Fourie), Noordhoff-Leyden, 1977.

24. Dantzenberg, J.H. and J.H. Zaat, "Quantitative Determination of Deformation by Sliding Wear", Wear, 23 (1973) 9-19.

25. Augustsson, G., "Strain Field Near the Surface Due to Surface Traction", S.M. Thesis, MIT, 1974.

26. Ruff, A.W., "Deformation Studies at Sliding Wear Tracks in Iron", Wear, 10 (1976) 59-74.

27. Duke, C.B., "Atomic Geometry and Electronic Structure of Solid Surfaces", Surface Effects in Crystal Plasticity (Eds. R.M. Latanision and J.T. Fourie), Noordhoff-Leyden, 1977.

28. Cahn, J.W. and J.E. Hillard, "Free Energy of a Non-Uniform System. I. Interfacial Free Energy", Journal of Chemical Physics, 28 (1958) 258-267.

3. GENERATION AND TRANSMISSION OF FORCE AT THE INTERFACE - THE GENESIS OF FRICTION

3.1 Introduction

One of the three fundamental aspects of tribology is the genesis of friction between sliding surfaces, as discussed in Section 1. Beginning in the early 1940's, serious research has been done to understand the friction mechanism. Although various theories have been advanced, the adhesion theory has dominated other theories[1-4]. However, the experimental results often deviate from the adhesion theory. The experimentally observed coefficients of friction are generally much larger than those predicted; the prediction is worse when the experiment is done in vacuum or in an inert environment which best simulate the conditions assumed in deriving the friction coefficient based on the adhesion theory. Further, the compatability argument for sliding surfaces, based on mutual solubility, cannot explain the large variations in the frictional behavior of metals with little difference in chemical solubility. Furthermore, there are some basic questions on adhesion, such as those discussed in Section 2, which cast doubt on the thesis that friction is entirely due to adhesion. It is clear that adhesion __does__ exist in some sliding situations, but the question is "how important is it in determining the frictional force?"

As discussed in Section 2, the essence of the adhesion theory is that asperities of sliding surfaces come in contact with opposing asperities and form welded junctions which must be sheared to satisfy the kinematic requirements. Therefore, the frictional force depends directly on the actual area of contact which is a function of the applied normal and tangential load. In order to explain the discrepancy between the experimentally measured and the theoretically predicted friction coefficients, the adhesion theory has relied on the argument that the real area of contact is larger when some of these junctions are under tensile loading, requiring a correspondingly larger force to shear the interface.

Because the adhesion theory of friction emphasizes the importance of adhesion between asperities, a great deal of attention in the past has been devoted to the role of surface energy and the mutual solubility of the contacting materials [1]. It has been argued that metals with greater solubility will more readily form strong junctions and thus have higher friction and wear coefficients. Many experimental results were presented in support of this argument. However, questions have been raised on the validity of the argument since surfaces are easily contaminated by chemisorption and physisorption and the chemical composition of the surface is different from that of the bulk [5]. A more realistic model appears to be that the frictional behavior is controlled by plowing, adhesion and asperity deformation [6].

The frictional force depends on the history of sliding. The frictional force undergoes significant changes during the early stages of sliding before reaching steady state frictional behavior. The time dependent nature of frictional behavior is a rich source of information in understanding frictional behavior. The difference between the static and kinetic coefficients of friction is wellknown but the time dependent nature of the kinetic coefficient of friction is less well known [7]. This phenomenon is very important because when the interface is lubricated, the time scale is so expanded that the important frictional phenomenon of interest between well lubricated surfaces, may in fact be confined to the early stages of the frictional behavior observed under the dry sliding conditions.

The frictional behavior of materials is important in tribology not only because of the frictional force between sliding surfaces of interest, but also because it generally affects the wear behavior[6,8]. The delamination wear is clearly affected by surface traction due to its effect on crack nucleation and propagation processes at the subsurface. Also, plowing by wear particles and asperity deformation affect both the wear process and frictional behavior.

This section describes a different theory to explain the frictional behavior of materials. It will be demonstrated that adhesion does not play a significant role at the onset of sliding in typical sliding situations and that the frictional force is generated as a consequence of asperity deformation, plowing by wear particles and adhesion. Furthermore, the coefficient of friction is not a given material property, because it depends on the mechanical properties of the opposing surface and the environment. Experimental results will first be summarized before presenting theoretical support for the postulated genesis of friction. Finally, a "Friction Space" concept is presented as a means of representing the frictional behavior of materials.

3.2 Typical Friction Tests and Experimental Observations of Frictional Behavior of Metals

In Section 1 the phenonemological aspects of friction were discussed. In studying the fundamental mechanisms of friction however, it is necessary to investigate beyond the general phenomenological aspects of friction in great detail. The results of such a study will be described in this subsection before presenting a friction theory in a later subsection.

A series of experiments were conducted at MIT in order to study the friction and wear behavior of various combinations of the following materials: Armco iron, AISI 1020, 1045, and 1095 steel. These iron based metals with differing carbon contents, have large differences in hard-phase concentrations and hardness. Listed materials were chosen to minimize chemical differences, although the surface of Armco iron is expected to be substantially different chemically, from the steel specimens because of the absence of interstitials carbon atoms in Armco iron.

Armco iron was recrystallized at 973° K for one hour. AISI 1020, 1045 and 1095 steels were austenitized at 1173° K for 15 minutes, oil quenched, and then tempered at 673° K for one hour to obtain a spherodized microstructure. The hardness and the volume fraction of cementite are listed in Table 3.1.

Material	Heat Treatment	Vickers Hardness (MPa)	Volume Fraction of Cementite
Armco Iron	973 K, 1 hr; air-cooled	980 ± 50	0.0004
AISI 1020 steel	Austenitized at 1173 K, 15 min; oil-quenched; 673 K, 1 hr; air-cooled	1710 ± 100	0.020
AISI 1045 steel	Spheroidized: 1173 K, 15 min; oil-quenched; 673K, 1 hr; air-cooled	4120 ± 130	0.067
AISI 1095 steel	Spheroidized: 1173 K, 15 min; oil-quenched; 673 K, 1 hr; air-cooled	6080 ± 350	0.142

TABLE 3.1 EXPERIMENTAL MATERIALS

Some tests were also conducted with OFHC copper, which was polished with 4/0 abrasive paper. These specimens were tested in air, sliding against AISI 1020 steel. The initial coefficient of friction, μ_i, was about the same as those obtained with steel iron specimens.

Samples of 6.35 mm in diameter were tested for friction and wear using crossed-cylinder geometry. The specimen (rotating cylinder) was rotated by the spindle of the lathe, and the slider (stationary cylinder) was held stationary in a holder attached to a lathe tool dynamometer which was mounted on the carriage of the lathe. Both normal and tangential forces were measured by a dynamometer-recorder assembly.

Tests were conducted in a purified argon atmosphere except for AISI 1020 steel where some samples were tested in air under both lubricated and unlubricated conditions. Water and light machine oil were used as lubricants. The experimental results were obtained under the following conditions: normal load of 1 kg (9.8 N), sliding speed of 0.02 m/s, total sliding distance of 36 m, and at room temperature.

Some of the specimens were sectioned along the sliding direction to measure the slope of the asperity by taking micrographs of the asperities since the asperities of the machined surfaces were orientation dependent.

Extremely well polished surfaces were slid against each other to investigate the nature of the surface damage after predetermined amounts of sliding. These surfaces were observed using scanning electron microscopy.

The friction and wear coefficients of the iron-carbon system are tabulated in Tables 3.2 and 3.3 respectively. There are several important results worth considering in detail. First, the coefficient of friction changes as a function of the distance slid, especially at the early stage of sliding. It usually has a low initial value and gradually increases until reaching a steady state value. After it reaches a maximum value, the friction coefficient sometimes drops down if the stationary slider is much harder than the moving specimen. The same pair of materials does not show the drop in the coefficient of friction when their roles are reversed. The initial coefficient of friction is always in the range of about 0.1 to 0.2 regardless of the materials tested and whether or not lubricants are used. Second, the steady state coefficient of friction and the wear rates are higher when identical metals are slid against each other than when a harder stationary slider is slid against a moving softer specimen. However, when a softer stationary slider is slid against a harder moving specimen, the steady state coefficients of friction are nearly the same as those of the identical materials sliding against each other. In this case, the wear rates of unidentical pairs of metals are much greater than those of identical metals.

These changes in the friction and wear behavior are related to the changes in the surface topography, as shown in Figure 3.1 and 3.2, which are the micrographs of the slider surface and the specimen surface, respectively. These figures show that when the stationary slider is harder than the specimens, the hard surface is polished to a mirror finish and the high spots of the softer surface also acquire the same mirror finish. When the material underneath the polished surface fails, new high spots are created, which become polished to a mirror finish again. This does not happen when the stationary slider is softer than the specimen or when identical metals are slid against each other. In these cases, many plowing grooves are observed and the surface always stays rough. From these experimental results the following observations are made:

1. The coefficients of friction vs. sliding distance (or time) may be summarized in Figure 3.3a, when identical metals

			Specimen (rotating cylinder)			
			Armco iron	1020 steel	1045 steel	1095 steel
Slider (stationary cylinder)	Armco iron	μ_i	0.13	0.20	0.24	0.20
		μ_s	0.71	0.75	0.69	0.76
		μ^*	----	----	----	----
	1020 steel	μ_i	0.18	0.20	0.13	0.12
		μ_s	0.55	0.68	0.57	0.65
		μ^*	0.80	----	----	----
	1045 steel	μ_i	0.16	0.17	0.17	0.12
		μ_s	0.52	0.53	0.71	0.69
		μ^*	0.77	0.71	----	----
	1095 steel	μ_i	0.17	0.17	0.14	0.17
		μ_s	0.51	0.54	0.58	0.67
		μ^*	0.76	0.73	----	----

μ_i = initial coefficient friction
μ_s = steady state coefficient of friction
μ^* = peak value of the friction coefficient

TABLE 3.2   FRICTION COEFFICIENTS

			Specimen (rotating cylinder)			
			Armco iron	1020 steel	1045 steel	1095 steel
Slider (stationary cylinder)	Armco iron	K_{sp}	46.0 ± 37.2	127.0 ± 45.0	331.0 ± 268.0	1210.0 ± 120.0
		K_{sl}	25.1 ± 17.5	6.94 ± 2.33	16.4 ± 14.0	17.1 ± 0.5
	1020 steel	K_{sp}	6.10 ± 2.75	143.0 ± 88.0	25.3 ± 5.0	42.3 ± 17.6
		K_{sl}	5.23 ± 2.60	85.4 ± 48.4	2.57 ± 2.27	1.80 ± 1.33
	1045 steel	K_{sp}	3.59 ± 1.69	15.6 ± 5.9	94.2 ± 17.8	565.0 ± 269.0
		K_{sl}	2.89 ± 2.37	13.9 ± 7.2	49.4 ± 7.5	60.0 ± 17.4
	1095 steel	K_{sp}	2.11 ± 0.89	8.83 ± 4.04	7.32 ± 5.10	15.1 ± 7.6
		K_{sl}	2.35 ± 0.95	4.97 ± 2.72	5.24 ± 3.78	14.7 ± 10.3

K_{sp} = wear coefficient of specimen; K_{sl} = wear coefficient of slider
All K values are to be multiplied by 10^{-4}.

TABLE 3.3   WEAR COEFFICIENTS

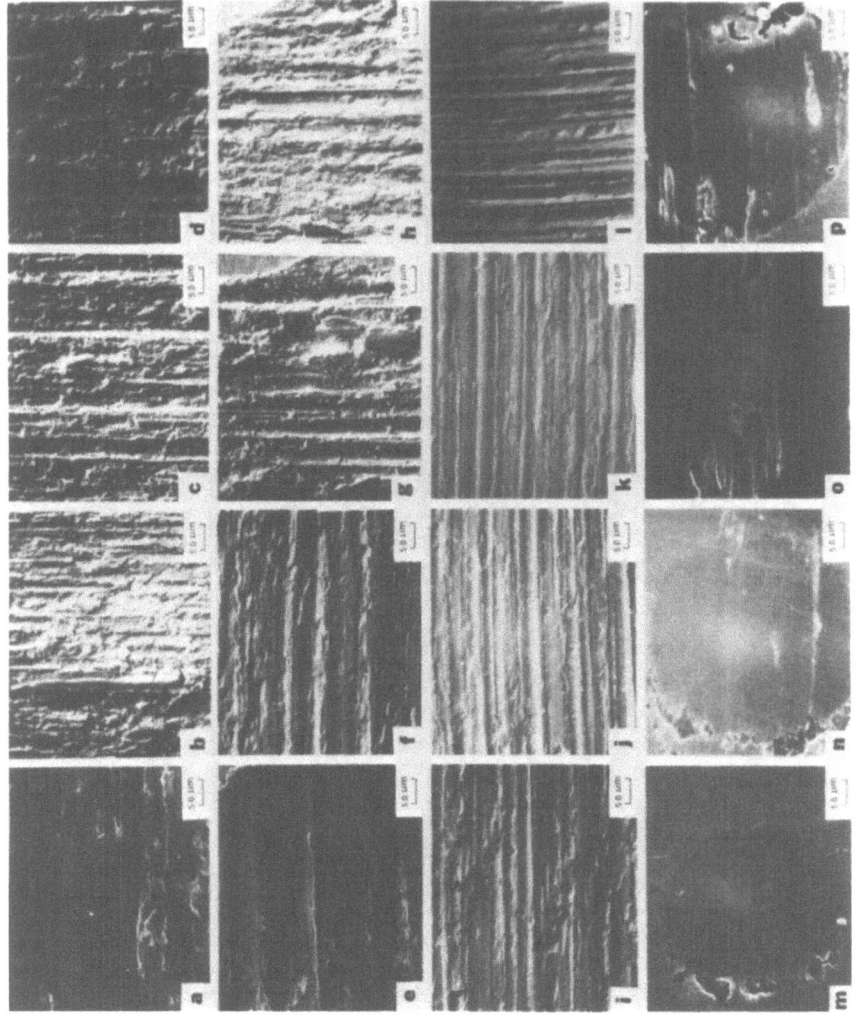

FIGURE 3.1 SCANNING ELECTRON MICROGRAPHS OF THE SURFACES
OF WORN SLIDERS
(A), (B), (C), (D) IRON ON IRON, 1020, 1045, AND 1095 STEEL, RESPECTIVELY;
(E), (F), (G), (H) 1020 ON IRON, 1020, 1045, AND 1095 STEEL, RESPECTIVELY;
(I), (J), (K), (L) 1045 ON IRON, 1020, 1045, AND 1095 STEEL, RESPECTIVELY;
(M), (N), (O), (P) 1095 ON IRON, 1020, 1045, AND 1095 STEEL, RESPECTIVELY

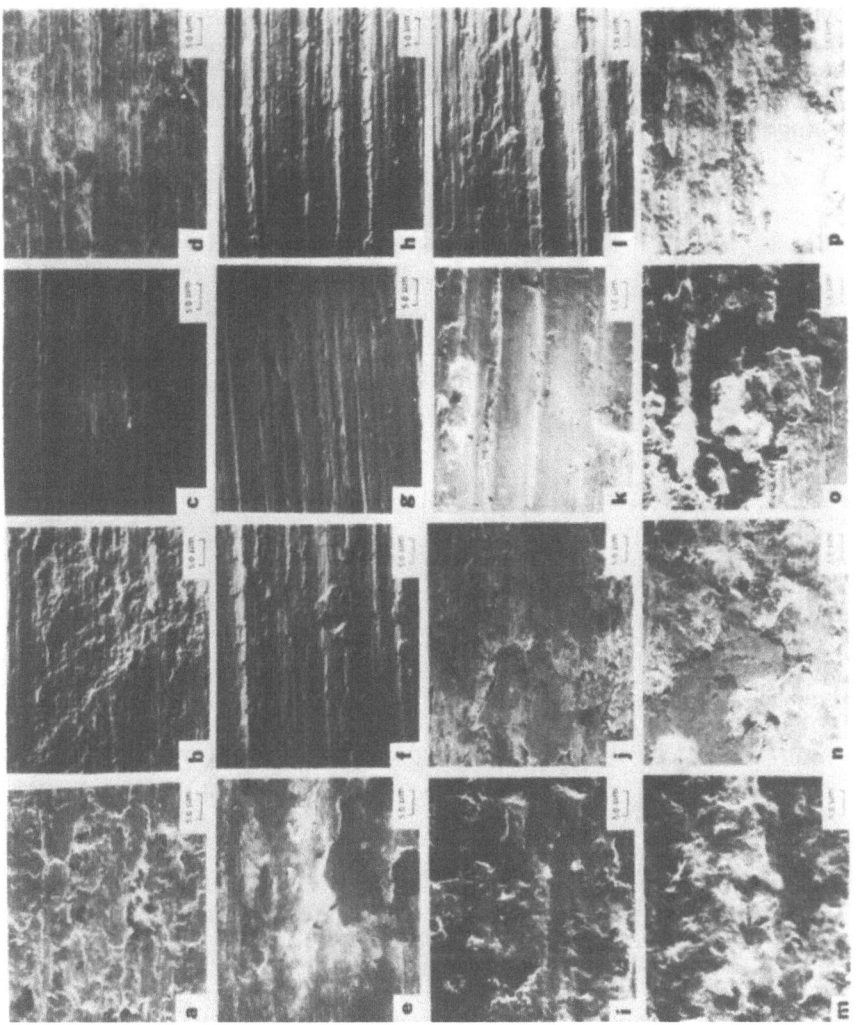

FIGURE 3.2 SCANNING ELECTRON MICROGRAPHS OF THE
SURFACES OF WORN SPECIMENS;
(A), (E), (I), (M) IRON ON IRON, 1020, 1045, AND 1095 STEEL, RESPECTIVELY;
(B), (F), (J), (N) 1020 ON IRON, 1020, 1045, AND 1095 STEEL, RESPECTIVELY;
(C), (G), (K), (O) 1045 ON IRON, 1020, 1045, AND 1095 STEEL, RESPECTIVELY;
(D), (H), (L), (P) 1095 ON IRON, 1020, 1045, AND 1095 STEEL, RESPECTIVELY

are slid against each other. The drop in the coefficient of friction in Figure 3.3b is associated with mutual polishing of the mating surfaces[9]. The behavior shown in Figure 3.3b results primarily when the hardness of the stationary slider is much greater than the moving specimen.

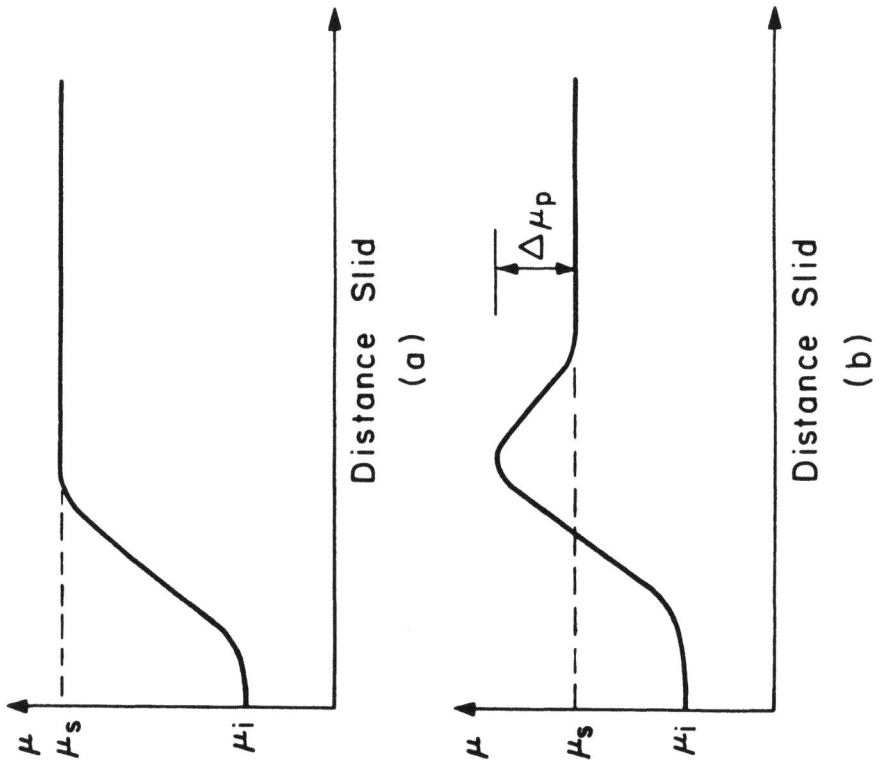

FIGURE 3.3   COEFFICIENT OF FRICTION VERSUS SLIDING DISTANCE:
(A) FOR AN ARMCO IRON SLIDER SLIDING AGAINST AN ARMCO IRON SPECIMEN ($\mu_i = 0.13$, $\mu_s = 0.71$);
(B) FOR AN AISI 1095 STEEL SLIDER SLIDING AGAINST AN ARMCO IRON SPECIMEN ($\mu_i = 0.17$, $\mu_s = 0.51$, $\Delta\mu_p = 0.25$)

2. When wear particles are brushed from the sliding interface, the coefficient of friction decreases to a low value and gradually reaches a steady state value again, as schematically illustrated in Figure 3.4. (The effect of wear particles on the friction coefficient was also reported by

Suh; Jahanmir and Abrahamson and more recently by Kuwahara and Masumoto[7, 10].)

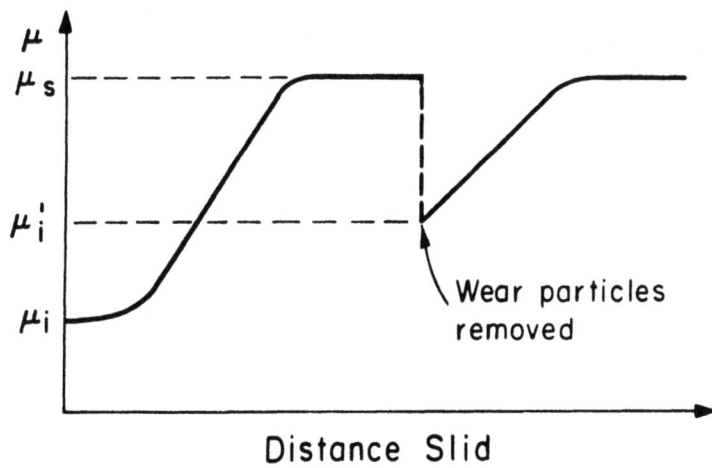

FIGURE 3.4   EFFECT OF REMOVING WEAR PARTICLES FOR AN ARMCO IRON SLIDER SLIDING AGAINST AN ARMCO IRON SPECIMEN ($\mu_i = 0.13$, $\mu_s = 0.71$, $\mu_i' = 0.4$)

3. The coefficient of friction can differ by as much as 0.2 even for the same pair of materials (which are chemically identical) depending on which is a stationary slider and which is a moving specimen, see Table 3.2.

4. The initial value of the kinetic coefficient of friction, μ_i, is in the neighborhood of 0.1 to 0.2 (but largely in the range of 0.12 to 0.17) for many materials tested, i.e., gold on gold, steel on steel, brass on steel, etc., and also regardless of whether or not lubricants are used[6, 9].

5. When the friction test is done with extremely well polished surfaces, plowing grooves are formed from the onset of testing.

6. Frictional behavior depends very much on experimental conditions.

3.3 Genesis of Friction

The experimental results clearly indicate that the observed friction coefficients cannot be explained in terms of the adhesion theory. The effect of entrapped wear particles and the existence of μ_i, which is independent of environmental conditions and materials tested, cannot be explained by the adhesion theory. The theory is further defined by the dramatic changes in the coefficient of friction when the role of the slider and the specimen is reversed.

Based on the experimental results discussed in the preceeding subsection, the following postulate was advanced to explain the genesis of friction between the sliding surfaces [6]:

"The coefficient of friction between the sliding surfaces is due to the various combined effects of asperity deformation, μ_d, plowing by wear particles and hard surface asperities, μ_p, and adhesion between the flat surfaces, μ_a. The relative contribution of these components depends on the condition of the sliding interface, which is affected by the history of sliding, the specific materials used, the surface topography and the environment."

In order to clarify the above postulate, the time dependent friction behavior of the materials tested, will be considered qualitatively by subdividing it into the following stages, see Figure 3.5.

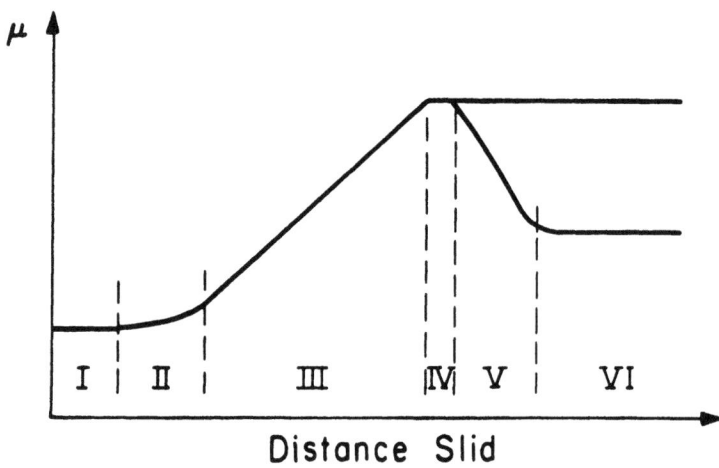

FIGURE 3.5 SIX STAGES IN THE FRICTIONAL FORCE VERSUS DISTANCE SLIDE RELATIONSHIP

a. Stage I -- In this early stage the coefficient of friction seems to be dictated by plowing of the surface by asperities. Adhesion does not play any significant role in this state due to the contaminated nature of the surface. The deformation of asperities does take place at the onset of sliding, which affects the static coefficient of friction. However, in Stage I, asperity deformation is not the major factor that determines the coefficient of friction, since they deform as soon as sliding commences and the surface is easily polished with the generation of new asperities in Stage I. Consequently, the coefficient of friction in this stage, μ_i, is largely independent of material combinations, the surface conditions, and the environmental conditions.

b. Stage II -- In this second stage, the frictional force begins to rise slowly due to increase in adhesion. When the interface is lubricated, Stage I persists for a long time and Stage II may not be present. The slope in Stage II can be steeper if the wear particles generated by the asperity deformation and fracture are entrapped between the sliding surfaces and proceed to plow the surfaces.

c. Stage III -- This stage is characterized by a steep increase in slope due to the rapid increase in the number of wear particles entrapped between the sliding surfaces as a consequence of higher wear rates. The slope is also affected by the increase in adhesion due to the increase in clean interfacial areas. The force required to deform the asperities will continue to contribute to the frictional force in this stage as long as surface asperities are present. The wear particles are generated when the process of wear particle formation by subsurface deformation, crack nucleation, and crack propagation postulated by the delamination theory or wear, is completed 11 Some of the wear particles get entrapped between the surfaces, causing plowing. The plowing will be greater when the wear particles are entrapped between metals of nearly equal hardness, because they will penetrate into both surfaces, preventing any slippage between the particle and the surface.

d. Stage IV -- This stage is reached when the number of wear particles entrapped between the interface remains constant. This occurs when the number of the newly entrapped particles equal the number of entrapped particles leaving the interface. The adhesion contribution to friction also remains constant is Stage IV. The asperity deformation continues to be important, since the wear by delamination creates new rough surfaces with asperities. However, in most cases asperity deformation is not as important as plowing since they deform readily and the frequency of new asperity generation is slow.

When two like metals are slid against each other or when the mechanism responsible for Stage V does not play a significant role, the coefficient between the two metals.

e. Stage V -- In some cases, such as when a very hard stationary slider is slid against a soft specimen, the asperities of the hard surface are gradually removed, creating a mirror finish as shown in Figure 3.6. In this case the frictional force decreases due to the decrease in plowing and asperity deformation. Plowing decreases since wear particles cannot anchor to a polished hard surface.

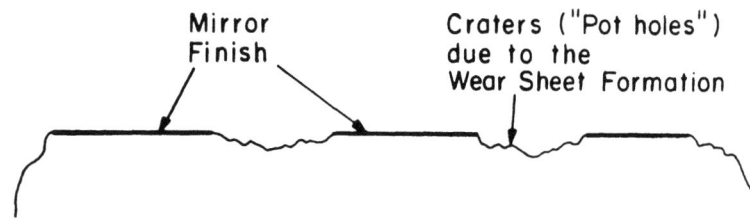

FIGURE 3.6 A HARD STATIONARY SURFACE POLISHED BY A SOFT SURFACE

f. Stage VI -- Eventually when the hard surface becomes mirror smooth to a maximum extent, the softer surface also aquires the same mirror finish and the frictional force levels off. The surfaces are never completely smooth since there are always "pot holes" due to the creation of delamination wear particles. These craters provide anchoring points for wear particles. When the hard surface is not stationary but moving against the softer surface, the hard surface remains rough, probably because polishing cannot take place due to geometric reasons. In this case, Stages V and VI are not present.

3.4 Analysis of the Friction Generating Mechanisms

The three basic mechanisms (i.e., asperity deformation, plowing, and adhesion) that are responsible for the generation of friction will be analyzed in this section. The asperity deformation determines the static coefficient of friction and also affects the dynamic coefficient of friction, since asperities are continuously generated due to delamination of wear sheets. However, the contribution of the asperity deformation to the dynamic coefficient of friction is not large relative to those by plowing and adhesion, since the asperities have to await the formation of delaminated wear particles which often requires a large number of cyclic loading by the asperities of the opposing surface. On the other hand, plowing takes place continuously whenever wear particles are entrapped between the sliding surfaces or when the asperities of the counterface plow in all cases when clean flat surfaces come into contact during steady state sliding. The relative magnitude of these components will be determined approximately by using the slip-line field.

3.4.1 Analysis of the Asperity Deformation, μ_d

Consider two representative asperities approaching each other, as shown in Figure 3.7. When these asperities come into contact with each other, they have to deform in such a manner that the resulting displacement field is compatible with the sliding direction and that the sum of the vertical components of the surface traction at the contracting asperities must be equal to the applied normal load. A possible slip-line field that satisfies the kinematic condition is given in Figure 3.8. The solution demands that the shear stress along OA be whatever is necessary to satisfy the condition that $\alpha = \theta$, which is mecessary to constrain the resulting deformation in the sliding direction. This would be possible even under the lubricated conditions if the interface, OA, is not perfectly smooth but rough enough to allow mechanical interlocking. The derivation of the normal and tangential force corresponding to the slip-line field shown in Figure 3.8 is given in the Appendix of this section.

The general solution is sketched in Figure 3.9. If it is assumed that the asperity deformation is the only phenomenon that takes place at the interface and is entirely responsible for the frictional force under a given load, the coefficient of friction due to asperity deformation μ_d varies from 0.39 to 1 as the slope of asperities increases from 0 to 45°. These values are closer to the static coefficient of friction than the dynamic friction coefficient measured during the early stage of sliding, i.e., Stage I[12].

In dynamic situations most of the normal load is carried by the entrapped wear particles and the flat contacts. Therefore, the actual contribution of the asperity deformation to the frictional force is expected to be a small fraction of the estimated value. This may be estimated indirectly from the experimental results.

It should be noted that the slip-line analysis done to determine μ_d is similar to Green's results which were discussed in Section 1[13]. However, in this case, μ_d, does not depend on adhesion since in the analysis the asperity

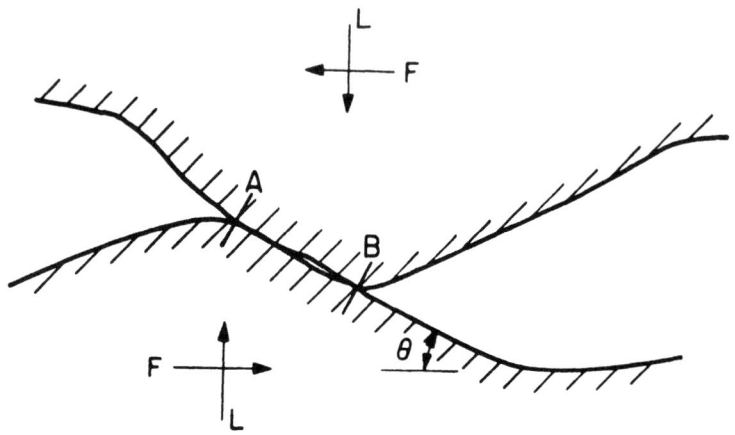

FIGURE 3.7   TWO INTERACTING SURFACE ASPERITIES

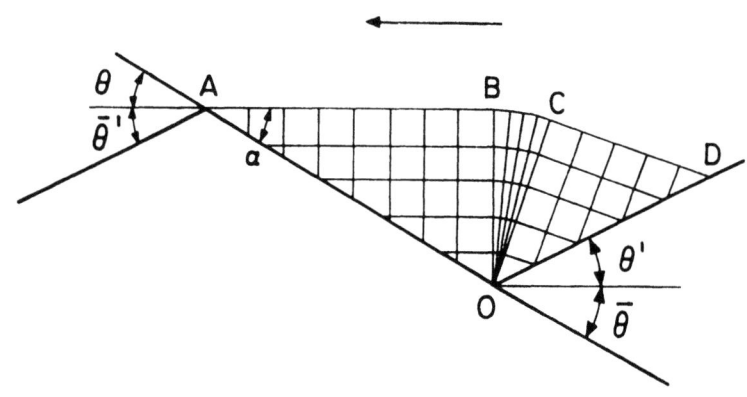

FIGURE 3.8   A GEOMETRICALLY COMPATIBLE SLIP-LINE FIELD. IT CAN BE SEEN THAT $\theta > \bar{\theta}$, $\theta' > \bar{\theta}$, AND $\theta = \alpha$

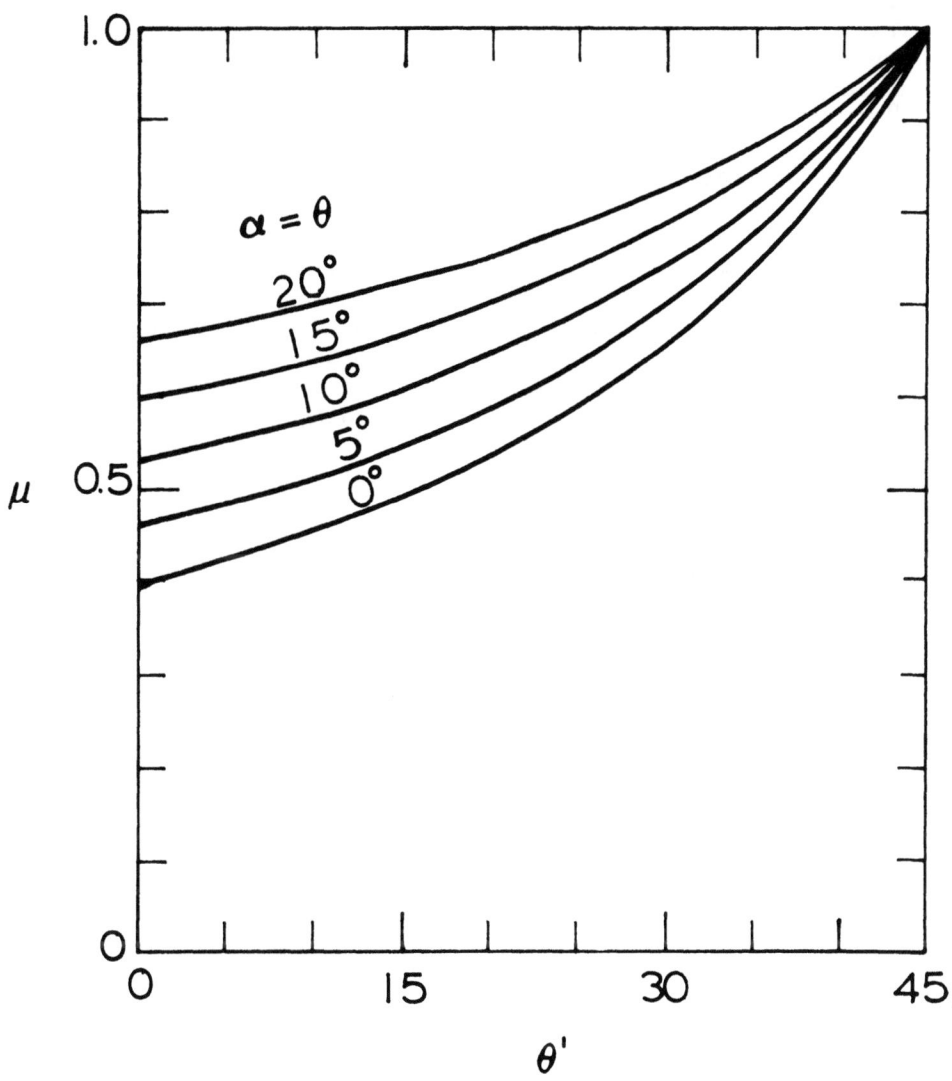

FIGURE 3.9   SLIP-LINE FIELD SOLUTION FOR FRICTION AS A FUNCTION OF THE SLOPE OF ASPERITIES

deformation was kinematically contrained. μ_d represents the frictional force due to the deformation of all interacting asperities. Before the onset of sliding between two surfaces, μ_d largely controls the static coefficient of friction.

3.4.2 Analysis of Adhesion Component of the Friction Coefficient, μ_a

A frictional force can arise due to the adhesion of two nearly flat surfaces. Unlike the deformation of asperities, this frictional force is a function of the adhesion between the two opposing surfaces. The adhesion force arises either due to the welding of two nearly flat portions of the surface or when the atoms are brought together in close proximity for interatomic interactions but without welding. The adhesion at the slopes of two interacting asperities has been included in deriving in μ_d as a subset of kinematically constrained deformation problems.

The specific experimental results presented in this section show that μ_a is not present (or is at least negligible) at the onset of sliding, probably due to the presence of contaminants on the surface. With the deformation of asperities and exposure of fresh new surfaces, the adhesion between nearly flat surfaces is expected to increase. The exact adhesion area cannot be determined a priori, since the applied normal load may also be carried out by interacting asperities and entrapped wear particles, although the limiting cases can be analyzed.

Consider two nearly flat surfaces coming into contact as shown in Figure 3.10. (Sometimes this type of contact is called a "rubbing" contact.) Depending on the nature of adhesion along the interface ED, the force required to move the rubbing surfaces with respect to each other varies. When there is no adhesion the force will be zero and when there is complete adhesion it will reach a maximum. The solution to this problem can be obtained again using the slip-line fields similar to that shown in Figure 3.10. The exact geometric shape of the slip-line field will depend on the boundary condition at ED. The solution sought can be adapted from the recent work of Challen and Oxley who derived an expression for the friction coefficient as [14];

$$\mu_a = \frac{A \sin \alpha + \cos (\cos^{-1} f - \alpha)}{A \cos \alpha + \sin (\cos^{-1} f - \alpha)} \quad (3.1)$$

where

$$A = 1 + \frac{\pi}{2} + \cos^{-1} f - 2\alpha - 2 \sin^{-1} \frac{\sin\alpha}{1-f}$$

f = strength of the adhesion at ED as expressed at a fraction of the shear flow strength of the softer material (3.2)

α = the slope of the hard asperity

For nearly flat surfaces α → 0. Therefore, Eqs. (3.1) and (3.2) reduce to

$$\mu_a = \frac{f}{A + \sin(\cos^{-1} f)} \quad (3.3)$$

$$A = 1 + \frac{\pi}{2} + \cos^{-1} f \quad (3.4)$$

It can be seen that μ_a varies from 0 to 0.39 as f changes from 0 to 1.

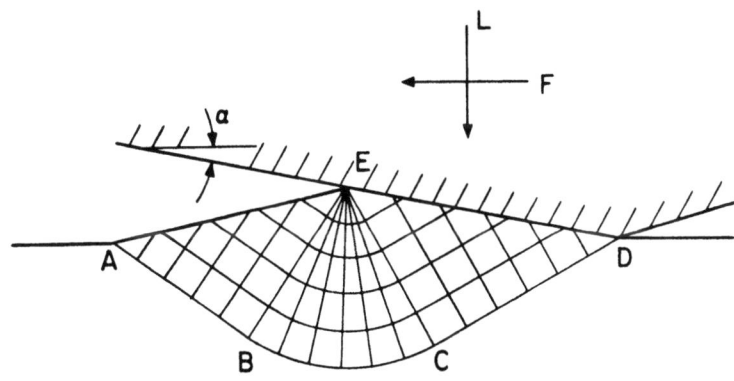

FIGURE 3.10 A SLIP-LINE FIELD FOR A RUBBING CONTACT

The friction coefficient determined by Eqs. (3.1) and (3.2) are based on the assumption that all the applied normal load is carried by the flat interfaces. However, since part of the normal load is also carried by purely elastic contacts, the interacting asperity junctions discussed in the preceeding section and the entrapped wear particles, μ_a, under typical sliding conditions, should be less than 0.4. The experimental results obtained with the hard AISI 1095 steel slider and the soft Armco iron specimen showed that the steady state coefficient of friction reached a value of 0.51 when both surfaces were polished smooth and thus the friction was caused primarily by adhesion. The agreement between the theory and the experiment is reasonable since asperity interactions and plowing by wear particles must have also contributed to the frictional force.

3.4.3 Analysis of Plowing, μ_p

The plowing component of the frictional force can be due to the penetration of hard asperities or due to the penetration of wear particles. The plowing due to wear particles is schematically illustrated in Figure 3.11. When two surfaces are of equal hardness, the particle can penetrate into both surfaces. As the surfaces move with respect to each other, grooves will be formed in one or both of the surfaces. When one of the surfaces move with respect to each other, grooves will be formed in one or both of the surfaces. When one of the surfaces is very hard and smooth, the wear particle will simply slide along the hard surface and no plowing can occur. However, when the hard surface is very rough, wear particles can anchor in the hard surface and plow the soft surface.

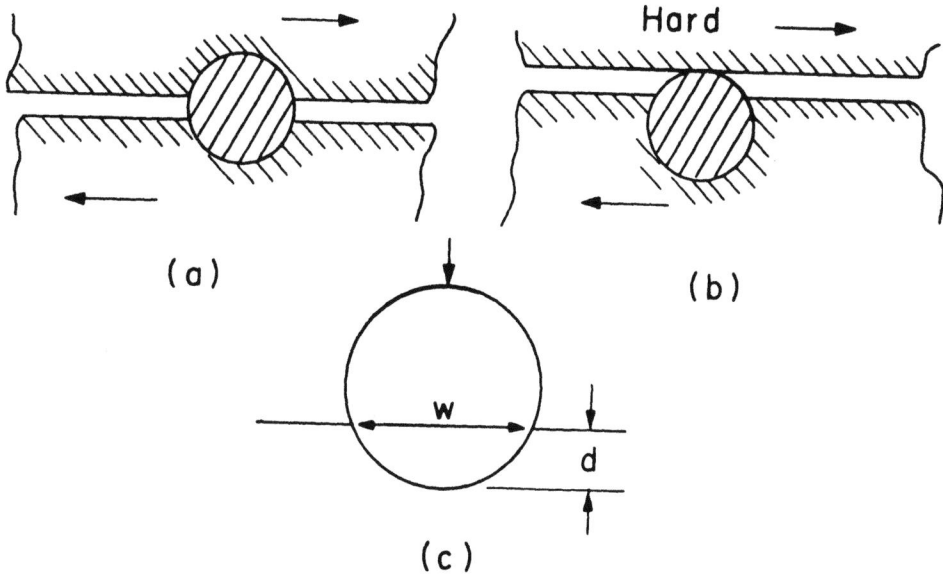

FIGURE 3.11 IDEALIZED MODEL OF WEAR PARTICLE INTERACTION BETWEEN TWO SLIDING SURFACES: (A) SURFACES OF EQUAL HARDNESS; (B) ONE SMOOTH VERY HARD SURFACE: (C) GEOMETRY OF THE WEAR PARTICLE

The friction due to plowing was investigated by Sin, et al., which showed that the contribution of plowing to the friction coefficient is very sensitive to the ratio of the radius of curvature of the particle to the depth of penetration[15]. The friction coefficient by plowing, μ_p, is given by:

$$\mu_p = \frac{2}{\pi} \left[\left(\frac{2r}{w}\right)^2 \sin^{-1} \frac{w}{2r} - \left(\frac{2r}{w}\right)^2 - 1 \right]^{1/2} \quad (3.5)$$

where w is the width of the penetration and r is the radius of curvature of the particle. The ratio of w/r, measured by sectioning the worn specimen is in the neighborhood of 0.8. Substituting this value into Eq. (3.5), the plowing coefficient of friction is found to be 0.2. This value is in the same range as the decrease in the friction coefficient observed by removing the wear particles from the Armco iron/Armco iron and the Armco iron/AISI 1095 steel interfaces, which were 0.31 and 0.16 respectively. The range of possible values of μ_p as a function of the ratio w/2r is shown in Figure 3.12.

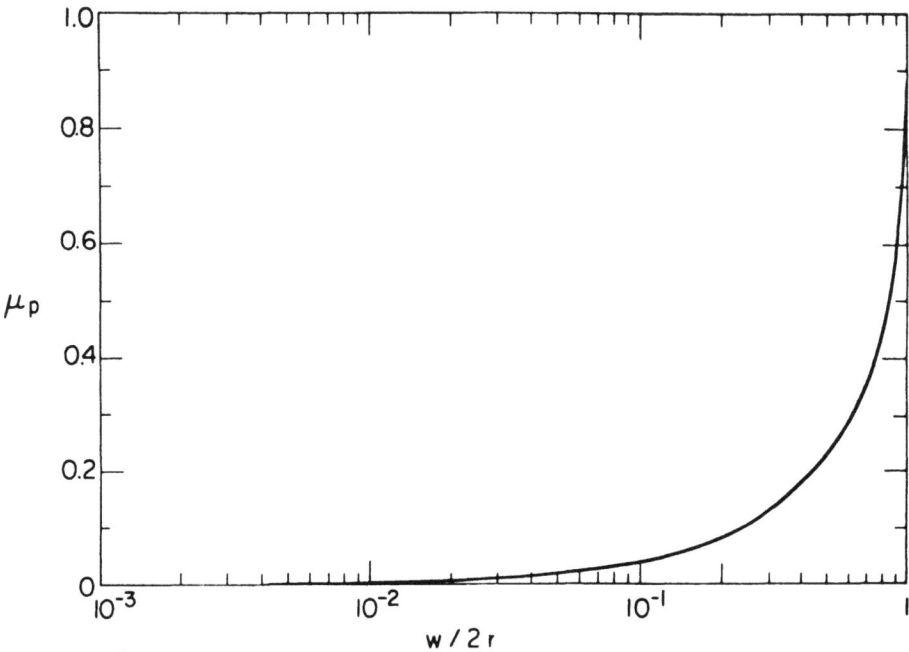

FIGURE 3.12 PLOWING COMPONENT OF THE FRICTION COEFFICIENT AS A FUNCTION OF THE RATIO OF THE WIDTH TO THE DIAMETER OF THE ENTRAPPED WEAR PARTICLE

Plowing not only increases the total frictional force and delamination wear, but also creates small wear particles which in turn affect the subsequent wear of sliding surfaces. Plowing action forms ridges along the sides of plowed grooves. When these ridges are deformed, flat, and subjected to repeated loading, some of these become loose wear particles with continued sliding. This is schematically illustrated in Figure 3.13.

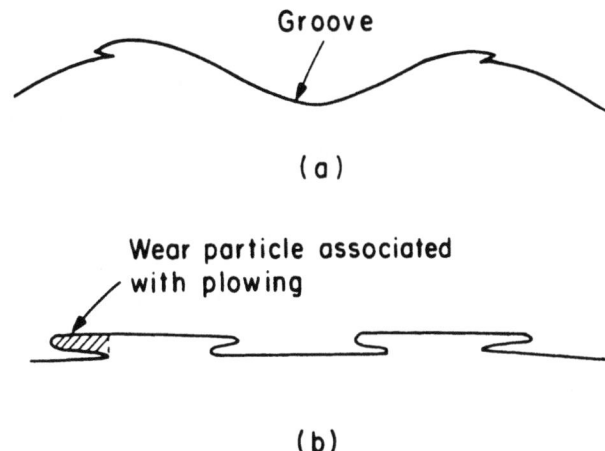

FIGURE 3.13 SCHEMATIC ILLUSTRATION OF WEAR PARTICLE FORMATION DUE TO PLOWING: (A) RIDGES FORMED ALONG THE SIDES OF THE PLOWED GROOVES; (B) FLATTENED RIDGES

3.5 Relative Contributions of μ_i, μ_a and μ_p to the Overall Friction Force

If the postulate for friction coefficient presented in this paper is correct, then the relative values of various friction components are as follows:

μ_d -- From 0.43 to 0.75, when the entire applied normal load is carried by typical surface asperities with a slope of $4°$ to $20°$. It appears that μ_d is responsible for the static coefficient of friction μ_i in Stage I. The reason μ_d is not a major factor in Stage I is that once the original asperities deform, asperity interactions cannot take place. This friction component can contribute partially to the steady state coefficient of friction if new asperities are continuously generated as a consequence of delamination wear process.

μ_a -- From about 0 to 0.4, depending on the nature of adhesion between the flat part of the interacting surfaces. The low value is for a well lubricated surface with light lubrication, while the high value is for identical metals sliding against each other without any surface contaminants and oxide layers.

μ_p -- From nearly 0 to 1.0, from a theoretical point of view, depending on the depth of penetration, but normally less than 0.4 in a typical situation. The high values are associated with two identical metals sliding against each other with deep penetration by wear particles, while the low value is obtained when either wear particles are totally absent from the interface or a soft surface is slid against a hard surface with a mirror finish.

The determination of the total friction coefficient in a given condition is complex. It is difficult to determine the relative contributions of μ_d, μ_a and μ_p to the total friction coefficient, because analyses for μ_d and μ_a were done assuming that the total normal load is carried by either asperities or flat contact areas. In real situations, the normal load will be apportioned among the asperity contacts, flat adhesion junctions, and the entrapped particles. However, it is quite plausible in many real situations, that each of these mechanisms which contribute to friction can take place sequentially rather than concurrently. Consider, for example, a flat junction and an asperity in contact as shown in Figure 3.14. When the flat areas first come into contact and form an adhesion junction, the analysis performed for μ_a is strictly valid. When the asperities also come into contact with further sliding, a large fraction of the normal load is still carried by the flat junctions due to their higher normal "stiffness," i.e., the force required to cause unit displacement along the vertical direction. In this case, in order to continue the sliding action the asperities will simply have to shear along the dotted line if the materials are identical, or along the crossed line if the top slider is much harder than the bottom slider. In this case, only a tangential force is required to deform the asperity plastically along the direction of sliding.

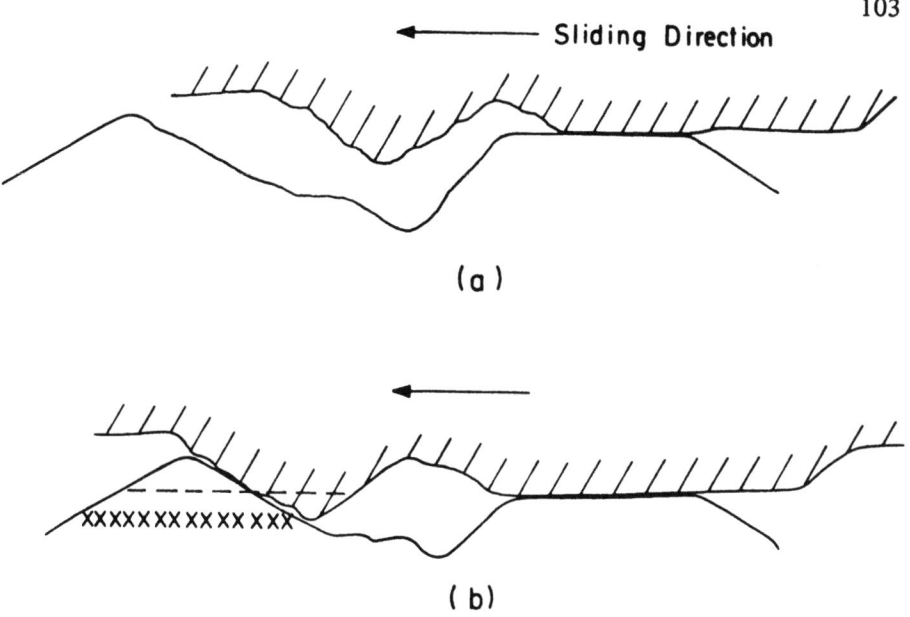

FIGURE 3.14 TWO SLIDING SURFACES IN CONTACT: (A) FIRST CONTACT OF FLAT SURFACES; (B) FLAT SURFACE CONTACT AND ASPERITY CONTACT

The coefficient of friction μ is represented in a "Friction Space" as a function of adhesion, plowing, and roughness as shown in Figure 3.15. The adhesion is expressed in terms of the nondimensional interfacial shear strength of the flat contacts which was defined by Eq. (3.2). The roughness is plotted in terms of the slope of the surface, while the plowing is given in terms of the ratio of the width of wear particle (or hard asperity) penetration to its radius. The lowest surface corresponds to the case of no asperities, i.e., $\theta = 0$, which forms the lower bound. The θ_i surface corresponds to the initial machined surface, while the θ^* surface represents the surface roughness when the peak of μ occurs. As the surface gets rougher and/or the number of the steady state asperities increase, μ will increase and the friction value will move in the Friction Space along the μ-axis. The friction surfaces θ^* is plotted from the actual experimental results obtained with Armco iron sliding against AISI 1095 steel. In this case, the asperity contribution to friction was 0.3.

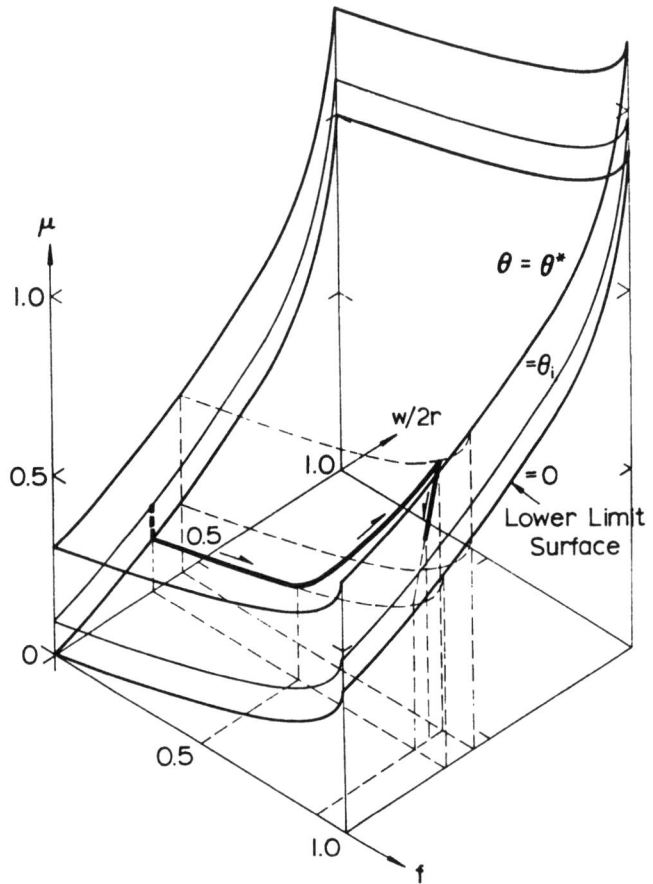

FIGURE 3.15 "FRICTION SPACE" SHOWING THE COEFFICIENT OF FRICTION AS A FUNCTION OF ADHESION BETWEEN FLAT CONTACTS, WEAR PARTICLE PENETRATION, AND SURFACE ROUGHNESS: $f = \tau_s/k$, WHERE τ_s IS THE SHEAR STRESS AT THE INTERFACE AND k IS THE SHEAR FLOW STRENGTH OF THE SOFT METAL; θ IS THE SLOPE OF ASPERITIES; $w/2r$ IS THE RATIO OF THE WIDTH OF ASPERITY PENETRATION TO THE DIAMETER OF THE PARTICLE

Although this figure is not precise, it gives a reasonable picture of what happens in a given situation. The paths of the friction coefficient change, shown in Figure 3.3(a) and 3.3(b), are shown in this "Friction Space," Figure 3.15. The friction is shown to start from the initial roughness plane θ_i to nearly flat surface, i.e., $\theta = 0$ and traces along the paths indicated in Figures 3.3(a) and 3.3(b). In many cases the θ_i surface may be above the θ^* friction space, depending on the initial surface finish relative to the steady state surface roughness.

The foregoing argument may be applied to the specific case of gold sliding against gold. When unlubricated gold specimens without any oxide layer are slid against each other, the frictional force is the sum of μ_a and μ_p or the sum of a fraction of μ_d, μ_a and μ_p, depending on the situation. It is most likely that the μ_a term will always be present since two flat surfaces in contact may be the most stiff system in supporting the normal load. For example, if the normal load is first borne by flat contacts only, friction will be entirely due to adhesion, μ_a, which will reach a maximum value. Then if asperities of the opposing surfaces come in contact, an additional frictional force will be required to deform the asperities. In addition to these frictional forces, the third frictional component may also affect the frictional behavior if the wear particles become wedged in between the sliding surfaces. Therefore, the coefficient of friction between gold on gold may be as high as 1.4 to 1.6 and fluctuate between a maximum and a minimum value.

Lubricated surfaces can have a coefficient of friction whose magnitude will be determined by the degree of plowing and asperity interaction. Lubricated surfaces, without any asperities, are found to have a coefficient of friction of approximately 0.04 for a hard surface and 0.12 for a soft iron surface. However, when wear particles are entrapped between the sliding surfaces, the plowing component of the frictional force can be present raising the friction coefficient.

In the past, the high friction coefficient between like metals has been explained in terms of greater adhesion due to their greater solubility[1]. However, the evidence presented in this paper shows that the so-called compatibility of metals is dictated more by their mechanical behavior rather than by their chemical behavior. This is quite reasonable since the diffusion rate at the typical sliding junctions at room temperature is so low that the solubility between the metals cannot account for the observed wear rates[16].

3.6 Concluding Remarks

In this section it is shown that the classical adhesion theory of friction cannot explain experimental results and that the frictional coefficient is not an inherent material property. The coefficient of friction depends very much on the sliding conditions, material combinations, and geometry. The coefficient of friction is composed of three components: that due to the deforming asperities, μ_d, that due to plowing by wear particles entrapped between sliding surfaces and hard surface asperities, μ_p, and that due to adhesion, μ_a. The contribution to the overall coefficient of friction by plowing and asperity deformation can be greater than that by adhesion. Typical values of μ_a can range from 0 to 0.4 and μ_p from 0 to 0.4 under typical conditions. However, μ_p can be as large as 1.0 when the depth of penetration by wear particles is large. The friction coefficient due to asperity deformation dictates the static friction coefficient can range from 0.43 to 0.75 depending on the slope of asperities.

The mechanisms responible for the genesis of friction presented in this section should be operative in the case of ceramics, thermosetting plastics, certain kinds of thermoplastics, and most metals. The frictional behavior of highly crystalline thermoplastics, such as teflon and high density polyethylene are affected also by the molecular orientation of molecules and their transfer to the counterface. The frictional behavior of crystalline thermoplastics will be treated in a separate section.

3.7 References - Section 3

1. Rabinowicz, E., "Friction and Wear of Materials," Wiley, New York, 1965. pp. 51-108.

2. Bowden, F.P. and D. Tabor, "Friction and Lubrication of Solids," Clarendon Press, Oxford, Part I, 1950, pp. 90-121, Part II, 1964, pp. 52 -86.

3. Ernst, H., and M.E. Merchant, "Surface Friction Between Metals - A Basic Factor in the Metal Cutting Process," Proceedings of the Special Summer Conference on Friction and Surface Finish, M.I.T. Press, Cambridge, 1940, pp. 76-101.

4. Shaw, M.C. and E.F. Macks, "Analysis and Lubrication of Bearings," McGraw Hill, 1949, pp. 457-461.

5. Buckley, D.H., "Definition and Effect of Chemical Properties of Surfaces in Friction, Wear, and Lubrication", Proceedings of the International Conference on the Fundamentals of Tribology, M.I.T. 1978, pp. 173-199.

6. Suh, N.P. and H.C. Sin, "The Gensis of Friction", Wear, 69(1981) 91-114.

7. Abrahamson, II, E.P., S. Jahanmir, and N.P. Suh, "The Effect of Surface Finish on the Wear of Sliding Surfaces," CIRP Ann. Int. Inst. Prod. Eng. Res., 24 (1975) 513-514.

8. Rabinowicz, E., Proceedings of the International Conference on Wear of Materials, St. Louis, Missouri, 1977, American Society of Mechanical Engineers, pp. 36-40.

9. Suh, N.P., H.C. Sin, M. Tohkai, and N. Saka, "Surface Topography and Functional Requirements for Dry Sliding Surfaces," CIRP Ann. Int. Inst. Prod. Eng. Res., 29 (1980) 413-418.

10. Kuwahara, K. and H. Masumoto, "Influence of Wear Particles on the Friction and Wear Between Copper Disk and Pin of Various Kinds of Metal," Journal of Japan Society of Lubrication Engineers, 25 (1980) 126-131.

11. Suh, N.P., "The Delamination Theory of Wear," Wear, 25 (1973) 111-124.

12. Kato, S., E. Mauri, A. Kobayashi, and T. Matsubayshi, "Characteristics of Surface Topography and Static Friction on Scraped Surface Slideway, Part I and Part II," Journal of Engineering for Industry, Trans. ASME, Vol. 103 (1980) pp. 97-108.

13. Green, A.P., "Friction Between Unlubricated Metals: A Theoretical Analysis of the Junction Model," Proceedings of Royal Society of London, A228 (19955) 191-204.

14. Challen, J.M. and P.L.B. Oxley, "An Explanation of the Different Regimes of Friction and Wear Using Asperity Deformation Models," Wear, 53 (1979) 229-243.

15. Sin, H.C., N. Saka, and N.P. Suh, "Abrasive Wear Mechanisms and the Grit Size Effect," Wear 55 (1979) 163-190.

16. Suh, N.P., "New Theories of Wear and Their Implications on Tool Materials," Wear, 1980, to appear.

3.8 Appendix - Section 3 - Slip-Line Field Solution for Sliding Contact

Figure 3.16 gives a possible slip-line field for the asperity contact between sliding surfaces. The interface AO between asperities and the stress free surface OD are both assumed to be straight with their directions defined by the angles θ and θ' measured from the sliding direction. From the figure it can be noticed that the slip-line ABCD is a β- line. Using the Hencky relations the stresses along the slip-line can be obtained.

At D, $\phi = \theta' + \frac{\pi}{4}$ and $p = k$ (1.A1)

Along AB, $\phi = \alpha - \theta + \frac{\pi}{2}$ and

$$P_{AB} = k(1 + \frac{\pi}{2} + 2\alpha - 2\theta - 2\theta')$$ (1.A2)

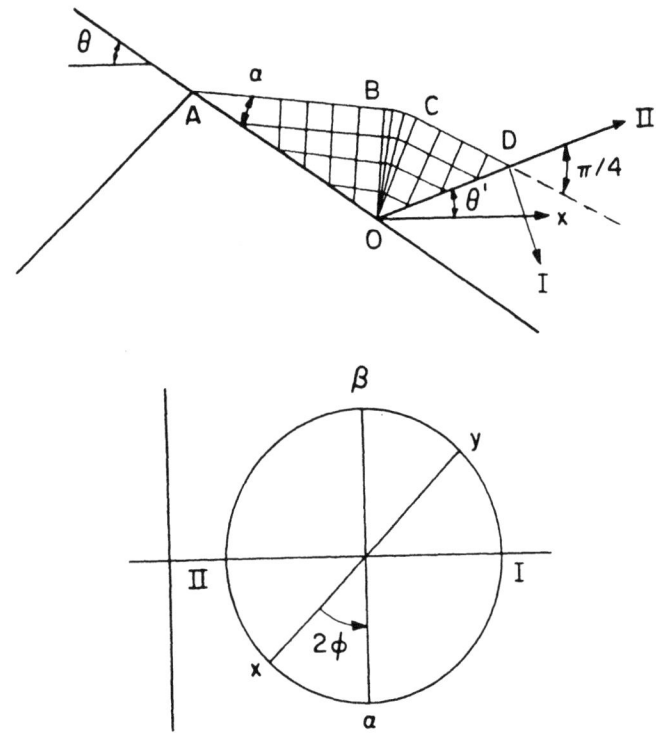

FIGURE 3.16 A SLIP-LINE FIELD FOR AN ASPERITY CONTACT

Isolating the junction along ABO as shown in Figure 3.17 we can find the resultant forces as

$$-F_y = (AB)p \cos(\theta-\alpha) - (AB)k \sin(\theta-\alpha)$$

$$+ (OB)k \cos(\theta-\alpha) - (OB)p \sin(\theta-\alpha) \quad (1.A3)$$

$$-F_x = F = (AB)p \sin(\theta-\alpha) + (AB)k \cos(\theta-\alpha)$$
$$+ (OB)k \sin(\theta-\alpha) + (OB)p \cos(\theta-\alpha)$$

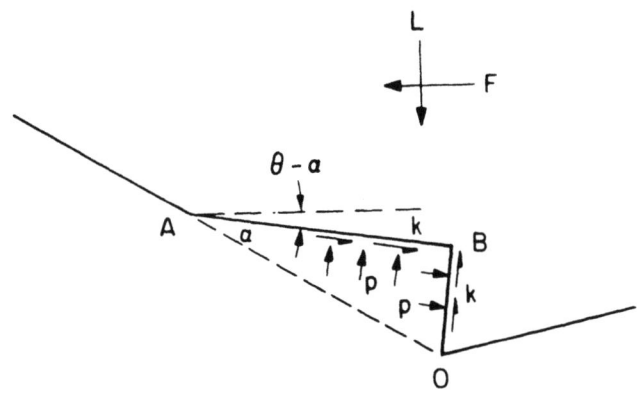

FIGURE 3.17   ISOLATION OF THE JUNCTION OF THE ASPERITY CONTACT ALONG ABO IN FIGURE 3.16

Using the geometric relations, L and F are expressed as

$$L = (AO)(p \cos\theta + k \sin(2\alpha - \theta)) \quad (1.A4)$$
$$F = (OA)(p \sin + K \cos(2\alpha - \theta))$$

Therefore, the coefficient of friction is

$$\mu = \frac{F}{L} = \frac{(1 + \frac{\pi}{2} + 2\alpha - 2\theta - 2\theta')\sin\theta + \cos(2\alpha-\theta)}{(1 + \frac{\pi}{2} + 2\alpha - 2\theta - 2\theta')\cos\theta + \sin(2\alpha-\theta)} \quad (1.A5)$$

When θ, θ', and α are small ($\theta, \theta', \alpha \to 0$)

$$\mu = \frac{1}{1+\frac{\pi}{2}} = 0.39$$

For several values of α, θ, and θ', the coefficient of friction is plotted in Figure 3.9.

If the junction does not weld along OA, the shear stress along OA will be much less than k. The interfacial shear stress is related to the angle α as $\tau = k \cos 2\alpha$. When there is no shear stress along OA, i.e., $\alpha = \pi/4$, the junction will slide along OA until the junction can deform under the influence of the normal load alone. During this sliding the coefficient of friction can be otained by substituting $\alpha = \pi/4$, into Eq. 1.A5 as

$$\mu = \frac{(1 + \pi - 2\theta - 2\theta') \sin\theta + \cos(\frac{\pi}{2} - \theta)}{(1 + \pi - 2\theta - 2\theta') \cos\theta + \sin(\frac{\pi}{2} - \theta)}$$

or

$$\mu = \tan \theta$$

which is the same expression as that derived from the roughness theory of friction.

The slip-line field solution derived above is a general solution. When the sliding occurs, the slip-line field should satisfy the kinematic condition, which corresponds to the case of the slip-line AB being parallel to the sliding direction. Therefore, α is equal to θ.

4. RESPONSE OF MATERIALS TO SURFACE TRACTION

4.1 Introduction

In Section 3 the origin and the magnitude of surface traction at asperity contacts were given. It was shown that the force acting on an asperity contact is due to one of the following forces: plowing, adhesion, and asperity deformation. The problem to be examined in this section is how the material responds to the external force applied at the interface between the sliding surfaces. In order to answer this question the responses of the surface layer will be treated as a continuum mechanics problem.

There are three basic responses to the external force:

1) plastic deformation of the surface layer, including the formation of ridges on the surface and the subsurface deformation,
2) subsurface crack and void nucleation
3) subsurface crack propagation

Each of these constitutes a separate class of mechanic problems and therefore will be treated individually in the following subsections.

For the purpose of analysis the material will be assumed isotropic and homogeneous. These are not necessarily good assumptions, because the surface becomes highly anisotropic and in-homogeneous during sliding. However, the need to simplify the mathematic analysis justifies these assumptions. On the other hand, these assumptions may not be too unreasonable in studying crack nucleation and propagation, since the amount of deformation at the subsurface where cracks nucleate and propagate is much less than the outermost layer of the surface.

It will be also assumed that the interfacial temperature rise is negligible because of the low sliding speed. This is a very reasonable assumption in the case of metals sliding at low speeds. In a later section on metal cutting, the case of high interfacial temperature rise will be considered.

4.2 Deformation of the Surface Layer

4.2.1 Formation of Surface Ridges

When plowing occurs, ridges form along the plowed groove regardless of whether or not chips are formed by the plowing action. These ridges become flattened as shown in Figure 4.1, which eventually fracture upon repeated loading. This process of wear particle formation has not been analyzed, although many experimental results support the qualitative view.

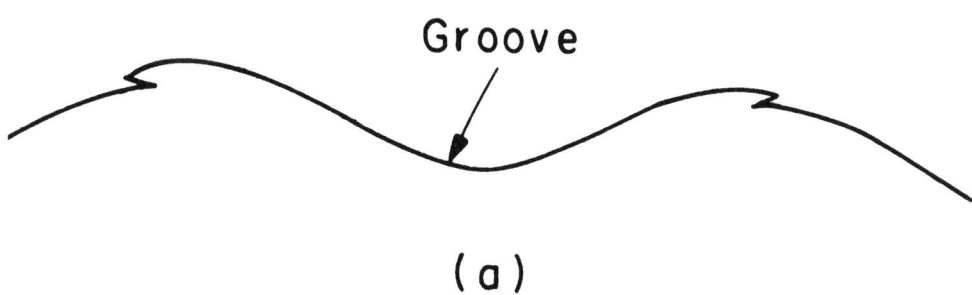

FIGURE 4.1

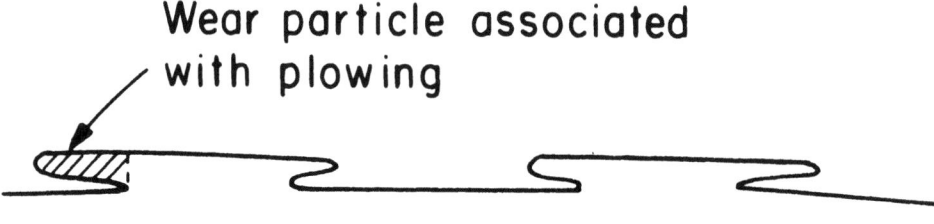

(b)

FIGURE 4.1   SCHEMATIC ILLUSTRATION OF WEAR PARTICLE FORMATION
             DUE TO PLOWING: (A) RIDGES FORMED ALONG THE SIDES
             OF THE PLOWED GROOVE; (B) FLATTENED RIDGES.

4.2.2 Deformation of a Semi-Infinite Elasto-Plastic Solid

The problem to be solved is the deformation of a semi-infinite elasto-plastic solid when it is loaded by moving asperities. The contact length at the asperity surface interface may be of the order of 10 to 100 µm and the distance between the asperity contacts may be 100 to 1000 µm apart. Therefore, the plastic deformation of the surface accumulates as the number of cyclic loading increases. The solutions to these problems can be obtained using the finite element method or using an approximate analytical/numerical technique. FEM solutions are possible for a variety of constitutive relationships but become very expensive to compute. The approximate technique used by Merwin and Johnson, Johnson and Jeffries, and Jahanmir and Suh will be outlined here[1, 2, 3]. This result will be compared to a FEM solution recently obtained at MIT.

When an elastic-plastic solid is loaded cyclically by moving asperities, residual stress and strain remain in the surface layer. This is of interest since these residual stress and strain fields may affect the subsequent crack nucleation and propagation processes. We will, therefore, show how the state of stress, due to the load applied by a moving asperity and the residual stress and strain field after repeated loading, can be determined. In the following subsections these results will be applied to solve crack nucleation and propagation problems.

Figure 4.2 shows the load exerted on a semi-infinite solid surface by an asperity. For convenience, it will be assumed that the asperity is stationary and the semi-infinite solid is moving with velocity U. The contact length is 2a. We will assume a plane strain condition, i.e., $\sigma_{zz} = 0$ and $\partial/\partial z = 0$, where the z axis is perpendicular to the plane of the paper. The stress distribution at the asperity contact will be assumed to be elliptic over the contact, which may be written as

$$\sigma_{xx} = \begin{cases} 2 q_0 \left[\frac{x}{a} - \left(\frac{x^2}{a^2} - 1 \right)^{1/2} \right] & \text{for } x \geq a \\ 2 q_0 \left[\frac{x}{a} + \left(\frac{x^2}{a^2} - 1 \right)^{1/2} \right] & \text{for } x \leq a \\ 2 q_0 \left[\frac{x}{a} - P_0 \left(1 - \frac{x^2}{a^2} \right)^{1/2} \right] & \text{for } x \leq a \end{cases} \quad (4.1)$$

$$\sigma_{yy} = \begin{cases} 0 & \text{for } -a \geq x \geq a \\ -P_0 \left(1 - \frac{x^2}{a^2} \right)^{1/2} & \text{for } |x| \leq a \end{cases} \quad (4.2)$$

$$\sigma_{xy} = \begin{cases} 0 & \text{for } -a \geq x \geq a \\ q_0 \left(1 - \frac{x^2}{a^2} \right)^{1/2} & \text{for } |x| \leq a \end{cases} \quad (4.3)$$

where σ's are stresses and the first subscript denotes the direction and the second subscript denotes the plane perpendicular to the axis. p_0 and q_0 are maximum normal stress and the maximum tangential stress at the contact.

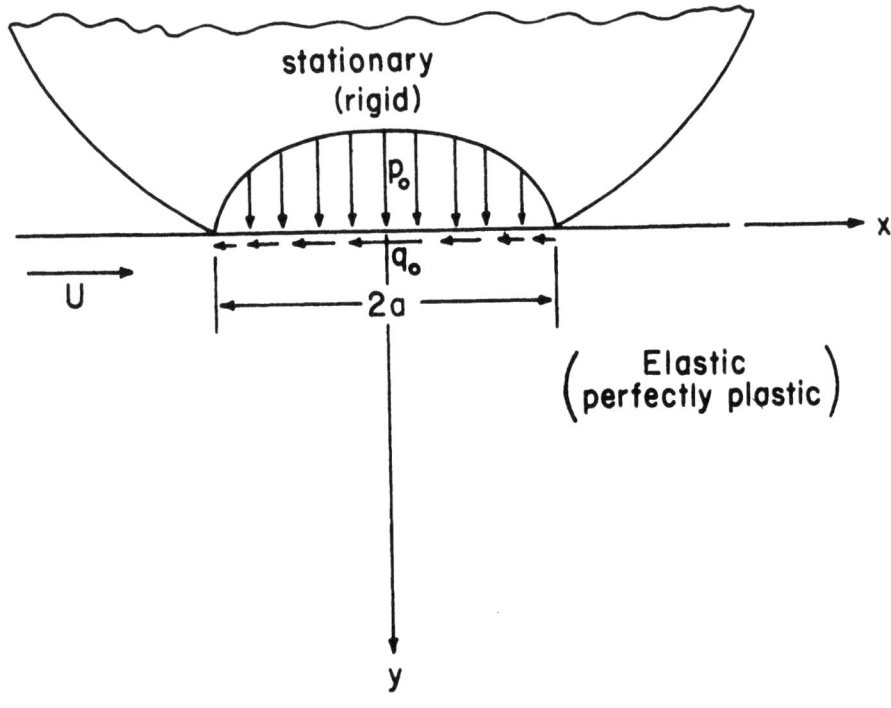

FIGURE 4.2 MODEL OF A CONTACT BETWEEN A STATIONARY RIGID ASPERITY AND A SLIDING ELASTIC-PERFECTLY PLASTIC PLANE.

1. Elastic case

We will first obtain a solution for an elastic solid which will then be used to obtain a solution for elasto-plastic solid. The governing equations are the equilibrium condition, geometric compatibility relations and the constitutive relations, which may, respectively, be written as:

$$\frac{\partial \sigma_{ij}}{\partial x_j} = 0 \qquad (4.4)$$

$$\varepsilon_{ij} = \frac{1}{2}\left(\frac{\partial u_i}{\partial x_j} + \frac{\partial u_j}{\partial x_i}\right) \qquad (4.5)$$

$$\sigma_{ij}' = \sigma_{ij} - \sigma = 2G(\varepsilon_{ij} - \varepsilon) = 2G\varepsilon_{ij}' \qquad (4.6)$$

where ε_{ij} is an incremental strain tensor; u's are displacement components, G is the shear modulus; σ_{ij}' and ε_{ij}' are deviatoric stress and strain components, respectively; and σ and ε are the hydrostatic stress and the dilatational strain defined as:

$$\sigma = \sigma_{ii}/3 \tag{4.7}$$

$$\varepsilon = \varepsilon_{ii} \tag{4.8}$$

When the subscripts repeat, summation is implied.

The elastic solution for this contact problem was first obtained by Smith and Liu[5]. The stresses below the surface due to the both tangential and normal loads are given as:

$$\sigma_{xx} = \frac{q_o}{\pi}\left[(2x^2 - 2a^2 - 3y^2)\psi + 2\pi\frac{x}{a} + 2(a^2 - x^2 - y^2)\frac{x}{a}\bar{\psi}\right]$$
$$- \frac{P_o}{\pi}y\left[\frac{a^2 + 2x^2 + 2y^2}{a} - \bar{\psi}\frac{2\pi}{a} - 3x\psi\right] \tag{4.9}$$

$$\sigma_{yy} = \frac{q_o}{\pi}y^2\psi - \frac{P_o}{\pi}y\left[a\bar{\psi} - x\psi\right] \tag{4.10}$$

$$\sigma_{xy} = \frac{q_o}{\pi}\left[(a^2 + 2x^2 + 2y^2)\frac{y}{a}\bar{\psi} - 2\pi\frac{y}{a} - 3xy\psi\right] - \frac{P_o}{\pi}y^2\psi \tag{4.11}$$

in which

$$\psi = \frac{\pi}{k_1}\frac{1 - \left(\frac{k_2}{k_1}\right)^{1/2}}{\left(\frac{k_2}{k_1}\right)^{1/2}\left[2\left(\frac{k_2}{k_1}\right)^{1/2} + \frac{k_1 + k_2 - 4a^2}{k_1}^{1/2}\right]} \tag{4.12}$$

$$\bar{\psi} = \frac{\pi}{k_1}\frac{1 + \left(\frac{k_2}{k_1}\right)^{1/2}}{\left(\frac{k_2}{k_1}\right)^{1/2}\left[2\left(\frac{k_2}{k_1}\right)^{1/2} + \frac{k_1 + k_2 - 4a^2}{k_1}^{1/2}\right]}$$

$$k_1 = (a + x)^2 = y^2$$

$$k_2 = (a - x)^2 + y^2$$

For plane strain the other stress components are given as:

$$\sigma_{zz} = \nu (\sigma_{xx} + \sigma_{yy}) \tag{4.13}$$

$$\sigma_{xz} = \sigma_{yz} = 0$$

where ν is the Poisson's ratio. In the calculation reported here a value of 0.3 is used.

2. Elasto-Perfectly Plastic Case

In order to find the stresses in the plastically deforming region and the cumulative plastic deformation after every passage of the asperity, it is necessary to trace the loading cycle of each point as it passes under the asperity. During plastic deformation it is assumed that the total strains remain identical with the elastic strains found from Eqs. (4.1) through (4.13) by applying Hooke's law. (The reasons behind this assumption will be discussed later.) The stresses at each point in the plastic region are then found by employing the incremental Prandtl-Reuss equations for an elasto-perfectly plastic material.

During plastic deformation the von Mises flow rule requires that the second invarient of the stress deviators, J_2', remains constant and equal to k^2, where k is the yield stress in shear. J_2 is defined as:

$$J_2 = \frac{1}{2} (\sigma_{ij}' \, \sigma_{ij}') \tag{4.14}$$

We may now proceed to find the stress-strain relations in the plastic region. The total strain rates, $\dot{\varepsilon}_{ij}{}^p$, are the summation of the elastic strain rates, $\dot{\varepsilon}_{ij}{}^e$, and the plastic strain rates, $\dot{\varepsilon}_{ij}{}^p$, or,

$$\dot{\varepsilon}_{ij} = \dot{\varepsilon}_{ij}{}^e + \dot{\varepsilon}_{ij}{}^p \tag{4.15}$$

The plastic strain rates can be found from the Prandtl-Reuss equations:

$$\dot{\varepsilon}_{ij}{}^p = \frac{\dot{W}^p}{2k^2} \sigma_{ij}{}' \qquad (4.16)$$

where $\dot{W}^p$ is the plastic strain energy rate,

$$\dot{W}^p = \dot{\varepsilon}_{ij}{}'^p \sigma_{ij}{}' = \bar{\sigma}\,\dot{\bar{\varepsilon}}^p \qquad (4.17)$$

By substituting (4.13) and (4.16) in (4.15) we obtain,

$$\dot{\varepsilon}_{ij}{}' = \frac{\dot{\sigma}_{ij}{}'}{2G} + \frac{\dot{W}^p}{2k^2} \sigma_{ij}{}' \qquad (4.18)$$

which can be rewritten as,

$$\dot{\sigma}_{ij}{}' = 2G\left(\dot{\varepsilon}_{ij}{}' - \frac{\dot{W}^p}{2k^2} \sigma_{ij}{}'\right) \qquad (4.19)$$

Since the plastic strain energy is much larger than the elastic strain energy,

$$\dot{W}^p \cong \dot{W} \qquad (4.20)$$

where $\dot{W}$ is the total strain energy rate. Therefore,

$$\dot{\sigma}_{ij}{}' \simeq 2G\left(\dot{\varepsilon}_{ij}{}' - \frac{\dot{W}}{2k^2} \sigma_{ij}{}'\right) \qquad (4.21)$$

where ε_{ij}, the total strain rate, was assumed to be identical with the elastic strains given by applying Hooke's law to (4.1) through (4.13). These relations apply during plastic deformation, as long as J_2 equals k^2 and $\dot{W}$ is positive. Otherwise, the elastic equations are used to find stresses.

In order to integrate (4.21), the time rates of change can be transformed to gradients with respect to x as follows:

$$\frac{d}{dt}(\sigma_{ij}{}'; \varepsilon_{ij}{}'; W) = U \frac{\partial}{\partial x}(\sigma_{ij}{}'; \varepsilon_{ij}{}'; W) \qquad (4.22)$$

where U is the steady sliding speed of the lower plane. Therefore, by substituting (4.22) in (4.21) the time rates of change will be replaced with x-derivatives and consequently U is cancelled out from the equations.

Since the strains are assumed to be known (i.e., equal to the elastic strains), Eq. (4.21) gives the gradients with respect to x of the stress deviators. The gradients are then integrated by using Runge-Kutta-Gill method to calculate the stresses[6]. The step-by-step numerical anaylsis for the stress cycle of a point at any depth y is as follows:

Starting from the elastic stress state at a position x_o where the stress at the point first satisfies the von Mises yield condition, on the entry side, the stress rates of change with x are found from Eq. (4.21). These stress rates are then used to predict the values of the stress components when the point has moved a small distance dx. Therefore, the state of stress of points at a constant depth is obtained as the point goes through the stress cycle.

The stresses which are found by the above procedure are inexact only to the extent that they do not satisfy the condition of equilibrium, since no attempt was made to maintain equilibrium during the loading cycle. The assumption of using strains identical to the elastic strains satisfies the compatibility condition. The stress boundary conditions (i.e., traction free surface outside the contact) were also satisfied. Therefore, the solution is only an approximation to the exact solution. However, it is possible to restore the condition of equilibrium at the end of the loading cycle, as discussed in the next subsection.

3. Residual Stresses and Strains

Had the loading been entirely elastic, the stresses would approach zero when the contact point approached $x = \infty$. However, as a result of plastic deformation each point must have a state of residual stress. The periodicity of the problem requires that residual stresses and strains be independent of x; i.e.,

$$(\sigma_{xx})_r = f_1(y)$$

$$(\sigma_{yy})_r = f_2(y)$$

$$(\sigma_{zz})_r = \nu (f_1 + f_2)$$

$$(\sigma_{xy})_r = f_3(y)$$

(4.23)

However, the equilibrium equations

$$\frac{\partial (\sigma_{xx})_r}{\partial x} + \frac{\partial (\sigma_{xy})_r}{\partial y} = 0$$

$$\frac{\partial (\sigma_{xy})_r}{\partial x} + \frac{\partial (\sigma_{yy})_r}{\partial y} = 0$$

(4.24)

require that

$$\frac{df_3}{dy} = 0 \quad \text{or} \quad f_3 = c_1$$

$$\frac{df_2}{dy} = 0 \quad \text{or} \quad f_2 = c_2$$

(4.25)

The boundary conditions of $\sigma_{yy} = 0$ and $\sigma_{xy} = 0$ at $x = \infty$ and $y = 0$, then give $f_2 = f_3 = 0$ at $y = 0$ or $c_1 = c_2 = 0$. Finally, the only possible state of residual stress is

$$(\sigma_{xx})_r = f(y)$$

$$(\sigma_{yy})_r = 0$$

$$(\sigma_{zz})_r = \nu f(y)$$

$$(\sigma_{xy})_r = 0$$

(4.26)

The numerical results of the preceeding subsection gives nonzero "pseudo-residual stresses" $(\sigma_{yy})_r$ and $(\sigma_{xy})'_r$ which violates equilibriium and stress boundary conditions. These conditions can be satisfied by permitting these stresses to relax elastically, which results in a state of residual strains. Carrying out this procedure, the residual strains at the end of a loading cycle are

$$(\varepsilon_{yy})_r = -\frac{(1 - 2\nu)}{2(1 - \nu)G} (\sigma_{yy})_r'$$

(4.27)

$$(\gamma_{xy})_r = -\frac{(\sigma_{xy})_r'}{G}$$

(4.28)

where $(\sigma_{yy})'_r$ and $(\sigma_{xy})'_r$ are the pseudo-residual stresses and $(\gamma_{xy})_r$ is the engineering shear strain. Furthermore, the residual stresses $(\sigma_{xx})_r$ and $(\sigma_{zz})_r$ become

$$(\sigma_{xx})_r = (\sigma_{xx})'_r - \frac{\nu}{(1-\nu)} (\sigma_{yy})'_r \qquad (4.29)$$

$$(\sigma_{zz})_r = (\sigma_{zz})'_r - \frac{\nu}{(1-\nu)} (\sigma_{yy})'_r \qquad (4.30)$$

Using the residual stresses, $(\sigma_{xx})_r$ and $(\sigma_{zz})_r$, as initial conditions in a repeated numerical integration of the last subsection, new values of residual stresses are calculated. This procedure is repeated until there is no further change in $(\sigma_{xx})_r$ and $(\sigma_{zz})_r$ (which corresponds to $(\sigma_{yy})'_r$ approaching zero). This condition is satisfied after 5 to 10 integration cycles. However, the residual shear strain per pass $(\gamma_{xy})_r$ approaches a constant value. Therefore, the steady state residual stress corresponds to the physical steady state of residual stresses and the residual shear strain corresponds to the physical steady state increment of shear strain for each passage.

4. The Procedure for Numerical Calculations

The preceeding method was programmed in DGC FORTRAN IV language and a Nova 2 minicomputer (which uses 32 bits for real numbers) and was used to perform the computations. The procedure is shown by a flow chart in Figure 4.3. The final state of stress and the residual stresses and strains calculated by the above procedure is an approximate solution for the case of steady state sliding (i.e., when the residual stress $(\sigma_{xx})_r$ and $(\sigma_{zz})_r$ become constant and the pseudo-residual stress $(\sigma_{yy})'_r$ becomes zero). Merwin and Johnson considered the integration cycles before steady state is reached to be the actual transient solution, but this is not necessarily correct since it is only after the last integration cycle that $(\sigma_{yy})'_r$ approaches zero.

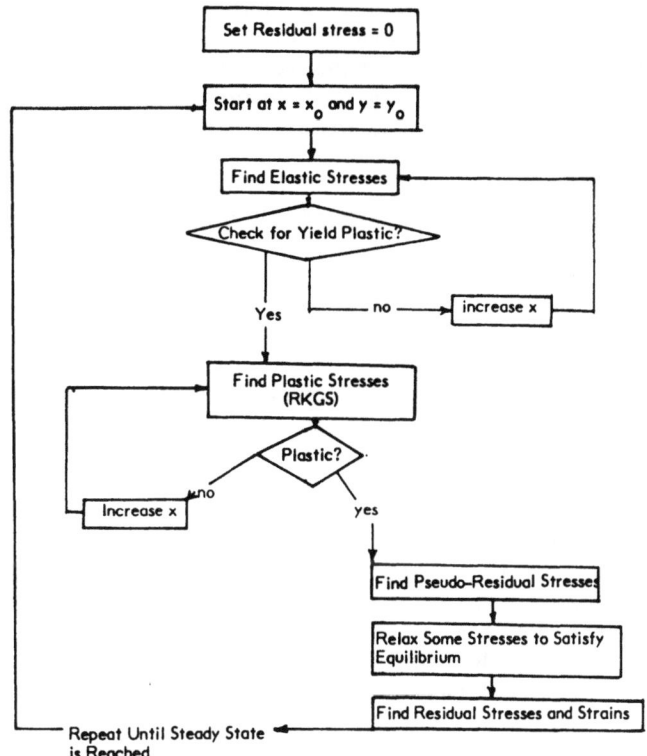

FIGURE 4.3 THE FLOW CHART FOR NUMERICAL CALCULATION OF THE STATE OF STRESS AND RESIDUAL STRESSES AND STRAINS UNDER A SLIDING CONTACT.

The assumption of allowing the total strains during plastic deformation to be identical with the elastic strains is reasonable for low friction coefficients (i.e., less than 0.5) since the region which continuously deforms plastically is contained by an elastic region around it. Therefore, the total strains cannot be much larger than the elastic strains. For larger friction coefficients, a boundary of the plastic region is at the surface, which may cause larger strain near the surface. Therefore, the solution becomes less exact at larger friction coefficients and near the free surface. However, it should still be reasonably deep below the surface near the elastic-plastic boundary.

The approximate solution found from the preceeding method has two types of instabilities. The stresses at the plastically deformed surface cannot be obtained due to the singularity of the elastic strain gradients at $x = a$. The solution also becomes unstable in that the residual stresses do not converge to steady state values very near the steady state elastic-plastic boundaries for low friction coefficients (lower than 0.5). However, there is no problem at a small distance from the elastic plastic boundary inside the plastic region.

5. Discussion of the Numerical Solutions

The Merwin and Johnson method was used to find the size of the plastically deforming zone under a contact, the state of stress, the residual stresses, and the residual strains during steady state sliding. The result for the applied normal stress $P_o = 4k$ and different tangential stresses ranging from $q_o = 0$ to $q_o = 4k$ is given in Figure 4.4. It should be noted that for zero friction, a state of shakedown is reached and the steady state deformation is purely elastic. The size of the plastic region increases with increasing friction coefficient. For friction coefficients smaller than 0.5, the plastic region is below the surface, whereas, at larger friction coefficients the plastic region extends to the surface. The large size of the plastic zone in front of the slider for large friction coefficient is surprising. Perhaps this is due to the fact that plowing was assumed for the stress boundary condition, but the displacements for plowing (i.e., raised material in front of the slider) were not considered. If the solution would allow a raised surface in front of the contact, stresses would be relieved below the surface in front of the contact, and the size of the plastic zone would decrease.

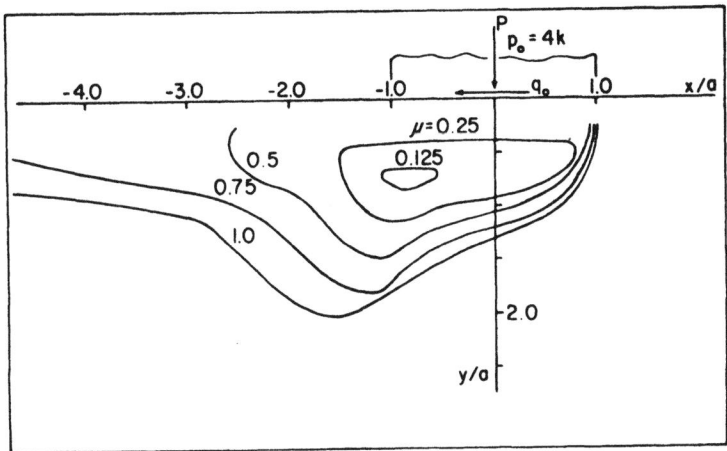

FIGURE 4.4   THE STEADY STATE PLASTIC DEFORMATION REGIONS IN AN ELASTIC-PERFECTLY PLASTIC MATERIAL UNDER A SLIDING CONTACT, FOR A MAXIMUM APPLIED NORMAL STRESS $p_o = 4k$ AND DIFFERENT FRICTION COEFFICIENTS.

The steady state σ_{xx}, σ_{yy} and σ_{xy} components of stress at various depths are given in Figures 4.5 to 4.7. (The coordinates are described in Figure 4.2.) It is noted that σ_{xx} is always compressive whereas σ_{yy} becomes tensile behind the contact and close to the surface. This tensile zone has also been found in obtaining a similar solution by a finite element method. The distribution of steady state residual stress $(\sigma_{xx})_r$ is given in Figure 4.8 for different friction coefficients. (As discussed earlier, the only possible residual stresses are $(\sigma_{xx})_r$ and $(\sigma_{zz})_r$.) Figure 4.8 shows that the residual stress is compressive and its magnitude, for large friction coefficients, is largest near the surface.

The steady state increment of shear strain, $(\gamma_{xy})_r$, per each passage of the asperity, is given in Figure 4.9. It is observed that the increment of shear strain is largest at the surface, for large friction coefficients, and its magnitude increases with increases in the friction coefficient. It should be noted that during steady state sliding, the only nonzero plastic strain is the shear strain which accumulates with an amount $(\gamma_{xy})_r$ after each passage of the slider.

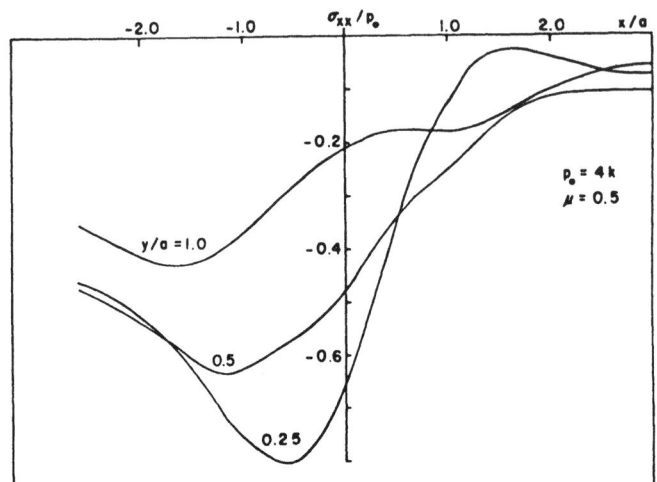

FIGURE 4.5 THE STEADY STATE σ_{xx} COMPONENT OF THE STATE OF STRESS AT DIFFERENT DEPTHS, NORMALIZED WITH RESPECT TO THE MAXIMUM APPLIED NORMAL STRESS p_o = 4k; FOR FRICTION COEFFICIENT μ = 0.5.

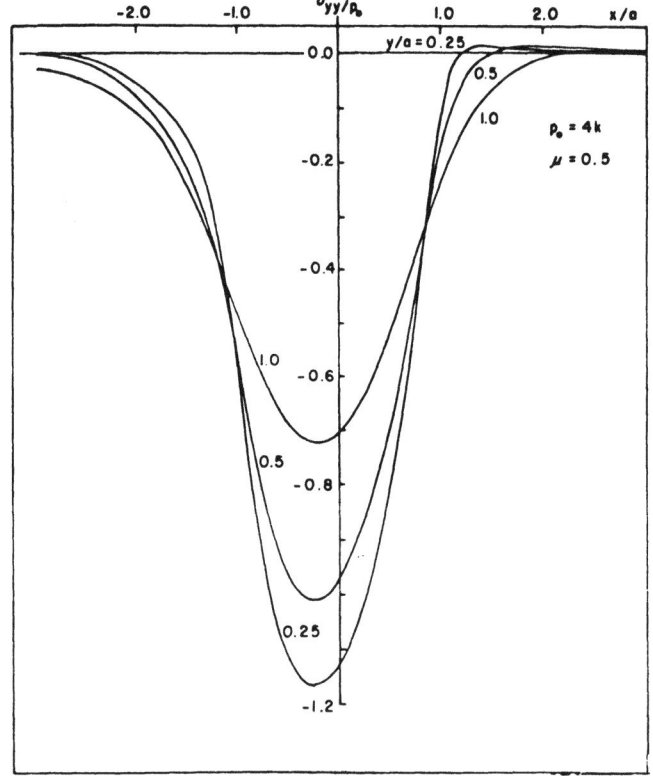

FIGURE 4.6 THE STEADY STATE σ_{yy} COMPONENT OF THE STATE OF STRESS AT DIFFERENT DEPTHS, NORMALIZED WITH RESPECT TO THE MAXIMUM APPLIED NORMAL STRESS p_o = 4k; FOR FRICTION COEFFICIENT μ = 0.5.

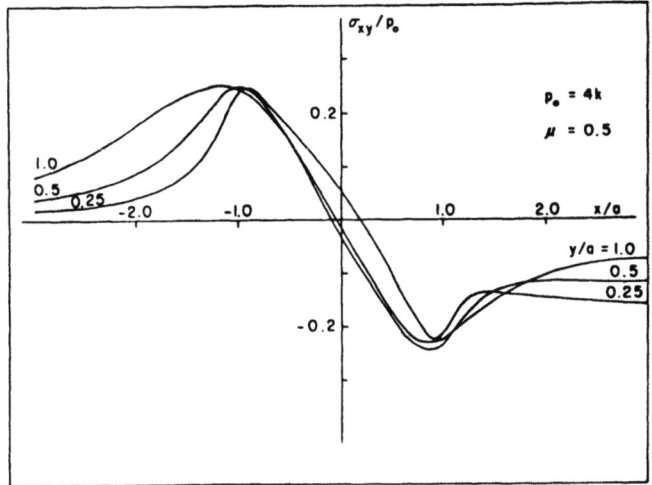

FIGURE 4.7 THE STEADY STATE σ_{xy} COMPONENT OF THE STATE OF STRESS AT DIFFERENT DEPTHS, NORMALIZED WITH RESPECT TO THE MAXIMUM APPLIED NORMAL STRESS $p_o = 4k$; FOR FRICTION COEFFICIENT $\mu = 0.5$.

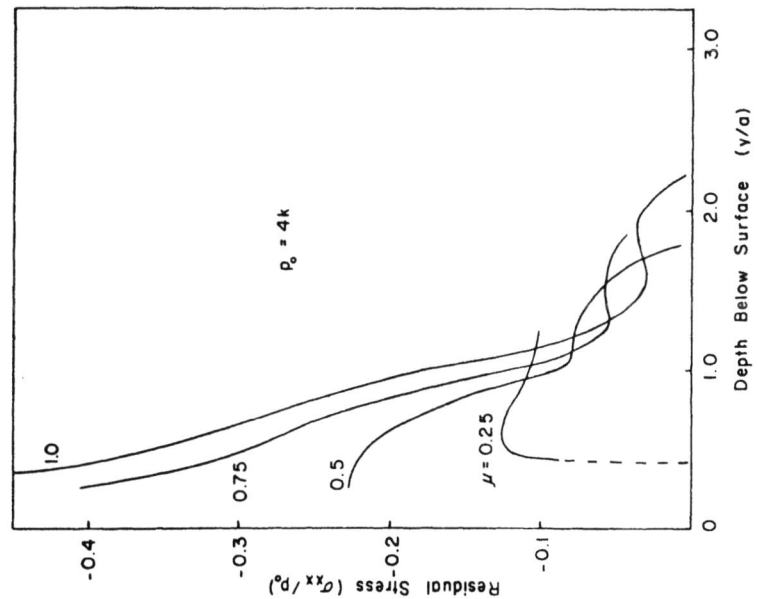

FIGURE 4.8 THE STEADY STATE RESIDUAL STRESS σ_{xx} FOR DIFFERENT FRICTION COEFFICIENTS, NORMALIZED WITH RESPECT TO THE MAXIMUM APPLIED NORMAL STRESS $p_o = 4k$.

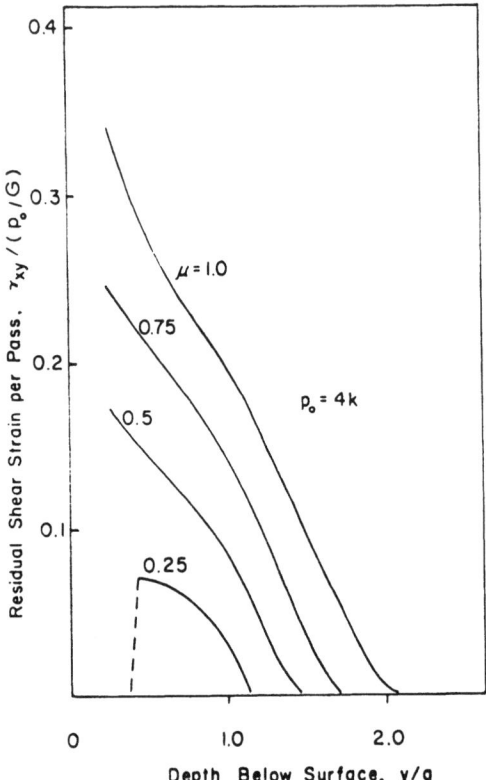

FIGURE 4.9   THE STEADY STATE RESIDUAL SHEAR STRAIN PER PASS, FOR DIFFERENT FRICTION COEFFICIENTS, NORMALIZED WITH RESPECT TO THE YIELD STRAIN IN PURE SHEAR AND THE MAXIMUM APPLIED NORMAL STRESS p_o = 4k.

4.2.3   Results of the Analysis Using FEM

The deformation of an elasto-plastic solid was also modeled and solved by using a finite element method[7]. The deformation of a semi-infinite, slightly work-hardening elasto-plastic solid was loaded by a moving asperity. The normal load was assumed to be 4k and the coefficient of friction 0.25. The material properties used were as follows: isotropic, slightly work-hardening (slope $d\sigma/d\varepsilon$ of the work-hardening region = 10^{-4} E where E is Young's modulus), E = 1.96×10^5 MPa = 2×10^4 kg/mm^2, ν = 0.28, σ_y = $\sqrt{3k}$ = 424 MPa = 43.3 kg/mm^2.

Figure 4.10 shows the plastically deformed region under a moving asperity during the first and fourth cycle, respectively. Under the moving load the material just in front of the load deforms plastically and part of the plastically deformed zone is much smaller than that under static load given in Figure 4.11. Furthermore, the repeated cyclic loading makes the plastic region smaller as shown in Figure 4.10(b). According to the numerical investigation of a rolling contact by Anand, the steady state deformation, after a few revolutions of a disk, would eventually reach a purely elastic state for a strain-hardening material, whereas elastic-perfectly plastic solids have elasto-plastic deformation at even high levels[8]. Therefore, the case shown in the figure may eventually reach a purely elastic state since it was assumed that the material is work-hardening.

The residual stress $(\sigma_{xx})_r$ is given in Figure 4.12 as a function of depth from the surface. It can be seen that the difference in $(\sigma_{xx})_r$ is almost negligible between third and fourth cycles. Also, the dominant residual stress is compressive in a direction parallel to the surface. The other components of residual stresses, $(\sigma_{yy})_r$ and $(\sigma_{xy})_r$ are very small. These two are the only components that can affect crack opening under sliding conditions. Therefore, it can be concluded that residual stresses do not affect crack propagation.

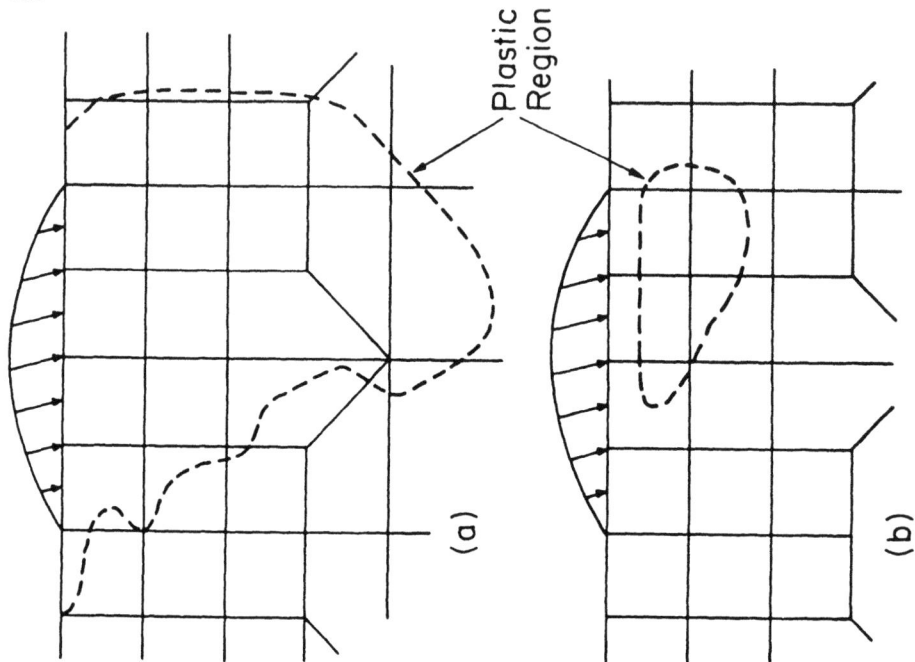

FIGURE 4.10 PLASTICALLY DEFORMED REGION UNDER A MOVING ASPERITY: (A) DURING THE FIRST CYCLE: (B) DURING THE FOURTH CYCLE.

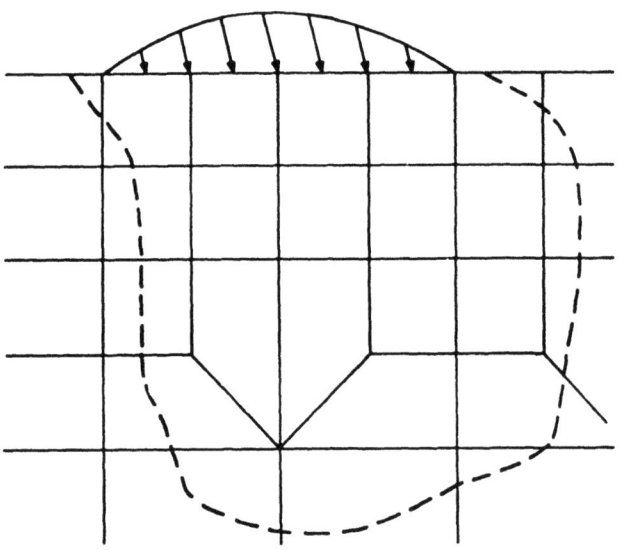

FIGURE 4.11 PLASTIC ZONE UNDER A STATIONARY ASPERITY.

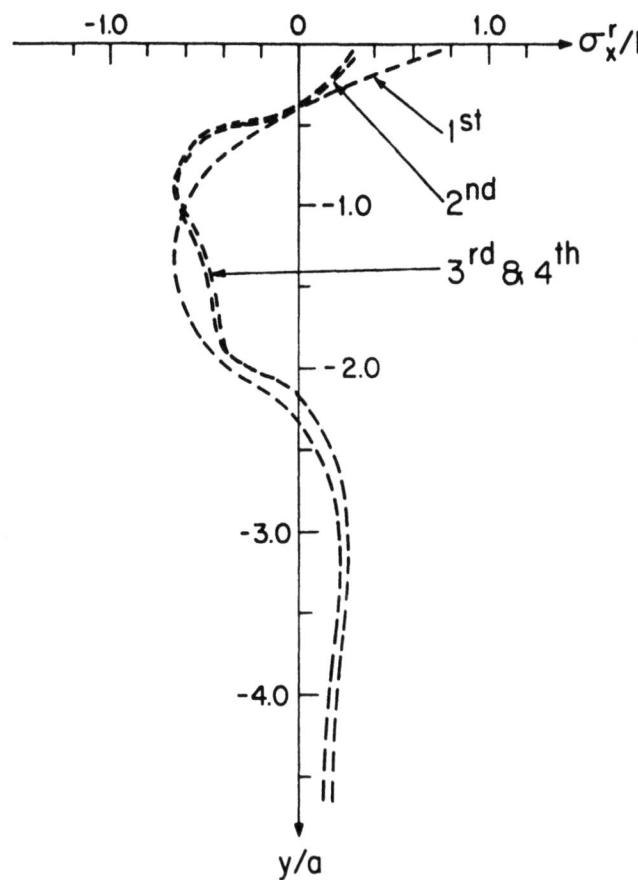

FIGURE 4.12 VARIATION OF RESIDUAL STRESS σ_x^r AS A FUNCTION OF DEPTH UNDER A MOVING ASPERITY.

These FEM results are in good agreement with the results obtained by Jahanmir and Suh using the technique developed by Merwin and Johnson[3, 1].

These theoretical results are also in agreement with the experimentally determined plastic strain fields which were obtained by measuring the grain deformation[9]. The plastic strain at the surface can be extremely large, especially on highly ductile copper, Figure 4.13.

4.3 Void and Crack Nucleation at the Subsurface

The failure of an initially flaw-free material proceeds in two sequential steps: (1) the initiation and nucleation of microcracks or microvoids and (2) crack propagation or void growth that leads to catastrophic failure. The first step of the failure process may not be an important process if flaws

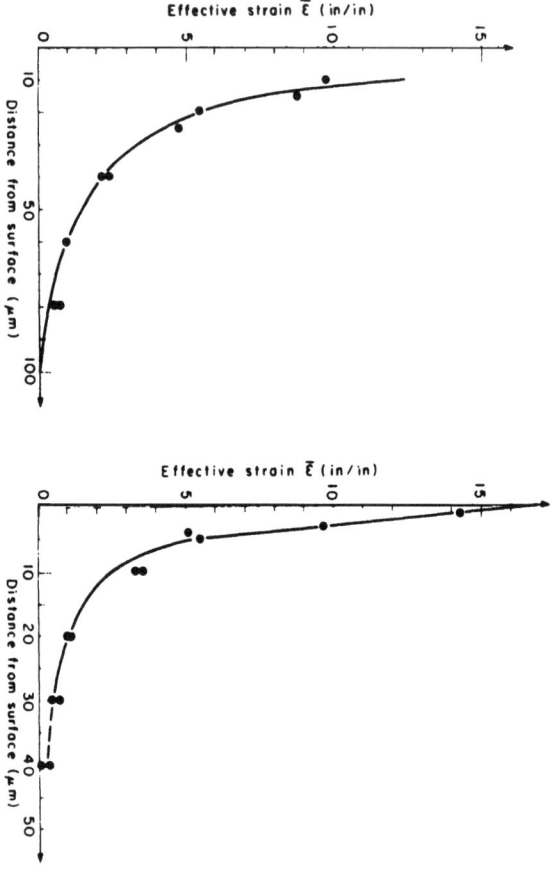

FIGURE 4.13   SUBSURFACE STRAIN VARIATION OBTAINED BY GRAIN SHAPE MEASUREMENTS: (A) COPPER SPECIMEN TESTED UNDER A NORMAL LOAD OF 2.1 kg AND AFTER A SLIDING DISTANCE OF 68 m IN ARGON; (B) AISI 1020 STEEL TESTED UNDER A NORMAL LOAD OF 2.4 kg AND A SLIDING DISTANCE OF 180 IN ARGON.

are initially present in the material. In this case, the
crack propagation may control the failure process. However,
in some materials, crack nucleation may determine the rate of
wear rather than crack propagation. In this subsection, the
mechanism of crack nucleation in a material without any flaws
will be analyzed in order to understand the origin of
deformation induced failure of elasto-plastic solids.

When the surface layer undergoes plastic deformation,
voids and cracks can nucleate. Cracks in two phase metals are
nucleated after repeated loading and consequent plastic
deformation of the surface layer due to the displacement
incompatibility between the hard particles and the matrix.
Experimental results are shown in Figure 4.14. Even in single
phase, metals cracks are present as shown in Figure 4.15.
Although the exact mechanism for crack and void nucleation in
single phase metals is not known, it has been suggested that
interactions of dislocations and the formation of dislocations
cells may be responsible [10 - 12]. It has also been suspected
that even in single crystals, crack nucleation may be a result
of displacement incompatibility between impurity inclusions
and the matrix[10]. Intersection of twins may also create
cracks[13].

The fact that the crack nucleation in two phase materials
is due to the displacement incompatibility between spherical
inclusions and the matrix can be illustrated as follows.
Consider a hard rigid spherical particle surrounded by an
elasto-plastic matrix which undergoes shear deformation. In
the absence of the hard inclusion the matrix will deform in
such a manner that an imaginary boundary will assume an
ellipsoidal shape as shown in Figure 4.16. However, in the
presence of the rigid sphere which is well bonded to the
matrix, the interface cannot assume the ellipsoidal shape
because of the geometric constraint imposed on the matrix by
the sphere. Consequently, normal stresses are developed at
the sphere-matrix interface. At location A the normal stress
is tensile, while at location B is compressive. When the
tensile normal stress at A exceeds the cohesive (or adhesive)
strength of the bonding at the particle-matrix interface, a
crack may nucleate, if the energy criterion is also satisfied.

The strength criterion may be expressed as

$$(\sigma_{kk})_{max} \geq \sigma_i \qquad (4.31)$$

where $(\sigma_{kk})_{max}$ is the maximum principal normal stress in
tension and σ_i is the ideal cohesive strength at the
interface. This is strictly a local criterion. When applied

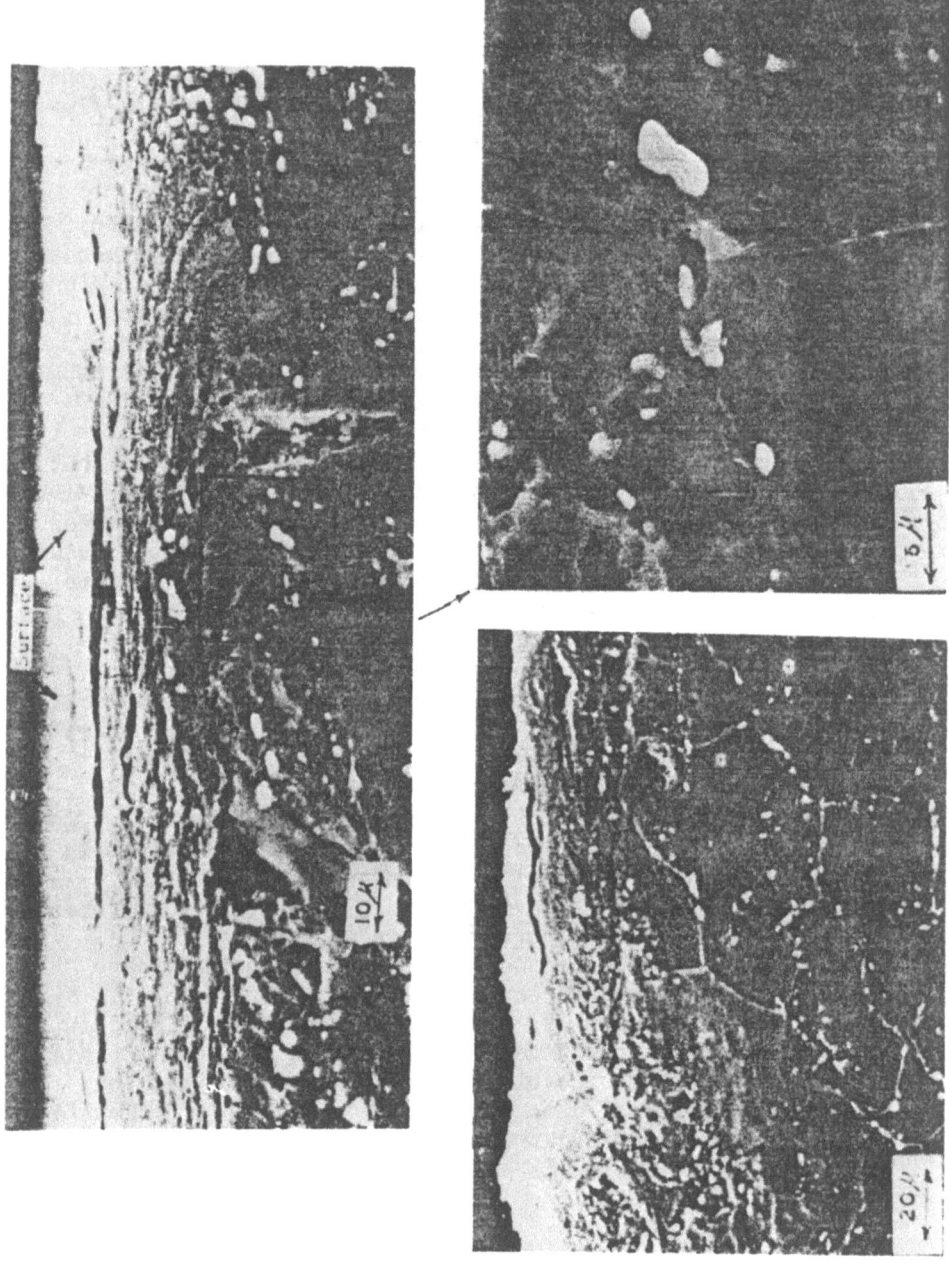

FIGURE 4.14 SUBSURFACE CRACKS AND DEFORMATION IN ANNEALED AISI 1020 STEEL.

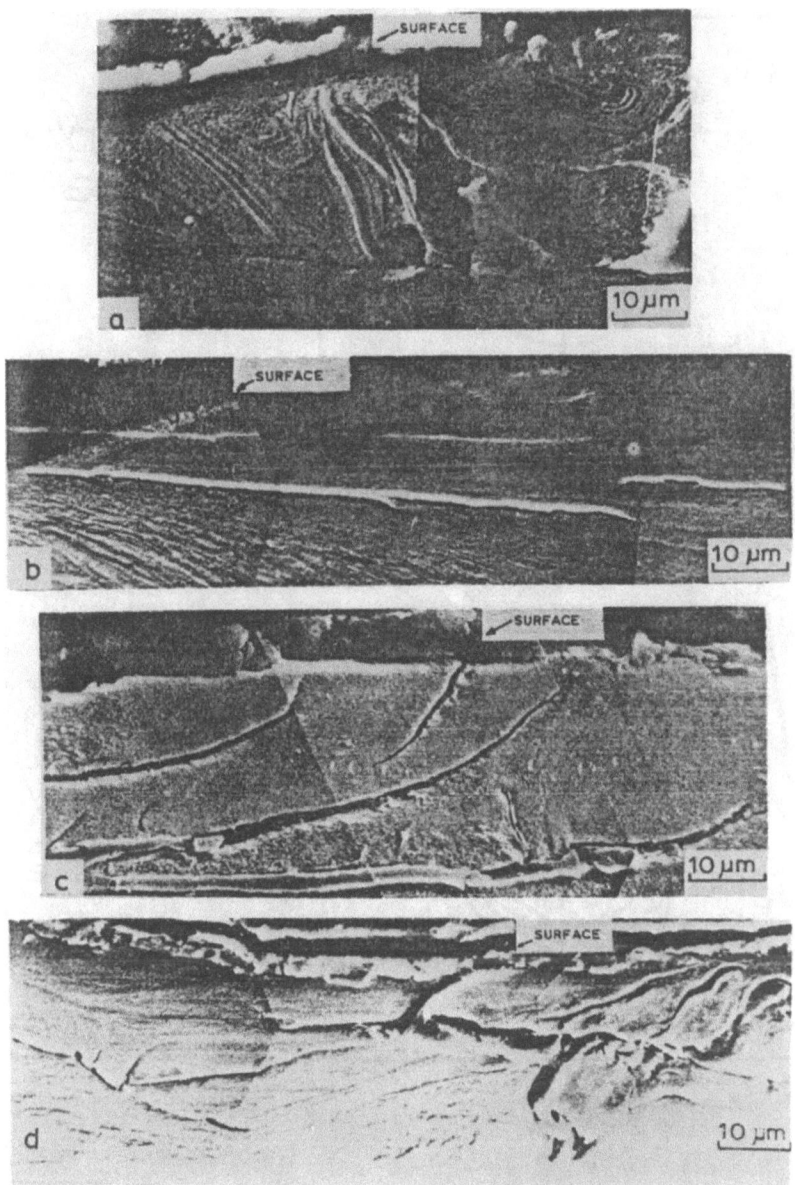

FIGURE 4.15 SCANNING ELECTRON MICROGRAPHS OF THE SUBSURFACE: (A) OFHC COPPER, (B) Cu – 5.7 at.% Sn, (C) Cu – 8.6 at.% Si and (D) Cu – 0.81 at.% Cr. THE NORMAL LOAD WAS 2 kg AT A SLIDING SPEED OF 2 m/min AND THE SLIDING DISTANCE WAS 200 m.

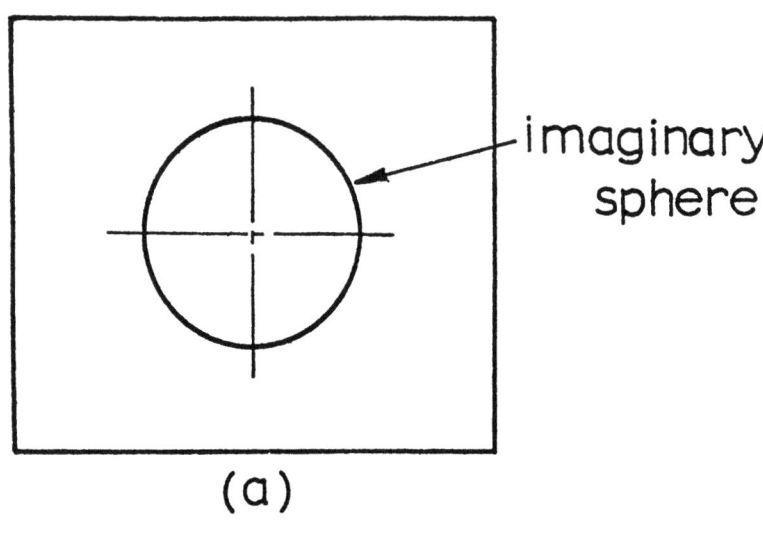

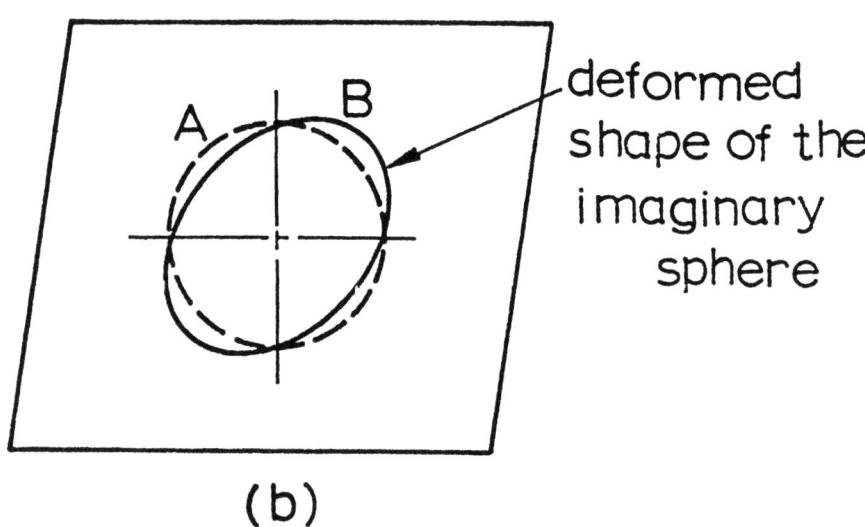

FIGURE 4.16 ILLUSTRATION OF THE DISPLACEMENT INCOMPATIBILITY BETWEEN THE MATRIX AND THE INCLUSION: (A) BEFORE DEFORMATION AND (B) AFTER DEFORMATION OF THE MATRIX.

to an inclusion filled material, the criterion itself bears no relationship to the size or shape of the inclusion. The shape of the inclusion will have to be considered in order to find the location of $(\sigma_{kk})_{max}$ and the relationship between the applied stress $(\sigma_{ij})_A$ and the interfacial stress $(\sigma_{kk})_{max}$. Once the stress concentration factor K is found, the stress criterion can be expressed in terms of the applied stress. In an elasto-plastic solid the interfacial stress concentration is limited by the flow strength of material $\sigma_y(\bar{\varepsilon}^p)$. For strain-hardening materials, the strain concentration around inclusions of different geometric shapes in inhomogeneous deformation fields are generally bound by two limiting idealizations of the plastic behavior of the material: a nonhardening rigid-plastic behavior and a linear hardening behavior with zero yield stress[14]. These results indicate that the interfacial stress, $(\sigma_{kk})_{max}$, at the surface of a cylindrical inclusion after yielding is bound by

$$\frac{3}{2}k \geq (\sigma_{kk})_{max} \geq 2k$$

where

$$(\sigma_{kk})_{max} = \sigma_{rr} - \sigma_t \qquad (4.32)$$

k = flow stress in shear

σ_t = hydrostatic tensile component of the applied stress

Since the limits on $(\sigma_{kk})_{max}$ are very close to each other and nearly equal to $\sigma_y(\bar{\varepsilon}^p)$, the interfacial radial stress on the cylindrical inclusion may be taken as

$$\sigma_{rr} = \sigma_y(\bar{\varepsilon}^p) + \sigma_t \qquad (4.33)$$

The energy criterion for void nucleation must be satisfied in addition to Eq. (4.31). When the inclusion filled material is subject to an external load, strain energy is stored in the elastic field within and around the inclusion. This strain energy will change as the elastic field changes during void nucleation. The energy in the matrix-inclusion system, E, should be sufficient to provide for the surface energy created by the void nucleation process. This may be expressed as

$$(E^s_{before}) - (E^s_{after}) \geq \Delta\gamma \cdot A \qquad (4.34)$$

where A is the surface area of the nucleated void (or crack) and $\Delta\gamma$ is the surface energy change during void nucleation. If the void nucleates at the matrix-inclusion interface,

$$\Delta\gamma = -\gamma_{M-I} + (\gamma_M + \gamma_I) \qquad (4.35)$$

where the subscripts M and I denote the matrix and inclusion, respectively, and γ the surface energy. Eq. (4.34) is inclusion size dependent, whereas the strength criterion is not. Figures 4.17 shows schematically both the strength and the energy criteria expressed in units of applied elastic strain as a function of the inclusion size. For inclusion size larger than d*, the energy criterion is always satisfied, whenever the strength criterion is reached. However, for inclusion sizes smaller than d*, satisfying the strength criterion does not necessarily guarantee the satisfaction of the energy requirement. Therefore, it may be argued that void nucleation will not be an energetically favorable process in this range of inclusion sizes.

In order to find σ_{rr} due to the accumulation of plastic shear strain, the solution of Eq. (4.33) is used. A point (x, y) is considered below the surface, subjected to a shear strain γ_{xy}. It is assumed that a rigid cylindrical particle is inserted at the center of a small volume element. The shear strain γ_{xy} develops an interfacial normal stress, σ_{rr}, around the particle, where σ_{rr} is a function of γ_{xy},

$$\sigma_{rr} = f(\gamma_{xy}) \qquad (4.36)$$

The shear strain γ_{xy} is the total accumulated strain and it is increased by increments of $(\gamma_{xy})_r$ after each asperity pass. Maximum σ_{rr} can be developed when

$$\gamma_{xy} = \gamma_y = \frac{k}{G} \qquad (4.37)$$

where γ_y is the yield strain in pure shear and G is the elastic shear modulus. The number of slider passages before maximum σ_{rr} is developed, can be found since the shear strain increment per pass, $(\gamma_{xy})_r$, is known from stress analysis. For an elasto-perfectly plastic solid, the maximum σ_{rr} due to accumulated plastic strain is equal to $\sqrt{3}k$ once the yield stress is exceeded. However, σ_{rr} depends on the position around the particle or,

$$(\sigma_{rr})_1 = \sqrt{3k} \sin 2\theta \qquad (4.38)$$

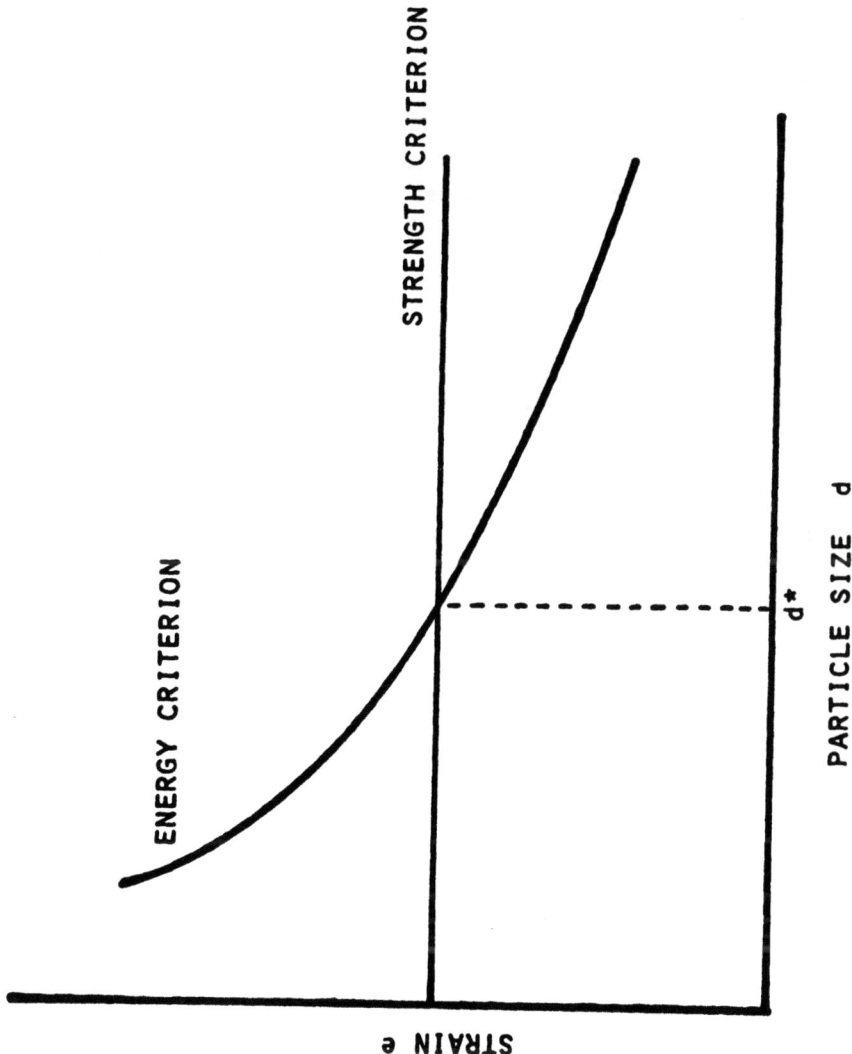

FIGURE 4.17 SCHEMATICS OF THE ENERGY AND STRESS CRITERION FOR VOID NUCLEATION.

where $(\sigma_{rr})_1$ is the normal interfacial stress due to the cumulative plastic deformation around a particle located at a depth y below the surface.

In order to find the maximum interfacial normal stress, σ_{rr}, due to the applied stress of the contact, Eq. (4.33) may be used. However, the stresses at each point must first be transformed to a state of maximum shear stress and hydrostatic stress by using Mohr's circle. Following the procedure of the last paragraph, it is assumed that the stresses act on an element and a particle is inserted at the center of the element, Figure 4.18. Therefore, using Eq. (4.33),

$$(\sigma_{rr})_2 = \sqrt{3}\tau_{max} \sin 2(\theta - \phi) + \sigma_h \qquad (4.39)$$

where τ_{max} is the maximum shear stress at the point (x, y), θ is the angle from the x-axis to the axis of the maximum positive shear stress, and σ_h is the hydrostatic stress at (x, y); $\sqrt{3}$ appears in the equation since the von Mises yield criterion is used.

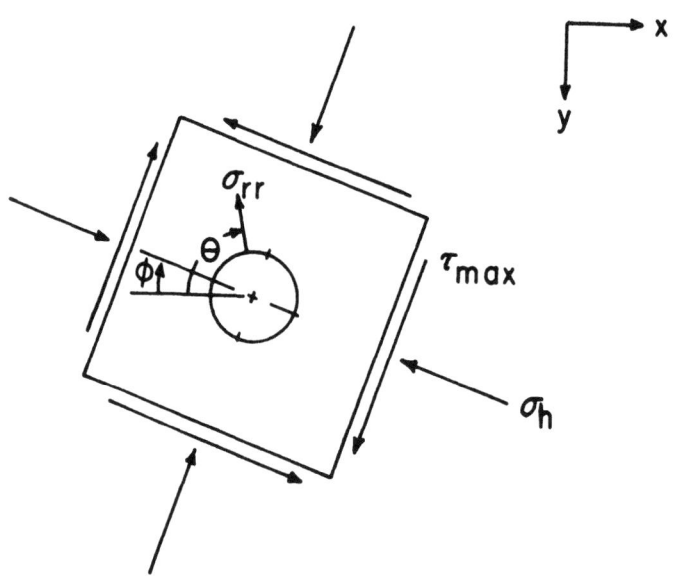

FIGURE 4.18 A RIGID CYLINDRICAL INCLUSION UNDER A GENERAL STATE OF STRESS.

The total σ_{rr} due to both the state of stress and the accumulation of plastic shear strain is found by adding Eq. (4.38) and Eq. (4.39), thus the maximum σ_{rr} which occurs at some angle $\theta = \theta_o$ can be obtained for each point (x, y)

$$(\sigma_{rr})_{max} = \sqrt{3}k \sin 2\theta_o + \sqrt{3\tau}_{max} \sin 2(\theta_o - \phi) + \sigma_h \quad (4.40)$$

The number of asperity passages before $(\sigma_{rr})_{max}$ can be developed at a given depth, can also be determined once the shear strain increment per pass is known at the respective depth.

Eq. (4.40) was solved numerically using Merwin and Johnson's method to calculate the state of stress and the residual stresses and strains. The contours of constant $(\sigma_{rr})_{max}$, found from Eq. (4.40), are plotted in Figures 4.19 to 4.21 for an applied normal stress at each asperity contact of 4k and friction coefficients of 1.0, 0.5, and 0.25. It should be noted that $(\sigma_{rr})_{max}$ is normalized with respect to k, the yield strength in shear, and all distances are normalized with respect to a, the half contact length. The figures show that $(\sigma_{rr})_{max}$ is compressive below the contact and attains its largest values in front of the slider, well below the surface.

If the particle-matrix bond strength is equal to 2k, Figures 4.22 to 4.24 show that the size of the region in which void nucleation is possible decreases with decreasing friction coefficient. The size of the void nucleation region is smaller for a stronger particle-matrix bond. It is not surprising to find that voids can only nucleate well below the surface ($\sigma_{rr} > 2k$), since the subsurface observations of wear samples in the last section showed that voids nucleate below the surface.

The minimum and the maximum depth of the void nucleation region is plottted in Figure 4.22, as a function of friction coefficient for the applied normal contact stresses, p_o and 4k and 6k. The range of depth for void nucleation increases with increasing friction coefficient for both applied normal stresses, but the voids can nucleate deeper below the surface for a larger applied normal stress. It is interesting to observe that at p_o = 6k and zero friction (which applies to a case of pure rolling), void nucleation is possible in a small region. However, at p_o = 4k and zero function coefficient, voids do not nucleate since the stresses shakedown to an elastic state during steady state condition.

The depth of the void nucleation region ($\sigma_{rr} > 2k$) is plotted in Figures 4.23 and 4.24 against the number of passes required to reach the maximum σ_{rr} at each depth, for different

140

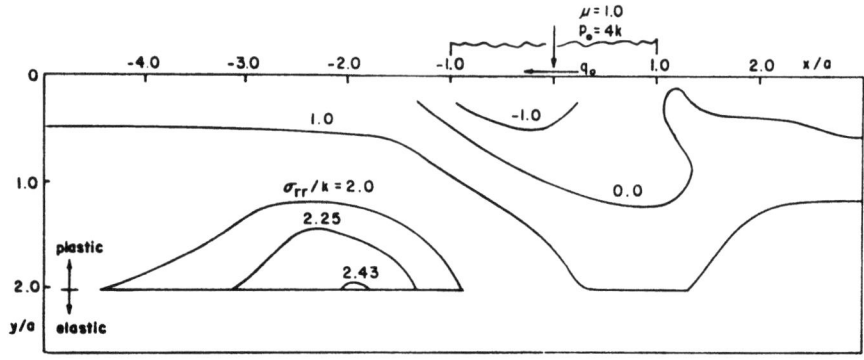

FIGURE 4.19 CONTOURS FOR σ_{rr} UNDER A SLIDING CONTACT, NORMALIZED WITH RESPECT TO k, THE SHEAR YIELD STRENGTH, FOR p_o = 4k and μ = 0.5.

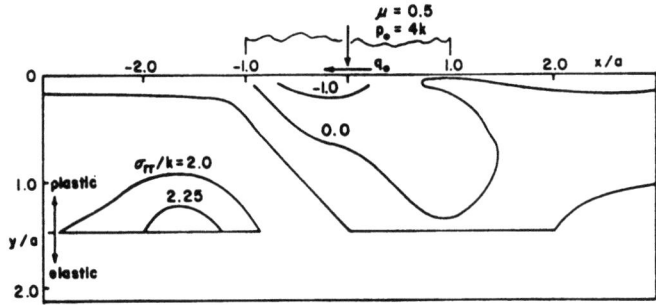

FIGURE 4.20 CONTOURS FOR σ_{rr} UNDER A SLIDING CONTACT, NORMALIZED WITH RESPECT TO k, THE SHEAR YIELD STRENGTH, FOR p_o = 4k AND μ = 0.25.

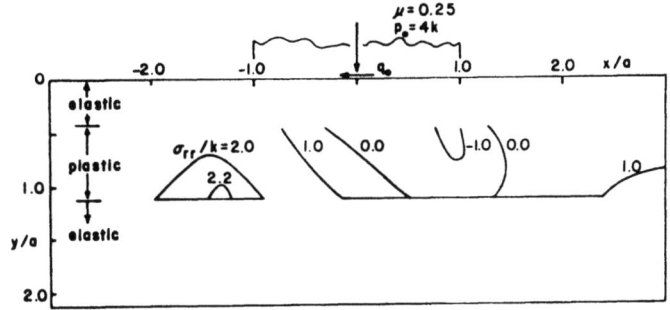

FIGURE 4.21   CONTOURS FOR σ_{rr} UNDER A SLIDING CONTACT, NORMALIZED WITH RESPECT TO k, THE SHEAR YIELD STRENGTH, FOR p_o = 4k AND μ = 0.25.

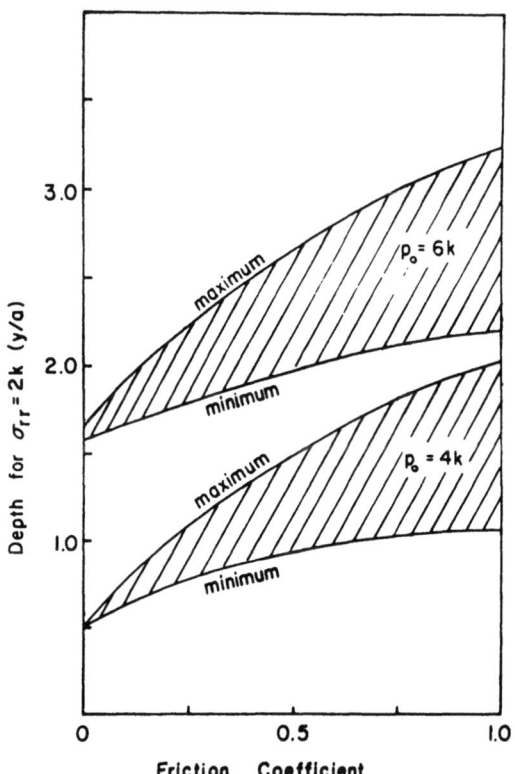

FIGURE 4.22   DEPTH OF VOID NUCLEATION REGIONS VERSUS FRICTION COEFFICIENT, FOR TWO DIFFERENT MAXIMUM APPLIED NORMAL STRESS p_o = 4k AND 6k.

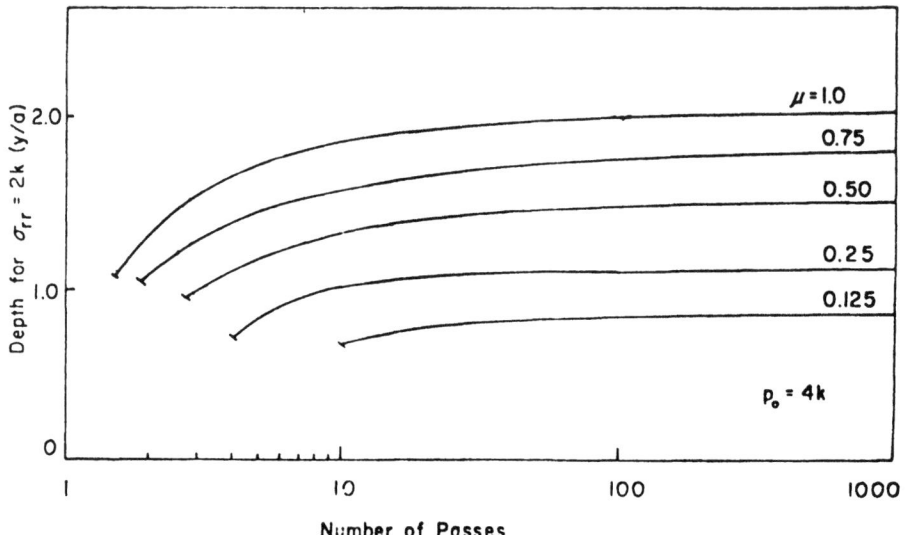

FIGURE 4.23   DEPTH OF VOID NUCLEATION REGIONS FOR DIFFERENT FRICTION COEFFICIENTS VERSUS THE NUMBER OF PASSES REQUIRED FOR VOID NUCLEATION, FOR A MAXIMUM APPLIED NORMAL STRESS, p_o = 4k

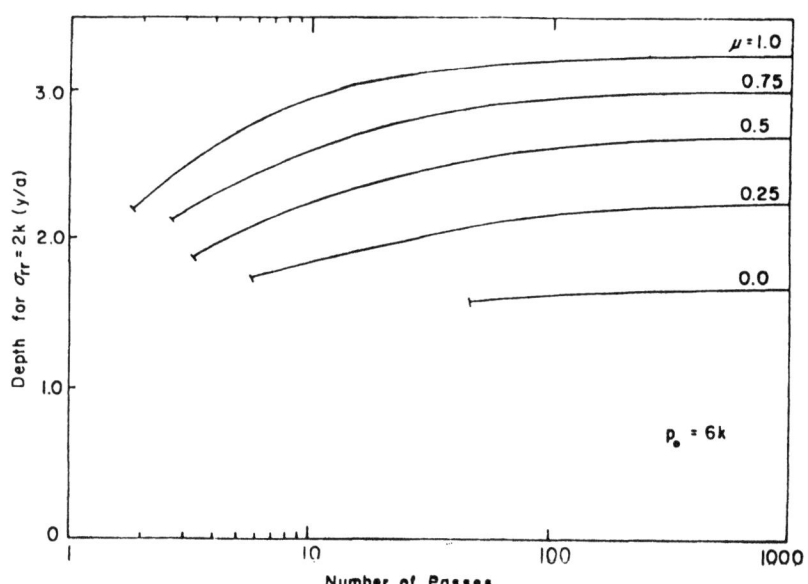

FIGURE 4.24   DEPTH OF VOID NUCLEATION REGIONS FOR DIFFERENT FRICTION COEFFICIENTS VERSUS THE NUMBER OF PASSES REQUIRED FOR VOID NUCLEATION, FOR A MAXIMUM APPLIED NORMAL STRESS p_o = 6k

friction coefficients and p_o of 4k and 6k. At a given depth, a smaller number of passes is required for void nucleation at larger friction coefficients. It should be noted that voids can nucleate at the minimum depths only after 1 to 10 passes; this implies that in many materials with hard inclusions, crack propagation may be the wear rate determining factor. It was estimated that about 10^5 asperity passes are required to create a wear sheet in AISI 1020 steel[15]. As the number of passes increases, the depth of void nucleation moves away from the surface.

It should be pointed out here that the addition of the interfacial normal stress at inclusions, due to the cumulative plastic deformation and due to the state of stress, violates the yield criterion. This violation occurs because an elastic-perfectly plastic material, σ_{rr}, cannot exceed the yield stress, i.e., $\sqrt{3}k$, and the solution may not be correct for $\sigma_{rr} \geq \sqrt{3}k$. However, if the voids nucleate when $\sigma_{rr} = \sqrt{3}k$, the stresses can relax and the yield criterion can be satisfied. The solution, however, does indicate that voids nucleate below the surface and that the depth and the number of passes depend on the friction coefficient and the applied normal load.

In the analysis of void nucleation, it was assumed that the hard particles are rigid. However, in real materials the particles have some elasticity, which would result in values of σ_{rr} smaller than the ones calculated by the above procedure. An exact analysis of the interfacial stress for elastic inclusions has been done for an elastic solid which may be adopted for an elasto-plastic solid[16, 17].

The criterion for void nucleation from large particles may be a combination of a local shear strain and a local interfacial tensile stress criterion. In the above analysis, it was assumed that the local stress criterion was sufficient. This assumption, however, may be a good approximation for equiaxial particles, but not for elongated particles. In the analysis of elongated particles, the local strain concentrations are large and void nucleation generally occurs by particle fracture. Therefore, as the particles become more elongated, the local shear strain criterion for void nucleation may become the dominating criterion.

4.4 Crack Propagation Due to Surface Traction

In the preceeding subsection, crack nucleation in an elasto-plastic solid was discussed in great detail. It was shown that cracks can nucleate at subsurfaces when large plastic deformation occurs, although the applied load by the asperity, creates compressive and shear stresses. It was also briefly stated that cracks nucleate at the surface of brittle solids, especially under rolling contacts. These cracks propagate perpendicular to the surface into the subsurface since the maximum normal stress at the surface is parallel to the surface, as shown in Figure 4.25. These cracks eventually change direction due to the changing direction of the maximum normal stress and the nature of the stress field. Various cases of crack propagation were illustrated by Suh[18].

In this subsection the crack propagation in an elasto-plastic solid under sliding conditions will be analyzed. As shown earlier, cracks are nucleated within a narrow zone below the surface due to the compressive and tangential load applied by an asperity. These cracks may propagate due to the repeated loading by moving asperities at the surface. These subsurface cracks will propagate at different rates, depending on where the cracks are located. The crack that propagates the fastest will control the wear process. Although these cracks grow little by little and reach a steady state crack length, the analysis is easier to perform when cracks have grown to a finite size.

When stresses are applied at the surface of a solid with a subsurface crack as shown in Figure 4.26, part of the crack is in a compressive region while a very small section of the crack is in a tensile region. These cracks always experience combined loading. The crack tip in the tensile regime experiences a crack opening-closing mode of loading (i.e., Mode I) as well as shear loading, whereas the crack in the compression region only experiences compressive and shear loading. The material around the crack tip in the tensile region is commonly in elastic state due to the unloading of the material with the passage of the asperity.

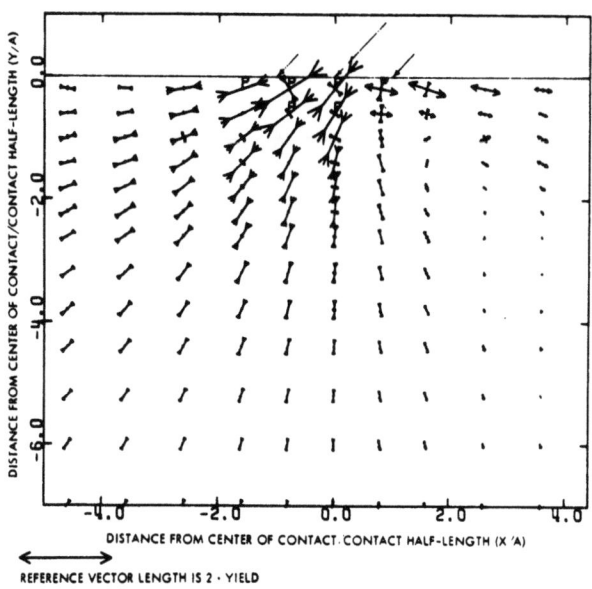

(a)

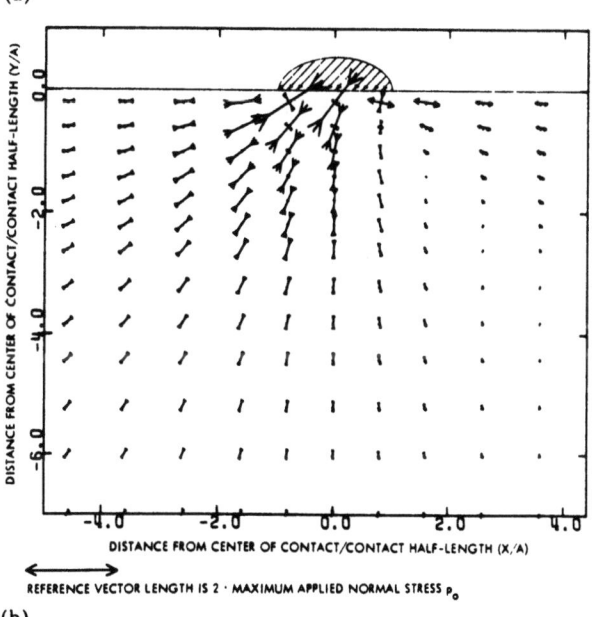

(b)

FIGURE 4.25 THE STRESS FIELD UNDER AN ASPERITY CONTACT
(COEFFICIENT OF FRICTION = 1). (LINES SHOW THE
DIRECTIONS AND MAGNITUDES OF PRINCIPAL STRESSES,
ARROWHEADS INDICATE SIGNS AND "p" INDICATES
PLASTICITY.) (A) AN ELASTOPLASTIC SOLID (FINITE
ELEMENT SOLUTION, APPLIED LOAD VECTORS NOT TO SCALE);
(B) AN ELASTIC SOLID (SOLUTIONS FOR AN ELLIPTICALLY
DISTRIBUTED LOAD, VECTORS NOT TO SCALE).

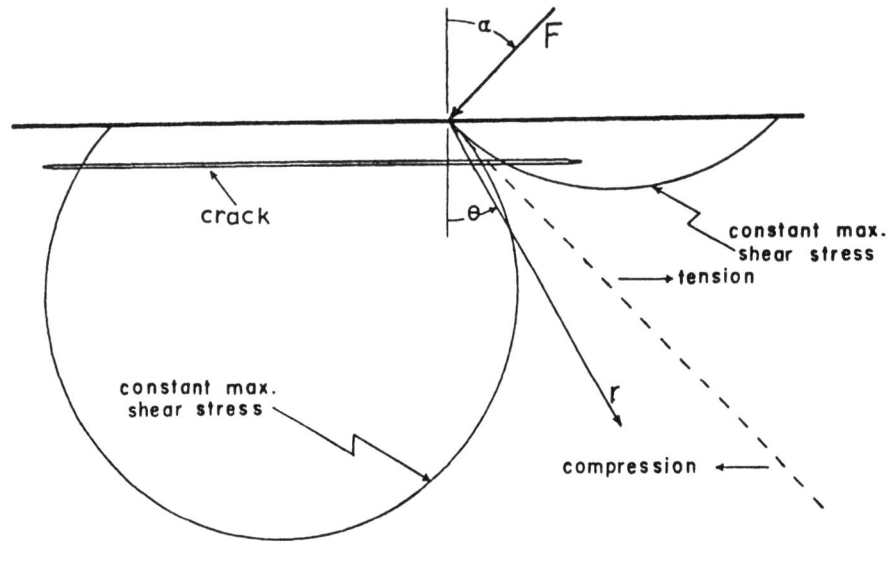

FIGURE 4.26   A POINT LOAD (LINE LOAD INTO THE PAPER, IN PLANE STRAIN) ON A HALFSPACE.

It has been well established in the fields of fracture mechanics and fatigue, that when cracks are loaded cyclically in Mode I, the crack propagation rate per cycle, dC/dN, correlates with the cyclic change in the stress intensity factor, ΔK_{II} as shown in Figure 4.27. The direction of the propagation in Mode I, correlates well with the direction of the maximum normal stress and also with the direction of minimum strain energy. When the change in the stress intensity factor per cycle is less than the threshold value, ΔK_{th}, cracks cannot propagate under Mode I loading.

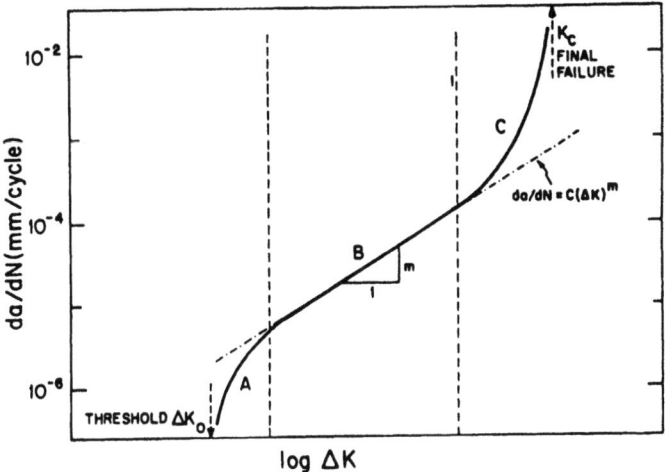

FIGURE 4.27 A TYPICAL PLOT OF THE CRACK EXTENSION PER CYCLE VS. THE LOGARITHM OF THE CHANGE IN THE STRESS INTENSITY FACTOR. THE PRIMARY MECHANISMS MAY BE DIVIDED INTO THREE REGIMES: REGIME A, NON-CONTINUUM MECHANISMS, SHOWING A LARGE INFLUENCE OF (I) MICROSTRUCTURE, (II) MEAN STRESS AND (III) ENVIRONMENT: REGIME B, CONTINUUM MECHANISMS (STRIATION GROWTH), SHOWING LITTLE INFLUENCE OF (I) MICROSTRUCTURE, (II) MEAN STRESS, (III) DILUTE ENVIRONMENT AND (IV) THICKNESS: REGIME C., "STATIC MODE" MECHANISMS (CLEAVAGE, INTERGRANULAR AND FIBROUS), SHOWING A LARGE INFLUENCE OF (I) MICROSTRUCTURE AND (II) THICKNESS BUT LITTLE INFLUENCE OF THE ENVIRONMENT.

Very little data are available on crack propagation under Mode II and Mode III loading. When there is combined loading of Mode I and Mode II (or III) type, there is very little known about the direction and the magnitude of crack propagation. It will be shown in this subsection that when the loading is both compressive and shearing, the crack propagation direction in an elasto-plastic solid is parallel to the τ_{max} direction.

4.4.1 Subsurface Crack Propagation in Linear Elastic Solids and Its Implications

For the case shown in Figure 4.23, Fleming and Suh analyzed the propagation of a subsurface crack parallel to the surface using a linear elastic fracture mechanics (LEFM) model[19]. This treatment was based on the fact that a subsurface crack is closed in the plastic region in front of the asperity contact, while the trailing end of the crack behind the moving asperity contact is subjected to tensile stress in the elastic region. This stress intensity factor, due to an elliptically distributed load at the asperity contact, was calculated using an approximate method based on weighting factors. The stress intensity factor for Mode I, computed in this manner, was very small being in the neighborhood of Δk_{th}, while the stress intensity factor for Mode II was about an order of magnitude larger than that for Mode I. Essentially, similar results were obtained by Hills and Ashelby, Rosenfield and Keer, et al, although they also computed the stress intensity factor in the compressive zone[20, 21, 22].

Recently, Sin and Suh have caluculated the stress intensity factor using finite element method for both an elastic solid and an elasto-plastic solid[23]. The FEM results for the elastic case show that:

1) The maximum values of K_{II} occur when the crack tips lie right below the asperity contact.
2) The closer the crack is to the surface, the larger is the maximum value of K_{II}.
3) K_{II} increases with crack length.

All these values of stress intensity factors for horizontal subsurfaace cracks are quite small. This is not surprising, in view of the fact that the largest stress component, due to the surface traction exerted by the asperity contact, is the normal stress parallel to the surface. Therefore, the largest stress intensity factor for an elastic solid is expected to be associated with cracks perpendicular to the surface.

Linear elastic fracture mechanics were applied to the subsurface problems as an approximation to the real elasto-plastic solid. Whether the linear elastic fracture mechanics (LEFM) approaches are appropriate depends on the plastic zone size at the crack tip. If this size is too big or comparable with such a dimension as the distance from the crack tip to the free surface, LEFM is no longer valid and plasticity plays a significant role. In this case, the local stress and strain history become important in determining the fracture criterion.

The estimate of the plastic zone size for Mode II loading is given by

$$r_p = \frac{1}{2\pi}(\frac{\Delta K_{II}}{k})^2 \qquad (4.41)$$

where ΔK_{II} is the stress intensity factor range in Mode II and k is the shear yield strength of the material. Tables 4.1 and 4.2 show the changes in stress intensity factor and estimated plastic zone sizes as a function of the depth of crack location for a small crack during one cycle of loading. Values for all parameters are the same as given before: k = 245 MPa = 25 kg/mm^2 and a = 10 μm. From these two tables, it can be seen that the plastic zone sizes estimated are comparable with the depth of crack and the stress intensity factor ranges calculated and are close to or less than the threshold intensity ΔK_{th}. Of course, there is no data on Mode II available. However, an approximate analysis indicates that the threshold intensity factor in Mode II is very close to that in Mode I[7].

There are two difficulties in accepting the linear elastic fracture mechanics approach to the subsurface crack problem. The cracks quite near the surface can have large values of ΔK_{II}, but they cannot propagate in a brittle manner due to the development of large plastic zones. On the other hand, cracks far below from the surface cannot grow since the stress intensity factor for these are much smaller than the threshold although the plastic zones are small and the plasticity restriction does not apply. Therefore, plasticity should come into the analysis of subsurface crack behavior under the moving asperity.

4.4.2 Crack Trajectory - Fracture Criteria in Mixed Mode

One of the major issues in studying crack propagation is how to predict the crack propagation direction. This is an appropriate topic to consider before the plasticity effects at the crack tip are further discussed.

Until recently, no suitable failure criterion could be found for the mixed mode fracture. Much of the attention has been given only to the problem of predicting the direction of crack extension when a body with cracks is simply loaded. There are two major criteria: maximum hoop stress and minimum strain energy density[24, 25]. The maximum hoop stress criterion states that crack growth will occur in a direction perpendicular to the maximum principal stress. On the other hand, the minimum strain-energy-density factor criterion postulates: (1) that the initial crack growth takes place in the direction along which the strain-energy-density factor possesses a stationary value, and (2) that crack initiation

d/a		μ : 0.25	0.5	1.0
0.3	Left	1.05	1.07	1.44
	Right	1.28	1.34	1.54
0.5	Left	1.08	1.11	1.15
	Right	1.35	1.40	1.49
1.0	Left	0.67	0.68	0.70
	Right	0.89	0.94	1.06

Note: a -- half length of the aperity contact
c -- length of the crack
d -- depth of the crack
μ -- friction coefficient

TABLE 4.1 STRESS INTENSITY FACTOR RANGE, K_{II} (MNm$^{3/2}$), AT LEFT AND RIGHT TIPS FOR A SMALL CRACK (c = 1/4α).

d/a		μ : 0.25	0.5	1.0
0.3	Left	0.96	1.00	1.48
	Right	1.44	1.56	2.08
0.5	Left	0.76	0.80	0.88
	Right	1.20	1.28	1.48
1.0	Left	0.12	0.12	0.12
	Right	0.20	0.24	0.32

Note: Nomenclature defined in Table 4.1

TABLE 4.2 RATIO OF PLASTIC ZONE SIZE TO DEPTH OF CRACK AT LEFT AND RIGHT TIPS FOR A SMALL CRACK (c = 1/4α).

occurs when the factor reaches a critical value. Comparison of the two criteria by several authors has shown that for tensile loading the differences between them are small. For compressive loading, however, not only do these two not agree, but neither criterion correlates well with physically observed behavior[26].

When these criteria are applied to the subsurface cracks in a compressive zone, they predict crack extension direction to be about 110° at the left (trailing) tip and about 70° at the right (leading) tip from the direction parallel to the surface, implying that crack extension occurs toward the surface at both tips. However, experimental results given in the next section show that the subsurface cracks grow parallel to the surface most of the time, before they become loose. McClintock has suggested, in the investigation of crack behavior in rail heads under rolling conditions, that cracks in a compressive field are most likely to grow in shear[27]. In fact, Forsyth has observed that fatigue cracks have two growth regimes[28]. In Stage I, cracks formed on the slip planes of the persistent slip bands, grow when they are most closely aligned with the maximum shear stress directions. In sliding wear, the slip planes tend to line up parallel to the surface maximum shear stress direction, and thus is likely to be the crack propagation direction.

In 2-dimensional deformation fields the stresses at the crack tip are expressed as[25]:

$$\sigma_{rr} = \frac{1}{2\sqrt{2\pi r}}[K_I \cos \frac{\theta}{2}(3 - \cos \theta) + K_{II} \sin \frac{\theta}{2}(3\cos\theta - 1)]$$

$$\sigma_{\theta\theta} = \frac{1}{\sqrt{2\pi r}} \cos \frac{\theta}{2} [K_I \cos^2 \frac{\theta}{2} - \frac{3}{2} K_{II} \sin \theta] \quad (4.42)$$

$$\sigma_{r\theta} = \frac{1}{2\sqrt{2\pi r}} \cos \frac{\theta}{2}[K_I \sin \theta + K_{II}(3 \cos \theta - 1)]$$

and the maximum shear stress τ_{max} is given by

$$(4.43)$$

τ_{max} will have maximum values when the conditions of $\partial \tau_{max}/\partial \theta = 0$ and $\partial^2 \tau_{max}/\partial \theta^2 > 0$ are satisfied. If we denote θ_m for the angle θ which satisfies the conditions and substitute into Eq. (4.42), then the stresses become $[\sigma_r]_{\theta_m}$, $[\sigma\theta]_{\theta_m}$, and $[\tau_r\theta]_{\theta_m}$ respectively. Thus, using the Mohr's circle transformation, the maximum of τ_{max} and the angle between the θ_m-direction and this maximum shear direction can be determined. When the distance r goes to zero, this direction ultimately becomes the direction of crack propagation.

If the above criterion is applied to the results obtained in the previous subsection for subsurface cracks, it predicts the angle to be between -5 and 5 degrees. These values are very small and therefore can practically be assumed zero, implying that cracks propagate in a plane coincident with the original cracks parallel to the surface.

4.4.3 The Crack Propagation Mechanism in Elasto-Plastic Solids

Linear elastic fracture mechanics are found to be useful in assessing the crack tip stress concentration and in determining the crack trajectory by mixed mode fracture criteria. However, the size restriction by plasticity considerations and the small stress intensity factors calculated (which is less than the threshold) suggest that the plastic fracture mechanics approach is required. Recently, Sin and Suh investigated this problem at MIT using the finite element method[23].

In this study ADINA (Automatic Dynamic Incremental Nonlinear Analysis, a finite element computer program for the static and dynamic displacement and stress analysis of solids, fluid-structure systems and structures) was used as in the elastic analysis case[29]. The material model used in this study was as follows: infinitesimal, material nonlinearity only[30].

The model used to calculate the elastic-plastic response under the moving load condition was the same as that for the elastic case. No dynamic effect was considered in the analysis. The material was assumed to be slightly work-hardening ($E_T = 10^{-4}$ E). The material properties used were as follows: isotropic, slightly work-hardening, $E = 1.96 \times 10^5$ MPa $= 2 \times 10^4$ kg/mm^2, $E_T = 19.6$ MPa $= 2$ kg/mm^2, $\nu = 0.28$, $\sigma_y = \sqrt{3}k = 424$ MPa $= 43.3$ kg/mm^2.

For the investigation of crack propagation only, a short crack (c = 1/4a) was used. Due to the prohibitively expensive computer cost, only a limited parameter study was conducted for the case of a = 10 μm, $p_o = 4k$, and μ = 0.25.

The problem was solved incrementally by moving the load step by step. For an accurate solution, the load increments per step should be sufficiently small. However, such a load step requires a large number of calculations that make the analysis very expensive. Therefore, larger load steps were used with iteration to obtain efficient and accurate solutions. The use of iteration can introduce some difficulties. The convergence process may be slow, requiring a large number of iterations that can be expensive. Also, some iterative methods do not converge for certain types of problems or large load increments. For details of numerical computation, the reader is referred to Reference (23). Figure 4.28 shows the development of the plastic zone along with the moving load step by step when this element is used. At the beginning, the shape of the plastically deformed zone is more or less the same as for the case of no crack inside the zone, except right around the crack tips. Due to the presence of a crack, the stress field changes significantly when the load moves over the crack. It can also be noticed that there are some spots inside the plastic region where unloading has taken place. With the repeated loading-unloading, it is expected that the overall plastic zone should become smaller as in the case of no crack.

Examination of the displacement of the nodal points on the crack surfaces shows that the upper surface initially slides forward and then slides backward. This is consistent with the result by McClintock who investigated the behavior of a subsurface crack in a rolling contact problem[27]. In Table 4.3, the relative sliding displacements of crack tip nodel points are shown for two different depths of crack location. It shows that the relative displacement increases with sliding, then decreases, and finally changes sign. The following can also be observed from the table: (1) the crack closer to the surface has large values, and (2) the left tip usually has large values of relative sliding displacements.

In Figures 4.29 and 4.30, the crack tip shear strains are plotted along the distance from the crack tip for different stages of the loading position as the asperity contact moves from left to right. The distribution of shear strain increases until it reaches a maximum value and then decreases. After it attains a minimum, it increases again. At very near the tip, the shear strain changes from positive to negative, and then to positive again. For a crack located at $d = 0.25a$, the shear strain can reach more than 120% at a point of 0.00625 μm away from the tip.

As shown in the preceeding subsection, the subsurface cracks are likely to grow in the direction parallel to the surface, even though it was based on elastic solutions. Also the experimental findings of surface texture development due to sliding strongly suggest that the cracks propagate along the slip planes nearly parallel to the surface[31, 32].

154

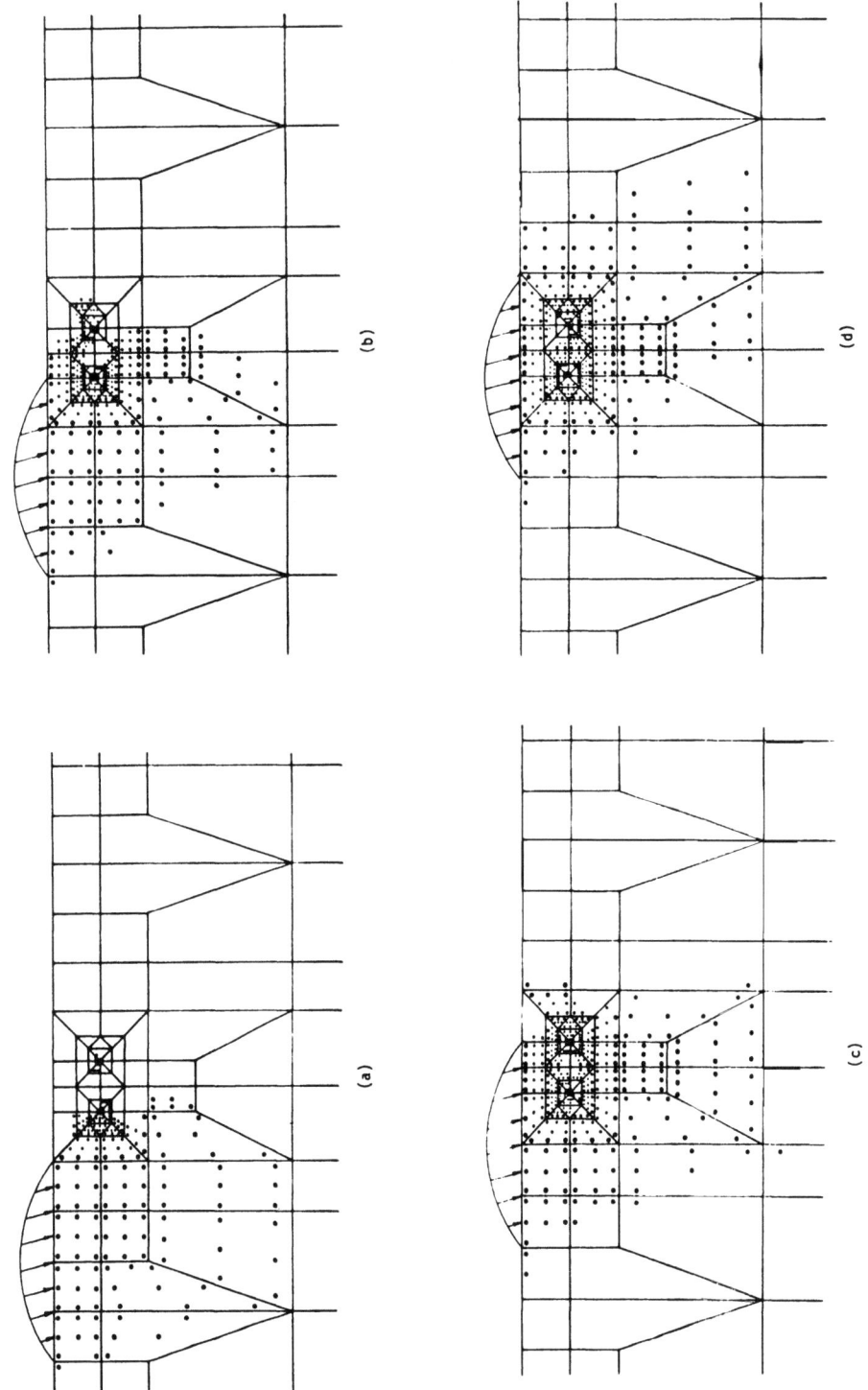

FIGURE 4.28

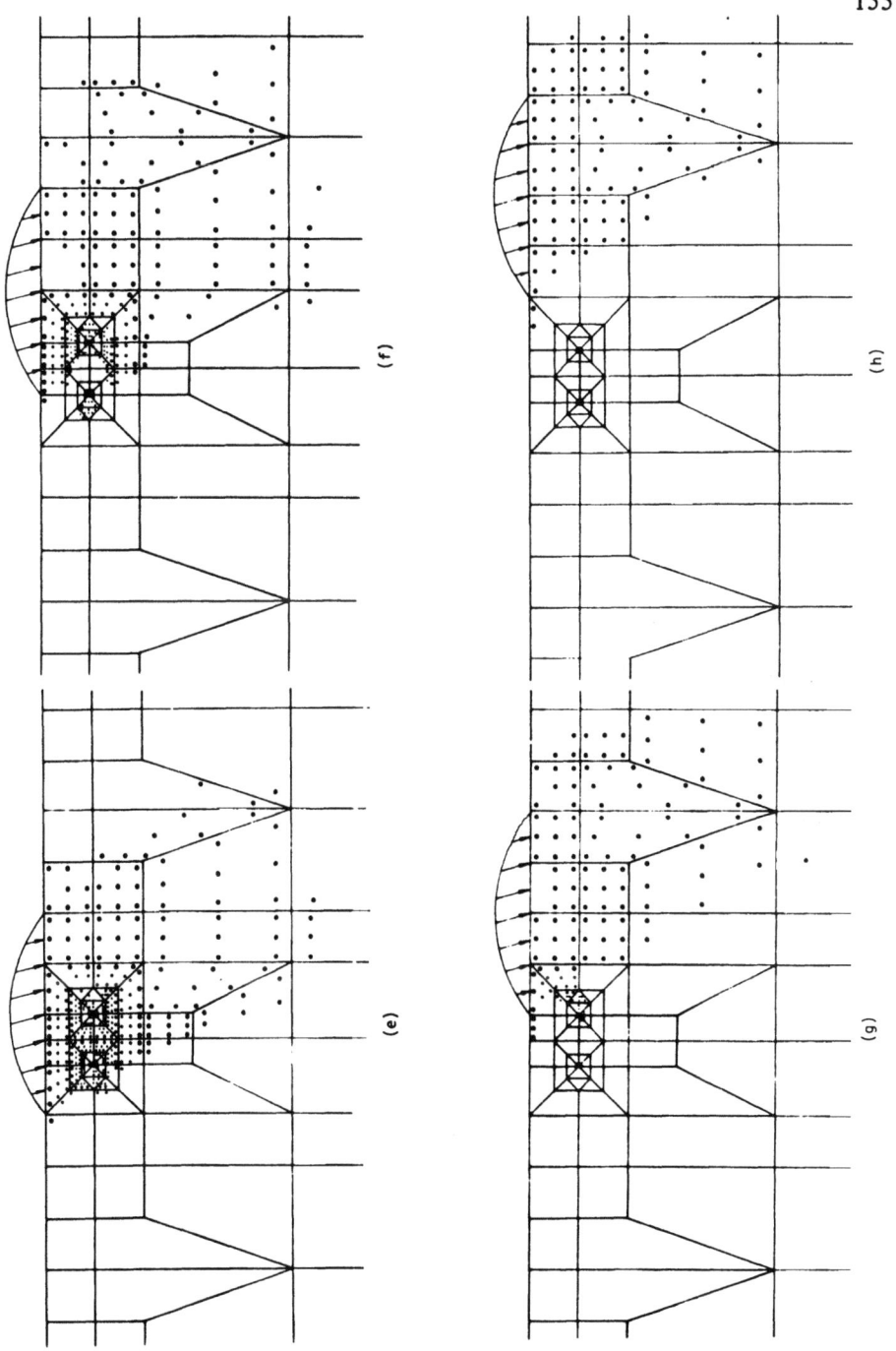

FIGURE 4.28 VARIATION OF PLASTICALLY DEFORMED ZONE AROUND A CRACK UNDER A MOVING ASPERITY: $a = 10$ μm, $d = 0.5\alpha$, $\mu = 0.25$, $p_o = 4k = 980$ MPa. DOTS INDICATE THE INTEGRATION POINTS.

(a) d = 0.5a

	a	b	c	d	e	f	g	h*
Right Tip	0.0009	0.0023	0.0033	0.0027	0.0015	-0.0009	-0.0008	-0.0001
Left Tip	0.0013	0.0030	0.0035	0.0028	0.0004	-0.0012	-0.0005	-0.00003

(b) d = 0.25a

	a	b	c	d
Right Tip	0.0003	0.0012	0.0077	0.0093
Left Tip	0.0002	0.0031	0.0141	0.0138

*Each step corresponds to the relative position shown in Fig. 4.28

TABLE 4.3 RELATIVE CRACK-TIP SLIDING DISPLACEMENT (μm)

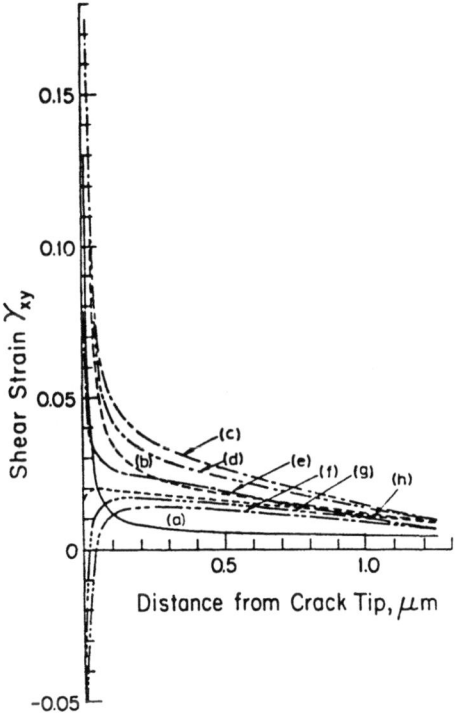

FIGURE 4.29 SHEAR STRAIN VERSUS DISTANCE FROM THE LEFT CRACK TIP WHEN PLASTIC ELEMENTS ARE USED. (A), (B) -- (H) CORRESPOND TO THE POSITION OF THE ASPERITY CONTACT RELATIVE TO THE CRACK TIP SHOWN IN FIG. 4.28

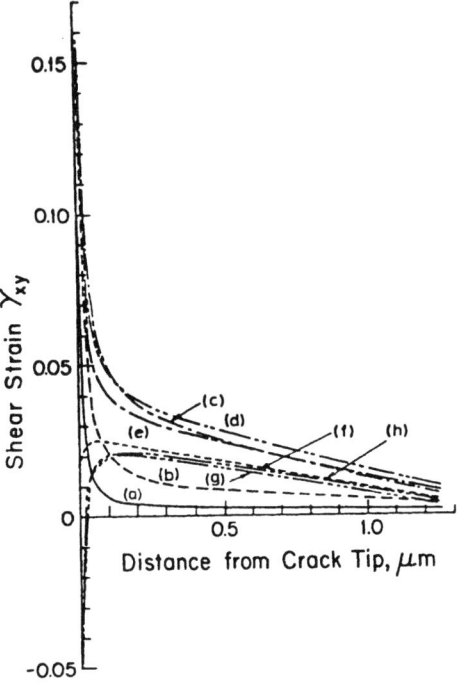

FIGURE 4.30 SHEAR STRAIN VERSUS DISTANCE FROM THE RIGHT CRACK TIP WHEN PLASTIC ELEMENTS ARE USED. (A), (B) -- (H) CORRESPOND TO THE POSITION OF THE ASPERITY CONTACT RELATIVE TO THE CRACK TIP SHOWN IN FIG. 4.28

The crack-opening displacement concept may be applied to the cracks in Mode II to determine the crack growth rate per cycle of loading. In this mode the relative sliding displacement occurs by means of slip due to crack tip deformation. If the maximum of this displacement is employed as a crack-tip sliding displacement (CTSD), denoted by ΔS, then the crack growth length, ΔC, may be expressed as

$$\Delta C = \Delta S - \Delta C_W \qquad (4.44)$$

where ΔC_W is the length of rewelding. As for ΔC_W, there is no satisfactory condition. According to Kikukawa et al, rewelding seems to be affected by environmental conditions only in Mode I[33]. In Mode II, they found that the length of rewelding, and therefore the crack growth length, does not differ discernably between the air and vacuum tests. Moreover, the ratio of $\Delta C/\Delta S$ was found to be nearly 0.16, which is fairly small in comparison with the ratio of $\Delta C/COD$ (0.55) for Mode I in their study.

As implied by Kikukawa et al's study, a crack can grow to the point where fracture really occurs, i.e., the fracture conditions are satisfied. These conditions may be addressed in terms of fracture strain, γ_f. If the material element at a distance r_f from the crack tip attains the fracture strain, actual fracture will occur up to this point. The strain γ_{xy} along the distance r from the crack tip can be written as

$$\gamma_{xy} = K_\varepsilon / r^m \qquad (4.45)$$

where K_ε is the strain intensity factor and m is the parameter which depends on the strain hardening exponent. For a nonhardening material and for the HRR field, m is unity. Therefore, r_f may be obtained by substituting γ_f for γ_{xy} in Eq. 4.45.

$$r_f = (K_\varepsilon / \gamma_f)^{-m} \qquad (4.46)$$

Figure 4.31 shows the shear strain distribution near the crack tip of cracks located at d/a = 0.25 and d/a = 0.5. If the contact length is 20 μm and if the fracture strain γ_f is assumed to be 0.4, the material within 0.0030 μm of the crack tip of the crack located at d = 5 μm will fracture, according to Figure 4.31. When the crack is located at d = 2.5 μm, the material within 0.02 μm of the crack tip will fracture.

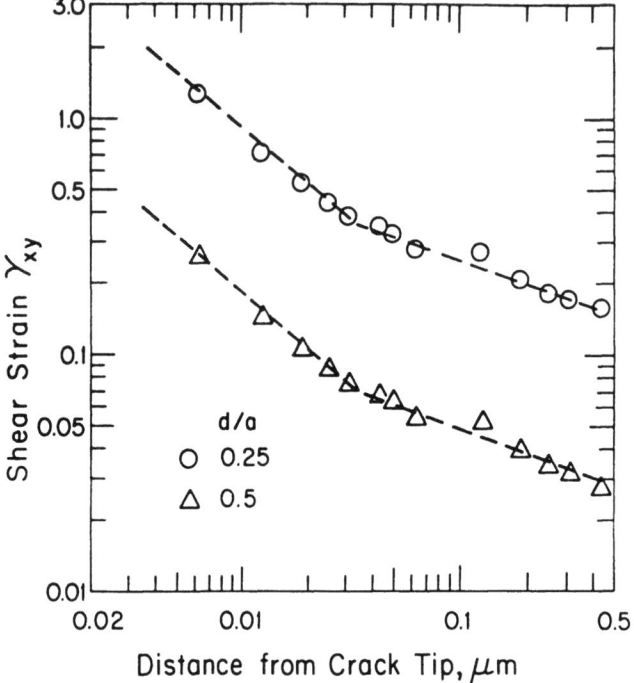

FIGURE 4.31 SHEAR STRAIN AS A FUNCTION OF DISTANCE FROM THE LEFT TIP FOR DIFFERENT DEPTHS OF CRACK LOCATION

4.4.4 Experimental Results

Crack propagation at the subsurface has been documented by Suh, et al[34]. Figures 4.32 and 4.33 show a typical micrograph of subsurface cracks in OFHC copper and iron solid solution, respectively. Figure 4.34 shows the early stages of crack propagation in annealed Fe-1.3%Mo. Cracks are not exactly parallel to the surface. This could be due to the loading condition, anistrophy of materials, and unique texture around second phases.

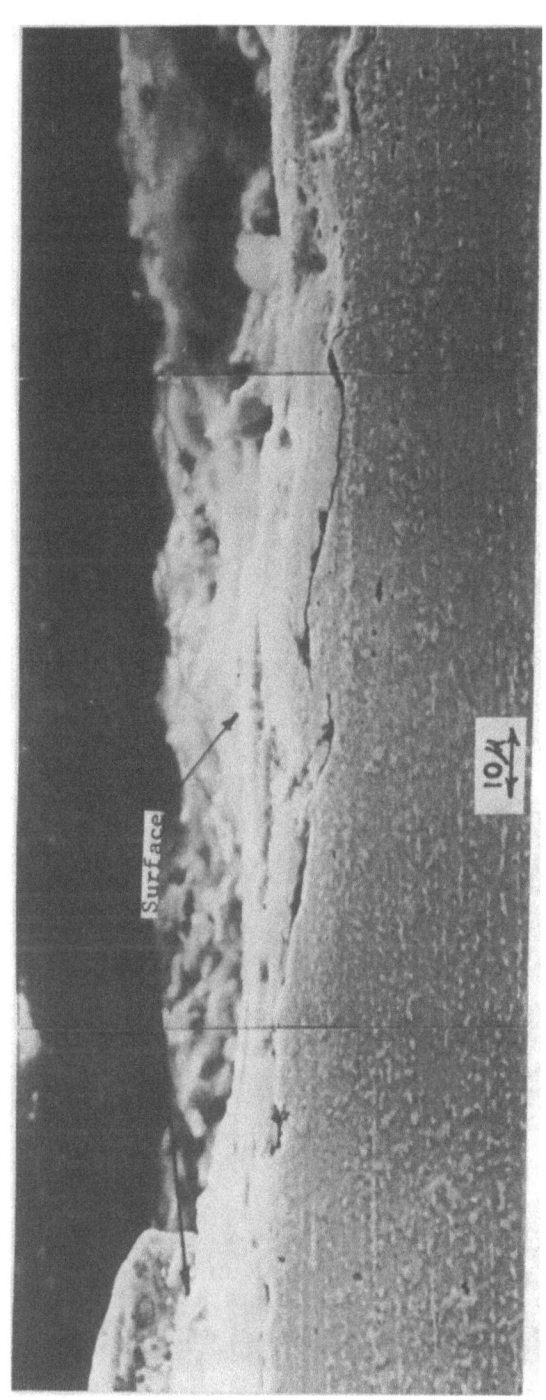

FIGURE 4.32 VOID AND CRACK FORMATION IN ANNEALED OFHC COPPER

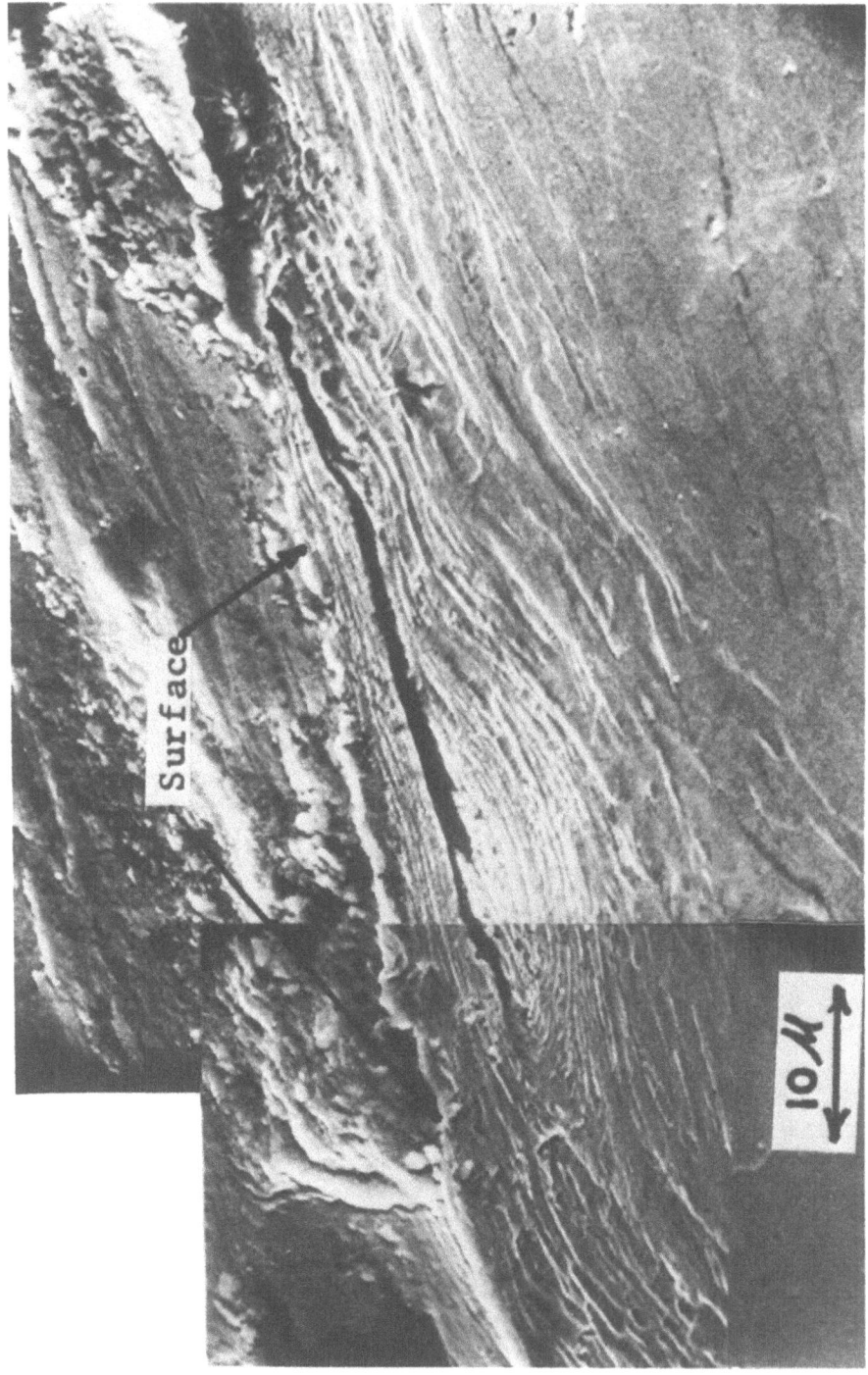

FIGURE 4.33 SUBSURFACE DEFORMATION AND CRACK FORMATION IN IRON SOLID SOLUTION

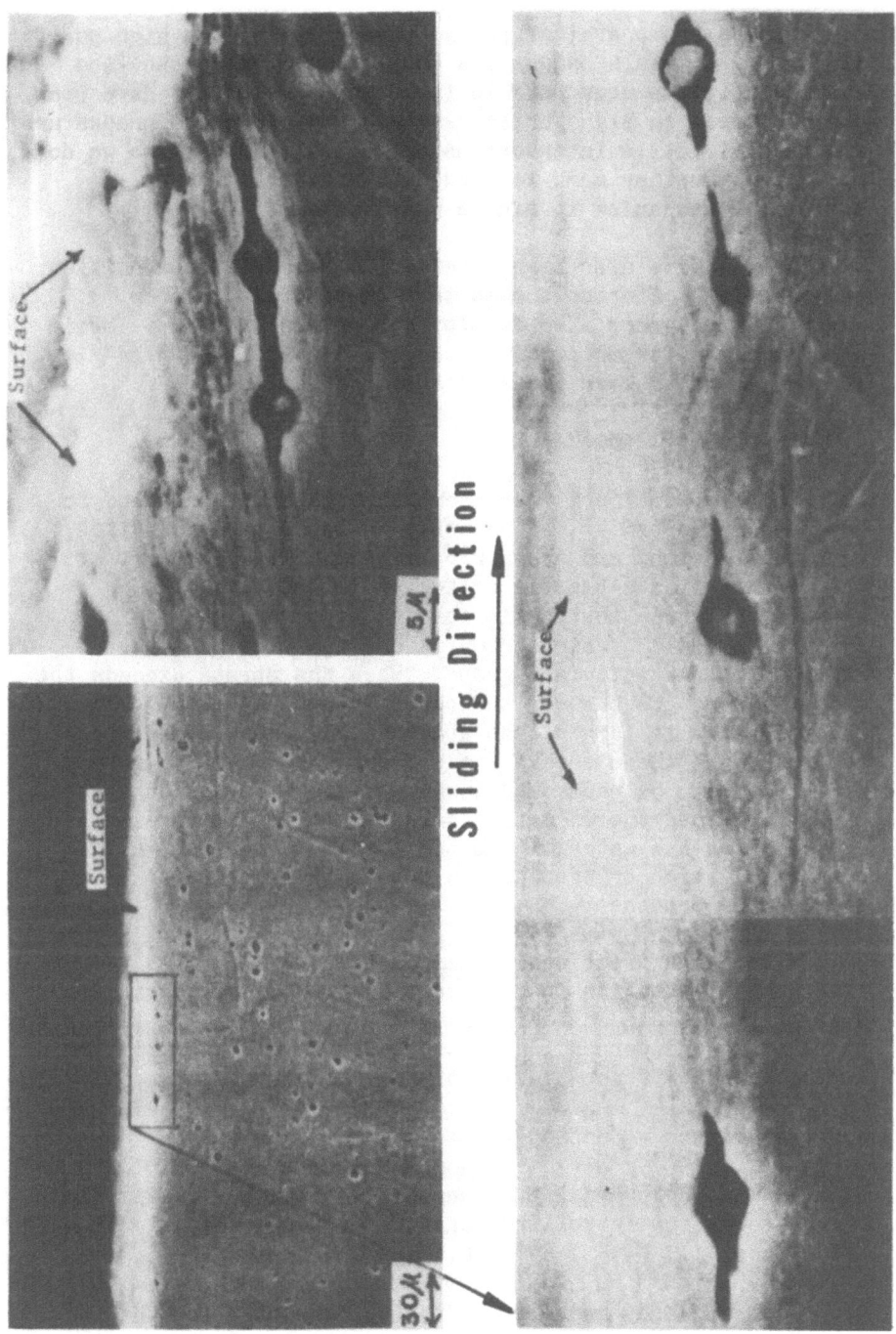

FIGURE 4.34 VOID FORMATION AROUND INCLUSION AND CRACK PROPAGATION FROM THESE VOIDS NEAR THE SURFACE IN ANNEALED Fe-1.3% Mo

Cracks are prevalent in two phase metals. In high purity single phase metals cracks are not a common occurrence and consequently the wear rate is low. However, cracks have been observed even in high purity metals. Whether these cracks are due to dislocation interactions or to small impurities we do not know. Further work is needed to understand crack nucleation mechanism in single phase metals.

Cracks have also been observed in polymers, composites, and ceramics. Clerico showed that chopped glass fiber reinforced polymers also develop subsurface cracks[35]. Swain also reported formation of subsurface cracks in alumina[36]. All these cracks lead to failure of the surface layer.

4.5 Concluding Remarks

In this section it was shown that materials respond to the forces exerted at the asperity contacts by deformation, crack nucleation, and crack propagation. It is shown that in the case of elasto-plastic solid, the plastic strain accumulates under the cyclic loading by moving asperities. With the plastic strain accumulation, the stress at the particle/matrix interface grows. When the stress exceeds the adhesive strength between the particle and the matrix and when the inclusion is larger than about 500°A, crack nucleation occurs. Once these cracks are present they can propagate which seems to be caused by large plastic deformation at the crack tip along the direction parallel to the surface. Depending on the material and the loading condition, either crack nucleation or propagation can be the wear rate controlling process.

Although a great deal of emphasis is given to elasto-plastic solids in sliding situations, cracks can also develop in brittle solids. In this case, cracks are likely to nucleate at and perpendicular to the surface. This mode of surface damage is not prevalent under sliding contact.

4.6 References - Section 4

1. Merwin, J.E. and K.L. Johnson, "An Analysis of Plastic Deformation in Rolling Contact," Proc. Institution Mech. Engrs. 177 (1963) 697-690.

2. Johnson, K.L. and J.A. Jefferis, "Plastic Flow and Residual Stresses in Rolling and Sliding Contact," Proc of Symposium on Fatigue in Rolling Contact, Inst. Mech. Engrs., (1963) 54-65.

3. Jahanmir, S. and N.P. Suh, "Mechanics of Subsurface Void Nucleation in Delamination Wear," Wear, 44 (1977) 17-38; also Jahanmir, S., Ph.D. Thesis, MIT, 1976.

4. Hertz, H., <u>Gesammtte Werke</u>, Vol. 1., Leipsig, Germany, 1895.

5. Smith, J.O. and C.K. Liu, "Stresses Due to Tangential and Normal Loads on an Elastic Contact with Application to Some Contact Problems," Journal of Applied Mechanics, Trans. A.S.M.E., 20 (1953) 157-166.

6. Gill, S., "A Process for the Step by Step Integration of Differential Equations in an Automatic Digital Computing Machine," Proc. Camb. Phil. Soc., 47 (1951) 90-108 and IBM, "Scientific Subroutine Package," Subroutine RKGS, 1975.

7. Sin, H.C., "Surface Traction and Crack Propagation in Delamination Wear," Ph.D. Thesis, MIT, 1981.

8. Anard, S.C., "Numerical Investigation of Stresses in the Inelastic Range in a Rolling Contact," Proc. Second Int. Conf. on Vehicle Structural Mechanics, Society of Automotive Engineers, 1977, pp. 121-127.

9. Augustsson, G., "Strain Field Near the Surface Due to Surface Traction," S.M. Thesis, MIT, 1974.

10. Suh, N.P., "An Overview of the Delamination Theory of Wear," Wear, 44 (1977) 1-16.

11. Argon, A.S., "Formation of Cavities From Nondeformable Second-Phase Particles in Low Temperature Ductile Fracture," J. Eng. Mat. Tech., Trans. ASME, (1976) 60-68.

12. Hirth, J.P. and D.A. Rigney, "Crystal Plasticity and the Delamination Theory of Wear," Wear, 39 (1976) 133-141.

13. McClintock, F.A., and A.S. Argon, "Mechanical Behavior of Materials," Addison-Wesley, Reading, MA, 1966.

14. Rhee, S.S. and F.A. McClintock, "On the Effects of Strain Hardening on Strain Concentrations," Proc. 4th Natl. Cong. Appl. Mech., ASME, (1962) pp. 10002.

15. Johanmir, S. and N.P. Suh, "Mechanics of Subsurface Void Nucleation in Delamination Wear," Wear, 44 (1977) 17-38.

16. Su, K.Y., "Void Nucleation in Particulate Filled Polymeric Materials and Its Implications on Friction and Wear Properties," Ph.D. Thesis, MIT, 1980.

17. Eshelby, J.D., "The Determination of the Elastic Field of an Ellipsoidal Inclusion, and Related Problems," Proc. Roy. Soc. A241 (1957) pp. 376-396.

18. Suh, N.P., "Wear Mechanisms: An Assessment of the State of Knowledge," Fundamentals of Tribology, (Ed. N.P. Suh and N. Saka), MIT Press, 1980.

19. Fleming, J.F. and N.P. Suh, "Mechanics of Crack Propagation in Delamination Wear," Wear, 44 (1977) 39-56.

20. Hills, D.A. and D.W. Ashelby, "On the Application of Fracture Mechanics to Wear," Wear, 54 (1979) 321-330.

21. Rosenfield, A.R., "A Fracture Mechanics Approach to Wear," Wear, 61 (1980) 125-132.

22. Keer, L.M., M.D. Bryant and G.K. Haritos, "Subsurface Cracking and Delamination," Solid Contact and Lubrication, A.S.M.E. Pulbication, A.M.D., (39) (1980) 79-95.

23. Sin, H.C. and N.P. suh, to be published.

24. Erdogan, F. and G.C. Sih, "On the Crack Extension in Plates Under Plane Loading and Transverse Shear," J. Basic Engr. 85 (1963) 519-527.

25. Sih, G.C., "Strain-Energy-Density Factor Applied to Mixed Mode Crack Problem," Int. J. Fracture, 10 (1974) 305-321.

26. Swedlow, J.L., "Criteria for Growth of the Angled Crack," Cracks and Fracture, ASTM STP 601, 1976, pp. 506-521.

27. McClintock, F.A., "Plastic Flow Around a Crack Under Friction and Combined Stress," Fracture 1977 (Eds. D.M.R. Taplin), Vol. 4, pp. 49-64, Pengamon, Oxford 1978.

28. Forsyth, P.J.E., "A Two Stage Process of Fatigue Crack Growth," Proc. Crack Prop. Symposium, Cranfield, Vol. 1, 1961, pp. 76-94.

29. Bathe, K.J., "ADINA - A Finite Element Program for Automatic Dynamic Incremental Nonlinear Analysis," Report 82448-1, Acoustics and Vibration Laboratory, MIT, 1975.

30. Bathe, K.J., "Static and Dynamic Geometric and Material Non-linear Analysis Using ADINA," Report 82448-2, Acoustics and Vibration Laboratory, MIT 1976.

31. Wheeler, D.R. and D.J. Buckley, "Texturing in Metals as a Result of Sliding," Wear, 33 (1975) 65-74.

32. Krause, H. and A.J. Demirci, "Texture Changes in the Running Surface of F.C.C Metals as the Result of Frictional Stress," Wear, 61 (1980) 325-332.

33. Kikukawa, M., M. Jono, and N. Adachi, "Direct Observation and Mechanism of Fatigue Crack Propagation," Fatigue Mechanisms, ASTM STP 675, 1979, pp. 234-253.

34. Suh, N.P., and Co-workers, "The Delamination Theory of Wear," Elsevier, 1977.

35. Clerico, M., "Sliding Wear Mechanisms of Polymers," Fundamentals of Tribology, (Eds. N.P. Suh and N. Saka), MIT Press, 1980.

36. Swain, M.V., "Microscopic Observations of Abrasive Wear of Polycrystalline Allumina," Wear, 35 (1975) 185-189.

5. SLIDING WEAR

5.1 Introduction

In Section 1, various wear processes have been briefly reviewed. It was stated that there are two classes of wear: one controlled by the mechanical behavior of materials and the other by the chemical behavior. One of these mechanical behavior controlled wear processes is sliding wear, which is often called inappropriately, adhesive wear in the literature. The purpose of this section is to examine various ways wear particles can be generated when two surfaces slide against each other. In this section, the discussion will be confined to those cases where the interfacial temperature rise is not high enough to cause wear processes controlled by the chemical behavior of materials.

Wear particles in elasto-plastic solids are generated during sliding by the following three mechanisms:

1) Asperity deformation -- Orginally existing surface asperities and the asperities generated due to the delamination wear can be removed upon deformation. They may be removed in a single asperity interaction or in multiple, repeated interactions.

2) Plowing -- Surfaces can be plowed by wear particles generated by asperity removal or by delamination. Hard particles from the environment can also plow the surface. Surfaces can also be plowed by hard asperities of the counterface. The consequence of plowing can result in the generation of chips by a cutting action or the formation of ridges which deform and get removed due to subsequent loadings.

3) Delamination -- Large wear particles are removed by the process of plastic deformation of the surface layer, and subsurface crack nucleation and propagation. This process of wear particle generation will be discussed in great detail.

Wear particles in brittle solids are generated by different mechanisms from those of elasto-plastic solids. They are normally produced by cracks running perpendicular to the surface starting from minute surface defects or subsurface defects very near the surface. These cracks then change direction and small particles are created.

Normally brittle solids, such as aluminum oxide and even glass, sometimes act as elastic-plastic solids when extremely small chips are cut from the surface. This size effect has been observed in many different situations as in grinding, sliding, and hydrostatic deformation of materials.

In Section 2 it was shown that the surface is highly heterogeneous if not contaminated. This fact was supported from experimental observations, and through quantum mechanical and thermodynamic reasoning. Therefore, it was emphasized that ashesion is normally difficult to achieve except when fresh new surfaces are exposed during asperity contacts.

In Section 3, the mechanism of crack nucleation was examined in depth, whereas in Section 4 crack propagation was investigated from a continuum mechanics point of view.

The purpose of this section is to describe the delamination theory of wear both for metals and polymers. This section will also present the means of reducing wear caused by sliding.

5.2 The Delamination Theory of Wear

5.2.1 Description of the Theory

The delamination theory of wear was introduced in 1972 in order to explain the wear of metals and other solid materials[1]. The theory has since been supported by a large number of experimental results[*,2]. The significance and implications of the mechanisms postulated by the delamination theory of wear can best be understood by considering the following questions:

1) Where does all the energy supplied by the external agent go?

2) Why and how does the coefficient of friction affect the wear rate?

3) Why do some hard metals wear faster than soft metals?

4) Why do most wear particles have an aspect ratio greatly different from unity?

5) Why does seizure occur?

6) How does the microstructure of metals affect the wear rate?

7) How do initial surface roughness and waviness influence the wear phenomenon?

The delamination theory of wear describes the following sequential (or independent if there are preexisting subsurface cracks) events which lead to loose wear sheet formation[1].

(1) When two surfaces come into contact, normal and tangential loads are transmitted through the contact points. Asperities of the softer surface are easily deformed and fractured by the repeated loading action, forming small wear particles. Hard asperities are also removed but at slower rates. A relatively smooth surface is initially generated, either when these asperities are deformed or when they are removed.

---

* The original paper is one of the most frequently cited papers in applied science according to the Institute for Scientific Information[3].

(2) The surface traction exerted by the harder asperities at the contact points, induces incremental plastic deformation per cycle of loading, accumulating with repeated loading. The increment of permanent deformation remaining after given cyclic loading is small compared with the total plastic deformation that occurs in that cycle due to the reversal of shear deformation.

(3) As the subsurface deformation continues, cracks are nucleated below the surface. Crack nucleation very near the surface cannot occur due to the triaxial state of compressive loading which exists just below the contact region.

(4) Once cracks are present (either by crack nucleation or from preexisting voids and cracks), further loading and deformation causes the cracks to extend and propagate, eventually joining with neighboring cracks. These cracks tend to propagate parallel to the surface at a depth governed by material properties and the state of loading. When the cracks cannot propagate because of the small stress concentration at the crack tip due to extremely small surface traction at the asperity contact, crack nucleation is the rate controlling mechanism.

(5) When the cracks finally shear to the surface, long and thin wear sheets delaminate. The thickness of the wear sheet is controlled by the location of subsurface crack growth, which is controlled by the normal and tangential loads at the surface.

A series of experimental studies conducted at M.I.T. have substantiated this theory, showing that the delamination process initiates when the subsurface plastic deformation causes the nucleation of voids, Figure 5.1[2]. With further deformation, these voids elongate and link up to form long cracks in a direction nearly parallel to the wear surface, Figure 5.2. At a critical length, these cracks shear to the surface, yielding a wear particle in the form of a long thin sheet as shown in Figure 5.3. The top surface of the wear sheet is generally smooth, while the fractured surface is rough sometimes showing dimples, Figure 5.4, and in some cases, shear failure surfaces. These experimental observations have been explained in terms of the theoretical models presented in the preceeding chapters.

5.2.2 A Model for Delamination Wear

A wear equation based on the delamination theory can readily be developed. For this purpose, consider a subsurface crack lying below the surface as shown in Figure 5.5. An

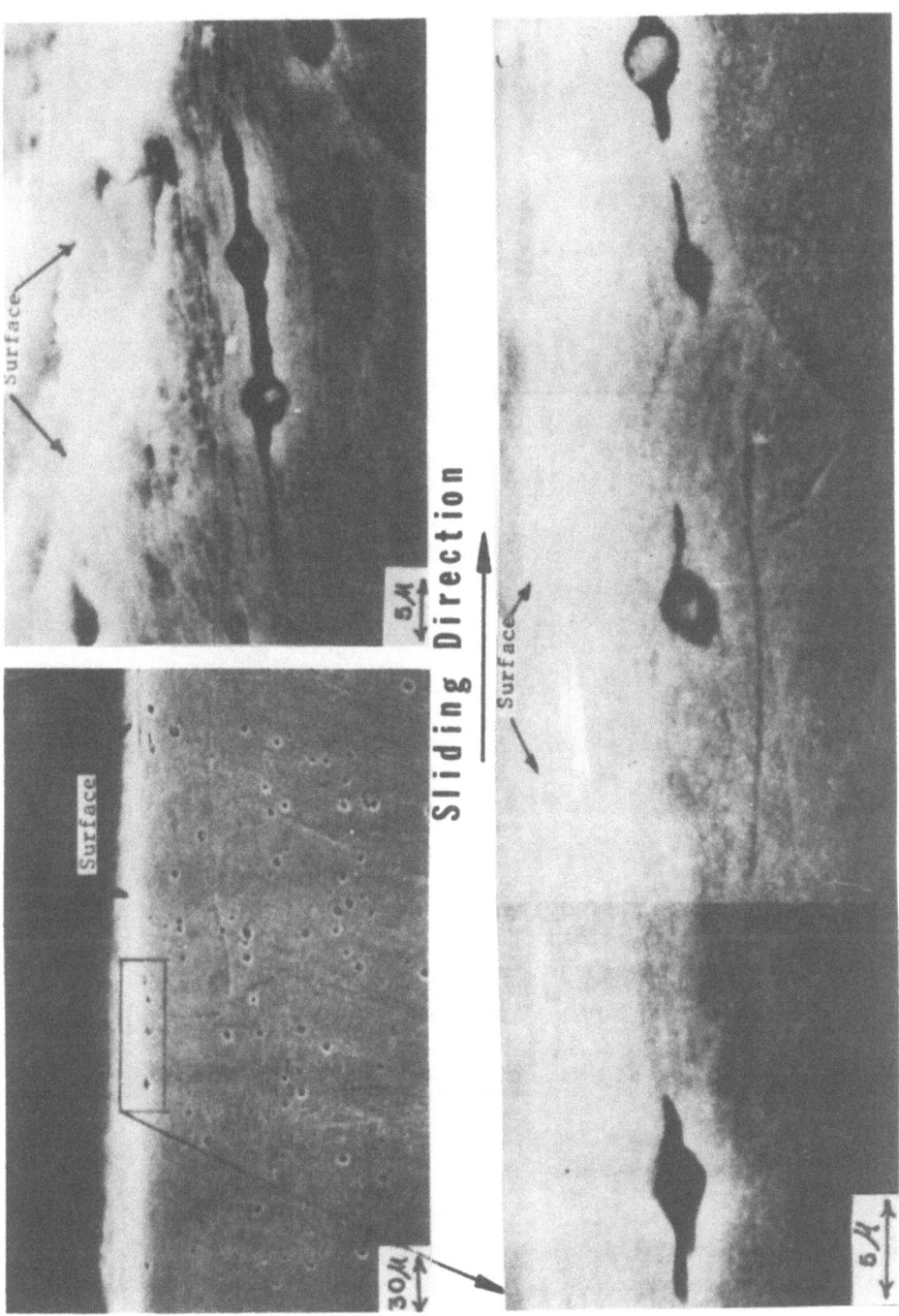

FIGURE 5.1 VOID FORMATION AROUND INCLUSIONS AND CRACK PROPAGATION FROM THESE VOIDS NEAR THE SURFACE IN ANNEALED Fe-1.3%Mo

FIGURE 5.2   SUBSURFACE DEFORMATION AND CRACK FORMATION IN
IRON SOLID SOLUTION

173

FIGURE 5.3 WEAR SHEET FORMATION IN IRON SOLID SOLUTION

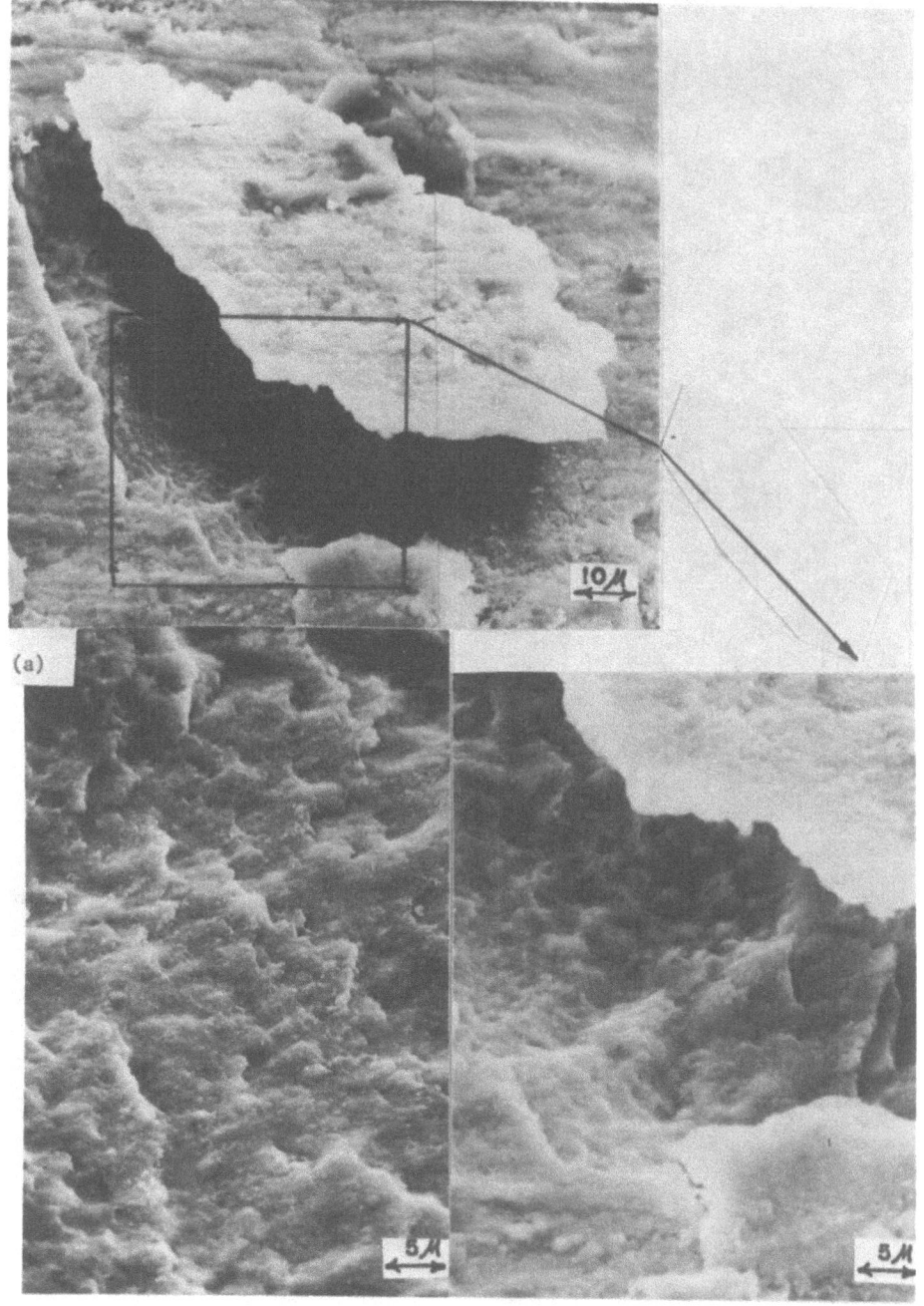

FIGURE 5.4 WORN SURFACE OF PURE IRON: (A) WEAR SHEET FORMATION;
(B) SHEAR DIMPLES BENEATH THE WEAR SHEET IN (A);
(C) DIMPLED APPEARANCE OF A WEAR CRATER

asperity is moving over the surface from left to right. The wear rate will be dictated by the crack propagation at both ends, L and R. The crack propagation for the ith cycle, ΔC_i, may be expressed as:

$$\Delta C_i = f(\mu, d, C, \text{material properties}) \qquad (5.1)$$

for both ends. If N is the total number of asperity passes required for removal of one layer, the volume V_1 for one crack of width, w, lying at a depth d is obtained as

$$V_1 = w \cdot d \cdot \sum_{i}^{N}(\Delta C_{L_i} + \Delta C_{R_i}) \qquad (5.2)$$

Therefore, the total volume, V, for one layer may be given by

$$V = N_c \cdot N_w \cdot w \cdot d \cdot \sum_{i}^{N}(\Delta C_{L_i} + \Delta C_{R_i}) \qquad (5.3)$$

where N_c is the number of cracks along the sliding direction and N_w is the number of wear sheets in the direction of contact width. Since $N_w \times w$ is in the order of the contact width L_w, the volume V becomes

$$V = N_c \cdot L_w \cdot d \cdot \sum_{i}^{N}(\Delta C_{L_i} + \Delta C_{R_i}) \qquad (5.4)$$

Let λ be the asperity contact spacing, ℓ_c the crack spacing, D the diameter of a specimen, and L the contact length as shown in Figure 5.6. Using this model the number N_c and N can be determined for a specimen and a slider, respectively.

Specimen:

Since the number of asperities per length of contact is L/λ, the number N is determined as

$$N = \frac{S}{\pi D} \frac{L}{\lambda} \qquad (5.5)$$

where S is the sliding distance required for removal of one layer. Also, N_c is given by

$$N_c = \pi D/\ell_c \qquad (5.6)$$

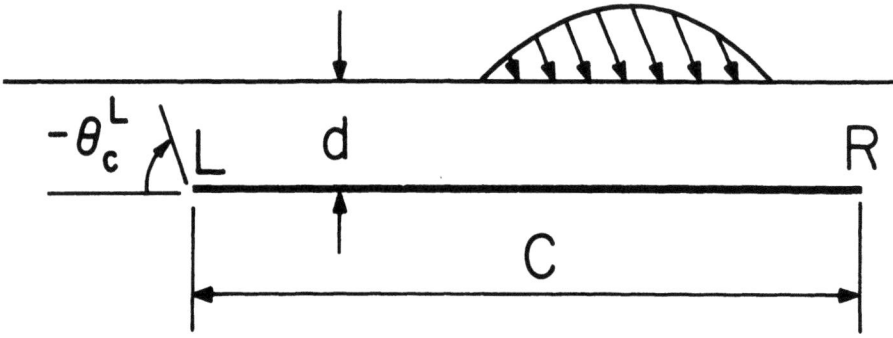

FIGURE 5.5 A SUBSURFACE CRACK UNDER A MOVING ASPERITY

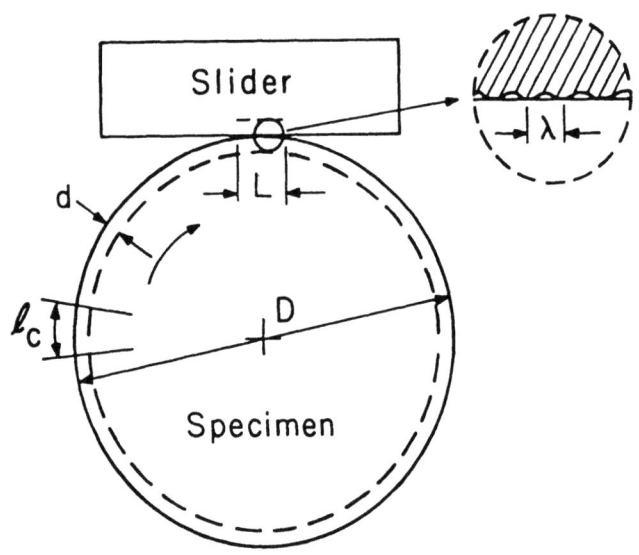

FIGURE 5.6 A MODEL OF WEARING SPECIMEN AND SLIDER

The wear rate is, by assuming $L_w = L$, finally obtained as

$$\frac{V}{S} = \frac{L^2 \cdot d \cdot (\Delta \bar{C}_L + \Delta \bar{C}_R)}{\lambda \cdot \ell_c} \qquad (5.7)$$

where $\Delta \bar{C}_L$ and $\Delta \bar{C}_R$ are the average crack propagation rates during N cycles.

<u>Slider</u>:

The numbers N and N_c are given by

$$N = S/\lambda \qquad (5.8)$$

and

$$N_c = L/\ell_c \qquad (5.9)$$

respectively. Therefore, the wear rate for a slider is obtained as

$$\frac{V}{S} = \frac{L^2 \cdot d \cdot (\Delta \bar{C}_L + \Delta \bar{C}_R)}{\lambda \cdot \ell_c} \qquad (5.10)$$

which is the same expression as given in Eq. 5.7 for a specimen. The wear coefficient K can be obtained by using Archard's equation as

$$K = \frac{3H \cdot L^2 \cdot d \cdot (\Delta \bar{C}_L + \Delta \bar{C}_R)}{W \cdot \lambda \cdot \ell_c} \qquad (5.11)$$

where H is the hardness of the material and W is the applied load.

Eqs. 5.7 and 5.11 show that the wear rate is directly proportional to the depth of crack and the average crack growth. The amount of crack growth, ΔC, depends on the depth, d. The results in Section 4 indicate that ΔC increase with decreasing d for a given friction coefficient of 0.25. For higher μ, the same tendency is expected.

ΔC depends on d because of shear strain concentration at crack tips. Under the moving load condition, it is highly likely that the plastic deformation is much more severe at the crack tips near the surface due to sliding between crack surfaces even for low friction. However, cracks are only observed at a finite distance below the surface, because they cannot nucleate very near the surface because of high triaxial state of compressive stress, as shown in Section 3. The actual location of the crack will also be determined by the location of second phase particles around which crack nucleation occurs preferentially.

For a longer crack, a much larger shear concentration is expected due to sliding of crack surfaces with respect to each other, which is affected by the shear traction between the crack surfaces. Therefore, the crack propagation rate will increase with increasing total crack length.

When the total crack length reaches a critical size for unstable crack propagation leading to fracture, the direction of crack propagation will change toward the surface, eventually forming a wear sheet. Fracture criteria for the combined mode of loading considered in Section 4, may provide a qualitative explanation for this. Let us consider a subsurface crack under a moving asperity shown in Figure 5.5. If the left tip, L, is in the tensile zone behind the load, then the maximum hoop stress and the minimum strain-energy-density factor criteria predict the crack growth of about $-70°$ at the left tip. When the critical condition is reached with the tensile component, the crack will be fractured to the surface. Micrographs of wear sheets about to delaminate show that subsurface cracks always reach the surface at the trailing edge of the crack[3].

Material properties such as strain-hardening property are the important factors affecting crack propagation, ΔC. Also, the anisotophy of the properties are expected to affect ΔC. Due to repeated loading effects of cyclic hardening, cyclic softening, and the Bauschinger effects are also important. However, it seems impossible at the present time to incorporate all these factors in the analysis because of the extreme complexity. The fracture strain is the simplest parameter that can be used in determining ΔC.

Experimentally, it is well known that the wear behavior of materials strongly depends on the coefficient of friction[4, 5]. A high shear strain distribution is expected ahead of extending cracks when the coefficient of friction is high. Therefore, the crack propagation rate is larger for higher friction coefficients. The exact relation between friction and wear cannot be given because of limited data. However, the elastic solutions obtained in Section 4, although they have already been shown to be inadequate, may be used to discuss the functional relation qualitatively.

In Figures 5.7 and 5.8, the maximum Mode II stress intensity factor for a given d are plotted as a function of the coefficient friction, μ,. It can be seen that the relation between K_{II} and μ is approximately linear for a given d. If we assume that Paris' crack growth law is still valid in this case, then the crack growth, ΔC, may be expressed as

$$\Delta C \sim (a\mu)^n \qquad (5.12)$$

where a is a constant. Since the wear volume is determined by the amount of the total ΔC accumulated, wear rates are a power function of friction. The exponent, n, is a constant which varies from 2 to 4 depending on material for ΔK values in stage II. This should give a rough idea about a functional relation between friction and wear, although the foregoing discussion is very approximate.

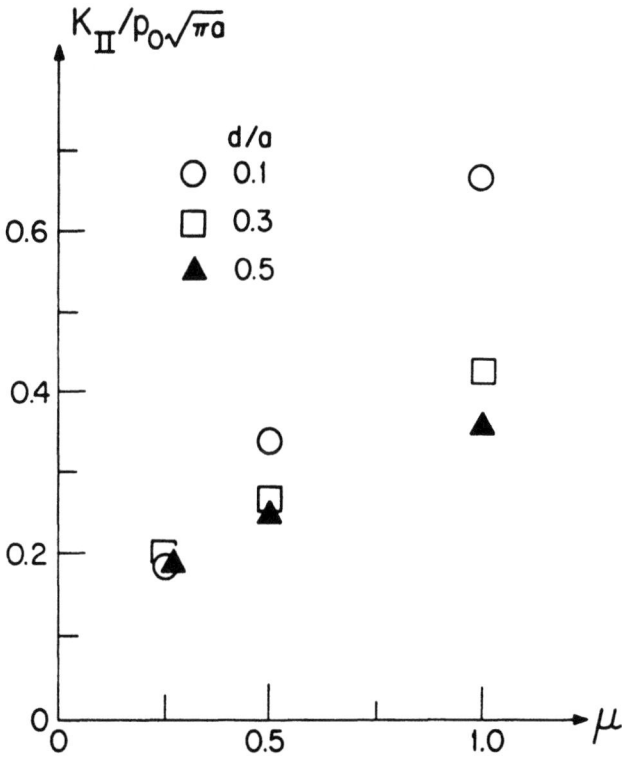

FIGURE 5.7   NORMALIZED ΔK_{II} AS A FUNCTION OF μ: c = 1/4a; LEFT TIP

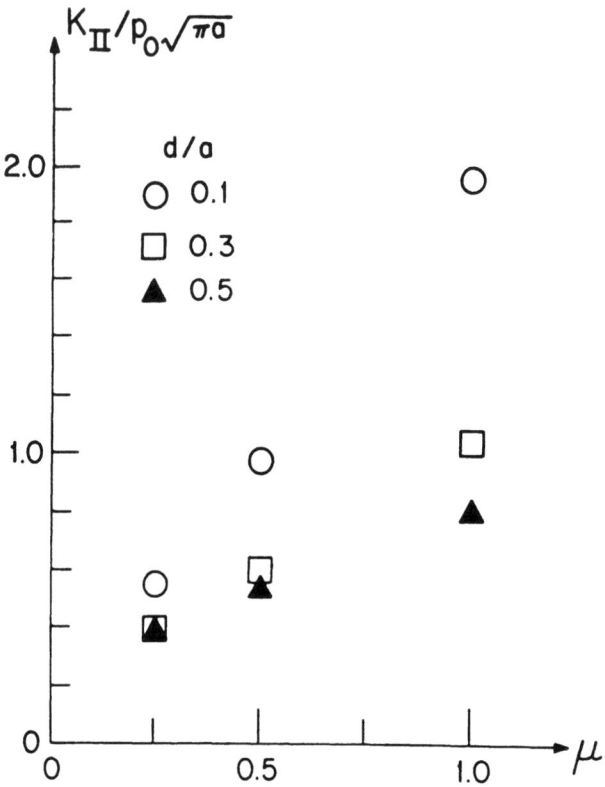

FIGURE 5.8   NORMALIZED ΔK_{II} AS A FUNCTION OF μ: $c = 3a$

5.2.3 Numerical Examples: Wear Coefficient Prediction

The wear model for delamination wear, i.e., Eqs. (5.7) and (5.11), may be used to predict wear coefficients for typical sliding wear situations. Two examples are given here.

Example 1:

The crack tip sliding displacement for $d = 5$ μm is 0.0035 μm (Table 4.3). If $\Delta \bar{C}_L = \Delta \bar{C}_R = 0.0035$ μm, and the apparent contact length $L = 2$ mm and both the asperity contact spacing, λ, and the crack spacing, ℓ_c, are assumed to be 100 μm, then

$$\frac{V}{S} = \frac{(2 \times 10^{-3})^2 (5 \times 10^{-6})(0.007 \times 10^{-6})}{(100 \times 10^{-6})(100 \times 10^{-6})} =$$

$$= 1.4 \times 10^{-11} \text{ m}^3/\text{m}$$

and for $H = 100$ kg/mm^2 and $W = 1$ kg,

$$K = \frac{3 \times 100 \times 10^6 \times 1.4 \times 10^{-11}}{1} = 4.2 \times 10^{-3}$$

Example 2:

For $d = 2.5$ μm, by taking $\Delta C_L = \Delta C_L = 0.014$ μm again from Table 4.3,

$$\frac{V}{S} = \frac{(2 \times 10^{-3})^2 (2.5 \times 10^{-6})(0.028 \times 10^{-6})}{(100 \times 10^{-6})(100 \times 10^{-6})}$$

$$= 2.8 \times 10^{-11} \text{ m}^3/\text{m}$$

and

$$K = 8.2 \times 10^{-3}$$

According to the experimental results, K is between 10^{-2} and 10^{-4} in most cases, which indicates that the model can predict wear behavior fairly well.

5.3 Microstructural Effects in Delamination Wear

The microstructure of metals affect the wear behavior of metals a great deal. Hardness and toughness are generally affected by the microstructure which in turn affect the wear properties. It has been well established that the hardness and the topography of the surface affect the number of asperity contacts and the size of each individual contact, in addition to the resistance to deformation of the surface layer[6,7]. As we have seen in Section 4, the toughness of metals is closely related to the crack propagation rate, which also effects the rate of wear sheet formation.

Several different aspects of the microstructure of metals affect the wear process: grain size; the properties, volume fraction, size, and distribution of second phase particles; and texture of the surface layer. Each of these aspects will be discussed in this subsection.

From the friction and wear point of view the ideal material that has a low coefficent of friction and a low wear rate is a hard material in which cracks cannot be nucleated and chemically inert. Such a material then must be single-phase without any impurities and tough. However, many engineering materials cannot be made both tough and hard. Hard materials generally have low toughness. Therefore, the choice of materials for tribological application must be made through compromise.

The wear of single phase polycrystalline metals was investigated at MIT[8-10]. OFHC copper, copper-chromium (0.58 and 0.81 at % Cr), copper-silicon (2.3 and 8.6 at % Si), and copper-tin (1.4, 3.4 and 5.7 at % Sn) alloys were tested. Table 5.1 gives the heat-treatment conditions, grain size, and chemical composition of these materials. Figure 5.9 shows typical microstructures of these materials. Figure 5.10 shows the Vickers hardness of the solid solutions as functions of the solute content. It can be seen that different elements produce different hardening for the same atomic content of solute. The effectiveness of tin increasing the hardness of copper should be noted.

Alloy	Composition		Recryscallization treatment		grain size
	Wt.%	At.%	Temperature(C)	Time (min)	(μm)
OFCH Cu	--	--	360	250	28
Cu-Sn	2.5	1.4	790	120	35
Cu-Sn	6.0	3.4	790	60	40
Cu-Sn	10.0	5.7	790	30	30
Cu-Si	1.0	2.3	850	300	75
Cu-Si	4.0	8.6	850	300	90
Cu-Cr	0.47	0.58	1070	5	450
Cu-Cr	0.66	0.81	1070	5	485

*The purity of the materials used was: OFHC 99.98%; Sn 99.89%; Si 99.9999%.

TABLE 5.1 EXPERIMENTAL MATERIALS: COPPER SOLID SOLUTIONS*

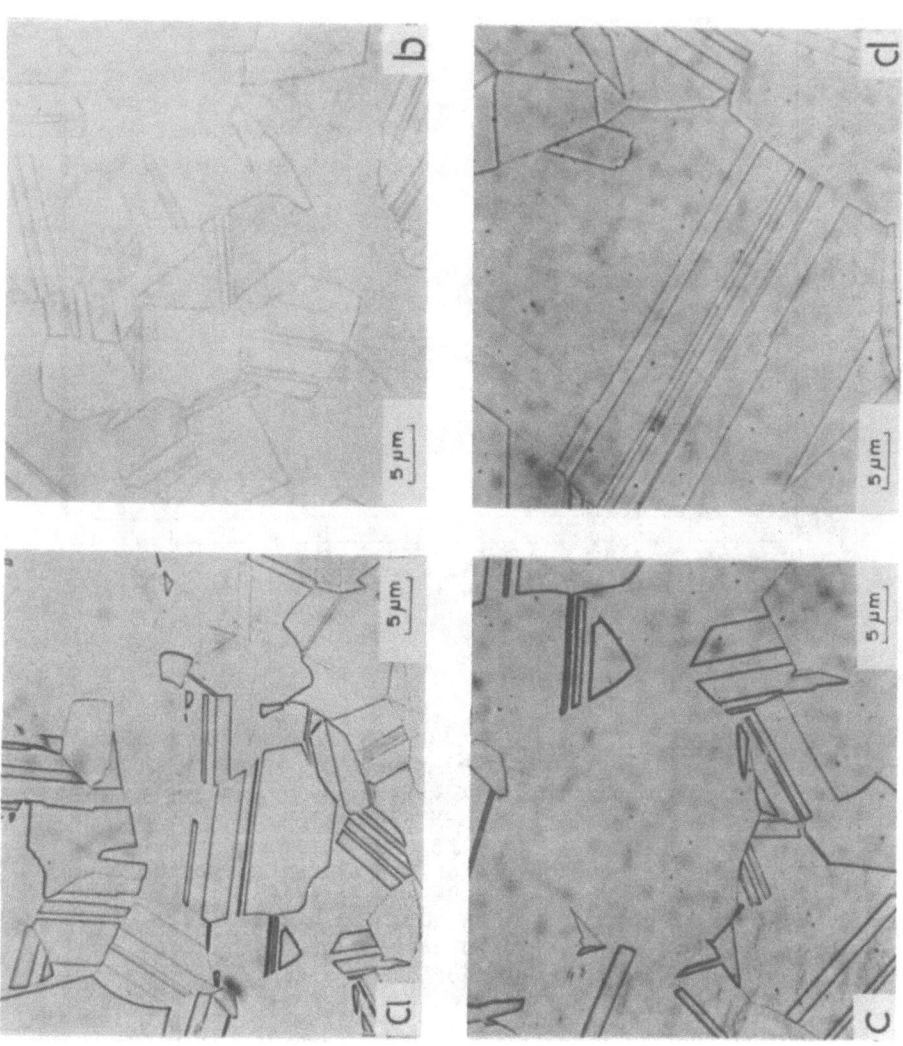

FIGURE 5.9 OPTICAL MICROGRAPHS SHOWING THE REPRESENTATIVE MICROSTRUCTURES OF SOLID SOLUTIONS: (A) OFHC COPPER, (B) Cu-5.7 at % Sn, (C) Cu-8.6 at % Si AND (D) Cu-0.81 at % Cr.

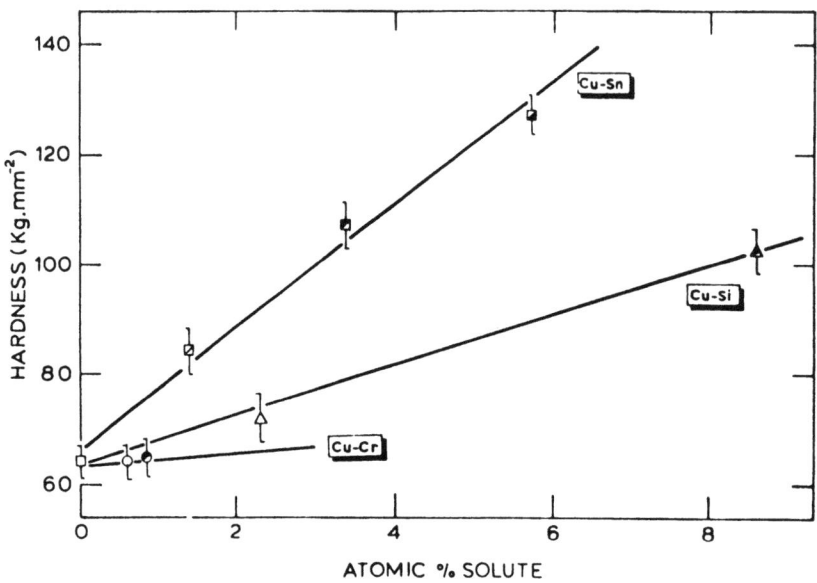

FIGURE 5.10 THE VICKERS MICROHARDNESS UNDER A 200 g LOAD VERSUS THE ATOMIC PERCENT OF SOLUTE. EACH POINT REPRESENTS AN AVERAGE OF FIVE MEASUREMENTS AND THE BARS REPRESENT THE STANDARD DEVIATION

An immediate consequence of the alloying was the change in the friction coefficient as shown in Figure 5.11. The harder Cu-Sn solution seems to have lower coefficients of friction than the Cu-Si solutions. Nevertheless, the difference in friction coefficient is not large enough to account for the large difference seen in wear rates which are shown in Figure 5.12. Figure 5.13 are SEM micrographs of some of the materials tested. The sliding direction in these micrographs is from the right to the left. The micrographs clearly show that delamination wear sheets have formed. It is also interesting to note that the top surface of the particles is very smooth. Figure 5.14 contains SEM micrographs of the subsurface of the specimens shown in Figure 5.13. The sliding direction is again from right to left. These micrographs indicate that cracks tend to propagate parallel to the surface and then extend to the surface. Although the angle of the crack, when the crack begins to propagate to the surface, is

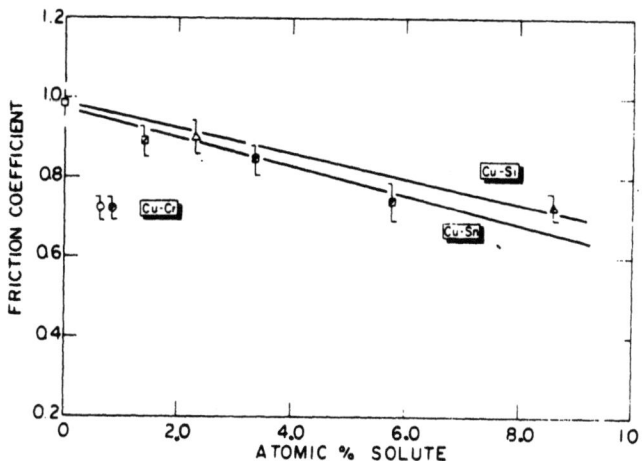

FIGURE 5.11 THE FRICTION COEFFICIENT AS A FUNCTION OF THE ATOMIC CONTENT OF SOLUTE. THE FRICTION COEFFICIENT WAS CALCULATED USING THE STEADY STATE TANGENTIAL FORCE. THE SLIDING SPEED WAS 2 m min^{-1}

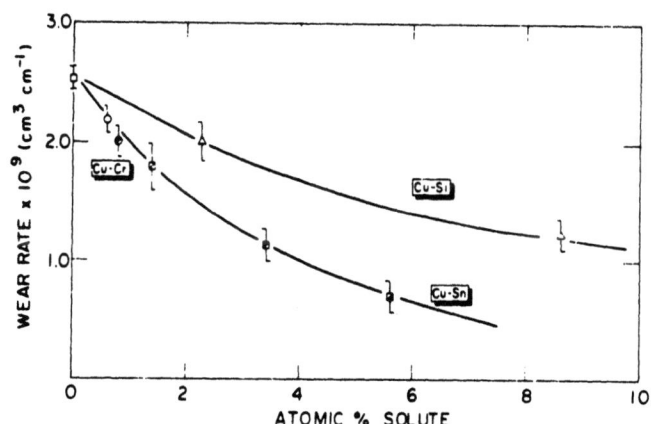

FIGURE 5.12 THE WEAR RATE AS A FUNCTION OF THE ATOMIC PERCENT OF SOLUTE. THE NORMAL LOAD WAS 2 kg AND THE DURATION OF THE TESTS WERE 100 MIN

FIGURE 5.13 SCANNING ELECTRON MICROGRAPHS OF WEAR TRACKS (A) OFHC COPPER, (B) Cu-5.7 AT % Sn, (C) Cu-8.6 AT % Si AND Cu-0.81 AT % Cr. THE NORMAL LOAD WAS 2 kg AT A SLIDING SPEED OF 2 m min^{-1} AND THE SLIDING DISTANCE WAS 200 m

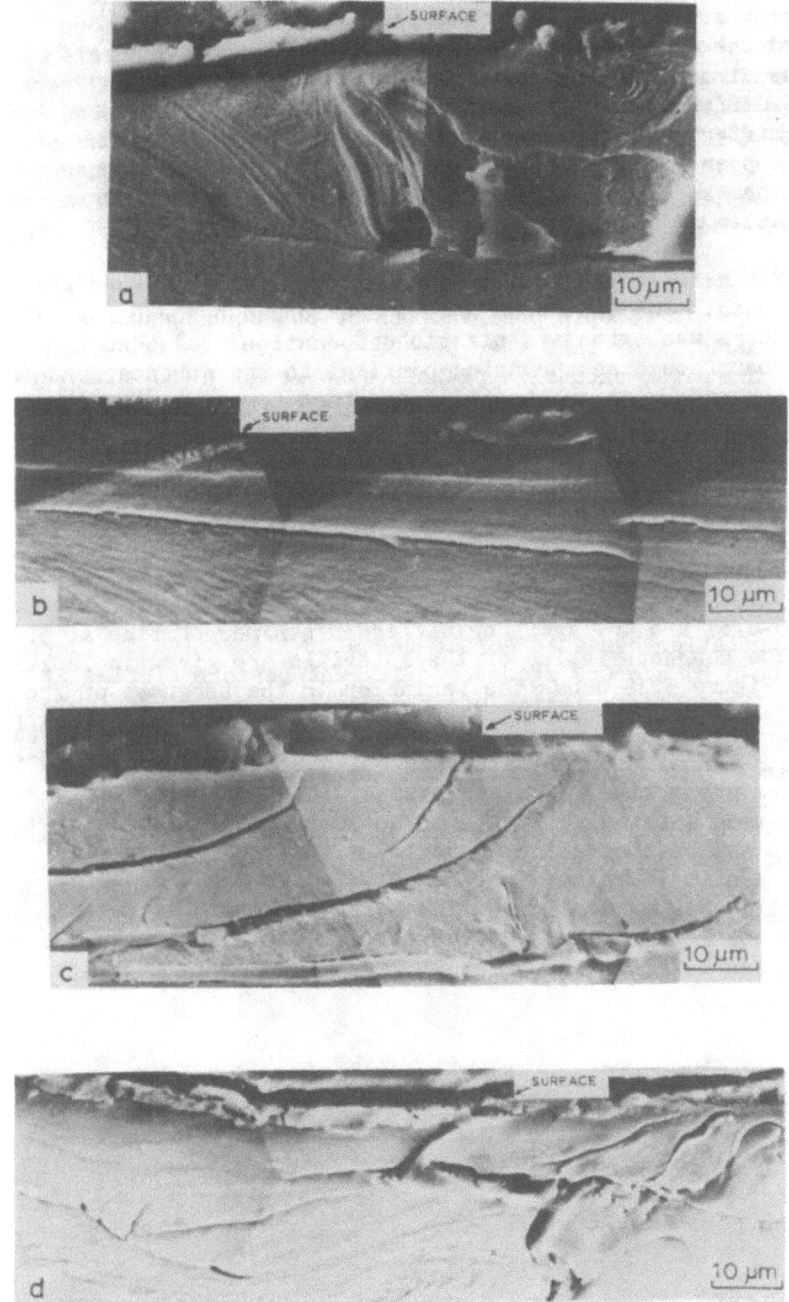

FIGURE 5.14 SCANNING ELECTRON MICROGRAPHS OF THE SUBSURFACE. THE MATERIALS AND TEST CONDITIONS ARE THE SAME AS IN THE PREVIOUS FIGURE

not always 70°, but it is interesting to note that most cracks change their direction as they approach the surface, more toward the direction perpendicular to the maximum normal stress direction. The depth at which cracks propagate seems to be different for specimens with different compositions. This difference is more striking between the micrographs of OFHC copper and Cu-5.7 a/o Sn solid solution; OFHC copper shows cracks propagating at a depth of the order of 50 μm, and the latter at a depth of the order of 15 μm.

The microscopic observations of the worn specimens show four facts: (a) that wear sheets are formed by cracking; (b) that there was extensive plastic deformation; (c) that there are subsurface cracks running parallel to the surface; and (d) that the cracks at the trailing end turn toward the surface. All of these observations are in accordance to the delamination theory of wear. These micrographs do not however, indicate how the cracks are initiated.

The effects of having a two phase structure on tribological behavior were investigated at MIT, using precipitation-hardened copper-chromium alloys (Cu-0.58 a/o Cr and Cu-0.81 a/o Cr) aged for different periods of time at 500° C[9]. The characteristics of the materials are given in Table 5.2. Figure 5.15 shows the variation in the hardness of the copper-chromium super-saturated solid solutions as a function of the aging time. The hardness of these materials initially increased with aging time and then decreased. The maximum value was reached after about 100 minutes of aging. The hardness is 65 kg/mm^2 for the solid solutions and 140 kg/mm^2 for the peak aged alloys. The aging time for the maximum hardness is different for the two alloys, whereas the maximum hardness is about the same.

Experimental materials

Parameter	Alloy	Aging time (min)		
		100	1 000	10 000
Volume fraction, $V_v \times 10^3$	Cu-0.58 Cr	5.19	5.25	5.31
	Cu-0.81 Cr	6.96	6.97	7.09
Mean free path λ (μm)	Cu-0.58 Cr	68.9	70.49	71.8
	Cu-0.81 Cr	51.84	53.02	53.00
Particle size d (μm)	Cu-0.58 Cr	0.54	0.55	0.58
	Cu-0.81 Cr	0.55	0.56	0.58

TABLE 5.2 EXPERIMENTAL MATERIALS

Figure 5.16 shows the friction coefficient and the wear rate as functions of the aging time. The friction coefficient is fairly constant for all treatment times for both alloys. The increase in hardness, resulting from the aging treatment, does not seem to affect the friction coefficient, probably because the hardness of the slider was much greater than the specimens. The wear rate initially decreases by a factor of three for both Cu-Cr alloys and then increases approximately linearly; the slope seems to be the same for both alloys. The minimum wear rate does not correspond to the maximum hardness. (In these figures the peak hardness is indicated by arrows A and B for Cu-0.58 Cr and Cu-0.81 Cr, respectively). Figure 5.16 also gives a plot of the wear coefficient as a function of aging time, which shows that it increases rapidly after 5 minutes of aging and then levels off asymptotically to a constant value.

As the microstructural differences between the two Cu-Cr alloys are characterized by the volume fraction and the mean free path of particles, the wear resistance (the inverse of wear coefficient) is plotted as a function of the volume

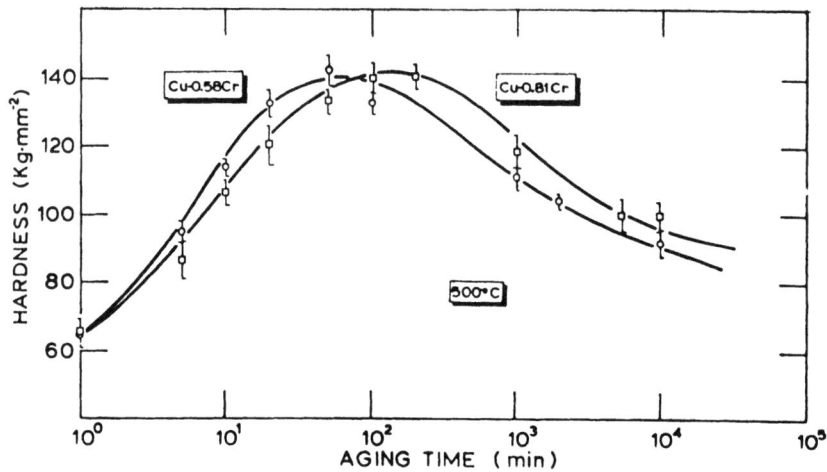

FIGURE 5.15   THE VICKERS MICROHARDNESS UNDER A 200 g NORMAL LOAD AS A FUNCTION OF AGING TIME. THE SPECIMENS WERE SUBJECTED TO AN AGING TREATMENT AT 500°C AND WERE WATER QUENCHED AT THE END OF THE TREATMENT

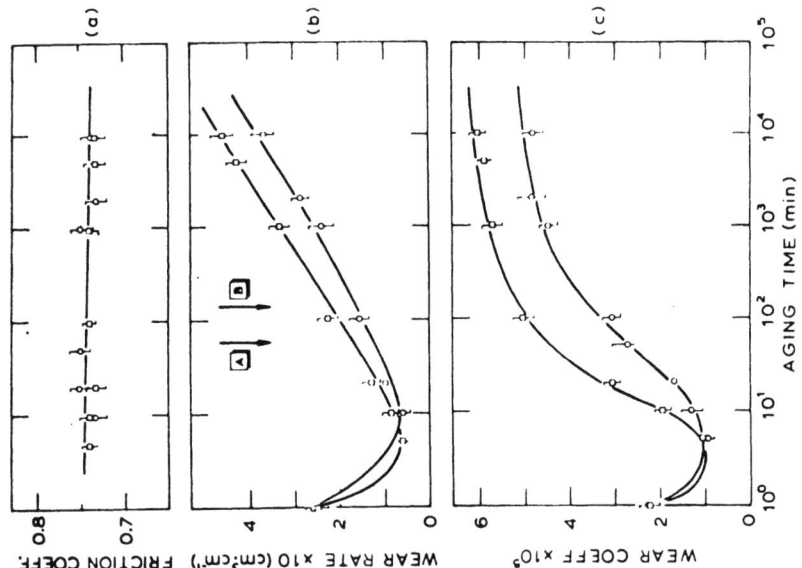

FIGURE 5.16   FRICTION AND WEAR PROPERTIES OF PRECIPITATION-HARDENED Cu-Cr ALLOYS AS A FUNCTION OF THE AGING TIME: (A) FRICTION COEFFICIENT, (B) WEAR RATE, AND (C) WEAR COEFFICIENT. THE NORMAL LOAD WAS 2 kg AND THE DURATION OF THE TESTS WERE 100 MIN AT A SLIDING SPEED OF 200 cm min^{-1}

fraction and the inverse of mean free path, as shown in Figure
5.17. The wear resistance decreases with an increase in the
volume fraction and with the inverse of the mean free path of
particles for the overaged alloys. These results are expected
since the crack nucleation cites increases with the number of
hard particles.

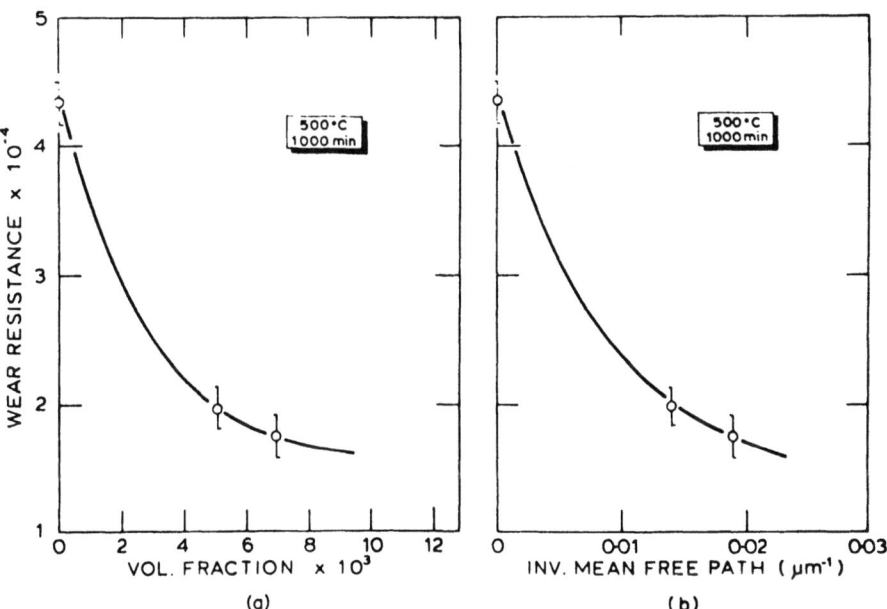

FIGURE 5.17 THE WEAR RESISTANCE (RECIPROCAL OF THE WEAR
             COEFFICIENT) VERSUS (A) THE VOLUME FRACTION AND
             (B) THE MEAN FREE PATH OF Cu-Cr ALLOYS FOR 1000 MIN
             OF TREATMENT

Figure 5.18 shows the micrographs of wear tracks of the precipitation-hardened alloys aged for 5 and 10,000 minutes. The sliding direction is from left to right. It can be seen that the surface details are similar to those shown earlier. The subsurface features for the same alloys are shown in Figure 5.19 where some second-phase particles can be seen in the overaged alloys. However, it is interesting to note that in Figure 5.19(a), subsurface cracks of the specimen aged for 5 minutes are very close to the surface, while for the specimen aged for 10,000 minutes, cracks are formed at a large depth. As an example of the morphology of wear particles, scanning electron micrographs of particles collected from the Cu-0.81 a/o Cr alloy aged for 10,000 minutes are shown in Figure 5.20. These particles are in the form of sheets and some have lamellar structure, possibly created by the presence of a large number of cracks above the crack that propagated the fastest.

These results with two phase metals show that the wear rate is also affected by the coherency of particles to the matrix. In the early stages of precipitation, the particles are coherent and therefore, the stress required to separate the particle from the matrix is large. Therefore, crack nucleation requires large amounts of subsurface deformation in order to develop sufficient interfacial stress between the matrix and the particle. Since the increased hardness decreases the deformation rate, the wear coefficient is decreased. The steep increase in wear coefficient after reaching a minimum value is due to the loss of coherency requiring less stress for decohesion of a particle from the matrix and is also due to the increased particle size. The deformation rate still decreases, but the deformation required for nucleation decreases even faster. Further, as the interparticle spacing decreases during the early stages of aging, cracks have to propagate smaller distances in order to join with neighboring cracks. This explains why the wear coefficient of the Cu-0.81% Cr alloy should be greater than that of the Cu-0.58% Cr alloy because the former has a larger volume fraction of second phase particles and possibly has a smaller interparticle spacing.

After the maximum hardness is reached, the volume fraction of particles does not increase with aging time anymore and coarsening of particles occurs, which increases the interparticle spacing. Crack nucleation at this point, tends to be relatively easy and the overall wear rate is controlled by the crack propagation rate. In this case, the crack growth and the interparticle spacing must be considered. Since the hardness does not change much for the overaged alloys, the number of asperities in contact must be nearly the

FIGURE 5.18 SCANNING ELECTRON MICROGRAPHS OF WEAR TRACKS OF THE PRECIPITATION-HARDENED Cu-0.81 AT % Cr ALLOY FOR AN AGING TIME OF (A) 5 MIN AND (B) 10000 MIN. THE SLIDING DIRECTION IS FROM LEFT TO RIGHT

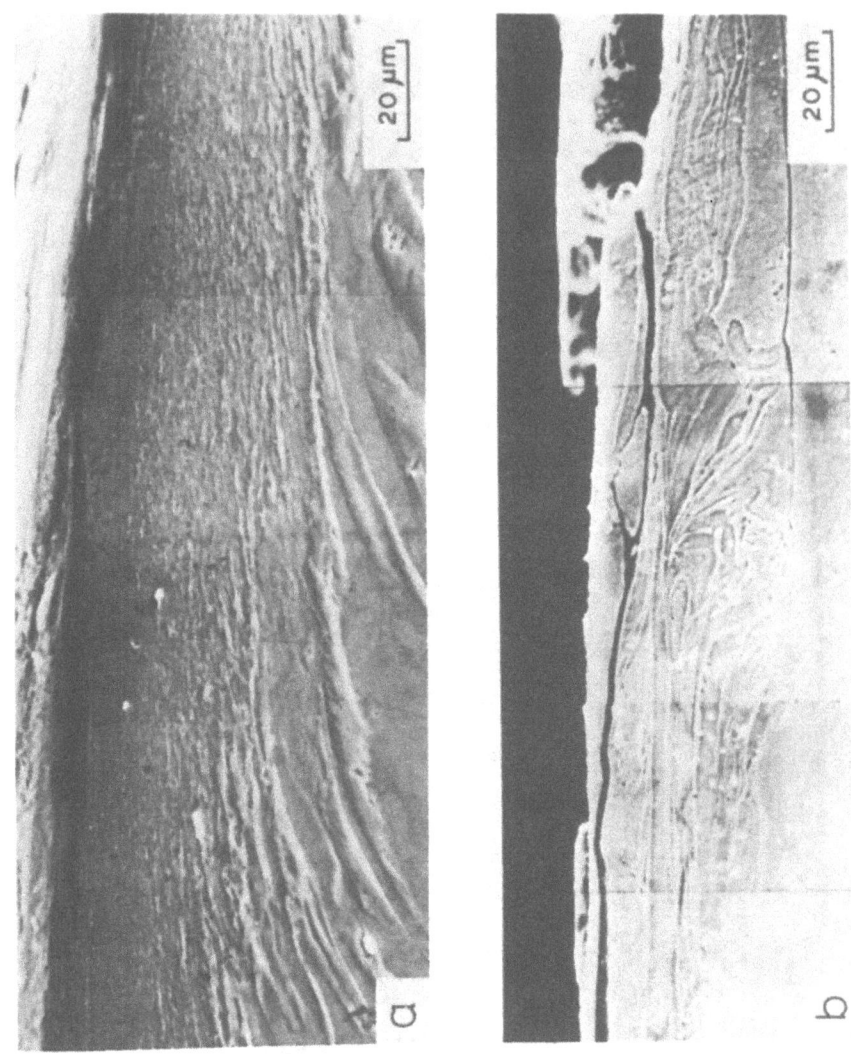

FIGURE 5.19   SCANNING ELECTRON MICROGRAPH OF THE SUBSURFACE OF
PRECIPITATION-HARDENED Cu-0.81 AT % Cr ALLOY AGED
FOR (A) 5 MIN AND (B) 10000 MIN

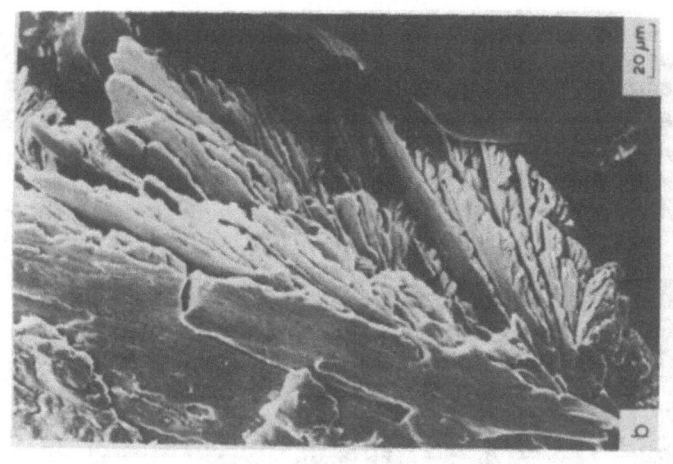

FIGURE 5.20 SCANNING ELECTRON MICROGRAPHS OF WEAR PARTICLES COLLECTED FROM WEAR TESTS ON THE SPECIMEN OF Cu-0.81 AT % Cr AGED FOR 10000 MIN

same and the number of asperities passing by will be about the same. Also since the matrices of both alloys are exactly the same, the crack growth rate must be the same to a first order approximation. Assuming that the material is delaminated in successive layers, the crack growth rate is calculated as the ratio of the mean free path to the number of cycles required to remove one layer. Such a calculation leads to a range of values between 4×10^{-4} and 6×10^{-4} µm/cycle for both alloys, and the wear rate depends basically on the mean free path of the interparticle spacing. Thus, when the overall wear process is controlled by the crack growth rate, the wear coefficient tends to level off toward asymptomatic values, since as the aging treatment is continued, the mean free path becomes roughly constant. This explains the leveling-off of the wear coefficient when the transition from crack nucleation rate controlled wear to crack propagation controlled wear occurs. The effect of the mean free path is also shown by the difference in the wear rate of Cu-0.81 a/o Cr and Cu-0.58 a/o Cr. As a higher volume fraction implies a smaller mean free path for the same particle size, the curve for Cu-0.81% Cr should be higher than that for Cu-0.58% Cr.

5.4 Sliding Wear of Polymeric Materials

5.4.1 Wear of Simple Phase Polymers

The basic thoughts involved in the delamination theory of wear can be applied to the wear of polymeric materials which have been used increasingly in bearing applications because of their low cost, excellent chemical resistance, and good wear properties.

Deformation properties of polymers have a profound influence on wear. Below the glass transition temperature, most amorphous polymers are relatively brittle. The sliding wear of glassy polymers is caused by crazing and brittle fracture when there is no thermal softening of the surface layer[11-15]. Semicrystalline polymers, which are usually used above the glass transition temperature and relatively ductile, show different wear behavior from the glassy polymers. Thick lump type wear debris or thin films are generated by plastic deformation of the surface layer in response to the applied normal and shear stresses[16-23, 19, 24-32]. Thin films are normally found on the counter surface face. The mechanism of the thin film transfer is not clear. The wear behavior depends upon the polymer properties and the sliding speed[19]. Thin films are formed only with highly linear crystalline polymers [such as polytetrafluoroethylene (PTFE), high density polyethylene (HDPE), polyoxymethylene (POM)], at slow sliding speed. The thickness of the film is of the order of 50 to 200 A and the molecular chains are highly oriented in the sliding direction[25].

Many different explanations have been advanced to explain the mechanism of the thin film transfer. Steijn explained the thin film transfer phenomenon as resulting from the formation of adhesion junctions and the drawing of thin fims across the sliding surface[27]. Tanaka et al, argued that the thin film transfer of PTFE is due to easy destruction of the special banded structure, but not due to the drawing of molecular chains[29]. Makinson and Tabor claimed that the easy slipping of crystalline slices causes the thin film transfer[19]. According to Briscoe et al, the behavior of the thin film transfer at a low sliding speed, appears to be connected with a smooth molecular profile and not with the crystallinity or band structure of PTFE[25]. In the case of the crystalline polymers, that transfer thick lumpy films rather than thin films, little orientation was found in the film although equally strong adhesion junction might have been present. It is more probable that the reorientation occurs prior to the thin film transfer[16]. Then the thin film is formed by shearing of the oriented layer, which is followed by molecular orientation possibly due to the elongation of molecular chain during the shearing process.

A careful examination of worn surfaces reveals very large plastic deformation of the surface and subsurface layer, as discussed in Section 4. The deformation and the deformation gradient are largest at the surface and decays rapidly away from the surface. In crystalline materials such as metal crystals, grains deform in such a manner that the slip planes align nearly parallel to the surface. Therefore, it is expected that the surface layer will shear easily and fracture parallel to the surface due to the alignment of the slip planes with the surface. In crystalline polymers with no bulky side pendant groups, the crystalline region of the polymer is expected to align parallel to the surface. Consequently, the crystalline platelets are expected to shear off parallel to the surface and elongate when the surface is plowed by the asperities of the opposing surfaces. The peeling process may be facilitated if there are preexisting cracks or if cracks are nucleated during the deformation process.

In addition to the deformation of the surface layer, the deformation and fracture resulting from the bulk deformation of the surface due, to the Hertzian contact load in glassy polymers, must be considered in establishing the criteria for wear. When only a normal load is applied at a contact between a sphere and a semi-infinite solid, the largest shear stress in the body occurs below the surface. The depth depends on the poisson ratio of the material and is 0.47 times the radius of the contact area (a) when poisson ratio is 0.3[30]. When normal and tangential loads are present at the asperity contact, the maximum shear stress is about 0.343 Po at y = 0.4a where Po is the normal stress at the center of contact area and y is the coordinate axis perpendicular to the contact area[31, 32]. In glassy polmers, fracture can be initiated from these locations of **maximum Hertzian shear stress**. Therefore, in order to prevent wear, it will be necessary to insure that the applied load does not exert a stress greater than a critical value at the point where the Hertzian shear stress is maximum.

Based on the foregoing arguments, a wear model for polymers can be postulated using the yield criteria, Figure 5.21. The wear model divides the wear conditions into two regimes: the thin film transfer case (Regime I) and the thick film or lump transfer case (Regime II). In Regime I, the relationship between the applied shear stress and the material shear yield strength are so that the thin film transfer occurs when

$$\tau_s \geq k_s$$

and (5.13)

$$\tau_{max} < k_b$$

where τ_s is the tangential component of the surface traction due to various mechanisms discussed in Section 3, τ_{max} is the maximum Hertzian shear stress, k_s is the shear yield strength of the surface layer, and k_b is the bulk shear strength. In this regime thin film transfer occurs. For example, since PTFE and HDPE have much lower shear yield strength at the surface than in the bulk, they can plastically deform only at the surface layer under typical sliding conditions and generate thin films by the mechanism discussed earlier. In Regime II, both the shear stress at the surface and the maximum shear stress can exceed the material strength. Therefore, bulky wear particles will be generated when the following two conditions are satisfied:

$$\tau_s \geq k_s$$

and (5.14)

$$\tau_{max} > k_b$$

As a result, plastic deformation may occur in the maximum shear region below the surface, generating lumpy wear particles.

The foregoing hypothesis has been checked by taking a polymer which belongs to Regime I and treating the surface so as to insure that the applied shear stress at the surface, τ_s, is less than the shear strength of the surface, k_s. This can be done by crosslinking the surface of polymers through helium plasma treatment.

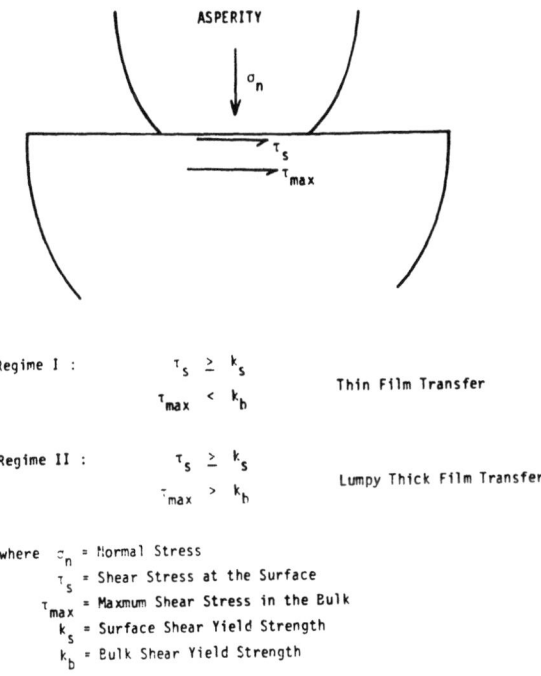

FIGURE 5.21 CONDITIONS FOR WEAR OF POLYMERS BY THIN FILM TRANSFER IN CRYSTALLINE POLYMERS & BY LUMPY THICK FILM TRANSFER

5.4.2 Wear of Polymeric Composites

Many composites are made of polymers reinforced with fibers. Many bearings and gears made of phenolics, nylon, polyurethane, teflon (PTFE), and acetal have been reinforced glass fibers and graphite fibers for improved wear resistance and dimensional stability. These composites with chopped fibers behave in such a manner that their tribological behavior can be predicted and improved based on the delamination theory of wear. Two of these studies will be briefly reviewed here.

Clerico tested nylon reinforced with randomly oriented chopped glass fibers[33]. She found that these composites delaminate by plastic deformation and subsurface crack nucleation and propagation. The wear particles collected were thin sheets.

Sung and Suh investigated the effect of fiber orientation on wear based on the delmaination theory of wear[34]. Three types of fiber reinforced plastics were studied. One was a unidirectionally oriented composite with graphite fibers and an epoxy matrix, fabricated from prepegs. The second composite was composed of unidirectionally oriented Kevlar-49 fibers in epoxy. The third composite material was a commercial grade bearing material (RT Duroid 5813), a polytetrafluoroethylene (PTFE) based material. This composite contained about 16 wt.% of planar oriented micro-glass fibers with two-thirds of the fibers being preferentially oriented in one direction. The ratio of fibers oriented in three orthogonal directions are approximately 2:1:0. The composition and the properties of these composites are given in Table 5.3.

Properties of composites

Unidirectional graphite fiber (Thornel 300)-epoxy (SP-228)

Specific gravity	1.58
Fiber volume	60%
Longitudinal tensile strength	200 klbf in^{-2}
Longitudinal tensile modulus	18 x 10^2 lbf in^{-2}
Compression strength	145 klbf in^{-2}
Transverse tensile strength	10 klbf in^{-2}
Transverse tensile modulus	1.2 x 10^1 lbf in^{-2}
Interlaminar shear strength	16 klbf in^{-2}

Unidirectional Kevlar-49 (DuPont)-epoxy

Specific gravity	1.33
Fiber volume	65%
Longitudinal tensile strength	223 klbf in^{-2}
Longitudinal tensile modulus	17 x 10^2 lbf in^{-2}
Interlaminar shear strength	9 klbf in^{-2}

Bidirectional glass microfiber-MoS$_2$-PTFE (RT Duroid[a])

Specific gravity	2.42
Glass fiber	16 wt.%
MoS$_2$	15 wt.%
Compression modulus[b]	158/183/130 klbf in^{-2}
Compression strength[b]	3.6/4.4/8.2 klbf in^{-2}

a Registered trademark of Rogers Corporation
b Three values are for the three orthogonal directions.

TABLE 5.3   PROPERTIES OF COMPOSITES

The experimental results show that the wear rate is the least when the fibers are placed perpendicular to the surface as shown in Figure 5.22 for the graphite fiber-epoxy composite. In this case, the friction was also the least as shown in Figure 5.23. Similar wear results were obtained with the Kevlar-epoxy composite, Figure 5.24, although the order of frictional force was changed. With the biaxially oriented glass micro-fiber - MoS$_2$ - PTFE composites the wear rate was again found to be the least when the largest fraction of fibers was oriented normal to the sliding plane Figure 5.25. These results indicate that when the fibers are normal to the surface, plastic deformation and crack propagation are both minimized, resulting in low wear rates.

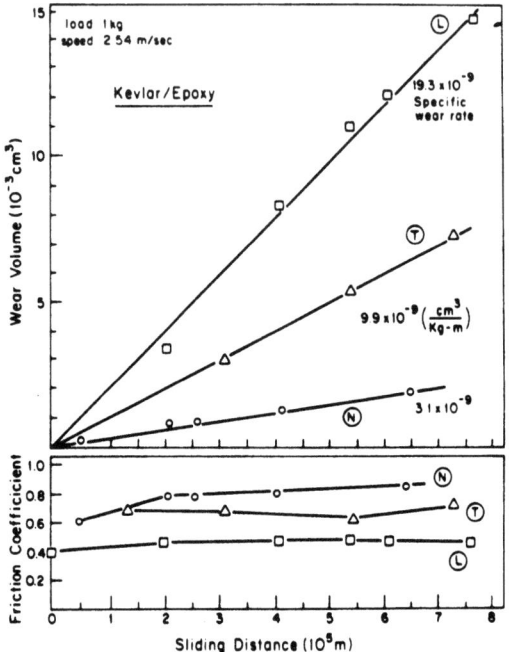

FIGURE 5.22 FRICTION COEFFICIENTS & WEAR VOLUME AS A FUNCTION OF SLIDING DISTANCE IN UNIAXIAL GRAPHITE FIBER-EPOXY COMPOSITE. SLIDING AGAINST 52100 STEEL, WITH FIBER ORIENTATIONS NORMAL LONGITUDINAL & TRANSFER TO THE SLIDING DIRECTION.

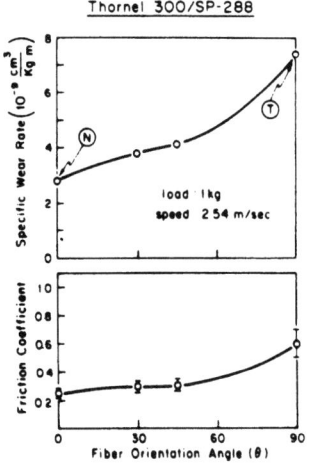

FIGURE 5.23 FRICTION COEFFICIENTS & SPECIFIC WEAR RATE AS A FUNCTION OF SLIDING DISTANCE IN UNIAXIAL GRAPHITE FIBER-EPOXY COMPOSITE SLIDING AGAINST 52100 STEEL, WITH VARYING FIBER ORIENTATIONS RANGING FROM NORMAL ($\theta=0$) TO TRANSVERSE ($\theta=90°$) TO THE SLIDING DIRECTION

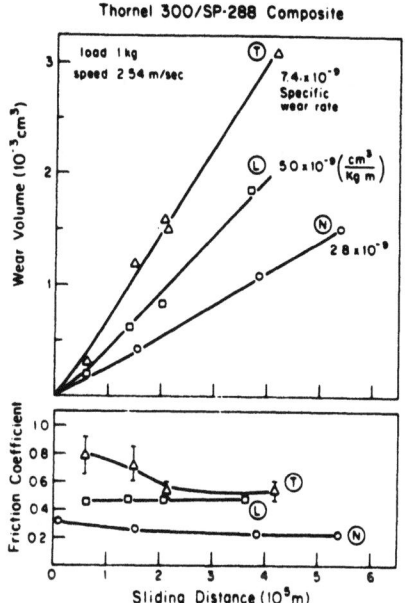

FIGURE 5.24 FRICTION COEFFICIENTS & WEAR VOLUME AS A FUNCTION OF SLIDING DISTANCE IN UNIAXIAL KEVLAR 49-EPOXY COMPOSITE SLIDING AGAINST 52100 STEEL WITH FIBER ORIENTATIONS NORMAL (N), LONGITUDINAL (L), AND TRANSVERSE (T), TO THE SLIDING DIRECTION

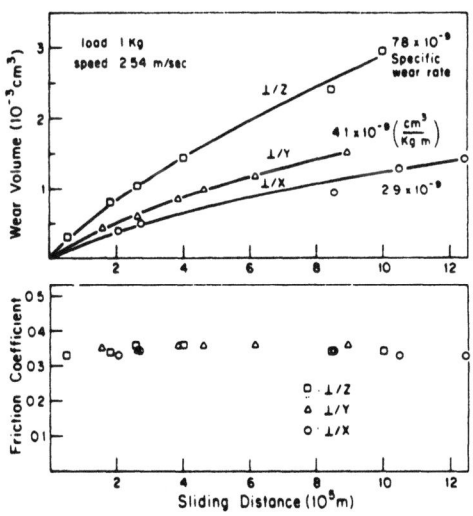

FIGURE 5.25 FRICTION COEFFICIENTS & WEAR VOLUME AS A FUNCTION OF SLIDING DISTANCE IN BIAXIALLY ORIENTED GLASS MICROFIBER MoS_2-PTFE COMPOSITE, SLIDING AGAINST 52100 STEEL WITH SLIDING PLANES NORMAL TO THREE ORTHOGONAL DIRECTIONS, X, Y, AND Z

5.5 Wear Minimization

A number of techniques have been devised to minimize the sliding wear of materials based on the ideas suggested by the delamination theory of wear. They are listed below without much elaboration. The details can be found in the references given.

1) Coating of the metal surface with a thin layer of soft metals. The thickness of the layer, which is very critical, is about 0.1 μm for nickel and 10 μm for gold. Such a thin layer prevents plastic deformation and thus, reduces wear[35].

2) Cross-linking of crystalline polymers to prevent the film transfer process by creating a strong layer on the surface so that the frictional force is always less than the strength of the surface layer[36].

3) Ion implantation of iron with titanium ion and nitrogen ion. This process creates a hard layer on the surface which prevents plowing, thus lowering the frictional force and the wear rate[37].

5.6 Concluding Remarks

In this section it was shown that elasto-plastic materials, be it a polymer or metal, wears by the delamination process. The delamination theory of wear explains the sliding wear process in detail and correctly. Wear of sliders is caused by plastic deformation, crack nucleation, and crack propagation.

Wear particles are also generated by asperity deformation and plowing, which accelerate the delamination process. However, in many cases the ultimate failure of the sliding surfaces is by the mechanisms postulated by the delamination theory of wear. This is because the dominant mechanism that controls the wear rate is the delamination mechanisms.

Sliding wear under high loads and speeds can also be caused by other mechanisms because of the high interfacial temperatures. This is a subject by itself.

5.7 References - Section 5

1. Suh, N.P., "The Delamination Theory of Wear," Wear, 25 (1973) 111-124.

2. Jahanmir, S., N.P. Suh, and E.P. Abrahamson, II, "Microscopic Observations of the Wear Sheet Formation by Delamination," Wear, 28 (1974) 235-249.

3. Jahanmir, S., "A Fundamental Study of the Delamination Theory of Wear," Ph.D. Thesis, Department of Mechanical Engineering, MIT, 1977.

4. Tohaki, M., "Microstructural Aspects of Friction," S.M. Thesis, Department of Mechanical Engineering, MIT, 1979.

5. Rabinowicz, E., "The Dependence of the Abrasive Wear Coefficient on the Surface Energy of Adhesion," Wear of Materials - 1977, ASME, New York, 1977, 36-40.

6. Gupta, P.K. and N.H. Cook, "Statistical Analysis of Mechanical Interaction of Rough Surfaces," Journal of Lubrication Technology, 94 (1972) 19-26.

7. Whitehouse, D.J., "The Effects of Surface Topography on Wear," Fundamentals of Tribology, (Eds. N.P. Suh and N. Saka), MIT Press, 1980, 17-52.

8. Pamies-Teixeira, J.J., N. Saka, and N.P.. Suh "Wear of Copper-Based Solid Solutions," Wear, 44 (1977) 65-75.

9. Saka, N., J.J. Pamies-Teixeira, and N.P. Suh, "Wear of Two-Phase Metals," Wear, 44 (1977) 77-86.

10. Saka, N., "Effect of Microstructure on Friction and Wear of Metals," Fundamentals of Tribology, (Ed. N.P. Suh and N. Saka), MIT Press, 1980, 135-170.

11. Puttick, K.E., L.S.A. Smith, and L.E. Miller, "Stress Field Round Indentation in Poly (Methyl Methacrylate)," J. Phys. D.; Appl. Phsy., 10 (1977) 617-632.

12. Billinghurst, P.R., C.A. Brookes, and D. Tabor, "The Sliding Processes as a Fracture-Inducing Mechanism," Proceedings of the Conference on the Physical Basis of Yield and Fracture, Oxford 1966, 253-258.

13. Matsushige, K., S.V. Radcliffe, and E. Baer, "The MechanicalBehavior of Polystyrene Under Pressure," J. Mat. Sci., 10 (1975) 833-845.

14. Van Den Boogaart, A., "Crazing and Characterization of Brittle Fracture in Polymers," Proceedings of the Conference on the Physical Basis of Yield and Fracture, Oxford (1966), 167-175.

15. Peterson, T.L., D.B. Ast, and E.J. Kramer, "Holographic Interferometry of Crazes in Polycarbonate," J. Appl. Phys. 45 (1974) 4220-4228.

16. Pooley, C.M. and D. Tabor, "Friction and Molecular Structure: The Behavior of Some Thermoplastics," Royal Soc. London, 329A (1972) 251-274.

17. Bowers, R.C., W.C. Clinton, and W.A. Zisman, "Frictional Properties of Plastics," Modern Plastics, Feb. (1954), 131-225.

18. Tanaka, K. and Y. Uchiyana, "Friction, Wear, and Surface Melting of Crystalline Polymers," Wear of Materials, ED. S.K. Rhee et. al., ASME (1977), 499-530.

19. Makinson, R.K. and D. Tabor, "The Friction and Transfer of Polytetrafluoroethylene," Proc. Roy. Soc. London, A281 (1964) 49-61.

20. West, C.H. and J.M. Senior, "Frictional Properties of Polyethylene," Wear, 19 (1972) 37-52.

21. James, D.I., "Surface Damage Caused by Polyvinyl Chloride Sliding on Steel," Wear, 2 (1958/59) 183-194.

22. Jain, V.K. and S. Bahadur, "Material Transfer in Polymer-Polymer Sliding," Wear of Materials, ASME (1977), 487-493.

23. Warren, J.H. and N.S. Eiss, Jr., "Depth of Penetration as a Predictor of the Wear, Polymers on Hard, Rough Surfaces," Wear of Materials, ASME (1977), 494-500.

24. Kar, M.K. and S. Bahadur, "Macromechanism of Wear at Polymer-Metal Sliding Interface," Wear of Materials, ASME (1977), 501-509.

25. Briscoe, B.J., C.M. Pooley and D. Tabor, "Friction and Transfer of Some Polymers in Unlubricated Sliding," Advances in Polymer Friction and Wear, Polymer Science and Technology Vol. 5A, Ed, H. Lee, Plenum Press (1974), 191-204.

26. McLaren, K.G. and D. Tabor, "The Friction and Deformation Properties of Irradiated Polytetrafluoroethylene (PTFE)," War, 8 (1965) 3-7.

27. Steijn, R.P., "The Sliding Surface of Polytetrafluoroethylene, on Investigation with Electron Microscope," Wear, 8 (1968) 193-212.

28. Sviridyonok, A.I., V.A. Bely, V.A. Smurigov, and V.G. Savkin, "Study of Transfer in Frictional Interaction of Polymers," Wear, 25 (1973) 301-308.

29. Tanaka, K., Y. Uchiyaman, and S. Toyooka, "The Mechanism of Wear of Polytetrafluoroethylene," Wear, 23 (1977) 153-172.

30. Timoshenko, S.P. and J.N. Goodier, "Theory of Elasticity," McGraw-Hill (1970), 409-414.

31. Poritsky, H., "Stresses and Deflections of Cylindrical Bodies in Contact with Application to Contact or Years and Locomotive Wheels," J. Appl. Mech., June (1950), 191-201.

32. Hamilton, G.M. and L.E. Goodman, "The Stress Field Created by a Circular Sliding Contact." J. Appl. Mech., June (1966), 371-376.

33. Clerico, M., "Sliding Wear of Polymeric Composites," Wear, 53 (1979) 279-301.

34. Sung, N.H. and N.P. Suh, "Effect of Fiber Orientation on Friction and Wear of Fiber Reinforced Polymeric Composites," Wear, 53 (1979) 129-141.

35. Saka, N., H.C. Sin, and N.P. Suh, "Prevention of Spline Wear by Soft Metallic Coatings," Final Report to DARPA and ONR, Contract No. N00014-76-C-0068, July, 1980.

36. Youn, J. and N.P. Suh, "Tribological Characteristics of Surface Treated Polymers," SPE ANTEC, (1981) 20-23.

37. Shepard, S.R. and N.P. Suh, "The Effects of Ion Implantation on Friction and Wear of Metals," ASLE/ASME paper to appear in Transactions, 1981.

6. CONCLUSIONS

Surface interactions are complex phenomena, requiring input from materials science, continuum mechanics, thermodynamics, and quantum mechanics. The surface is always highly contaminated because of the high free energy of the surface. The atoms of the contaminants are well mixed with the substrate atoms, again to lower the free energy.

Therefore, adhesion cannot occur readily under quasi-static conditions at low temperatures.

Friction is caused by plowing, adhesion, and asperity deformation. Therefore, it depends strongly on environment. Through proper design of the interface and materials, friction can be controlled. It was shown that friction must be considered in the Friction Space, since it is not an invariant in the temporal frame of tribological applications.

Finally, the response of materials are shown to be plastic deformation, crack nucleation, and crack propagation. Based on detailed considerations of each one of these processes, the delamination theory of wear was introduced. It is shown that although there are many different wear particle generating mechanisms, the delamination process is the rate determining process.

7. ACKNOWLEDGEMENTS

I am grateful to the financial support provided by the Defense Advanced Research Projects Agency and the Office of Naval Research over the years to conduct research on the Delamination Theory of Wear. Much of the work reported here was done through the partnership between my students, colleagues and myself. I am particularly indebted to Drs. S. Jahanmir, N. Saka, and H.C. Sin for their contributions to our research efforts. Thanks are also due Mrs. Margaret McDonald for her patience in typing the manuscript.

MATERIALS IN TRIBOTECHNICAL APPLICATIONS

A.W.J. de Gee
Metaalinstituut TNO, Apeldoorn, The Netherlands

1. INTRODUCTION

When set to the task of discussing the role of materials in tribology, one is immediately faced with the need to decide whether to follow a "theoretical model approach" or a "systems approach"[1,2]. The first approach assumes a consistent behavior of materials under a variety of application conditions, the second allows for appreciable deviations in behavior from one application to the other.

Without rejecting the "model approach" all together (see below), in the present case the "systems approach" is chosen, if only because of the fact that the wear rate of a particular, well defined material may differ by several decades from one application (i.e. set of conditions) to the other. In the following, two examples are given of cases in which relatively minor changes in conditions result in a considerable - and a priori unpredictable change in tribological behavior.

The first example of the essential unpredictability of tribological processes is shown in Figure 1. This figure shows the pronounced influence of a change in dimensions of contacting bodies on their wear rates[3]. In the present case this is caused by a pronounced difference in friction induced surface temperature, as a result of the change in dimensions. Actually, the wear process changes radically from one geometry to the other. If a pin with R = 2 mm is combined with a ring with R = 40 mm, the wear mechanism is that of adhesive wear. If, on the other hand, a pin with R = 4 mm is combined with a ring with R = 20 mm, the wear mechanism is that of oxidative wear.

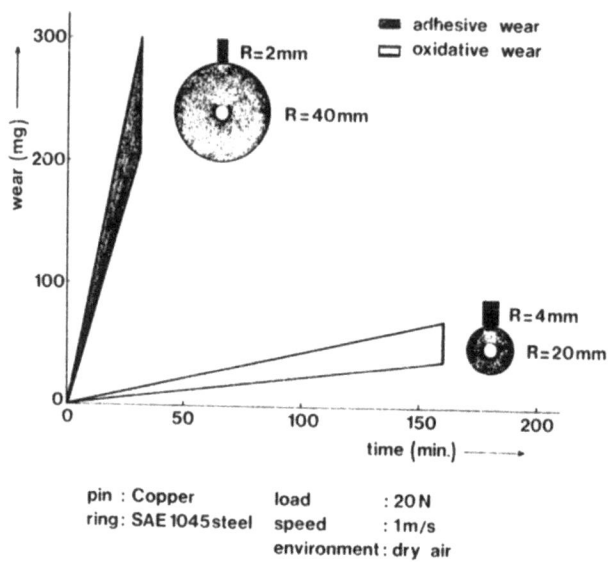

FIGURE 1   THE INFLUENCE OF A CHANGE IN DIMENSIONS OF CONTACTING BODIES ON THEIR WEAR RATES

The second example concerns the influence of the chemical composition of the bearing material and that of the lubricant cover gas on the wear of Cu 6 Sn alloys, Figure 2, used in combination with shafts of, respectively, carburized steel AISI 5120 and through hardened steel AISI D3[4]. Copper-tin alloys usually contain small quantities of phosphorous (up to 0.1%) as a result of the fact that the melt is deoxidized with "copper-phosphorus" (copper with approximately 10% Cu_3P). If oxygen from the air penetrates into the melt, phosphorous is oxidated preferentially, thereby protecting the metals against oxidation. The phosphorous oxide, thus formed, can be found in the slag. If desired, phosphorus-free alloys can be obtained by melting together pure metals under vacuum (no oxidation of the metals), by exposing a phosphorous containing melt for some time to the air, thereby removing the surplus of phosphorous by oxidation or by depositing the bearing material on a steel substrate by a process of metal spraying under controlled oxidation conditions. In the first instance the resulting alloy contains neither phosphorous nor tin oxide. In the latter two cases however, a very small amount of tin oxide is formed, which - after solidification - is present in the alloy in the form of finely divided crystals with a diameter from 0.1 to 0.5 µm.

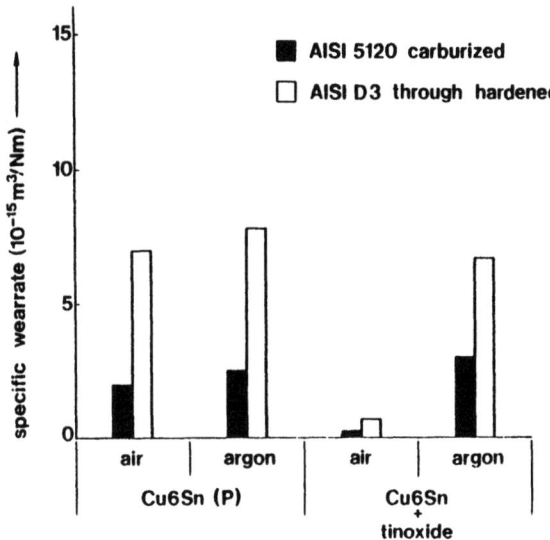

FIGURE 2 THE INFLUENCE OF THE CHEMICAL COMPOSITION OF BEARING MATERIAL AND LUBRICANT COVER GAS ON THE WEAR OF Cu 6 Sn ALLOYS

Laboratory tests with a tribometer as well as full scale bearing tests showed that these tin oxide containing materials behave exceptionally well, provided that the environment contains oxygen[5]. This effect is shown in Figure 2. The explanation of the pronounced differences in behavior is that tin oxide catalyses the formation of surface active, polymolecular products ("friction polymers") in the contact zone. Such polymer formation is possible only by virtue of the fact that the lubricant contains suitable "building stones." In the present case, such "building stones" are free radicals, which are formed in the lubricant by reaction with oxygen from the cover gas. Figure 2 also shows that the nature of the steel also has a pronounced effect on the wear of the Cu Sn alloy (although the surface roughness of the shafts was the same in both cases). Still the nature of the shaft does not affect the phosphorous/tin oxide effect.

There would not have been the remotest possibility of predicting a priori the above effects on the basis of theoretical models.

Although a strong case for a systems approach can be built upon the above evidence, this does not mean that theoretical modeling would not be worthwhile and - in fact - a necessary activity in tribology. Actually, it is only by virtue of such models that effects as discussed above can be explained. Also, theoretical models may show how far observations can be extrapolated to "related systems" (for instance: if the mechanism by which lauric acid functions in boundary lubrication is understood, see Figure 3, one can safely predict that stearic acid will have similar effect, although it will still be hard to predict the minimum concentration at which such effect manifests itself for the first time). An obvious area in which the principle of "extrapolation by virtue of theoretical modeling" is applied widely is that of improvement of the properties of a particular class of materials, as for instance deformation - martensite forming manganese steels[5].

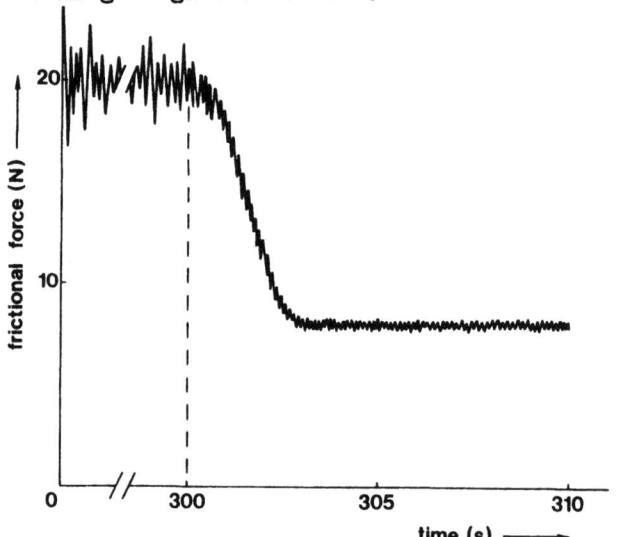

FIGURE 3   EFFECT OF ADDITION OF LAURIC ACID ON THE FRICTION OF BRONZE AGAINST STEEL

LUBRICATION CONDITION: BOUNDARY

300 s AFTER BEGINNING OF THE TEST 1 cm^3 of LAURIC ACID ($C_{11}H_{23}COOH$) IS ADDED TO THE LUBRICANT BATH (1000 cm^3 SAE 10 BASE OIL).

Having chosen for a systems - (application directed) approach for a discussion of "materials in tribology", practically meaningful application areas are to be defined.

In the present case a choice has been made for the following areas:

- lubricated plain bearings
- rolling element bearings, gears, cams and cam followers
- dry bearings
- earth moving, mining and dredging equipment.

The above four application areas cover a considerable percentage (80% at a rough estimate) of the technical areas where tribology is important. In addition, the tribological aspects of surface treatments and coatings are covered separately, as such treatments and coatings may be applied in widely different application areas.

In order to promote a unified approach towards the behavior of materials in the above application areas, a brief survey of wear mechanisms and their interrelation precedes the detailed discussion of the separate areas.

1.1 Wear Processes

Wear is usually defined as the undesirable, progressive loss of substance from the operating surface of a body, occurring as a result of relative motion at the surface. It can be due to the removal of material as a result of frictional heating (wear by melting or evaporation) or to the continuous removal of reaction products (e.g. oxides) from the surface (chemical, corrosive or oxidative wear). In all other cases, wear is the result of local overstressing of one or both of the contacting bodies, due to normal and tangential (friction) forces. This may be (but is not necessarily) accompanied by transfer of material from one body to the other. When transfer occurs, the process is identified as one of adhesive wear; otherwise the term delamination wear is applicable. Finally, if the action of an abrasive is involved, the designations abrasion or abrasive wear are used.

If tangential forces are low, as is usual in rolling element bearings, the contacting surfaces may still suffer from surface fatigue, leading to wear as a result of cyclic variations in the normal force. If cyclic loading in a direction perpendicular to the surfaces is due to impact, the term impact wear is used. Finally, in erosion a solid surface suffers from attack by solid particles carried in a fluid. Depending upon the angle of attack, either abrasion or impact wear may predominate.

Figure 4 gives a survey of the different wear processes that may occur under conditions of sliding motion between two contacting bodies and shows how these relate to friction.

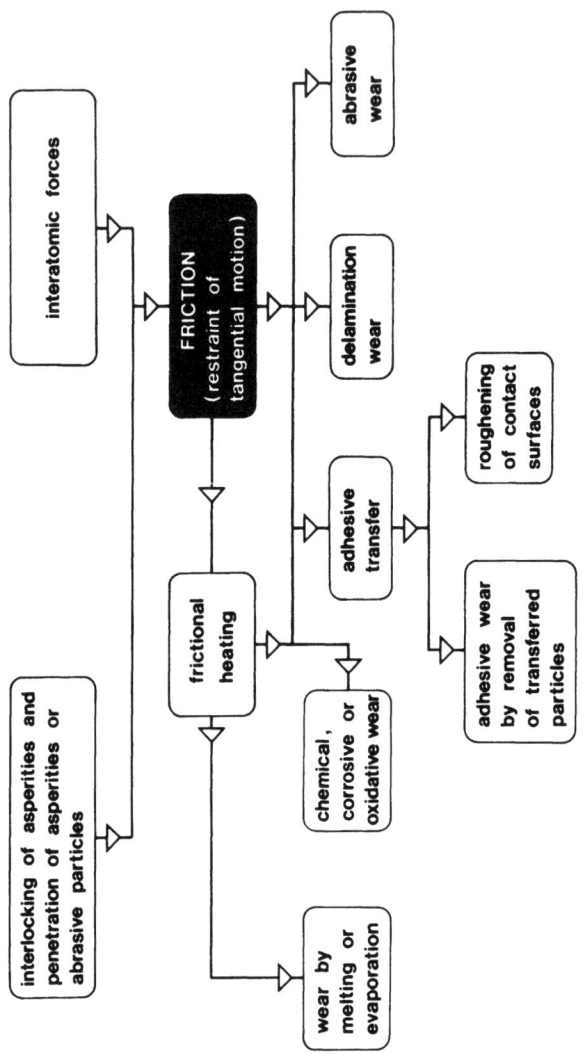

FIGURE 4   SURVEY OF WEAR PROCESSES THAT MAY OCCUR UNDER CONDITIONS OF SLIDING MOTION BETWEEN TWO CONTACTING BODIES AND THEIR RELATION TO FRICTION

A consequence of the fact that, usually, wear is caused by overstressing of the surface zone of materials due to normal and tangential forces, is that friction (as resulting from adhesive forces and plowing) and the dynamic strength properties of the materials, that are subjected to friction, are equally important in determining the wear rate. Although from this it seems reasonable to adhere to the statement "no wear without friction" it is also clear that in general there will be no unambiguous relation between friction and wear. This is because of the fact that - from one material combination to the other - a difference in friction may or may not be balanced by a difference in dynamic strength properties.

2. MATERIALS FOR LUBRICATED PLAIN BEARINGS

2.1 Non-Tribotechnical Versus Tribotechnical Quality Criteria

In selecting materials for use in lubricated plain bearings, one usually applies non-tribological as well as tribological selection criteria. The first group comprises hardness (and the related properties compressive strength and deformability), the thermal properties heat conductivity, specific heat, maximum admissible surface temperature and coefficient of thermal expansion, the electrical properties (i.e. in particular the resistance against electrical discharge pitting) and, last but not least, the price as related to availability, machineability, etc. A general property of these non-tribological criteria is that they are - at least in a first approximation - system-independent. A non-tribological criterion which does depend strongly on the systems properties (i.e. in particular the environment and the temperature) is the resistance of the bearing material against corrosive action of the lubricant (particularly important if an "unconventional" liquid as, for instance, water is used).

In contrast to the above, quality criteria of a typical tribological nature are always heavily system-dependent. This means, among other things, that in practice one can never characterize a single bearing material as such. Instead one should always characterize the combination of bearing material, lubricant and journal material. This applies in particular if unconventional (nonferrous) journal materials and/or unconventional lubricants are to be used.

International cooperative work under the auspices of the International Standards Organization, ISO, has led to the definition of four main categories of tribological selection criteria, namely:

- behavior under conditions of boundary lubrication
- behavior under conditions of surface fatigue
- behavior under conditions of (fluid) erosion
- behavior under conditions of cavitation erosion.

2.2 Tribological Quality Criteria

Figure 5 gives a survey of the criteria which, together, determine the tribological behavior of bearing materials and their interrelation. In the following paragraph, Figure 5 is discussed in detail.

When considering behavior under conditions of boundary lubrication (A), one distinguishes between behavior under normal conditions of lubricant supply, (very) low speed and low thermal loading (A1), and the behavior under conditions of interrupted lubricant supply, nominal speed and high thermal loading (A2). The first set of conditions is encountered in plain bearings which normally function under conditions of hydrodynamic lubrication, but which are run at speeds far below nominal speed (as, for instance, during starting and stopping). Under such conditions, the development of frictional heat is negligible and the bearing runs under psuedo-isothermal conditions. The latter situation occurs in bearings in which the supply of lubricant stagnates while the bearing runs at nominal speed. Under such conditions, there is considerable development of frictional heat and the temperature of the bearing rises rapidly. In both situations, A1 and A2, the lubricant may or may not be contaminated with abrasive particles ("dust"). This leads to situations A1-1, A1-2, A2-1, and A2-2. In all these cases, the behavior of the combination of bearing material - lubricant - journal material can be characterized with three parameters, namely the coefficient of friction (a), the process roughness or, better, the contact parameter (b), and the wear rate (c).

A major difference between situations A1 and A2 is that in situation A1 the system rapidly attains equilibrium, the characteristic parameters coefficient of friction, process roughness and wear rate attaining constant values. In situation A2 on the contrary drastic transition effects are bound to occur. In this case it is not so much the high values of the characteristic parameters after reaching equilibrium which are important, but much more the time which elapses before a transition in one or more of the characteristic parameters occurs. Accordingly, problems associated with stagnation of the lubricant supply at nominal speed are usually solved by application of a suitable monitoring system, for instance based on continuous measurement of the bearing temperature. On the basis of this philosophy a better material-lubricant combination means a combination with a longer delay time.

Further particulars on determination and use of coefficient of friction, process surface roughness (or contact parameter) and wear rate can be found in References 7, 8, and 9.

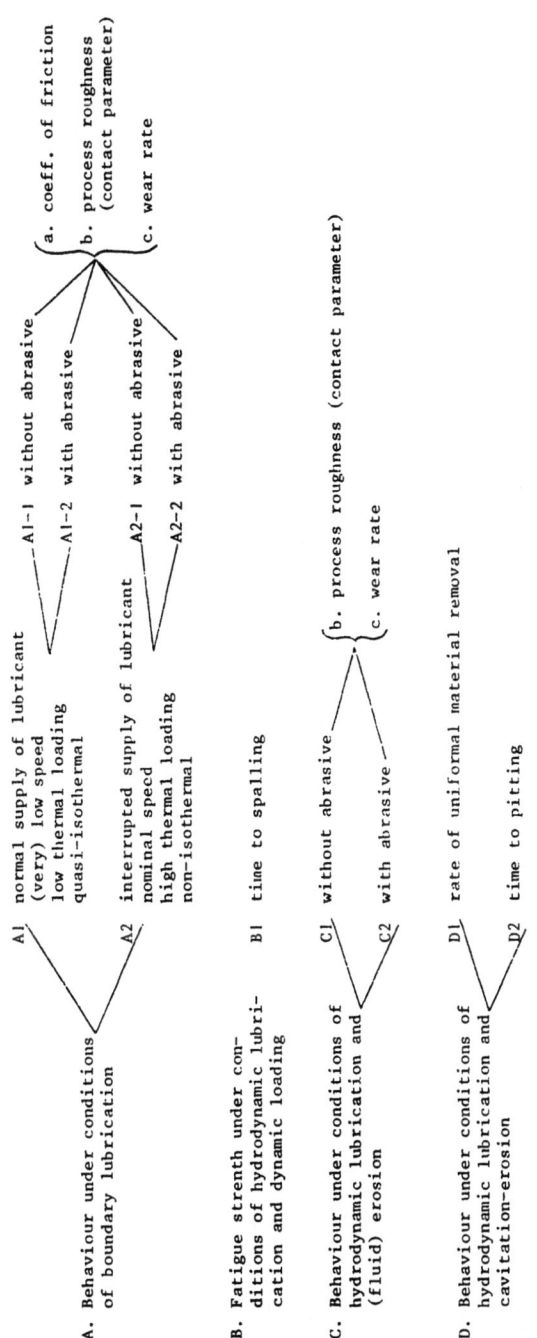

FIGURE 5 SURVEY OF THE CRITERIA WHICH DETERMINE THE TRIBOLOGICAL BEHAVIOR OF BEARING MATERIALS AND THEIR INTERRELATION

2.3 Typical Journal Bearing Materials and Their Performance Characteristics

Materials, particularly suited for application in plain bearings, should be able to carry the load on the bearing (no gross plastic deformation), should have an adequate fatigue life, should be readily deformable to accommodate misalignments, and should be able to embed dust particles. Some of these properties are conflicting. This applies for instance for fatigue life versus embedability. The first generally requires application of the bearing material in thin layers (the normal forces are then effectively carried by the hard steel substrate) while embedability, in particular of coarser particles, requires a considerable layer thickness.

According to the above, material combinations which are suitable for use in plain bearings, which occasionally have to function under conditions of solid-to-solid contact, preferably should possess low values of coefficient of friction, process surface roughness, and wear rate as well as adequate (fatigue) strength properties.

Metallic bearing materials, which more or less fulfill these requirements, usually contain elements which either do not form solid solutions with iron, the main constituent of most journals, or which form brittle intermetallic compounds with iron. The first fact is connected with the widely used, though controversial, rule which states that the tendency towards adhesion is the lowest for pairs of metals with almost zero mutual solubility, i.e. for metals which are the least compatible in a metallurgical sense. This point of view, which has been defended most strongly by Rabinowicz, has led to the formation of "compatibility chart" for various metal combinations, derived from binary diagrams of the respective elements[10].

In weighing the arguments of supporters and opponents of the "solubility rule," it should first be pointed out that strong forces between atoms of different types frequently lead to the formation of intermetallic compounds, thereby reducing rather than increasing solid solubility. In that case there will be no correlation between solid solubility and tendency towards adhesion. Further, it should be noted that the evidence which led to the design of the compatibility chart, without exception, concerns technical, that is, contaminated systems, in which oxide films and - in some cases - lubricants were present. This implies that, under the relevant conditions, contact between the virginal metal sufaces was largely inhibited by interfacial layers. This seems to be a crucial factor because experiments in a vacuum showed beyond doubt that very clean surfaces of virtually insoluble metals such as silver and iron easily form strong adhesive junctions[11]. The above leads to the conclusion that the solubility rule is of practical consequence only for (lightly) contaminated surfaces, for which the interatomic forces are reduced to an appreciable extent.

The formation of brittle, intermetallic compounds in the friction interface can be a beneficial factor because, if such compounds are formed, the adhesive junctions tend to fracture in the original interface, with the result that adhesive material transfer from one sliding component to the other and - ultimately - adhesive wear does not occur. Obviously, a necessary requirement for brittle compound formation is that a certain amount of interfacial diffusion occurs. In practical systems one can hardly imagine a situation in which instantaneous diffusion will take place to any measurable extent during a single contact cycle. In such cases a certain amount of preliminary transfer will take place. Upon subsequent contacts frictional heating will facilitate interfacial diffusion with the result that brittle interfacial compounds can be formed, the transferred material remains attached for a small number of contact cycles only, and the wear particle, which forms ultimately, remains very thin.

Good examples of elements which do not form (solid) solutions with iron are lead and silver. The first element is a well-known component in lead based white metals and in leaded bronzes; the second metal can be used as such in the form of thin plating. Examples of brittle compound forming elements are tin and antimony. Again these are used frequently as main alloying components in bearing materials, most notably in white metals or babbitts, of which the lead babbitt Pb15Sb10Sn and the tin babbitt Sn12Sb are important representatives.

In many bearing materials, the alloying element with the favorable tribological properties, is present in the form of a separate, soft phase. This is, for example, the case in copper-lead, bronze-lead, and aluminum-tin bearings. In the first two cases, lead forms the second, soft phase; in aluminum-tin alloys this is the case for tin. Under load the softer phase is partly extruded and the interface becomes covered with a thin layer of lead or tin, respectively.

Obviously, in bearing materials, the ability of a metal surface to form suitable boundary lubricant films by reaction with the lubricant is an equally important feature. An example of this has already been given in the introduction (friction polymer formation, catalysed by finely divided tin oxide).

Similar, although not always equally strong effects are frequently found with other materials.

As stated above, plain bearings may also suffer from wear due to the presence of hard abrasive particles in the lubricant. This wear takes the form of (usually very mild) erosion if the effective width, d, of the abrasive particles is smaller than the minimum thickness of the lubricant film, h_{min}. However, if $d > h_{min}$ and in particular under conditions of boundary lubrication, more or less severe abrasive wear occurs. Under such conditions the power of the bearing surface to embed foreign particles (i.e. its embeddability), the surface hardness in the fully work-hardened condition and the elasticity of the journal are important criteria. Further, surface active components in the lubricant may play an important role in reducing the cutting action of angular abrasive.

Although the above properties (i.e. embedability, hardness, elasticity, and "lubricity") may be determined separately, Working Group 6 of Subgroup 2 of ISO/TC123 has recommended to adopt a "systems approach", i.e. to study the reaction of the system bearing material - lubricant - journal material to introduction of abrasive particles in the lubricant[7]. In principle this can be done in a laboratory apparatus, for instance of the "pin-on-ring" type, as well as in a real journal bearing. In both cases, the main criterion is the degree of damage (scratch formation) to the journal surface. This can be characterized adequately by means of Talysurf tracings as shown in Figure 6 for four classes of materials I - IV. In a specific test it was found that the white metals Pb Sb15Sn 10 and SnSb12 fall in class I, characterized by the fact that the surface condition of the steel mating surface remains virtually unaffected[12]. With Ag and CuSn6Pb10, class II, some local roughening of the surface of the ring occurs. Next, CuSn6Pb10 (P) and CuPb20 form class III, with appreciable roughening of the ring surface and finally AlSn20 and CuSn8 fall in class IV, characterized by severe scratching of the contact surfaces.

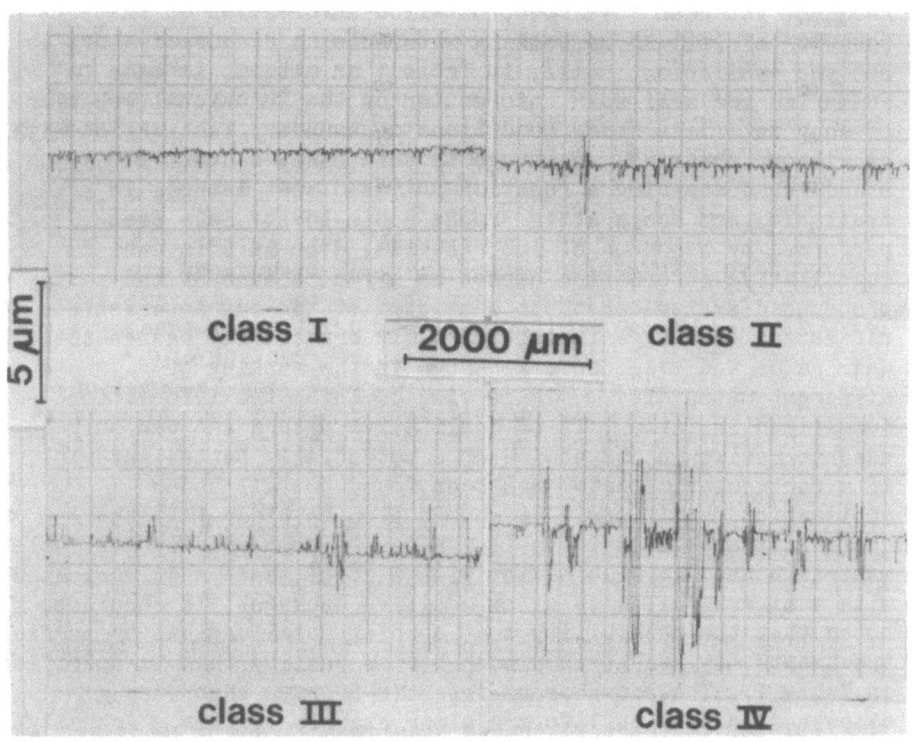

FIGURE 6 TALYSURF STYLUS TRACING OF STEEL SHAFTS, RUN AGAINST DIFFERENT BEARING MATERAILS UNDER CONDITIONS OF BOUNDARY LUBRICATION IN THE PRESENCE OF ABRASIVE DUST

As the mating surface, steel SAE 1045, and the lubricant were kept constant, differences in behavior must be due to differences in embeddability. The results show that this embeddability apparently does not correlate in a simple way with hardness. In fact the hardness of the AlSn20 alloy and that of the white metals were the same, while the embeddability apparently differed considerably. Another striking feature is the significant difference in behavior between the phosphorus-free and the phosphorus containing, CuSn6Pb10 alloy. In a lubricant without abrasive, this difference is explained in terms of friction polymer formation. The present results permit the conclusion that such polymers also considerably mitigate the scratching action of hard particles in the lubricant.

If only because of the fact that for the greater majority of criteria, shown in Figure 5, generally accepted test methods

have not yet been developed, detailed information on the
behavior of well-known bearing materials as discussed above is
not yet available. Still, in Table 1 an attempt is made to
summarize the available information on the functional behavior
of such materials under conditions of boundary lubrication with
or without abrasive. Although in this case, a draft ISO
standard, describing a function-oriented test method, is
available, and cooperative "round robin tests" have been
performed by a number of laboratories, even in this case
quantitative information cannot be given because of the
pronounced system-dependent character of the characteristic
parameters (a change in lubricant may drastically change the
wear rates and even cause a change in the ranking of
different materials)[7, 9]. In view of this, the information
summarized in Table 1 is of qualitative nature and can only be
used to assist in the choice of a potentially better material,
if a particular material has been found to fail under
application conditions. It should also be clear that the
qualifications given in Table 1 refer to commercially available
materials used in combination with a steel shaft and lubricated
with a mineral oil with a low additive package. It should be
noted that the possibility that special metallurgical or surface
treatments may result in a much better quality than is specified
in Table 1. The phosphorous-free tin-bronzes that were
discussed in Section 1 form a clear example of this. Actually,
the behavior of these materials under conditions of boundary
lubrication is so superior that, in the second column of Table
1, they would easily score "+++"; however, they are not
commercially available.

3. MATERIALS FOR ROLLING ELEMENT BEARINGS, GEARS, CAMS AND TAPPETS

3.1 Similar Characteristics

Rolling element bearings, gears, cams and tappets possess
a number of similar characteristics as shown schematically in
Figure 7. They are:

- Counterformal two body contact situation, leading to
  relatively high (Hertzian) contact pressures.
- Surfaces usually made from hardened steel.
- Elastohydrodynamic lubrication, characterized by the fact
  that the lubricant film formation is caused by
  hydrodynamic pressure built-up in the lubricant as well
  as by elastic deformation of the contact surfaces. As a
  result of such deformation, the thickness of the
  lubricant film is considerably larger than follows from
  "classical" hydrodynamic theories, which assume
  undeformable surfaces.

Material	behaviour under conditions of boundary lubrication	
	without abrasive particles[1]	with abrasive particles[2]
Pb-Babbitt (10 μm layer)	++	+
Pb-Babbitt (100 μm layer)	++	+++
Sn-Babbitt (10 μm layer)	+++	+
Sn-Babbitt (100 μm layer)	+++	+++
Silver (10 μm layer)	++	±
Silver (100 μm layer)	++	+
Aluminium-Tin	+	+
Copper-Lead	++	++
Sn-Bronze[3]	+	±
Sn Pb - Bronze	+	++
Al-Bronze[4]	+	+

Notes

1) Rough estimate, based on data from different sources, concerning coefficient of friction, wearrate and process surface roughness (contact parameter); c.f. Fig. 5

2) Rough estimate, based on data from different sources, concerning damage to the steel shaft (wear and scratching)

3) See also the note in the text on phosphorus-free tin bronzes

4) Can only be used in combination with a hardened, for instance carburized, shaft.

Explanation of signs

+++ excellent
++ good   } in most technical applications
+ fair
± unsatisfactory for a number of applications.

TABLE 1 FUNCTIONAL BEHAVIOR OF SOME WELL-KNOWN BEARING MATERIALS, OPERATING UNDER CONDITIONS OF BOUNDARY LUBRICATION

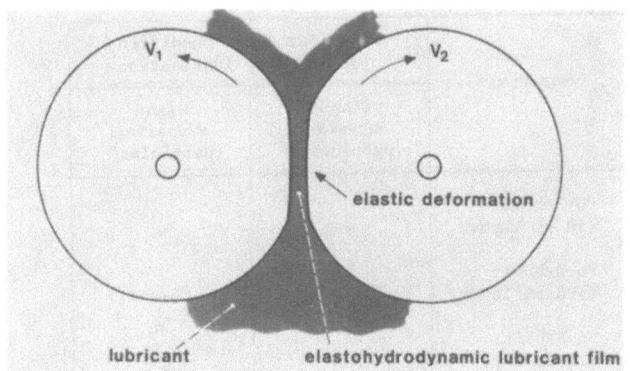

FIGURE 7 SIMILAR CHARACTERISTICS OF ROLLING ELEMENT BEARINGS, GEARS, CAMS, AND TAPPETS

In spite of the above similarities, there are also pronounced differences in tribo-technical behavior between rolling element bearings and rolling cams on the one hand and sliding cams on the other hand, while gears occupy an intermediate position. The main difference in this respect is that, in rolling element bearings and in rolling cams, the percentage of slip (sliding motion) usually is of the order of 1%. The same is true for gear teeth at the rolling circle. However, at top and bottom of gear teeth, up to 40% slip may occur, while sliding cams function with a slip percentage of 100%.

This difference in type of motion has important consequences for the tribological behavior of the materials, in particular because coefficients of friction in pure rolling are on the order of $f = 0.001 - 0.01$, while under conditions of sliding motion, in the presence of a lubricant, coefficients of friction are on the order of $f = 0.01 - 0.1$. As frictional energy is dissipated, mainly in the form of heat, this difference in friction results in the fact that the contact process in rolling element bearings and rolling cams nearly always proceeds at ambient temperature, while in sliding cams relatively high "friction induced temperatures" can occur. This difference in "thermal loading" has important consequences for the formation of an elastohydrodynamic lubricant film and - in connection with that - for the occurrence of different wear mechanisms. In rolling element bearings and rolling cams, functioning under "ideal" conditions (no misalignment, viscosity of the lubricant carefully chosen, no abrasive dust in the lubricant), one usually finds that the life of the components is limited by the occurrence of spalling (surface fatigue), caused by the cyclic variations in normal force. The same holds for gear teeth at the rolling circle. However, at top and bottom of gear teeth and in sliding cams, the relatively high friction induced temperatures frequently interfere with lubricant film formation, as a result of which corrosive, abrasive, or adhesive wear may occur in an early stage of use.

3.2 Surface Fatigue/Spalling

In general, the resistance against surface fatigue of components of rolling element bearings and rolling cams increases with increasing hardness. However, as fatigue cracks are usually intitiated at sites in the material where high local stresses occur, local defects and impurity inclusions play a role of major significance. This aspect of fatigue induced spalling of rolling elements has been covered extensively by Scott et al[13,14,15]. As an example, Figure 8 shows the relation between lifetime under conditions of rolling contact and the (calculated) minimum film thickness/roughness ratio, h/R, for two cylinders (initial line contact), made of maraging steel ASTM A-538, grade C, lubricated with SAE 10W base oil. Two different manufacturing methods of the steel were applied, i.e. vacuum melting and electroslag remelting[16]. In the present case the minimum film thickness, h, was calculated with the equation of Dowson and Higginson for inital line contacts[17]:

$$h = \frac{1.6 \alpha^{0.6} (\eta_0 u)^{0.7} (E')^{0.03} R^{0.43}}{F_N^{0.13}} \quad (1)$$

in which   is the pressure exponent of the viscosity, defined by:

$$\eta = \eta_0 e^{\alpha p}$$

η_0 = dynamic viscosity of the lubricant at p = 1 atm.
$u = \frac{1}{2} (v_1 + v_2)$, in which v_1 and v_2 are the surface speeds of the cylinders

$$E' = \frac{E}{1-v^2}$$

$R = \frac{R_1 R_2}{R_1 + R_2}$, in which R_1 and R_2 are the radii of curvature of the cylinders.

In the example given in Figure 8, R_1 = 74 mm and R_2 = 77 mm. The rotation frequency of the cylinders was 50 s^{-1}, so that v_1 = 23.24 m/s and v_2 = 24.18 m/s. The percentage of slip (sliding motion) thus amounted to:

$$\frac{v_1 - v_2}{\frac{1}{2}(v_1 + v_2)} \cdot 100\% = 4\%$$

The composite roughness parameter R was calculated with:

$$R = (R_{m_1}^2 + R_{m_2}^2)^{\frac{1}{2}}$$

in which R_{m_1} and R_{m_2} are the standard deviations in the distribution of roughness heights (r.m.s values) of cylinder 1 and cylinder 2, respectively.

In calculating h with Equation (1), it was assumed that during a test the viscosity of the oil was constant, i.e. that the temperature did not change significantly as a result of friction. As the percentage of slip was as low as 4%, this seemed a reasonable assumption. Variations in h/R were realized by changing the oil bath temperature, T, by which the value of η_o could be varied between $9.5 \cdot 10^{-3}$ Ns/m^2 (at T = 50°C) and $4.0 \cdot 10^{-3}$ Ns/m^2 (at T = 90°C).

The R-values were calculated for the new, as machined surfaces. Thus, the h/R values in Figure 8 refer to the lubrication condition directly upon the application of normal force. During running-in, the highest roughness peaks on the ring surfaces wear away, with the result that, upon running-in, h/R increases appreciably.

At each h/R ratio that was applied (i.e 0.7, 1.0, 1.2 and 2.5), ten pairs of vacuum melted rings and ten pairs of electroslag remelted rings were tested. The "L_{50} -lifetimes," given in Figure 8, were defined as the time which elapsed until five ring combinations in each group of ten, were damaged as a result of spalling fatigue. Figure 9 shows typical spalling damage which, in the present case, occurred suddenly after $7 \cdot 10^6$ revolutions.

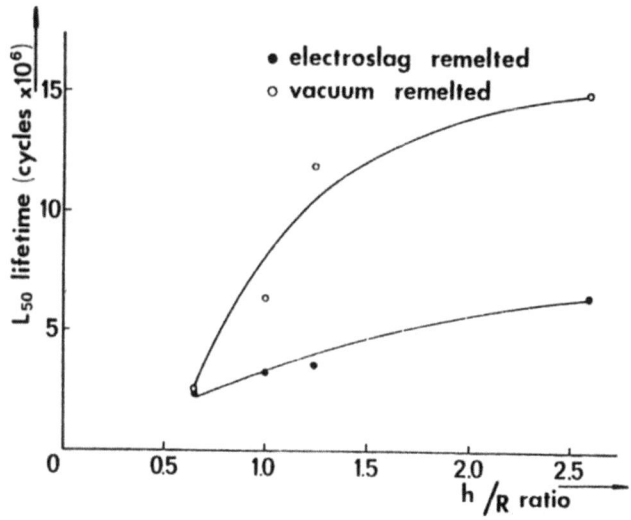

FIGURE 8 THE RELATION BETWEEN L_{50} LIFETIME AND (CALCULATED) MINIMUM FILM THICKNESS/ROUGHNESS RATIO $\frac{h}{R}$ FOR CONTACTING CYLINDERS (INITIAL LINE CONTACT OF MARAGING STEEL ASTM A-538, LUBRICATED WITH SAE 10 W BASE OIL

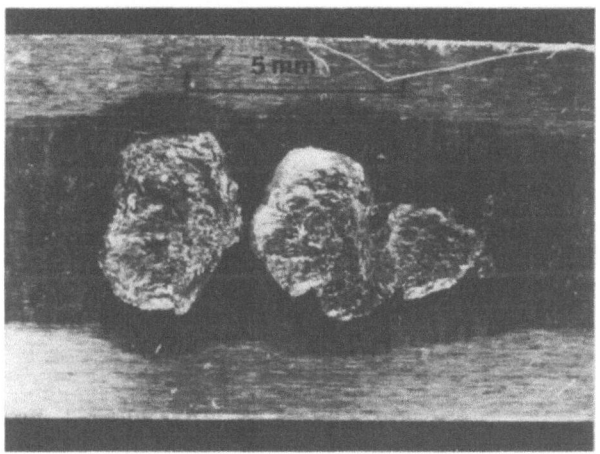

FIGURE 9 SPALLING DAMAGE

Figure 8 shows that for both steel batches the lifetime decreases with decreasing h/R ratio. This is due to the fact that the lower the value of h/R, the more roughness peaks penetrate the elastohydrodynamic lubricant film, in particular during the running-in period. During such penetration, the surfaces are locally damaged which induces the initiation of fatigue cracks from the surface into the bulk of the material. Figure 8 also shows that at h/R > 0.6 the vacuum melted steel has a significantly longer lifetime than the electroslag remelted steel. The reason for this difference is that the vacuum melted batch contains less inclusions than the electroslag remelted batch.

At h/R > 2.5, crack initiation at the surface becomes a rare event. In fact fatigue failure in this lubrication regime is nearly exclusively due to crack initiation below the surface, in which stress concentrations around inclusions play an important role.

Metallographic examination of bodies, subjected to high cyclic loading, frequently shows the existence of "white etching zones" which frequently turn out to be prone to crack formation. Typical white etching zones, from which cracks originate, are shown in Figure 10. Metallographic work showed that the white etching zones in maraging steel consist of a soft modification of the material (hardness 300 HV instead of 560 HV), but with the same chemical composition as the surrounding matrix. It was also found that precipitation hardening as 480°C completely restores the original structure and hardness of the material.

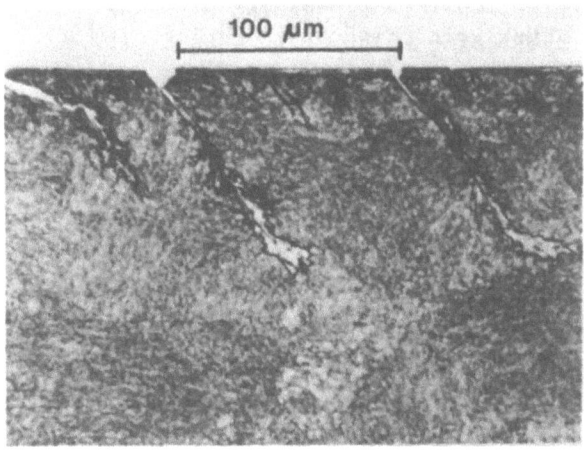

FIGURE 10   WHITE ETCHING ZONES FROM WHICH CRACKS ORIGINATE

As stated above, the pronounced decrease in lifetime at decreasing h/R ratio, is due to minor damage to the surfaces during running-in under conditions of asperity contact. A similar unfavorable effect can be due to corrosive attack of the steel surfaces. In this respect, it can be particularly dangerous if the lubricant contains water. Through a mechanism of preferential adsorption of water on the steel surface, followed by capillary condensation in tiny cracks formed during machining of the surfaces and local expansion, dangerous cracks can be formed on the surface which can lead to premature failure of the components, "infant-mortality."

Because of the pronounced influence of inclusions on spalling fatigue, very pronounced differences in lifetime can exist between different batches of steel with the same nominal chemical composition. An example of such a batch effect, dating back to the late sixties, is shown in Figure 11[18]. This shows the L_{10} lifetimes of the inner rings of 23 different groups of geometrically identical ball bearings, which were manufactured in an identical way, from different batches of vacuum degased ball bearing steel AISI 52100. The different batches were produced by seven different steel manufacturers. The L_{10} lifetime is defined as the time which elapses until 10% of the bearings in each group has been damaged by spalling fatigue. It can be seen that from one manufacturer to the other, pronounced differences occur. In particular, the results obtained with the materials from manufacturer E illustrate quite clearly that apparently identical materials may behave quite differently under conditions where surface fatigue predominates as a wear mechanism. It should be noted that since 1970 the situation has improved considerably. This is mainly due to the fact that steel to be used for rolling element bearings is now produced very carefully, thus keeping the impurity content and in particular, the distribution of the impurities in the material constant.

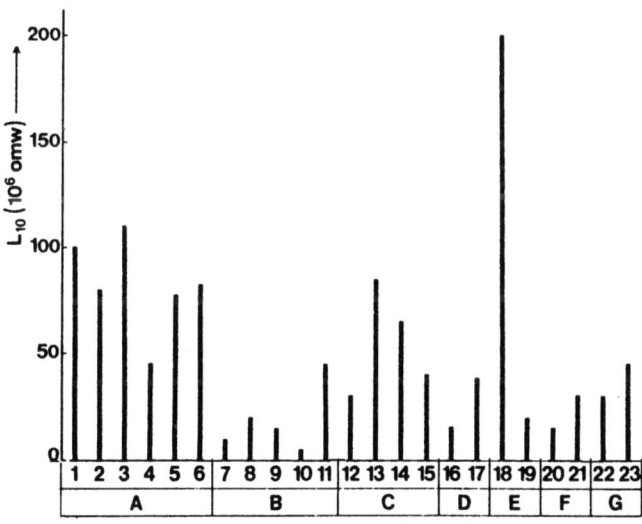

FIGURE 11 THE EFFECT OF STEEL QUALITY (BATCH EFFORT) ON THE ENDURANCE (L_{10} LIFETIME) OF BALL BEARINGS (FROM REF. 17)

1, 2, 3 .... 23: batch numbers
A, B, ... G: steel manufacturers

3.3 The Behavior of Real Rolling Element Bearings

Provided that rolling element bearings are assembled in accordance with the directions of the manufacturer, that a lubricant of adequate viscosity is used and that the lubricant does not contain water or dirt, the lifetime of the bearing is limited exclusively by subsurface (spalling) fatigue. Under such conditions the well-known Lundberg-Palmgren equation applies:

$$L_{10} = (\frac{C}{P})^k \cdot 10^6 \qquad (2)$$

C is called the basic dynamic load capacity, defined as the force on the bearing under which 90% of the bearings has a lifetime of more than 1 million revolutions. P is the equivalent dynamic bearing load, which can be found by adding the radial force, Fr, and the axial force, Fa, according to:

$$P = XF_r + YF_a \qquad (3)$$

in which X and Y are constant factors for a particular type of bearing. The exponent k in Equation (2) is equal to about 3.

For each desired value of the L10-lifetime and for a known value of P, Equation (2) yields a value for C, on the basis of which, for each type of rolling element bearing, the necessary dimensions (diameters) of the bearing can be found in catalogues provided by the various rolling element bearing manufacturers.

The Lundberg-Palmgren Equation yields values for the L_{10}-lifetime, i.e. the time which elapses until 10% of the bearings in a larger group have suffered from spalling fatigue. Obviously, there may be cases in which one desires to obtain information on the time to lower failure percentages (e.g. L2-values). To that purpose, the L_{10} -values found, according to Equation (2), can be multiplied with the appropriate a_x factors given in Table 2, for failure percentages of 5, 4, 3, 2, and 1, respectively.

x	a_x
10	1.00
5	0.62
4	0.53
3	0.44
2	0.33
1	0.21

TABLE 2 CALCULATION OF TIME TO FAILURE PERCENTAGES $\leq 10\%$, ACCORDING TO: $L_x = a_x \cdot L_{10}$

It goes without saying that the Lundberg-Palmgren Equation cannot yield a prediction for the lifetime of one particular bearing. It only predicts the "average endurance" of a large amount of seemingly identical bearings. Even then one should be careful, because the Lundberg-Palmgen equation only applies under ideal conditions. Faulty assembly, abrasive dust in the lubricant, etc. can cause disasterous and essentially unpredictable reductions in lifetime.

3.4 Counterformal Contacts under Conditions of Sliding Friction

If counterformal surfaces are in contact under conditions of sliding friction (as in sliding cams and tappets), and elastohydrodynamic lubricant film can still be built up, provided that the conditions are suitable. However, because an appreciable amount of frictional heat is generated under conditions of sliding contact, the system is this case is certainly not isothermal by nature.

It has been shown that the lubrication condition of sliding counterformal steel contacts can be described in the form of a transition diagram, which defines the lubrication condition as a function of normal force F_N, sliding speed v and oil bath temperature, T[19, 20, 21, 22, 23]. Such diagrams apply equally well to ball-against-ball, ball-against-cylinder, crossed cylinder contacts, and also, they apply to virginal as well as to run-in surfaces. A cross section at constant T, for a completely oil submerged point contact, shows three regions: Figure 12; region I, in which friction is low and wear is nil or very low, region III, where severe wear and scuffing occur and an intermediate region II, characterized by a (temporarily) high friction but mild wear. It is assumed that in the three regions the following lubrication mechanisms apply:

region I: (partial) elastohydrodynamic lubrication

region II: boundary lubrication

region III: unlubricated contact (although the specimens are still fully submerged in the lubricant).

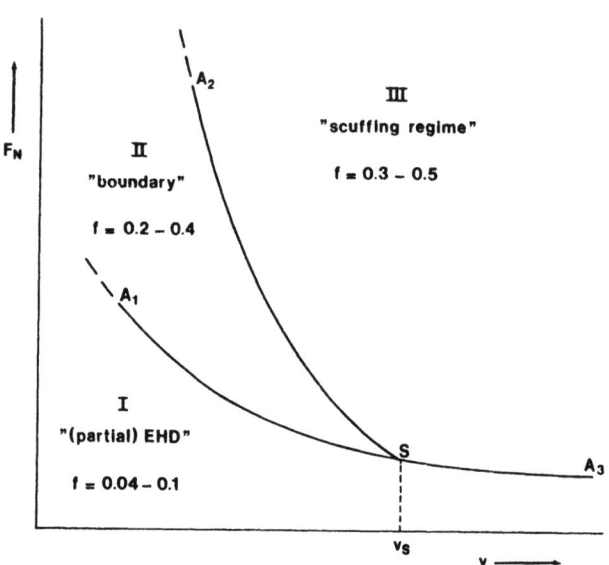

FIGURE 12 CROSS SECTION AT CONSTANT TEMPERATURE OF A TRANSITION DIAGRAM, APPLICABLE TO COMPLETELY OIL SUBMERGED SLIDING POINT CONTACTS

The lower curve A1-S-A3 has been called the "primary transition curve". It is believed to be continuous, point S merely being the intersection between curves A1-S-A3 and A2-S. Early work on the subject as well as recent research show that the location of the curve A1-S-A3 depends on viscosity. This is the main reason for assuming that in region I a thin elastohydrodynamic lubricant film keeps the surfaces apart[24, 25, 26]. The pronounced decrease in load-carrying capacity at increa sing speed of sliding is supposed to result from the effect of frictional heating in the contact zone, causing an appreciable decrease in effective lubricant viscosity[24]. Upon running in regions II or III, reaction (oxide) layer formation may cause a secondary transition towards a stable regime with very little wear. This regime is characterized by a coefficient of friction, $f = 0.1$ and a specific wear rate $< 0.1 \cdot 10^{-6}$ mm^3/Nm. If this regime is reached after functioning for a certain period of time in region II, the component normally is still serviceable. In fact, region II is probably identical to the region of "incipient scuffing," which is sometimes found in actual machine components[27]. On the contrary, when functioning in region III, the component usually becomes unserviceable long before protective films have formed. Figure 13 shows characteristic friction force - time records, obtained at testing ball-against-cylinder contacts, Figure 14, at v = 0.5 m/s in marine diesel engine oil of 60°C under air cover, and at three different values of normal force, F_N.

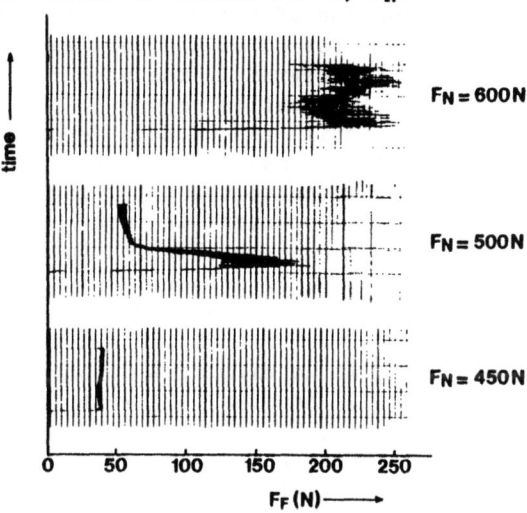

FIGURE 13   CHARACTERISTIC FRICTION FORCE - TIME RECORDS FOR BALL-AGAINST-CYLINDER CONTACT, LUBRICATED WITH MARINE DIESEL ENGINE OIL OF 60°C UNDER AIR COVER

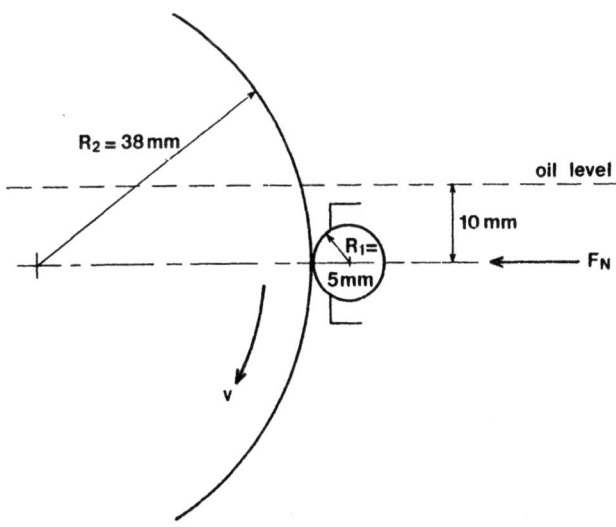

FIGURE 14   BALL-AGAINST-CYLINDER GEOMETRY

At normal forces F_N = 450 N, 500 N, and 600N, the system runs in, respectively, regions I, II, and III. Figure 13 clearly shows that in region II the process is regenerative by nature, friction decreasing rapidly to its f = 0.1 level. In this particular example, this took about 25 seconds. Figure 15 shows characteristic parts of the worn surfaces of the test specimens after termination of the given experiments. From the widths of the wear scars formed on the ball surfaces it can be seen that, in lubrication regime I, the wear rate of the pin is very small indeed. In lubrication regime II the wear rate is much higher, athough it still can be considered acceptable for most practical applications, mainly in view of the fact that the occurrence of wear is limited to the brief high friction period. In lubrication regime III, however, the wear rate is extremely high, which is completely unacceptable in practice. The corresponding ring surfaces, after termination of the tests, are also shown in Figure 15. It can be seen that, at F_N = 450 N (region I), the ring surface remains virtually undamaged. Actually, in most cases it is difficult to trace the wear track. At F_N = 500 N (region II), however, the wear track is clearly visible with the naked eye. Microscopic observation reveals that, in this region, adhesive material transfer occurs at a very minor scale. Finally, at F_N = 600 N (region III) severe adhesive wear effects are found upon inspection of the ring surface. This condition is similar to that of gear teeth which suffered from severe scuffing.

As stated earlier, the primary transition curve A1-S-A3, Figure 12, is assoicated with collapse of a partial EHD lubricant film. The location of the secondary transition curve A2-S, however, is probably associated with a metallurgical transformation in the steel, occurring within a narrow range of contact temperatures of the order of 500 - 600°C

Elastohydrodynamic action being responsible for the partial separation of the surfaces in region I, one would expect the load-carrying capacity of the EHD film, the curve A1-S-A3, to decrease with decreasing speed in the very low speed region (effect not shown in Figure 12). Recent research showed that this curve does indeed show a maximum, occurring in the speed range of v = 0.001 - 0.003 m/s.

It goes without saying that the actual location of the transition curves in the transition diagram depends on a great number of factors, of which the most important are:

- the viscosity-temperature behavior of the lubricant
- the radii of curvature in the contact area and the local Hertzian pressure which results from these radii

235

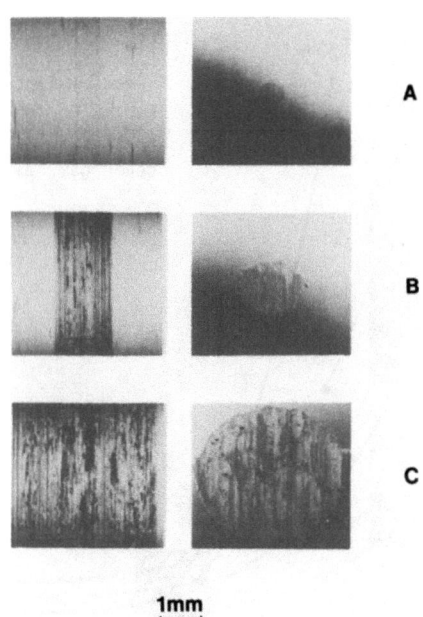

1mm

FIGURE 15 CHARACTERISTIC PARTS OF THE WORN SURFACES
OF TEST SPECIMENS, PRODUCED IN,
RESPECTIVELY, REGIONS I, II, AND III
OF THE TRANSITION DIAGRAM (FIGURE 12)

- the roughness of the contacting surfaces, the chemical composition, and the degree of aeration of the lubricant
- the chemical composition and structure of the steel.

As an example, Figure 16 shows the effect on the transition diagram of adding the lubricant additive zincdialkyldithiophosphate (ZDP) to a mineral base oil. It can be seen that addition of ZDP results in a shift of the total diagram to higher F_N-values, which means that the load-carrying capacity of the system increases over the whole range of experimental conditions.

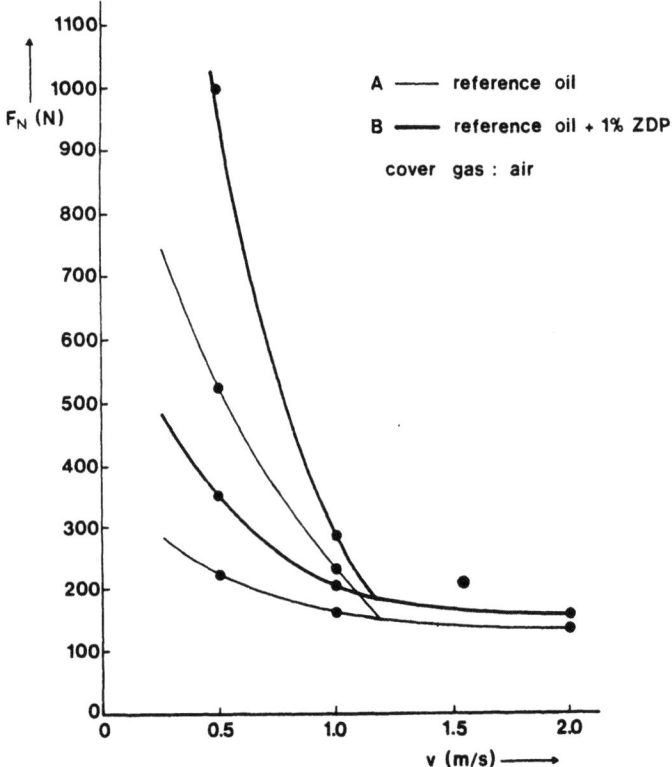

FIGURE 16 THE EFFECT OF ZINCDIALKYLDITHIOPHOSPHATE (ZDP) ON THE TRANSITION DIAGRAM FOR OIL SUBMERGED SLIDING POINT CONTACTS

Results obtained with different types of steel in mineral baseoil without additives under air cover, are shown in Figure 17. Three types of steel were used; steel AISI E 52100 with 1.5 wt.% chromium, steel AISI A2 with 5.6 wt.% chromium, and steel AISI D2 with 12 wt.% chromium. Each steel was subjected to three different heat treatments, which resulted in percentages of retained austenite (R.A.) of 2-4%, 17-21%, and 55% respectively.

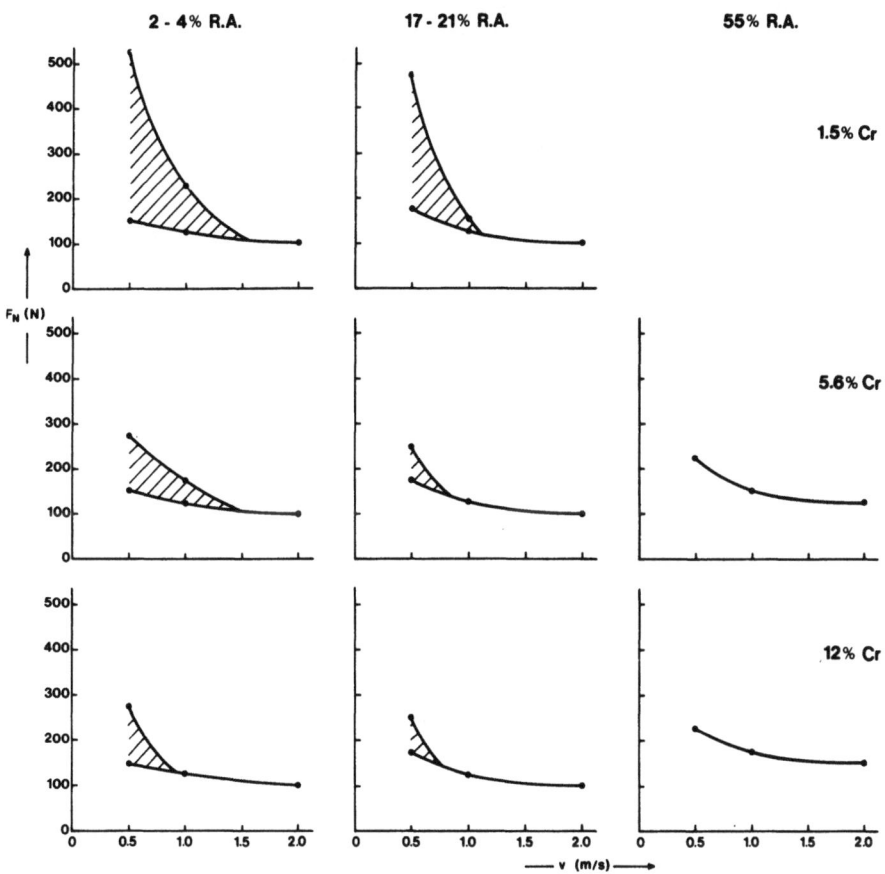

FIGURE 17   TRANSITION DIAGRAMS FOR STEELS WITH DIFFERENT PERCENTAGES OF CHROMIUM AND RETAINED AUSTENITE (R.A.)

Figure 17 shows that the chromium content of the steel has little or no effect on the load-carrying capacity of the partial EHD film (i.e. the location of the primary transition curves in the different diagrams). On the other hand, the chromium content is found to influence quite strongly the second primary transition (i.e. from the region of boundary lubrication or incipient scuffing to the scuffing region). In fact, at v = 0.5 m/s, the load-carrying capacity decreases from about 500 N for steel E 52100 to about 275 N for steels A2 and D2. The differences in heat treatment, as manifest in the R.A. percentages, are found to be of a dualistic nature. First, the load-carrying capacity of the partial EHD film shows a tendency to increase with increasing R.A. percentage. Secondly, the load-carrying capacity of the boundary film tends to decrease with increasing R.A. percentage. As a result of these effects, the importance of regime II decreases appreciably with increasing percentage of alloying elements as well as with increasing percentage of retained austenite. This is borne out quite clearly by the diagrams shown in Figure 17.

Although a first primary transition (collapse of the EHD lubricant film) has been observed in many different systems, severe wear and scuffing in region III only occurs if untreated steel surfaces are used. Surface treatments (most notably salt bath nitriding) may result in a relatively mild wear process, even in region III. In such a case, wear remains mild until the protecting surface layer has been worn away. This aspect of steel-steel interaction is discussed in Section 6.

4. MATERIALS FOR DRY BEARINGS

4.1 General Aspects

Dry running, unlubricated bearings should not be used without special reason; if at all possible (and economically feasible), lubrication should be applied. The main reason for this is that with a few exceptions, only lubricated bearings, and obviously bearings in which the journal is kept in position by magnetic forces, can function completely without wear. Probably the only exceptions to this rule are lightly loaded rolling element bearings (although without lubrication it will be hard to avoid some cage wear), lightly loaded plastic bearings (as can be found, for instance, in office equipment), and bearings operating in vacuum and lubricated with molybdenum-disulphide. In such systems, the no-wear running period may sometimes surpass the desired lifetime. In lightly loaded plastic bearings in which the effects of contact with the mating component, usually a steel journal, are limited to elastic deformation of the surface asperities, the no-wear period is limited by the occurrence of micro-fatigue effects, as have been described by Bayer et al [28].

Also in dry running rolling element bearings the no-wear period is limited by fatigue. However, in contrast to the situation with plastic bearings, the damage after the first occurrence of fatigue effects in rolling element bearings usually is of such intensity that further use in the "wear period" is usually impossible. The behavior of bearings, lubricated with molybdenum-disulphide, is discussed in detail in subsection 4.4.

Reasons for choosing a dry running, unlubricated bearing, even in cases where a certain amount of wear may be expected, can be listed as follows:

1. The bearing has to function at high or low temperature, realistic boundaries being +250°C and -50°C.

2. The bearing has to function in vacuum.
   As far as this point is concerned it should be realized that nowadays lubricants with extremely low vapor pressure (10 - 50 Pa at 200°C) are commercially available.

3. The environment in which the bearing has to function must be kept extremely pure, so that leakage of lubricant cannot be allowed.
   Although, in principle, very efficient lubricant seals are available, dry running bearings are, for this reason, frequently applied in the textile industry, domestic appliances, etc.

4. The bearing has to function in a process fluid with nonlubricating (perhaps even degreasing) properties. Notwithstanding the presence of a fluid, one can, in these cases, often consider the bearing as "running dry."

5. The price of the equipment in which the bearing is to be used and/or the possibilities for maintenance are such that application of a lubricated system is virtually impossible.
   Again, this situation is found frequently in the field of domestic appliances; still in such equipment the grease lubricated spiral groove bearing is slowly gaining ground[29].

4.2 Materials

Materials for dry running, unlubricated bearings are frequently built-up on the basis of the following components which in some cases, may also be used as such:

- plastics (thermoplastic polymers and thermosetting resins)
- carbon and graphite
- metals with food running properties against iron such as silver, antimony and lead; see subsection 2.3
- molybdenum-disulphide MoS_2.

Common examples of materials which belong to one or more of these categories are described by Lancaster in the "Tribology Handbook"[30]. Data, taken from this reference, which are very useful in deciding which material to apply in a specific practical application, are reproduced here in the form of Tables 3 and 4.

Table 3 contains information on plastic based materials. Although important differences in properties may occur from one material to the other, they have certain features in common:

- a relatively low maximum allowable surface temperature
- a relatively high coefficient of thermal expansion
- a very low heat conductivity.

Material		$\bar{p}_{max}$ MN/m^2	T_{max} °C	$\frac{\Delta L}{L}$ $10^{-6}/°C$	λ $\omega/m\ °C$
thermoplastics	nylon, acetal	10	100	100	0.24
thermoplastics + fillers	fillers: - MoS$_2$ - PTFE - glass - graphite, etc.	15-20	150	60-100	0.24
PTFE + fillers	fillers: - glass - bronze - mica - carbon - metals	2-7	250	60-100	0.25-0.5
thermosets + fillers	phenolics, epoxies + asbestos, textiles, PTFE	30-50	175	10-80	0.4

$\bar{p}_{max}$ = maximum value static projected bearing pressure

T_{max} = maximum service temperature

$\Delta L/L$ = coefficient of thermal expansion

λ = heat conductivity

TABLE 3 SOME CHARACTERISTIC PROPERTIES OF PLASTIC BASED DRY BEARING MATERIALS (AFTER LANCASTER, REF. 29)

These facts are to be taken into account quite seriously in each practical design situation.

The characteristic properties of carbon or metal based materials are summarized in Table 4. Here we see that in contrast to the situation with plastic based bearing materials (summarized in the lower row of Table 4), these materials are characterized by a relatively high maximum allowable surface temperature, a generally quite modest thermal expansion and a relatively excellent heat conductivity. From this, one rapidly gathers that these materials are particularly suited for high temperature applications.

Material		$\bar{P}_{max}$ MN/m^2	T_{max} °C	$\frac{\Delta L}{L}$ 10^{-6}/°C	λ w/m °C
carbon-graphite	may contain resin	1-3	500	1.5-4	10-50
carbon-metal	metal: Cu, Ag, Sn, Pb, Sb	3-5	350	4-5	15-30
metal-solid lubricant	bronze-graphite-MoS$_2$ Ag-PTFE	30-70	250-500	10-20	50-100
"plastics" (summary of Table 3)	frequently with fillers	2-50	100-250	10-100	0.24-0.5

$\bar{P}_{max}$, T_{max}, $\Delta L/L$ and λ: see legend Table 3

TABLE 4 SOME CHARACTERISTIC PROPERTIES OF CARBON OR METAL BASED DRY BEARING MATERIALS (AFTER LANCASTER, REF. 29)

The importance of maximum allowable service temperature and heat conductivity is clearly illustrated by Figure 18, which describes experience acquired in designing a refrigerating plant. Figure 18 shows the behavior of three different bearing materials, each based on a combination of polytetraflourethylene (PTFE) and bronze, used in combination with journals of stainless steel and copper, respectively, and used at an environmental temperature of -20°C. Bearing materials as well a journal materials show pronounced differences in heat conductivity. Bearing materials A and B possess a PTFE matrix, filled with bronze powder, while bearing material C possesses a bronze matrix, filled with PFTE. As a consequence of their structure, materials A and B have a relatively low heat conductivity, 4 to 10 mW/cm K, respectively. In material C, the fact that bronze forms the matrix guarantees a relatively high heat conductivity of 200 mW/cm K.

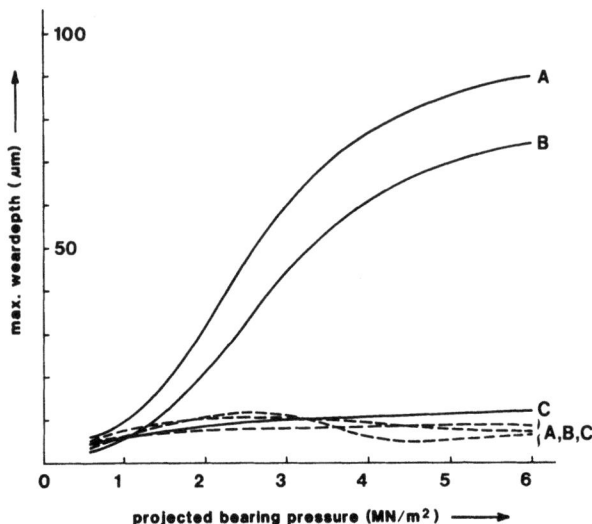

FIGURE 18   WEAR DEPTH AS A FUNCTION OF PROJECTED BEARING
PRESSURE FOR THREE PTFE-BRONZE COMPOSITES, A, B
AND C, USED IN COMBINATION WITH JOURNALS OF,
RESPECTIVELY, STAINLESS STEEL (FULL LINE) AND
COPPER (DASHED LINES)

Figure 18 shows that if materials A and B are used in combination with stainless steel which has a rather poor heat conductivity, a pronounced increase in wear depth occurs in the range of bearing pressures between 1 and 3 MN/m^2. This is caused by a pronounced increase in surface temperature of the bearing as a result of dissipation of frictional heat. Clearly, the heat conductivity of material C is so high that this phenomenon is not found if that material is used. When using copper as journal material, a material with a relatively high heat conductivity, none of the materials A, B, or C show the pronounced increase in wear depth that is found with materials A and B in combination with a stainless steel journal. This example clearly illustrates the important role of heat conductivity in the tribo-technical behavior of bearing materials based on plastic. The total energy that dissipates per unit of projecting bearing surface, as a result of friction (Q'), is given by

$$Q' = f \cdot p \cdot v \qquad (4)$$

in which f is the coefficient of friction, p is the projected bearing pressure, and v is the speed of sliding. A "maximum allowable pv-value" $(pv)_{max}$ is sometimes specified for a particular bearing material. If only because of the fact that f may depend quite strongly on v as well as p, this can be misleading, in particular in the case of plastic based materials. Instead, the use of p against v curves is recommended[30]. For a number of plastic based materials such curves are given in Figure 19.

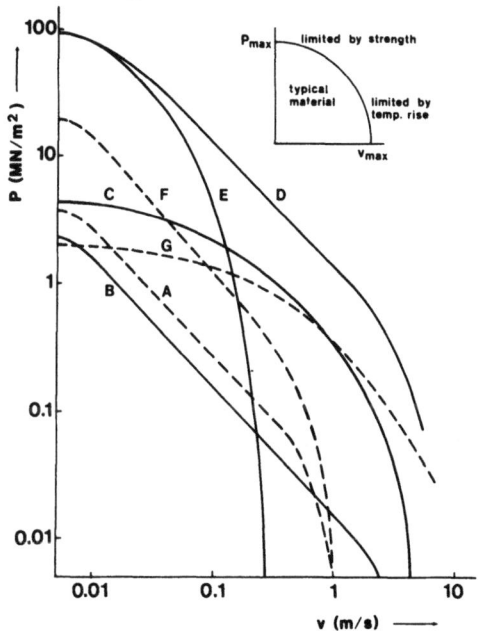

FIGURE 19   P-V CURVES FOR A NUMBER OF PLASTIC BASED BEARING MATERIALS (AFTER LANCASTER, REF. 29)

Clearly, information as given in Figure 19 is based on some form of general concensus as to what wear rates are acceptable for practical application. More precise information on the wear rates to be expected when using dry running bearings is hard to give because of the pronounced systems dependency of such data. As a first approximation, information on the specific wear rates of characteristic dry bearing materials can be found in Reference 31. Figure 20, taken from this reference, relates the specific wear rate, expressed in mm^3 material removed per meter sliding distance, per N normal load, for a number of materials to the projected bearing pressure p. This figure relates to situations in which either the heat production (low sliding speed) or the possibilities for heat conduction are such that "high" temperatures in the bearing are avoided. Figure 20 shows that, even within one "family" of materials, the specific wear rate may easily differ by a factor of 5-10. Therefore application oriented tribo-metry will generally be necessary if one wishes to obtain more precise information[5].

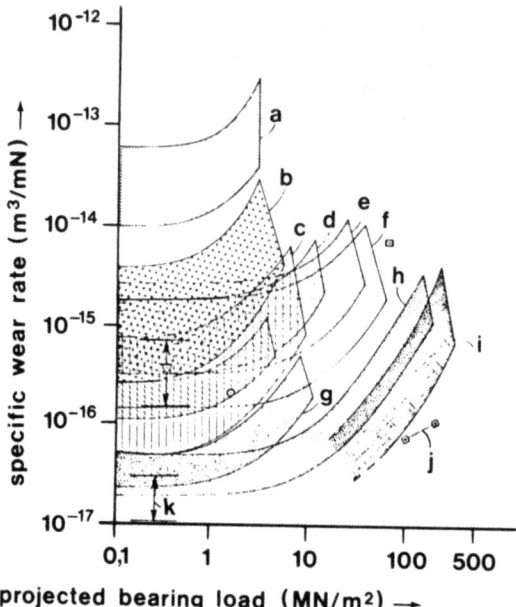

A : PTFE (unfilled)
B : carbon-graphite (including metal and thermoset impregnated types)
C : filled PTFE's
D : filled nylons and acetals
E : reinforced phenolic resin laminates with solid lubricant fillers
F : filled polyimides and p-oxybenzoyl polymers
G : Epoxy-bonded bronze / Pb filled PTFE
H : PTFE / Pb in porous bronze on a metal backing (available wear depth ≈ 0.04 mm)
I : Woven PTFE / glass fibre / metal back (available wear depth ≈ 0.25 mm)
J : PTFE fibre / high strength CF filament wound bearings
K : High modulus CF reinforced thermoset (pin-on-disc results).

FIGURE 20 SPECIFIC WEAR RATE AS A FUNCTION OF PROJECTED BEARING LOAD FOR A NUMBER OF "DRY BEARING MATERIALS" (AFTER CREASE, REF.30)

4.3 The Effect of Surface Roughness and Material Transfer on the Wear of Plastics

Apart from the pronounced effect of friction induced surface temperatures, the roughness of the mating journal surface and possible changes in that roughness, exert a pronounced influence on the tribological behavior of plastic bearing materials. Figure 21, taken from Reference 32, shows this for six different plastics. From the upper part of Figure 21, it can be seen that for all materials, the coefficient of friction, f, goes through a minimum, the location of which

varies with the nature of the material. The reason for this
effect is that at low values of journal roughness, adhesive
forces in the interface predominate (the materials "glue
together"), while at higher roughness values the machining
forces increase considerably, due to the chiselling action of
the asperities of the steel journal surface. The lower part of
Figure 21 shows that although for all materials, wear increases
with increasing journal roughness, the intensity of the effect
differs considerably from one material to the other. The wear
of PTFE depends very strongly and the wear of nylon (PA-66) only
very slightly on roughness.

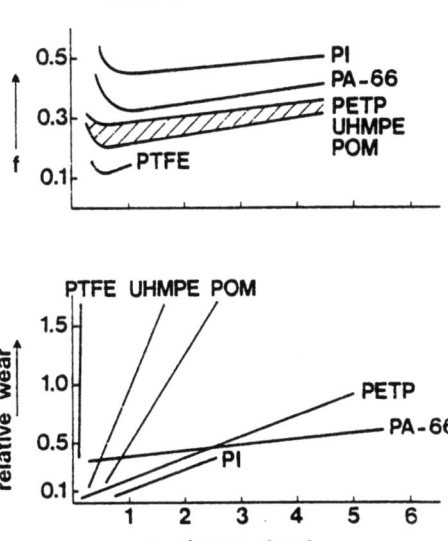

FIGURE 21 FRICTION AND WEAR OF A NUMBER OF UNFILLED PLASTICS AS
A FUNCTION OF THE ROUGHNESS OF THE MATING SURFACE
(AFTER ERHARD AND STRICKLE, REF. 31)

A complicating factor in discussing the effect of roughness
of the mating surface on wear of plastic bearing materials is
that the material of the bearing may transfer to the steel
journal, thus masking the original roughness pattern. This
effect has been studied and discussed comprehensively by
Lancaster[33]. Figure 22, taken from this reference, shows 14
different plastic bearing materials in sliding contact with
steel, the ratio, (λ), between the wear rates, characteristic
for a wear process with transfer and the wear rates,
characteristic for a wear process without transfer, as a
function of the strain at fracture. Figure 22 shows that the
ratio, Ψ, decreases appreciably with increasing ductility
(increasing strain at fracture). For the relatively ductile
plastics, 6-14, the formation of a transferred layer on the

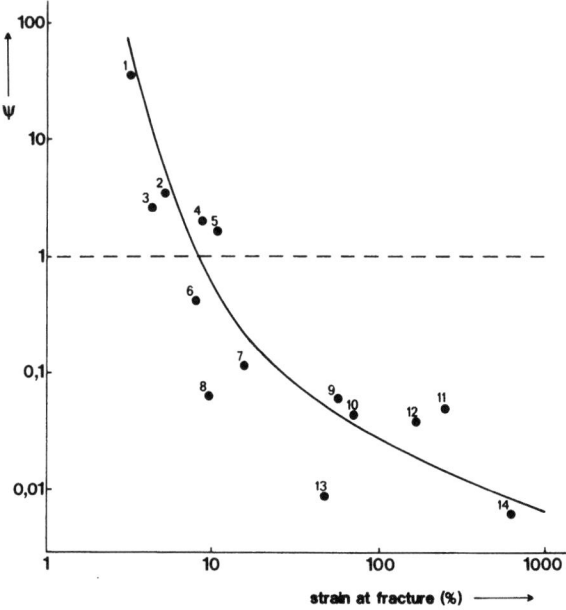

1 : epoxy-828
2 : polystyrene
3 : polyvinylidenechloride
4 : polyether-17449
5 : polymethylmetacrylate (perspex)
6 : polytetrafluorchloroethylene
7 : acrylonitril-butadienestyrene colepolymer
8 : polycarbonate
9 : nylon 11
10 : nylon 6.6
11 : polytetrafluorethylene (teflon)
12 : polypropylene
13 : polyacetale
14 : polyethylene

R_m-value shaft : 0.08 µm
ambient temperature : 25°C.

FIGURE 22 RATIO Ψ BETWEEN THE WEAR RATES, CHARACTERISTIC FOR A WEAR PROCESS WITH TRANSFER AND THE WEAR RATES, CHARACTERISTIC FOR A PROCESS WITHOUT TRANSFER FOR 14 DIFFERENT BEARING MATERIALS IN SLIDING CONTACT WITH STEEL AS A FUNCTION OF THE DUCTILITY (STRAIN AT FRACTURE) (AFTER LANCASTER, REF. 32)

steel surface is favorable, because this reduces the roughness of the mating surface and thereby the micro-machining action of the asperities. For the relatively brittle plastics, 1-5, a transferred layer with good adherence to the journal surface is not formed with the result that, in this case, it is favorable to prevent transfer.

In the case of Lancaster's experiments, the transfer situation was realized by traversing several times over the same track in the unlubricated condition. A wear process without plastic transfer was realized by introducing a suitable lubricant, preventing hydrodynamic effects by using a suitably low viscosity.

In adapting the information of Figure 22 to practical tribological situations, it should be kept in mind that, with plastics, a decrease in temperature can result in a sudden drastic decrease in ductility at the so-called "embrittlement temperature." Figure 22 shows that as a result, the wear behavior of such a plastic can change drastically.

4.4 The Effect of Fillers

It is, again, Lancaster who described conprehensively the effect of fillers on the behavior of plastic bearing materials[33].

Addition of molybdenum-disulphide, MoS_2, or polytetrafluroethylene, PTFE, in either elastomers or resins, usually leads to transfer of the filler to the mating metal surface. This results in an appreciable decrease in coefficient of friction and thus, in reduction of frictional heat. The ultimate effect on wear cannot easily be predicted because the reduction in friction is usually counterbalanced by a reduction in mechanical strength of the plastic. According to Lancaster, the strength reducing effect of PTFE is less pronounced than that of MoS_2.

Another well-known addition to plastic bearing materials consists of carbonfiber. Such fiber gives additional mechanical strength to the plastic, may be instrumental in conducting the frictional heat from the surface, and may also favorably reduce the roughness of the mating steel surface. If the fibers possess a graphite-like structure, a transferred film can be formed, in the same way as described above for MoS and PTFE. The formation of such transferred films can lead to a considerable reduction in wear rate. An example of this is shown in Figure 23, again taken from Reference 33. It can be seen that the initially relatively high wear rate decreases appreciably as a result of the gradual formation of a transferred graphite film on the mating steel surface. Addition of water results in removal of the transferred film and in a sudden, drastic increase in wear rate.

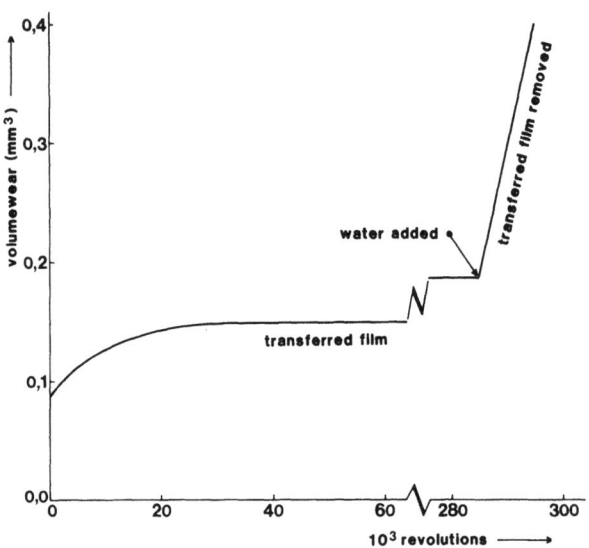

FIGURE 23   EFFECT OF THE FORMATION OF A TRANSFERRED FILM OF GRAPHITE ON THE WEAR OF EPOXYRESIN, FILLED WITH NONPOLISHING CARBON FIBRE (AFTER LANCASTER, REF. 32)

Carbonfibers with a cubic structure (non-graphitic) do not form a transferred film on the mating steel surface. They have, however, a polishing action, as a result of which the roughness of the mating surface decreases gradually with time. This also has a favorable effect on wear of the filled plastic, as shown in Figure 24, again taken from Reference 33. Figure 24 shows the volume wear of PTFE, filled with, respectively, non-polishing carbon fibers (A) and polishing carbon fibers (B), as a function of the number of revolutions and for three different initial roughnesses of the mating steel surface. It turns out that, when using the nonabrasive fiber, the wear rate increases considerably with increasing surface roughness of the steel. This is in accordance with the information from the lower part of Figure 21. However, if the plastic filled with polishing fibers is used, the volume of wear is found to be nearly independent of the initial surface roughness of the mating steel surface. Actually, the differences which are still found are due to the fact that it takes some time for the roughness of the steel surface to attain its equilibrium value. From the slope of curve 1 in part A and curves 1, 2, and 3 in part B, it can be concluded that the c.l.a.1 value of the polished steel surface is approximately 0.04 µm.

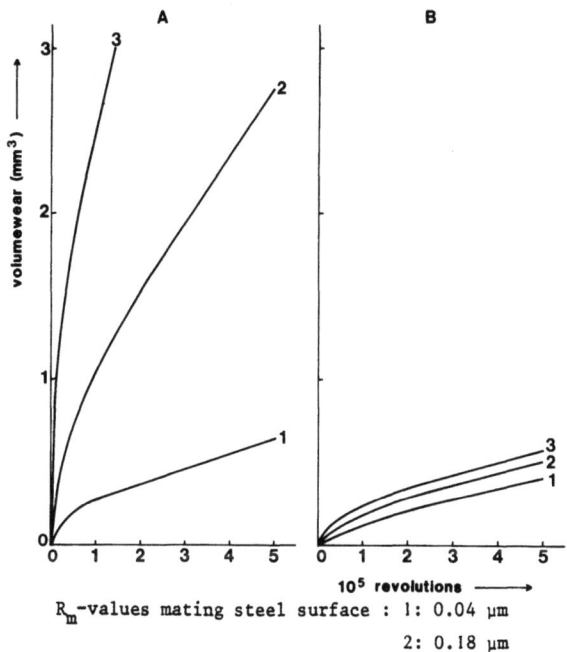

R_m-values mating steel surface : 1: 0.04 μm
2: 0.18 μm
3: 0.5 μm

FIGURE 24   WEAR OF PTFE FILLED WITH, RESPECTIVELY, NONPOLISHING
(A) AND POLISHING (B) CARBON FIBERS AS A FUNCTION OF THE
NUMBER OF REVOLUTIONS OF THE JOURNAL FOR THREE DIFFERENT
INITIAL ROUGHNESSES OF THE MATING STEEL SURFACE

The favorable effect of polishing is even more pronounced in the case of water lubricated epoxyresin-carbonfiber composites, because water in any case prevents the formation of transferred films, Figure 23.

Figure 25 shows the relationship between the wear rate and the initial surface roughness of the mating steel surface for the following three cases:

a) composites with polishing carbonfiber in contact with stainless steel
b) composites with non-polishing carbonfiber in contact with stainless steel
c) composites with polishing carbonfiber in contact with hard chromium.

Figure 25 shows that the polishing action in case (a) leads to a situation in which the specific wear rate is virtually independent of the initial roughness of the mating steel surface. Clearly, a similar effect is not found if a composite with non-polishing carbonfiber is used, case (b). The same is true in case (c), because, in that situation, the polishing fiber is not able to polish the hard chromium, as a result of which the effect of the initial roughness of the steel surface persists throughout the entire lifetime of the component.

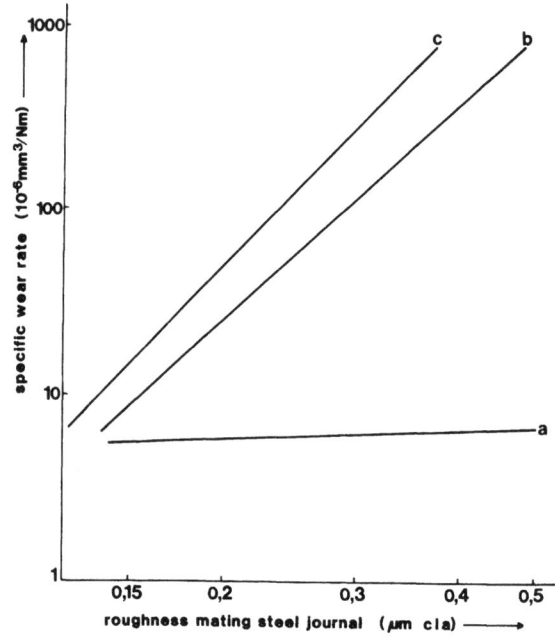

FIGURE 25 RELATION BETWEEN THE WEAR RATE AND THE INITIAL SURFACE ROUGHNESS OF THE MATING STEEL SURFACE FOR THE FOLLOWING CASES:

A. COMPOSITES WITH POLISHING CARBON FIBER IN CONTACT WITH STAINLESS STEEL
B. COMPOSITES WITH NONPOLISHING CARBON FIBER IN CONTACT WITH STAINLESS STEEL
C. COMPOSITES WITH POLISHING CARBON FIBER IN CONTACT WITH HARD CHROMIUM

(AFTER LANCASTER, REF. 32)

It is not accidental that in tests, the results of which are shown in Figure 25, stainless steel or chromium were used. If ordinary steel had been used, corrosion might have led to an appreciable change in surface roughness. In practice such corrosive attack of the mating steel surface usually leads to a reduction in roughness and thus, to a reduction in wear of the plastic bearing material. However, in some cases, depending on the corroding agent, the roughness may increase, with an accompanying unfavorable increase in wear rate.

Lancaster studied the wear of epoxyresin-non-polishing carbonfiber composites in different environments. The results are shown in Figure 26. The lowest wear rate is found in dry condition because a transferred graphite film is formed. On the other hand, the highest wear rate is found in water, because water prevents the formation of a transferred film, while a decrease in roughness of the mating surface by either polishing or chemical etching (corrosion) does not occur. When using seawater, some "chemical polishing" occurs, because the stainless steel which is used as mating surface, is not completely stable in seawater. This results in a considerable reduction in wear rate of the plastic. This effect is enhanced considerably by using solutions of coppersulphate or ironchloride, which produce a stronger etching effect on stainless steel. The organic liquids, "mineral oil" and "diester", have no corrosive action with respect to stainless steel. In this case, the considerable reduction in wear rate as compared to that in water, is to be ascribed to the formation on the steel surface of a transferred film consisting of a mixture of the resin and the lubricant.

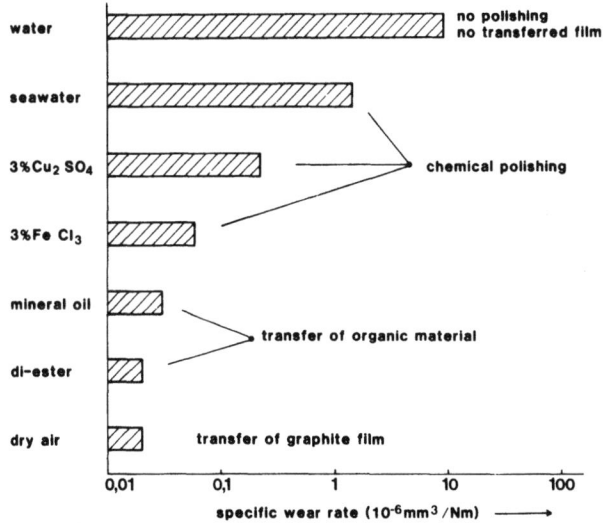

FIGURE 26 WEAR RATE OF EPOXYRESIN NONPOLISHING CARBON FIBER COMPOSITES IN DIFFERENT ENVIRONMENTS (AFTER LANCASTER, REF. 32)

It has already been mentioned that corrosive attack of the metal surface can lead to an increase in surface roughness and, thereby, to an increase in wear of the plastic bearing material. The same holds for fillers with a pronounced action. If the individual particles of the filler are relatively big, the roughness of the mating steel surface may increase rather then decrease as a result of contact with the bearing material, resulting in an increase in wear rate of the plastic. This may happen if glass is used as a filler, in particular if the mating surface is relatively soft (aluminum, bronze, titanium, etc.).

From the above, it will be clear that the tribological behavior of plasti-metal combinations is rather complicated. In fact function-oriented tribo-metric testing sill frequently be necessary if one wants to find the best material for a particular application[5].

4.5 Plastic-Plastic Combinations

In the above subsection, the behavior of plastic-metal combinations was discussed. Thereby the important effect of frictional heating of the plastic in the contact zone was emphasized. This is even more important in the case of plastic-plastic combinations, which generally have a very poor heat conductivity. Thus, they can only be used in very lightly loaded and/or very slowly moving mechanisms. Published information on the behavior of plastic-plastic combinations is very scarce. From research performed at TNO, the results of which as yet are unpublished, the following general rules can be derived:

- The greater the difference in chemical composition and structure of the contacting surfaces, the lesser the wear will be. This is another illustration of the "compatibility rule" which was discussed for metal-metal combinations in subsection 2.3.
- The presence of fillers with polishing effects in one of the contacting surfaces, invariably leads to a considerable increase in wear rate of the mating surface. In this case, plastic-plastic combination, this also holds for polishing carbonfibers.
- Contrary to the situation with plastic-metal combinations, introduction of a liquid will always lead to a decrease in wear rate because, in this case, a decrease in adhesion and material transfer will always be favorable.

4.6 Solid Lubricants

As stated in the introduction of this chapter solid lubricants (i.e. in particular molybdenum-disulphide, MoS_2, and graphite) frequently form an important component in materials for dry bearing applications. As such, dry lubricant films, formed by vigorous rubbing-in of suitably pretreated metal surfaces may already result in a dry running bearing with favorable performance characteristics. Further resin or waterglass bound films find widely spread applications and finally, graphite of MoS^2 may be added to lubricating grease.

An important argument for applying a solid lubricant is the very good thermal stability which can be realized with some systems. Actually, dry MoS$_2$ powder in neutral or reducing environments may function adequately up to temperatures of 800°C while a lubricant as calciumfluoride with ceramic binder may function adequately up to 1200°C.

If the solid lubricant is not replenished during use, the applicability is limited by the endurance of the initially applied lubricant film. The following is based on Reference 34, updated if necessary.

4.7 Main Properties of MoS$_2$ and Graphite

Molybdenum-disulphide as well as graphite have a hexagonal crystal structure, as shown for graphite in Figure 27 and for MoS$_2$ in Figure 28. In both cases the hexagonal structure leads to the formation of platelike crystals with a thickness to length ratio on the order of 1 : 1000. In pure MoS$_2$, the attractive forces between neighboring sulphur planes are very low, with the result that shear is easy. On the other hand, the load-carrying capacity in a direction normal to the hexagonal plane is very high. This explains the extremely favorable friction properties of MoS$_2$. Unfortunately, impurities of different kinds in particular oxygen or water hinder the free movement of the hexagonal movement crystals. This is entirely different with graphite, which shows low friction only if an adequate amount of "impurity molecules" is adsorbed on the carbon lattice. In this respect water vapor is found to be extremely favorable. Obviously, this basic difference in response to "impurities" has important consequences for the practical application of MoS$_2$ and graphite. For example, in a vacuum molybdenum-disulphide will show an excellent performance, while graphite will fail utterly as a lubricant. On the other hand, in very humid environments, (relative humidity > 80%) and under conditions of frictional contact, molybdenum-disulphide will hydrolyze spontaneously, leading to the formation of molybdenum-disulphide, molybdenum-trioxide, sulphur-dioxide, and free sulphur. Under such conditions, graphite functions fully adequately. The above hydrolysis of MoS$_2$, is one of the most convincing examples of "tribo-chemical reactions", as in vitro, it is impossible to decompose MoS$_2$, even by boiling with a concentrated solution of hydrochloric acid.

As far as the load-carrying capacity of the film is concerned, MoS$_2$ is undoubtedly highly superior to graphite. In

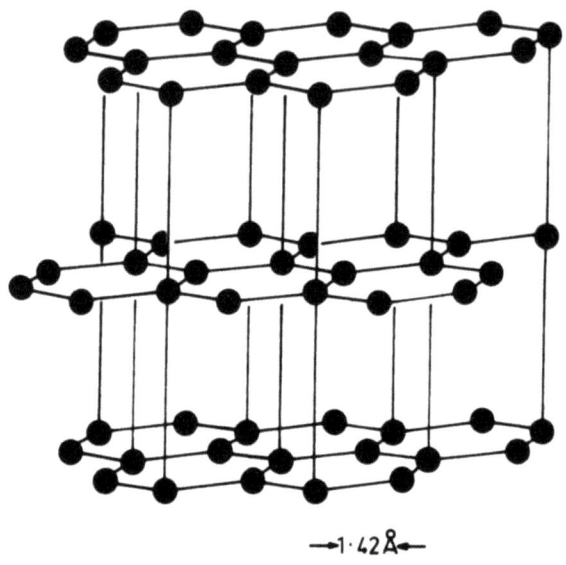

FIGURE 27   CRYSTAL STRUCTURE OF GRAPHITE

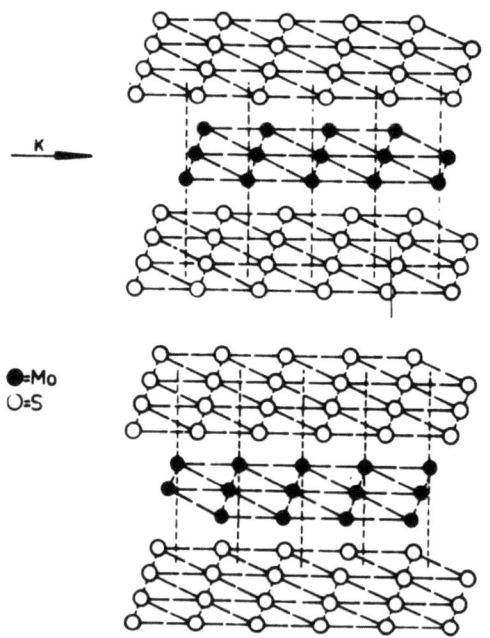

FIGURE 28   CRYSTAL STRUCTURE OF MOLYBDENUM-DISULPHIDE

fact, an MoS_2 film, formed from dry powder will carry any load which does not lead to gross plastic deformation of the substrate. The same is by no means true for graphite. In MoS_2 lubricated systems, one will usually find a decrease in coefficient of friction with increasing normal force. For a lubricant film formed from dry MoS_2 powder, this behavior is illustrated in Figure 29. It can be seen that the relative humidity of the atmosphere is an important parameter in the f/F_N relation. This is probably due to the fact that the friction behavior of MoS_2 improves with decreasing humidity. With increasing normal force, the heat which is generated in the friction zone, increases leading to an increase in surface temperature of the lubricant film. At a constant relative humidity of the surrounding atmosphere, this leads to a decrease in coefficient of friction with increasing normal force. Figure 29 shows that very low coefficients of friction can be obtained with MoS_2, of the order of $f = 0.03$. Even under optimum conditions, the coefficients of friction, measured with graphite as a lubricant, are higher of the order of $f = 0.1$. As is to be expected, on the basis of the friction mechanism, described above, an increase in normal force will, in this case, lead to a usually moderate increase in f.

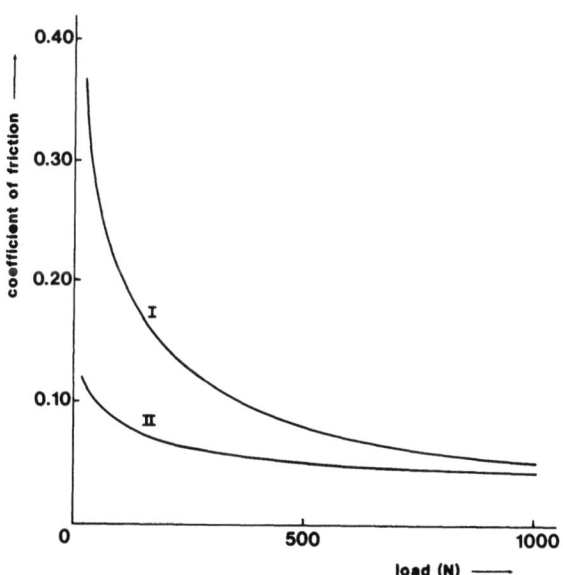

FIGURE 29   RELATION BETWEEN COEFFICIENT OF FRICTION AND NORMAL FORCE FOR MOLYBDENUM-DISULPHIDE POWDER FILMS IN DIFFERENT ENVIRONMENTS

  I : AIR OF $25°C$ AND 50% RELATIVE HUMIDITY
  II : AIR OF $25°C$ AND 0% RELATIVE HUMIDITY

If the solid lubricant is supplied continuously from the matrix of the bearing material (as, for instance, in the case of plastic - MoS_2 composites), the endurance of the lubricant film does not come into play. However, in cases where one relies on the performance of a rubbed-in and burnished film on a steel substrate, the endurance usually is the determining factor in a practical application.

As far as molybdenum-disulphide is concerned, the best performance is obtained if the contact surfaces are made of a steel with HV = 200-300 and a steel with HV = 700-900, respectively. The MoS_2 powder is rubbed on the harder surface, after pretreatment by wet blasting with quartz or Al_2O_3 to a final roughness of approximately 0.1-0.3 µm c.l.a. An even better performance/endurance is found if the substrate is finished by phosphating. If solid lubricant films are formed by mechanical rubbing-in, which is a usual and dependable application procedure, the layer thickness in the fully compressed state depends on the rubbing-in time. Thus, the time necessary to form a layer with a compressed thickness of 1 m will vary considerably from one powder to the other. For MoS2, in two modifications (i.e. average particle sizes of 30 µm and 3 µm, respectively), the related (also hexagonal) compound tungsten-disulphide WS_2 and for graphite, this rubbing-in time is given in Table 5. It can be seen that the rubbing-in time indeed varies considerably from one powder to the other, a fact which should be taken into serious account if such powders are to be applied in practice. Table 5 also shows the endurances measured under a normal force of 200 N and at a speed of sliding of 0.25 m/s in, respectively, dry air and nitrogen with approximately 100 ppm oxygen. It runs out that the endurances of 1 µm thick layers of, respectively, coarse and fine MoS_2, is the same (i.e. 18 hours in dry air and more than 100 hours in nitrogen). Thus, the difference between coarse and fine powdered MoS_2, which is frequently claimed on the basis of practical experience, is entirely due to differences in rubbin-in time. Table 5 also shows the relative inferiority as far as the endurance of rubbed-in films is concerned of WS_2 and graphite. In accordance with the friction mechanism described above, the endurance of a graphite film in nitrogen is virtually nil, while under such conditions, WS2 still performs moderately well. Because of the ease of layer formation during rubbing-in, WS_2 excels at relatively short rubbing-in times. This is shown in Table 6 which lists the endurance in dry air for layers formed during 1 minute rubbing-in procedures. This table shows that at such short rubbing-in times, coarse MoS_2 particles do not function at all.

POWDER	PARTICLE SIZE (μm)	RUBBING-IN[1] TIME (min)	ENDURANCE [2] (h)	
			DRY AIR	NITROGEN WITH 100 ppm OXYGEN
MoS_2	10-50	200	18[3]	> 100
MoS_2	1- 5	20	18[3]	< 100
WS_2	1-25	1	2[3]	22
GRAPHITE	1-10	20	4[4]	< 1

TABLE 5   ENDURANCE OF SOLID LUBRICANT FILMS, FORMED BY RUBBING-IN WET-BLASTED STEEL SUBSTRATES WITH DRY POWDER TO A LAYER THICKNESS (COMPRESSED) OF 1 μm

1. TIME NECESSARY TO FORM A LAYER WITH A THICKNESS OF 1 μm (IN THE FULLY COMPRESSED STATE)

2. TEST CONDITIONS: NORMAL FORCE: 200 N
   SLIDING SPEED: 0.25 m/s

3. DRY AIR

4. AIR WITH 60% REL. HUM.

RUBBING-IN PROCEDURE; NEOPRENE SPONGE UNDER 14 N FORCE

particle size (μm)	powder	endurance in air[1] (h)
10-50	MoS$_2$	< 0.1 [2]
1- 5	MoS$_2$	1 [2]
1-25	WS$_2$	2 [2]
1-10	graphite	< 0.1 [3]

TABLE 6 ENDURANCE OF SOLID LUBRICANT FILMS, FORMED BY RUBBING-IN WET-BLASTED STEEL SUBSTRATES WITH DRY POWDER DURING 1 MINUTE

1) TEST CONDITIONS: SEE LEGEND TABLE 5
2) DRY AIR
3) AIR WITH 60% REL. HUM.

The unfavorable effect of oxygen on the endurance of MoS$_2$ is once more shown in Figure 30. In a morphological sense, the intake of oxygen manifests itself by the formation of a membraneous like layer with a metallic luster. Such layers, which are to be considered as intermediate degradation products of MoS$_2$ powder films, suffer from dynamic blister formation. Figure 31 shows a magnified picture of a part of a blistered MoS$_2$ film. The blisters which can be seen have formed directly after the passage of the slider. Upon subsequent slider passage, they would have been flattened only to form again after repeated slider passage. Figure 31 clearly shows a crack formed in the central blister in a direction perpendicular to the direction of sliding. Such cracks ultimately lead to descaling of relatively large parts of the MoS$_2$ membrane.

5. MATERIALS FOR MINING, EARTH MOVING AND DREDGING EQUIPMENT

In mining, earth moving, and dredging equipment three wear processes predominate; abrasive wear, erosion, and impact wear.

5.1 Abrasive Wear

Situations of pure abrasive wear are frequently encountered in cases where a component moves through a loosely packed soil or where a load of abrasive material (for instance sand) is moved with respect to a guiding surface.

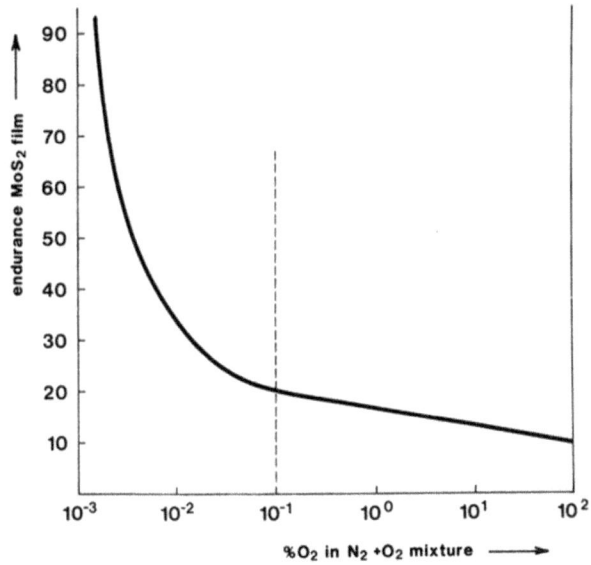

FIGURE 30   THE EFFECT OF OXYGEN ON THE
ENDURANCE OF MoS_2 POWDER FILMS

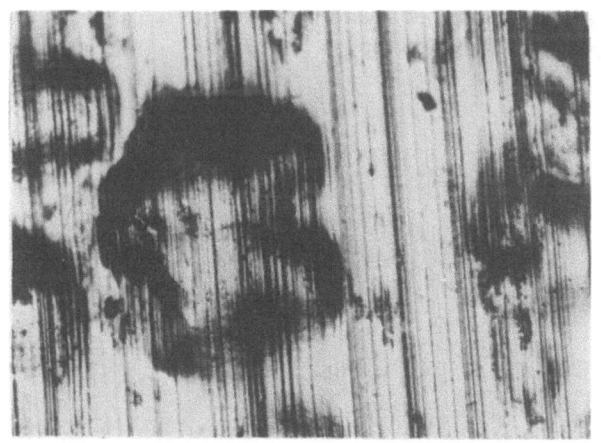

FIGURE 31   BLISTERS IN AN MoS_2 POWDER FILM

Although, from a systems point of view, the abrasive wear mechanism is relatively simple or in any case well defined, it has taken a considerable period of time before a clear insight in the material factors which control the tribological behavior of materials under conditions of abrasion could be developed. Quite recently, Torrance developed an expression relating the abrasion resistance of pure metals or alloys to their hardness and Young's moduli[35]. His expression seems to agree well with a wide range of published experimental results. The wear resistance under conditions of pure abrasion is usually measured in tests in which a small metal cylinder or pin is weighed and then pressed under a standard load against a moving abrasive surface, generally a standard grade of abrasive paper. The abrasive is slowly traversed beneath the pin in such a way that the pin is always rubbed by fresh abrasive. The pin may describe a spiral on the abrasive disc or follow a helix on an abrasive drum. The test is stopped when the pin has traversed a standard distance. It is then reweighed and the volume, v_1, lost during the test, is calculated from the weight loss. The test is then repeated using a reference material and the relative wear resistance, β_1, of the metal is expressed as $\beta = v_r/v_1$; where v_r = the volume loss of the reference material. Use of a reference material is made, because differences in quality of the abrasive (paper) may result in appreciable differences in the absolute wear amounts of the test materials. If β_1 is plotted against the Vickers hardness for a series of metals, a remarkable fact emerges; all the pure metals and annealed alloys lie on a straight line passing through the origin, but alloys, hardened by heat treatment, lie beneath this line on straight lines of lower gradient. This is shown in Figure 32, from Reference 36.

Abrasion/Abrasive Erosion
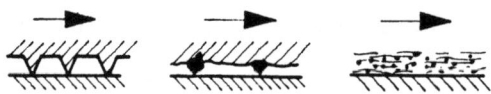

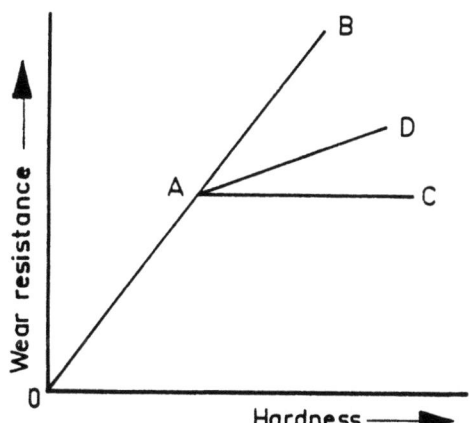
FIGURE 32 RELATION BETWEEN WEAR RESISTANCE AND HARDNESS FOR
PURE METALS (LINE O-A-B), HEAT TREATED STEELS (LINE
A-D) AND COLD WORKED MATERIALS (A-C)(AFTER KRUSCHOV, REF.35)

The anomalous behavior of heat treated alloys has been explained partly by Richardson, who confirmed that hardening a metal by cold working, has no effect on its abrasion resistance[37]. He also observed that the surface hardness of an abraded metal is considerably higher than its bulk hardness, because of the intense plastic strain which is induced by abrasion. He therefore suggested that the abrasion resistance of a metal correlates better with its hardness in a heavily worked condition than with its bulk hardness. Although experiments showed this to be the case, heat treated alloys were still found to fall on lines of lower gradient than pure metals.

A second approach was used by Buttery and Archard, who attempted to simulate the action of an abrasive grit by plowing scratches in the surfaces of various metals with a Vickers indenter[38]. By careful measurements made on Talysurf traces taken across the scratches, they tried to find how much of the scratch volume was pushed into sidewalls and how much was removed as wear debris. On the basis of these results, they suggested that an abrasive grit removes metal more effectively from a heat treated alloy than from a pure metal, in this way explaining in a semi-quantitative way, the lower than expected abrasion resistance of heat treated materials.

The above results were criticized by Johnson, who pointed out that although there is a tendency for metal, displaced by an indenter, to be ploughed plastically into sidewalls, it is possible for some of the displacement to be accommodated by a small elastic distortion of metal adjacent to the plastic zone[39]. Torrance took up this argument and developed the following relation for the total volume of material, removed by an abrasive[35]:

$$v = \frac{b\,Fs}{H\tan^2\theta}\left(1 + \frac{KH}{E}\right) \tag{5}$$

in which b = the fraction of the groove volume removed as wear debris; F = the force applied to the test specimen; s = the length of the grooves ploughed by the abrasive particles; H = the Vickers indentation hardness of heavily worked material (i.e according to Richardson); θ = the semi-apical angle of the (supposedly conical) particles; E = the Young's modulus of the surface that undergoes abrasive wear; and K = a factor which in a first approximation can be considered as a constant for a wide variety of metals and θ-values. Since the wear resistance, β_i, is defined as:

$$\beta_i = v_r/v_i$$

Equation (5) gives:

$$\beta_i = \frac{H_i}{H_r}\frac{1 + KH_r/E_r}{1 + KH_i/E_i} \tag{6}$$

in which the index, i, refers to the material under test and the index, r, to the reference material.

Although Torrance's equation is based on many assumptions and approximations, it correlates surprisingly well with experimental results, as is shown in Reference 35. In this reference the measured values of the wear resistance, β_i, are correlated with the values of β_i calculated with Eq. (6) for a large number of tests performed by Richardson, Moore and Mutton, and Watson[37, 40, 41]. It is shown that, despite the many simplifying assumptions, Eq. (6) gives a much better prediction of the relative abrasion resistance than was hitherto possible.

As stated in subsection 1.1, friction and wear do not usually correlate if different materials are compared in a single test. This applies equally well for abrasion as for other tribological situations. In abrasive wear, as occurring in the cutting of tightly packed soil in a dredging operation, one particular construction material may come in contact with soils of different composition, surface structure, etc. In that case, the chances of finding workable correlations between friction and wear are much better. Figure 33, taken from Reference 42, shows that this is indeed the case. Correlations, as shown in Figure 33, open the possibility of monitoring cutter wear by means of power loss measurements. It was found that application of this principle in dredging practice actually resulted in improved efficiency in the use of cutter teeth.

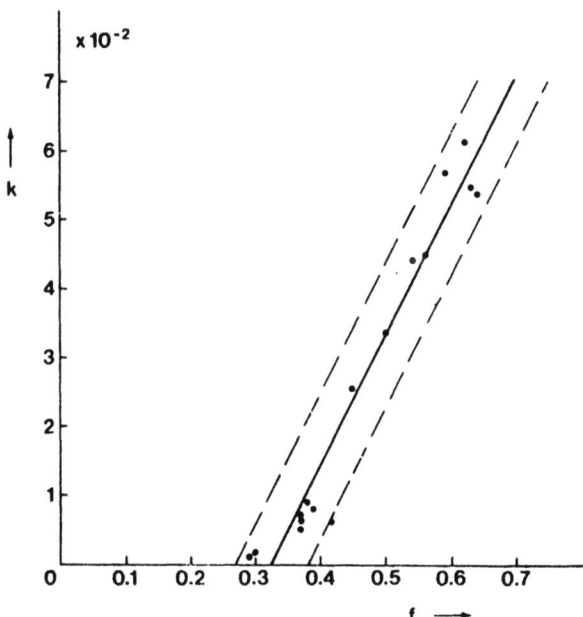

FIGURE 33 CORRELATION BETWEEN WEAR COEFFICIENT k AND COEFFICIENT OF FRICTION f, WHERE k = (VOLUME OF WEAR) (HARDNESS) / (LOAD) (SLIDING DISTANCE). ———, LINEAR REGRESSION CURVE, ----, CONTROL LIMITS AT TWO STANDARD DEVIATIONS TO EITHER SIDE OF THE REGRESSION CURVE. 95% OF THE DATA ARE EXPECTED TO LIE WITHIN THESE CONTROL LIMITS.

5.2 Impact Wear

Wear Phenomena associated with impact are described comprehensively by Engel[43]. In mining or earth moving operations, impact wear can be due to collisions between metal components and hard rocks. Under those conditions, the resistance of the material against impact induced surface fatigue appears to be of crucial importance. This necessarily implies that under rock-moving conditions, the materials should be hard and ductile, a state of affairs which is, in fact, desirable in the entire field of wear, but which is not obtained easily. In the application area under discussion, it is in particular the manganese steels which have emerged as wear resistant materials with unique properties. In fact, the favorable wear behavior of these materials, the "Hadfield steels", is due to the fact that they are open to the formation of deformation-martensite as a result of mechanical work. Recently, the situation has been summed up by Zum Gahr et al[6]. The authors show beyond a doubt that there may be pronounced differences between steels from a single "family", only by virtue of the fact that some can form deformation martensite and others cannot. Actually this is a fine example of a case in which, within a particular class of materials, a theoretical model based on metallurgical knowledge may adequately predict the field behavior of new alloys.

5.3 Erosion

Erosion is loss of material from a surface due to contact with particles which are carried in a fluid. Practical cases in which the lifetime of components is limited by erosion are encountered in application areas which range from dredging operations (cold; seawater) to gas turbine applications (hot; corrosive gasses).

In erosion the mechanisms of abrasion and impact wear usually occur simultaneously, the relative importance of each of these mechanisms depending on the predominant angle of impact, α, under which the eroding particles hit the wearing surface. Obviously this complicates the picture quite considerably, meaning that a systems approach is even more appropriate in the case of erosion than in the case of pure abrasion or pure impact wear. Another factor of considerable influence is formed by the presence of the medium which carries the particles, which may be either gas or liquid and which may exert an independent corrosive action on the wearing surface. Other important parameters determining the behavior of materials under conditions of erosion are the particle concentration, the nature and distribution of the eroding particles, the speed, and the temperature.

For the case of aluminumoxide particles in an air stream attacking aluminum and plain carbon steel surfaces, it has been attempted successfully to set up theoretical models for the relation between erosion and angle of attack, α[44, 45]. Figure 34 summarizes the results of this work in terms of erosion versus α curves for extremely ductile (1), extremely brittle (2) and "ordinary" materials (3). From Figure 34 it can be seen that when working with ordinary engineering materials, as for instance steels, one finds a maximum in the erosion curve, which usually falls in the range $20° < \alpha < 40°$. The precise location of the maximum, however, depends quite strongly on the properties of the wearing surface and on the precise experimental conditions, which limits the predictive power of the theoretical model quite severely. In fact it is not likely that a general erosion-theory, which covers the effects of all influencing factors (including corrosion), will ever be developed. For that reason, for the majority of new technical applications, one has to rely on simulative testing in selecting suitable materials[46].

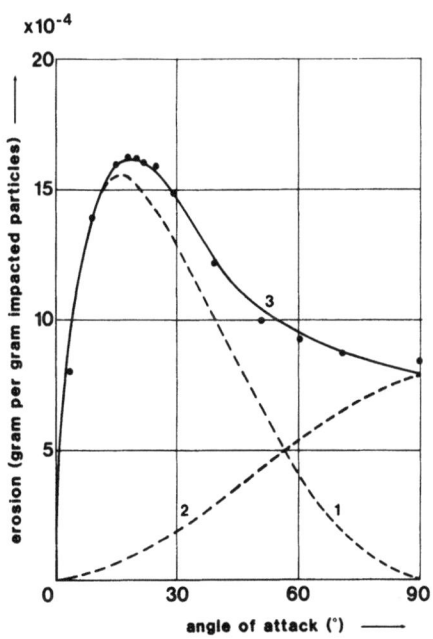

FIGURE 34   EROSION VERSUS ANGLE OF ATTACK CURVES FOR DIFFERENT TYPES OF ERODING MATERIAL (AFTER NEILSON & GILCHRIST, REF. 44)

1: IDEAL DUCTILE MATERIAL (THEORETICAL CASE)
2: EXTREMELY BRITTLE MATERIAL (FOR EXAMPLE, GLASS)
3: ORDINARY ENGINEERING MATERIAL (IN THIS CASE ALUMINIUM, ERODED BY 210 µm ALUMINIUMOXIDE PARTICLES IN AN AIR STREAM AT V = 125 m/s.

A typical result, obtained in a current research program, performed at the TNO Laboratory, in which different materials are compared with respect to their behavior under conditions of erosion in particle-seawater slurries, is shown in Figure 35. This figure shows the relative wear rates, β_i, for "rubber" and "polybutene", calculated by relating their volume losses to that of plain carbon steel C22 (Vickers hardness 150). Actually, if the volume loss of the reference steel exceeded that of the test material, a factor β_{i+} was calculated according to:

$$\beta_{1+} = \frac{v_{steel}}{v_{test\ mat.}}$$

If the volume loss of the test material exceeded that of the reference material, a factor β_{1-}, was calculated according to:

$$\beta_{1-} = \frac{v_{test\ mat.}}{v_{steel}}$$

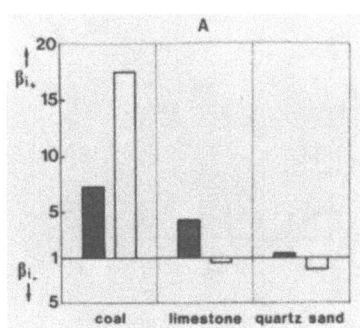 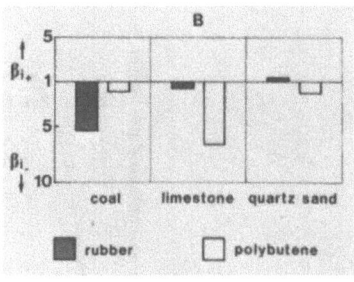

FIGURE 35   RESULTS OBTAINED IN TESTING "RUBBER" AND "POLYBUTENE" IN, RESPECTIVELY, COAL-SEAWATER, LIMESTONE-SEAWATER AND QUARTZSAND-SEAWATER SLURRIES.

THE PICTURE CONTAINS RELATIVE WEAR RATES, OBTAINED BY COMPARING THE WEAR OF THE TEST MATERIALS WITH THE WEAR OF PLAIN CARBON STEEL C22 (150 HV).

A: RESULTS OBTAINED UNDER CONDITIONS OF ELECTROCHEMICAL CORROSION

B: RESULTS OBTAINED UNDER CONDITIONS WHERE ELECTRO-CHEMICAL CORROSION IS INHIBITED EFFECTIVELY.

In the present tests, the abrasive agents were ground coal, limestone, and quartz sand, each with an average particle diameter about 600 μm. The other test parameters were: concentration of abrasive, 19 vol.%; angle of attack, α : 80°; slurry speed, 4 m/s; and exposition time, 6 h. Part A of Figure 35 shows results obtained under conditions in which (for steel

C22) electrochemical corrosion could occur; pieces of other metal attached to the test specimen. Part B of Figure 35 shows results obtained with electrically isolated test specimens. It can be seen that in quartz sand/seawater mixtures, the results obtained under "corrosion conditions" (part A of Figure 34), are the same as those obtained under "noncorrosion conditions" (part B of Figure 34), with rubber performing slightly better and polybutene perfoming worse than steel C22 by a factor of approximately 2. Apparently in quartz sand/seawater slurries, electro-chemical corrosion effects do not contribute significantly to the process of material removal. Figure 34 also shows that this is quite different in coal/seawater and limestone/seawater mixtures where the results obtained under "corrosion conditions" differ considerably from those obtained under "noncorrosion" conditions.

Figure 35 only gives an impression of the situation at particular values of angle of attack α, particle size (distribution), and abrasive concentration. Variations in the values of each of these parameters considerably influence the results.

Results as shown in Figure 35 can be applied in practice by first performing tests under real operating conditions in prototype equipment made from the reference material (in this case steel C22), and then applying materials with high β_{i+} values at locations which are found to suffer from excessive wear. Such selective application of better, but generally also more expensive, materials may contribute significantly to the overall performance of the installation, paying due regards to economic considerations.

In the present case, it would not be wise to apply rubber or polybutene coatings for use in coal or limestone slurries if, in practice, electrochemical corrosion can effectively be suppressed (in quartz sand, however, application of rubber might improve the situation somewhat). If, however, electrochemical corrosion also plays an important part in the practical application, the use of rubber or polybutene (in coal-seawater slurries) or rubber (in limestone-seawater slurries) will considerably improve the situation.

6. THE TRIBOLOGICAL ASPECTS OF SURFACE TREATMENTS AND COATINGS

Thermal and thermochemical treatments of surfaces and coating techniques become increasingly important because of two reasons:

- the demands on machines and equipment in relation to exposition to hostile environments and high temperatures increase,
- the need for cheaper and more readily available construction materials becomes more pressing.

Actually when applying a high quality surface treatment or a coating to a relatively simple substrate material, it may be possible to reconcile the above requirements effectively.

By now, a large number of coatings and surface treatments are available, which have as a rule, been developed and exploited independently. Thus it is difficult, if not entirely impossible, to compare the features of the different processes on a really objective basis. Still, this section tries to provide a rough guide for process selection, which should be followed by function-oriented tribometry and careful weighing of the tribological as well as the economical factors, if the best possible solution for a given practical application is to be found[5].

Much of the information contained in this section is derived from a multi-client study, performed by the International Research and Development Corporation, Ltd[47]. However, experiences gained at TNO and information from other literature sources are included as well.

6.1 Selection Criteria

Although the "wear resistance" undoubtedly is the most important criterion in selecting a suitable surface treatment or surface coating, other properties will be more or less important, depending on the details of the practical application. Actually, in each application, each selection criterion should be allotted a "weighing factor." In addition to "wear resistance," the following criteria may come into play:

- layer thickness
- workpiece temperature during treatment
- porosity
- hardness
- ductility
- adherence
- price
- corrosion resistance
- "fatigue strength"
- maximum service temperature

The layer thickness should be adapted to the requirements of the practical application. As an example, in the case of precision

measuring equipment, one will generally apply highly wear
resistant treatments or coatings of relatively small thickness,
because the admissible amount of wear usually is very small. On
the other hand, in equipment for the dredging industry one will
usually apply surface welded coatings of considerable thickness
(up to 20 mm). A quantitative example of the interaction wear
rate/layer thickness/type of application is given in the
following.

Table 7 shows data relevant to surfaces treated by
gasnitriding and hard chromizing, both processes belonging to
the family of "thermochemical treatments." Under the present
conditions of load, speed, lubrication, and temperature, the
wear rate of the gasnitrized surfaces was 0.2 µm per day and
that of the surfaces treated with chromium 0.02 µm per day
(Table 7, column 3). The wear rate of the substrate material
was 50 µm per day in both cases. The effective layer thickness
of the gasnitrided surfaces was 300 µm and that of the surfaces
treated with chromium, 20 µm. From the data, it follows that at
a maximum allowable depth, Δh_{max} of 300 µm, the liftime of the
gasnitrided surface, t_{max}, is about four years. On the other
hand at a maximum allowable wear depth, Δh_{max} of 20 µm, t_{max}
reduces to only three months (0.27 year). In the case of
surfaces treated with the hard-chromizing process, the maximum
allowable wear depth is irrelevant, because in both cases the
lifetime of the layer is 2.7 years, being the time necessary to
remove the chromium diffusion layer by wear.

From this example we see that if the requirements are:

- lifetime   : 4 years
- maximum wear depth : 300 µm

gasnitriding is the appropriate process.

If, on the other hand, the requirements are:

- lifetime   : 2 years
- maximum wear depth : 20 µm

the hard-chromizing process is to be chosen.

process	layer thickness (μm)	wearrate (μm/day)		t_{max} (year)	
		layer	substrate	Δh_{max} = 300 μm	Δh_{max} = 20 μm
gasnitriding	300	0.2	50	4.1	0.27
chromizing	20	0.02	50	2.7	2.7

TABLE 7 LIFETIME t_{max} OF, RESPECTIVELY, GASNITRIDED AND CHROMIZED SURFACES FOR TWO VALUES OF THE MAXIMUM ALLOWABLE WEAR DEPTH Δh_{max}

Although in reality, the situation may be much more involved than is suggested in Table 7, if only because more processes are included in the comparison, the example shows that data on the wear rate alone are not sufficient to determine which process is most suitable for a particular application. The workpiece temperature during performance of the surface treatment or the application of the surface coating is important because a relatively high temperature may lead to severe distortions of the workpiece. In general, there will be a tendency to apply the high temperature treatments only in cases where a certain degree of distortion is acceptable. If this is not the case, one has to reckon with additional costs of machining of the surface treated layer or coating and also with possible undesirable local changes in layer thickness as a result of machining. The porosity of a coating can influence its mechanical strength and thereby its wear resistance. Also pores, which connect the coating surface with the substrate surface, may transmit a corrosive environment which may lead to interface corrosion. Obviously hardness and ductility are mutually related properties. As stated previously for most applications, one wishes to obtain a high hardness and a high ductility. In practice this usually means that an acceptable compromise has to be found. The importance of the criteria "adherence" and "price" is self-evident. Corrosion resistance, "fatigue strength" and maximum service temperature are typical system-dependent properties. It depends much on the type of application, i.e. the composition of the environment, the temperature, the loading pattern, etc., whether special requirements as to these criteria are to be met.

From the above, it will be clear that a surface treatment or surface coating which behaves excellently under one particular set of conditions may fail utterly under other conditions. Thus commercial claims regarding "universally

applicable coatings" should be treated with a healthy
scepticism.

6.2 A Survey of Surface Treatment and Surface Coating Techniques

The (commercially) available surface treatment and surface
coating techniques which are available to the designer of wear
resistant surfaces are summarized in Figures 36 and 37
respectively. It can be seen that one may choose from a great
amount of tehnical possibilities. Actually, the situation is
even more involved because in certain cases many competitive
commercial processes are on the market which may or may not lead
to essential differences in quality. Obviously, the scope of
this presentation does not permit a detailed discussion of each
of the individual processes mentioned in Figures 36 and 37.
Instead Table 8 provides for each of the main categories
information on the following criteria:

- maximum layer thickness, that can be obtained within
  reasonable limits of economic procedure
- maximum temperature of the workpiece during treatment
- maximum hardness that can be obtained.

Further, Table 8 lists for each of the main categories some
typical examples of materials suitable for treatment or coating
materials and also some typical applications.

If ranges are given for the maximum values of layer
thickness and workpiece temperature or hardness, this indicates
that different processes, which belong to one and the same
category, yield different values for these parameters. For
instance in the category "thermal hardening", the process of
"flame hardening" yields a hardening depth up to 5000 μm; laser
hardening, on the other hand, produces a maximum hardening depth
of some 500 μm. In some cases the differences thus found are
even more extreme. For instance, the case within the group or
"thermochemical treatments" is cited. Here the "chromizing"
treatment yields a maximum layer thickness of 20 μm, while
"carburizing" may easily yield layers with a maximum thickness
of 4000 μm. It should finally be noted that the examples as to
coating materials and characteristic applications are by no
means exhaustive. In fact the more recently developed processes
such as, certain vapor deposition methods and electrochemical
treatments, followed by thermal diffusion may find wider
applications in the near future[48, 49].

Table 8 shows that, in principle, a confusingly large
amount of different surfaces can be obtained, which makes the
ultimate choice for a given practical application difficult.

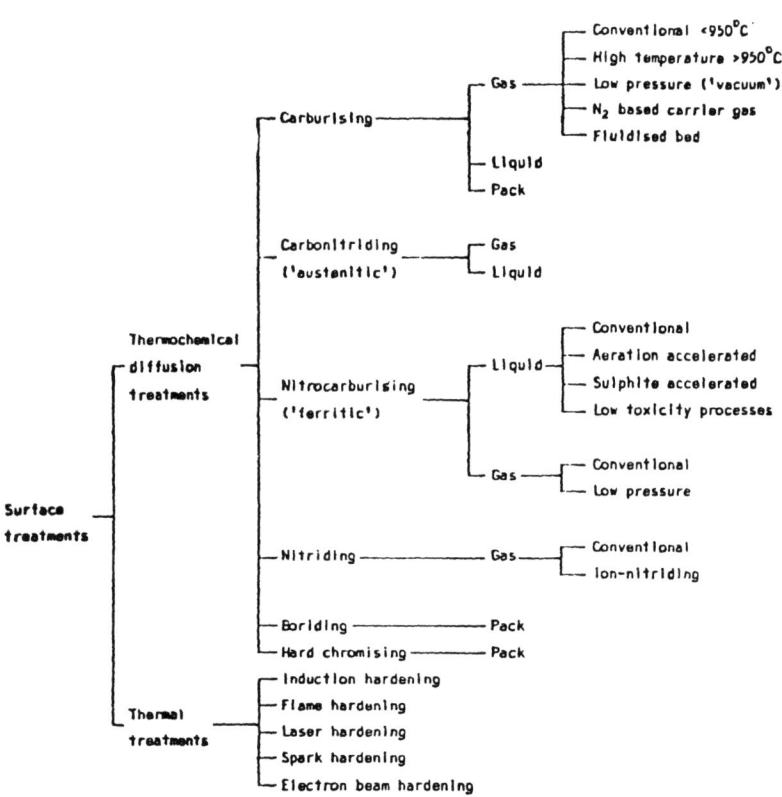

FIGURE 36 SURVEY OF SURFACE TREATMENT TECHNIQUES

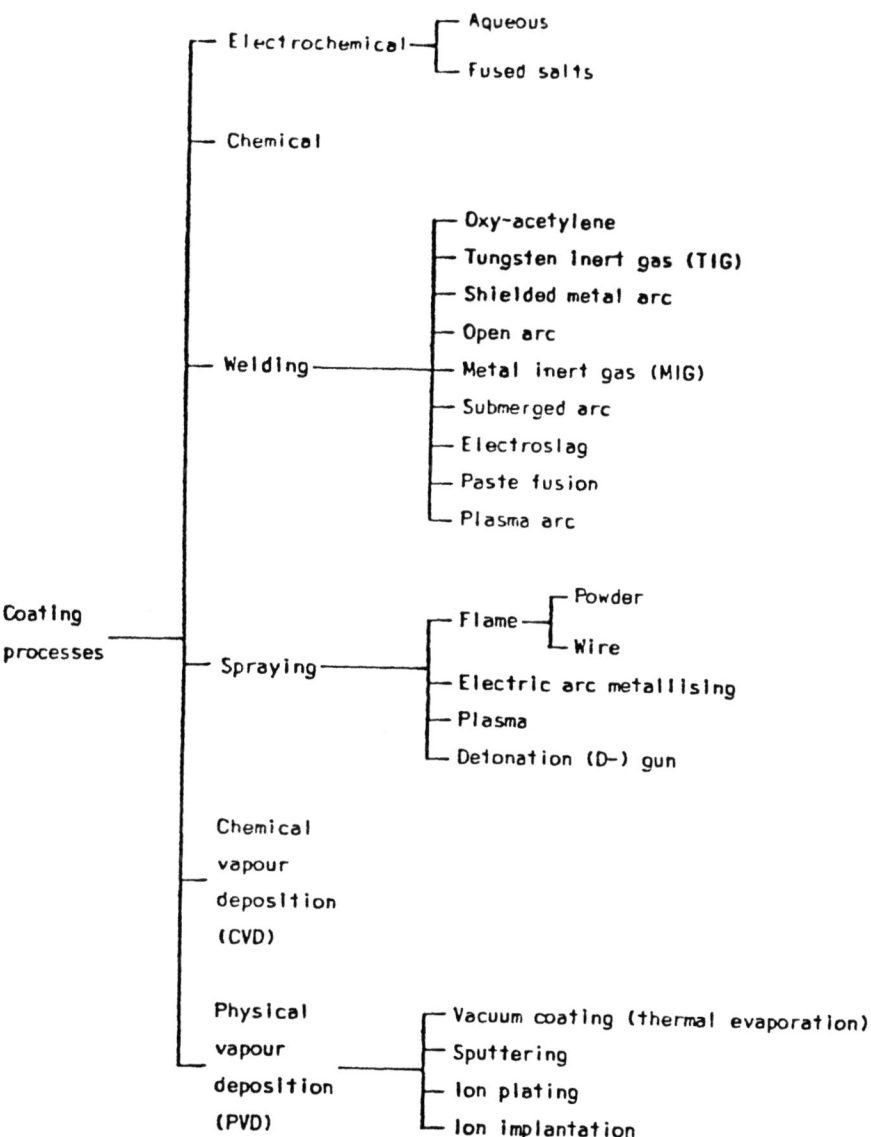

FIGURE 37   SURVEY OF SURFACE COATING TECHNIQUES

process	maximum thickness coating (μm)	maximum temperature workpiece (°C)	some typical examples of coating materials	maximum hardness	some characteristic applications
thermal hardening	500 – 6000	100 – 1000	Steel (0,3% – 0,5% C)	500 – 900	automobile engine components
thermochemical treatments	20 – 4000	400 – 1000	Steel + C, N, B, Cr or V	500 – 2000	gears, roller bearing races, crankshafts
electrochemical treatments	50 – 500	50	Ag + Re Cr Cr_3C_2 / Co Al_2O_3 on Al (anodizing)	150 1000 2500/500 (1000)	rotary switches cylinder liners diesel engine piston rings plastic injection moulds
chemical treatments	5 – 50	50 – 200	NiP or NiB $MHPO_4 / M_3(PO_4)_2$ (phosphating)	950 380	casting dies, thread guides gears, tappets, cam shafts
welding	20000	1400	12% Mn steel martensitic steels Co-alloys WC/Ni	600 900 700 2000/700	earth and rock engaging equipment; dredging cutter heads
spraying	1000	100 – 1100	13% Cr steel + oxides Mo + oxides NiAl Co-alloys WC/Co or Cr_3C_2/NiCr Al_2O_3 or Cr_2O_3	350 400 350 700 2000/1000 2500	shafts, valve seats, dies
chemical vapour deposition (CVD)	10	1000	TiC, TiN, W_2C, WC, Al_2O_3, etc.	3000	metal cutting tools wire drawing dies
physical vapour deposition (PVD)	5 – 150	20	TiC, TiN, W_2C, WC, B_4C, Si_3N_4, etc. Al, Pb, In, W C, N	4000 100 – 1000 1500	metal cutting tools compressor blades ball bearings, shafts wire drawing dies

TABLE 8   A SURVEY OF SURFACE TREATMENT AND SURFACE COATING TECHNIQUES

The issue is further complicated by the fact that, in a particular application, surfaces which have been treated by different processes may well be combined. That this can be very beneficial is illustrated in Figure 38. This shows that the total amount of volume wear, produced in 20 hours testing under the conditions specified, differs significantly for combinations of nitrocaburized (NC) and carburized (C) surfaces, with the latter combination producing about twice the amount of wear as the first combination. Remarkably enough, the volume wear of both surfaces, but in particular that of the carburized surfaces, but in particular that of the carburized surfaces, is reduced considerably if combinations or carburized and nitrocarburized surfaces are made. In fact Figure 38 shows that if in particular the shaft, item B in Figure 38, is to be protected, the combination carburized shaft --- nitrocarburized mating surface yields an approximately 25 times better performance than a combination carburized shaft - carburized mating surface. In a way this is still another example of the "compatability effect" that was also described in subsection 2.3 for combinations of metals and in subsection 4.5 for combinations of plastics.

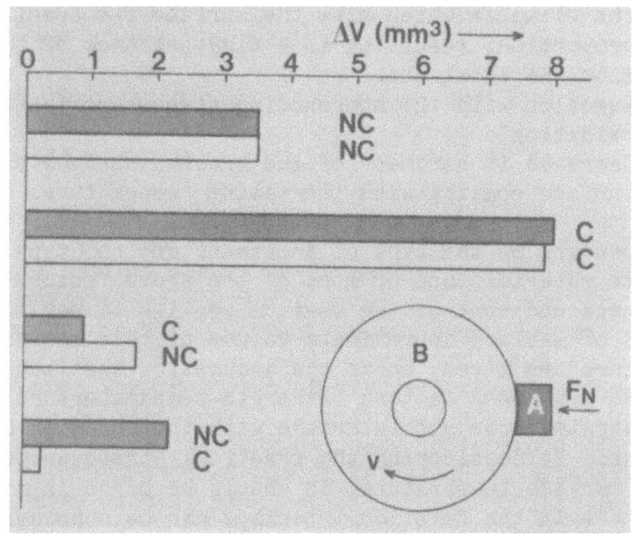

FIGURE 38 VOLUME WEAR MEASURED WITH A PIN (A) AND RING (B) TEST RIG, AN ILLUSTRATION OF THE "COMPATIBILITY EFFECT"

NC: NITROCARBURIZED LAYER
C : CARBURIZED LAYER
NORMAL FORCE: 1000 N, SPEED OF SLIDING: 0.2 m/s,
        EXPERIMENT
DURATION: 20 HOURS, ENVIRONMENT: DRY AIR OF $200^{\circ}C$

6.3 Thermochemical Processes

In Table 9 the main properties and functional characteristics of the thermochemical treatments carburizing, carbonitriding, nitrocarburizing, nitriding, boriding, and chromizing are summarized. For comparison, the same properties are given for laserhardening (category "thermohardening"), hard chromium plated surfaces (category "electrochemical processes") and WC/Co coatings (category "spraying"). Typical performance characteristics of the various treatments and coatings are the maximum surface temperature, the fatique strength, the wear resistance under conditions of abrasive and adhesive wear, and the corrosion resistance. The maximum service temperature of a component which has been surface treated or coated depends on a number of different factors, some of which are system oriented. These are:

- Strength/temperature relation of the substrate.
  As a result of softening of the substrate the component as a whole may become inoperable, even if the coating itself still functions adequately.
- Diffusion into the substrate material of one or more of the elements which give the surface its special properties, resulting in a disappearance of the coating into the substrate.
- Reaction with the surrounding atmosphere (usually oxidation).
- Decrease in hardness of the treated surface or the surface coating with increasing temperature.

Depending on the type of treatment and the type of substrate material, one or more of the above factors will predominate and control the maximum service temperature. In column 5 of Table 9 approximate values of this maximum servie temperature are given, under the assumption that the first of the above mentioned factors (strength-temperature relation of the substrate) does not interfere with the function of the component. In considering the practical consequences of the maximum service temperature, it should be borne in mind that the temperature in the friction interface can be considerably higher than the environmental temperature, because of frictional heating effects. Thus in deciding whether a particular treatment or coating can be applied the real surface temperature should be estimated.

Because of the very strongly system oriented character of the properties "fatigue strength", "wear resistance," and "corrosion resistance," it is not possible to give for these properties quantitative data (although, in treatment selection for a well-defined practical application, function oriented

process	thickness coating (μm)	maximum temperature workpiece (°C)	maximum hardness HV	"fatigue strength"	"wear resistance" abrasive	"wear resistance" adhesive	"corrosion resistance"	maximum service temperature (°C)
laser hardening	500	100	500 – 900	++	+	+	+│	200
carburizing	250 – 4000	850 – 950	700 – 900	+++	+	+	+│	200
carbonitriding	50 – 750	750 – 900	600 – 850	++	+	++	+│	200
nitrocarburizing	20*, 1000**	570	500 – 650*	++	+	++(+)	+│	200
nitriding	400 – 600	500 – 525	800 – 1050	+	++	+++	+	500
boriding	25 – 100	850 – 1000	1000 – 2000	+	+++	+(+)	+	500
chromising	10 – 20		1500 – 2000	+	+++	+++	+++	600
chromium plating	2 – 250	50	1000	–	++	+++	+++	500
WC/Co plasma spraying	500	250	2000/1000	+│	++	++	+	500

*thickness and hardness of compound layer on mild steel
**total depth of diffusion zone

TABLE 9   IMPORTANT PROPERTIES OF SOME SURFACE TREATED OR COATED SURFACES

tribometry may produce such data). As far as the "fatigue strength" is concerned, Table 9 shows that the thermochemical treatments as well as laser hardening (and "thermohardening" in general) lead to an increase in fatigue strength of the surface. This is because during the formation of martensite in the surface zone (as occurs during thermal hardening) as well as by diffusion of carbon, nitrogen, borium, or chromium in the surface, compressive stresses are built up in the surface zone, which counteract the formation of microcracks. Table 9 shows that, in particular, carburizing leads to an appreciable increase in "fatigue strength" of the surface. As far as this point is concerned two reservations should be made:

- If the surface treated components are finished by grinding, which is usually the case after carburizing, carbonitriding, and nitriding, the grinding process may, but not necessarily does, cause tensile stresses in the surface which may nullify the favorable effects of the surface treatment itself.

- If possible, a sharp transition between surface treated zone and substrate material should be avoided, because this may lead to spalling phenomena. This is in particular the case if the surface treated component is to be applied under conditions of cyclic loading, as in rolling element bearings. In any case it should be avoided that the location of the zone with maximum stress coincides with the transition area.

Table 9 shows that plasma sprayed WC/Co layers usually do not influence the fatigue strength of the component. Hard chromium plating, on the other hand, may lead to a significant decrease in fatigue strength because cracks in the chromium layer may initiate cracks in the substrate. This can be remedied by applying an intermediate layer, acting as crack barrier (for instance a thin nickel-chromium layer). However, this leads to a considerable increase in costs.

As far as the "wear resistance" is concerned one should distinguish between the resistance against "abrasive wear" and that against "adhesive wear". If abrasive wear predominates, hardness and elastic modulus are the predominating factors. As far as adhesive wear is concerned, diminishing friction is of equal importance. In this respect the nitrocarburizing, nitriding and chromizing processes distinguish themselves favorably from the other processes. Table 9 also shows that, as far as "wear resistance" is concerned, hard chromium plating and WC/Co plasma spraying in general are a match for the thermochemical process. Table 9 finally shows that the corrosion resistance of the treated surfaces is generally similar or somewhat better than that of the untreated substrate. Conforming to expectations, chromium containing surfaces, obtained by chromizing or chromium plating of surfaces, generally show an excellent corrosion resistance because of the intrinsic favorable properties of chromium. The corrosion resistance of WC/Co sprayed layers depends on their porosity. If porosity is very low (c.f. 50) or if a porous layer is sealed with a resin, a very good resistance against corrosion can be obtained.

6.4 CVD and PVD Processes

In the seventies chemical vapor deposition (CVD) and physical vapor deposition (PVD) have rapidly gained a foothold in the world of surface coating techniques. In chemical vapor depsoition a very dense, hard, and well-adhering coating is formed on the surface of a substrate, heated to a temperature of approximately 1000°C, by reaction of gaseous compounds[48]. The most frequently applied materials are listed in column 4 of Table 7; suitable substrate materials are steels, cermets, and carbides. The layer thickness is limited to approximately 10 μm, because unacceptable stresses build up in thicker layers. The equipment necessary for performing the CVD process consists in principle of a vaporizer in which reactive gasses are added to a carrier gas, a container in which the substrate to be coated is placed and in which the chemical reactions occur and a heat supply system. The primary reaction partners are usually liquid. The carrier gas, usually hydrogen, is led through the liquids during which it becomes saturated with the relevant vapors. Heating takes place by induction or radiation.

A typical example is the formation of titaniumcarbide via the reaction:

$$TiCl4 + CH4 \rightarrow TiC + 4HCL$$

in which hydrogen is used as a crrier gas. At atmospheric pressure, dense TiC layers are formed on the substrate at temperatures in the range of 900°- 1200°C, at a rate of layer formation of approximately 5 μm/hour. The deposition temperature can be lowered to 700°C by working under reduced pressure of approximately 0.1 kPa. The structure of the TiC coating tends to be columnar and this leads to easy crack propagation and consequent lowering of transverse rupture strength of the substrate, a decrease of 40% having been measured with cemented carbides. Attempts to reduce this effect, by the use of catalysts which increase the number of nucleation sites on the substrate, have been successful[52]. Up till now, the use of CVD coated surfaces has virtually been restricted to tool tips. However, in the late seventies a number of interesting tribological applications became known[53]. These include rolling element bearings, sliding gyroscope bearings, and measuring equipment.

A typical advantage of the CVD process is its excellent "throwing power", by which an irregularly shaped component can be covered on all sides with a layer of uniform thickness.

Physical vapor deposition (PVD) processes are always performed at low pressure, the vapor to be deposited being formed by thermal evaporation or electrical emission ("sputtering") of a suitable material, in some cases in the presence of a chemically active gas. The desired coating forms by condensation at the substrate surface. All PVD processes have in common that one needs relatively expensive equipment for their performance which is also energy intensive. On the other hand, the coating quality of some PVD processes is superior. In particular this holds for the newer process of ion-implantation[54]. In this process ions are accelerated by voltages of 100 - 150 kV and are focused on the substrate to be coated. By the high velocity with which the ions hit the substrate, they can penetrate to a depth of approximately 1 μm. This yields a very good adhesion. A typical example is dies for wire drawing made of WC, the endurance of which was increased by a factor of 10 by ion implantation of carbon.

7. REFERENCES

1. Salomon, G., "Recent Advances in the Application of Systems Thinking in Tribology," lecture presented at the eighth meeting of the International Research Group on Wear of Engineering Materials, Saint-Ouen (Paris), May 1981; to be published (text can be obtained from the Technical Secretariat, c/o TNO, P.O. Box 541, 7300 AM Apledoorn, The Netherlands).

2. Czichos, H., "Tribology, a Systems approach to the Science and Technology of Friction, Lubrication and Wear," Elsevier, Amsterdam, 1978.

3. Begelinger, A. and de Gee, A.W.J., "Synopsis of Results from an International Co-operative Wear Programme," Lubrication Engineering 26 (1970), 56-63.

4. Vaessen, G.H.G. and de Gee, A.W.J., "Boundary Lubrication of Bronzes - Metallurgical Effects," Transactions ASLE 16 (1973) 203-207.

5. De Gee, A.W.J., "Selection of Materials for Tribotechnical Applications - The role of Tribometry," Tribology International, August 1978, 233-239.

6. Bauschke, H.M., Hornbogen, E, and Zum Gahr, K.H., "Abrasive Wear of Austenitic Steels," Aeitschr. fur Metallkunde 72 (1981), 1-13 (in German).

7. ISO/DP7148 "Testing of the Tribological Behavior of Bearing Materials for Oil Lubricated Applications; Part I: Coefficient of Friction and Wear Rate under Conditions of Boundary Lubrication and Ample Supply of Lubricant to the Friction Couple," Copies can be obtained from: H. Tepper, Deutsches Institut fur Normung e.V., Kamekestrasse 8, D-5000 Koln 1, Germany.

8. De Gee, A.W.J., "Selection of Materials for Lubricated Journal Bearings," Wear 36 (1976), 33-61 and Wear 42 (1977), 251-261.

9. Habig, K.H., Borszeit, E., and de Gee, A.W.J., "Friction and Wear Tests on Metallic Bearing Materials for Oil Lubricated Bearings," to be published (Wear 1981).

10. Rabinowicz, E., "The Influence of Compatability on Different Tribological Phenomena," Transactions ASLE 14 (1971), 206-212.

11. Buckley, D.H., "Fretting in Aircraft Systems," AGARD Conference Proceeedings Nr. 161 (1975), 13/1-13/15.

12. Begelinger, A. and de Gee, A.W.J., "Abrasive Wear of Bearing Materials," Proceedings 2nd Eurotrib Conference, Dusseldorf, October 1977, Band I.

13. Scott, D., "The Effect of Steel Making, Vacuum Melting and Casting Techniques on the Life of Rolling Bearings," Vacuum 19 (1969), 167-169.

14. Scott, D., and McCullagh, P.H., "The Role of Nitrogen Content on the Rolling Contact Fatigue Performance of EN31 Ball Bearing Steels," Wear 25 (1973), 339-344.

15. Scott, D., "New Materials for Rolling Mechanisms," Wear 43 (1977), 71-87.

16. Vaessesn, G.H.G. and de Gee, A.W.J., "Rolling Contact Fatigue of Maraging Steels of Different Production History, Influence of Film Thickness/Roughness Ratio," Paper C7, Second Symposium on Elastohydrodynamic Lubrication, I. Mech. E., Leeds, 1972.

17. Dowson, D. and Higginson, G.R., "Elastohydrodynamic Lubrication," International Series in Material Science and Technology, Vol. 23, Pergamon Press, 2nd Edition, 1977.

18. McCool, J.I., "Load Ratings and Fatigue Life Prediction for Ball and Roller Bearings," Transactions ASME, Journal of Lubrication Technology, January 1970, 16-22.

19. Salomon, G., "Failure Criteria in Thin Film Lubrication," Wear 36 (1976), 1-6.

20. Czichos, H., "Failure Criteria in Thin Film Lubrication; Investigations of Different Stages of Film Failure," Wear 36 (1976), 13-17.

21. Begelinger, A. and de Gee, A.W.J., "On the Mechanism of Lubricant Film Failure in Sliding Concetrated Steel Contacts," Transactions ASME, Journal of Lubrication Technology 98 (1976), 575-579.

22. Begelinger, A., de Gee, A.W.J. and Salomon, G., "Failure of Thin Film Lubrication - Function-Oriented Characterization of Additives and Steels," Transactions ASLE 23 (1980), 23-34.

23. Begelinger, A. and de Gee, A.W.J., "Failure of Thin Film Lubrication - The Effect of Running-In on the Load Carrying Capacity of Thin Film Lubricated Concentrated Contacts," Transactions ASME, Journal of Lubrication Technology 103 (1981), 203-211.

24. Begelinger, A. and de Gee, A.W.J., Thin Film Lubrication of Sliding Point Contacts of AISI 52100 Steel," Wear 28 (1974), 103-114.

25. Czichos, H., "Failure Criteria in Thin Film Lubrication; the Concept of a Failure Surface," Tribology International 7 (1974), 14-20.

26. Begelinger, A. and de Gee, A.W.J., "Failure of Thin Film Lubrication - A Detailed Study of the Lubricant Film Breakdown Mechanism," to be published in Wear.

27. Fowle, T.I., "Gear lubrication: Relating Theory to Practice," Lubrication Engineering 32 (1976), 17-34.

28. De Gee, A.W.J. and Vaessen, G.H.G., "A Note of Bayer and Ku's Model for Zero Wear," Wear 18 (1971), 492-496.

29. Muyderman, E.A., "Grease Lubricated Spiral Groove Bearings," Tribology International, June 1979, 131-137.

30. Lancaster, J.K., "Dry Rubbing Bearings," in Tribology Handbook, ed. M.J. Neale, Butterworth, London, 1973.

31. Crease, A.B., "The Wear Performance of Rubbing Bearings - Improved Data for Design," Paper 10, Proceedings 3rd Leeds-Lyon Symposium, Leeds, I. Mech. E., London, 1976.

32. Erhard, G. and Strickle, E., "Gleitelemente aus thermoplastischen Kunststoffen," Kunststoffe 62 (1972), 1-9 (in German).

33. Lancaster, J.K., "Dry bearings: A Survey of Materials and Factors Effecting Their Performance," Tribology International 6 (1973), 219-251.

34. Salomon, G., Begelinger, A., van Bloois, F.I., and de gee, A.W.J., "Characterization and Tribological Properties of MoS_2 Powders and Related Chalcogenides," Transactions ASLE 13 (1970), 134-147.

35. Torrance, A.A., "The Correlation of Abrasive Wear Tests," Wear 63 (1980), 359-370.

36. Krushcov, M.M., "Resistance of Metals to Wear by Abrasion as Related to Hardness," Proc. Conf. on Lubrication and Wear, 1957, I. Mech. Eng., London, 46-55.

37. Richardson, R.C.D., "The Maximum Hardness of Strained Surfaces and the Abrasive Wear of Metals and Alloys," Wear 10 (1967), 353-382.

38. Buttery, T.C. and Archard, J.F., "Grinding and Abrasive Wear", Proc. Inst. Mech. Eng., London, 185 (1970 - '71), 537-542.

39. Johnson, K.L., Discussion of Ref. 37, Proc. Inst. Mech. Eng., London, 185 (1970 - '71), D 205.

40. Moore, M.A., "The Relationship Between the Abrasive Wear Resistance, Hardness, and Microstructure of Ferritic Materials", Wear 28 (1974), 59-68.

41. Mutton, P.J. and Watson, J.D., "Some Effects of Microstructure Abrasion Resistance of Metals," Wear 48 (1978), 385-398.

42. De Gee, A.W.J., "A Note on the Relation Between Friction and Wear," Wear 65 (1981), 397-398.

43. Engel, P.A., "Impact of Materials," Elsevier, Amsterdam 1976.

44. Bitter, J.G., "A Study of Erosion Phenomena," Wear 6 (1963), 169-190.

45. Neilson, J.H. and Gilchrist, A., "Erosion by a Stream of Solid Particles," Wear 11, 1968, 111-122.

46. De Bree, S.E.M., Begelinger, A. and de Gee, A.W.J., "A Study of the Wear Behavior of Materials for Dredge Parts in Water-Sand Mixtures," Proc. Third International Symposium on Dredging Technology, BHRA, Bordeaux, March 1980, 299-314.

47. Arthur, G., Birch, D., Michie, G.M., Moorhouse, P., and Wells, T.C., "Wear Resistant Surfaces - A Guide to Their Production, Properties and Selection," Report International Research and Development Co. Ltd., England, 1979.

48. Powel, C.F., Oxley, J.H., and Blocker, J.M., "Vapor Deposition," John Wiley and Sons, New York, 1966.

49. Caubet, J.J., and Gregory, J.C., "Thermal and Chemico-Thermal Treatments of Non-Ferrous Materials to Reduce Wear," Tribology International 4 (1971), 8-14.

50. Habig, K.H., "Verschleiss und Harte von Werkstoffen" ("Wear and Hardness of Materials"), Carl Hanser Verlag, Munich-Vienna, 1980.

51. Van Nederveen, H.B., Verburgh, M.B., and Houben, J.M., "The Densification of Plasma Sprayed Coatings by Subsequent Hot Isostatic Pressing," paper 51, Proc. 9th International Spraying Conference, The Hague, May 1980.

52. Hintermann, H.E., Gass, H., and Lindstrom, J.N., "Nucleation and Catalysed Growth of TiC, Produced by CVD on Cemented Carbide," Third International Conference on CVD, 1972, 352-360.

53. Hintermann, H.E., Menoud, C., and Maillat, M., "Friction and Wear Characteristics of SiC, TiC and TiN Hard Coatings Rubbing Against Each Other," Lecture Presented at the 8th Meeting of the IRG on Wear of Engineering Materials, Saint-Ouen, May 1981, to be published in Wear.

54. Hartley, M.E.W., "Ion Implantation and Surface Modification in Tribology," Wear 34 (1975), 427-438.

SURFACES

M. J. EDMONDS
BRIGHTON POLYTECHNIC

1. INTRODUCTION

Surfaces are at the heart of the multidisciplinary subject of Tribology. Without an accurate assessment and characterization of surfaces, tribological research cannot progress. In this chapter the author has outlined some of the important aspects concerned with surface technology - ranging from the composition of a surface, specification, terminology, characterization, selection and production, to some of the advances that have taken place in recent years.

The discussion closes with a short account of possible future trends in surface technology and the likelihood of in-process and on-line measurement within the next decade.

1.1 Brief Historical Review

Interest in surfaces and the effects of movement of one surface relative to another, almost certainly stretches back in history to the stone age, Figure 1. Although this may perhaps be disputed by some scholars on the grounds of lack of positive archaeological evidence, it is difficult to visualize primitive man manipulating heavy objects other than by rolling them on logs, thus creating the earliest type of plain bearing. We can also appreciate him rubbing a piece of wood or bone perhaps against a stone, in order to sharpen an edge during the manufacture of tools or weapons. As he progressed, we can imagine him using water or animal fats to lubricate the surfaces to make them easier to operate.

FIGURE 1   "TRIBOLOGY?   THAT'S THE STUDY OF TRIBES ISN'T IT!!!

We have to wait many centuries though before any positive documentary evidence is available to support ideas and practices. Indeed, the earliest known detail is that credited to Leonardo da Vinci (1452 - 1519) who, as far as we are aware, was the first person to develop and record ideas of the basic concepts of friction[1].

Interest in surfaces today, although far removed from primitive man, results from the evolution of the rolling bearing, or to use the modern word which embraces all such matters - TRIBOLOGY.

It is really to the study of the rolling bearing that we have to seek for our origins in the subject. Archaeologists believe that the ancient Egyptians and Greeks made considerable use of the principle of rolling bearings. A piece of a bearing believed to have been devised by the Greek Diades (330 BC) is still in existence and was used as part of a battering ram. Fragments of materials found in Lake Nemi - Italy, in 1928, again support the idea that rolling bearings were used more than 2,000 years ago. The period of the Industrial Revolution in Britain and Europe inevitably made a major impact into the field of rolling contact. A notable exhibit in the Norwich Museum of an iron ball thrust bearing, with many design characteristics of modern bearings, made its appearance about 1780.

In 1787 one of the earliest British patents, number 1580, was awarded to Mr. John Garrett of Gloucester for the arrangement of various types of rolling elements to form bearings. Several interesting developments occurred in the years that followed; and in the middle part of the eighteenth century, with the advent of the bicycle, patents for a variety of rolling applications became abundant.

In 1881, Heinrich Hertz published in Germany, a treatise destined to become a classic, on the deformation of curved elastic bodies in contact[2]. This work provided the first mathematical background to analysis involving surfaces. Apart from a few papers presented to learned societies, there appeared to be very little literature available directly concerned with Surface Technology until about 1930 when a paper by R. E. W. Harrison appeared, which was concerned with a survey of Quality Standards[3]. Shortly after, a paper by Firestone, Durbin, and Abbott was published concerning tests for smoothness of machined finishes[4]. The instrument they used to provide surface profile measurements was an optically amplified stylus system in which a lever arrangement was used to record on photographic paper. A year later in 1933, Abbott and Firestone published a paper which may be regarded as the forerunner for all the subsequent surface analysis work that has taken place[5]. By accepting the fact that no surface is truly flat but made up of an irregular array of hills and valleys, they suggested that if an imaginary mean plane is drawn through the surface profile irregularities then it is possible, by drawing parallel planes on either side of the mean plane, to produce a curve of statistical features of the surface texture.

Figure 2 illustrates a typical surface profile showing the depth versus bearing area relation. The curve is now known as the "Abbott bearing area curve," and demonstrates the fact that when two surfaces are interfaced, the real area of contact supporting the surfaces is only a small percentage of the nominal area. This concept will be discussed in greater detail later, since it is an important aspect of surface technology.

To a large extent, until about 1940, much of the analytical interest in surfaces seems to have remained within the disciplines of physics, chemistry and metallurgy. Engineers at this time were more concerned with methods of production of surfaces, and the design of machines that produced them.

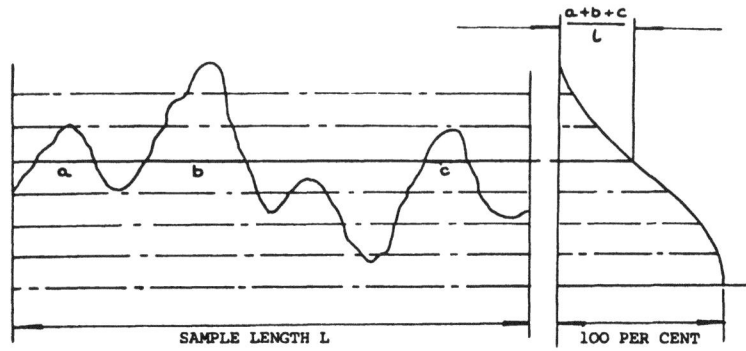

FIGURE 2 THE BEARING AREA CURVE

However, as technology advanced, it became imperative that engineers cross the disciplinary boundaries and gain a greater fundamental understanding of the surfaces and materials they were dealing with.

The period of the second world war produced the next significant advances in the technology. These advances resulted from the fact that aircraft engines had to be developed at an ever increasing rate, be kept running, and respective engine mechanical failures reduced to a minimum. Led by Dr. R. E. Reason, a team of engineers in Leicester set out on a research and development program to develop an instrument capable of quantitative assessment of surface texture, and suitable for use alongside production engineering workshops[6]. Hitherto, all previous assessment attempts were mainly confined to laboratories. Standards of manufacture were generally accomplished by a "go and no go" gauge arrangement. If a component looked good, felt good (tactile testing), and worked, then the process of its manufacture was repeated. Such a system proved extremely difficult, particularly when it became necessary to subcontract work out as the need arose to increase component production rate. Some means had to be established to indicate the type of surface finish specified by the designer, in order to supplement tolerance specifications. This led to the "centre-line-average" technique, or Ra value as it is known today, based on the assessment instruments developed by R. E. Reason and his team.

This brings us up to the early 1950's. One other major advance of special significance connected with surface technology must be noted, and that is the work of Bowden and Tabor, the principal tribophysicists of the day[7]. In 1950 they published their classical work, The Friction and Lubrication of Solids, Part I. Many important topics, vital to the study of Tribology are covered in their text, ranging from area of contact between solids, friction, and adhesion to the influence of liquid films.

During the past thirty years, many advances have taken place in the study of Surface Technology, and the remainder of this chapter will be concerned with outlining various features that have emerged. To quote Dr. Tom Thomas; "the number of papers published on matters relating to Surface Technology is now doubling each year." This illustrates the wide range of interests associated with the topic, and its place in modern technological systems. The most comprehensive survey on Surface Technology is, to the best of the author's knowledge, that of Thomas and King[8]. This book contains 651 references together with abstracts set in chronological order from 1921 to 1976, and it provides the most useful literature survey on the topic so far in print.

2. COMPOSITION

To the layperson, a surface probably represents simply the outside face of a body, but it is in fact much more complicated than that. A surface must be recognized as an amalgam consisting of several zones which grow organically out of its parent material having both chemical and physical characteristics dependent on the composition of the material itself and the environment in which it is situated.

The base of the surface amalgam consists of a zone of work-hardened material on top of which is a microcrystalline structure. This is termed the BIELBY layer, and for surfaces which have been derived by means of a machining operation, is formed as a result of the melting and flow of the surface molecular layers. The layers form due to hardening and subsequent quenching as they cool against the underlying material. Figure 3 shows a cross-section of this very complex structure. The basic structure is further complicated by the addition of chemical reactions with the atmosphere, contaminants of dust particles, and molecular films absorbed from the environment. In Figure 4, orders of magnitude of depth are given for the various structural zones, plotted on a log scale. very great and the surface will yield plastically. Penetration

From values such as these, it is clear methods of production not only influence the surface finish, but must also have an effect on the subsurface. A very interesting paper by Wallace shows the condition of subsurface material after different finishing processes, and appears to substantiate the figures quoted by Halling and Barwell[10, 9, 20].

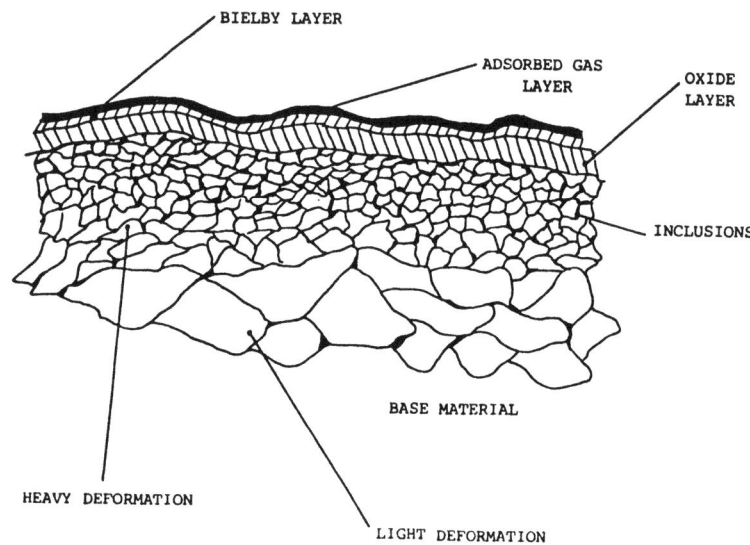

FIGURE 3 SURFACE COMPOSITE

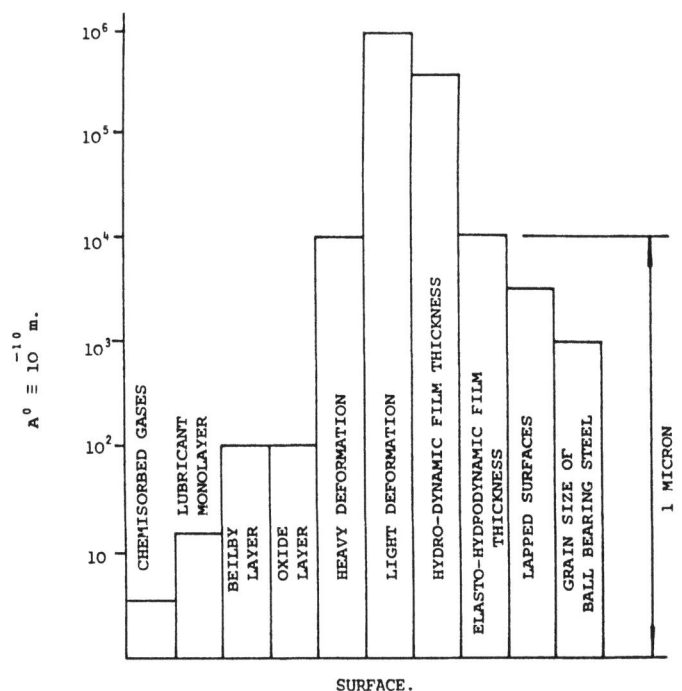

FIGURE 4 TYPICAL ORDERS OF MAGNITUDE OF SURFACE FEATURES

3. HARDNESS

The ability of a surface to resist bulk surface deformation in the plastic mode is referred to as hardness. This is usually measured as a result of penetration of the body by a diamond shape in the form of an inverted pyramid. As the diamond impinges on the surface, stress at the point of entry will be will continue until the diamond has penetrated sufficiently to have interacted with sufficient material to produce an opposing force which will bring it to rest. It is usual to quote, as a hardness figure, the value of the applied load divided by the area of the resulting permanent impression.

Values of hardness have been found to correspond closely with a value of three times the yield stress in compression. Likewise, when two surfaces are brought together, the highest peaks (asperities) on the one surface will come into contact with asperities on the other surface and the intensity of load initially will be very great. It may be expected that the asperities will first yield elastically, but in most metals the elastic component is almost negligible, and the plastic mode appears to predominate [11]. The scale of action will, however, be much smaller than that of the conventional hardness test.

The hardness, expressed as a stress, is normally referred to as the flow stress of the material. If work- hardening is ignored, and the material assumed to be uniform, then it can be shown that the true area of contact will be reached when the applied load equals the product of the true area of contact and the flow stress[12].

4. SURFACE TEXTURE MEASURMENT

Many methods for surface texture measurement both contacting and noncontacting, have been devised over the past thirty years[8]. Generally speaking, non-contacting methods involve optical systems, and contacting methods utilize a fine pointed stylus. A very comprehensive survey, which includes comparisons of all systems developed up to 1942, was published by Schiesinger and more recently by Schneider[13, 17]. The major advantage with optical techniques is the fact that no surface damage is incurred, whereas traversing a stylus over a surface often results in a very fine scratch. Despite this situation, the most widely used instrumentation employs a stylus. This utilization is not only a function of convenience, but principally because more quantitative detail can be obtained. Nevertheless, much qualitative information can be derived from optical systems, particularly electron microscopes. Table 1, below, gives some idea of the resolution obtainable from a variety of techniques.

Technique	Resolution (μm)	
	Lateral	Vertical
Profilometric Instruments	1.3-2.5	0.005-0.25
Electron Microscope	0.005	0.0025
Optical Microscope	0.25-0.35	0.18-0.35
Interference Microscope	0.25	0.25
Oblique Sectioning	0.25	0.025

TABLE 1   SURFACE RESOLUTION

4.1 Stylus Systems

The early stylus instruments used a phonograph needle; however, modern systems enploy a truncated diamond pyramid of tip radius approximately 2.5 μm[15, 16]. The angle between the faces is 112°, and the dimensions of the rectangular flat about 3 μm x 8 μm. The shorter edge is arranged to be parallel with the direction of motion. The applied load is of the order of 0.7 mN. Variations of styli are available for special applications, down to a tip radius of 0.1 μm.

Vertical movement of the stylus is recorded electrically via an L.V.D.T. (linear variable differential transformer), with the displacement amplified. The electrical signal from a displacement-sensitive transducer can provide three kinds of information: (1) an analogue record, i.e. a profile graph on a chart recorder; (2) a digital record of closely spaced ordinates of the profile graph, which can be recorded on punched or magnetic tape for subsequent processing in a computer; or (3) meter indications of numerical values which are descriptive of the profile.

It should be pointed out that the graphical description of a profile should not be taken too literally, because magnification factors on the instruments yield up to 10,000 times vertical displacement relative to the horizontal displacement. What appears to be a close spike-like profile in the compressed state would yield, if plotted on a 1:1 ratio, a gentle undulating surface, characteristic of moorland rather than say a profile view of the Alps. By virtue of this, penetration into the valleys by the stylus is not such a critical factor as might be assumed. Although, if measurements of abrasive surfaces such as the assessment of grit in grinding wheels are considered, then the problem could be more acute. If such an assessment as this is desired, it would seem more practical to approach the problem a different way utilizing the electron microscope.

4.2 Electron Microscope

As mentioned earlier, the S.E.M. (scanning electron microscope) does offer an alternative means of surface geometry assessment. Magnifications up to 5,000 x are easy to relate to ordinary hand-held magnifiers. If it is desired to simply look at a surface at a reasonable magnification, using 'Y' modulation can be useful. This means operating the S.E.M. such that an isometric picture is built-up, by scanning the surface in a similar fashion to tracing a profile using a stylus system. Figure 5 illustrates the view of a medium ground surface using this technique, together with its magnified S.E.M. image, the angle being set at 45°.

5. SPECIFICATION

No surface exists in nature, or man-made, which can be described as being truly planer. The reason for this is that all matter is made up of atoms, and for such a state to exist the center of all atoms, or ions in the surface layer, would have to lie in the same plane. No element appears to be able to conform to this requirement. The nearest element which might qualify is very carefully cleaved mica. In the vast majority of cases, the surface of a solid has irregularities, although in some cases only atomic in scale, but frequently very much larger. For machined surfaces, we can take it for granted that whatever method of preparation has been applied, the surface will always have irregularities of some form or another. The measurement and adequate assessment of surfaces is of vital importance to all engineering and science systems. Friction, wear, contact deformation, reflection, heat transfer, electrical and thermal conductance, tolerances and fits are all influenced by and depend on a comprehensive characterization of the micro- and macro-geometrical texture. Many papers have now been published on each of the above mentioned topic areas; subjects ranging from the assessment of the surface of ships hulls to the surface of lip seals[18, 19]. As already mentioned, Thomas and Kings book offers a wide range of references together with an excellent literature review; and readers are advised to use this as their starting point for any particular area of interest.

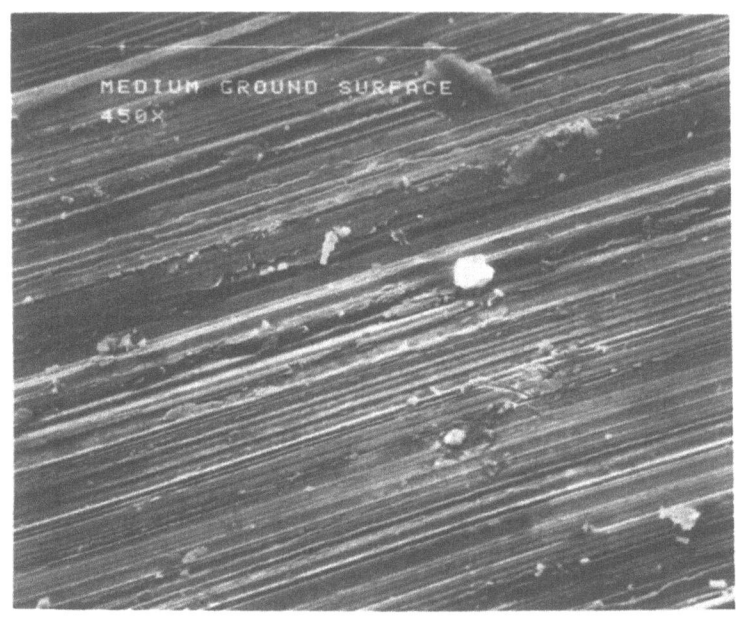

FIGURE 5(A) S.E.M. SCAN OF A MEDIUM GROUND SURFACE

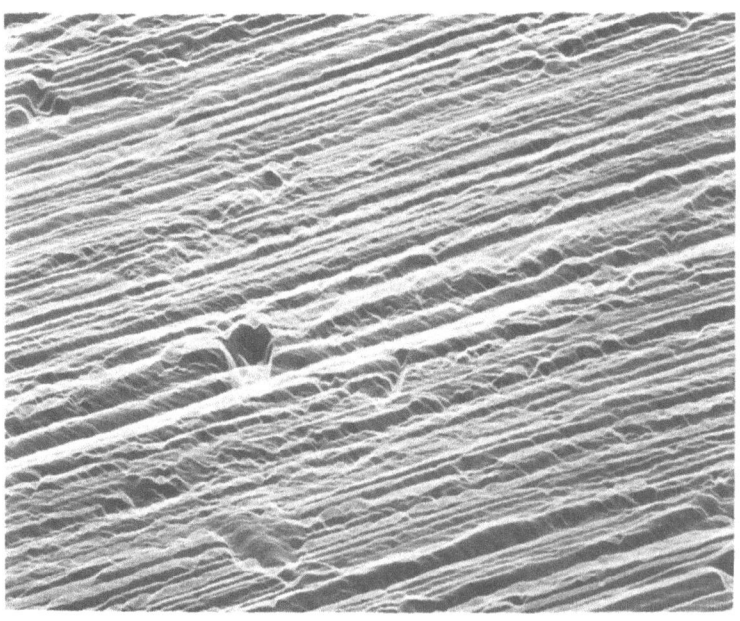

FIGURE 5(B) S.E.M. SCAN OF 'Y' MODULATION

5.1 Terminology

The geometric texture of machined surfaces is controlled by characteristics of the finishing process by which the surface is produced. On a microscopic scale, irregularities are evident, and roughness is formed by fluctuations in the surface of short wavelengths (termed microroughness). The form is characterized by hills (known as asperities) and valleys of varying amplitudes and spacing, see Figure 6. Orders of magnitude of measurement are either microinches or micrometers (micron), (1 µm = 1 x 10^{-6}m). In addition, on many surfaces a larger wavelength of roughness can also be observed called waviness, which is referred to as macroroughness. Normally this is measured in mm. Other features which also exist in a surface in the form of undulations, and exhibit much larger wave-lengths, are errors of form. Again, these may be measured in mm. Such features as these are generally caused by vibrations of the machine, work piece, or the cutting tool.

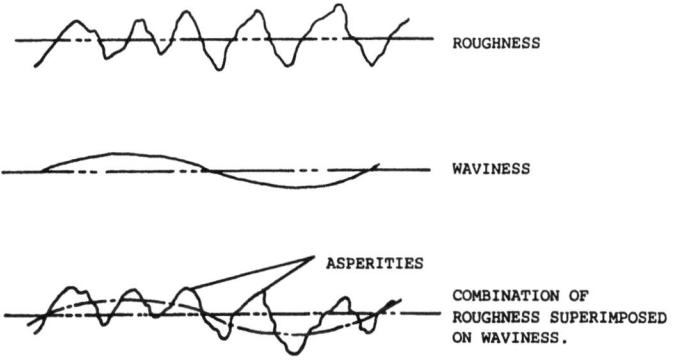

FIGURE 6   COMPONENTS OF A SURFACE

The distribution of the asperities over a surface can be either directional or homogeneous in all directions, depending on the method of preparation. Surfaces produced by shaping, milling, or planning for example have a definite pattern or lay. But those produced by lapping, electropolishing, etc., show an isotropic distribution in all directions.

5.2 Symbols Used in Surface Analysis (ANSI B461 - 1978)

5.2.1 Center-Line Average

The universally recognized parameter of roughness has the symbol Ra. Formerly, this was called the center-line average, C.L.A. (British), and the Arithmetic Average, A.A., (America). It is the arithmetic mean of the departure, "y", of the profile from the theoretical mean line, and is normallly determined as the mean result of several consecutive sampling lengths, "L", see Figure 7.

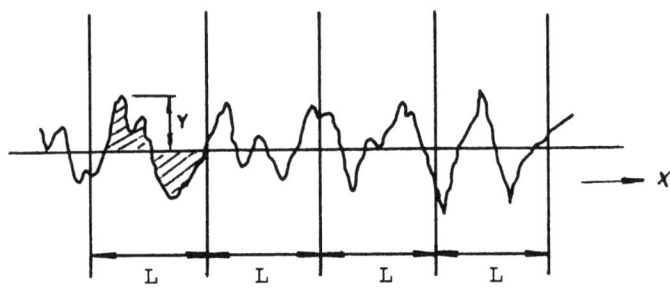

FIGURE 7   CENTER-LINE AVERAGE

5.2.2   Root Mean Square

The Root Mean Square value, R.M.S., is no longer used in the context of surface texture measurement, but is used in statistical analysis. It was formerly given the symbol Rq

where $Rq = \sqrt{\frac{1}{L} \int_0^L Y^2(x)dx}$

In statistical analysis the symbol used is "σ".

5.2.3   Ten Point Height

Rz is the ten point height, and is the average distance between the five highest peaks and the five deepest valleys within the sampling length and measured perpendicular to it, see Figure 8.

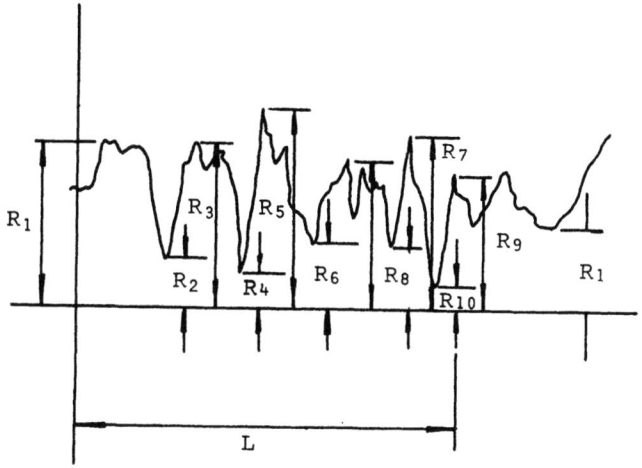

FIGURE 8   TEN-POINT HEIGHT

5.2.4 R_t, R_{max} and R_{tm}

R_t is the maximum peak to valley height within the assessment length. R_{max} is the maximum peak to valley height within a sampling length L. Since the value can be affected by a spurious scratch or a particle of dust on the surface, it is more usual to use the average, "R_{tm}", of five consecutive sampling lengths, see Figure 9.

where $R_{tm} = \dfrac{R_{max_1} + R_{max_2} + R_{max_3} + R_{max_4} + R_{max_5}}{5}$

$$R_{tm} = \frac{1}{5} \sum_{i=1}^{i=5} R_{max_1}$$

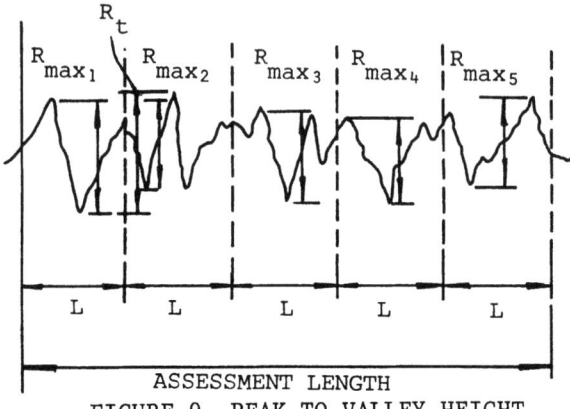

FIGURE 9 PEAK TO VALLEY HEIGHT

5.2.5 R_p and R_{pm}

R_p is the maximum profile height from the mean line within the sampling length. R_{pm} is the mean value of R_p determined over 5 sampling lengths, see Figure 10.

$$\text{where } R_{pm} = \frac{R_{p_1} + R_{p_2} + R_{p_3} + R_{p_4} + R_{p_5}}{5} = \frac{1}{5} \sum_{i=1}^{i=5} R_p$$

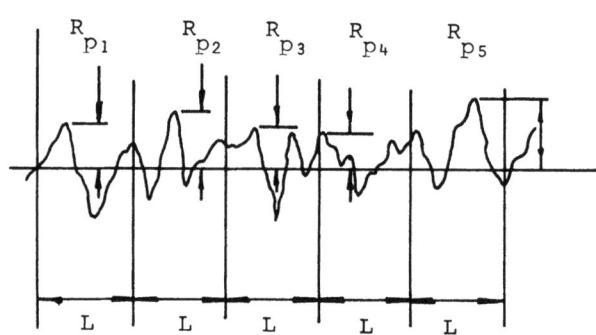

FIGURE 10 MAXIMUM PROFILE HEIGHT

5.2.6 P_c

Peak count is the number of peak/valley pairs per inch projecting through a band width, b, see Figure 11.

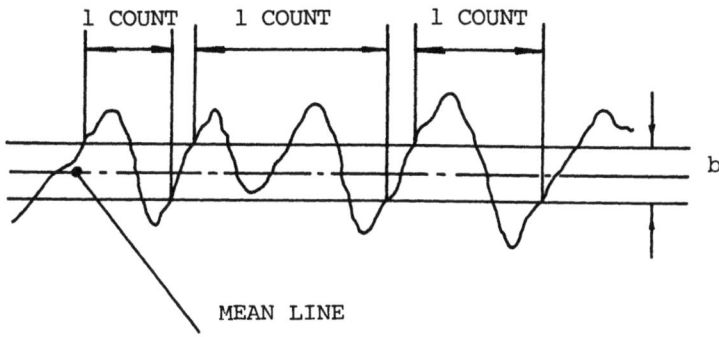

FIGURE 11   PEAK COUNT

5.2.7 Average Wavelength

Because no single parameter can adequately represent the complex patterns found on a surface, other parameters were introduced, but all are derived principally from amplitude, and as such are relatively unaffected by the spacing of irregularities. Universal acceptance of these symbols and their meaning has not, as far as the author is aware, been totally agreed upon. In Europe, the symbol, λa is used which is not derived from amplitude but from spacing. This is termed the average wavelength (λa) and is now used in conjunction with Ra. Thus, the average roughness and the openness or closeness of the surface texture can be specified.

Some typical values of Ra and λa are shown in Table 2.

Process	Ra (μm)	λa (μm)
Turning	1.7	100
Fine turning	0.3	37
Grinding (flat)	0.95	77
Circumferential Grinding	0.55	33
Single tooth milling	2.2	300
Multiple tooth milling	5.1	1300
Fine cut planning	5.1	440

TABLE 2   Ra VERSUS λa VALUES

From the table we can see how two surfaces prepared by different machine processes can produce a similar value of Ra, and yet have a value of λa grossly different. Clearly it is necessary to specify both parameters and not simply rely on just the Ra value alone.

5.3 Cut-Off

The question of "cut-off" is another feature which should be added to the specification of a surface. Since the surface texture can include roughness of different crest spacings which may be superimposed on one another, it is necessary to choose a sampling length, L, sufficiently long enough to enable the various crest spacings to be accommmodated. Electrical integrating instruments have now been devised such that the meter cut-off values are made equal to the sampling lengths. When the meter is switched to the desired meter cut-off, the pick-up used to follow the profile of a surface gives the average results from several consecutive sampling lengths.

Table 3 shows a selection of suitable cut-offs which are normally used for a variety of common manufacturing processes.

Manufacturing Process	Typical roughness height (Ra) μm	Range of Peak Spacings (mm)				
		0.08	0.25	0.8	2.5	8.0
Super Finishing	0.05 - 0.2	*	*	*		
Lapping	0.05 - 0.4	*	*	*		
Honing	0.1 - 0.8		*	*		
Grinding	0.1 - 1.6		*	*	*	
Diamond turning	0.1 - 0.4		*	*		
Turning	0.4 - 6.3			*	*	
Boring	0.4 - 6.3			*	*	*
Broaching	0.8 - 3.2			*	*	
Milling	0.8 - 6.3			*	*	*
Shaping	1.6 - 12.5			*	*	*

TABLE 3   SUITABLE CUT-OFF VALUES

Various limiting features are apparent when selecting the appropriate cut-off value such as the finite dimensions of the stylus tip, the inherent noise level of the instrument, and the type of skid nose piece which may be added to the instrument. It is therefore advisable to seek guidance from the appropriate instrument manuals before proceeding to take surface measurements.

5.4 Surface Designation

The cost of machining to achieve set tolerances is largely dependent on the skill of the operator, the quality and type of machine tool, and the material. If we add to this just the basic surface texture requirement, Ra, then clearly the cost of production must increase. A graph showing relative costs plotted against surface texture is shown in Figure 12. The cost of failure of a component which can in turn affect other components, and which can lead to catastrophic failure, must be viewed relative to the cost of producing the component in the first place. It is therefore necessary to first decide the need, then to assess the value, and finally to communicate very specifically the necessary information to all those people involved in the manufacture of any particular component or system. In order to do this, the design engineer must suppply very specific information relating to the surface finish desired. Ideally, a detailed engineering drawing should contain information on the maximum and minimum Ra value, the maximum waviness height, the appropriate cut-off, the lay, and maximum roughness spacing.

At first sight, the above detail may appear to present an appreciable burden to both the detail draughtsman and the machine operator. However, the problem can be overcome reasonably well using the existing technique of a tick, "√", to indicate where machining has to be performed. Based on ANSI Y 1436 - 1978, Reference 21, which is almost in agreement with I.S.O. 1320, the following format, Figure 13, has been agreed upon.

In practice, not all areas on a machine component need to be treated with such rigor. Only those parts which require precision machining and close tolerances such as slideways, shaft fittings, bearings, etc., or areas where contacting surface are concerned, need the full specification. Other than this, the Ra value is generally sufficient. But how does the machine operator interpret the values? By far and away the best method is simply experience, together with visual examples and the old and surprisingly sensitive fingernail test. With the newer instruments coming onto the market, sample specimens can be taken and assessed fairly rapidly. Unfortunately, this often involves removing the component from the machine tool, and it is particularly time consuming and difficult to reset, should the component not meet the requirements.

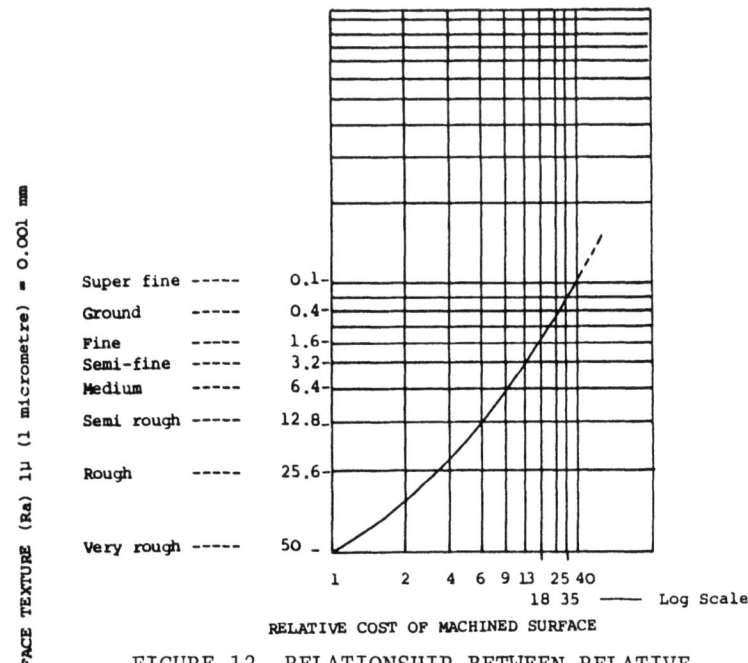

FIGURE 12   RELATIONSHIP BETWEEN RELATIVE
COST TO SURFACE TEXTURE

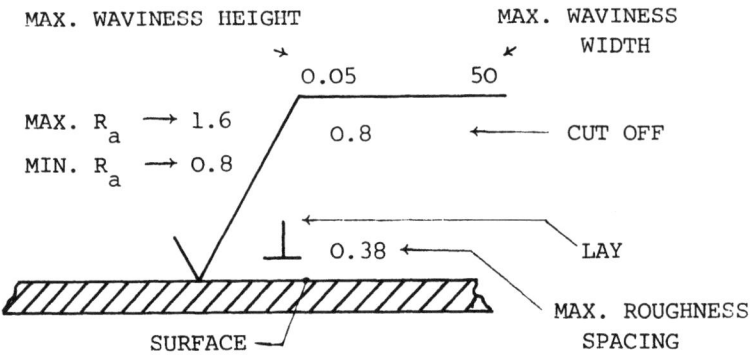

FIGURE 13   SURFACE TEXTURE SYMBOLS

```
        ALL VALUES ARE IN mm EXCEPT RA = µm
        SURFACE MAY BE PRODUCED BY ANY METHOD
      ─ MATERIAL REMOVAL BY MACHINING IS REQUIRED:
        (I.E., MATERIAL MUST BE PROVIDED FOR THAT PURPOSE)
.025m ─ MATERIAL REMOVAL ALLOWANCE (mm)
        MATERIAL REMOVAL PROHIBITED
```

5.5 Replication

In an effort to overcome the problem mentioned above, considerable work has now been done on surface replication[22, 23]. This is a technique which permits a plastic replica to be made of the area of interest which can be assessed away from the machine. Depending on the type of surface, very good correlations can be obtained between the original surface and the replica. In some cases a 95% accuracy has been claimed. The major problem is the greater the adhesion of the replica to the surface the more difficult it is to remove it. However, the better the adhesion, the more faithful is the reproduction. A compromise, as in most cases in engineering, has so far to be accepted.

6. CHARACTERIZATION

6.1 The Statistical Parameters

The principal production parameters for describing machine surfaces have been outlined in the preceeding subsections. In 1968 a new technique was introduced by Peklinik by applying spectral analysis to the study of surface topography[24]. This technique has probably produced the most significant advance in the subject in the past decade. Based on the mathematical treatment of natural phenonmena, namely the motion of ocean waves, mathematician Lonquet-Higgins produced a mathematical treatise which laid the foundations of what today has become one of the most powerful tools in the study of signal processing[25]. In recognizing that machined surfaces are made up of random and periodic components analogous to noise and vibration signals, surface scientists have developed and utilized the technique of spectral analysis with a considerable degree of success in recent years[26, 27, 28, 29, 30, 31]. The most significant advance, and the one which has created the most valuable contribution, is the Auto-correlation function, together with its Fourier-transfrom, the power spectral density function. These mathematical functions aim to separate the random and periodic components that occur in many surfaces generated by machining processes. Separation of these components is possible by analyzing the two- dimensional profile of a surface.

6.2 The Auto-Correlation Function

The analysis required to obtain the Auto-correlation function of a surface consists of making comparisons between a given length of profile and the same profile displaced in the length domain by increasing amounts. At each stage, ordinates in the overlapping portions of each profile are multiplied and their sum is reduced to an average value. Figure 14 illustrates the situation.

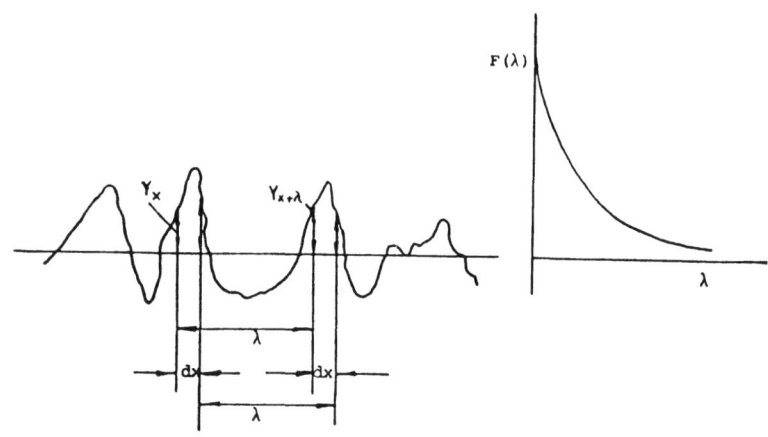

FIGURE 14   THE AUTO-CORRELATION FUNCTION

Consider the points Yx and Yx+λ on the profile separated by a distance λ. The product of their heights is Yx, x+λ. If the length λ is moved a distance dx along the profile then its ends define a new set of points Yx + dx, Yx + dx + λ, which will yield another height product. By repeating this procedure for the whole length of the profile, the following parameter can be defined:

$$\sigma^2_{YY}, \lambda = \frac{1}{L - \lambda} \int^{L-\lambda} Y_x, Y_x + \lambda \qquad (1)$$

where σ is the R.M.S. value.

If λ is small in comparison with the average dimensions of the peaks then the products Yx, Yx + λ will, in the main, be mostly finite and positive; thus, Equation (1) will be finite and positive. If λ is large in comparison with the average dimensions of the peaks, the products are also likely to be of either sign, and their sum will therefore be small in magnitude. By plotting σ^2_{yy}, λ against λ, the function decays from a finite positive value to nearly zero, and the rate of decay can then be used as a measure of the spacing of the surface texture. This function is called Auto-covariance, which for zero delay is the variance, i.e. σ^2_{yy}, o = σ^2.

If both sides of the equation are normalized by dividing by σ^2, the Auto-correlation coefficient is thus obtained from:

$$F\lambda = \frac{1}{\sigma^2 (L-\lambda)} \int_0^{L-\lambda} Y_x, Y_x + \lambda^{dx} \qquad (2)$$

Since this function has the properties associated with the correlation coefficient of discrete statistics, i.e. its numerical value can vary between ± 1, the variation with delay length λ is the Auto-correlation function.

An Auto-correlation function of 1.0 indicates complete agreement, whilst - 1.0 indicates complete disagreement. A value of zero shows that no correlation exists.

Figure 15 illustrates two typical surface profiles that might be produced by differenct processes, together with their corresponding Auto-correlation functions. The shape of Figure 15(A) is somewhat periodic in form and typical of a surface that might be produced by end-milling or by shaping, operating across the lay. The plot of the Auto-correlation function follows a similar periodic form. The general decay indicates a decrease of correlation as the sampling interval, ℓ, increases, and the oscillation component indicates the inherent periodicity of the profile. Figure 15(C), on the other hand, is random and typical of a ground surface. The rate of decay of the Auto-correlation function, Figure 15(D), is fairly rapid, and some periodicity can also be seen. Again, this illustrates that the profile has been obtained operating across the lay. The initial decay is characteristic of the surface component, and the mean distance between the points where the function becomes zero is an indication of the half-wavelength of the surface waviness. At this point, the surface ordinates in the construction of the curve are statistically independent, and the function may be approximated by:

$$F(\lambda) = e^{-x/\alpha^1} \qquad (3)$$

where α^1 is the correlation distance.

Peklenik suggested that most engineering surfaces can be categorized by one of five Auto-correlation functions, which range from linear to sinusoidal functions to the preceeding exponential function, Equation (3)[24].

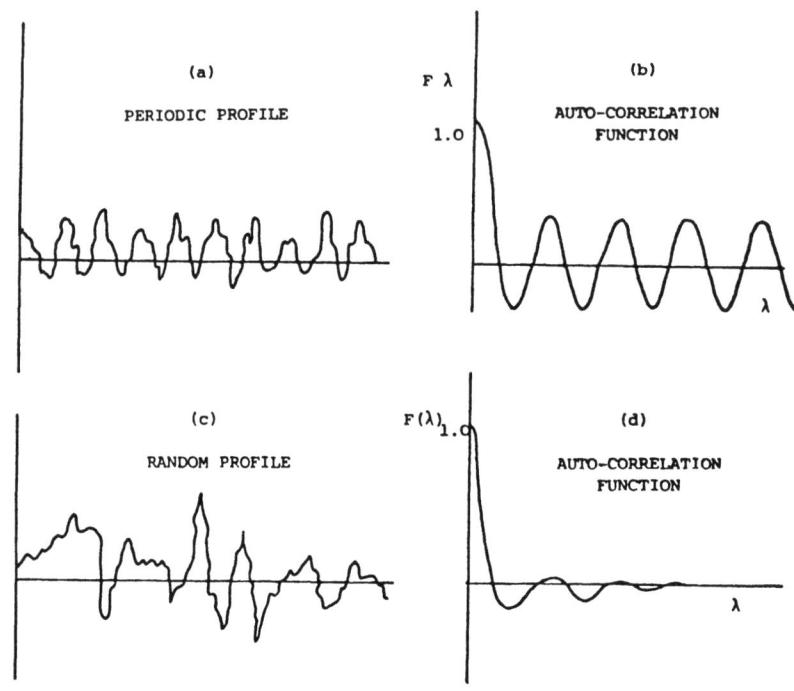

FIGURE 15   TYPICAL SURFACE PROFILES AND THEIR RESULTING AUTO-CORRELATION FUNCTIONS

6.3 The Power Spectral Density Function

As previously mentioned, the Fourier transform of the Auto-correlation function is termed the power spectral density of the discrete wavelengths contained in the profile variation, that is to say, it reveals the dominant frequencies in the profile[32, 33]. Large wavelength periodic components appear as spikes in the P.S.D. curve and, generally speaking, for a random signal with a wide spectral density distribution, the Auto-correlation function exhibits a high rate of decay. The P.S.D., $P(\omega)$, and Auto-correlation function are related by:

$$P(\omega) = 4 \int_0^{L-\lambda} \sigma^2 YY, \lambda \quad \cos 2\pi \lambda f \; \lambda d \ldots \quad (4)$$

The curve of $P(\omega)$, Figure 16, is the power spectrum due to the fact that its instantaneous value at a given frequency is proportional to the mean square of the signal at a given frequency, and hence its power. The area under the power spectrum curve is the total power present at all the frequencies, i.e. the mean square value of the whole signal, or the square of the r.m.s. roughness.

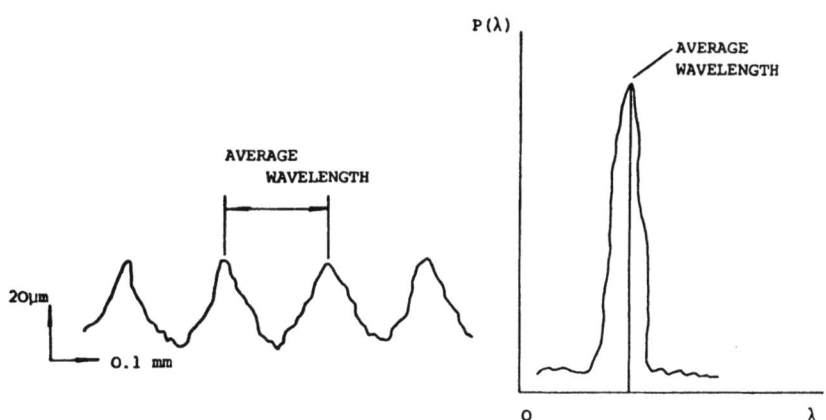

FIGURE 16   THE POWER SPECTRAL DENSITY FUNCTION

Any surface with a periodic component will have well defined peaks in its surface profile, which will also show up in the power spectrum curve. Depending on the number of ordinates chosen for a given length of profile, the P.S.D. curve will reveal not only the predominant frequency but also, in order of magnitude, any underlying frequencies or harmonics. This is particularly so for machined surfaces typically end-milled, turned, etc. However, for surfaces which have been produced by grit-blasting, where an appreciable periodic component is not apparent, i.e. when the surface is random rather than deterministic and the height distribution is Gaussian, the frequencies may be suppressed.

Generally, for most surface specifications, the Auto-correlation length can only be determined from the Auto-correlation function. Also, any periodic variation in the profile appears as a similar periodic variation in the curve of the auto-correlation function versus ordinate separation. In the P.S.D. function, any periodic function in the underlying surface appears as discontinuities. Nayak and Nara proposed that waviness and roughness should be discussed in terms of bandwidths of the spectrum, rather than of fixed wave lengths[34,35]. They proposed that the previous deterministic view of small sinusoids, namely "roughness" imposed on large sinusoids of fixed wavelengths, i.e. waviness or errors of form, should be replaced by a picture of overlapping bands of wave lengths from a continuous spectrum. From this it follows that the size, shape, and height of the surface texture also forms a continuous distribution depending on the spectrum. Thomas and Sayles adopted this approach and proposed a theory to predict the total roughness of a specimen of any size from a standard profilometer measurement[36]. Their theory suggests that any two isotropic surfaces produced by ordinary engineering processes, no matter by what means, will have identical power spectra if their measured roughness is the same. This is because, as they suggest, no profilometer analysis would be able to distinguish between a surface roduced by blasting with very small beads and one lapped with coarse particles of abrasive materials to the same finish. To a certain extent, this is in agreement with the statistical theory of surface generation to Williamson[36].

A typical example of an engineering surface is shown in Figure 17.

6.4 The Height Distribution of Surface Texture

The Abbott and Firestone bearing area curve has already been mentioned. In statistical terms, this is the cumulative distribution of the ordinate distribution curve, and can be written as:

$$F(z) = \int_{\infty}^{\infty} \psi(z) \, dz. \qquad (5)$$

Where z refers to the heights of the ordinates in the profile measured from the center-line, and $\psi(z)$ is the probability density function of the distribution of these heights. In practice, the derivation is achieved by taking measurements of z_1, z_2, etc. at some discreet interval and summing the number of ordinates at any given height level. Computer programs are now available for performing this task, and a typical curve for a grit blasted surface is shown in Figure 18. The straight portion of the graph indicates a Gaussian distribution, and is characteristic of a surface whose texture is uniformly distributed. For a more detailed discussion of this, readers are referred to Williamson's paper, Reference 28, or Halling, Reference 9.

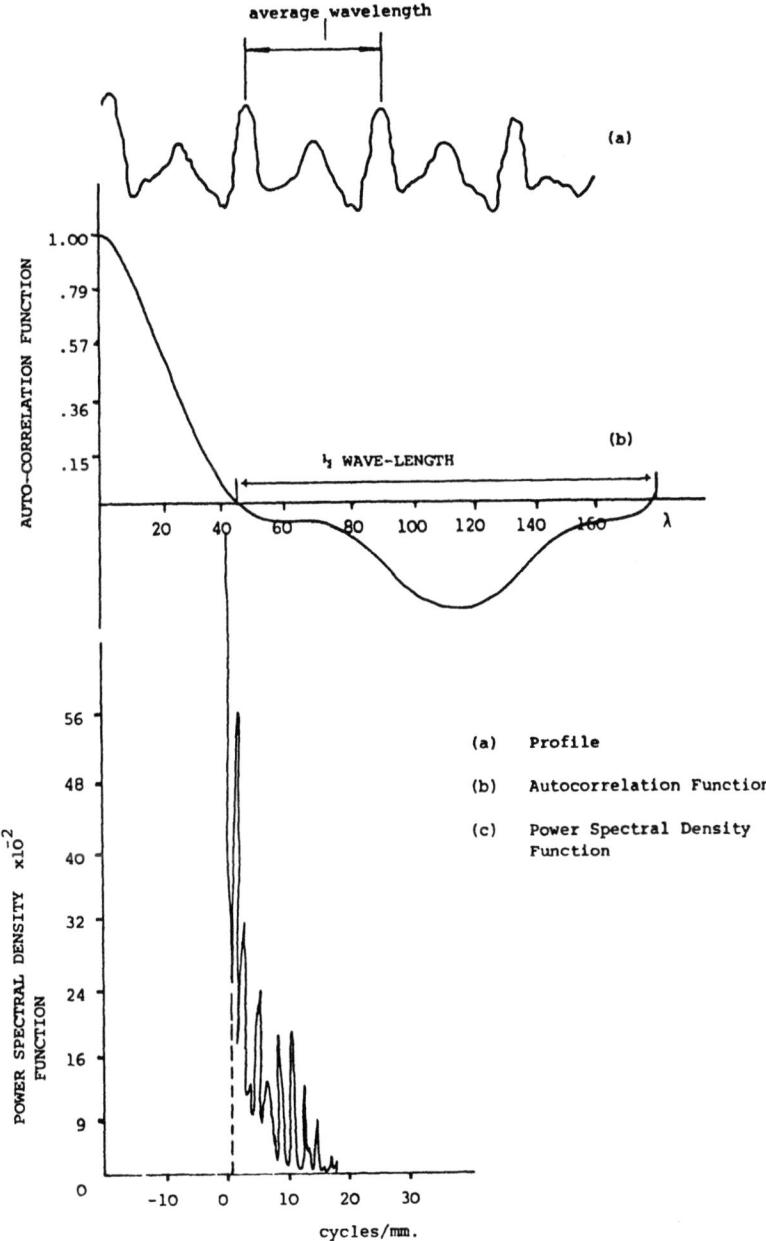

FIGURE 17 TYPICAL CHARACTERISTIC CURVES FOR AN END-MILLED SURFACE

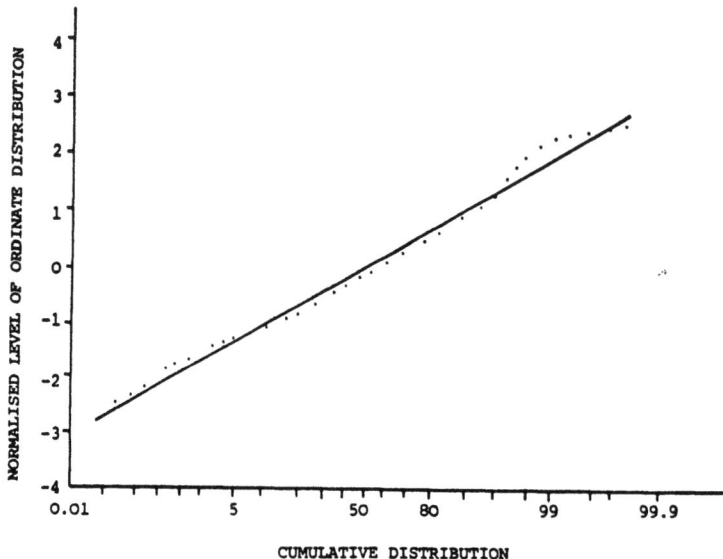

FIGURE 18   TYPICAL ORDINATE DISTRIBUTION CURVE FOR A
GRIT-BLASTED SURFACE

Although several common surface preparations produce near-Gaussian distributions, many do not. It is therefore necessary to define some other statistical parameters for measuring the various forms.

6.5   Moments, Skewness and Kurtoris

6.5.1   Moments

The nth moment of the distribution curve $\psi(z)$, about the mean is defined by:

$$M_n = \int_{-\infty}^{\infty} z^n \psi(z) \, dz. \qquad (6)$$

Thus the R.M.S. value $\equiv \sigma \equiv \left[\int_{-\infty}^{\infty} z^2 \psi(z) \, dz\right]^{1/2}$

$$= \left[2\text{nd moment of } \psi(z)\right]^{1/2}$$

also the C.L.A. $\equiv 2 \int_{0}^{\infty} z\psi(z) \, dz.$

$$= 2 \times \{\text{the 1}^{st} \text{moment of } 1/2 \; \psi(z)\}.$$

Clearly the 1st moment of the whole of $\psi(z)$ about the mean is zero, from which the center line of the distribution can be obtained.

For a Gaussian distribution curve the nth moment is given by:

$$M_n = \frac{1}{\sigma(\pi)^{1/2}} \int_{-\infty}^{\infty} z^n e^{-z^2/2} dz \qquad (6)$$

If n is odd the term vanishes, which it would for a symmetrical curve, and if n is even then

$$M_n = \frac{n!}{2^{n/2}(n/2)!} \sigma^n \qquad (7)$$

Thus the 2nd moment becomes just σ^2 which is the variance.

6.5.2 Skewness

Skewness is a measure of the departure of a distribution curve from symmetry and is defined by:

$$S = \frac{\int_{-\infty}^{\infty} z^3 (\psi)(z) dz}{\sigma^3} \qquad (8)$$

$$= \frac{3^{rd} \text{ moment of } \psi(z)}{\sigma^3}$$

If the curve is a Gaussian distribution symmetrical about the axis then the skewness would be zero.

If the curve is either positive or negative then the shape of the skew would follow a shape similar to that shown in Figure 19.

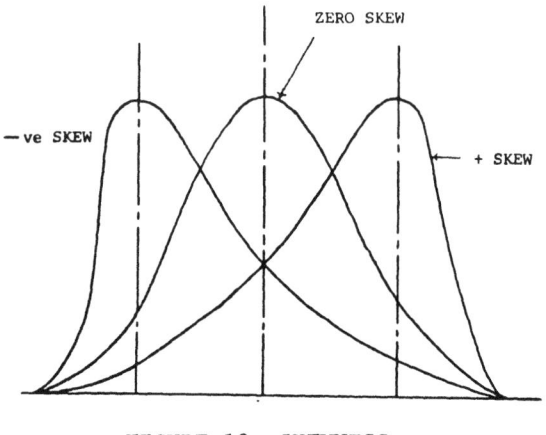

FIGURE 19 SKEWNESS

6.5.3 Kurtosis

This indicates the spikiness in the surface texture and is measured by the hump on the distribution curve. It is defined by:

$$k = \int_{-\infty}^{\infty} \left| \frac{z^4 \psi(z) \, dz}{\sigma^4} \right.$$

$$= \frac{4^{th} \text{ moment of } \psi(z)}{\sigma^{-4}} \qquad (9)$$

For a Gaussian distribution $k = 3$, obtained by substitution in Equation (7), i.e.

$$k = \frac{1}{\sigma^4} \frac{4 \times 3 \times 2}{2 \times 2 \times 2} \sigma^4 = 3$$

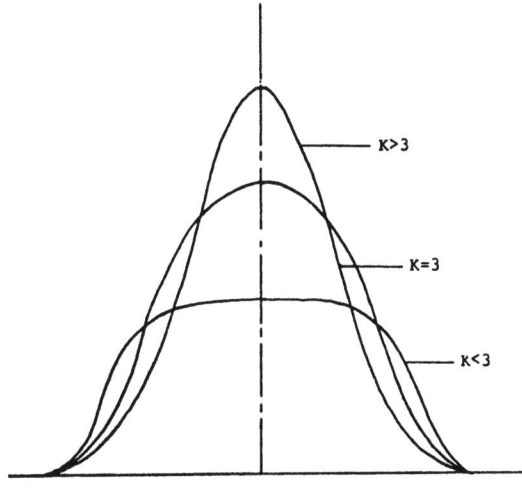

FIGURE 20 KURTOSIS

If k is less than 3 the curve is called platykurtic, and if k is greater than 3 the curve is called lepto-kurtic.

7. TOPOGRAPHY, AND REAL AREA OF CONTACT

So far the discussion has centered around the profile of a surface and the analysis concerned with two- dimensional surface topography. But topography, as the name implies, is three-dimensional; it has length, breadth, and depth.

The first successful attempt at describing a surface three dimensionally by means of a contacting method, namely a stylus, and utilizing multiple parallel tracing can be attributed to Williamson in 1968[28]. Since then, other workers have reported similar systems depending on the application to which their interests are directed, i.e. static, rolling, or sliding contact[38, 39, 40, 41, 42]. The major problems and limitations imposed on these systems were in creating an arbitrary flat datum, and the size of an area compatible with the amount of information that could be analyzed due to the vast data storage and data handling problems that such systems incur. An additional problem was the time element to physically record the vast amount of information. Advances in computer technology during recent years have greatly assisted in this problem, and future workers will be able to consider analyzing surface areas larger than have been possible hitherto.

Figure 21 shows a 2.5 mm² isometric view of a rough turned surface of roughness Ra = 20.64 µm. The horizontal axis is a 100 x 100 data point array, and the vertical axis magnified to 255 data points, with the peak to valley height being 54 µm.

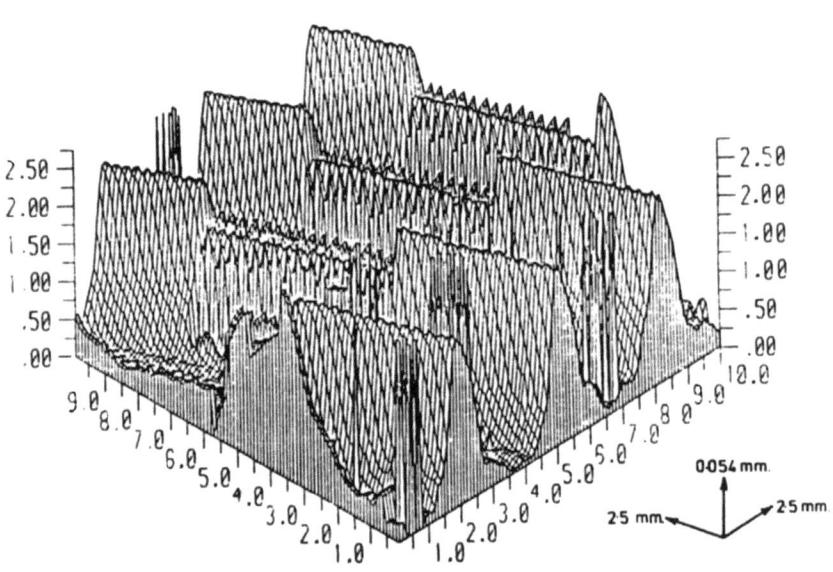

FIGURE 21   ISOMETRIC VIEW OF A ROUGH TURNED SURFACE

The machining marks are clearly visible indicating the undulating nature of the surface topography. Throughout the central portion, the asperities appear suppressed across the ridges, which suggests that possibly some flaw may have existed in the machine tool, or alternatively that over that portion of the surface the area had been damaged. Subsequent optical examination of the actual surface did in fact reveal a damaged portion - a small indentaion about 1 mm wide and about 10 µm deep. Other areas of the surface appeared to be satisfactory, which indicated that the cutting tool was not the cause. The important feature revealed by this form of analysis is that the technique is a useful means of detecting and quantifying detail of small orders of magnitude. "Wear" is a typical example where the application could be useful, particularly if the instrumentation has relocation facilities[37, 38].

Figure 22 shows a contour map of the surface with eight different contour levels, equally divided between 0 and 255 data points where 20.65 μm = 255 data points. Again, the undulating nature of the surface is clearly revealed. The darker areas, where it appears that contours coalesce, can also be seen as sharp "spikes" in the isometric view. The explanation for this is that, due to the very rough nature of the surface, the stylus response could not cope with the speed of travel as it scanned the surface and "bounced" in this area. Subsequent tests on other types of machine surfaces did not reveal spikes of this nature, which suggests that speed of traverse is a necessary consideration, and particularly so if analysis is undertaken on very rough surfaces.

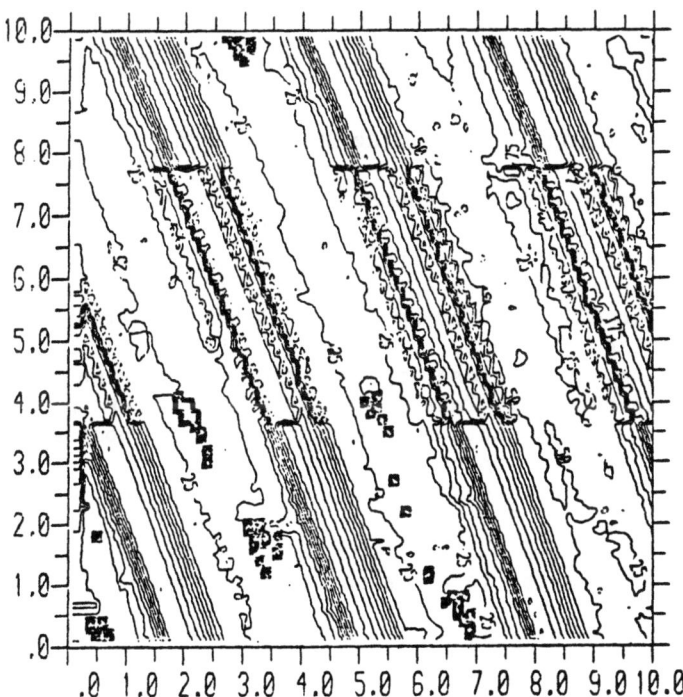

FIGURE 22   CONTOUR PLOT OF A ROUGH TURNED SURFACE

7.1 Real Area of Contact

When one solid is placed in contact with another solid the dominant asperities, which may be situated on the peaks of the waves of both solids initially coming into contact, suffer deformation and carry on deforming until the load is finally supported by areas of discrete contacts. Any increase in loading, such as bolting or rivetting, causes further deformation of asperity contacts, and also may extend to the waviness in the surface. Most of the asperities deform plastically, but elastic deformation of asperities can occur at the edges of the microscopic contact areas where the stresses are relatively low. The induced stresses in the surface waves however, are always much smaller than those in the individual asperities, and generally speaking wave deformation is normally regarded as being elastic. Several other factors influence the pattern of contact, namely the magnitude and time dependence of the applied force, the stiffness and hardness of the superficial material, and the likelihood of contaminant layers or interfacial fluids.

In many engineering applications it is necessary to pass a heat flux, or transfer electrical energy from one solid to another, and it is the true area of contact that usually dictates the thermal and electrical resistance which occurs at the interface.

7.2 Deformation of Contacting Surfaces Under Load

Greenwood and Williamson proposed in their classic paper, "Contact of Nominally Flat Surfaces," that most surfaces do not undergo a transformation from elastic to plastic deformation as the load increases, but deform plastically even under the lightest loads[11]. By using a statistical technique to characterize a surface, they showed that engineering surfaces exhibit, to a very close approximation, a Gaussian distribution of surface heights. This led them to the belief that not only were the laws of friction dependent upon the topography of the surface, but more importantly they were able to postulate, for a Gaussian surface, a parameter which would indicate the mode of deformation.

This parameter is known as the plasticity index, β

where $\beta = \dfrac{E'}{M} \left| \dfrac{\sigma'}{\rho_s} \right|^{1/2}$ (10)

and $\dfrac{1}{E'} = \dfrac{1 - \nu^2}{E_1} + \dfrac{1 - \nu^2}{E_2}$ (11)

and $\frac{1}{\rho} = \frac{1}{\rho_1} + \frac{1}{\rho_2}$ (12)

and $\sigma' = (\sigma_1^2 + \sigma_2^2)^{1/2}$ (13)

The deformation was considered to be plastic for all but the lightest loads, when β>1, and elastic under all but the most extreme loading conditions, if β<0.6. In the limited range, 0.6<β<1, an increase in load could change the contact from elastic to plastic. The majority of surfaces investigated by Greenwood gave mainly plastic deformation. The value of β was formed from calculations based upon the onset of plastic deformation, but for a single contact there is something like a 200 fold range of load (which implies approximately a 50 fold range of the deformation depth) between the onset and what Tabor has termed "full plasticity"[7]. Therefore, if severe plastic deformation is being considered, it seems likely that much higher values of β may be required. This view is shared by Whitehouse and Archard who proposed that the values of β under estimated the probability of plastic flow, because the curvature of the peaks increase with increasing asperity height[44]. Thus the highest peaks, which are those involved in contact, have a smaller radius than the total population of peaks.

7.3 Sampling Interval

Very little experimental data is available relating sampling interval to mean asperity curvature, $\bar{\rho}$, which is a necessary parameter in accurate determination of theoretical predictions. It has been suggested that for surfaces which possess a Gaussian distribution of surface heights, a three point analysis should be used to obtain the correct value of $\bar{\rho}$[43]. The value taken should be that value which corresponds to a sampling interval of 1/3 of that resulting from the main frequency of the surface structure, (i.e. 1/3 of the sampling interval corresponding to an Auto-correlation function of +0.1). Some surface structures may possess more than one principal frequency, in which case it would be necessary to take a value of 1/4 or even possibly 1/5 of the value corresponding to 0.1. However, a sampling interval of about 1/3 seems to be appropriate in most cases.

In general, the plasticity index is significant only if it is applied to the main long wavelength structure of the surface. Whitehouse and Archard argue that if values of $\bar{\rho}_s$ corresponding to the smallest scale structure is used, then the derivation of the plasticity index becomes invalid because of the deformation of adjacent asperities[44].

7.4 Estimation of Real Contact Area

Archard has shown that when elastic deformation occurs the actual contact area and load are related by the equation

$$A_a = KW^m \tag{14}$$

where K is a function of the assumed surface structure and the elastic constants of the material[45]. Analysis based on different mathematical models show, Figure 23, that m lies between 2/3 (for contact between a smooth sphere and a smooth flat) and 1 (the value approached by more complex models which more nearly simulate the roughness and waviness of real surfaces). As the complexity of the mathematical models increases, the number of contact areas similarly become more near directly proportional to the load with the size of each contact area becoming less dependent on load.

The theory developed by Greenwood and Williamson for elastic contact takes into account two material properties; the hardness and modulus of elasticity, and three topographic parameters; the mean radius of asperities, their surface density, and the spread of mean height[11]. The theory led to expressions of total real contact area, the number of micro-contacts, the load, and the (electrical) resistance between two contacting surfaces, in terms of separation of their mean planes. In agreement with experimental evidence, the theory indicates that the number of micro-contacts and the real area of contact depend only on the load and not on the nominal contact pressure. Further, it also indicates that the separation of the surfaces is not very sensitive to the pressure, the separation of similar surfaces being approximately equal to the center line average of roughness. The fact that the ratio of real contact area to load is nearly constant for elastic contact leads to the concept of an "elastic hardness" which can be used for predicting the real contact area for given loads, just as the conventional hardness is used when plastic deformation is assumed.

The authors assumed that the heights of the asperities were represented by a well defined continuous distribution function. By measuring several metal surfaces, they found that even though the entire height distribution might not be represented by any known function, the uppermost peaks formed a reasonable good approximation to a Gaussian distribution. In other words, if the contact load is not very high, the Gaussian distribution of heights is certainly a good appproximation.

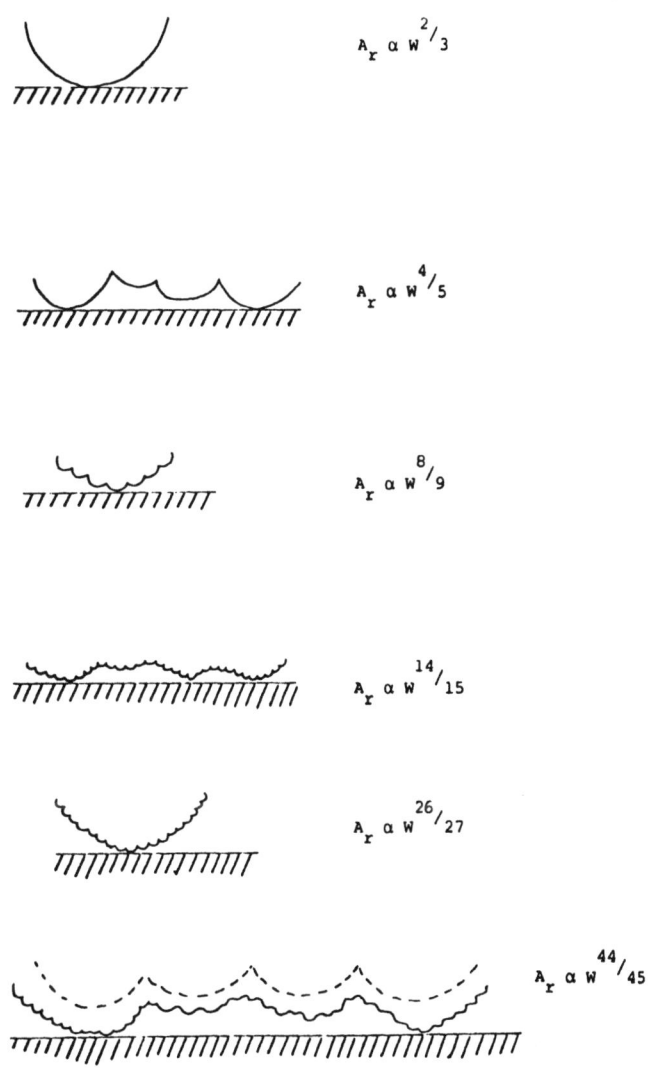

FIGURE 23 ELASTIC SURFACE MODELS

The next significant advance on the mechanism of surface contact was made by Tsukizoe and Hisakado[46, 47]. They considered the contact between two surfaces, one of which had a Gaussian distribution of surface heights. In order to predict the number of contacts and the average contact spot radius, they had to make the following assumptions about the shape and distribution of surface asperities:

i) The distribution curve obtained from the profile curve had a normal distribution.
ii) The surface was isotropic.
iii) The surface contained a large number of asperities in the form of cones of equal base angle.
iv) Plastic deformation occurred at the contacts. Hence, no interference occurred between asperities, and the ratio of the real area of contact, A_r, to the nominal area of contact A_N, was given by:

$$\frac{A_r}{A_N} = \frac{P}{P_m} \qquad (15)$$

or

$$\frac{A_r}{A_N} = \frac{W}{A_N P_m} \qquad (16)$$

where P = applied nominal pressure

P_m = flow pressure of the softer material

W = applied normal load.

Based upon the above assumptions, Tsukizoe and Hisakado deduced expressions for the separation, the real area of contact, the number of contact points, the average radius of the contact points, and the distribution of the radii of the contact points. A similar analysis was recently presented by Jones[43], which is based upon the work of Tsukizoe, Hisakado and Thomas[46, 47, 48].

Tsukizoe and Hisakado obtained an expression for the mean plane separation, $t = (u/\delta)$, and the nominal and real areas of contact expressed in terms of the probability of contact for the normal distribution;

$$\Phi(t) = \left| 1 - \frac{2W}{A_{N/P_m}} \right| \int_0^\infty \phi(t) \, dt \qquad (17)$$

where $\Phi(t) = \int_0^t \phi(t)\, dt$ (18)

They assumed that the surface was isotropic and that the distribution curve obtained from the profile curve of the surface was normal[46];

$$\phi(t) = \frac{1}{\sqrt{2\pi}}\, e^{-t^2/2} \quad (19)$$

where t is the normalized plane separation given by $t = u/\sigma$, u is the mean plane separation, and, σ, the r.m.s. roughness for the two surfaces, is obtained from

$$\sigma = (\sigma_1^2 + \sigma_2^2)^{1/2} \quad (20)$$

where σ_1 and σ_2 are the r.m.s. roughness of the profile curves for surfaces 1 and 2 respectively.

They also assumed plastic deformation and hence constant material flow pressure.

The real area was estimated from Equation (16) It was assumed that the flow pressure was almost load independent and there was no work-hardening of the conical asperities which have a small base angle.

Finally, the population of micro-contact bridges formed between mating surfaces may be expressed in terms of probability functions.

Figure 24 shows the distribution of sizes of micro-contacts as a function of dimensionless mean plane separation, t, (and hence of the imposed load) for contacts between nominally flat isotropic surfaces containing Gaussian distributions of surface heights[46, 47]. The data is presented in dimensionless form, micro-contact radii being normalized using a normalized mean absolute surface slope, ψ. The plotted function is of the form

$$f(a'_1) = \frac{\left[(t + a'_1)^2 - 1\right] \phi(t + a'_1)}{t\, \phi(t)} \quad (21)$$

where $a'_i = \pi^\Psi a_{i/2}$

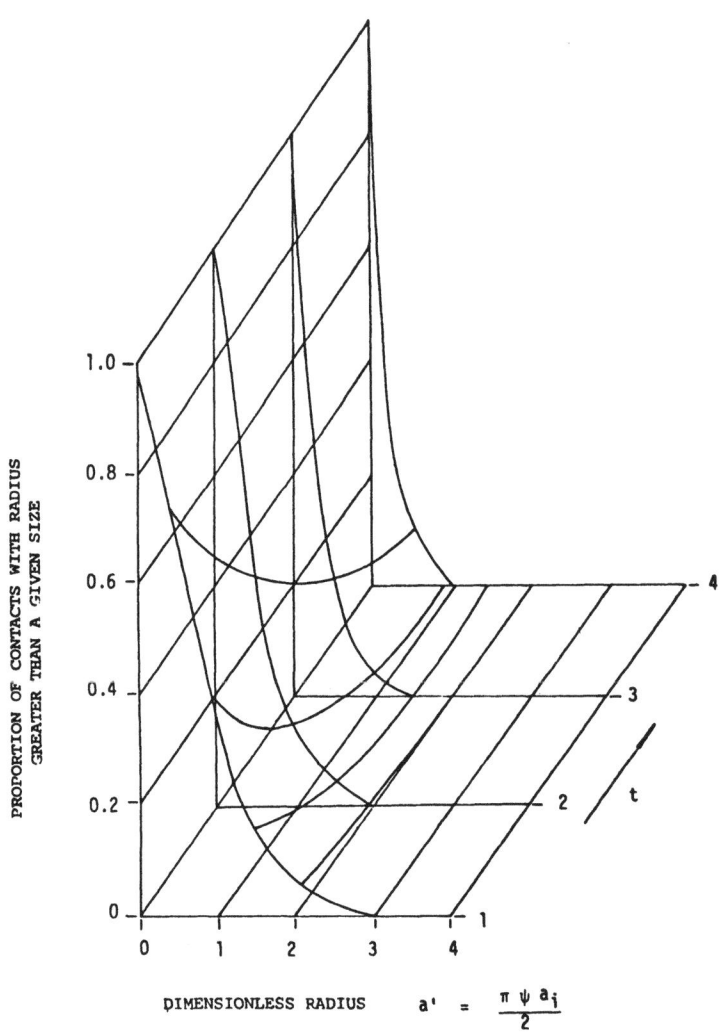

FIGURE 24 PROPORTION OF MICRO-CONTACTS WITH RADIUS GREATER THAN A GIVEN SIZE

Tsukizoe and Hisakado presented the theoretical relationships relating load with mean plane separation, mean contact spot size and number of contact spots, graphically in dimensionless form[46, 47]. From this it was shown that as the load increased, the average contact spot size increased slowly while the number of contacts increased quite rapidly. Thus, the

total area of contact would probably be increased by the
increase in the number of contacts.

Tsukizoe and Hisakado tested their hypothesis
experimentally using similar and dissimilar materials produced
by different machining processes. They concluded that:

i) The relation between load and separation given by
    Equation (17) was reasonable for contact between
    normally distributed surfaces.
ii) The real area of contact under the applied load was
    independent of the surface roughness. The
    experimental results confirmed the assumption that
    deformation of the metal at the contact was plastic.
iii) The number of contact points under the appplied load
    was influenced by the degree of roughness and was
    proportional to $\tan^2/\sigma^2$.
iv) Over a wide load range ($10^{-4} < P/P_m < 10^{-2}$ MNm^{-2}) the
    mean contact radius increased by a factor of 1.6
    while the number of contacts increased by a factor of
    42.2. Therefore, the total contact area increased
    mainly as a result of the increase in the number of
    contacts.

In general, the theoretical predictions and experimental
results compared quite favorably. For a dimensionless loading,
$3 \times 10^{-3} < P/P_m < 2 \times 10^{-2}$, the discrepancy for the main plane
separation and load was quite small. O'Callaghan and Probert
assumed that when two surfaces are in contact, the contacting
regions deform in an ideal plastic manner and the true contact
between a rough surface and a flat surface can be regarded as
that due to a number of small indentations[49]. Hence the true
contact pressure equals the micro-indentation hardness:

i.e. $\quad A_r = P/M$ \hfill (22)

where M is the Vickers micro-hardness of the softer surface.

Thomas and Probert extended the theory of Tsukizoe and
Hisakado and developed the following dimensionless relationship
for number and mean size of contacts[48];

$$\frac{N_a}{\bar{z}^2} = \frac{\pi t \, \phi(t)}{8} \qquad (23)$$

and $\bar{a}\psi = \dfrac{2}{\pi t}$ \hfill (24)

The dimensionless mean plane separation, t, was linked to the real area of contact, A_r, through the normal probability integral[48].

$$A_r = \tfrac{1}{2} - \Phi(t) \hfill (25)$$

Subsituting Equation (22) into Equation (25) yields

$$\dfrac{P}{M} = \tfrac{1}{2} - \Phi(t) \hfill (26)$$

Thus the number of contacts spots per unit nominal area, N_a, and their mean radius a, can be related to the loading pressure, P. Subsequently the real area per unit area of contact may be obtained from expression

$$A_r = N_a \pi (\bar{a})^2 \hfill (27)$$

The analysis required the measurement of r.m.s. roughness, o mean asperity slope, $|\psi|$, and micro-hardness, M.

In general, it is extremely difficult to measure the total number of micro-contact spots in practice. Even at very high objective magnification ranges, the resolution of the lens limits the lower size of spots that can be measured. However, it is possible to measure size distribution of the larger micro-contact spots to a high degree of accuracy. In order to compare experimental data with theoretical predictions it is common to use the latter in cumulative form[43, 50].

Figure 25 shows the proportion of contact area supported by contacts of radius greater than a given size, plotted against t and a'. From Figures 24 and 25 it is clear that the disadvantage of not being able to measure the smaller micro-contact spots, is less pronounced for contact area than for the number of contact spots.

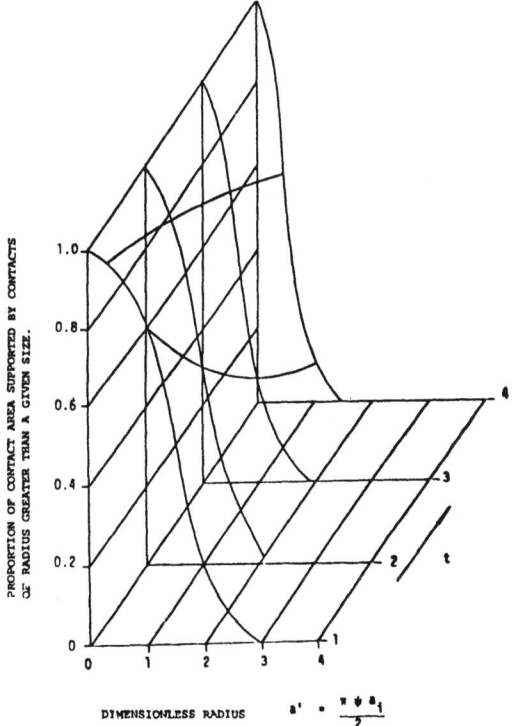

FIGURE 25   PROPORTION OF CONTACT AREA SUPPORTED BY CONTACTS OF RADIUS GREATER THAN A GIVEN SIZE

O'Callaghan and Probert showed that the analysis of Tsukizoe and Hisakado only applies when[49]:

$$2\Phi(t) = 1 - \phi(t)/t \tag{28}$$

i.e. when P/M < 0.3.

When P/M > 0.3 the distance between adjacent asperities becomes too small for assuming that the deformation of the surface in contact was ideally plastic and no interference occurred between asperities.

By using the arithmetic mean radius, $\bar{a}$, in Equation (24), the real area of contact would be underestimated by approximately 23.5%. The correct parameter should therefore be the r.m.s. spot radius $\bar{\bar{a}}$ [43]. Equation (24) therefore becomes:

$$A_r = N_a \pi (\bar{\bar{a}}) \tag{29}$$

where $\bar{\bar{a}} = 1.43\ \bar{a}$ (30)

For soft, rough surfaces in contact with harder smooth surfaces, the contact areas are formed by deformation of asperities. The correct hardness for this mode of deformation would be approximated by the Mallock hardness. The Mallock hardness is defined as the flow pressure of a single asperity under load.

The deformation of single asperities under load has been studied by Thomas et al, References 51 and 52. Uppal and Probert have shown that the flow pressure of the single asperity varied with applied load[52]. Thomas found that the flow pressure for single asperity models was several times the Vickers hardness for deformation ratios greater than 0.8. The latter also found variation of the asperity flow pressure with varying asperity slopes. An important factor shown by Uppal and Probert was that when an idealized asperity was deformed, material from the apex was displaced to the shoulders[52]. For multi-asperity models this would result in asperity interaction and yet another parameter influencing the flow pressure of material in contact.

O'Callaghan and Probert stated that despite the existence of many theories to predict values for the parameters describing static contact, there is very little direct experimental data to corroborate the various estimates, and thus it is difficult to quantify the overall effect of the usual simplifying assumptions which are listed below[49]:

i) Large scale surface undulations are neglected.
ii) No interference between adjacent asperity bridges occur. In fact, contact bridge interaction ensues under much lower loads than has been previously suspected.
iii) Contact regions are circular. This applies only at very low loads, and is then only an approximation.
iv) The distribution of surface heights is Gaussian.
v) Asperities may be described as right circular cones.
vi) Contacting asperities deform in an ideal plastic manner. This implies that as loading proceeds all the material within intersections of the surfaces is nullified. It follows that the contact behavior at all loads may be predicted from material properties and geometries measured prior to loading. However, since matter is conserved, the material from the peaks must enter the valleys in a surface. The result of neglecting this displaced material is that

the calculated values of the mean plane separation under load are less than the true values.

7.5 Image Analysis

As previously mentioned, it is extremely difficult to measure contacts very accurately; but one method developed by the author, which has proved quite fruitful, utilizes a Quautimet Image Analyzer, Figure 26[41]. This instrument, which was originally designed to size and count flaws in a metal structure, consists of a microscope through which a specimen is viewed, and a videcon head which transfers the image to a viewing screen via a computer. The intensity of light reflected from the surface of the specimen can be adjusted at selected levels and, depending on the desired mode of operation, detects area fractures in the image, their chord length, their perimeter, and their number and size distribution.

The adopted technique consisted of pressing a hard (steel) machined surface against an optically flat soft (copper) surface under a known load, and upon separation, placing the impressed surface and the copper-surface, under the microscope. The imprint of the asperities are clearly visible even under low magnification. It is then relatively easy to count the number of asperities using the Image Analyzer by taking radial counts across the surface.

Figure 27 shows the spiral effect of a turned surface with the Analyzer set in the perimeter mode. The contact features are those produced by pressing a turned stainless steel surface against a flat copper surface under a load of 2.2 MN_m^{-2}. The contact features away from the center of the specimen, again of turned surface under a load of 18 MN_m^{-2}, are shown in Figure 28. Figure 29 presents contact features of a grit-blasted surface are shown under a load of 2.2 MN_m^{-2}. The analyzer automatically records the data, which can then be fed directly to a teletype printer for subsequent analysis or other data processing devices.

8. PRODUCTION OF SURFACES

A variety of techniques are available for the production of surfaces, several of which are listed in Table 4. Typical values of Ra that should be achieved by each technique are included (providing the tools are in good condition). The latter point, regarding the quality of the tool is particularly important since a blunt, worn, or damaged tool can result in severe degradation of the workpiece in the form of induced stresses and high temperature gradients, which can severely affect the surface integrity of the material as well as the

FIGURE 26 QUANTIMET IMAGE ANALYZER

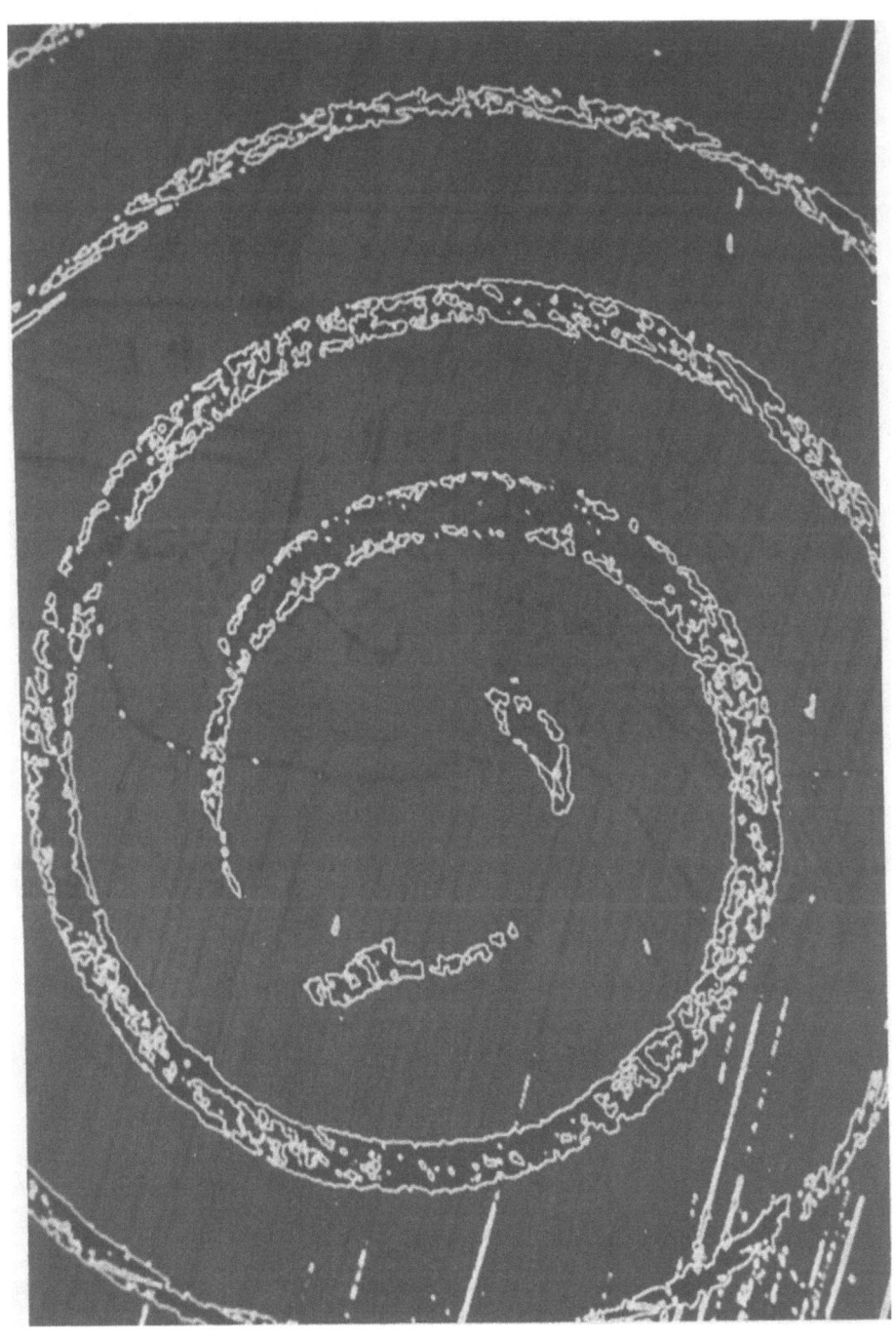

FIGURE 27   CONTACT SPOTS OF A TURNED SURFACE SHOWING SPIRAL EFFECT

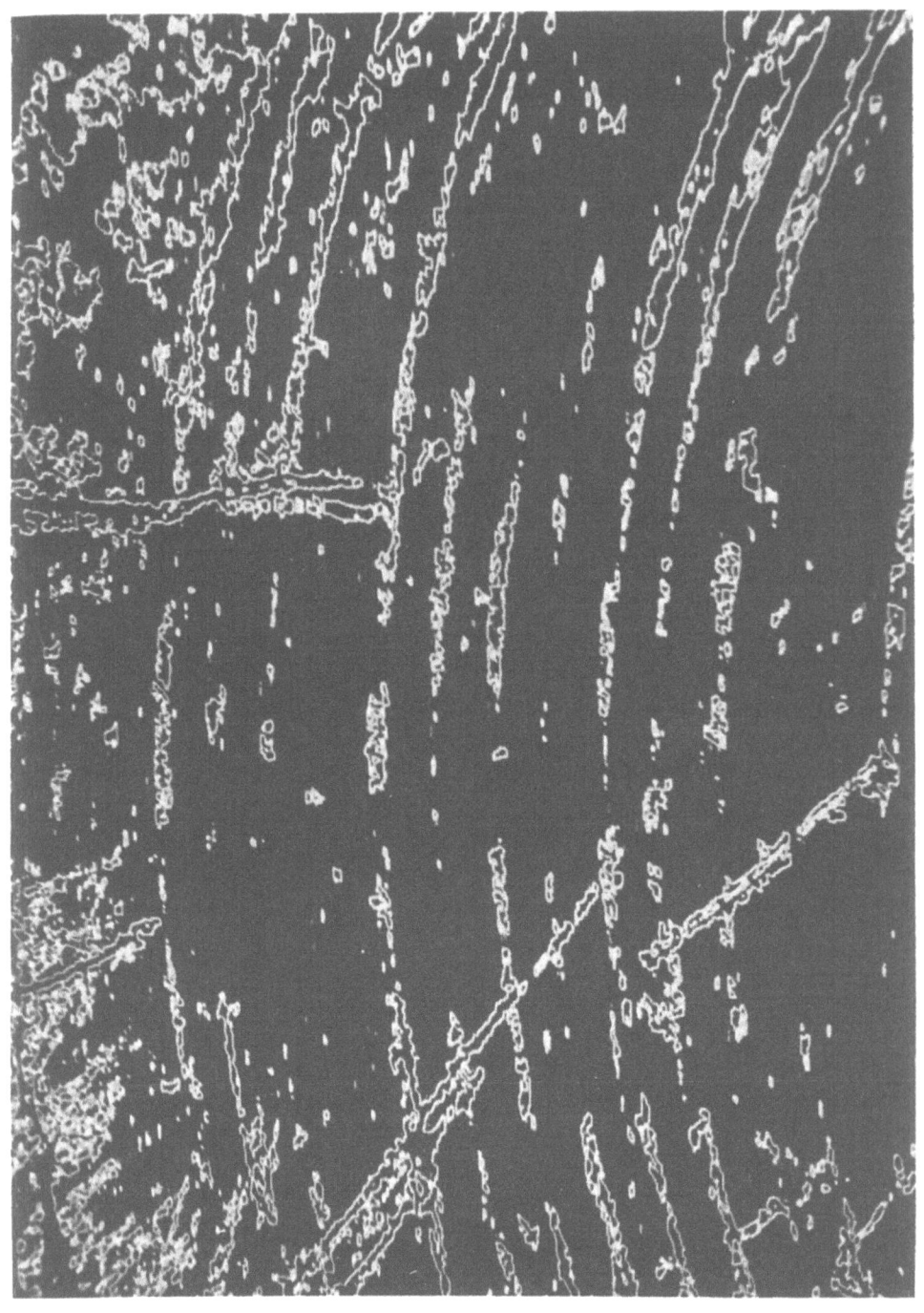

FIGURE 28 CONTACT SPOTS OF A TURNED SURFACE AWAY FROM THE CENTER OF THE SPECIMEN

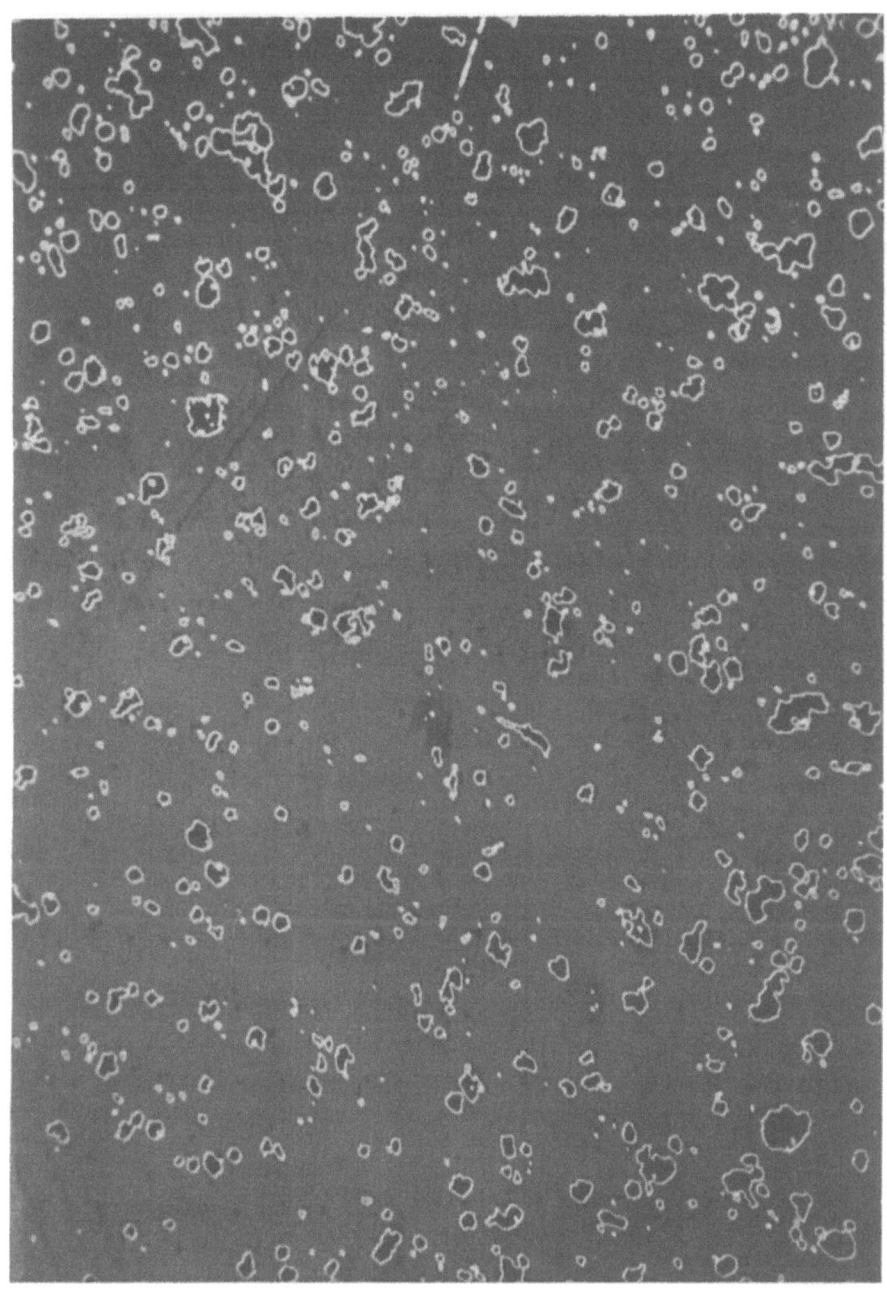

FIGURE 29 CONTACT SPOTS OF A GRIT-BLASTED SURFACE

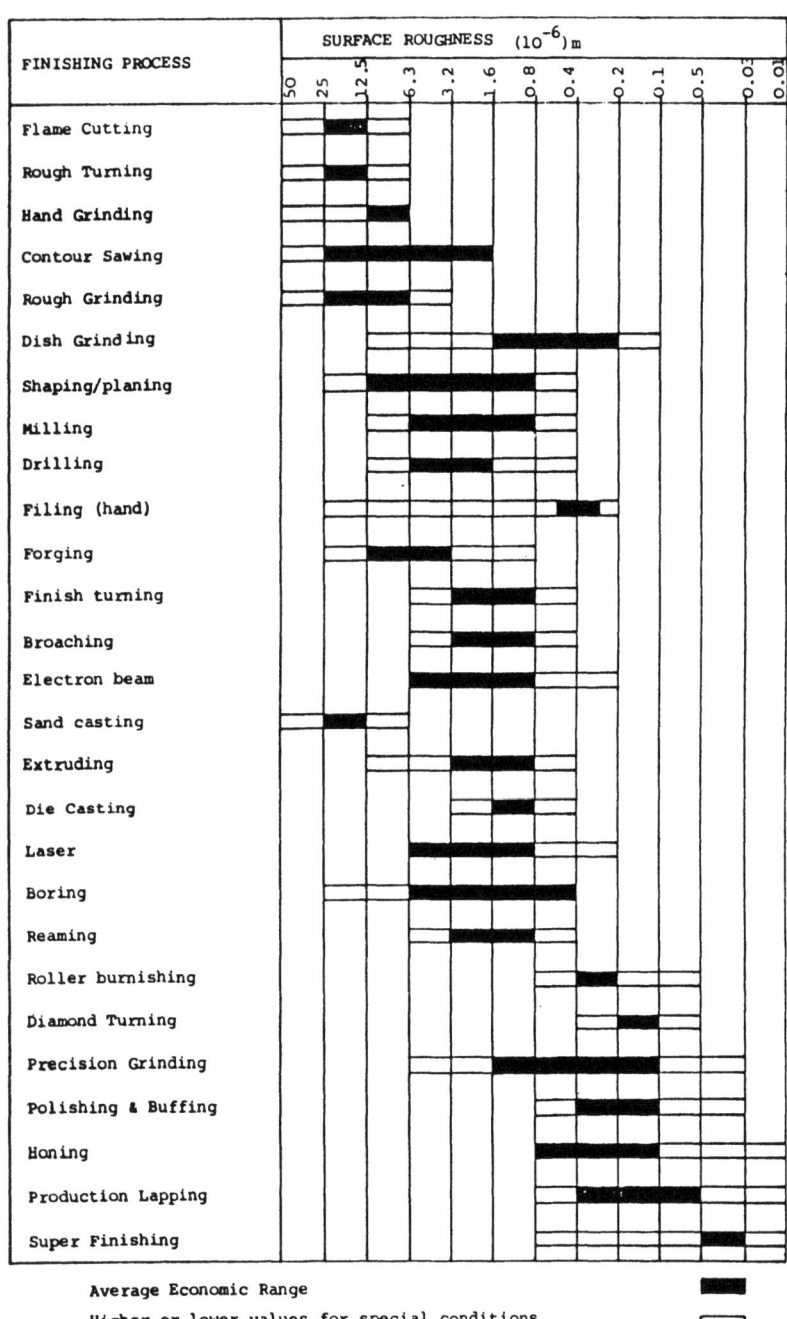

TABLE 4 SURFACE ROUGHNESS PRODUCED BY COMMON PRODUCTION METHODS

surface finish. The need therefore, to control the quality of surface texture, is of paramount importance and is normally a functional requirement of a production engineer. Control is generally instituted not so much in seeking superlative finishing of components, but rather to secure a surface texture of known type and roughness which from experience, has proven to be the most suitable to provide long life, fatigue resistance, maximum efficiency, and functional requirement at the lowest possible cost. The benefits derived are many and include such things as reduction in vibration, wear, and energy consumption.

8.1 Surface Integrity

Although surface finish is of vital importance in any tribological situation, it is of little value, if in creating the desired finish, the surface integrity has been altered to such an extent that the component fails as a result of the stresses it is likely to encounter. Considerable research has been carried out in recent years to establish working criteria on the effects of fatigue, micro-cracks and surface irregularities which are often likely initiation sites for failure. In addition, metallurgical and mechanical alterations, as a result of excessively high removal rates, can reduce the overall quality of the component.

Essentially, the term surface integrity is the study which embraces the interrelationships of metallurgy, machinability, and mechanical testing. For a very good and comprehensive review of all such matters, readers are referred to Section 18 of ASTM-E673 (1980); "Surface Analysis Definitions and Terminology," or alternatively, Surface Integrity Encyclopedia (Special unpublished collection of data and effects from specific material-process combinations), maintained for reference by the Machinability Data Center, Cincinnati, Ohio 45209.

8.2 Surface Roughness as a Result of Geometric and Kinematic Reproduction of Tool Point

A machined surface is created as a result of geometric and kinematic reproduction of the tool, and is generally dictated by both the kind of material and the tool point shape. The tool point shape is characterized by the following:

i) the nose radius,
ii) the leading cutting edge angle and the auxiliary cutting edge angle.

The factor which characterizes the tool point and workpiece kinematics, and which influences surface roughness, is the feed per tool point. In the case of single-tool turning for example, this is equivalent to the feed per revolution. The resulting swarf material is either of the form of chips or spiral formation.

Considerable research has taken place in this area and various theoretical treatments have been developed, to try and correlate predictable surface finish relative to tool shape, speed of cut, feed, etc[53, 54]. Generally, the theoretical values calculated according to reproduction formula are smaller than the real mean total height of surface irregularities primarily due to the influence of physical and dynamic phenomena. Features such as built-up-edge formation, friction between the cut surface and the tool point, vibrations, and embedding of particles of material being machined, or of the tool point, all influence the effect of surface finish. Tables have been produced in order to aid the machine operator and a wealth of information is available relating cutting speeds and material composition[55]. In addition, some very useful nomograms have also been produced which offer an approximate selection of cutting conditions to achieve a given surface finish[56]. The presence of numerous variables, however, make it impossible to give any precise answers, and thus the skill and experience of the technical operators is still the best criteria.

9. SELECTION OF SURFACES

The selection of surfaces for production purposes is in many respects a matter of subjective judgement based on experience of previous similar systems. As previously mentioned, cost is one of the governing factors in the manufacturing process as shown by the exponential form of the graph in Figure 12. Considerable information has been gained from various surveys covering characteristics of most commonly employed metal forming processes. These include such features as:

i) Dimensional accuracy
ii) As produced surface finish
iii) Machine-finish allowance
iv) Maximum and minimum component size and section thickness
v) Minimum production total
vi) Ideal angle, form of undercut, holes, etc.
vii) Selection of alternative types of material.

Set out in tabular form, comprehensive detail of the above characteristics can be found in the Engineering Design Guides, Series No. 29; The Selection of Materials, by Waterman, Reference 57. These guides are based on information derived from the Materials Optimizer of the Fulmer Research Institute. Other useful sources of information can be found in References 58, 59, and 60.

For the purpose of this short discussion, Tables 5 and 6 are presented by courtesy of the Editor of the Materials Optimizer, Fulmer Research Institute Ltd. These tables illustrate some of the practical features which must be taken into account, concerning the selection of surfaces.

10. CONCLUSION

This chapter has been written essentially as an overview of a very complex discipline within the field of Tribology. Many detailed areas have, by necessity, been omitted, but it is hoped that the author has been able to convey an appreciation of the subject, and that it will be helpful to those readers who may future, and in what way will surface measurement trends progress? The answer almost certainly lies in-process and on-line measurement, as recently discussed by Young, Vorburger and Teague[61]. Some of the techniques developed in recent years, using non contacting methods, and in particular optical systems, yield the most promising techniques. Laser optics are already now well established, and the data-handling capabilities of modern micro-computers, which can respond rapidly to change, particularly in an area scan, offer advantages to the surface technologist unheard of only a decade ago. Other areas which might be explored are non contacting capacitive transducers, again linked to the computer, and possibly air-jet transducers. Statistical analysis of surfaces will undoubtedly be performed operating in the double integral domain, and no doubt new parameters which more accurately describe a surface, will be developed. Techniques recently reported by Thwaite concerning optical Fourier transforms open the doors to exciting new developments[62].

The need for surface technologists to pursue their activities is paramount, because the development of automated quality control through in-process inspection, linked on-line with process computer systems, will form an integral part of manufacturing assemblies within the next decade.

11. NOTATION

a	radius of a micro-contact spot	m
a^1	dimensionless micro-contact spot radius	
$\bar{a}$	arithmetic micro-contact spot-radius	m

	Type of contact	Examples	Type of wear	Remarks	Materials for use
Movement between two components in contact	Conformal (area contact)	Plain bearings, piston rings in cylinders, electrical contacts	Adhesive wear, surface fatigue, some abrasive wear and fretting	Surfaces usually need some running-in	Soft materials operating with hard materials
	Counterformal (concentrated contact)	Rolling bearings, gears, cams and tappets, electrical contacts	Surface fatigue, some adhesive wear and abrasive wear	Adhesive wear occurs only when there is a substantial amount of sliding	Hard with hard
	Conformal and counterformal	Plain bearings, piston rings in cylinders, rolling bearings, gears, cams and tappets, electrical contacts	Running-in	Involves mutual adjustment of the surfaces of the two components	Surface treatments and coatings or special surface finishes
Movement between a component and a mass of material	Particulate solids rubbing over a surface	Excavator buckets, sandblast apparatus, and other particulate-materials handling plant	Abrasive wear	The type of abrasive wear experienced depends mainly on the contact pressure	Hard (sliding contact), elastic (bouncing contact)
	Particulate solids or liquids in a fluid over a surface	Equipment pumping abrasive slurries, turbine blades in wet steam, aircraft leading edges in rain	Particle-impact erosion	Only becomes noticeable at high impact velocities and/or high particle densities	Hard (low impact angles), elastic and ductile (high impact angles)
	Stream of fluid flowing over a surface	Ship propellors, pipes and ducts	Cavitation erosion	Requires low local pressure to initiate the process. Corrosion can occur	High ultimate resilience characteristic

TABLE 5 CONDITIONS CAUSING WEAR (MATERIALS OPTIMIZER FULMER RESEARCH INSTITUTE 1974)

Material	Machinability index
Ball-bearing steel (BS 970 534A99)	30
Inconel (77.5% Ni-16.0% Cr-6.5% Fe)	35
Phosphor bronze (9.5% Cu-5.0% Sn-0.2%P)	40
Stainless steel (18% Cr-8% Ni)	45
Nickel steel (3.5% Ni.0.3% C)	50
Nickel-chrome steel (0.45%C)	50
Pearlite cast iron	50
Wrought iron	50
Monel (70% Ni-30% Cu)	55
Structural steel	60
Copper ¼-hard rolled	60
Aluminium bronze (5% Al)	60
Chrome-molybdenum steel (BS 970 830M31)	65
Cast steel (0.35%C)	70
Chrome steel (free cutting)	70
Cast copper	70
Brass α	80
Cast iron (soft)	80
Free-cutting mild steel	100
Malleable cast iron (ferritic)	120
Free cutting α or β brass	200–400
Aluminium (half hard)	300–1500
Magnesium (6.5% Al)	500–2000

Machinability index is only a very rough guide to the relative ease or difficulty of machining different materials. The figures may be taken to represent ease of machining in terms of speed of metal removal, higher values denoting easier or faster machinability.

TABLE 6 MACHINABILITY INDEX OF SOME COMMON METALS
(MATERIALS OPTIMIZER)
(FULMER RESEARCH INSTITUTE 1974)

$\bar{\bar{a}}$	RMS mean micro-contact spot radius	m
$e^{\alpha\lambda}$	decay of exponential function	
t	normalized mean plane separation	
y	profile height ordinate	m
$y(x)$	surface profile	mm
$y(x+\lambda)$	height of a surface profile at coordinate $(x+\lambda)$	mm
A_r	real area of contact	m²
A_n	nominal area of contact	m²
A_1	real area of contact per unit nominal contact area ($= A_r/A_n$)	
E	elastic modulus of a material	Nm⁻²
E^1	effective elastic modulus of a contact	Nm⁻²
$F(\lambda)$	Auto-correlation function	
M	Vickers micro-hardness	Nm⁻²
N	number of micro-contacts	
N_a	number of micro-contacts per unit area	m⁻²
P	applied pressure (W/A)	Nm⁻²
$P(W)$	power spectral density	
P_m	flow pressure of a single asperity	Nm⁻²
W	applied loading	N
W^*	dimensionless loading factor $(= W/\sigma_s^2 M)$	

α^1	Correlation distance	m
β	plasticity index	
λ	sample interval	m
σ	RMS roughness	m
σ	RMS roughness of a surface diameter d.	m
ρ	radius of curvature	m
ν	Poissons ratio	
μ	microscopic	10^{-6} m
ψ	surface slope	radius
$|\psi|$	mean absolute surface slope	radius
ϕt	normal probability density function	

$$= \frac{e^{-t^2/2}}{\sqrt{2\pi}}$$

$\Phi(t)$ Normal probability integral

$$\int_0^t \phi(t)\, dt$$

12. REFERENCES

1. DaVinci, L., Codice Atlanticus, Milan, British Museum Library.

2. Hertz, H., The Contact of Elastic Bodies, Miscellaneous Papers (London Macmillan, 1896) 146 - 168, 173 - 183.

3. Harrison, R.E.W., "A Survey of Surface Quality Standards and Tolerance Costs Based on 1929-30 Precision Grinding Practice," Trans A.S.M.E. 53, 11 - 25 (1931).

4. Firestone, F.A., Durbin, F.M., and Abbott, E.J., "Test for Smoothness of Machined Surfaces," Metal Progress 21, 57-9 (1932).

5. Abbott, E.J., and Firestone, F.A., "Specifying Surface Quality," Mech. Engng., 55, 569-72 (1933).

6. Reason, R.E., "Surface Finish and Its Measurement," J. Inst. Prod. Engrs., 23, 347-72 October (1944).

7. Bowden, F.P., and Tabor, D., "The Friction and Lubrication of Solids, Part I." Book published by Oxford University Press U.K. (1950).

8. Thomas, T.R., and King, M., "Surface Topography in Engineering - A State-of-the-Art Review and Bibliography," Guy, N.G. (ed) (B.H.R.A.) Fluid Engineering, Beds., U.K.

9. Halling, J. (ed), "Principles of Tribology," Macmillan Press Ltd. (1978).

10. Wallace, D.A., "Surface Finish," S.A.E. Trans. (1940).

11. Greenwood, J.A., and Williamson, J.B.P., "Contact of Nominally Flat Surfaces," Proc. Roy. Soc. A295, 300-19 (1966).

12. Greenwood, J.A., "The Area of Contact Between Rough Surfaces and Flats," Trans. A.S.M.E., Ser (F)., J. Lubrication Technology, 89, 1, 81 - 91 (1967).

13. Schlesinger, G., "Surface Finish," Inst. of Prod. Engrs. Report, 231 (1942).

14. Bennet, H.E., and Porteous, J.O., "Relation Between Surface Roughness, and Specular Reflectance at Normal Incidence," J. Opt. Soc. AM., 51, 123-9 (1961).

15. Abbott, E.J., and Goldschmidt, E., "Surface Quality," Mech. Engng. 59, 813 - 825 (1937).

16. Barash, M.M., "Measuring the Finish of Rough Surfaces," Int. J. Mach. Tool Des. Res. Vol. 3, 97 - 100 (1963).

17. Schneider, E.J., "Surface Finish Testing - The Present State-of-the-Art and its Future," S.A.E. Paper 700144, 6 (1970).

18. Chaplin, P.D., "The Analysis of Hull Surface Roughness records," European Shipbuilding, 16, 40-7 (1967).

19. Thomas, T.R., Holmes, C.R., McAdas, H.T., and Bernard, J.C., "Surface Features Influencing the Effectiveness of Lip Seals: A Pattern - Recognition Approach," S.M.E. Paper 1075-128, 16 (1975).

20. Barwell, F.T., "Bearing Systems - Principles and Practices," Oxford University Press (1979).

21. Surface Analysis Definitions of Terminology Relating to. ASTM-E673 (1980).

22. Pearson, J. and Hopkins, M.R., "Plastic Replicas for Surface - Finish Measurement," J. Iron and Steel Inst., 67 - 70 (May 1948).

23. Young, A.P., and Clegg, B.H., "Replica Method for Examining Surface Profiles," Rev. Sci. Instrum. 30, 444-6 (June 1959).

24. Peklenik, J., "New Developments in Surface Characteristic and Measurements by Means of Random Process Analysis," Proc. Inst. Mech. Engrs. London Part 3K, 182, 108-26 (1967).

25. Lonquet-Higgins, M.S., "Statistical Properties of an Isotropic Surface Random Surface," Phil. Trans. Royal Soc., A250, 157 - 174 (1957).

26. Bendat, J.S., and Piersol, A.G., "Measurement and Analysis of Random Data," Wiley (1966).

27. Greenwood, J.A., "Surface Measurement, Experimental Methods in Tribology," Proc. Inst. Mech. Engrs. London, Part 3G vol. 182, 1 - 6 (1968).

28. Williamson, J.B.P., "Microtopography of Surfaces, In Properties and Metrology of Surfaces," Proc. Inst. Mech. Engrs. London, Park 3K, 182, 21 - 30 (1967).

29. Whitehouse, D.J., "Some Ultimate Units on the Measurement of Surfaces Using Stylus Techniques, Measurement and Control," 8, 147 - 151 (1975).

30. Whitehouse, D.J., "Modern Trends in the Measurement of Surfaces," Rev. M. Mec. 21, 19 - 28 (March 1975).

31. Thomas, T.R., "Recent Advances in the Measurement and Analysis of Surface Microgeometry," Wear (33) 2, 205-33 (July 1975).

32. Wilen, J.E., "Characterization of Cylinder Bore Surface Finish, A Review of Profile Analysis," Wear, Vol. 19, 143 - 162 (1972).

33. Radhakrishnan, V., "Statistical Behavior of Surface Profiles," Wear, Vol. 17, 259 - 267 (1971).

34. Nayak, P.R., "Random Process Model of Rough Surfaces," Trans. Amer. Soc. Mech. Eng. Vol. 93, No. 2, 398 - 407 (1971).

35. Nara, J., "About the Standardization and Spectral Measurement of Surface Waviness," Annals of International Committee for Prod. Eng., Research Vol. 9, No. 3, 687 - 693 (1971).

36. Williamson, J.B.P., Pullen, J., Hunt, R.T., "True Shape of Solid Surfaces," In Surface Mechanics Ed F.F. Ling Book Pub. ASME, New York, 24 - 35 (1969).

37. Williamson, J.B.P. and Hunt, R.T., "Relocation Profilometers," J. Phys. E. Sci. Instrum., 1, 7, 749-52 (July 1968).

38. Grieve, D.J., Kaliszer, H., and Rowe, G.W., "A Normal Wear Process Examined by Measurements of Surface Topography," ANN. C.I.R.P., 18, 585-92 (1970).

39. Peklenik, J., and Kubo, M., "A Basic Study of a Three Dimensional Assessment of the Surface Generated in a Manufacturing process," Ann. C.I.R.P., 16, 257-65 (1968).

40. Thomas, T.R., and Sayles, R.S., "Mapping a Small Surface," Journal of Physics. E. Sci, Inst., Vol. 9, 855 - 861 (1976).

41. Edmonds, M.J., "Effects of Surface Configurations on Thermal Energy Transfers Across Pressed Contacts," Ph.D. Thesis, Cranfield Inst. of Tech. (1978).

42. Williams, A., and Idrus, N., "Detection and Measurement of Damage to Surfaces After Static Contact Loading," Proc. Int. Conf. Metrology and Properties of Eng. Surfaces, Leicester Poly, UK., 281 - 291. (April 1979).

43. Jones, A.M., O'Callaghan, P.W. and Probert, S.D., "Prediction of Contact Parameters From Topography of Contacting Surfaces," Wear (31), 89 - 107 (1975).

44. Whitehouse, D.J., and Archard, J.F., "The Properties of Random Surfaces of Significance in Their Contact," Proc. Roy. Soc., A 316, 97 - 121 (1970).

45. Archard, J.F., "Elastic Deformation and the Laws of Friction," Proc. Roy. Soc. 243 A, 190 - 205 (1957).

46. Tsukizoe, T., and Hisakado, T., "On the Mechanism of Contact Between Metal Surfaces - The Penetrating Depth and Average Clearance (Pt. 1)," Trans. A.S.M.E. Ser. D.J., Basic Engng. 873, 666 - 674 (1965).

47. Tsukizoe, T., and Hisakado, T., "On the Mechanism of Contact Between Metal Surfaces, Pt. II the Real Area of Contact, and the Number of Contact Points," Trans. A.S.M.E. Ser. F. 3 Lubr. Tech. 90, 1, (81 - 88) 1968.

48. Thomas, T.R., and Probert, S.D, "Establishment of Contact Parameters from Surface Profiles," Br. J. Appl. Phys., Ser. 3.3, 277 - 291 (1970).

49. O'Callaghan, P.W., and Probert, S.D., "Effects of Static Loading on Surface Parameters," Wear (24), 133 - 145 (1973).

50. Edmonds, M.J., Jones, A.M., O'Callaghan, P.W., and Probert, S.D., "Prediction and Measurement of Thermal Contact Resistance," Wear (50), 299 - 319 (1978).

51. Thomas, T.R., Uppal, A.H., and Probert, S.D., "Hardness of rough surfaces," Nature Physical Science, Vol. 229, No. 3, 86 - 87 (1971).

52. Uppal, A.J., and Probert, S.D., "Deformation of Single and Multiple Asperities on Metal Surfaces," Wear (20), 381 - 400 (1972).

53. Radford, J.D., and Richardson, D.B., "Production Engineering Technology," 3rd Edition, McMillan (1981).

54. Kaczmarck, J., "Principles of Machining - By Cutting Abrasion and Erosion," Peter Peregrinns Ltd. (1976).

55. Schubert, P.B. (Ed.), "Machinery's Handbook," 21st edition, Industrial Press. Inc. (1979).

56. Kachmarck, J., "Technological Possibilities of Obtaining Good Surface Finish," Czasopismo Technique 4 1961.

57. Waterman, N.A., "The Selection of Materials," Engineering Design Guides No. 29, O.U.P.

58. ASTM Standards ASTM-E673 Section 18 (1980).

59. Surface Integrity Encyclopedia (Special unpublished collection of data and effects from specific material - process combinations). Maintained for reference by the Machinability Data Center, Cincinnati, Ohio.

60. British Standards. BS. 4500. ISO Limits and Fits Part I: General, Tolerances and Deviations (1969), Also BS 1134, Part 1 and 2 (1972).

61. Young, R., Vorburger, T., and Teague, C., "In-Process and On-Line Measurement of Surface Finish," Annals of CIRP. Vol. 29, 1 (1980).

62. Thwaite, E.G., "The Direct Measurement of the Power Spectrum of Rough Surfaces by Optical Fourier Transformation," Wear, 57, 71 - 80 (1979).

If you have any concerns about our products,
you can contact us on
ProductSafety@springernature.com

In case Publisher is established outside the EU,
the EU authorized representative is:
**Springer Nature Customer Service Center GmbH
Europaplatz 3, 69115 Heidelberg, Germany**

Printed by Libri Plureos GmbH
in Hamburg, Germany